CONCEPTS IN INHALATION TOXICOLOGY

CONCEPTS IN INHALATION TOXICOLOGY

Second Edition

Edited by

Roger O. McClellan
Chemical Industry Institute
of Toxicology
Research Triangle Park, North Carolina

Rogene F. Henderson
Inhalation Toxicology Research Institute
Lovelace Biomedical and Environmental Research Institute
Albuquerque, New Mexico

CRC Press
Taylor & Francis Group
Boca Raton London New York

CRC Press is an imprint of the
Taylor & Francis Group, an **informa** business

CRC Press
Taylor & Francis Group
6000 Broken Sound Parkway NW, Suite 300
Boca Raton, FL 33487-2742

First issued in paperback 2019

© 2009 by Taylor & Francis Group, LLC
CRC Press is an imprint of Taylor & Francis Group, an Informa business

No claim to original U.S. Government works

ISBN-13: 978-1-56032-368-6 (hbk)
ISBN-13: 978-0-367-40158-0 (pbk)

Contents

PART 3: BASIC BIOLOGY OF THE RESPIRATORY TRACT

PART 4: DISPOSITION OF INHALED MATERIAL

PART 5: EVALUATION OF THE RESPOSES OF THE RESPIRATORY TRACT TO INHALED TOXICANTS

            Inflammation and Fibrosis   471
            Kevin E. Driscoll

            Introduction   471
            Cytokines   472
            Role of Cytokines in Pulmonary Inflammation
            and Fibrosis   476
            Cytokines as Biomarkers of Lung Toxicity   495
            Conclusion   496
            References   496

Chapter 17 Applications of Behavioral Measures to
            Inhalation Toxicology   505
            Bernard Weiss and Alice Rahill

            Introduction   505
            Measures of Aversive and Irritant Properties   506
            Ozone   506
            Behavioral Toxicity of Volatile Organic Solvents   511
            Inhalation as a Source of Exposure to
            Neurotoxic Metals   520
            Chemical Sensitivity Syndromes   524
            Summary   528
            Acknowledgments   528
            References   529

Chapter 18 Noncarcinogenic Responses of the
            Respiratory Tract to Inhaled Toxicants   533
            *Donald L. Dungworth, Fletcher F. Hahn, and
            Kristen J. Nikula*

            Introduction   533
            Factors Affecting Anatomic Location of Damage   534
            Nose   537
            Larynx   546
            Acute Responses of the Lung to Inhaled
            Toxicants   546
            Chronic Responses of the Lung to Inhaled
            Toxicants   550
            Correlation between Acute and Chronic
            Responses   555
            Exposure Concentration-Time-Response
            Relationships for Reactive Gases   559
            Time-Response Relationships for Particles   563

PART 6: RISK ASSESSMENT

Contributors

David E. Bice, Ph.D.
Inhalation Toxicology Research
   Institute
Lovelace Biomedical and
   Environmental Research Institute
Albuquerque, New Mexico 87185

James A. Bond, Ph.D.
Chemical Industry Institute of
   Toxicology
Research Triangle Park, North Carolina
27709

Yung-Sung Cheng, Ph.D.
Inhalation Toxicology Research
   Institute
Lovelace Biomedical and
   Environmental Research Institute
Albuquerque, New Mexico 87185

Alan R. Dahl, Ph.D.
Inhalation Toxicology Research
   Institute
Lovelace Biomedical and
   Environmental Research Institute
Albuquerque, New Mexico 87185

Kevin E. Driscoll, Ph.D.
Miami Valley Laboratories
The Proctor & Gamble Company
Cincinnati, Ohio 45239

Donald L. Dungworth, B.V.Sc., Ph.D.,
   M.R.C.V.S., Dip. A.C.V.P.
Professor of Veterinary Pathology,
   Emeritus
University of California, Davis
Davis, California 95616

Aron B. Fisher, M.D.
Institute for Environmental Medicine
University of Pennsylvania School of
   Medicine
Philadelphia, Pennsylvania 19104

Fletcher F. Hahn, D.V.M., Ph.D.
Inhalation Toxicology Research
   Institute
Lovelace Biomedical and
   Environmental Research Institute
Albuquerque, New Mexico 87185

Rogene F. Henderson, Ph.D.
Inhalation Toxicology Research
   Institute
Lovelace Biomedical and
   Environmental Research Institute
Albuquerque, New Mexico 87185

Julia S. Kimbell, Ph.D.
Chemical Industry Institute of
   Toxicology
Research Triangle Park, North Carolina
27709

Joe L. Mauderly, D.V.M.
Inhalation Toxicology Research
    Institute
Lovelace Biomedical and
    Environmental Research Institute
Albuquerque, New Mexico 87185

Roger O. McClellan, D.V.M.
Chemical Industry Institute of
    Toxicology
Research Triangle Park, North Carolina
    27709

Michele A. Medinsky, Ph.D.
Chemical Industry Institute of
    Toxicology
Research Triangle Park, North Carolina
    27709

Frederick J. Miller, Ph.D.
Chemical Industry Institute of
    Toxicology
Research Triangle Park, North Carolina
    27709

Owen R. Moss, Ph.D.
Chemical Industry Institute of
    Toxicology
Research Triangle Park, North Carolina
    27709

Kristen J. Nikula, D.V.M., Ph.D.
Inhalation Toxicology Research
    Institute
Lovelace Biomedical and
    Environmental Research Institute
Albuquerque, New Mexico 87185

Robert F. Phalen, Ph.D.
Community and Environmental
    Medicine
University of California
Irvine, California 92717

Shankar B. Prasad, M.B.B.S.
Southern California Air Quality
    Management District
Diamond Bar, California 91765

Alice Rahill, Ph.D.
University of Rochester Medical Center
Rochester, New York 14642

Richard B. Schlesinger, Ph.D.
Institute of Environmental Medicine
New York University Medical Center
Tuxedo, New York 10987

M. B. Snipes, Ph.D.
Inhalation Toxicology Research
    Institute
Lovelace Biomedical and
    Environmental Research Institute
Albuquerque, New Mexico 87185

Bernard Weiss, Ph.D.
University of Rochester Medical Center
Rochester, New York 14642

Brian A. Wong, Ph.D.
Chemical Industry Institute of
    Toxicology
Research Triangle Park, North Carolina
    27709

Hsu-Chi Yeh, Ph.D.
Inhalation Toxicology Research
    Institute
Lovelace Biomedical and
    Environmental Research Institute
Albuquerque, New Mexico 87185

Foreword

Almost 50 years ago, when I arrived at the University of Rochester with a master's degree in Chemistry hoping to become a toxicologist, there was no formal program in toxicology there or elsewhere. Professor Harold Hodge, the first chairman of pharmacology at Rochester, had encouraged me to come to Rochester to undertake my Ph.D. in pharmacology because toxicology at that time was largely treated as an extension of pharmacology. Professor Hodge was already recognized as a national leader in the field of toxicology and later was to become the Society of Toxicology's first president. Moreover, he had a large biological research laboratory under his aegis that was carrying out a toxicology program sponsored by the U.S. Atomic Energy Commission (AEC). So my career in toxicology began.

In the late 1940s, there not only was a national demand for studies of the hazardous properties of radionuclides such as the alpha-emitters (for example, radon and plutonium) but also for a number of stable chemicals (for example, beryllium oxide and the rare earth oxides) for which toxicological information was seriously lacking. From a research perspective, the availability of radioisotopes for use as tracers was making its initial impact on toxicology and on biomedical sciences generally. It was a special time; both the needs and the methods of research were receiving a major impetus by these early postwar priorities of the AEC.

Among Dr. Hodge's colleagues were Drs. Herbert Stokinger, William Neuman, Aser Rothstein, Henry Blair, Frank A. Smith, J. Newell Stannard, and Charles LaBelle. Kenneth Lauterbach, Sid Laskin, Herbert Wilson, Leonard Leach, and over a hundred technicians and graduate students provided technical support. This outstanding group of educators and researchers had already developed a major inhalation facility for exposing animals, for they appreciated that the principal toxicological risks presented by these radioactive and nonradioactive chemicals were based on their existence as dusts in the workplace. My plans to become a toxicologist were modified somewhat to become an inhalation toxicologist.

In those days, the technology of inhalation toxicology was virtually un-developed, so each experimental study required the construction of aerosol generators, exposure units, and samplers to fit its needs. The analytical requirements were similarly custom-made. For very hazardous dusts such as beryllium oxide (BeO), special enclosures were developed with nose-only units. I recall my first days working under the direction of Charles LaBelle at his automatic exposure unit developed for rabbits, in a laboratory that always had some suspicious white powder on work surfaces. His work soon revealed the exceptional toxicity of high-fired BeO and contributed to the active controversy as to whether berylliosis was the devastating lung disease that it was thought by some to be.

Professor Hodge allowed his graduate students some time to learn the ropes of research while taking preclinical courses in the medical school and serving as graduate assistants in the pharmacology course for second-year medical students. The G.I. Bill gave me free tuition but limited me to earn less than $200 per month to support my family. On the brighter side, in the medical school I was to learn from and soon afterward associate with the likes of Professors Wallace Fenn, Arthur DuBois, Hermann Rahn, and Arthur Otis, perhaps the most formidable collection of respiratory physiologists ever assembled into one faculty. Thomas Mercer was also a graduate student and colleague of mine; we spent hours together shooting the breeze about aerosol and inhalation exposure problems while shooting at water roaches with rubber band-propelled paper clips.

Down the hall from LaBelle's lab in the Annex was the main chamber room, which was to remain busy even into the late 1950s. It was always filled with mice, rats, dogs, and monkeys being exposed to dusts, mainly uranium compounds. Electron microscopes soon appeared on the scene and were applied to aerosols, as were molecular filters. Laskin's early cascade impactor sampler became available commercially and so did the Wright Dust-feed Mechanism from England. As the physical scene changed, so did I—trying to comprehend the small but precious collection of books on the shelf: *Mechanics of Aerosols* by Fuchs, *Micromeritrics* by Della Valle, *Smoke: A Study of Aerial Dispersion Systems* by Whytlaw-Gray, *Industrial Dust* by Drinker and Hatch, and *Dust Is Dangerous,* by C.N. Davies. And, of course, I nearly memorized our limited but equally precious collection of the pioneering research papers of A. M. Van Wijk and H. S. Patterson, I. B. Wilson and V. K. LaMer, C. E. and J. H. Brown, A. M. Baetjer, M. W. First and L. Silverman, T. F. Hatch, E. J. King and G. Nagelschmidt, W. Findeisen, and C. N. Davies.

During our doctorate years, we graduate students learned a lot about being plumbers, glass blowers, lathe and drill press operators, animal caretakers, colloidal chemists, radiation biologists, and aerosol physicists by imitating our peers and by trial and error, always managing to do it the hard way. We also learned to spend a minimum of four to six hours per week in the library and to emulate the accomplishments of predecessors and contemporaries in the inhalation field. How lucky we were! My first research presentation, like that of most graduate students, was at the Federation of American Biologists annual meeting in Atlantic City, where one was given all of 10 minutes to speak! Describing several years of research in 10 minutes was a humbling experience! After I became a faculty member in 1951, the American Industrial Hygiene Association (AIHA) annual meeting became the premier place to present research accomplishments. There one was allowed up to 40 minutes!

There were very few journals in the United States at that time that were particularly interested in publishing papers on aerosol and inhalation studies. For the most part, two national journals carried virtually all of the experimental studies on dust deposition and retention and on animal toxicology. These were the *AIHA Journal* and *Health Physics*.

One of the early miracles of becoming an inhalation toxicologist was being asked by Professors Ted Hatch and Paul Gross to review the galleys of their new monograph *Pulmonary Deposition and Retention of Inhaled Aerosols*, which was published in 1964. To peruse their remarkable monograph 30-something years later is a rewarding exercise for any inhalation toxicologist.

In the 1950s and 1960s, deposition and retention measurements were becoming increasingly feasible without the use of costly serial-sacrifice procedures because radioisotopically labelled dusts could be utilized. Nevertheless, such studies did not seek to be quantitative, mainly because the use of first-order kinetics to describe dust retention did not require it. Then, of course, there also were the uncertainties associated with external measurements of dust deposited within respiratory structures. At New York University in 1955, Albert and Arnett demonstrated the practicality of using human subjects for determining dust retention. Those interested in modeling dust deposition in the human airways, such as Landahl and coworkers, were stimulated and encouraged by the new capability of comparing model predictions to experimental measurements in human subjects.

The inhalation studies begun at the University of Rochester during the wartime 1940s set the stage for the expansion of research during the 1950s and 1960s and for Rochester gaining an international reputation as one of the early centers for inhalation studies in humans and experimental animals. From those early days until now, the graduates, postgraduates, visiting faculty, and regular faculty of Rochester who made major contributions to inhalation toxicology generally and to the development of other inhalation research centers across the nation include Tom Mercer, Bob Thomas, Bill Bair, Lou Casarett, Otto Raabe, Bob Phalen, Bruce Boecker, Herbert Stokinger, James Scott, Charles Yuile, Dick Cuddihy, Owen Moss, Bean T-H Chen, Jaroslav Vostal, Douglas Craig, Bruce Stuart, Bernard Greenspan, John Morris, Sidney Soderholm, David Velasquez, John Ballou, Bruce Lehnert, Bart Dahneke, Justin Postendorfer, Werner Stöber, Newell Stannard, Juraj

Ferin, Jack Finkelstein, Mark Utell, and Günter Oberdörster. Many of the aforementioned scientists were important to the development and productivity of the Inhalation Toxicology Research Institute in Albuquerque that Roger McClellan so effectively directed for many years and where Rogene Henderson is still conducting internationally acclaimed research on lung responses to inhaled materials. Several are currently associated with Roger McClellan and contribute to the inhalation toxicology capability of the Chemical Industry Institute of Toxicology (CIIT) at Research Triangle Park, NC. In diverse ways, many of these scientists have contributed to *Concepts in Inhalation Toxicology*.

As toxicology emerged as an independent biomedical science, curricula and research activities developed in the United States at a remarkable rate. In many universities, industries, and government laboratories, additional centers for inhalation studies began or expanded at, for example, Kettering Laboratory, General Motors, University of Pittsburgh, Harvard School of Public Health, Lovelace ITRI, New York University, Johns Hopkins University, Los Alamos Scientific Laboratory, DuPont, and Research Triangle Park (CIIT and EPA). Sources of research funding for inhalation toxicology studies steadily increased as the National Institutes of Health, U.S. Environmental Protection Agency, National Institute of Environmental Health Sciences, Food and Drug Administration, Department of Defense, National Institute of Occupational Safety and Health, and many industries offered grants and contracts. Of course, there also was the impetus provided by the concurrent enactment of the national toxicology program, the Clean Air Act, the Toxic Substances Control Act, the Occupational Safety and Health Act, and other legislation.

While it would be incorrect to conclude that Rochester played more than a minor role in the phenomenal growth of the field of inhalation toxicology, Rochester, like all other centers of toxicological investigations, was greatly affected by these events, and it profited from the remarkably interactive research environment that prevailed.

It is gratifying that the first edition of *Concepts in Inhalation Toxicology* has been so widely accepted. Its success is partly a testimonial to the experts assembled by Drs. Henderson and McClellan. Each of these experts has had a long and illustrious research career, has played an important leadership role in his or her particular specialties within the field, and has remained actively productive in the field. It is also a reflection of the book's excellent coverage of basic concepts and quantitative approaches and to the comprehensive treatment of evaluations of respiratory responses to inhaled particles and gases.

This second edition updates the 1989 edition and adds significantly to the discussion of risk assessment of inhaled toxicants, a topic Dr. McClellan has actively pursued. In a field that is expanding so rapidly in scope, knowledge, and sophistication, this second edition will be welcomed by all serious students and practitioners of inhalation toxicology, whether their perspective is hands-on, regulatory, or managerial. For me, the progress that has occurred since the days when the principal challenge was to develop the elementary technology of conducting inhalation research is both satisfying and overwhelming. Now we are researching the interactions of particles with respiratory tract tissues, cells, cellular organelles—the molecular constituents of respiratory system structure and function. What an exciting time! Presently, we have inhalation toxicologists who specialize in macrophage function, cytokine networking, immunomodulation, genetic coding, mucous rheology, mechanisms of oxidant injury, and lung growth, and the possibilities continue to grow.

If those entering the field of inhalation toxicology research had to choose only one book on the general subject, a book that would allow them to savor the complex and stimulating advances in the field, in my judgment there is no better choice than the second edition of *Concepts in Inhalation Toxicology*.

P. E. Morrow

Preface

The field of inhalation toxicology is growing rapidly. The general public, industry, organized labor, and governmental regulatory agencies are increasingly concerned that the air we breathe in our homes, workplaces, and outdoor environment not impair human health. Periodically, the Lovelace Inhalation Toxicology Research Institute (LITRI), an internationally recognized leader in the field, has offered tutorial workshops on the basic concepts of inhalation toxicology. The first edition of *Concepts in Inhalation Toxicology* was based on the fourth workshop in the series. The first edition has been used as background material for other workshops and university courses in advanced toxicology and as a reference work for scientists engaged in the conduct of research with airborne materials and the assessment of their risks to human health. Since the first edition was so well received, we were encouraged to prepare a second edition incorporating the latest developments in the field.

The purpose of the second edition is to again convey to the reader the key concepts in the field of inhalation toxicology. The book is directed toward scientists who conduct, manage, evaluate, or interpret research work in the field of inhalation toxicology. The book is intended to be especially useful to two groups of individuals. The first group comprises those scientists without previous experience in inhalation toxicology who are planning to initiate work in this field in their own laboratories. Our goal is to introduce those scientists to the basic concepts of inhalation toxicology so that they can bring their own specialized expertise from other fields to bear in resolving important scientific questions about airborne materials. The second group comprises the increasing number of scientists, engineers, and other analysts who are involved in risk assessment. These individuals, who typically have varied educational backgrounds and no direct research experience in inhalation toxicology, are frequently called on to assess the risks of airborne materials. The increase in such risk assessment activities is not surprising since inhalation is a major avenue by which toxic materials enter the body. Our goal is to introduce both groups to the special nuances and concepts of inhalation toxicology that will enhance the validity of their risk assessments. If you are not included in either of these groups, we hope that you will still find the book useful in guiding your own endeavors.

The book is divided into six parts. After an introductory overview of the field of inhalation toxicology, the three chapters in Part 2 review the essentials of a good exposure system and the procedures for generating and characterizing exposure atmospheres, including both gases and particles. In Part 3, three chapters describe the basic morphological and biochemical characteristics of the respiratory tract. Part 4 contains four chapters on dosimetry, or the deposition and clearance of inhaled particles and gases and factors affecting the disposition of inhaled compounds. The fifth and largest part deals with the evaluation of responses to inhaled materials. It contains seven chapters describing approaches in the fields of histopathology, respiratory function, immunology, biochemistry, and behavioral science for evaluation of responses to inhaled materials. Finally, Part 6 is a chapter on methods of risk assessment for inhaled toxicants.

The editors thank the staff of the Chemical Industry Institute of Toxicology and the Lovelace Inhalation Toxicology Research Institute and colleagues from other organizations whose substantial efforts made this second edition possible.

Roger O. McClellan
Rogene F. Henderson

Preface to the First Edition

The field of inhalation toxicology is rapidly growing. There is an increasing interest on the part of the general public, industry, and governmental regulatory agencies that the air we breathe in our homes, in our workplaces, and in our outdoor environment not impair the health of people. Periodically, the Lovelace Inhalation Toxicology Research Institute (LITRI), an internationally recognized leader in the field, has been pleased to offer tutorial workshops on the basic concepts involved in inhalation toxicology. The present volume is based on the fourth of this series of workshops, which was cosponsored by an outstanding group of organizations whose members have a strong interest in improving the quality of inhalation toxicology research. The cosponsors, in addition to the Lovelace Inhalation Toxicology Research Institute, were the Lovelace Medical Foundation, the U.S. Department of Energy, the Toxicology Study Section and the Safety and Occupational Health Study Section of the National Institutes of Health, the Office of Exploratory Research of the U.S. Environmental Protection Agency, and the Inhalation Specialty Section of the Society of Toxicology.

The purpose of this book is to convey to the reader the basic concepts that are important in the field of inhalation toxicology. The book is directed toward scientists who either conduct, manage, evaluate, or interpret research work in the field of inhalation toxicology. It is intended that the book be especially useful to scientists who have not previously worked in inhalation toxicology and who are now planning to initiate work in the field in their own laboratories.

The book is divided into five major sections. After an introductory overview of the field of inhalation toxicology, three chapters review the essentials of a good exposure system and how one generates and characterizes exposure atmospheres, including both gases and particles. This section is followed by two chapters describing the basic morphological and biochemical characteristics of the respiratory tract. The next section contains three chapters on dosimetry—the deposition and clearance of inhaled particles and gases and the factors that affect the disposition of inhaled compounds. The fourth and largest section deals with evaluation of responses to inhaled materials and contains eight chapters describing approaches in the fields of histopathology, respiratory function, immunology, biochemistry, and behavioral science for evaluation of the response to inhaled materials. Finally, the last section is a chapter on methods of risk assessment for inhaled toxicants.

The editors thank the numerous members of the Lovelace Inhalation Toxicology Research Institute staff whose untiring efforts made both this book and the workshop on which it is based possible.

Roger O. McClellan
Rogene F. Henderson

Part One

Introduction

Chapter One

An Introduction to Inhalation Toxicology

Roger O. McClellan

Toxicology has traditionally been defined as the science or knowledge of poisons. Within this framework, inhalation toxicology can be viewed as the science or knowledge of inhaled poisons or toxicants. In a broad sense, the specialty of inhalation toxicology is concerned with

1   the physical and chemical characteristics of material in the air,
2   the basic biology of the respiratory tract,
3   the deposition and retention of inhaled materials in the body and their interactions with critical biological units, and
4   the way in which such interactions with the respiratory tract and other systems produce disease.

The major elements of inhalation toxicology are depicted schematically in Fig. 1, with mechanistic linkages between exposure, dose, and response. Each of these interrelated areas is addressed in *Concepts in Inhalation Toxicology, Second Edition.* Knowledge of these concepts provides a basis for evaluating potential health risks of airborne materials as well as for identifying related research needs. These concepts are critical to the design, conduct, and interpretation of specific experiments. Finally, the concepts are an

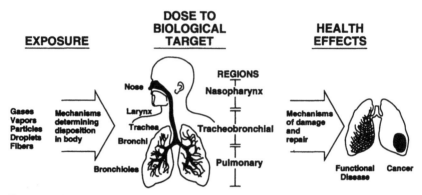

Figure 1 Schematic representation of the major elements of inhalation toxicology oriented toward understanding the toxicity of airborne materials.

essential consideration in integrating new experimental results with our current knowledge to yield the best possible assessment of the health risks of airborne materials.

Three topics will be considered in the introduction: the basis for concerns about the toxicity of airborne materials, the use of both basic science and risk assessment orientations in inhalation toxicology, and the organization and contents of this book.

CONCERN ABOUT AIR POLLUTION
A Historical View

There has been concern for many years—indeed, centuries—about the role of airborne materials in disease. This concern, which undoubtedly preceded recorded history, may have coincided with man's use of fire for cooking and heating and later for manufacturing. In the Western world, concern for air pollution accelerated following introduction of coal as an energy source upon Marco Polo's return from the technologically more advanced societies of the Far East. As early as the 13th century, there was concern over coal smoke and odor in London, and the Queen of England reportedly moved from London to Nottingham because of the insufferable smoke. In 1661, John Evelyn published *Fumifugium: Or The Inconvenience of the Air and Smoake of London Dissipated,* which drew attention to the problem of air pollution.

In the early years of the industrial revolution, industry was dependent on wood as a fuel and was therefore located near forests. This restriction served as a throttle on industrial development. The development of effective steam engines in the 18th century provided a major impetus for the use of coal as a fuel. With coal, the pace of industrial activity increased, and factories developed near sources of coal and along the waterways that served to transport the coal efficiently. Increasingly, the smoke and ash from coal-fired boilers in factories, power plants, and locomotives became of concern. Interestingly, this concern about air pollution was stimulated by impairment of visibility and soiling as well as effects on health. In addition, the damage to vegetation from emissions produced in the smelting of sulfide ores gave clear evidence of yet another effect of air pollution. Air pollution problems were recognized in many industrial countries in the late 19th century, but the problem was probably nowhere as apparent as in London. The high levels of pollutants combined with the notorious London fog created a serious problem and is reputed to have given rise to the term *smog*, a contraction of smoke and fog.

In the 1900s, a series of incidents drew attention to air pollution as a serious and even acute problem. In December 1930, static atmosphere conditions in the Meuse Valley of Belgium resulted in pollution that caused over 60 deaths and hundreds of illnesses. In December 1952, smog in London resulted in about 4000 deaths, principally among the infirm, the old, and those with respiratory diseases. The effects were not restricted to humans. Acute respiratory symptoms were also reported in cattle at the Smithfield Club livestock show. The effects may have been most severe in the heavyweight prize cattle, who were well cared for and had their litter and excreta promptly removed. Other cattle that were not as well cared for apparently fared better, perhaps because of some neutralizing benefits of ammonia from the excreta on the effects of sulfur oxides and sulfuric acid. Another major smog episode in London during December 1962 resulted in 340 deaths. In October 1948, a particularly calm and stable meteorology in Donora, Pennsylvania, resulted in a marked increase in levels of oxides and inorganic sulfates and increased morbidity and mortality from respiratory effects.

Despite early concerns about whether airborne materials could cause disease, relatively little research was conducted prior to World War I to improve our knowledge of the relationship between the air pollution and disease. The unfortunate use of poisonous gases in World War I provided a stimulus for developing a better understanding of how these specific agents produced disease. Unfortunately, the meager literature indicates that research in this area was not substantial or very long-lasting. In retrospect, the lack of research related to airborne toxicants is probably not surprising, considering the generally low level of biomedical research activity in the first third of the 20th century. The strong preventive medicine orientation that exists today toward diseases of both occupational and environmental origin had not yet developed.

There were a few exceptions to this general lack of concern for the health hazards of airborne materials. Perhaps the most notable exceptions were the efforts of Dr. T. F. Hatch at the University of Pittsburgh and Drs. C. K. and P. Drinker at Harvard University, who initiated pioneering studies in the related fields of inhalation toxicology and industrial hygiene in the 1920s. Their work focused on concern for airborne materials causing occupational diseases. As an aside, *Industrial Dust: Hygienic Significance, Measurement, and Control* by Drinker and Hatch (1936) and *Pulmonary Deposition and Retention of Inhaled Aerosols* by Hatch and Gross (1964) are classic references in the field.

If the pre–World War II era can be considered the prenatal period of inhalation toxicology, then World War II can certainly be considered the era of rapid postnatal growth and development. The stimulus for this development came from concern for health effects being caused by airborne materials of three markedly different types: chemical and biological warfare agents, radioactive materials, and automotive exhaust emissions.

Concern for airborne chemical and biological agents had its roots in the use of airborne materials for warfare during World War I. This tragic experience gave rise to strong sentiment during World War II that the use of such agents not be repeated and that the United States and its allies be prepared to defend against their use. This gave impetus to military-sponsored research to better understand how airborne materials of either a chemical or biological nature produce disease.

The issue of potential health effects of radioactive materials was brought to the forefront by the Manhattan Project, established by the United States government to develop the atomic bomb. Very early, an awareness developed of potential occupational exposure to some of the key materials used in the project. These materials included the uranium used both as the starting material for controlled nuclear fission and in the manufacture of atomic bombs; plutonium, the man-made heavy element resulting from capture of neutrons by uranium, which was also used in manufacturing bombs; and the radionuclides that were by-products of controlled nuclear fission. Plutonium, an alpha-emitting radionuclide, was of particular concern because it was believed to concentrate in the skeleton and cause bone cancers, as observed in luminous-dial painters who had ingested the alpha-emitter radium in the early 1900s. With uranium, plutonium, and the fission products, the potential for inhalation exposures of workers was recognized, and there was well-justified concern for the respiratory tract. Research and related stringent standards and control procedures were effective in minimizing disease from occupational exposures to radioactive materials.

In contrast to the concern for plutonium and other man-made radionuclides, there was a failure to fully appreciate the role of exposures to radon, a radioactive gas, and its daughter products in causing lung cancer in mines in central Europe with high levels of radon. Some of these mines had been used to provide the pitchblende ore used by the Curies and others as a source of radium. The failure to fully appreciate the linkage resulted in a repeat of history, with an excess of lung cancers observed in uranium miners in the United States in the 1950s and later. Ironically, the problem is still with us in the 1990s, when concern has shifted to even lower levels of exposure to radon and its daughters in the indoor environment.

The medical efforts of the Manhattan Project were under the capable leadership of Dr. Stafford Warren, a radiologist who had been at the University of Rochester prior to World War II. When concern arose for the toxicity of uranium, he turned to his former Rochester colleagues, Drs. H. E. Stokinger and H. C. Hodge, for assistance. Their collaboration led to development of the University of Rochester Atomic Energy Project, which did pioneering work on the inhalation toxicology of uranium and other materials of interest in the Manhattan Project. The impact of the Rochester activities extended well beyond the knowledge acquired on specific agents to substantive contributions to our general knowledge in inhalation toxicology and toxicology generally and the associated education of many people who have been major contributors to advances in the field. For example, Louis J. Casarett, coeditor with John Doull of the first edition of the classic reference *Toxicology: The Basic Science of Poisons,* received his Ph.D. from the University of Rochester and spent his early research career working in the Atomic Energy Project. As an aside, the reference work is now in its fourth edition (Amdur, Doull, and Klaassen, 1991).

Beyond the kind of institutional impacts noted above, the atomic era ushered in by World War II had an additional major impact in its emphasis on quantitation. With the use of radioactive materials and their precise measurement, it became possible to quantitate the amount of a radioactive material in the air and the fraction deposited and retained in the body and to relate the observed biological effects to various measures of dose. Some of the elements such as plutonium and strontium were of concern specifically because they had radioisotopes (^{239}Pu and ^{90}Sr). For these radionuclides, the use of a retained and effective dose orientation was quite natural and in keeping with dosimetric developments in the interrelated fields of radiation biology, radiation protection, and radiology. The same approach was extended to radioactively labeled organic compounds using radioisotopes such as ^{3}H, ^{14}C, and ^{32}P. Readers who have an interest in the history of developments in the radiation field, including aspects of inhalation toxicology, are referred to the monumental reference work *Radioactivity and Health: A History* prepared by J. Newell Stannard (1988), an early toxicologist who spent a scientific career at the University of Rochester before beginning a second career in advising and writing. The impact of quantitative capability on inhalation toxicology derived from the radiation field is readily apparent in a number of chapters in this book.

The post–World War II era brought with it an increased awareness of air pollution arising from another source—motor vehicles. Perhaps nowhere was this more the case than in the Los Angeles basin with its marked increase in population, industry, and cars. As concern increased for the Los Angeles smog in the late 1940s and 1950s, the name "Bay of Smokers," which was coined in 1542 by Juan Rodrigues Cabrillo for San Pedro Bay, took on a contemporary tone. In the early 1950s, Professor A. J. Haagen-Smit of the California Institute of Technology discovered the critical nature of the interactions among oxides of nitrogen and hydrocarbons from vehicle exhaust and sunlight in producing ozone and other photochemical oxidants as key components of Los Angeles smog. Findings such as those of Haagen-Smit serve as a reminder of the extent to which inhalation toxicology links the physical and chemical sciences to the life sciences.

Since World War II, there has been a steadily increasing awareness of the importance of understanding the health effects of airborne materials and of limiting occupational and environmental exposures, thereby minimizing the potential for induction of disease related to such exposures. Unfortunately, this awareness has been periodically stimulated with incidents that heighten our concern. For example, the recognition of a linkage between the occurrence of hemangiosarcomas of the liver, a rare cancer in workers occupationally exposed to vinyl chloride, served to emphasize that inhalation exposure can result in effects other than respiratory disease. The accidental release of a large quantity of methyl isocyanate from a pesticide-manufacturing plant in Bhopal, India, in December 1984 caused deaths of over 2000 individuals living near the plant who inhaled the airborne toxicant. This accident underscores the extent to which industrial activities can have an impact beyond plant boundaries. One result of the Bhopal accident has been increased concern for the effects of accidental releases from a wide range of facilities in which large quantities of many different chemicals are stored and used.

Concern for the toxicity of airborne materials has resulted in a number of federal and state laws to regulate sources of airborne materials and minimize their impacts on health and the environment. In the United States, the major pieces of relevant federal legislation are the Occupational Health and Safety Act administered by the Occupational Safety and Health Administration of the Department of Labor; Mine Safety and Health Legislation administered by the Mine Safety and Health Administration within the Department of Labor; and the Clean Air Act, Toxic Substances and Control Act, Comprehensive Environmental Remediation Compensation and Liability Act (Superfund), and Resource Conservation and Recovery Act administered by the Environmental Protection Agency. These pieces of legislation, as well as the concern of corporations and organized labor for the well-being of workers and the safety of products, have provided a major impetus for inhalation toxicology research.

Morbidity and Mortality of Respiratory Disease

Concern for the toxicity of airborne materials is reinforced by recent vital statistics on respiratory disease in the United States. A large portion of respiratory disease is associated with cigarette smoking. Indeed, the influence of cigarette smoking is so substantial that it complicates attempts to establish the role of other occupational or environmental factors in producing respira-

Table 1 Impact of Acute Respiratory Conditions in the United States, 1993

Condition	Number[1]	Bed days[1] (per 100 persons per year)	Restricted activity days[1]	Lost work days[2]
Common cold	26.8	25.9	67.1	21.2
Other upper respiratory infections	11.3	13.4	37.8	11.8
Influenza	52.2	104.4	194.7	83.3
Acute bronchitis	4.7	9.4	23.7	9.2
Pneumonia	2.0	10.2	18.6	3.9
Other respiratory conditions	2.0	5.2	10.2	3.2
Total	99.0	168.5	352.1	132.6

[1]All ages.
[2]All ages 18 years and over, currently employed persons.
(Adapted from Vital and Health Statistics, 1994)

tory disease. Nonetheless, it is appropriate to consider the overall statistics with a view that some undefined portion of respiratory disease is attributable to factors other than cigarette smoke that act alone or in combination with cigarette smoke.

Statistics on the occurrence of acute and chronic respiratory disease in the United States in 1992 are presented in Tables 1 and 2. The high prevalence of these conditions takes on particular significance in terms of impact when the percentage of medically attended and millions of days of restricted activity and lost workdays are considered, as can be done for the acute conditions.

Table 2 Prevalance of Chronic Diseases of the Respiratory Tract for All Ages, United States, 1993

Condition	Number (per 1000 persons)
Chronic bronchitis	54.3
Asthma	51.4
Hay fever or allergic rhinitis without asthma	93.4
Chronic sinusitis	146.7
Deviated nasal septum	7.0
Chronic disease of tonsils or adenoids	11.0
Emphysema	7.6
Total	371.4

(Adapted from Vital and Health Statistics, 1994)

The impact on society of medical costs and lost productivity is, without question, measured in the tens of billions of dollars.

The impact of respiratory diseases is also reflected in mortality statistics (Table 3). Of the 15 leading causes of death, 3 involve the respiratory system: cancer, chronic obstructive pulmonary disease, and pneumonia and flu. Collectively, these diseases currently accounted for about 15% of the deaths in the United States. These statistics do not include deaths due to diseases of other organ systems that result from inhaled materials, a situation that must be considered, as noted earlier with the vinyl chloride example.

Statistics on cancer deaths give clear insight into the prominent contribution of lung cancer to mortality and the role of inhalation of carcinogenic agents (Fig. 2a,b). In reviewing the changing pattern of cancer in U.S. males (Fig. 2a), the steady increase in rate of lung cancer since the 1930s that finally reached a plateau in the 1990s stands out. This increase is generally attributed to lung cancer induced by cigarette smoking after a long latent period. The situation for females (Fig. 2b) appears more favorable on first glance. However, closer examination reveals that the pattern of increasing lung cancer rate for females is simply lagging behind that for males by about 30 years.

Table 3   Mortality for Leading Causes of Death in the United States, 1990

Rank	Cause of death	Death rate per 100,000 population	Percentage of total deaths
1	Heart diseases	289.5	33.5
2	Malignant neoplasms	203.2*	23.5*
3	Cerebrovascular diseases	57.9	6.7
4	Accidents and adverse effects	37.0	4.3
5	Chronic obstructive pulmonary diseases	34.9	4.0
6	Pneumonia and influenza	32.0	3.7
7	Diabetes mellitus	19.2	2.2
8	Suicide	12.4	1.4
9	Chronic liver disease and cirrhosis	10.4	1.2
10	Human immunodeficiency virus infection	10.1	1.2
11	Homicide and legal intervention	10.0	1.2
12	Nephritis, nephrotic syndrome, nephrosis	8.3	1.0
13	Septicemia	7.7	0.9
14	Atherosclerosis	7.3	0.8
15	Certain conditions originating in the perinatal period	7.1	0.8
	All other causes	116.9	13.5
	All causes	863.8	100.0

*Approximately 30% of deaths caused by malignant neoplasms are due to lung cancer.
(From Vital Statistics of the United States, 1994)

The rate for lung cancer in women is now approaching that for breast cancer, which had been the leading cause of cancer deaths in women for 40 years. The observed increase in lung cancer, which began in the 1960s, is generally thought to be related to women who began smoking during World War II and in subsequent years when they entered the work force in large numbers.

As noted, the vast majority of lung cancer is believed to be associated with cigarette smoking. What is not well established is the extent to which cigarette smoking interacts with other factors in the induction of lung cancer. For some situations such as uranium mining (with exposure to radon and its radioactive progeny) and exposure to asbestos, evidence of an interactive effect is overwhelming (Whittemore and Millan, 1983; Lubin and Steindorf, 1995; Selikoff and Hammond, 1979). This interactive effect may be illustrated by considering the statistics for asbestos and cigarette smoke exposure (Table 4). The lung cancer rate for smokers without asbestos exposure was increased about 10-fold over that of nonsmokers without asbestos exposure. The lung cancer rate of nonsmoking, asbestos-exposed individuals was increased about fivefold over that of individuals who were nonsmokers and not exposed to asbestos. Likewise, the lung cancer rate of smokers who were exposed to asbestos was increased about fivefold over nonsmokers exposed to asbestos. The smokers exposed to asbestos had a lung cancer rate over 50-fold greater than individuals who were neither cigarette smokers nor exposed to asbestos. One of the challenges for the field of inhalation toxicology is to understand the pathogenesis of lung cancer induced by smoking and exposure to agents such as asbestos or radon and its progeny, either alone or in combination. Exposure of the general public to asbestos has been of concern for some time.

Concern has recently mounted for exposure of the general public to both cigarette smoke (as a result of passive exposure) and radon and its progeny as well as their interactions. The 1986 Report of the U.S. Surgeon General (U.S. Department of Health and Human Services, 1986) and the 1986 Report of the National Academy of Sciences, National Research Council (NAS,

Table 4   Lung Cancer Death Rates Associated with Cigarette Smoking and Asbestos Exposure (Standardized Mortality Rates shown as Lung Cancers/10^5 Persons)

No work with asbestos and no cigarettes	11.3		
		5×	
Work with asbestos and no cigarettes	58.4		11×
			53×
No work with asbestos and cigarette smoking	122.6		
		10×	
Work with asbestos and cigarette smoking	601.6	5×	

(Adapted from Selikoff and Hammond, 1979, JAMA 242:458, 1979. © 1979 American Medical Association.)

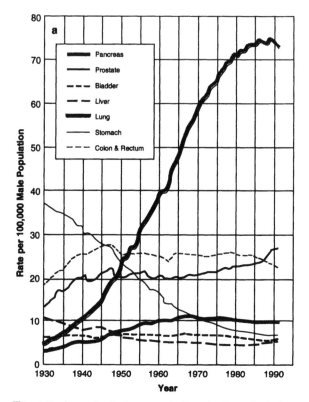

Figure 2 Age-adjusted cancer death rates for selected anatomic sites for the United States population, 1930–1991. Age-adjusted to the 1970 U.S. standard population. (a) Males and (b) females. (Adapted from Wingo et al., 1995, CA—Cancer J Clin 45:22–23, 1995, with permission from J. B. Lippincott Company, Philadelphia, PA.)

1986) both concluded that environmental tobacco smoke (ETS), or passive cigarette smoke, was a risk factor for induction of lung cancer and other respiratory diseases. In my view, these conclusions with ETS are strengthened when the evidence is viewed from an exposure-response perspective and active cigarette smoking is viewed as a high-exposure data point that anchors the upper end of the exposure-response relationship for its closely related toxicant, ETS. The reader interested in the issue of health effects of ETS is referred to a review by Samet (1992).

The case for the risk of lung cancer from residential exposure to radon and its progeny is also controversial. In this case, the issue is quantitative extrapolation from uranium miner exposures to the lower and in some cases very much lower exposures in the residential environment (Lubin and Steindorf, 1995). Using data on the distribution of radon levels in U.S. single-family homes (Nero et al., 1986) and risk factors derived from a meta-analysis of miner data, it is estimated that about 15,000 lung cancers may be

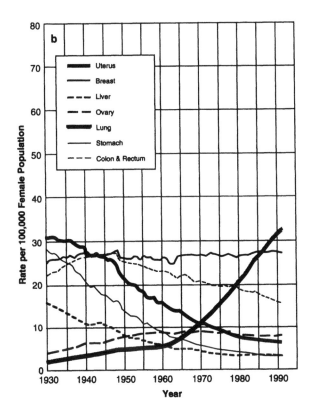

attributed to residential radon with an uncertainty range of 6000–34,000 lung cancer deaths per year (Lubin et al., 1994).

An awareness of the total prevalence of cancer and functional disorders of the respiratory tract as well as the known associations between specific airborne agents and respiratory disease serves as a warning that newly developed chemicals may have potential for producing such diseases. This gives impetus to a proactive approach to inhalation toxicity studies and associated risk assessment evaluations so that potentially hazardous materials can be identified early and exposure to them appropriately controlled. This proactive approach places a premium on being able to estimate risks from laboratory animals and *in-vitro* cell and tissue studies in the absence of human data.

ORIENTATIONS TO INHALATION TOXICOLOGY

Inhalation toxicology can be viewed as having several complementary orientations. As will be noted, the different orientations are probably more a reflection of the intentions and interests of the individuals conducting the research or using the data generated than major differences in the actual research.

A Basic Science Orientation

A basic science, or subject matter, orientation to inhalation toxicology is depicted in Fig. 1. This orientation emphasizes the importance of understanding (1) the physical and chemical characteristics of what is in the air and (2) the normal biology of the respiratory tract as a basis for understanding how materials in the air interact with the respiratory tract and how these interactions progress via delivered dose to critical biological units to the pathogenesis of respiratory disease. This approach has been used to guide inhalation toxicology research at the Inhalation Toxicology Research Institute, the Chemical Industry Institute of Toxicology, and other institutions that have mounted broad-based, multidisciplinary team efforts to understand the toxicity of airborne materials.

A schematic representation for assessing the toxicity of inhaled fibers is depicted in Fig. 3 (McClellan and Hesterberg, 1994), which is an expanded version of Fig. 1. This figure contains more detail than Fig. 1 and helps one appreciate the fact that these schemes are only general models of complex overall process(es). As techniques become more sophisticated with time, it is expected that more detailed models will be created that give added insight into the critical steps in the processes by which airborne toxicants can cause disease. The present models convey both deterministic and probabilistic linkages for various steps between exposure, dose to critical biological units, and response. In my opinion, an important development in future models will be the role of repair processes. These processes are expected to be of major importance in

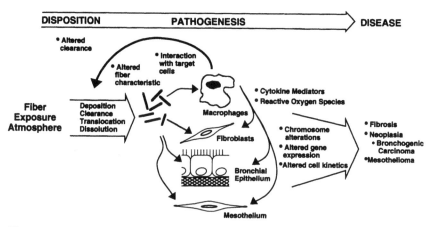

Figure 3 A schematic representation of an expanded exposure-dose-response approach to understanding the human health risks of exposure to fibers. From McClellan and Hesterberg (1994).

determining the nature of exposure-response relationships at low levels of exposure, where the prevalence of disease associated with airborne toxicants blends with the prevalence of disease of spontaneous or unknown origin.

This basic science orientation also emphasizes the extent to which successful research in inhalation toxicology brings together the physical and biological sciences. The quality of any study depends on both kinds of inputs. If inadequate attention is given to the characterization of either the airborne toxicant or the respiratory tract, then it is unlikely that the study can be meaningfully interpreted.

A Risk Orientation

Inhalation toxicology has a very pragmatic orientation to developing information that can be used to assess and limit the health risks of airborne materials. In recent years, this orientation for inhalation toxicology, and indeed for toxicology in general, has become more formalized. A major stimulus to this formalization has been the move toward increased quantitation of health risks and documentation of the basis for the quantitation.

Landmark documentation of the multiphased approach to research, risk assessment, and risk management was developed by a committee of the National Academy of Sciences (NAS) and the National Research Council (NRC) (National Academy of Sciences, 1983). More recently, another NAS-NRC Committee has reviewed the methodologies used to assess health risks of airborne hazardous materials (National Academy of Sciences, 1994). The interrelationships among research, risk assessment, and risk management are illustrated in Fig. 4, which is taken from an appendix to the 1994 NAS-NRC report authored by McClellan and North (1994). Each of these activities has a role in the overall process. Research provides essential information for doing a

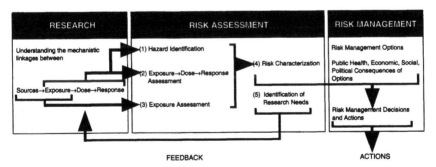

Figure 4 A schematic representation of the linked research-risk assessment-risk management paradigm (McClellan and North, 1994). (Reprinted with permission from Science and Judgment in Risk Assessment. © 1994 National Academy of Sciences. Courtesy of the National Academy Press, Washington, DC.) (Adapted with permission from Risk Assessment in the Federal Government: Managing the Process. © 1983 National Academy of Sciences. Courtesy of the National Academy Press, Washington, DC.)

scientifically valid risk assessment. The risk assessment provides a structured approach to integrating scientific information on risks so that it can be considered with other information in making risk management decisions. An equally important role of the risk assessment process, which is usually not recognized, is the identification of information gaps. Research that addresses these information gaps will serve to reduce the uncertainty in risk assessment. Risk assessment is therefore a potentially very powerful tool for research planning.

Major contributions of inhalation toxicology to the risk assessment process are in the hazard identification and exposure-dose-response elements of the process. It is increasingly recognized that information in these areas must be developed using multiple approaches: epidemiological studies, investigations with whole animals, and research using macromolecular, cellular, and tissue systems. This information must ultimately be joined with exposure assessment information to yield a characterization of risk.

SOURCES OF INFORMATION

Typically, our interest is in the effects of toxic agents on humans. Thus, if exposed human populations are available, the opportunity for epidemiological studies must be fully exploited. Unfortunately, even when exposed human populations are available, there may still be major obstacles to acquiring adequate information for developing quantitation risk assessment without using data from other sources. For example, population size may be inadequate, good exposure data may not be available, the effect of confounding factors such as cigarette smoking may mask subtle effects of other toxicants, the effects of other toxicants are rarely unique, and there is a long time interval between exposure and occurrence of some effects such as cancer. Human exposures may not have occurred with some materials, especially those newly synthesized, thereby precluding epidemiological studies.

In a few instances, the conduct of controlled human-exposure studies may be ethical. Such studies are ethical when there is a high degree of confidence that the material of interest is not capable of producing irreversible health effects such as cancer at the exposure levels being studied. The basis for the conduct of such studies is further reinforced if the material is a common pollutant being studied at exposure levels similar to those found in the ambient environment. Examples are studies with ozone, oxides of nitrogen, or acidic aerosols. Controlled human-exposure studies are also ethical when a material such as a pharmaceutical agent is intended for therapeutic application in humans. In addition to studying specific agents to obtain information on those agents, other studies may be done to provide insight into generic processes. Examples of this kind are apparent in the chapter describing the deposition and retention of inhaled particles in humans.

In the absence of fully adequate data for risk assessment or for understanding the mechanisms of action of toxicants, *in-vitro* studies using molecules, cells, or tissues from people or laboratory animals and *in-vivo* studies using laboratory animals are necessary.

A wide array of specimens is available for use in *in-vitro* studies, including tissues or cells taken directly from the respiratory tract or cell lines derived from the respiratory tract. The two best examples are tracheal-bronchial epithelium or macrophages. Such specimens can be used as the source of specific cells or enzymes that can be studied intact. Such studies may extend from evaluating the metabolic capacity of the cells or tissues to assessing specific effects such as cytotoxicity, phagocytic capacity, mutagenicity, or cell-transforming ability of specific agents. *In-vitro* studies are typically of short duration, can be conducted under rigorously controlled conditions, and can be used to assess a relatively large number of experimental variables than can be economically studied in the intact animal. The study of cells or tissues outside the body removes them from the complex homeostatic environment of the animal. This is an advantage in allowing the evaluation of certain functions in isolation. On the other hand, the results of such studies must be interpreted with special care relative to their significance in the whole animal or person.

In *in-vivo* studies with laboratory animals, the animals are essentially used as surrogates for humans. In doing so, it is fully recognized that no single laboratory animal species serves as an ideal surrogate for humans. In interpreting studies conducted with rats, for example, it is important to recognize that the rat is *not* a miniaturized Homo sapiens. All such studies have to be interpreted with the recognition that each species has certain characteristics similar to and others that are different from those of humans. The challenge is to take advantage of both the similarities and differences in estimating responses in humans.

ORGANIZATION OF THE BOOK

The book is organized into five sections in addition to the introductory section (Part One). In a sense, these five sections closely parallel the various elements of inhalation toxicology depicted in Fig. 1. Part Two is concerned with what is in the air, and Part Three reviews the basic biology of the respiratory tract. These sections provide the essential underpinnings for considering the disposition of inhaled materials, the topic covered in Part Four. Part Five proceeds logically to an evaluation of responses of the respiratory tract to inhaled toxicants. The last section (Part Six) attempts to integrate these various topics by illustrating their use in assessing the health risks of inhaled toxicants.

The three chapters on inhalation exposure in Part Two illustrate the importance of strong inputs from physical science and engineering into inhalation toxicology. These disciplines provide the essential scientific basis for the generation and characterization of airborne atmospheres and the delivery of these atmospheres in a reproducible manner to experimental subjects. It has been noted, only partially in jest, that a successful inhalation toxicologist must combine some of the skills of both plumbers and ventilation engineers. This is apparent in considering the chapter by Cheng and Moss, who review the various exposure system options that are available and emphasize the use of an integrated systems approach.

The complementary chapters by Wong and by Moss and Cheng in Part Two review the approaches available for generation and characterization of test atmospheres containing gases and vapors or particles, respectively. These chapters emphasize the extent to which the generation and characterization of each test atmosphere poses unique challenges. In short, there is no universal generation and characterization system that can be taken off the shelf and put into use when a test atmosphere of a new material is being evaluated to determine its toxicity.

The three chapters in Part Three on the basic biology of the respiratory tract provide some of the essential information that must be considered in assessing the toxicity of airborne materials. In a sense, this section is concerned with describing the target tissue, cells, and macromolecules for the toxicants delivered with the methodologies described in Part Two. Phalen, Yeh, and Prasad describe the basic morphology of the respiratory tract with special attention given to its physical dimensions. The morphology is essential information for understanding the disposition of inhaled materials, especially particles or fibers. Thus, this chapter is an essential prelude to the later chapters that consider the deposition of inhaled particles and clearance processes. In Chapter 6 of this section, Fisher discusses the important metabolic properties of the lung. In Chapter 7, Dahl calls attention to the substantial metabolic activity of various regions of the respiratory tract. This metabolic capability is critical for olfaction and detoxification of inhaled materials. In some cases, the metabolic activity can also serve to increase the toxicity of an inhaled chemical by producing a toxicologically more active metabolite.

Part Four considers the disposition of inhaled material. It begins with a chapter by Schlesinger, who considers the deposition and clearance of particles. Schlesinger emphasizes the role of aerodynamic and diffusional properties of particles in influencing their deposition. Further, he calls attention to important species differences in particle deposition and retention. In Chapter 9, Snipes extends this theme by reviewing the long-term retention of particles and presents models of the predicted patterns of accumulation of inhaled particles of various solubilities in different species. This presentation provides a strong argument for considering the retained lung burden of material rather than exposure concentration when scaling exposure-dose-response relationships between species.

In Chapter 10, Miller and Kimbell address the uptake of inhaled reactive gases. The parameters of solubility and reactivity are shown to have a significant role in determining the regional distribution of inhaled gases. Significant species differences in uptake of specific gases are illustrated, again emphasizing the importance of understanding these differences in extrapolating from laboratory animals to humans.

In Chapter 11 of Part Four, Bond and Medinsky note the range of factors that may influence the disposition of inhaled organic compounds. Particular attention is given to the influence of particle association, which has been shown to result in prolongation of the clearance half-time for some polycyclic aromatic hydrocarbons when associated with carbonaceous particles.

In Part Five, seven chapters illustrate the broad range of responses of the respiratory tract to inhaled material. The section starts with a chapter by Hahn, who reviews the carcinogenic responses of the respiratory tract. He illustrates the various types of tumors that are produced in laboratory animals by a wide range of toxic materials. In Chapter 13, Mauderly describes the assessment of pulmonary function and how it is affected by inhaled materials. The chapter emphasizes the extent to which detailed functional evaluations can be carried out in laboratory animals similar to what is done with humans. This is followed by a chapter by Bice, who describes immunologic responses of the respiratory tract to inhaled materials. In Chapter 15 of Part Five, Henderson describes how biomarkers obtained by lavage of the respiratory tract can be used to evaluate the health status of the respiratory tract. Bice's strong cellular orientation is carried forward in Chapter 16 by Driscoll, who describes the role of cytokines, produced by cells within the lung, in mediating inflammatory and fibrotic responses in the lung. In the next chapter, Weiss and Rahill address a topic that has received only limited attention by experimenters, the use of behavioral changes as an end point for inhaled toxicants. This lack of attention is surprising, since almost all of us are aware from personal experience of instances in which airborne materials influenced our own behavior. Undoubtedly, this area will receive more attention in the future as collaboration is established between inhalation toxicologists and behavioral scientists. In the last chapter, Dungworth, Hahn, and Nikula review the noncarcinogenic responses of the respiratory tract, emphasizing morphological changes.

The last section (Part Six) of the book is my chapter on the broad topic of risk assessment. It reviews the main steps in the risk assessment process and shows how information from inhalation toxicology studies is integrated with information from other sources to develop risk assessments for airborne materials. The chapter illustrates how data from diverse sources provide useful information for assessing human health risks from exposure to airborne materials.

SUMMARY

This chapter has considered three topics:

1 the basis for concern about the toxicity of airborne materials,
2 the use of both a basic sciences and a risk assessment orientation to inhalation toxicology, and
3 the contents of this book.

The first portion of the chapter emphasizes the historical origins of inhalation toxicology. History clearly shows that the roots of this specialty derive from concern for how airborne materials influence human health. This historical background is reinforced by consideration of contemporary health statistics, which indicate the extent to which diseases of the respiratory system account for substantial human morbidity and mortality. The role of cigarette smoking as a factor in respiratory disease, and especially in lung cancer, is increasingly recognized. The roles of other airborne toxicants in causing respiratory disease, either alone or in combination with cigarette smoke, are much less clear.

The second portion of this introductory chapter calls attention to the use of a risk assessment orientation for identifying research needs and for integrating information from multiple sources. It draws attention to the use of information from multiple approaches such as epidemiological studies, investigations with whole animals, and research using macromolecular, cellular, and tissue systems.

The third portion of the chapter reviews the organization and contents of the book. The discussion of the first chapters in the book underscores the extent to which inhalation toxicology is dependent upon bringing together input from the physical and biological sciences. With introduction of the biology, an additional theme emerges—the importance of a comparative species orientation in understanding the disposition and effects of inhaled materials. In considering the effects of inhaled materials, it is readily apparent that we are concerned with an array of responses from reversible morphological and functional changes to irreversible alterations, including the development of cancer. A further theme that emerges is the importance of quantitation in measuring exposures, determining the dose of toxicant to critical macromolecules and cells, assessing responses, and developing relationships among exposure-dose-response that are suitable for assessing the human health risks of exposure to airborne materials.

ACKNOWLEDGMENTS

The author gratefully acknowledges the many useful interactions with his colleagues at the Chemical Industry Institute of Toxicology in recent years and earlier at the Inhalation Toxicology Research Institute. Their contributions to the development of the concepts presented in this chapter and volume are immeasurable.

REFERENCES

Amdur, MO, Doull, J, Klaassen, CD (eds): Casarett and Doull's Toxicology: The Basic Science of Poisons, Fourth Edition. New York, NY: Pergamon Press, 1991.

Dinker, P, Hatch, T: Industrial Dust: Hygienic Significance, Measurement, and Control. New York, NY: McGraw-Hill, 1936.

Hatch, TF, Gross, P: Pulmonary Deposition and Retention of Inhaled Aerosols. New York, NY: Academic Press, 1964.

Lubin, JH, Boice, JD, Jr, Edling, C, Hornung, RW, Howe, G, Kunz, E, Kusiak, RA, Morrison, HI, Radford, EP, Samet JM, Tirmarche, M, Woodward, A, Yao, SY, Pierce, DA: Lung Cancer and Radon: A Joint Analysis of 11 Underground Miner Studies. Report No. 94-3644. U.S. National Institutes of Health, Besthesda, MD, 1994.

Lubin, JH, Steindorf, K: Cigarette use and the estimation of lung cancer attributable to radon in the United States. Radiat Res 141:79–85, 1995.

McClellan, RO, Hesterberg, TW: Role of biopersistence in the pathogenicity of man-made fibers and methods for evaluating biopersistence: A summary of two round-table discussions. Environ Health Perspect 102(Suppl 5):277–283, 1994.

McClellan, RO, North, DW: Making full time use of scientific information in risk assessment (Appendix N-2). In Science and Judgment in Risk Assessment, Committee on Risk Assessment of Hazardous Air Pollutants, National Research Council. National Academy Press, Washington, DC, 629–640, 1994.

National Academy of Sciences, National Research Council, Committee on the Institutional Means for Assessment of Risks to Public Health: Risk Assessment in the Federal Government: Managing the Process. Washington, DC: National Academy Press, 1983.

National Academy of Sciences, National Research Council, Committee on Passive Smoking 1986: Environmental Tobacco Smoke: Measuring Exposures and Assessing Health Effects, Washington, DC, National Academy Press, 1986.

National Academy of Sciences, National Research Council: Science and Judgment in Risk Assessment. Washington, DC, National Academy Press, 1994.

Nero, AV, Schwehr, MB, Nazaroff, WW, Revzan, KL: Distribution of airborne radon-222 concentrations in U.S. homes. Science 234:992–997, 1986.

Samet, JM: Environmental Tobacco Smoke in Environmental Toxicants: Human Exposures and Their Health Effects, edited by M Lippman, New York, NY: Van Nostrand Reinhold, pp. 231–265, 1992.

Selikoff, IJ, Hammond, EC: Editorial: Asbestos and smoking. JAMA 242:458–459, 1979.

Stannard, JN: Radioactivity and Health: A History (Baalman, R, Jr, ed.). Richland, WA: Pacific Northwest Laboratory, 1988.

U.S. Department of Health and Human Services. The Health Consequences of Involuntary Smoking. Washington, DC: U.S. Government Printing Office, 1986.

Vital and Health Statistics: Current Estimates from the National Health Interview Survey, 1993. DHHS Publication No. (PHS) 95–1518, 1994.

Vital Statistics of the United States 1990. Volume II—Mortality. U.S. Department of Health and Human Services, 1994.

Whittemore, AS, Millan, A: Lung cancer mortality among U.S. uranium miners: A reappraisal. JNCI 71:489–499, 1983.

Wingo, PA, Tong, T, Bolden, S: Cancer statistics, 1995. CA 45:8–30, 1995.

Part Two
Inhalation Exposure

Chapter Two

Inhalation Exposure Systems

Yung-Sung Cheng and Owen R. Moss

INTRODUCTION

Inhalation is a major route through which material enters the body. Inhalation experiments using laboratory animals are carried out under controlled conditions to assess the toxicity of aerosols, gases, and vapors. Chapters 2–4 of this monograph describe physical and engineering aspects of the inhalation exposure. This chapter will discuss components of inhalation exposure systems, exposure chambers and their performance, environmental conditions, and facility requirements. The design and performance of exposure chambers have been reviewed by Drew and Laskin (1973), Willeke (1980), Leong (1981), MacFarland (1983, 1987), Phalen (1984), and Salem (1987). In this chapter, we will briefly review various systems and focus on the design and operation of exposure systems being used for inhalation toxicity studies.

Type of Inhalation Exposure

Inhalation experiments are performed for various purposes including toxicity and pharmacokinetic studies of airborne materials and airway challenge tests for drugs and allergens. Different endpoints are used to determine the inhalation toxicity of airborne materials. The LC_{50} study uses lethality, whereas a bioassay may use carcinogenicity, histopathology, hematology, biochemistry, and pulmonary function. Their selection dictates the inhalation protocol in terms of exposure duration, exposure concentrations, and number of animals required. Pharmacokinetic experiments are used to study deposition, retention, distribution, and metabolism of inhaled materials. For such studies, it is essential to accurately account for the amount of material inhaled, deposited, and retained. Animal breathing patterns are often monitored on-line, so that the exposure system must be specifically designed to provide these measurements. The airway challenge (bronchoprovocation) procedure is used to measure acute airway response to bronchoactive drugs or allergens; for this procedure, animals are usually housed in a plethysmograph which is used as the inhalation chamber. The animal is exposed to a series of progressively increasing concentrations of test material until positive indication of increased airflow resistance in the lung is measured (Cropp et al., 1980; Gerlach et al., 1989).

The duration of an inhalation exposure is determined by the type of protocol. An acute exposure is usually a single, short-term exposure, ranging from a few minutes to several hours to relatively high concentrations of material. A typical LC_{50} protocol consists of one, 4- to 6-hour exposure; a repeated exposure protocol is usually 14 or 28 days; a subchronic study is usually 13 weeks; and a chronic study lasts from 1 year to the life span of the exposed animal, typically 24 exposure months (EPA, 1982; OECD, 1981; Huff et al., 1986).

Subchronic and chronic exposures are either intermittent (6 to 7 hours/day; 5 days/week) or continuous (20 to 24 hours/day; 7 days/week). The intermittent mode simulates occupational exposures, whereas the continuous mode simulates environmental exposures. Generation, delivery, and control systems for intermittent exposures are often simpler than those in the continuous systems. Food and water are usually withheld during intermittent inhalation exposures. In continuous exposures, food and water are provided in the exposure chamber, so contamination of food with the test material is inevitable. An interval between exposures is usually needed for animal care and maintenance; however, some systems allow animal care during exposure (Thomas, 1965). Because of the long exposure duration in continuous studies, the system is often unattended and therefore requires sophisticated automatic monitoring and operation.

Test Materials

Airborne test materials include gases or vapors, liquid droplets or solutions, and dry powder or fibers. Generation and characterization of the material depend upon its physical form. Most materials used in inhalation exposures are single compounds, but they can be a mixture of two or more compounds. Complex mixtures, such as automobile exhaust (Cheng et al., 1984), coal-liquid distillation products (Springer et al., 1986), and cigarette smoke (Baumgartner and Coggins, 1980; Chen et al., 1992), require a specially designed system for generating, delivering, and characterizing the test atmosphere.

Radioactive and nonradioactive test materials are also used in inhalation experiments. Aerosols containing radionuclide are of environmental concern, and radiolabeled compounds are frequently used in pharmacokinetic studies to trace and quantify the fate of inhaled materials. Preparation, generation, and detection of radioactive exposure atmospheres require special techniques (Newton et al., 1980; Bond et al., 1986), and the exposure systems must be designed to minimize release of the materials.

EXPOSURE SYSTEMS

Inhalation exposure systems are designed and operated to provide test atmospheres to the breathing zone of laboratory animals or human volunteers. Based on how the test material is delivered, there are three basic types of exposure systems: static, with no aerosol flow in the chamber; recirculating, with a closed-loop; and dynamic, with a single-pass flow-through system.

Static Exposure System

In this system, the test atmosphere is produced by introducing a finite amount of material into a closed exposure chamber. The static system has two advantages: (1) consumption of test material is minimal, and (2) continuous generation of aerosols and gases is not required. The disadvantages of this system are depletion of oxygen concentration, rising temperature and decreasing airborne concentration. These factors limit the practical exposure duration to about 1 hour. The decay of toxicant concentration in the chamber is due to the deposition or absorption on the surfaces of exposure apparatus (Chen et al., 1992; Hefner et al., 1975), and deposition in the experimental objects.

Static exposure systems have been improved (Bolt et al., 1976; Lutz and Schlatter, 1977; Filser and Bolt, 1981). A glass desiccator has been used as the exposure chamber with soda lime placed inside the chamber to absorb exhaled CO_2 (Fig. 1). Fresh O_2 is available on demand, and the chamber temperature can be reduced by placing the chamber in a water bath (Lutz and Schlatter, 1977). By using an all-glass exposure system, the decay of test vapors and gases in the absence of animals has been held to a minimum for some chemicals (Bolt et al., 1976). Approximately first-order decay has been attained for other chemicals, which are absorbed by the soda lime (Filser and Bolt, 1979). Under these conditions, exposures in static chambers have been

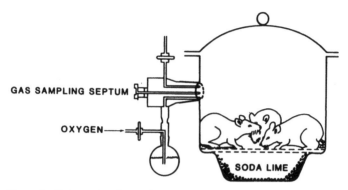

Figure 1 Schematic of a static exposure system using a glass desiccator (from Filser and Bolt, 1981, with permission).

conducted for up to 30 hours. These improved systems have been used extensively to study the uptake and metabolism of organic vapors and gases.

Recirculating Exposure System

A recirculating exposure system is a static exposure system with a circulating loop that remove waters and CO_2 and adds fresh O_2 (Jaeger et al., 1977; Anderson et al., 1979). With the loop, loss of gases or vapors in the system have been slightly higher than in the improved static system (Jaeger et al., 1977; Anderson et al., 1979). This type of recirculating system is similar to the improved static system in that toxicant cannot be added after the exposure begins. Neither the static nor the recirculating system maintains a steady concentration in the chamber; therefore, they are most useful in pharmacokinetic studies using radiolabeled gases and vapors. After correcting for material, the decay curve of the toxicant concentration can be analyzed with physiologic models to obtain detailed information on absorption, distribution, and metabolism (Anderson et al., 1980; Filser and Bolt, 1981).

Dynamic recirculating systems in which test material can be replenished have been described (Morken, 1955; Paustenbach et al., 1983; Newton et al., 1992). These systems maintain a steady concentration with minimum test material. Fig. 2 depicts such a system including a generator of radon gas, an aging chamber to allow ingrowth of radon progeny, an exposure chamber, and a filter and reconditioning system (Newton et al., 1992). A dedicated PC is used to monitor and control the operation of the system.

Dynamic Exposure System

In a dynamic system, the test atmosphere is continuously delivered to and exhausted from the animal chamber in a flow-through manner; test material

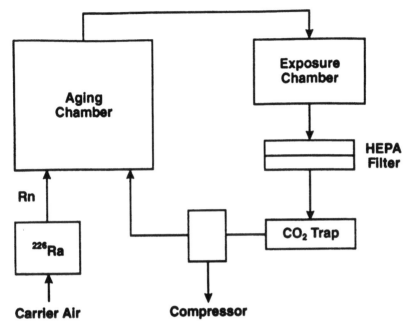

Figure 2 Schematic of a recirculating exposure system (from Newton et al., 1992, with permission).

is not recirculated. After an initial buildup of the toxicant, a stable chamber concentration is maintained if the generation rate remains constant and is usually reproducible on a daily basis once conditions have been optimized. Prediction of this equilibrium concentration requires accurate information on generation rate, loss of test material in various parts of the system, and flow rates (Fraser et al., 1959).

In this system, temperature and relative humidity (RH) are well regulated, oxygen is automatically replenished, and CO_2, ammonia, and other vapors are removed continuously. This system can be used for both toxicological and pharmacokinetic studies of aerosols and gases. The dose delivered is easier to estimate from the steady airborne concentrations, and the environmental conditions in the chamber allow long-term studies with a large number of animals. However, a dynamic exposure system consumes larger quantities of test material than either the static or recirculating systems. Pumps, fans, compressed air, and vacuum source are also required.

A dynamic exposure system (Fig. 3) consists of six essential components:

1   Generation   The generator produces aerosol, gas, or vapor from the test material. Various generating techniques are described in Chapter 3 (aerosols) and Chapter 4 (gases and vapors).

2   Treatment   A test atmosphere from the generator is frequently

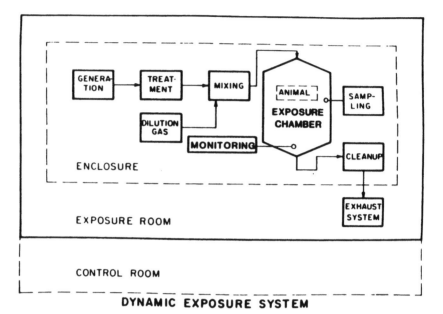

DYNAMIC EXPOSURE SYSTEM

Figure 3 Components of a dynamic exposure system.

heated to ensure complete evaporation. High temperatures are sometimes used to facilitate chemical decomposition, for example, to form metal oxide particles (Raabe et al., 1971; Kanapilly et al., 1978). Aerosols generated from powder dispersion and wet nebulization techniques are often highly charged (Chow and Mercer, 1971; Yeh et al., 1988). Removing the charge is important because highly charged particles tend to deposit within the exposure system before reaching the animals. If inhaled, charged particles may differ from an electrically neutral aerosol through altered deposition patterns and efficiency in the respiratory tract. Radioactive sources (for example, Kr-85, Am-241, and Ni-63) can reduce the charge carried by the aerosol (see Chapter 4).

   3 Dilution   Filtered dilution air is added to the aerosol or vapor stream to achieve the desired concentration and reduce the temperature of the test material after heat treatment. The dilutor should produce a well-mixed test atmosphere in the delivery lines.

   4 Chamber   The chamber houses the animals to be exposed and provides a means of delivering the test atmosphere to the breathing zone. It also provides a stable environment that meets the animal welfare standard for safe housing conditions.

   5 Exhaust Clean-up   Exhaust from the chamber may contain hazardous or radioactive materials that must be diluted with clean air to environmentally safe levels or removed before being exhausted to the atmosphere. HEPA filters are generally used to remove particles; activated charcoal or water sprays are used to remove vapors and gases. Oxidizers or incinerators are sometimes used to convert hazardous vapors into inert gases.

6 Monitoring Devices   Monitors measure the test atmosphere for concentration, chemical composition, and particle-size distribution (of aerosols), as well as flow rate, temperature, and relative humidity in the exposure chamber.

Safety Considerations for Inhalation Systems

The most important consideration in designing an inhalation system is to ensure a safe working environment for laboratory personnel and to prevent the discharge of pollutants to the outside atmosphere. Exposure of the operator and animal-care staff to toxicants must be minimized by placing part of exposure system (for example, the chamber and generator), or the entire system in a ventilated enclosure such as a fume hood or a glovebox. Pressure gradients are maintained between the chamber, chamber enclosure (if one exists), and laboratory in which it is operated so that leakage from the chamber is prevented (Fig. 3). For highly toxic materials, a separate control room with readouts from instruments monitoring the exposure system can further reduce the potential for human exposure. Workers who must enter the laboratory may be required to wear respirators, eye protection, special clothing, gloves, and shoe covers.

EXPOSURE CHAMBERS

Chambers have been designed in various sizes and shapes (Fraser et al., 1959; Jemski and Phillips, 1965; Drew and Laskin, 1973; MacFarland, 1983; MacFarland, 1987). There are two basic types of design: whole-body and nose/head-only chambers. In the whole-body chamber, the animals are immersed in the exposure atmosphere, whereas in the nose- or head-only exposure chamber, primarily the head or nose is exposed. As expected, deposition of aerosols or vapors on animal fur is higher in a whole-body chamber (Wolff et al., 1982). Skin contamination may increase deposition and absorption of test material through ingestion (Langård and Nordhagen, 1980; Chen et al., in press). Animal handling is minimized in the whole-body exposure if the animals are housed in the chamber; nose-only exposures are generally more labor intensive if repeated exposures are required. In selecting a particular chamber for the study, the species, material availability, exposure duration, toxicity of material, and animal monitoring requirements should be considered. Table 1 lists the capacities of whole-body and nose-only chambers that are currently used.

Animal Species and Number

The species and numbers of animals to be used in a study determine the size and numbers of exposure chambers required. Chambers are designed to accommodate various species, including rodents, dog, and primates. The

Table 1 Capacity of Inhalation Exposure Chambers

Name	Type	Animal Capacity		Volume (m³)	Flow Rate (L min⁻¹)
		Mice	Rats		
Hinners	Whole-body vertical flow		16	0.32	100–200
H-2000	Whole-body vertical flow	180	72	1.0	300–600
H-1000	Whole-body vertical flow	360	144	1.7	150–300
Ferin & Leach	Whole-body horizontal flow		72	3.0	1600–3200
Active Dispensive	Whole-body diagonal flow		30	0.15	30–110
Lovelace	Nose only	80/96	80/96	0.01	10–40
Cannon	Nose only	40	40	0.0025	10–40

number of animals is determined by the choice of endpoints to be evaluated, whether one or both sexes are used, and considerations of statistical analyses. For a life-span study, the survival curve of the test animals and spontaneous incidence of certain diseases should also be considered.

Amount of Toxicant Available

The amount of toxicant available and its cost may be major considerations in the choice of a chamber. In a dynamic exposure system, total consumption of the test material is directly proportional to the concentration and flow rate through the chamber. Only a small portion of the test material is inhaled and deposited in the animals; most test material is not used unless it is reclaimed. The chamber volume and flow rater for a nose/head-only chamber are one order of magnitude smaller than a whole-body exposure chamber accommodating the same number of animals and thus requires less test material.

Duration of Exposure

In a whole-body chamber, animals are usually housed in individual cages and can be exposed to the toxicants continuously or intermittently for long periods of time under well-regulated environmental conditions. On the other hand, a restraining device is needed to keep animals in the proper position in a nose/head-only chamber; therefore, this approach is most useful for single or repeated exposures of up to several hours. With experienced personnel, a nonoffensive test material and careful animal care, subchronic and even chronic nose-only exposures of up to 2 years have been carried out (Smith et al., 1981).

Animal Monitoring

Monitoring physiological conditions (such as lung functions) and obtaining clinical chemistry samples from the test animals are essential in inhalation studies. Respiratory parameters are important endpoints for toxicity and sensitization tests; they are also required to estimate the deposition of material in the respiratory tract. Absorption rate of test material into the circulatory system is a key measurement in pharmacokinetic studies. Plethysmographs are often used to monitor respiration during exposure, and catheters are used to take blood samples during some studies using nose-only exposure chambers. It is usually more difficult to monitor animals in whole-body exposure systems.

Whole-body Exposure Chambers

Small Chambers   Whole-body exposure chambers have been used for the past 100 years (Fraser et al., 1959; MacFarland, 1983). Early chambers were made of commercially available containers such as jars, desiccators, columns, cylinders, and aquaria from glass, lucite, or other transparent material for easy observation. Glass battery jars were mounted horizontally to form a small chamber (Spiegl et al., 1953; P'an and Jegier, 1970). Modified versions of jar-type chambers are still part of recirculating exposure systems (Anderson et al., 1979; Paustenbach et al., 1983). Desiccators require minimum modification for exposing small animals (Comstock et al., 1962; Lutz and Schlatter, 1977; Jaeger et al., 1977), and are used most often as static exposure chambers (Bolt et al., 1976; Filser and Bolt, 1981). Lucite cylinders with top and bottom end caps have been used as portable exposure chambers (Laskin and Drew, 1970; Riley, 1986). Animals have been loaded and unloaded through the top cap and the chamber opened in a vertical-flow fashion. Large aquaria (390 liters) are also used as exposure chambers (Barrow and Steinhagen, 1982).

These chambers are inexpensive, easy to construct, and available in many different sizes. Most of these systems are custom-made in individual laboratories, with little information available on the chamber performance. They are useful for exposing a few rodents to gases and vapors when distribution in the chamber is uniform (Barrow and Steinhagen, 1982; Riley, 1986). However, distribution of aerosols within the chamber is usually not as uniform as it is for gases or vapors.

Large-Scale Chambers   Large-scale, whole-body exposure chambers are usually made of stainless steel because it is easy to clean and is resistant to corrosion. Many chambers are box-shaped with a cubic or hexagonal cross section. A few chambers are spherical or hemispherical and have minimum surface-to-volume ratio (Thomas, 1965; Stuart et al., 1969). Chambers have largely been designed based on the requirement for good aerodynamic performance for aerosols and for sufficient capacity to expose large numbers of animals. Chambers with vertical flow, horizontal flow, and diagonal flow have been developed.

An early design is the University of Rochester chamber (volume, 1.3 m³) with a hexagonal cross section (Fig. 4). This chamber has a tangential upper inlet, an upper cone to allow even distribution, an exposure section, and a lower cone for exhaust. It was designed in 1950 to provide uniform mixing within the exposure volume during chronic exposures of animals to aerosolized material (Leach and Spiegl, 1958; Leach et al., 1959). It was originally used for simultaneous exposure of 4 monkeys, 8 dogs and 40 rats using two levels of cages. Animals could be observed easily with windows that formed the chamber walls.

Vertical-flow chambers with cubic exposure sections, tangential inlets, and pyramid-shaped upper and lower sections (Fig. 5) have been described by Hinners et al., 1968. Four chambers with volumes from 0.32 to 2.8 m³ were designed. A similar chamber of rectangular shape was designed by Laskin et al., 1969; it had a straight inlet and 1.3 m³ volume. The Hinners and Laskin chambers are easier to fabricate than the Rochester chamber and appear to have similar performance characteristics.

The Hinners and Laskin chambers are still used, although with some modifications, including larger volumes (Carpenter and Beethe, 1978; Schreck et al., 1981). Exposure chambers constructed at Dow Chemical Co. were designed as small rooms, having a door on each end, one leading to a clean area and the other exiting to the contaminated corridor. These "room" chambers have a vertical-flow design, with volumes of about 18 m³, large enough to hold six racks of animal cages, which were rolled in and out of the chambers every day. The chamber walls were made of cement block coated with epoxy (MacEwen, 1978).

The major disadvantage with Hinners- or Rochester-type chambers is that no catch pans were used to separate animals in multiple levels because of concern for uniform mixing of test material throughout the chamber volume. Without catch pans, animals on lower levels are contaminated by excreta from above. In subchronic and chronic studies, additional animal holding facilities are required, and animals must be transported daily to exposure chambers.

A vertical-flow chamber has been developed using catch pans to improve aerosol uniformity inside the chamber (Moss et al., 1982) (Fig. 6). The total volume of the chamber is 2.3 m³ of which animals and caging occupy 1.7 m³. There are three levels of caging; each level is split into two tiers which are offset from each other and from the chamber walls. Drawer-like, stainless-steel-wire cage units, composed of individual animal compartments, are sus-

Figure 4 University of Rochester hexagonal exposure chamber.

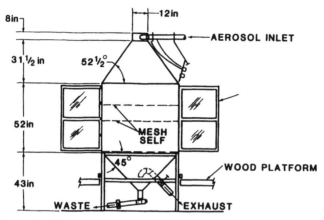

Figure 5 Hinners exposure chamber (from Hinners et al., 1968, with permission).

pended in the space above each tier and away from the wall. Stainless-steel catch pans are suspended below each cage unit. The chamber was designed using flow visualization techniques in a scale model to determine how to best maintain uniform aerosol distribution throughout the chamber when the catch pans are in position (Fig. 6). Incoming air is diverted by a distribution plate under the flow entrance so that the air flows along the inner surface of the top transition piece to the vertical walls. An aerosol or gas is directed toward the cage unit by free-standing eddies formed by the catch pan. Each stationary catch pan acts like a propeller blade to distribute the test material uniformly (Griffis et al., 1981; Cheng et al., 1984, 1989; Yeh et al., 1986). Animals are housed in individual compartments of cages. The space require-

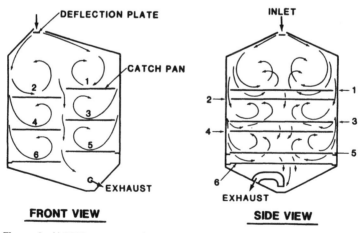

Figure 6 H-2000 exposure chamber.

ments for different species of rodents (Table 2) follow the guidelines published by the National Institutes of Health (1985). This chamber (H-2000 Lab Products, Inc., Maywood, NJ) provides enough space for simultaneous exposure of 144 rats (body weight less than 400 g), 360 mice, or 60 guinea pigs. A smaller chamber (H-1000, Lab Products, Inc., Maywood, NJ) with half the capacity is also available.

Catch pans are also used in horizontal-flow chambers with multiple-tier cages. Fig. 7 shows such a chamber with a rack holding 24 cages (capacity is 1 to 3 rats per cage) in 4 tiers (Ferin and Leach, 1980). A manifold connects four tubes with the aerosol generator; a perforated plate (diffuser) in front of the cages produces laminar uniform flow in the horizontal direction. An evaluation of performance showed uniform aerosol distribution throughout the chamber. Other horizontal flow chambers for rodents and dogs have been reported (Carpenter and Beethe, 1980; Hemenway and MacAskill, 1982). Heinrich et al. (1985) described a large inhalation facility having rows of large exposure rooms (12 m³) with horizontal flow design.

Most vertical-flow and horizontal-flow chambers have conical or pyramidal sections for entrance and exhaust to provide uniform aerosol flow. This design increases the chamber volume but is more expensive to fabricate. The H-1000 was simplified with a flat top and bottom. A compact rectangular chamber (0.21 m³ volume) was designed using manifolds at the top and bottom to distribute test material to 10 rats or 40 mice (Marra and Rombout, 1990). Several chambers can be stacked up to better use the floor space in an exposure room. An active-dispersion, diagonal-flow system that uses a

Table 2   Minimum Space Recommendations for Rodents

Animals	Weight (g)	Floor Area (cm²)	Height (cm)
Mice	< 10	39	12.7
	10–15	52	12.7
	15–25	77	12.7
	> 25	96	12.7
Rats	< 100	110	17.8
	100–200	148	17.8
	200–300	187	17.8
	300–400	258	17.8
	400–500	387	17.8
	> 500	451	17.8
Hamsters	< 60	65	15.2
	60–80	84	15.2
	80–100	103	15.2
	> 100	123	15.2
Guinea Pig	≤ 350	387	17.8
	> 350	651	17.8

NIH (1985).

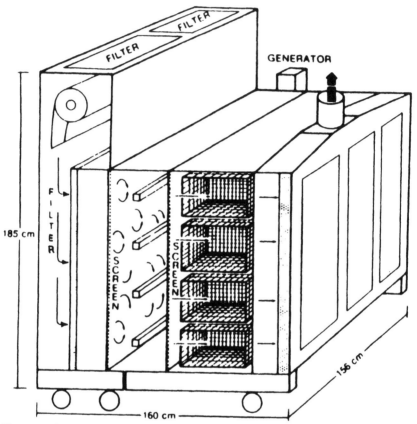

Figure 7 A horizontal flow exposure chamber (from Ferin and Leach, 1980, with permission).

rectangular box (Leong et al., 1981), is shown in Fig. 8. The aerosol flows into the upper corner of the chamber and exits at the diagonally opposite, lower corner. A wide-angle, positive-pressure air nozzle placed in the aerosol entrance disperses the aerosol throughout the chamber.

Head/Nose-only Chambers

Head- or nose-only exposure chambers were originally designed for experiments where reduced skin contamination was important, when a small-sized chamber was desirable or when the test material was in short supply. The Henderson (1952) apparatus, consisting of a horizontal, cylindrical tube with several animal ports, was frequently used for exposing rodents to airborne pathogens to study infectious respiratory diseases (Jemski and Phillips, 1965). Vertical chambers of cylindrical or hexagonal shapes with a single ring of 6

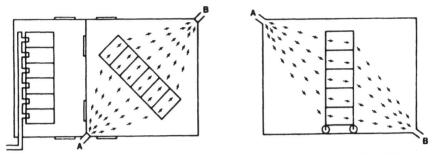

Figure 8 A diagonal flow exposure chamber (from Leong et al., 1980, with permission).

to 8 rodent ports around their peripheries were used to study radioactive materials (MacEwen et al., 1961; Thomas and Lie, 1963); a multi-tiered version capable of exposing 48 mice was also designed (Lane, 1959). Later, designs of multiple-level exposure chambers were improved by considering the uniform aerosol distribution.

Multi-Tiered Rodent Chambers In large, nose-only exposure chambers for rodents, the uniformity of aerosol concentration is an important issue. It is also desirable to avoid having animals rebreathe the exhalation of other animals. One way to improve the uniformity of the aerosol is to divide a rectangular chamber into numerous vertical channels with precise control of the aerosol flow rate through each channel (Walsh et al., 1980; Andre et al., 1989) as shown in Fig. 9. However, even with multiple levels of animal ports, the animals still rebreathe the exhalations from the animals below. A rectangular chamber addressing these problems was designed for simultaneous exposure of 88 rodents (Raabe et al., 1973). This chamber has been further

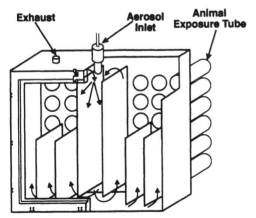

Figure 9 A nose-only exposure chamber with vertical channels (from Andre et al., 1989, with permission).

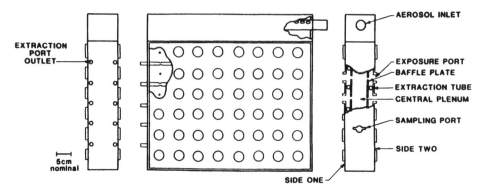

Figure 10 Schematic of a 96-port rectangular nose-only exposure chamber and a photograph of 80-port chamber inside a glove box (from Yeh et al., 1990).

modified to accommodate 80 and 96 rats or mice (Hoover and Mewhinney, 1985; Yeh et al. 1990) with the same basic design as shown in Fig. 10. The aerosol, introduced at the top of chamber through a horizontal tube with holes facing upward, enters the center compartment of the chamber separated by two baffle plates with holes facing each animal port. The aerosol is drawn through the baffle plates past the animals' noses and is withdrawn

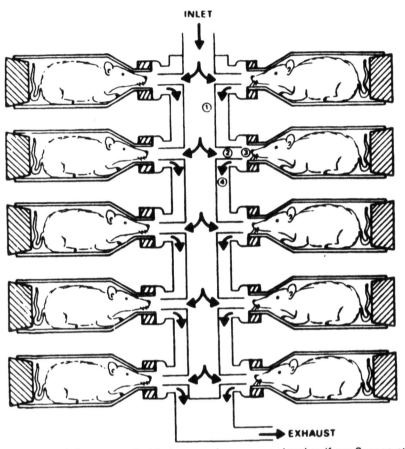

Figure 11 A flow-past cylindrical nose-only exposure chamber (from Cannon et al., 1983, with permission).

from the chamber at each exposure port through extraction tubes. This process ensures that each animal breathes a fresh aerosol of uniform concentration.

Another design provides a separate aerosol supply and exhaust in a vertical concentric cylinder with multiple levels of rodents in a ring arrangement (Cannon et al., 1983). Fig. 11 shows the schematic of the flow-past chamber, where the aerosol flows down the inner cylinder for delivery to individual animal ports through a tube, and the expired air is drawn through the outer cylinder. The Cannon design is similar in principle to the nose-only chamber for cigarette smoke exposure designed by Baumgartner and Coggins (1980). The Cannon design for 52 animals (Lab Products, Maywood, NJ) and a smaller chamber with the aerosol delivered from the bottom (Inhalation Systems Westwood, NJ) are commercially available. An 80-port flow-past cham-

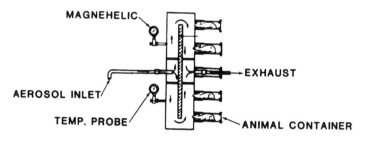

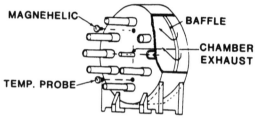

Figure 12 A disc-shape nose-only exposure chamber (from Leach et al., 1984, with permission).

ber with improved aerosol uniformity and easy access to the animals has been designed by Hemenway et al. (1990). A third design for uniform aerosol distribution is the disc-shaped exposure chamber shown in Fig. 12. The test atmosphere is drawn into the center of the chamber, forced toward the edge of the baffle past the noses of the animals, and out through the central outlet (Leach et al., 1984). The symmetrically placed animals (up to 8) are exposed to a uniform concentration and not subjected to exhalations.

Exposure Chambers for a Single Animal   Nose-only exposure chambers have been tailored to suit different studies. Many systems are custom-made in the laboratory, especially those accommodating only a few animals or designed specifically for humans and large animals. Smaller units, with one or more animal ports, are usually equipped with respiratory monitoring devices for accurate measurement of particle deposition or gas uptake. Several systems for measuring aerosol deposition in humans use either a pneumotachograph (Stahlhofen *et al.,* 1980), or inductive plethysmograph as a vest encircling the rib cage (Blanchard and Willeke, 1983). Similar systems for dogs, monkeys and humans are also described by Lippmann (1980). An apparatus for continuous measurement of respiration and vapor uptake in rats uses a body plethysmograph and on-line gas chromatograph (Dahl et al., 1987; Fig. 13). Dogs and monkeys are often under anesthesia during exposure (Dahl et al., 1990; Snipes et al. 1991), and can be housed in a ventilation box to control the breathing patterns (Fig. 14).

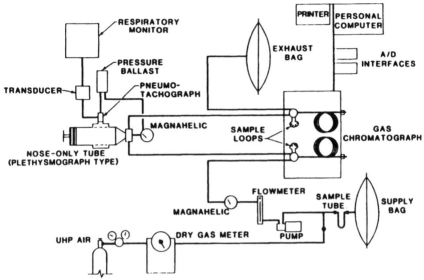

Figure 13 An exposure system for determining exact dose of vapor to the rat (from Dahl et al., 1987, with permission).

Animal Tubes and Respiratory Measurements Many animal holders or restraining devices have been developed to keep the nose or head in position to receive the test atmosphere and to minimize contamination of the remainder of the body (Fig. 15). Plastic tubes with openings at one end for the nose and at the other end for stoppers are widely used. The shapes of these holders include a machined 60° cone (Lippmann, 1980), a reduced circular section with conical internal shape (Raabe et al., 1973), and a molded conical shape (Cannon et al., 1983); they are designed to fit a corresponding animal port in the exposure chamber. Because these tubes cover the body of the animal,

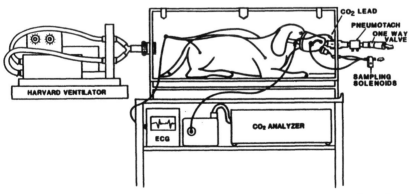

Figure 14 A nose-only vapor exposure system with ventilation for dogs (from Snipes et al., 1991, with permission).

CHAMBER

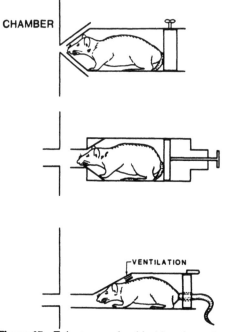

Figure 15   Tube type animal holders for rodents.

they may induce thermal or other stresses. Thin-walled tubes of a size appropriate to the species exposed should be selected to avoid the stress caused by a tight fit or suffocation, which can occur when the tube is large enough so that the animal can turn around. To help dissipate body heat, perforated tubes (Chang et al., 1983), or wire-mesh restrainers have been used (Landry et al., 1983; Griffith and Standafer, 1985); the latter requires a seal around the animal's neck. An exposure tube that has an inner wire-mesh restraining tube can be used alone during most of the exposure period (Landry et al., 1983). An aluminum tube can be placed over this assembly when respiratory measurements require a plethysmograph.

Exposure systems for large animals usually consist of a mask and some arrangement to keep the animal in the position for exposure (Bair et al., 1969; Stuart et al., 1969; Boecker et al., 1964). A large head-only exposure chamber with attached animal holding boxes has been used for simultaneous exposure of 10 dogs to radon daughters (Stuart et al., 1969). It was designed to keep dogs in a comfortable, seated position. A flow of clean air, separate from the test material, was circulated through the holding boxes.

In many inhalation studies, it is necessary to know the amount of material deposited and retained within the respiratory tract of each animal. This information can be obtained by measuring the difference in the amount of material an animal inhales and exhales. A special face mask with a 2-way valve to collect aerosol samples in the inspiratory and expiratory air and a pneumotachograph to measure the respiratory rates were developed in a dog nose-only exposure system by Vocci et al. (1956). A series of developments followed in the monitoring of physiological functions in inhalation exposures with nose-only systems.

The whole-body plethysmograph has been used to obtain pertinent respiratory data, including tidal volume, respiratory frequency, and minute volume. The subject is housed in a plethysmograph chamber so that only the nose or head region is exposed. Figure 16 shows a body plethysmograph for rats (Medinsky et al., 1985; Mauderly, 1986), that is similar in size to standard nose-only tube and fits in the animal port of a Lovelace multi-port, nose-only chamber. Tidal volume is measured by pressure changes in the apparatus, caused by the volume changes of respired air. Respiratory rates are obtained by counting the number of pressure pulses with time. The body plethysmograph method is frequently used, especially with rodents where it can be substituted for an animal holder in a multiple-port, nose-only exposure chamber (Thomas and Lie, 1963; Murphy and Ulrich, 1964; Boecker et al., 1964). Generally, no anesthesia is required, and the animal's response is similar to that of other animals which are not monitored in the same exposure unit.

EVALUATION OF EXPOSURE CHAMBERS

Four questions should be asked before an exposure starts:

1 Does air leak into or out of the chamber?
2 How much is test material lost in delivery lines?
3 Is the concentration stable in the chamber? and
4 Is the test material dispersed uniformly throughout the chamber?

Leak Test

Chambers often leak around doors, lids, entrances, outlets, and sampling ports. Leaks are especially common where tubes housing rodents are connected in nose-only exposure chambers with multiple ports. Air leaks dilute toxicant concentration and may alter distribution of the test material. Concentration changes can result in overestimating uptake of test material in a static or recirculating system. Large leaks affect the balance of flow in a dynamic system and may also affect the safety of the operator. Slightly negative pressure (1 to 5 cm H_2O) is usually maintained in the chamber to prevent the test atmosphere from leaking into the exposure room.

After the exposure system is assembled, the chamber must be checked for leaks. A simple approach is to measure the rate at which air leaks into a chamber under a slight vacuum (Moss, 1983; Mokler and White, 1983). Such a vacuum (e.g., 5 cm H_2O) is then set, and the chamber is sealed. The change in vacuum or pressure difference, Δ_t, between the chamber and the ambient air is measured as a function of time. A slow increase of pressure indicates an inward leak that can often be expressed as an exponential function:

$$\Delta_t = \Delta_0 \exp(-kt), \tag{1}$$

where Δ_0 is the initial or set pressure difference, t is time (min), and k is the decay constant (min^{-1}). The leak rate (liters/min) in a chamber of volume, V (liters), operated at pressure difference, Δ, and ambient pressure, P_a, can be estimated from the decay constant:

$$\text{leak rate} = kV\frac{\Delta}{P_a}. \tag{2}$$

These equations were derived from the ideal gas law and the assumption that the rate of change of pressure difference is directly proportional to the leak rate. An example of such data for three H-2000 chambers (chamber volume = 2.3 m^3) is shown in Fig. 17. The calculated leak rates were 0.23, 0.63, and 6.7 L min^{-1}, respectively, if the chamber were operated at 2 cm H_2O negative in Albuquerque, NM (620 mm Hg ambient pressure). Because the chamber cannot be made 100% leak-proof, an acceptable leak rate, such as 2% of total flow rate, must be established.

    If the measured leak rate is unacceptable, then leaks should be located and sealed. The chamber is slightly pressurized (1 to 2 cm H_2O), and a soap solution "painted" on possible leak points. The bubbles produced by escaping air pinpoint the problem areas. An ultrasonic detector achieves the same result by registering high-frequency noise from leaking air. Other methods involve introducing a trace gas in the chamber or using smoke as a tracer.

Loss of Material

The delivery line and internal surface of the chamber may lower the aerosol concentration because of absorption, adsorption, chemical reaction or deposition of particles (Silver, 1946). The rate of loss is specific for the test compound and the surface materials. For example, small surfaces of rubber in gaskets or rubber stoppers can markedly lower the concentrations of certain organic vapors or gases (Silver, 1946; Bolt et al., 1976). Aerosol particles can deposit on the surfaces by sedimentation, diffusion, and electrostatic charge. Sedimentation and diffusion deposition are functions of particle size (Chen et al. 1992), and electrostatic deposition is a function of charges carried by particles. For dynamic exposure systems, the turbulence caused by the flow or fans within the chamber would also affect the deposition (Okuyama et al., 1986; Cheng et al., 1992).

    The magnitude of any surface effect depends on the quantity of surface exposed per unit volume and is governed by the shape and size of the chamber and presence of additional interior surfaces. A spherical chamber has the smallest surface:volume ratio; however, a cubical chamber is more easily constructed. The surface/volume ratio of either chamber decreases as chamber size increases. Baffles, racks, animal cages and pans, lights, and fans within the chamber increase the surface:volume ratio.

Because of their additional surface, subjects within a chamber also affect the concentration. Aerosol concentrations often drop from values expected from test runs when animals are placed in a chamber. Silver (1946) suggested that the volume of animals should not be more than 5% of in whole-body chamber to avoid excessive reduction of aerosol concentration.

Chamber Concentration

The concentration in static or recirculating exposure chambers decreases continuously because no additional test material is introduced. The amount absorbed by the animals can be estimated by the concentration decay curve after correcting for decay in the chamber without animals (Filser and Bolt, 1979; Anderson et al., 1980).

In a dynamic exposure system, the chamber concentration of test material rises rapidly when the system is first turned on, then slowly approaches a theoretical equilibrium as described by the follow equation (Silver, 1946):

$$C = C_0 [1 - \exp(-\frac{F}{V} t)], \tag{3}$$

where:

C = concentration,
C_0 = equilibrium chamber concentration (= generator minute output $\times$ (1−losses)/F),
F = total flow through the chamber,
V = chamber volume,
t = time.

When the generator is stopped, the concentration decreases exponentially from the equilibrium value, C_0 according to the equation:

$$C = C_0 \exp(-\frac{F}{V} t). \tag{4}$$

The concentration-time characteristics of the chamber are most usefully expressed by stating the time, t_x, required to attain a given percentage $x\%$ of the equilibrium concentration as shown by:

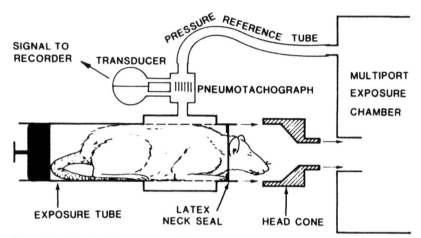

Figure 16 A body plethysmograph for rats connected to a nose-only exposure chamber (from Medinsky et al., 1985, with permission).

$$t_x = k \frac{V}{F}, \tag{5}$$

where $k = -\ln [(100 - x)/100]$. For example, for t_{90}, $k = 2.303$, and for t_{99}, $k = 4.605$. Values of t_{90} or t_{99} can be obtained during the buildup and decay of the chamber concentration and, theoretically, should be the same.

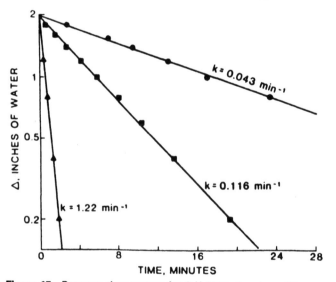

Figure 17 Pressure-time curves for 3 H-2000 chambers. The upper line is for a new chamber, the middle line is for a chamber that has been used for 18 months and the lowest line is a chamber with deteriorated gaskets on the doors (from Mokler and White, 1983, with permission).

Equations 3–5 are valid only if: (1) the output of the generator is stable and the test material thoroughly mixed before entering the chamber, and (2) mixing within the chamber is rapid and thorough (Moss, 1983). When both criteria are satisfied, t_{90} will be the same, regardless of the sampling point, and will agree with the theoretical value predicted by equation 5. However, these two conditions are difficult to achieve in practice. If condition 1 is not achieved, at least two airstreams will enter the chamber; one relatively clean and the other having a higher concentration of test material. The two streams will be mixed within the chamber. There will be zones continuously diluted by relatively clean air and zones that are continuously fed by air having a higher aerosol concentration. The mean concentration between sampling points equals the target equilibrium concentration, C_0, but the variation between sampling points would be large and stable. This should not happen if the test atmosphere is thoroughly and uniformly mixed before entering the chamber, even if the chamber has very poor internal mixing characteristics.

When uniform mixing is not achieved, zones of little or no air movement, i.e., stagnant zones, are created. The equilibrium concentration throughout the chamber will still be the same and will be independent of sampling position, provided thorough mixing in the inlet occurs and enough time has passed. The t_{90} will not be the same throughout the chamber, however. The concentrations at sampling points within the stagnant zones will change at a slower rate, hence longer t_{90}. Comparison of measured values of t_{90} with the theoretical value of t_{90} indicates the degree of uniform mixing.

The stability of aerosol concentration is readily measured with a real-time monitor such as a RAM-1 (MIE Inc., Billerica, MA) or a MIRAN infrared spectrometer (Foxoboro Wilkes, Norwalk, CT) for vapors. A typical concentration-time curve in a dynamic exposure chamber (Fig. 18) consists of three parts.

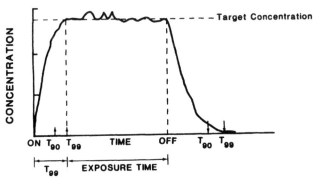

Figure 18 Concentration time curve in a dynamic exposure chamber.

1 the buildup before concentration reaches 99% (t_{99}) of the equilibrium concentration,

2 the flat portion of exposure time, T, and

3 the fall portion of t_{99} minutes when the concentration falls to approximately 1% of the equilibrium concentration. The rising and falling portions of concentration curve are theoretically equal. The duration of exposure at the target concentration is then $T + t_{99}$, which should be used to calculate the product of concentration and time. In practice, $T + t_{90}$ is also used, because t_{90} is often easier to measure than t_{99} in chambers with nonideal mixing characteristics. When the flat portion, T, is very much larger than t_{99}, the rising and falling curves may be neglected and the exposure considered as a square-wave function.

Uniformity of Chamber Concentration

After chamber concentration is established, the uniformity of test atmosphere in the chamber should be evaluated. This is especially important in multiple-level chambers. Lung burdens of animals are proportional to the concentration maintained in the breathing zone (Raabe et al., 1973; Griffis et al., 1981; Cannon et al., 1983; Hemenway et al., 1983); therefore, an uneven distribution of toxicant will result in highly variable exposure concentrations received by animals in the same exposure group. In a study involving several concentrations, high variability within groups may result in overlapping of lung burdens between groups.

The uniformity or degree of mixing within a chamber has been measured by flow visualization, dynamic flow tests, and sampling from many different points within the chamber (Moss, 1981). In the flow visualization method, smoke or a dye tracer is injected into the chamber, and the tracer is followed visually or recorded by camera or video. Flow patterns, mixing and "dead spots" can be identified. However, the information is qualitative unless computer modeling techniques are used. To design and modify chambers (Moss et al., 1982; Holmberg et al., 1980) it is more practical to use small-scale models that are dynamically similar to a full-scale chamber. Dye is sometimes injected into the water used in these models for detecting mixing conditions (Moss, 1981).

Dynamic flow tests consist of continuously monitoring a bolus of tracer gas injected into the airstream as it passes the inlet and exhaust ports of the chamber. The mean time, t_i, of the concentration versus time plot is calculated for the sampling point at the entrance ($i = 1$) and exhaust ports ($i = 2$). Chamber mixing is evaluated by calculating the percent dead space, %DS:

$$\%DS = 100(T_m - V/F)/T_m, \tag{6}$$

where:

T_m = measured residence time = t_2-t_1,
F = total flow through the chamber,
V = chamber volume.

The percent dead space gives a measure of uniformity of mixing within the chamber (Hemenway et al., 1982). A high percent dead space indicates that stagnant zones exist and that a fraction of the inlet air is being shunted through the chamber.

Flow visualization or dynamic flow methods are especially useful in design and modification of chambers, but these tests do not provide information on test material concentrations at different locations. Such information is obtained by sampling at different locations in the chamber, using the test material at target concentrations (point test). The conventional way is to sample simultaneously and to use the coefficient of variation (standard deviation/mean) of these samples as an indication of the degree of uniform mixing. The following steps should be considered in carrying out the procedure:

1   Select appropriate sampling locations.   Tests in a large, multi-tier, whole-body chamber should include samples from different levels and depths (Yeh et al., 1986). For nose-only exposure chambers of more than 10 animal ports, random selection of 4 to 6 locations will minimize the number of samples (Yeh et al., 1990).

2   Calibrate flow rates and the instruments used. Use low sample flow rates (0.5 to 3 L min^{-1}) to minimize alteration of the mixing pattern. Results obtained using high sampling rates may not represent the actual operation.

3   Sample all locations simultaneously and calculate the coefficient of variation.

If more than 10 locations are selected or if the number of sampling locations is greater than the available number of samplers, then a series of samples should be taken simultaneously. Each set should consist of samples from a fixed reference point and several from other points. Concentrations taken over the same time period are normalized by the reference sample, and all normalized concentrations from several sampling periods are analyzed together.

A second method of making point tests has been reported by Decker et al. (1985). In this method, samples are taken sequentially in different locations of a chamber using a real-time aerosol mass monitor such as a RAM-1 or RAM-S unit (MIE Inc., Billerica, MA). The procedure is as follows:

1   Choose a reference and other sampling locations.
2   Measure the concentration in the reference location.
3   Sample other locations (half the total number).
4   Measure the concentration again in the reference location.
5   Sample the remaining locations.

6   Measure the concentration in the reference location the third time.

7   Calculate the total variation, the coefficient of variation of samples in different locations including the reference point (one sample for each location).

8   Calculate temporal variation, the coefficient of variation of the three reference samples.

9   The spatial variation is calculated as:

$$CV^2_{\text{spatial}}(\%) = CV^2_{\text{total}}(\%) - CV^2_{\text{temporal}}(\%) \,. \tag{7}$$

Equation 7 is valid under the assumption that spatial and temporal variations of chamber concentrations are independent. The sequential procedure is simpler, less time-consuming, and less interruptive to the exposure than the simultaneous procedure. Therefore, it could be used during the exposure with animals in the chamber. These two methods for chamber uniformity showed comparable results if the spatial variation was below 15%; at higher values, the sequential procedures gave much larger values (Cheng et al., 1989). The real-time monitor must be carefully calibrated with the test material if meaningful results are to be obtained (Cheng et al., 1988; 1989). The toxicant concentrations should be maintained at a relatively stable level during the procedure, because the accuracy of equation 7 in estimating spatial variation is reduced when the temporal variation is large.

Several factors may influence the mixing pattern of a exposure chamber: (1) chamber design, (2) mixing of the test atmosphere before it reaches the chamber, (3) chamber integrity, and (4) aerosol particle size. Table 3 lists the degree of spatial uniformity of aerosols in currently used large nose-only and whole-body exposure chambers. It is reasonable to expect that a large exposure chamber will have a spatial variation of about 5–10% for 1-μm particle-size aerosols. This variation can be kept below 5% for several well-designed chambers; for other chambers, the variation can be reduced.

Table 3   Degree of Spatial Uniformity in Exposure Chambers

Chamber Type	Aerosol Size MMAD	Spatial Variation CV (%)	Reference
96 Port Nose-only	1 μm	2.8–5.4	Yeh et al. (1990)
40 Port Nose-only	1 μm	6	Cannon et al. (1983)
27″ Hinners	1.1 μm	1.1–1.6	Yeh et al. (1990)
H-2000	1.6 μm	5	Cheng et al. (1989)
	3.1 μm	13	
H-1000	1.6 μm	8	Cheng et al. (1989)
	3.1 μm	46	
Horizontal Flow	1.0 μm	8–10	Ferin and Leach (1980)
Horizontal Flow	2–3 μm	34–42	Hemenway et al. (1983)
Diagonal Flow	3 μm	10–12	Leong et al. (1981)

Uniform mixing of a test atmosphere at the chamber inlet is essential if uniform distribution is to be expected. Inlet design is one important consideration in chamber design. Baffle panels, screens, cones, deflectors, and manifolds have been used to improve mixing in the chamber inlets. A relationship between the shape of the flow inlet and chamber mixing has been reported (Carpenter and Beethe, 1980). If dilution air is introduced, then the degree of mixing of the clean air and toxicant flows at the inlet profoundly affects the degree of mixing throughout the chamber. Yeh et al. (1986) have examined three types of in-line dilutors in order to obtain a uniform distribution of aerosol concentration in whole-body exposure chambers. A laminar flow element has been used in a flow-past, nose-only exposure chamber to increase the uniformity (Hemenway et al., 1990). Fans inside the chamber near the entrance are also useful in improving aerosol mixing (Jemski and Phillips, 1965; Walsh et al., 1980; Cheng et al., 1989). A large, whole-body exposure chamber with rotating animal cages has been used to even out the aerosol concentration (Strong and Walsh, 1992).

Fig. 19 shows spatial variations in a H-2000 exposure chamber for different aerosols reported in the literature (Moss et al., 1982; Griffis et al., 1981; Cheng et al., 1984; 1989; Henderson et al., 1984; Yeh et al., 1986). The spatial variation increases with particle size, because larger particles do not follow the streamlines of the flow, and losses due to impaction and sedimentation change the concentration inside the chamber. On the other hand, gases and smaller particles closely follow airflow streamlines. Also shown in the same figure are data obtained with increased mixing by placing a small box fan underneath the inlet. The spatial variation for a H-2000 chamber was reduced to below 5% for 1.6 and 3.1 μm particles (Cheng et al., 1989). The

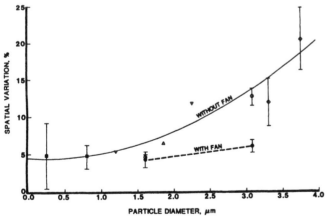

Figure 19 Spatial variations of aerosol concentrations in H-2000 exposure chambers as a function of aerosol particle size. Data are mean and standard deviation from (●) Moss et al. (1982); (■) Cheng et al. (1984), (▲) Griffis et al. (1981); (▼) Yeh et al. (1986); and (♦) Henderson et al. (1984).

spatial variation in a smaller H-1000 chamber was reduced more dramatically by using a similar fan (Cheng et al., 1989).

If the concentration distribution is not acceptable, rotating animals inside the chamber is an effective way to average out the lung burdens achieved (Hemenway et al., 1983). Animals, or cage units holding more than one animal, can be rotated in a repeated-dose or longer study on a weekly or other regular basis. Animal rotation is not useful for single, acute exposure.

ENVIRONMENTAL CONDITIONS WITHIN EXPOSURE CHAMBERS

Environmental factors inside an exposure chamber can influence the health and response of animals. Temperature, RH, airflow, light, noise, and ammonia levels may be inappropriate if proper controls are not maintained. Animals inhaling a test material may have diminished capacity to maintain homeostasis in an unusual environment. It is necessary to hold environmental variables to a narrow range so that the biological effects of the test material can determined rather than the effect of environmental factors (Rao, 1986). The concern for environmental factors in an exposure system is similar to that for conditions in an animal room.

Temperature

Most laboratory animals used in inhalation studies are homoiothermic, so they typically maintain a stable internal body temperature. To do so, heat production and loss from the body must be controlled. An ambient temperature around the test animals lower than the body temperature helps to transfer heat away from the animals. The ranges of chamber temperatures recommended for rodents are given in Table 4. Temperatures can influence severity,

Table 4 Recommended Temperature and Relative Humidity Ranges for Rodents in Exposure Chambers

Rodent	Temperature (°F)			R.H. (%)		
	Comfort Range[a]	Recommended		Recommended		
Rat	75–84	72–78[b]	68–75.2[c,d]	40–70[b]	40–60[c]	30–70[d]
Mouse	79–88	72–78[b]	68–75.2[c]	40–70[b]	40–60[c]	30–70[d]
Hamster	70–79	72–78[b]	68–75.2[c]	40–70[d]	40–60[c]	30–70[d]

[a]Weihle (1971).
[b]NIH (1985).
[c]EPA (1982).
[d]OECD (1981).

duration, nature, and variance of the toxic response of test chemicals (Weihe, 1971). Extremes of temperatures in a chamber could result in mortality.

Laboratory animals lose heat by radiation, convection, and evaporation. In a dynamic whole-body exposure chamber, the heat produced by test animals is transferred to the chamber walls, then to the exposure room by radiation, and to airflow through the chamber by convection and evaporation (Berstein and Drew, 1980). Therefore, the flow rate and temperature of intake air and the temperature of the room containing the exposure chamber are two factors available for the temperature control (Fig. 20). A large number of animals in a chamber tends to raise the chamber temperature; therefore, the room temperature needs to be cooler. As the primary means of heat removal from the chamber (70–95%), a temperature gradient of about 3 to 5°C between the chamber and room temperature is needed to maintain H-2000 and H-1000 exposure chambers at the desired temperatures when fully loaded with animals. However, there are situations when the room is too cold for smaller rodents such as mice and hamsters. For example, if separate

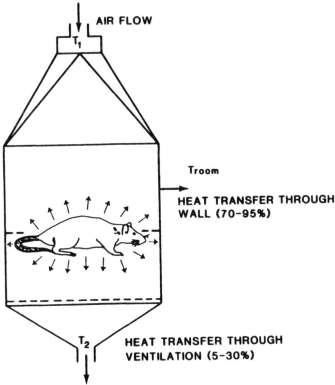

Figure 20 Schematic of heat transport in a dynamic whole-body exposure chamber.

exposure chambers housing rats and mice are in the same room, the room temperature optimal for rat chambers may be too cold for mice.

In nose-only chambers, rodents are usually confined inside exposure tubes with little or no air circulating around the body. The exposure tube prevents heat transfer by evaporation and convection, leaving radiation heat transfer from tube walls as the only effective means of heat removal. Some animals, rats and mice for example, regulate temperature with their tails. If a thin-walled "tail extension tube" is added to the back seal of the nose-only exposure tube, or if the tail is allowed to hang free (Fig. 15), the animal can lower its body temperature during exposure by over 2°C under some room conditions. It is critical to maintain a low temperature around the tubes to prevent overheating of the test animals. If many animals are loaded in multiple-port chambers such as Lovelace 96-port or Battelle 40-port chambers (Cannon et al., 1983), heat transfer from the animal tubes is not efficient. A greater temperature gradient between the chamber and room, circulation of cool air around the chamber (Sterner et al., 1991), and/or the addition of "tail tubes" are needed to facilitate adequate control of body temperature.

Relative Humidity

The RH in the chamber air is important for heat balance. The RH influences feed consumption, behavior, disease, and chemical effects on animals (Rao, 1986). At a constant room temperature of 21°C, rats at low RH consume more food than at high RH, and mice are more active at low RH than at high RH (Clough, 1982). High RH promotes the production of ammonia in rodent cages, and a high concentration of ammonia increases the susceptibility of animals to infections, especially to respiratory pathogens (Clough, 1982). The recommended RH range for rodents in animal facilities as well as exposure chambers is 30–40% minimum to 60 to 70% maximum (Table 4).

To maintain appropriate RH in a chamber, the air intake should be conditioned and humidified by a steamer or evaporative humidifier. Fig. 21 shows an example of RH readings in a H-2000 whole-body exposure chamber where 144 rats were housed in the chamber. Air with 35–40% RH was delivered to the chamber 24 hours a day. The excreta trays were cleaned, and cageboard was replaced at about 7 AM and 4 PM. The RH increased between cleanup times. This cycling of RH in the chamber shows the effects of animals and animal-care activities.

Air Ventilation

Air quality and flow rate must be maintained through ventilation of an exposure chamber. Filtered and conditioned air should be used to dilute the test material during exposure and nonexposure periods to maintain environmen-

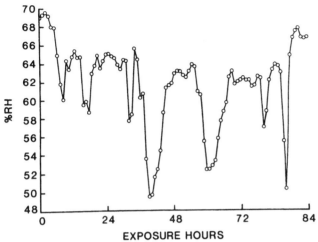

Figure 21 Change of relative humidity measured in a H-2000 chamber housing rats.

tal conditions that are free of trace gases, dusts, and microorganisms. Air movement in an exposure chamber also helps to maintain proper temperature and RH. Thus, a conditioned air supply (30–40% RH and about 17 to 20°C) should be used.

If animals are to be kept in whole-body exposure chambers 24 hours per day during the study, then ammonia levels in the chamber should be maintained at a proper level. The effects of ammonia production arising from animal excreta in exposure chambers have been investigated (Barrow and Dodd, 1976; Malanchuk et al., 1980; Weissman et al., 1980; Higuchi and Davies, 1993). Concentration of ammonia in a chamber increases directly with time and number of animals and inversely with air flow. Also, extraneous compounds may be formed in exposure chambers by reaction of ammonia and test material (Campbell et al., 1969; Barrow and Dodd, 1976; Malanchuk et al., 1980). Exposure of rats for 4 to 6 weeks to 25 to 250 ppm of ammonia increases the susceptibility of animals for infectious diseases (Broderson et al., 1976). To maintain ammonia concentrations at 2.5 ppm or below, cageboard should be changed daily and chambers and cages washed weekly (Weissman et al., 1980). Bacteriostatic cageboards are recommended as they reduced the production of ammonia. A H-2000 chamber has been modified with an air flow manifold through the excreta pans in order to maintain ammonia concentration below 25 ppb (Higuchi and Davies, 1993).

Sufficient air flow is required to supply oxygen for animals in the exposure chamber. The flow rates in whole-body exposure chambers are usually kept at 10 to 15 (chamber volume) air changes per hour. Because of the large chamber volume to animal volume ratio, these flow rates supply enough air

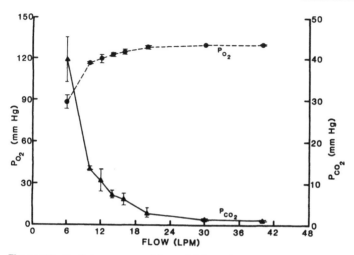

Figure 22 Partial pressure of O_2 and CO_2 as function of flow rate in a 80-port rectangular nose-only exposure chamber.

for animals. On the other hand, this air change rate does not provide enough air for animals in a nose-only chamber. Measured minute volumes for rats housed in nose-only tubes increases when the flow rate per rat through the apparatus approaches the normal rat minute volume, and returns to normal when the flow rate doubles (Landry et al., 1983; Dahl et al., 1987). We have also measured increased CO_2 and reduced oxygen concentrations when the flow rate was reduced from 48 to 6 L min^{-1} with 72 F344 rats in a 80-port nose-only exposure chamber (Barr et al., 1987). Fig. 22 shows that when the total flow rate was reduced to about 1.5 times the minute volume per animal (22 L min^{-1}), the partial pressure of the oxygen started to decrease. The CO_2 concentration in the chamber continued to increase with reduced flow rates. However, no abnormal behaviors were observed. Thus, we recommend that the flow rate through a nose-only exposure chamber should be at least 1.5 times the amount:

$$Q(\text{L min}^{-1}) \geq 1.5 \times \text{animal number} \times \text{minute volume.} \qquad (8)$$

Values of minute volumes for unsedated, unrestrained rodents and dogs (Fairchild, 1972; Mauderly, 1974; Mauderly et al., 1979), and rodents in exposure tubes (Landry et al., 1983; Mauderly, 1986), are available. When the flow rate in an 80-port nose-only exposure chamber was cut off, the oxygen partial pressure was depleted to the terminal value of 46 mm Hg in merely 11 minutes (Barr et al., 1987). Thus, the flow rate or oxygen level in the nose-only exposure chamber should be monitored continuously. The absolute minimum alarm limit for the oxygen level should be 18%.

FACILITIES

Exposure chambers must be operated in a facility with adequate control of temperature, humidity, clean supply air, clean compressed air, exposure control, and data collection in addition to protecting animals from outside agents during normal animal care and exposure operation procedures (McClellan et al., 1984).

Temperature

Temperature in each room housing exposure chambers must be adjustable and independent of that in other rooms in the facility.

Relative Humidity

Relative humidity in the exposure rooms should be controlled, building wide, at a fixed level, independent of season. In addition, the RH should be controlled for each exposure system. A dual air-supply system can be used for the chambers; one line is set at 70 or 80% RH, the other at 30 to 40% RH. A valve at the top of each chamber can be used to blend these two airstreams in order to maintain constant humidity. An automated valve, linked with a humidity detector through a computer-controlled circuit, will maintain constant RH in the chambers regardless of changes in room RH.

Room Air Movement

Room air movement should be 10–15 air changes per hour, with the air balance in the room set so that pressure is either positive or negative with respect to outer hallways and the rest of the building; a single-pass system is best. If a major portion of the room air is recirculated, an adequate cleanup system must be in place, e.g., scrubber, electrostatic precipitator, or high-efficiency filters.

Exposure Air

Exposure air should be supplied to the rooms containing chambers or atmosphere generation systems. Dry, particle-free, compressed air is used in most aerosol and vapor generation systems, and in chamber dilution systems. The required flow rate depends on the size and number of chambers.

Exposure Vacuum

Exposure vacuum should be supplied to the rooms containing chambers or atmosphere generation systems. This vacuum should be in a line sized to allow a flow rate of at least 50 L min^{-1} per chamber.

Exhaust

Exhaust from the chambers should be diluted following pre-cleaning techniques specific to each experiment. The exhaust system from each exposure room should allow simple installation of temporary cleanup devices such as scrubbers or absorption devices in addition to high-efficiency filters.

Exposure Rooms

Exposure rooms should be designed to provide a clear separation between work related to animal care and work related to maintaining exposure concentrations. Several room designs have been discussed by McClellan et al., (1984). One approach is to place all exposure chamber services above the level of the top of the chamber, either on the wall or in the ceiling. Shelves or walkways suspended over the chambers should be installed for control equipment, and should be accessible for cleaning or storage during periods when there are no inhalation toxicology projects on line. Movement of personnel and equipment in the facility and through each exposure room should be controlled to protect the animals on study and contain any hazardous material.

Exposure Control and Data Collection

Exposure control and data collection should be as automated as practical. A distributed control system that generates daily electronic or paper reports is an efficient approach to automate a facility over the course of several inhalation toxicology studies. The system can be expanded and improved as each study is added. Minimal control of such a system should include monitoring and adjustment of chamber air flows, temperature, RH, and pressure drop (to determine the presence of leaks). Computer-controlled exposure systems have been described (Blanchard and Willeke, 1983; Dahl et al., 1987; Snipes et al., 1991).

Specific requirements for design, construction, or modification of an inhalation toxicology research facility will vary depending on the scale, work flow, and type of research proposed. The general guidelines above combined with the specific culture of the organization and staff using the facility should result in an efficient and cost-effective environment for conducting the high-technology toxicology studies for inhaled material.

SUMMARY

The purpose of this chapter is to present an overview of inhalation systems, especially those pertinent to inhalation toxicity studies using laboratory animals. The information is intended to enhance understanding of the basic principles in operating these systems and may contribute to the selection of appropriate equipment suitable for individual studies. Comprehensive references cited in the chapter are listed here.

ACKNOWLEDGMENT

This research was supported by the U.S. Department of Energy, Office of Health and Environmental Research, under Contract No. DE-AC04-76EV01013 at the Inhalation Toxicology Research Institute, Albuquerque, NM.

REFERENCES

Andre, S, Charuau, J, Rateau, G, Vavasseur, C, Metivier, H: Design of a new inhalation device for rodents and primates. J. Aerosol Sci 20:674–656, 1989.

Anderson, ME, Gargas, ML, Jones RA, Jenkins, LJ: The use of inhalation techniques to assess the kinetic constants of 1,1-dichloroethylene metabolism. Toxicol Appl Pharmacol 47:395–409, 1979.

Anderson, ME, Gargas, ML, Jones, RA, Jenkins, LJ: Determination of the kinetic constants for metabolism of inhaled toxicants *in vivo* using gas uptake measurements. Toxicol Appl Pharmacol 54:100–116, 1980.

Bair, WJ, Porter, NS, Brown, DP, Wehner, AP: Apparatus for direct inhalation of cigarette smoke by dogs. J Appl Physiol 26:847–850, 1969.

Barr, EB, Cheng, YS, Mauderly, JL: Determination of oxygen depletion in a nose-only exposure chamber. In: Annual Report of the Inhalation Toxicology Research Institute, LMF-120, pp. 72–75. Springfield, VA: NTIS, 1987.

Barrow, CS, Steinhagen, WH: Design, construction and operation of a simple inhalation exposure system. Fundam Appl Toxicol 2:33–37, 1982.

Baumgartner, H, Coggins, CRE: Description of a continuous-smoking inhalation machine for exposing small animals to tobacco smoke. Beitr Tabakforschung International 10:169–174, 1980.

Bernstein, DM, Drew, RT: The major parameters affecting temperature inside inhalation chambers. Am Ind Hyg Assoc J 41:420–426, 1980.

Blanchard, JD, Willeke, K: An inhalation system for characterizing total lung deposition of ultrafine particles. Am Ind Hyg Assoc J 44:846–856, 1983.

Boecker, BB, Aguilar, FL, Mercer, TT: A canine inhalation exposure apparatus utilizing a whole-body plethysmograph. Health Phys 10:1077–1089, 1964.

Bolt, HM, Kappus, H, Buchter, A, Bolt, W: Disposition of [1,2-C] vinyl chloride in the rats. Arch Toxicol 35:153–162, 1976.

Bond, JA, Benson, JM, Sun, JD, Hylarides, MD: Use of radiolabeling in environmental and toxicological research. In: Hazard Assessment of Chemicals, edited by J Saxena, pp. 29–60. Washington, DC: Hemisphere, 1986.

Broderson, JR, Lindsey, JR, Crawford, JE: The role of environmental ammonia in respiratory mycoplasmosis of rats. Am J Pathol 85:115–130, 1976.

Campbell, KI, Crider, WL, Knott, JM, Malanchuk, M: Alien constituents in experimental atmospheres. In: Proceedings of 5th Annual Conference on Atmosphere Contamination in Confined Spaces, pp. 297–306, 1969.

Cannon, WC, Blanton, EF, McDonald, KE: The flow-past chamber: An improved nose-only exposure system for rodents. Am Ind Hyg Assoc J 44:923–928, 1983.

Carpenter, RL, Beethe, RL: Cones, cone angles, plenums and manifolds. In: Proceedings of Workshop on Inhalation Chamber Technology, edited by RT Drew, BNL 51318, pp. 21–30. Springfield, VA: NTIS, 1978.

Carpenter, RL, Beethe, RL: Airflow and aerosol distribution in animal exposure facilities. In: Generation of Aerosols, edited by K Willeke, pp. 459–473. Ann Arbor, MI: Ann Arbor Science, 1980.

Chang, JCF, Gross, EA, Swenberg, JA, Barrow, CS: Nasal cavity deposition, histopathology, and cell proliferation after single or repeated formaldehyde exposure in B6C3F$_1$ mice and F344 rats. Toxicol Appl Pharmacol 68:161–176, 1983.

Chen, BT, Bechtold, WE, Mauderly, JL: Description and evaluation of a cigarette smoke generation system for inhalation studies. J Aerosol Med 5:19–30, 1992.

Chen, BT, Yeh, HC, Cheng, YS: Evaluation of an environmental reaction chamber. Aerosol Sci Technol 17:9–24, 1992.

Chen, BT, Benz, JV, Finch, FL, Mauderly, JL, Sabourin, PJ, Yeh, HC, Snipes, MB: Effect of exposure mode on amounts of radiolabeled cigarette particles in lungs and gastrointestinal tracts of F344 rats. Inhal Toxicol (in press).

Cheng, YS, Yeh, HC, Mauderly, JL, Molker, BV: Characterization of diesel exhaust in a chronic inhalation study. Am Ind Hyg Assoc J 45:547–555, 1984.

Cheng, YS, Barr, EB, Benson, JM, Damon, EG, Medinsky, MA, Hobbs, CH, Goehl, TJ: Evaluation of a real-time aerosol monitor (RAM-S) for inhalation studies. Fundam Appl Toxicol, 10:321–328, 1988.

Cheng, YS, Barr, EB, Carpenter, R.L., Benson, JM, Hobbs, CH: Improvement of aerosol distribution in whole-body inhalation exposure chambers. Inhal Toxicol, 1:153–166, 1989.

Cheng, YS, Su, YF, Chen, BT: Plate-out rates of radon progeny and particles in a spherical chamber. In: Indoor Radon and Lung Cancer: Reality or Myth? edited by FT Cross, pp 709–729. Richland, WA: Battelle Press, 1992

Chow, HY, Mercer, TT: Charges on droplets produced by atomization of solutions. Am Ind Hyg Assoc J 32:247–255, 1971.

Clough, G: Environmental effects on animals used in biomedical research. Biol Rev 57:487–523, 1982.

Comstock, EG, Rue, RR, Gast, JH: A system for exposure of mice to an atmosphere containing carbon particles. Am Ind Hyg Assoc J 23:88–90, 1962.

Cropp, GJA, Bernstein, IL, Boushey, HA, Hyde, RW, Rosenthal, RR, Spector, SL, Townley, RG: Guidelines for bronchial inhalation challenges with pharmacologic and antigenic agents. In: American Thoracic Society News, pp. 11–19. Spring 1980.

Dahl, AR, Gugliotta, TP, Hanson, RL, Mauderly, JL, Rothenberg, SJ: A method for the continuous measurement of respiration and vapor uptake in rats. Am Ind Hyg Assoc J 48:505–510, 1987.

Dahl, AR, Bechtold, WE, Bond, JA, Henderson, RF, Mauderly, JL, Muggenburg, BA, Sun JD, Birnbaum, LS: Species differences in the metabolism and disposition of inhaled 1,3-butadiene and isoprene. Environ Health Perspect 86:65–69, 1990.

Decker, JR, Moss, OR, Goehl, TJ: Method of Determining Uniform Distribution Test Compound in Inhalation Bioassay Exposure Chambers. Presented in 1985 American Association for Aerosol Research Annual Meeting, Nov. 18–22, Albuquerque, NM, 1985.

Drew, RT, Laskin, S: Environmental inhalation chambers. In: Methods of Animal Experimentation, edited by WI Gay, pp. 1–41. New York: Academic Press, 1973.

EPA: Health Effects Test Guidelines. EPA 560/6-82-001, PB82-232984, 1982.

Fairchild, GA: Measurement of respiratory volume for virus retention studies in mice. Appl Microbiol 24:812–818, 1972.

Ferin, J, Leach, LJ: Horizontal airflow inhalation exposure chamber. In: Generation of Aerosols, edited by K Willeke, pp. 517–523. Ann Arbor, MI: Ann Arbor Science, 1980.

Filser, JG, Bolt, HM: Pharmacokinetics of halogenated ethylenes in rats. Arch Toxicol 42:123–136, 1979.

Filser, JG, Bolt, HM: Inhalation pharmacokinetics based on gas uptake studies. Arch Toxicol 47:279–292, 1981.

Gerlach, RF, Medinsky, MA, Hobbs, CH, Bice, DE, Bechtold, WE, Cheng, YS, Gillett, NA, Birnbaum, LS, Mauderly, JL: Effect of four-week repeated inhalation exposure to unconjugated azodicarbonamide on specific and non-specific airway sensitivity of the guinea pig. J Appl Toxicol 9:145–153, 1989.

Griffis, LC, Wolff, RK, Beethe, RL, Hobbs, CH, McClellan, RO: Evaluation of a multi-tiered inhalation exposure chamber. Fundam Appl Toxicol 1:8–12, 1981.

Griffith, RB, Standafer, S: Simultaneous mainstream-sidestream smoke exposure systems. II. The rats exposure system. Toxicology 35:13–24, 1985.

Hefner, RE, Watanabe, PG, Gehring, PJ: Preliminary studies on the fates of inhaled vinyl chloride monomer in rats. Ann NY Acad Sci 246:135–148, 1975.

Heinrich, U, Pott, F, Mohr, U, Stöber, W: Experimental methods for the detection of the carcinogenicity and/or cocarcinogenicity of inhaled polycyclic-aromatic-hydrocarbon-containing emissions. In: Carcinogenesis, Vol. 8, pp. 131–146. New York: Raven Press, 1985.

Hemenway, DR, Carpenter, RL, Moss, OR: Inhalation toxicology chamber performance: A quantitative model. Am Ind Hyg Assoc J 43:120–127, 1982.

Hemenway, DR, MacAskill, S: Design, development and test results of a horizontal flow inhalation toxicology facility. Am Ind Hyg Assoc J 43:874–879, 1982.

Hemenway, DR, Sylvester, D, Gale, PN, Vacek, P, Evans, JN: Effectiveness of animal rotation in achieving uniform dust exposures and lung dust deposition in horizontal flow chambers. Am Ind Hyg Assoc J 44:655–658, 1983.

Hemenway, DR, Jakab, GJ, Risby, TH, Sehbert, SS, Bowes, SM, Hmieleski, R: Nose-only Inhalation system using the fluidized-bed generation system for coexposures to carbon black and formaldehyde. Inhal Toxicol 2:69–89, 1990.

Henderson, DW: An apparatus for the study of airborne infection. J Hyg 50:53–68, 1952.

Henderson, RF, Cheng, YS, Dutcher, JS, Marshall, TC, White, JE: Studies on the Inhalation Toxicity of Dyes Present in Colored Smoke Munitions. Final Report, Inhalation Toxicology Research Institute, 1984.

Higuchi, MA, Davies, DW: An ammonia abatement system for whole-body small animal inhalation exposures to acid aerosols. Inhal Toxicol 5:333–343, 1993.

Hinners, RG, Burkart, JK, Punte, CL: Animal inhalation exposure chambers. Arch Environ Health 16:194–206, 1968.

Holmberg, RW, Moneyhun, JH, Dalbey, WE: An exposure system for toxicological studies of concentrated oil aerosols. In: Inhalation Toxicology and Technology, edited by BKJ Leong, pp. 53–64. Ann Arbor, MI: Ann Arbor Science, 1981.

Hoover, MD, Mewhinney, CJ: Evaluation of a 96-port nose-only exposure chamber. In: Annual Report of the Inhalation Toxicology Research Institute, LMF-114, pp. 41–44. Springfield, VA: NTIS, 1985.

Huff, JE, Haseman, JK, McConnell, EE, Moore, JA: The National Toxicology Program, toxicology data evaluation techniques and long-term carcinogenesis studies. In: Safety Evaluation of Drugs and Chemicals, edited by WE Lloyd, pp. 411–447. New York, NY: Hemisphere, 1986.

Jaeger, RJ, Shoner, LG, Coffman, L: 1,1-Dichloroethylene hepatotoxicity: Proposed mechanism of action and distribution and binding of ^{14}C radioactivity following inhalation exposure in rats. Environ Health Perspect 21:113–119, 1977.

Jemski, JV, Phillips, GB: Aerosol challenge of animals. In: Methods of Animal Experimentation, edited by WI Gay, Vol. 1, pp. 274–341. New York, NY: Academic Press, 1965.

Kanapilly, GM, Tu, KT, Larsen, TB, Fogel, GR, Luna, RJ: Controlled generation of ultrafine metallic aerosols by vaporization of an organic chelate of the metal. J Colloid Interface Sci 65:533–547, 1978.

Landry, TO, Ramsey, JC, McKenna, MJ: Pulmonary physiology and inhalation dosimetry in rats: Development of a method and two examples. Toxicol Appl Pharmacol 71:72–83, 1983.

Lane, WB: Investigating inhalation hazard using synthetic radioactive fallout. Air Pollut Cont Assoc J 9:102–104, 1959.

Langård, S, Nordhagen, AL: Small animal inhalation chambers and the significance of dust ingestion from the contaminated coat when exposing rats to zinc chromate. Acta Pharmacol 46:43–46, 1980.

Laskin, S, Drew, RT: An inexpensive portable inhalation chamber. Am Ind Hyg Assoc J 31:645–646, 1970.

Laskin, S, Kuschner, M, Drew, RT: Studies in pulmonary carcinogenesis. In: Inhalation Carcinogenesis, edited by MG Hanna et al., CONF-691001, pp. 321–352. US AEC, Springfield, VA: NTIS, 1969.

Leach, LJ, Spiegl, CJ: An animal inhalation exposure unit for toxicity screening. Am Ind Hyg Assoc J 19:66–68, 1958.

Leach, LJ, Spiegl, CJ, Wilson, RH, Sylvester, GE, Lauterbach, KE: A multiple chamber exposure unit design for chronic inhalation studies. Am Ind Hyg Assoc J 20:13–22, 1959.

Leach, CL, Oberg, SG, Sharma, RP, Drown, DB: A nose-only inhalation exposure system for generation, treatment and characterization of formaldehyde vapor. Am Ind Hyg Assoc J 45:269–273, 1984.

Leong, BKJ: Inhalation Toxicology and Technology. Ann Arbor, MI: Ann Arbor Science, 1981.

Leong, BKJ, Powell, DJ, Pochyla, GL, Lummis, MG: An active dispersion inhalation exposure chamber. In: Inhalation Toxicology and Technology, edited by BKJ Leong, pp. 65–76. Ann Arbor, MI: Ann Arbor Science, 1981.

Lippmann, M: Aerosol exposure methods. In: Generation of Aerosols, edited by K Willeke, pp. 443–458. Ann Arbor, MI: Ann Arbor Science, 1980.

Lutz, WK, Schlatter, CH: A closed inhalation chamber for quantitative metabolism studies of volatile compounds with small laboratory animals. Toxicol Lett 1:83–87, 1977.

MacEwen, JD, Urban, ECJ, Smith, RJ, Vorwald, AJ: A new method for massive dust exposures by inhalation. Am Ind Hyg Assoc J 22:109–113, 1961.

MacEwen, JD: Nonconventional systems, Chapter 2. In: Proceedings of Workshop on Inhalation Chamber Technology, edited by RT Drew, BNL-51318, pp. 9–16. Springfield, VA: NTIS, 1978.

MacFarland, HN: Design and operational characteristics of inhalation exposure equipment: A review. Fundam Appl Toxicol 3:603–613, 1983.

MacFarland, HN: Designs and operational characteristics of inhalation exposure equipment. In: Inhalation Toxicology, pp. 93–120. New York: Dekker, 1987.

Malanchuk, M, Barkley, NP, Contnur, GL: Interference of animal source ammonia with exposure chamber atmospheres containing acid particulate from automobile exhaust. J Environ Pathol Toxicol 4:265–276, 1980.

Marra, M, Rombout, JA: Design and performance of an inhalation chamber for exposing laboratory animals to oxidant air pollutants. Inhal Toxicol 2:187–204, 1990.

Mauderly, JL: Influence of sex and age on the pulmonary function of the unanesthetized Beagle dog. J Gerontol 29:282–289, 1974.

Mauderly, JL, Tesarek, JE, Sifford, LJ, Sifford, LJ: Respiratory measurements of unsedated small laboratory mammals using nonbreathing valves. Lab Anim Sci 29:323–329, 1979.

Mauderly, JL: Respiration of F344 rats in nose-only inhalation exposure tubes. J Appl Toxicol 6:25–30, 1986.

McClellan, RO, Boecker, BB, Lopez, JA: Inhalation toxicology: Consideration in the design and operation of laboratory. In: Concepts in Toxicology, edited by F Homburger, pp. 170–189. Basel, Switzerland: Kagar AG, 1984.

Medinsky, MA, Dutcher, JS, Bond, JA, Henderson, RF, Mauderly, JL, Snipes, MB, Mewhinney, JA, Cheng, YS, Birnbaum, LS: Uptake and excretion of [^{14}C]methyl bromide as influenced by exposure concentration. Toxicol Appl Pharmacol 78:215–225, 1985.

Mokler, BV, White, RK: Quantitative standard for exposure chamber integrity. Am Ind Hyg Assoc J 44:292–295, 1983.

Morken, DA: A Radon Exposure Unit for Small Animals. UR-307, University of Rochester, Atomic Energy Report, 1955.

Moss, OR: Comparison of three methods of evaluating inhalation toxicology chamber performance. In: Inhalation Toxicology and Technology, edited by BKJ Leong, pp. 19–28. Ann Arbor, MI: Ann Arbor Science, 1981.

Moss, OR, Decker, JR, Cannon, WC: Aerosol mixing in an animal exposure chamber having three levels of caging with excreta pans. Am Ind Hyg Assoc J 43:244–249, 1982.

Moss, OR: Sampling in exposure chambers. In: Air Sampling Instruments for Evaluation of Atmospheric Contaminants, edited by PJ Lioy, 6th Edition. pp. J2–J5 Cincinnati, OH: ACGIH, 1983.

Murphy, SD, Ulrich, CE: Multi-animal test system for measuring effects of irritant gases and vapors on respiratory function of guinea pigs. Am Ind Hyg Assoc J 25:28–36, 1964.

NIH: Guide for Care and Use of Laboratory Animals. Publication No. 85–23, Bethesda, MD: National Institute of Health, 1985.

Newton, GJ, Kanapilly, GM, Boecker, BB, Raabe, OG: Radioactive labeling of aerosols: Generation methods and characteristics. In: Generation of Aerosols, edited by K Willeke, pp. 399–426. Ann Arbor, MI: Ann Arbor Science, 1980.

Newton, GJ, Cuddihy, RG, Yeh, HC, Boecker, BB: Design and performance of a recirculating radon-progeny aerosol generation and animal inhalation exposure system. In: Indoor Radon and Lung Cancer: Reality or Myth? edited by FT Cross, pp. 709–729. Richland, WA: Battelle Press, 1992.

OECD: Guideline for Testing of Chemicals. Organization for Economic Cooperation and Development, Paris, France, 1981.

Okuyama, K, Kousaka, Y, Yamamoto, S, Hosokawa, T: Particle loss of aerosols with particle diameters between 6 and 2000 nm in stirred tank. J Colloid Interface Sci 110:214–223, 1986.

P'an, AYS, Jegier, Z: A simple exposure chamber for gas inhalation experiments with small animals. Am Ind Hyg Assoc J 31:647–649, 1970.

Paustenbach, DJ, Carlson, GP, Christian, JE, Born, BS, Rausch, JE: A dynamic closed-loop recirculating inhalation chamber for conducting pharmacokinetic and short-term toxicity studies. Fundam Appl Toxicol 3:528–532, 1983.

Phalen, RF: Inhalation Studies: Foundation and Techniques. Boca Raton, FL: CRC Press, 1984.

Raabe, OG, Kanapilly, GM, Newton, GJ: New methods for the generation of aerosols of insoluble particles for use in inhalation studies. In: Inhaled Particles III, edited by WH Walton, pp. 3–18. Surrey, England: Unwin Brothers, 1971.

Raabe, OG, Bennick, JE, Light, ME, Hobbs, CH, Thomas, RL, Tillery, MI: An improved apparatus for acute inhalation exposure of rodents to radioactive aerosols. Toxicol Appl Pharmacol 26:264–273, 1973.

Rao, GN: Significance of environmental factors on the test system. In: Managing Conduct and Data Quality of Toxicology Studies, pp. 173–185. Princeton, NJ: Princeton Scientific, 1986.

Riley, A: A new approach to the construction of small animal inhalation chambers: Design and evaluation. Am Ind Hyg Assoc J 47:147–151, 1986.

Salem, H: Inhalation Toxicology: Research methods, applications and evaluation. New York, NY: Dekker, 1987.

Schreck, RM, Chan, TL, Soderholm, SC: Design, operation and characterization of large volume exposure chambers. In: Inhalation Toxicology and Technology, edited by BKJ Leong, pp. 29–52. Ann Arbor, MI: Ann Arbor Science, 1981.

Silver, SD: Constant flow gassing chambers: Principles influencing design and operation. J Lab Clin Med 31:1153–1161, 1946.

Smith, DM, Ortiz, LW, Archuleta, RF, Spalding, JF, Tillery, MI, Ettinger, HJ, Thomas, RG: A method for chronic nose-only exposures of laboratory animals to inhaled fibrous aerosols.

In: Inhalation Toxicology and Technology, edited by BKJ Leong, pp. 89–106. Ann Arbor, MI: Ann Arbor Science, 1981.

Snipes, MB, Spoo, JW, Brookins, LK, Jones, SE, Mauderly, JL, Orwat, TB, Stiver, JH, Dahl, AR: A method for measuring nasal and lung uptake of inhaled vapor. Fund Appl Toxicol 16:81–91, 1991.

Spiegl, CJ, Leach, LJ, Lauterbach, KE, Wilson, R, Laskin, S: A small chamber for studying test atmospheres. AMA Arch Ind Hyg and Occup Med 8:286–288, 1953.

Springer, DL, Miller, RA, Weimer, WC, Ragan, HA, Buschbom, RL, Mahlum, DD: Effects of inhalation exposure to a high-boiling (288–454°C) coal liquid. Toxicol Appl Pharmacol 82:112–131, 1986.

Stahlhofen, W, Gebhart, J, Heyder, J: Experimental determination of the regional deposition of aerosol particles in the human respiratory tract. Am Ind Hyg Assoc J 41:285–298, 1980.

Sterner, RT, Johns, BE, Crane, KA, Shumake, SA, Gaddis, SE: An expensive humidifying and cooling system for inhalation chambers. Inhal Toxicol 3:139–144, 1991.

Strong, JC, Walsh, M: A facility for studying the carcinogenic and synergistic effects of radon daughters and other agents in rodents. In: Indoor Radon and Lung Cancer: Reality or Myth? edited by FT Cross, pp. 731–740. Richland, WA: Battelle Press, 1992.

Stuart, BO, Willard, DH, Howard, EB: Uranium mine air contaminations in dogs and hamsters. In: Inhalation Carcinogenesis, edited by MG Hanna et al., CONF-691001, pp. 413–427. US AEC, Springfield, VA: NTIS, 1969.

Thomas, AA: Low ambient pressure environments and toxicity. Arch Environ Health 11:316–322, 1965.

Thomas, RG, Lie, R: Procedures and Equipment Used in Inhalation Studies on Small Animals. Lovelace Foundation, LF-11, Albuquerque, NM, 1963.

Vocci, FJ, Krackow, EH, Swann, HE, Eipper, JE, Ballard, TA: An apparatus for quantitative studies of inhaled aerosol by dosimetric procedure. Air Pollut Cont Assoc J 6:69–71, 1956.

Walsh, M, Pritchard, JN, Black, A, Moores, SR, Morgan, A: The development of a system for the exposure of mice to aerosols of plutonium oxide. J Aerosol Sci 11:467–474, 1980.

Weihe, WH: The significance of physical environment for the health and state of adaptation of laboratory animals. In: Defining the Laboratory Animal. Inst Lab Animal Resource, pp. 353–378. Washington: Natl Acad Sci, 1971.

Weissman, SH, Beethe, RL, Redman, HC: Ammonia concentrations in an animal inhalation exposure chamber. Lab Animal Sci 30:974–980, 1980.

Willeke, K: Generation of aerosols and facilities for exposure experiments. Ann Arbor, MI: Ann Arbor Science, 1980.

Wolff, RK, Griffis, LC, Hobbs, CH, McClellan, RO: Disposition and retention of 0.1 μm $^{67}Ga_2O_3$ aggregate aerosols in rats following whole body exposures. Fundam Appl Toxicol 2:195–200, 1982.

Yeh, HC, Newton, GJ, Barr, EB, Carpenter, RL, Hobbs, CH: Studies of the temporal and spatial distribution of aerosols in multi-tiered inhalation exposure chambers. Am Ind Hyg Assoc J 47:540–545, 1986.

Yeh, HC, Carpenter, RL, Cheng, YS: Electrostatic charge of aerosol particles from a fluidized bed aerosol generator. J Aerosol Sci 19:147–151, 1988.

Yeh, HC, Snipes, MB, Eidson, AF, Hobbs, CH, Henry, MC: Comparative evaluation of nose-only versus whole-body inhalation exposure for rats-aerosol characterization and lung deposition. Inhal Toxicol 2:205–221, 1990.

Chapter Three

Generation and Characterization of Gases and Vapors

Brian A. Wong

INTRODUCTION

The accurate administration of a test compound to a laboratory animal is crucial to a toxicity study. For an inhalation toxicity study, the test compound (either a gas or an aerosol) is introduced into the atmosphere of the chamber housing the test subject. A gas test atmosphere is produced by obtaining the test compound in a suitable form, in a gaseous or a liquid state, or as a chemical precursor, and then generating and diluting the gas to proper concentration to be admitted into the exposure chamber. The chamber exposure atmosphere is characterized for the compound of interest to ensure that the concentration is at the desired level. That characterization information can then perhaps be used as feed back to make adjustments in the generation system to maintain the proper concentration. This chapter will describe various ways of producing a gas or vapor atmosphere, present some examples, and then discuss some of the instrumentation and equipment commonly used to generate and characterize a gaseous exposure atmosphere. For a more detailed review of gas generation methods, see Nelson (1992).

In the gaseous state of matter, individual molecules interact very little with each other, but constantly move and collide with each other and the walls of the containment vessel. A gas expands to fill the shape of its container, and will continuously expand if released. A gas may be liquefied by subjecting it to high pressures and low temperatures. The critical temperature of a gas is the temperature above which it is impossible to liquefy, regardless of the applied pressure. Therefore, a compound with a critical temperature below room temperature cannot exist as a liquid at room temperature. The first section of Table 1 lists some common gases with low critical temperatures.

A compound with a critical temperature above room temperature, but with a boiling point below room temperature, can be subjected to pressure and liquefied. These compounds will readily vaporize when released to atmospheric pressure. The second part of Table 1 lists some compounds with boiling temperatures below room temperature and critical temperatures above room temperatures. These compounds are generally available as liquefied gases stored under their own vapor pressure.

A compound with a boiling point above room temperature is, by definition, a liquid at room temperature. However, in equilibrium with the liquid, there are molecules of the compound in the gas state (vapor). These gas molecules exert a vapor pressure that increases with increasing temperature. Tables of temperature and vapor pressure are listed in the CRC Handbook of Chemistry and Physics (Lide, 1994), the Matheson Gas Data Book (Braker and Mossman, 1980), the Properties of Gases and Liquids (Reid et al., 1977) and numerous other sources. The third section of Table 1 lists some compounds with boiling temperatures above room temperature. These compounds are obtained as liquids.

CONCENTRATION UNITS

Prior to discussing the methodology for generating atmospheres of these three classes of gases and vapors, the units to describe the amount or concentration of gas in an atmosphere must be established. It is common practice to describe concentration as "ppm" (parts per million), or volume of gas per million volumes of total gas.

$$1 \text{ ppm of Benzene gas} = (0.000001 \text{ L Benzene}/ 1 \text{ L Total gas}) \times 10^6 \quad (1)$$

The total gas includes the benzene, but for relatively low volumes of the gas of interest (< 5000 ppm), the carrier gas is the main constituent and comprises the total volume of gas. A concentration of 1 ppm is equivalent to 1 μl gas/liter of air, or 1 ml of gas/m^3 of air. Concentration may also be reported as ppb (parts per billion) or percent (%).

Mass concentration units (weight per volume, or mg/m^3) may also be used. The conversion from ppm (C_v) to mg/m^3 (C_m) is:

$$C_m \text{ (mg/m}^3) = C_v(\text{ppm}) \frac{M}{22.4} \frac{P}{P_0} \frac{T_0}{T} \text{ (g/L)} \quad (2)$$

Table 1 Physical Properties of Selected Gases and Liquids

Name	Formula	MW	T_c °C	P_c Atm	T_b °C	ref
Hydrogen	H_2	2.016	−239.9	12.80	−252.8	1
Nitrogen	N_2	28.013	−146.9	33.54	−195.8	1
Carbon monoxide	CO	28.010	−140.2	34.5	−191.5	1
Oxygen	O_2	31.9988	−118.6	49.77	−183.0	1
Carbon dioxide	CO_2	44.011	31	72.85	−78.4	1
Hydrogen chloride	HCl	36.461	51.4	81.5	−85.0	1
Dichlorodifluoromethane	CCl_2F_2	120.914	111.8	40.71	−28.9	1
Ammonia	NH_3	17.031	132.4	111.3	−33.5	2
Formaldehyde	CH_2O	30.026	134.8	65	−19.2	2
Chlorine	Cl_2	70.906	144	76.1	−+34.05	1
Vinyl chloride	C_2H_3Cl	62.499	151.5	56.8	−13.8	1
1,3-butadiene	C_4H_6	54.092	152.0	42.7	−4.4	1
Sulfur dioxide	SO_2	64.063	157.6	77.81	−10.0	1
Ethylene oxide	C_2H_4O	44.054	195.8	70.97	−10.5	1
Menthanol	CH_4O	32.042	239.4	79.9	64.6	2
Vinyl acetate	$C_4H_6O_2$	86.091	251.8	43	72.8	2
Chloroform	$CHCl_3$	119.378	263.2	54.0	61.1	2
Benzene	C_6H_6	78.114	288.9	48.3	80.1	2
Styrene	C_8H_8	104.152	373.8	39.4	145.1	2
Water	H_2O	18.015	374.1	217.6	100.0	2

MW = Molecular Weight
T_c = Critical Temperature
P_c = Critical Pressure
T_b = Boiling Point at atmospheric pressure
References:
1 Braker and Mossman (1980).
2 Reid et al. (1977).

where the factor 22.4 is the molar volume at $T_0 = 273K$ and $P_0 = 760$ mmHg, T and P are the local temperature and pressure of $C(mg/m^3)$, and M is the molecular weight. The conversion from mg/m³ to ppm is

$$C_v(ppm) = C_m(mg/m^3) \frac{22.4}{M} \frac{P_0}{P} \frac{T}{T_0} (L/g) \tag{3}$$

with the units as before (Moss, 1994).

Other relationships and concentrations may be calculated using the ideal gas law and other laws of chemistry dealing with the behavior of gases.

Dilution

A generation system may require a dilution step to achieve the proper concentration. The diluted concentration is calculated as:

$$C_2(\text{ppm}) = C_1(\text{ppm}) * Q_1(\text{L/min}) / Q_2(\text{L/min}) \tag{4}$$

where C_2 is the diluted concentration, Q_2 is the total flow after dilution, and C_1 is the input concentration and Q_1 is the inlet flow. Fig. 1 shows a schematic of a dilution system. For very low target concentrations, several dilution steps may be required. However, uncertainties in controlling flows for each of the dilution steps multiply, leading to low precision at the final output. Multiple dilutions are usually not recommended. See Moss (1994) for further discussion of dilution systems. This equation can also be used to determine the theoretical or nominal exposure chamber concentration, C_2, where Q_2 is the total flow through the chamber, C_1 is the generator output concentration, and Q_1 is the generator flow.

Liquid Vaporization

When a volume of liquid (V_l) is vaporized, the volume of gas (V_g) produced is:

$$V_g \,(\text{ml}_g) = V_l \,(\text{ml}_l)\, \rho\, (\text{g/ml}_l)\, \frac{22.4}{M}\, \frac{P_0}{P}\, \frac{T}{T_0}\, (\text{L/g}) \times 10^3 \,(\text{ml}_g/\text{L}) \tag{5}$$

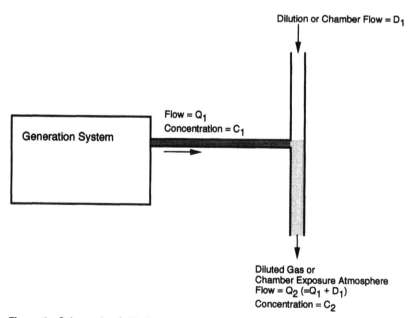

Figure 1 Schematic of dilution system or chamber inlet system.

where ρ is the liquid density.

The vaporization rate (Q_l) to achieve a target concentration (C_v) can be determined based on equations (5) and (4):

$$Q_l \text{ (ml}_l\text{/min)} = \frac{C_v(\text{ppm})}{10^6} \, Q_2 \text{ (L/min)} \, \frac{1}{\rho \text{ (g/ml}_l)} \, \frac{M}{22.4} \, \frac{P}{P_0} \, \frac{T_0}{T} \text{ (g/L)} \qquad (6)$$

The following equation provides the nominal concentration based on the liquid vaporization rate and chamber airflow (Q_2):

$$C_v \text{ (ppm)} = \frac{Q_l \text{ (ml}_l\text{/min)}}{Q_2 \text{ (L/min)}} \, \rho \text{ (g/ml)} \, \frac{22.4}{M} \, \frac{P_0}{P} \, \frac{T}{T_0} \text{ (L/g)} \times 10^6 \qquad (7)$$

The use of some of these equations will be provided in examples later in this chapter.

BASIC CONSIDERATIONS

When a target concentration has been established, there are other basic considerations related to safety and the practicality of generating the test atmosphere. These include:

Inflammability Limits Many gases, particularly organic gases, when mixed with oxygen in the proper proportions, will react readily at elevated temperatures. Thus, a flame or spark in the mixture can result in a fire or explosion. Lower and upper inflammability limits, usually given as the percent of gas in either air or oxygen, define the concentration range that can sustain a chain-type reaction. Many references such as the CRC Handbook of Chemistry and Physics (Lide, 1994) have tables of the upper and lower inflammability limits. Material Safety Data Sheets provided by chemical manufacturers may also contain this information. All proposed studies should be carefully checked to ensure that the generated concentrations are well below the lower inflammability limits. Note that if a concentrated gas is diluted, the gas may pass through the inflammability limits during dilution, rendering a certain volume of the generation equipment susceptible to explosions. An inert gas such as nitrogen, with no oxidizer present, may be used as a diluent, to avoid explosive situations. To avoid sparks, metal tubing and parts should be grounded to prevent static charge buildup.

Reactivity The gas or vapor may react with materials in the generator, tubing conveying the gas or vapor, chamber structural materials, or with other gases that may be present in the chamber.

• Generator   The materials of construction of the generator may react with the liquid or gas being generated. Materials such as glass, Teflon or stainless steel that are non-reactive with most chemicals are typically used in generation systems. Occasionally overlooked are the materials used as seals for various pieces of equipment such as flow meters or valves, which are usually constructed of a polymeric material such as buna-n or viton rubber. These seals may be very efficient at adsorbing organic compounds, leading to a reduced concentration, or reduced sealing capacity. For example, an ethylene oxide atmosphere reacted with the viton seals in a mass flow controller, causing the controller to stick. Using seals made of kalrez® solved the problem.

• Tubing   The tubing used to convey the test compound may also react. For example, styrene vapor may be absorbed by tygon tubing in the generation system.

• Exposure Chamber   Stainless steel is a common material for chamber construction. However, stainless steel is subject to corrosion from chlorine or sulfur dioxide-containing atmospheres (Hinners et al., 1968). Plastics (especially acrylics) may also be used because of their transparency and ease of machining (MacFarland, 1983). However, they may adsorb significant amounts of organic vapors, and may lead to clouding of the transparent plastic or to a lower than expected gas concentration.

• Other Gases   Ammonia, produced from the activity of urease bacteria on urine and feces, can react with chlorine to produce chloramines (Barrow and Dodd, 1979). Ammonia can also react with acid aerosols and gases (Higuchi and Davies, 1993).

Typically, the materials of choice for construction of the generation system are stainless steel, glass, and Teflon®, for inertness to most chemicals. However, a chemical needs to be checked for compatibility with all materials in the generation system.

Maximum Achievable Concentration   For a compound that is a liquid at room temperature and pressure, the compound's vapor pressure sets an upper limit to the gas concentration that can be achieved. This vapor pressure can be used in the generation process to produce a vapor at a known concentration. However, operating too near the vapor pressure of a vapor can lead to condensation of the vapor with small changes in temperature or pressure (such as might be encountered when passing through an orifice or constriction in the delivery line).

Oxygen Concentration   The operator should take care not to reduce the oxygen concentration below 19% (Snellings and Dodd, 1990) in the chamber, especially if the gas of interest is introduced by an inert carrier gas.

Relative Humidity   The relative humidity may affect the gas concentration, especially if the gas reacts with water, or if the detector is sensitive to water concentration.

Operator Safety   Safety of personnel operating the generation equipment and personnel handling the animals being exposed is of utmost importance. Exposure to the compound of interest must be minimized at all times, during the generator development, testing, and operation. Where possible, failsafe devices and procedures should be installed and used in the generation system. For example, in the system (described later) of Miller et al. (1980), electrical equipment such as heaters and the liquid pump are connected to pressure-activated switch. If the compressed air system should fail, the electrical components connected to the switch would automatically shut off. Safety features should be put into place to permit diversion of the gas flow from the chamber or shut off the gas generation equipment quickly. Monitoring of the area for the test compound should be conducted during exposures to ensure that there are no leaks in the generation system. Care should be taken that the chamber is clear of the compound before personnel are permitted to enter the chamber, or remove animals from the system.

After consideration of these safety and feasibility aspects, the actual methodology for generation of the gaseous test compound can be selected.

GENERATION

The method selected to generate a gas or a vapor will depend on several factors, including the target concentration, physical properties of the material, availability of the material from commercial sources in desired forms, and the techniques used by previous workers to generate the same or similar materials. This next section will provide examples of the various techniques used to generate gases and vapors.

Generation of Gases   Compounds available as gases (as exemplified by the first section of Table 1) may be generated from a cylinder of the pure gas, as shown in Fig. 2. A pressure regulator is attached to the tank (whose pressure may be greater than 2000 psig when full) to reduce and regulate the pressure. A flow controller (rotameter or mass flow controller) is then used to control the flow of gas into the chamber air supply. Compounds in the other sections of Table 1 may be available as mixtures of gases diluted in an inert gas, with the maximum concentration of these mixtures determined by the compound vapor pressure. Pure gases and standard mixtures and custom mixtures are readily obtainable from commercial gas suppliers.

Example—Himmelstein, et al. (1994) used a cylinder of 15,000 ppm butadiene in nitrogen that was subsequently diluted into the chamber air to generate target concentrations of 62.5, 625, and 1250 ppm butadiene. Gas flows of 62.5, 625, and 1250 ml/min were required for a total flow of 15 L/min through the chamber. The concentration was monitored with an infrared spectrometer.

Generation of Vapors   Vapors are generated by the controlled vaporization of liquids and pressurized liquids at a steady rate into the chamber air

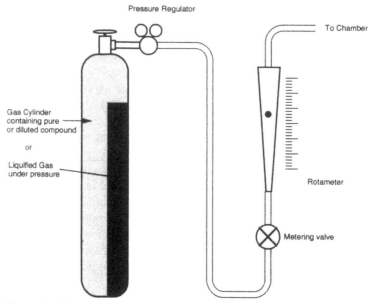

Figure 2 Schematic of a system to generate a gas from a gas cylinder. The pressure regulator reduces the pressure of the gas going to the metering valve and flow meter (a rotameter is shown). The figure also shows that a cylinder containing a liquefied gas can be used in a similar manner.

supply. Critical factors that affect the stability and predictability of the generation system include the efficiency of vaporization of the gas and the accuracy of the flow control of the liquid being vaporized. Efficiency of vaporization is enhanced by (1) increasing the mixing of the liquid vapors with the carrier gas, (2) increasing the surface area of the liquid and (3) heating the liquid. Metering pumps can be used to improve the accuracy of the liquid flow control.

A simple technique is to "mix" an inert carrier gas with the liquid to be vaporized. The mixing can occur by injecting the liquid directly into the carrier gas stream, or by bubbling the carrier gas through the liquid (Nelson, 1992).

Example—Fig. 3 shows a 10 gallon stainless steel pressure vessel that was partially filled with chloroform. At equilibrium, a chloroform vapor pressure of 171.4 mmHg at 22.2°C and a local barometric pressure of 752 mmHg would produce a static concentration of about 228,000 ppm. Nitrogen was bubbled through the liquid at a calculated rate of 260 ml/min and carried chloroform vapors into the chamber inlet air flow of 2000 L/min to produce a chamber concentration of 30 ppm. Concentration could be adjusted by changing the flow of nitrogen. Warheit et al. (1992) used a similar system to generate a 100, 540, or 2200 ppm atmosphere of HCFC-122 (1,1-difluoro-1,2,2-trichloroethane).

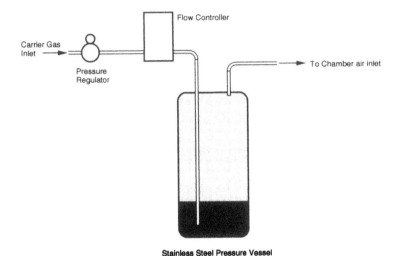

Stainless Steel Pressure Vessel

Figure 3   Generation system bubbling carrier gas through a liquid. The compound to be vaporized is stored in a stainless steel pressure vessel. A flow meter (mass flow controller) meters the carrier gas that bubbles through the liquid. Vapors are carried out to the chamber.

A similar system such as one described by Snellings and Dodd (1990) can be used for compounds that are liquefied under pressure, or systems where very low concentrations are required. The liquid is placed in a stainless steel pressure vessel and pressurized with an inert gas (usually nitrogen). The composition of the vapor/inert gas mixture is controlled by the vapor pressure of the liquid test material. This mixture is driven by the pressure through a flow meter to the chamber.

Example—A 1 ppm benzene atmosphere was generated by nitrogen pressurizing a stainless steel vessel containing liquid benzene. A pressure of 10 psig produced gas concentration of about 63000 ppm of benzene in the pressure vessel. A mass flow controller metered the resultant benzene vapor/nitrogen mixture at 32 ml/min into the chamber inlet air flow of 2000 L/min.

Example—A 100 ppm ethylene oxide atmosphere was generated by adding nitrogen gas to a dual valve cylinder (Fig. 4). Ethylene oxide is normally removed from the cylinder by adding nitrogen in one valve and pressurizing the cylinder to 50 psig. That pressure is used to push the ethylene oxide liquid through a liquid eductor tube connected to the second valve. The connections were reversed, and nitrogen was added into the valve connected to liquid eductor tube. The other valve that drew from the head space was connected to a mass flow controller that metered out the resulting mixture. At 20°C, the ethylene oxide vapor pressure is 1095 mmHg. Adding nitrogen to make a 40 psig mixture results in a gas mixture of about 388000 ppm EtO. This mixture at a flowrate of 515 ml/min will produce a chamber concentration of 100 ppm at 2000 L/min.

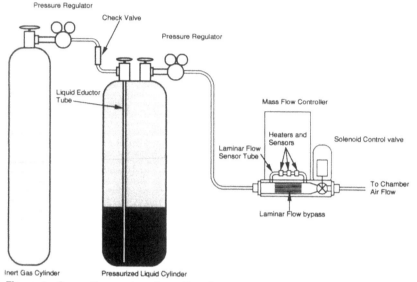

Figure 4 Generation system for a liquefied gas. The liquefied gas is stored under pressure. An inert gas is added to the liquefied gas cylinder through the liquid eductor tube. The vapor mixture exits through the second valve and is metered through a flow meter (mass flow controller) to the chamber.

Alternatively for gases liquefied under pressure, the greater than atmospheric vapor pressure can be used push liquid or vapor through the generation and delivery system.

Example—Pullinger et al. (1979) inverted a 1,3 butadiene cylinder and used the vapor pressure to push the liquid into a distribution system. At the end of the distribution system, the butadiene passed through a heat exchanger and was vaporized. The vapors were then metered into the exposure chambers to produce nominal concentrations of 700, 1000, 2000, 3200, and 8000 ppm into 1000 L/min.

Example—Snellings et al. (1982) heated an ethylene oxide cylinder to 35°C with a temperature controlled bath. The vapor pressure was used to push the ethylene oxide gas through a stainless steel manifold to valves and flow meters. The EtO vapor was then metered into exposure chambers at nominal concentrations of 10, 33, and 100 ppm. Figure 2 is a schematic of this type of system.

Vaporization of the liquid may be enhanced by increasing the surface area over which the liquid is flowing. Often times, the vaporizing section utilizes a counter-current flow system in which a carrier gas flows in the opposite direction as the liquid flow. The combination of counter current flow, and increased liquid surface area greatly enhances vaporization.

Example—Potts and Steiner (1980) described a counter-current system using a multi-plate distillation column. A low-volatility liquid was continuously pumped in at the top of the distillation column. As the liquid flowed down across the plates of the column, air flowed counter-current, entraining vapor. The distillation plates increased the liquid surface area exposed to the air flow.

As the liquid is vaporized, the liquid and surroundings will cool from the latent heat of vaporization, altering the vapor pressure. To counteract such changes, the liquid and the vaporization area temperature can be controlled by immersion in a constant temperature bath. Alternatively, the vaporization rate can be increased by adding heat to the vaporization area. Heat tapes, or heating jackets may be used to add controlled heating to the vaporization system.

Example—In a modification to the system previously described, Potts and Steiner (1980) encased the distillation column in a water jacket thermostatted a few degrees below room temperature. This system controlled the vaporization rate more precisely, and provided a more stable concentration.

Example—A 25% w/w solution of aqueous glutaraldehyde was pumped into a rotating evaporator column by a metering pump. Heated compressed air was passed up through the column, entraining glutaraldehyde vapors. The rotating column spread the liquid over the entire column surface, enhancing the vaporization of glutaraldehyde. Fig. 5 shows a schematic based on this generator (Gieschen et al., 1991). This system is similar to a system described by Weigel et al. (1991), in which a commercial rotary evaporator is used to vaporize a liquid. A water bath is used to maintain the liquid in the rotating flask at a constant temperature. Nitrogen gas carries the vapors into condensing section, where the vapor is cooled, and higher boiling impurities can be condensed out (Fig. 6).

In these previously mentioned systems, the rate of transfer of vapors to the gas stream is controlled by the rate of vaporization from the liquid. Excess, unvaporized liquid is collected. The vaporization rate is sensitive to the local temperature and to the saturation of the air or carrier gas. A "nominal" concentration is difficult to determine, since there is no way to measure or estimate the mass flux of vapor into the air stream. If, instead of having excess liquid, all of the liquid entering the system is vaporized, then the liquid flow rate controls the vaporization rate. The gas concentration can be controlled by metering the flow of liquid to control the vaporization rate, and a "nominal" or theoretical concentration can be determined.

A syringe pump, in which a syringe plunger is slowly depressed by a motor drive, can be used to accurately control liquid flow to be vaporized. This system usually cannot generate large concentrations for long periods of time, however schemes do exist that allow one syringe to be filled simultaneously as another is being emptied, allowing continuous operation (Nelson, 1992).

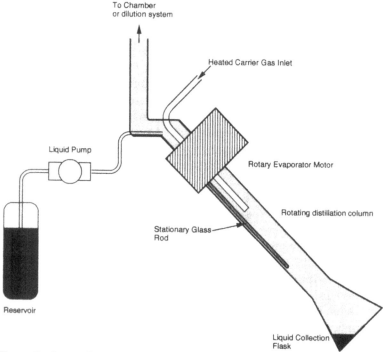

Figure 5 Generation system based on Gieschen et al. (1991). A rotary evaporator motor is used to rotate a glass evaporator column. A liquid pump introduces the compound which flows down the wall of the evaporator column and is spread across the wall by a glass rod. Heated air is introduced to evaporate the liquid and carry the vapors out to the chamber.

    A piston pump is commonly used to introduce the liquid into a vaporizing section. This pump is valveless design that uses a reciprocating and rotating piston to alternately suction and discharge a liquid. A pulsing flow of liquid is produced. If the pulsing period is short relative to the residence time in the vaporization area and the chamber, the pulsations will not be noticeable. Piston pumps are capable of very low flows and different materials of construction are available for chemical compatibility.

    A peristaltic pump can also be used to meter liquid flow. A peristaltic pump uses rollers attached to a rotor that compress a length of flexible tubing. As the rollers move, they force liquid through the tube. The liquid flowrate is controlled by a combination of the size of tubing used and the rotational speed of the rotor. Various types of tubing can be used, to ensure chemical compatibility.

    Example—Miller et al. (1980) describe a system that utilizes a piston pump to meter liquid into the long end of a J-shaped glass tube assembly. Glass beads in the long end of the J-tube greatly increase the surface area of

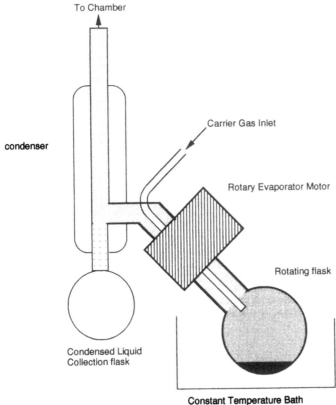

Figure 6 Generation system based on Weigel et al. (1991). A rotary evaporator system is used. Liquid is contained in the rotating round-bottom flask that is maintained in a constant temperature bath. Carrier gas is introduced to carry the vapor into a condensing section. The vapor is cooled, and any less volatile contaminants are condensed out. The remaining vapor mixture exits to the chamber or dilution system.

the liquid. Heated air flows in through the short end of the J, counter-current to the liquid flow. Maples and Dahl (1993) used a heated J-tube to vaporize various alkene and alkene oxides. Tilbury et al. (1993) used a similar system, where the J-tube was stainless steel rather than glass, and the J-tube itself was heated rather than the carrier air, to vaporize unleaded gasoline (Fig. 7). In this study, unleaded gas at nominal concentrations of 67, 292, or 2056 ppm into 340 to 390 L/min airflow was generated. Because unleaded gas is a complex mixture, vaporization of the liquid must be complete to avoid fractionation of the mixture (in which the more volatile materials go into the exposure atmosphere, and the less volatile materials remain in the liquid phase).

Example—Decker et al. (1982) developed a generator that consisted of

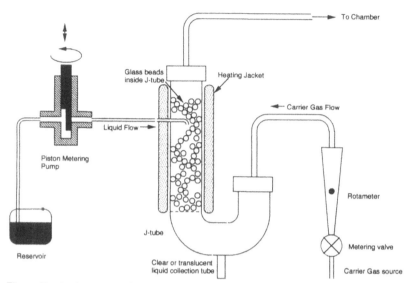

Figure 7   J-tube generation system based on Miller et al. (1980). Liquid is pumped into the glass-bead filled long end of a J-shaped tube. Carrier gas, in a counter current flow, entrains vapor and carries it out to the chamber or dilution system. The J-tube may be heated as shown here, or the carrier gas may be heated as described by Miller et al.

a cylindrical stainless steel assembly covered by a glass fiber wick. The liquid to be vaporized was pumped through a tube that contacted the wick. The liquid was drawn into the wick and vaporized into the surrounding air stream. A heater inside the stainless steel cylinder aided the vaporization process. This generator was designed to be placed directly into the chamber inlet air flow duct. A modification of this system was used to generate 2-methoxy ethanol at 400 ppm (see Fig. 8).

Example—Carpenter et al. (1975) and Snellings and Dodd (1990) used a spiral grooved glass tube as the vaporization section. Liquid was introduced at the top of the glass tube, and flowed down the inner wall of the grooved section. The grooves increased the surface area. The glass tube was electrically heated with resistance wire to aid vaporization. The system was designed to have the total chamber air flow counter-current to the liquid. Isopropanol was generated at concentrations of 100, 500, 1500, or 5000 ppm for a 13 week exposure using this system (Burleigh-Flayer, et al., 1994).

The liquid surface area can be increased by nebulizing the liquid into small droplets. The small droplets evaporate quickly, and the resultant vapor is introduced into the chamber air inlet stream. Some nebulizer designs allow the liquid consumption to be controlled.

Example—Vapor phase nitric acid (HNO_3) was generated by nebulizing

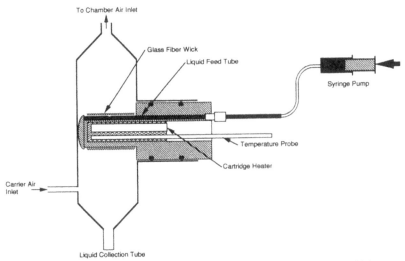

Figure 8   A wick generator system based on Decker et al. (1982). Liquid is pumped through a feed tube into a glass fiber wick. As air flows past the wick, vapor is entrained. Vaporization can be enhanced by heating the assembly with the cartridge heater.

a dilute aqueous solution of HNO_3 (Nadziejko et al., 1992). Target concentrations were 0.25 and 1.00 mg/m³.

Example—Bogdanffy et al. (1994) generated vapors of vinyl acetate by metering liquid vinyl acetate into a nebulizer. The droplets were mixed with filtered air and evaporated to produce concentrations of 50, 200, and 600 ppm of vinyl acetate vapor.

For low concentrations, a gas permeation system can be used. This system is based on the predictability of a gas diffusing through a permeable wall or a capillary tube (Fig. 9). The diffusion rate is dependent on the temperature, permeable wall properties or capillary dimensions and physical properties of the gas. A permeation device containing the test compound is placed in an oven, where the diffusing gas is carried into a diluent gas stream. Because the diffusion rates are small, and the devices carry only a limited amount of material, this system is generally used only for small exposure systems with relatively low total airflows.

Example—Schlesinger et al. (1994) used certified permeation devices to generate atmospheres containing 50, 150, or 450 µg/m₃ of HNO_3.

Direct Generation   A gas may be generated by a chemical reaction from precursor chemicals. This method of generation is specific to the desired gas. Two examples of this method of generation are ozone and formaldehyde. Ozone is a very reactive gas and cannot be stored for any length of time.

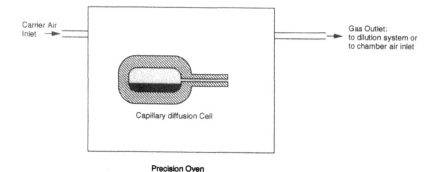

Precision Oven

Figure 9 Generation system based on permeation devices. The device shown is based on the diffusion of gas through a capillary tube. Other devices are based on a tube made of a permeable material (typically polymeric). Liquid inside the tube diffuses through the tube material at a controlled rate determined by oven conditions. The diffusing vapors are entrained into carrier air flow through the oven.

Consequently, it must be generated as it is being used. Ozone is generated by passing oxygen through an ultraviolet generator or an electric discharge system (Nelson, 1992).

   Example—Schlesinger et al. (1994) generated an atmosphere of 0.15 ppm ozone using an ultraviolet ozone generator.

   Formaldehyde gas can be generated from the thermal decomposition of paraformaldehyde, a solid polymer of formaldehyde. The paraformaldehyde is held in a beaker inside a sealed stainless steel container, and heated in a well-regulated oven. Carrier gas (nitrogen) is passed through the stainless steel container, where it mixes with formaldehyde vapors, and is carried out to the chamber (Fig. 10). By varying oven temperatures, beaker sizes and chamber air flows, formaldehyde concentrations can be varied from 6 to 15 ppm (Chang et al., 1983).

   Nelson (1992) describes systems to generate mercury, hydrogen cyanide, chlorine dioxide and other specialized chemicals.

GAS CONTROL

A gas generation system requires the control of pressure and flow as indicated in the schematics. A gas flows from a region of high pressure to a region of low pressure. Therefore, adjusting the pressure gradient or the area of the conducting tubing controls the flow of gas.

Pressure Regulation

The collision of gas molecules with the walls of the container exerts a force on the walls. This force is measured as pressure, and is defined as a force per unit area. Some of the common units of pressure are pounds per square inch

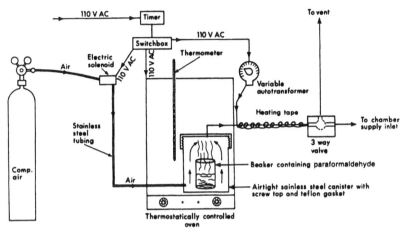

Figure 10 Formaldehyde generation system. Solid paraformaldehyde is slowly decomposed to formaldehyde vapors which are entrained into the carrier gas stream. (Chang et al., 1983, Fig. 1 Reprinted with permission from Academic Press).

(psi) or newtons per square meter (Pascal). Pressure may also be expressed in atmospheres (atm) relative to the atmospheric pressure at sea level. Another common pressure unit is the length of column of mercury or water that exerts a force equal to the surrounding air pressure. So, at sea level, 0°C:

$$1 \text{ atm} = 14.7 \text{ psi} = 1.013 \times 10^{-5}\text{Pa} = 760 \text{ mmHg} = 29.92 \text{ inHg}$$

It has been common practice with psi units to use two points of reference—absolute vacuum and atmospheric pressure. When measured with respect to absolute vacuum, pressure is reported as psia. When measured compared with the atmospheric pressure, units are reported as psig, or gauge pressure. Obviously, the gauge pressure is subject to variations in the local atmospheric pressure due to weather changes. Also, the gauge pressure varies from location to location, because of altitude. Barometric pressure is usually reported in mmHg, while small pressures, such as the static pressure differential between chamber and room may be reported in inH_2O.

Pressure regulators are used in dynamic situations to take a gas from high pressure to a lower pressure. Pressure regulation is important in maintaining a constant flow as gas flow is controlled by the pressure differential upstream and downstream of the flow metering device. If either the upstream pressure or the downstream pressure changes, the gas flow will change. Pressure regulators are essential for reducing internal pressures near 2000 psig in gas cylinders to more usable pressures around 20 psig and maintaining this lower pressure as the cylinder pressure decreases as the gas is used. As a cylinder empties, the reduced pressure may change slightly from the original set point at high cylinder pressure. The measured reduced pressure should be monitored over the "life" of the cylinder, and adjusted, if necessary, to maintain a constant pressure. Pressure regulators are also used to reduce air pressure from compressed air systems.

Flow Monitoring and Control

In order to assure proper operation, determine dilution ratios or estimate chamber concentrations, the volumetric flow of gas must be monitored and controlled. Many devices used to accomplish these tasks will be described briefly in this section. Commonly used units for gas flow are cubic centimeters or milliliters per minute (cc/min or ml/min) and liters per minute (L/min).

• Orifice   One technique to monitor or control flow is to place a restriction in the ducting carrying the air and measure the pressure differential (or pressure drop) between the upstream side and the downstream side of the restriction. The restriction can be a simple orifice in a flat plate, a tapered restriction (venturi), or a bundle of capillary tubes. (Hering, 1989). Knowledge of the pressure drop, relative cross-sectional areas of the duct and the constriction and the gas density is used to compute the flow rate. Typically, the computed flow rate is used to select the proper size restriction. A calibration graph of flow rate vs. pressure drop is used to determine actual flow rates.
• Critical Orifice   If the pressure differential is large enough, the flow through the orifice reaches a limit because it cannot exceed the speed of sound. When the flow reaches sonic velocity, the flow becomes critical, so that the device is known as a critical orifice. These devices are widely used because downstream pressure only has to be held below a certain level.
• Valves   Valves are used to control gas flow by providing a variable passage for the gas to flow through. Valves can be a simple on/off system, such as a ball valve or toggle valve. They can also be used to regulate flow by providing a variably-sized passage for the gas to flow through. Other valves can be used to divert gas flow from a source to one direction or another (three-way valves). Valves may be manually controlled or electrically or pneumatically operated.
• Rotameters   In a rotameter, gas flows upward through a vertical tapered tube. A float inside the tube is suspended at a point where the drag force exerted by the gas flowing past equals the weight of the float (see the rotameter schematic in Fig. 2). A scale is etched on the outside of the tube to provide readings for calibration curves. Rotameters are typically constructed from a block of acrylic plastic or a glass tube. The range of a rotameter can be altered by using floats of differing densities. Rotameters come in a variety of sizes and flow ranges, ranging from a few cc/min to hundreds of L/min. Rotameters can be equipped with a valve to control flow.
• Mass Flow Meters   The flow of gas is directed through a flow bypass section. This section can be a laminar flow element, a tube constriction, or other. A portion of the flow is diverted around the bypass section into the flow sensor section. The sensor section consists of some combination of heating and temperature sensing elements. As the gas flows through the sensing

tube, the gas is heated. The temperature differential between the inlet and outlet of the sensor is proportional to the flow and the heat capacity of the gas. The flow meter in Fig. 4 shows a schematic of a mass flow meter. Differences in heat capacities between gases can be compensated for electronically so that the meter can be direct reading for any gas.

  • Mass Flow Controllers   Mass flow controllers link a mass flow meter with a control valve (Fig. 4). A signal or voltage is set, and the control valve adjusts the gas flow to maintain the setpoint. Mass flow controllers offer the advantages of excellent repeatability and stability. They are easy to link to computers for process control and automation.

Calibration   Most flow monitoring devices require calibration against some kind of standard. For higher flows, a dry gas meter (similar to the meters used to measure natural gas usage in homes) is commonly used to measure the volume of gas that flow through the device. A timer is used in conjunction with the dry gas meter to measure flow rates. Also commonly used are soap-bubble meters. A burette can be modified to permit the flow of gas to carry a soap film up through the burette. The user must measure the time required for the soap film to traverse a known volume. Soap bubble meters are commercially available where the soap film travel is opto-electronically timed to provide a direct readout of flow rate for the user. Recently, a frictionless piston meter has become commercially available that is easy to use and covers a wide range of flows. The reader is referred to Nelson (1992) and Hering (1989) for further information on calibration of air flows.

CHARACTERIZATION OF GAS CONCENTRATION

It may be possible to calculate a nominal concentration from the liquid flow rate and the dilution gas flow rates, but several factors may cause the actual concentration to deviate from the theoretical. These include inaccuracies in the calibration of the liquid flow or air flow devices, variations in temperature, or adsorption of the gas of interest onto generator or chamber surfaces. Therefore, inhalation studies require the measurement of gas concentration of the exposure atmosphere. Additionally, comparison of the nominal versus the measured concentration can give insight into the quality of the generation system or the presence of previously undetected leaks in the system.

The instruments most commonly used to characterize the exposure concentration in inhalation studies are the gas chromatograph (GC) and the infrared spectrometer (IR). A wide range of gases and vapors can be analyzed by both instruments.

Infrared Spectrometry   Molecules can absorb wavelengths of infrared radiation that are characteristic of the bonds between its constituent atoms. Hence, the infrared spectra can be used to identify molecular species. The concentration of a gas can be determined by measuring the amount of absorption of a selected wavelength of light as it passes through a cell containing the gas. In the infrared spectrometer, a monochrometer or a filter is used to select a certain wavelength of light. The light beam may be reflected off mirrors several times to form a long path length beam through the cell containing the test atmosphere. It then is measured by a sensor. The output is given in absorbance units, which can then be related to concentration when properly calibrated. Examples of the use of IR spectrometry in inhalation studies is provided by Chang et al. (1983), Bogdanffy et al. (1994), and Himmelstein et al. (1994).

Typically, a sample of the chamber atmosphere is drawn through the gas cell until the reading stabilizes. The concentration is determined from concentration vs. absorbance units calibration graph. Water has some absorbance bands in the infrared, and changes in relative humidity may interfere with the measurement of low gas concentrations in those bands.

Gas Chromatography   Gases can be drawn through a column that contains a packing material with differing affinities for the gases being separated. The gases require different times to pass through the column and are identified by their retention times within the column. A detector at the end of the column provides a measure of the amount of gas that is present. The gas chromatograph instrument consists of an injection system, a column and a detector, all housed within an oven to define the temperature conditions. The detectors that are commonly used on gas chromatographs are flame ionization, photoionization, thermal conductivity, and electron capture detectors. The GC is typically used by continually drawing a sample from the exposure chamber. From this line, a volume of the atmosphere is drawn into a sample loop and then pulled into the column. The GC is programmed to quantify only the material found at the proper retention time. Usually, column and operating conditions of the GC are selected to have a short retention time for the compound of interest. Therefore, the analysis time is short, and the measure concentration is a near real-time measurement. Maples and Dahl (1993), Tilbury et al. (1993), Bogdanffy et al. (1994), Warheit et al. (1992), Kari et al. (1993), and Carpenter et al. (1975) used GC techniques to measure exposure chamber concentration.

There are other instruments, commonly used for environmental sampling, that may find applicability to gas characterization for inhalation studies. One is the photoionization detector (PID). This detector is used on GC's, but it can also be used without a chromatographic column. The user loses the benefit of the separation of gases, but if only one gas is being introduced into the chamber, separation may not be required. A photoacoustic IR analyzer uses the absorbance of infrared radiation by molecules coupled to a unique detector. The selected wavelength of IR radiation is pulsed at a certain frequency. The temperature of the gas absorbing the radiation will increase, causing a pressure increase. The pulsing pressure is detected by microphones, with the strength of the acoustic signal proportional to the concentration of gas present.

Some gases have specific monitors or detectors. Schlesinger et al. (1994) modified an oxides of nitrogen chemiluminescence analyzer to measure nitric acid (HNO₃). That study and Nadziejko et al. (1992) also used an ultraviolet light absorption ozone monitor to measure ozone in the test atmosphere. Other electrochemical sensor and monitors specific for certain gases are available and may be found in industrial hygiene and environmental sampling literature.

Batch Methods   These methods require that a batch of the exposure atmosphere be sampled and held in a storage system for later analysis. A sample of the exposure atmosphere may be drawn through an impinger or bubbler at a known flow rate for a certain period of time. The gas becomes dissolved in the impinger fluid, or may react with a chemical in the fluid. The impinger fluid is later analyzed by appropriate methods for the total gas content. Knowledge of the gas amount and the total volume sampled provides the gas concentration. Barrow and Dodd (1979) used impingers containing a sulfuric acid solution to collect ammonia for later analysis via a colorimetric technique. Nadziejko et al. (1992) sampled an atmosphere containing nitric acid and adsorbed HNO_3 vapors onto a nylon filter. The filter was extracted into an aqueous media and analyzed for nitrate ion on an ion chromatograph. Katz (1977) provides methods for analyzing air samples for a variety of chemicals.

Batch methods have the disadvantage of a time delay from when the sample is taken to when results of the analysis are ready. The time delay prevents close control of the concentration during the exposure as immediate feedback is not available. For this, real-time or near real-time techniques are required.

Calibration   Gas characterization instruments require periodic calibration to assure the reliability of their analysis. The various techniques are roughly grouped as static or dynamic calibration. One excellent static system is a premixed gas cylinder with a certificate of analysis from a commercial gas supplier. A specified gas concentration may be ordered, and the vendor will provide an analysis of the gas mixture. Custom calibration mixtures of different concentrations can be used to generate a calibration curve. Alternatively, gas calibration mixtures may be prepared in Tedlar® or Teflon® bags. A known volume of carrier gas or air can be admitted into the bag. A gastight syringe is used to inject a known volume of the test compound gas or liquid into the bag. Contents of the bags can then be processed through a GC or IR spectrometer to generate a calibration curve.

An IR spectrometer with a gas cell is easily calibrated by connecting a bellows pump to the inlet and outlet of the gas cell, recirculating the contents. A known volume of liquid or gas is injected into this closed loop and the absorbance units plotted against the known concentration. A series of injections is made to obtain the calibration curve. Prior to the calibration curve, a spectra can be obtained to determine the wavelength of maximum absorption.

The gas permeation system, previously mentioned in the generation section, can also be used to dynamically generate a known gas concentration for calibration purposes. Moss (1994) and Nelson (1992) provide more detailed descriptions of calibration methods.

AUTOMATION AND COMPUTER CONTROL

The following is an example of a standard procedure for the daily operation of an inhalation exposure:

1 Prior to the scheduled start of the exposure, turn on the generator heaters and other systems that require a warm up period.

2 After warm-up, start the generator system with the gas flowing to an exhaust system.

3 When the generator has stabilized and the exposure system is ready, start the flow of gas through the exposure system. Begin timing the exposure.

4 Use the analytical instrument to monitor the concentration of the gas in the exposure system. Readings are usually taken at predetermined intervals, such as once per hour or sooner. Any information that would indicate a problem with the analytical instrumentation should be recorded immediately in a daily operational log. Such information might be important if readings are unusual.

5 Based on the measured concentrations, adjust the generator if necessary to achieve the target concentration. Record any settings for the generation and analytical system, changes to any settings, and any other pertinent readings.

6 At the end of the exposure time period, switch the flow of gas from the exposure chamber to exhaust and begin to shut down the generator.

7 Gather the data and compute daily averages. Concentration readings taken within the exposure chamber ramp up time (chamber t_{99}) should not be included in daily averages. Any other readings should be included, unless it can be demonstrated that the analytical instrument was not operating properly.

8 Once the system has been shut down, preparations should be made for the next exposure.

Additional steps such as running a calibration standard through the analytical instrument may also be incorporated, depending on need.

The proliferation of data logging systems, computer data acquisition systems, and computer control systems has made automated data recording and control of inhalation studies readily achievable. Analytical instruments with an analog or "chart recorder" output can be easily connected to a computer data acquisition system to record concentrations automatically (step 4 above). These data can then be automatically tabulated and averaged over each day. With additional programming knowledge, an instrument with an RS-232 communications port can be connected up to a computer system to transmit data for automatic collection.

Computer systems such as process control systems or building management systems can be used to automate the exposure process. These systems can be programmed to switch on devices such as heaters, solenoid valves, and mass flow controllers at certain times, automating steps 1, 2, 3, 6, and 8 above. The data from the analytical instrument can be compared with the target concentration and control signals sent to the generation system (for example) to adjust a liquid pump rate or carrier gas flow (step 5). Such a system, programmed to accomplish many of the steps outlined above, releases the operator from tedious and time-consuming data gathering. However, such a system still requires careful attention from the operator to ensure that the experiment is proceeding properly. Additionally, several safeguarding steps should be programmed into the system to handle unlikely, but major events. Occurrences such as a power outage, fan breakdown, analytical instrument failure (lamps burning out, flame detector going out), the failure of a mass flow controller, or accidental entry by laboratory personnel can jeopardize an exposure. Programs that will take appropriate action such as switching gas flow to exhaust, shutting off the gas generation system, or issuing an alarm to appropriate personnel should be built into the automated system.

SUMMARY

A variety of methods are available to generate gases and vapors for inhalation studies. These include the direct generation and dilution of gases stored in cylinders, the controlled vaporization of liquids to gases, and the subsequent dilution and transport to the exposure chamber. The gas concentration may be characterized by gas chromatography, infrared spectrometry, or other instrumentation.

REFERENCES

Barrow, CS, Dodd, DE: Ammonia production in inhalation chambers and its relevance to chlorine inhalation chambers. Toxicol Appl Pharmacol 49:89–95, 1979.

Bogdanffy, MS, Dreef-van der Meulen, HC, Beems, RB, Feron, VJ, Cascieri, TC, Tyler, TR, Vinegar, MB, Rickard, RW: Chronic toxicity and oncogenicity inhalation study with vinyl acetate in the rat and mouse. Fundam Appl Toxicol 23:215–229, 1994.

Braker, W, Mossman, AL: Matheson Gas Data Book, Secaucus, NJ: Matheson Gas Products, 1980.

Burleigh-Flayer, HD, Gill, MW, Strother, DE, Masten, LW, McKee, RH, Tyler, TR, Gardiner, T: Isopropanol 13-week vapor inhalation study in rats and mice with neurotoxicity evaluation in rats. Fundam Appl Toxicol 23:421–428, 1994.

Carpenter, CP, Kinkead, ER, Geary, DL, Sullivan, LJ, King, JM: Petroleum Hydrocarbon Toxicity Studies. I. Methodology. Toxicol Appl Pharmacol 32:246–262, 1975.

Chang, JCF, Gross, EA, Swenberg, JA, Barrow, CS: Nasal cavity deposition, histopathology, and cell proliferation following single and repeated formaldehyde exposures in B6C3F1 mice and F-344 rats. Toxicol Appl Pharmacol 68:161–176, 1983.

Decker, JR, Moss, OR, Kay, BL: Controlled-delivery vapor generator for animal exposures. Am Ind Hyg Assoc J 43:400–402, 1982.

Gieschen, AW, Greenspan, BJ, Westerberg, RB, Goehl, TJ, Roycroft, JH: Generation, monitoring and concentration verification of ppb concentrations of glutaraldehyde for inhalation studies. Abstract 334, Toxicologist 11:105, 1991.

Hering, SV: Air Sampling Instruments for evaluation of atmospheric contaminants, 7th ed., Cincinnati, OH: American Conference of Governmental Industrial Hygienists, 1989.

Higuchi, MA, Davies, DW: An ammonia abatement system for whole-body small animal inhalation exposures to acid aerosols. Inhalation Toxicol 5:323–333, 1993.

Himmelstein, MW, Turner, MJ, Asgharian, B, Bond, JA: Comparison of blood concentrations of 1,3-butadiene and butadiene epoxides in mice and rats exposed to 1,3-butadiene by inhalation. Carcinogenesis 15:1479–1486, 1994.

Hinners, RG, Burkart, JK, Punte, CL: Animal inhalation chambers, Arch Environ Health, 16: 194–206, 1968.

Kari, FJ: NTP Technical report on toxicity studies of glutaraldehyde administered by inhalation to F344/N rats and B6C3F1 mice. National Toxicology Program Toxicity Report Series 25. U.S. Department of Health and Human Services, 1993.

Katz, M, ed.: Methods of Air Sampling and Analysis, 2nd Ed., Washington, DC: Amer. Public Health Assoc., 1977.

Lide, DR, ed.: Handbook of Chemistry and Physics, 75th edition. Boca Raton, FL: CRC Press, Inc., 1994.

MacFarland, HN: Designs and operational characteristics of inhalation exposure equipment— A Review. Fundam Appl Toxicol 3:603–613, 1983.

Maples, KR, Dahl, AR: Levels of epoxides in blood during inhalation of alkenes and alkene oxides. Inhalation Toxicol 5:43–54, 1993.

Miller, RR, Letts, RL, Potts, WJ, McKenna, MJ: Improved methodology for generating controlled test atmospheres. Am Ind Hyg Assoc J 41:844–846, 1980.

Moss, OR: Calibration of gas and vapor samplers, in Air Sampling Instruments, 8th ed., edited by S. Hering, Cincinnati, OH: American Conference of Governmental Industrial Hygienists, 1994.

Nadziejko, CE, Nansen, L, Mannix, RC, Kleinman, MT, Phalen, RF: Effect of nitric acid vapor on the response to inhaled ozone. Inhalation Toxicol 4:343–358, 1992.

Nelson, GO: Gas Mixtures, Preparation and Control. Chelsea, MI: Lewis Publishers, Inc., 1992.

Potts, WJ, Steiner, EC: An apparatus for generation of vapors from liquids of low volatility for use in inhalation toxicity studies. Am Ind Hyg Assoc J 41:141–145, 1980.

Pullinger, DH, Crouch, CN, Dare, PRM: Inhalation toxicity studies with 1,3-butadiene - 1. Atmosphere Generation and Control. Am Ind Hyg Assoc J 40:789–795, 1979.

Reid, RC, Prausnitz, JM, Sherwood, TK: The Properties of Gases and Liquids, New York: McGraw-Hill, 1977.

Schlesinger, RB, El-Fawal, HAN, Zelikoff, JT, Gorczynski, JE, McGovern, T, Nadziejko, CE, Chen, LC: Pulmonary effects of repeated episodic exposures to nitric acid vapor alone and in combination with ozone. Inhalation Toxicol 6:21–41, 1994.

Snellings, WM, Dodd, DE: Inhalation studies In: Handbook of In Vivo Toxicity Testing, edited by Arnold, DL, Grice, HC and Krewski, DR, San Diego, CA: Academic Press, Inc., 1990.

Snellings, WM, Zelenak, JP, Weil, SC: Effects on reproduction in Fischer 344 rats exposed to ethylene oxide by inhalation for one generation. Toxicol Appl Pharmacol 63:382–388, 1982.

Tilbury, L, Butterworth, BB, Moss, O, Goldsworthy, TL: Hepatocyte cell proliferation in mice after inhalation exposure to unleaded gasoline vapor. J Toxicol Env Health 38:293–307, 1993.

Warheit, DB, Carakostas, MC, Frame, SR: 2-Week inhalation toxicity study with HCFC-122 in rats. Inhalation Toxicol 4:81–93, 1992.

Weigel, RJ, Clark, ML, Westerberg, RB, Decker, JR, Goehl, TJ: Use of a rotary evaporator system to generate high purity vapors from reactive liquids for inhalation studies. Abstract 333, Toxicologist 11:105, 1991.

Chapter Four

Generation and Characterization of Test Atmospheres: Particles and Droplets

Owen R. Moss and Yung-Sung Cheng

INTRODUCTION

A toxicologist responsible for the conduct of inhalation exposures to solid and liquid particles must also be able to review the relevant aerosol generation and characterization related to the study. This brief discussion of the basic physicochemical properties of particles as well as of the basic principles of aerosol generation and monitoring instruments is intended to be a starting point for choosing instrumentation for exposure control and documentation. The reader is assumed to be a toxicologist, biologist, or engineer who is performing or evaluating exposures of animals to airborne material.

The control and documentation of inhalation exposures require understanding of the basic properties of particles, including use of the log-normal distribution in describing particle size. Techniques in the generation of airborne particles include condensation, dry dispersion, and wet dispersion of liquids and solids. Characterization of airborne particles may be by number, mass, surface area, particle size, chemical composition, and even fractal dimension. Reviews by Willeke and Baron (1993) and Cohen (1995) are the two most recent overviews of the basic properties, generation, and characterization of aerosols.

BASIC PROPERTIES OF PARTICLES

The basic property of particle size, measured as both physical and dynamic size, influences the control and output in particle generation, the accuracy of sampling, and the approach to characterization (Fig. 1). Physical size is related to particle geometry. For example, volume-equivalent diameter (d_v) refers to the diameter of a sphere having the same density and mass as the particle in question. For fibrous particles, both width and length are required for physical characterization.

Dynamic size of an airborne particle includes aerodynamic and mobility diameters which are measured by using instruments with detection or collection capabilities based on the inertia or mobility of the particle. Inertial instruments provide measures of the aerodynamic diameter (d_{ae}) or diameter of a unit density sphere $(\rho_0 = 1 \text{ g/cm}^3)$ having the same settling velocity as the particle in question. Instruments that measure inertia include the horizontal

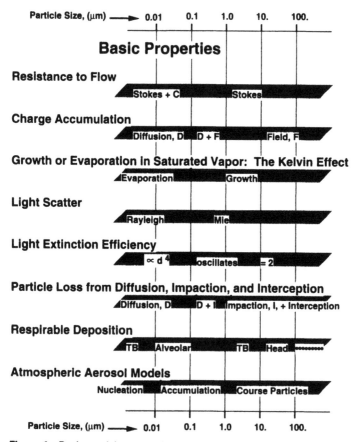

Figure 1 Basic particle properties that influence generation, sampling, and characterization.

elutriator, cascade impactor, cyclone, electric single particle aerodynamic relaxation time analyzer, aerosol particle sizer, aerosizer, and aerosol centrifuge. The electrical aerosol analyzer and diffusion battery measure the respective mobility equivalent diameter (d_{me}) or diameter of a sphere having the same dynamic mobility (particle velocity/resistance force) as the particle in question.

The relationship between the aerodynamic and mobility equivalent diameter and the volume equivalent diameter can be expressed as follows:

$$d_v = d_{ae} \left[\frac{\rho_0 C(d_{ae})}{\rho_1 C(d_v)} \right]^{0.5} \tag{1}$$

$$d_v = d_{me} \left[\frac{C(d_v)}{C(d_{me})} \right] \tag{2}$$

The C term in Equations 1 and 2 is a correction for particle slip ("C" for Cunningham; see Willeke and Baron, 1993; Hinds, 1982). Particles less than 1 μm in diameter slip through the air according to their size and the mean free path length of the air molecules. These equations are valid for spherical particles or isometric particles of near-spherical shape. For irregular-shaped particles, more complex relationships must be used (for example, Cheng, 1986; Asgharian and Yu, 1989).

Resistance to Flow

One of the first practical applications for calculating the settling of particles was made by Sir Isaac Newton in a study on the properties of cannonballs moving through air. These particles were so large that the major resistance to their movement in air was the displacement of air in their path. This resistance force was directly proportional to the square of the velocity and the square of the particle diameter $[f_r \approx (v^2)(d^2)]$.

In this century, Stokes became interested in the laminar flow of any liquid or gas around a falling particle. Under such a flow regime (the Stokes regime) the resistance to motion was directly proportional to the velocity and to the diameter of the particle $[f_r \approx (v)(d)]$. Later, researchers noticed that particles fell faster than predicted by Stokes as they became smaller. As particle size approaches that of the spaces between air molecules, the resistance to movement decreases. In fact, small particles slip through the air instead of falling through a continuous fluid medium, and the slip factor C was derived to account for this difference. To accurately predict the terminal settling velocity, the resistance force calculated according to Stokes was divided by C (Equations 1 and 2) (Willeke and Baron, 1993; Hinds, 1982; Mercer, 1973).

Charge Accumulation

Particles accumulate positive or negative charges through static electricity, through random collisions with unipolar ions (such as air ions or electrons), or by becoming charged in an electromagnetic field, where unipolar ions move to the surface of the particle along field lines. Static electrification can also take place as particles are separated and formed from bulk material. Because static electrification is difficult to control, it is not used for deliberately charging particles. Diffusion charging is the method used most often to charge particles less than 0.1 μm in diameter, while field charging is used primarily for charging particles larger than 10–15 μm in diameter.

Growth or Evaporation in Saturated Vapor: The Kelvin Effect

Small liquid particles tend to evaporate in saturated vapor, whereas large particles in the same saturated vapor tend to grow. The evaporation of small droplets in saturated vapor conditions, known as the Kelvin effect, is only seen for particles less than 0.1 μm in diameter.

Light Scatter

For particles less than 0.08 μm in diameter, Rayleigh described the scatter of incident light from one particle. He assumed that, at any instant, the electromagnetic field from the incident light was uniform over the particle surface. When the electromagnetic field is not uniform over the surface of the particle, the light scatter must be described by the complex relations derived by Mie. Mie scattering (Fig. 1) occurs when the particle diameter is equal to or larger than the wavelength of light (0.4 μm for violet, 0.5 μm for green, and 0.7 μm for red).

Light Extinction Efficiency

Light extinction is the loss in intensity of an incident beam of parallel light as it passes through a cloud of particles. This loss in intensity is due both to absorption of light by the particles and to scatter of light from particle surfaces. The scattered portion of the light composing the beam is no longer parallel and therefore not measured by a detector aligned to the original beam. Extinction efficiency, a measure of this loss, is the ratio between

   1   the radiant power in the light beam that is lost by scattering and absorption and
   2   the radiant power incident on an individual particle.

For particles 8 μm or larger in diameter, the light extinction efficiency is relatively independent of the particle diameter and has a value of approximately 2. For particles less than 0.1 μm in diameter, the light extinction efficiency is directly proportional to the fourth power of the particle diameter (that is, d^4). The extinction efficiency oscillates as a function of increasing diameter for particles that have diameters between 0.1 and 8 μm (Fig. 1).

Particle Loss from Diffusion, Impaction, and Interception

Diffusion, impaction, and interception play a dominant role in the removal of particles in sampling lines or in branching airways such as the lung. Particles less than 0.1 μm in diameter are collected mainly by diffusion to the walls. Those with diameters greater than 1 μm are collected mainly by impaction and, if sufficiently large, by interception (Fig. 1). Care must be taken to avoid or to account for losses that can occur when sampling aerosols from an exposure chamber or from the environment.

Respirable Deposition

Particles with either a physical diameter less than 0.03 μm or an aerodynamic diameter greater than 5 μm are most likely to deposit in the tracheobronchial region. Particles with physical and aerodynamic diameters between these sizes tend to deposit mainly in the alveoli. Particles with an aerodynamic diameter greater than 10 μm tend to deposit in significant amounts in the head or naso-oro-pharyngeal airways. In Fig. 1, the bar labeled respirable deposition region shows regions in the lung where particles of the indicated size are most likely to deposit.

Atmospheric Aerosol Models

The particulate component of the atmosphere over large urban areas is composed of a nucleation mode, an accumulation mode, and a coarse mode (Fig. 1). A significant portion of the accumulation mode is generally believed to form by coagulation of particles in the nucleation mode, while the coarse mode appears to be composed of particles from sources that are distinct from the sources of the other two modes. Most of the mass of particles in the urban atmosphere is either in the accumulation or the coarse mode. However, the combined surface area—the major cause of reduced visibility—is due mainly to particles 0.1 μm to 2 μm in diameter, which compose the accumulation mode.

PARTICLE SIZE

The log-normal distribution can be used to describe particle size, even though it is not always the best-fitting curve for the particle size data. This particular skewed distribution (Fig. 2) has the advantage of being completely described by just two parameters—the median diameter and the geometric standard deviation (spread in the distribution). These two parameters allow some basic and relatively simple calculations to be made for estimating the count, surface area, and mass distributions of any given sample of particles once the two parameters of any one of those distributions have been measured. Attempting to fit the measured particle size distribution with a single or bimodal log-normal distribution function is often worthwhile when the measured particle diameter must be converted to the diameter of interest for risk analysis or must be converted for estimation of inhaled dose.

Properties of the Log-Normal Distribution

Along the characteristically skewed distribution curve shown in Fig. 2 are indicated diameters that are normally encountered in inhalation toxicology and aerosol physics literature. From left to right, they are as follows:

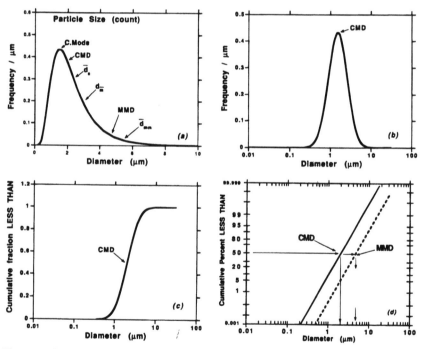

Figure 2 Fraction of the total count per micron for a log-normally distributed sample. a   Skewed distribution for a linear scale of both size (on the abscissa) and normalized frequency (on the ordinate). b   The apparent normal distribution when a log scale of size (on the abscissa) is used. c   Typical S-shaped curve of a log normal sample for a linear scale of cumulative frequency (on the ordinate) and log scale of size (on the abscissa). d   Typical straight-line fit when the cumulative frequency on the ordinate is presented as a probability scale.

Count mode diameter (C_{Mode})—is the most frequent particle diameter.

Count median diameter (*CMD*)   is generally used to describe a log-normal distribution. It is the diameter of a particle that is both larger and smaller than half the particles sampled.

Count mean diameter $(\bar{d}_c)$,   or the average particle diameter, is calculated by first multiplying each measured diameter, d_i, by the number, n_i, of particles having that diameter. The count mean diameter is the quotient of the sum of these products, $\Sigma n_i d_i$, divided by the total number of particles sampled, $N_T = \Sigma n_i$.

Diameter of average mass $(d_{\bar{m}})$ is another average particle diameter related to the total mass of particles sampled. The mass of the particle of average mass multiplied by the total number of particles sampled equals the total mass. The total mass of particles sampled is the sum of the product of the single-particle mass, m_i, calculated for each measured diameter, d_i, by the number, n_i, of particles having that diameter, $M_T = \Sigma m_i n_i$. The average mass is the quotient of this number divided by the total number of particles, $N_T = \Sigma n_i$. The diameter of average mass is calculated by assuming a sphere and applying the density of the material to convert from this average mass to volume and then to diameter.

Mass median diameter (**MMD**)   is the diameter of the particle having a mass that is both larger and smaller than the mass of half the particles sampled.

Mass mean diameter $(\bar{d}_{mm})$ is an average particle diameter, calculated by first multiplying each measured diameter, d_i, by the cumulative mass, $m_i n_i$, of all particles having that diameter. The mass mean diameter is the quotient of the sum of these products, $\Sigma m_i n_i d_i$, divided by the total mass, $\Sigma m_i n_i$, of the particles sampled.

Methods of Plotting Log-Normal Distribution

The basic frequency plot of a log-normal distribution having a geometric standard deviation of 2 is shown in Fig. 2a. The scale of the ordinate is normalized to frequency (or number) per micron. This type of graph is also used in analyzing cascade impactor data.

Basic Frequency (Linear-Linear)   In Fig. 2a, the independent variable is the diameter, d, given on a linear scale in microns. The dependent variable is the frequency—in this case, the frequency per micron of particle diameter.

Frequency (Linear-Log)   If Fig. 2a is replotted as in Fig. 2b with the abscissa changed from a linear to a log scale, the plot of the distribution takes on the characteristic bell shape of a normal distribution.

Cumulative Frequency (Linear-Log)   The curve becomes S-shaped when the ordinate is changed to show cumulative frequency (less than a given size).

Cumulative Frequency (Probability-Log)   If the cumulative frequency of the curve in Fig. 2c is replotted with the ordinate written as a probability scale, the cumulative frequency curve for the log-normal distribution becomes a straight line (Fig. 2d). Moreover, the 50% intercept of this line is the median diameter (in the case shown, CMD), and the slope is the geometric standard deviation (GSD).

The geometric standard deviation is the quotient of the median diameter divided by the diameter indicated by the intercept of the 15.87% line and the cumulative frequency curve: GSD = $(CMD)/(d_{15.87})$ in the example shown in Fig. 2. In this example, the GSD can also be calculated as the quotient of the diameter indicated by the intercept of the 84.13% line and the cumulative frequency curve divided by the median diameter: GSD = $(d_{84.13})/(CMD)$.

A very useful property of the log-normal distribution is that if the data shown in Fig. 2 are replotted with the mass distribution (mass frequency or fraction, instead of the count frequency) on the ordinate, the slope, GSD, of the line in Fig. 2d is the same as for the count frequency distribution. The new cumulative frequency line will merely shift to the right, as shown.

The usefulness of the log-normal distribution is that any of the other diameters can be calculated from the equation shown in Fig. 3 once the GSD and one of the diameters discussed above are measured.

$$\frac{d_x}{CMD} = e^{b(\ln(GSD))^2}$$

d_x	b
C.Mode	-1.0
CMD	0.0
$\bar{d}_c$	0.5
$d_{\bar{m}}$	1.5
MMD	3.0
$\bar{d}_{mm}$	3.5

Figure 3   Relation between count mode (C. Mode), count median (CMD), count mean $(\bar{d}_c)$, average mass $(d_{\bar{m}})$, mass median (MMD), and mass mean diameters $(\bar{d}_{mm})$ of a log-normal particle size distribution.

Bimodal Distributions

In inhalation toxicology and aerosol physics sampling programs, the measured cumulative distribution often does not form a straight line on a log-probability plot (Fig. 2d). The distribution may be the summation of two separate log-normal distributions such as those shown in Fig. 4. When the two log-normal distributions do not significantly overlap (Fig. 4a), the cumulative distribution curve may have a characteristic inverted S shape, like the sample in Fig. 5. The two distributions can be separated from the combined cumulative distribution curve by an iterative process that begins with estimating the fraction, f, of small particles (smalls), estimating the best-defined distribution (either smalls or bigs), estimating the second distribution, and iterating the process for a best fit to the data (Fig. 6). A general rule in this process is to initially work with an exponential, power, or polynomial curve fit to the original data points.

Estimating the Fraction of Smalls   The first estimate of the fraction, f, of the bimodal distribution composed of a separated distribution of smalls is

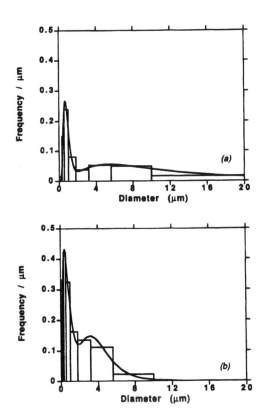

Figure 4 Two examples of bimodal, log-normal size distributions. a The distributions do not significantly overlap. b The two distributions do overlap.

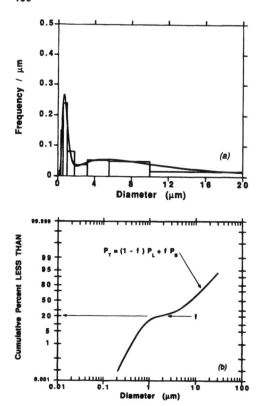

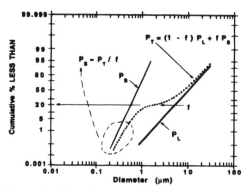

Figure 5 Characteristic S-shaped curve for the cumulative summation of two, non-overlapping, log-normal distributions.

made from the inflection or horizontal point in the cumulative distribution curve as plotted on log-probability paper (Figs. 5 and 6). This fraction is adjusted in the iterative calculations to bring the cumulative frequency curve of the combined distributions closer to the original data. The two distributions composing the sampled distribution can be expressed by:

Figure 6 Bimodal distribution. Estimate the distribution of small particles and then calculate the distribution of large particles.

$$P_T = (1 - f)P_L + fP_S \tag{3}$$

where P_T is the combined, cumulative distribution function, P_L is the cumulative distribution function for the distribution of large particles (P_L will range from 0%–100%), and P_S is the cumulative distribution function for the distribution of small particles (P_S will also range from 0%–100%).

Estimating the Best-Defined Distribution   When plotted on log-probability paper, one end of the S-shaped distribution will, in general, be better defined than the other in terms of a straight-line segment extending toward one extreme of the probability scale. In this region of the probability scale, the distributions may be relatively far apart in size. For example, if the tail of the curve toward the small-particle diameters is better defined (Fig. 5), a particle size region can be chosen that is below the region where the distribution of large particles contributes to changes in the cumulative percentage curve.

For an estimated fraction, f, of smalls (Fig. 6), there is a range of diameters where P_L is approximately 0, so that Equation 3 is simplified:

$$P_T = fP_S \tag{4}$$

In this region of the curve, the points belonging to the distribution of small particles, independent of the distribution of large particles, can be estimated by dividing P_T by f:

$$P_S \approx P_T/f \tag{5}$$

If the reverse is true, and the tail of the cumulative distribution curve toward the large particle diameter is better defined, then there is a range of diameters where P_S is approximately 100%. Therefore in this region of the curve, the points belonging to the distribution of big particles, independent of the distribution of small particles, can be estimated by subtracting 100 f from P_T and dividing the result by $(1 - f)$:

$$P_L \approx (P_T - 100\, f)/(1 - f) \tag{6}$$

Estimating the Second Distribution   When one of two distributions is drawn as a straight line on the probability paper, the other distribution can be estimated by applying the original relationship for P_T (Equation 3).

Iterating the Process for a Best Fit to the Data   The key variables in the fitting process are the estimate of f, the slope (GSD), and location (CMD, for the example in Fig. 5, 6, and 7) of the first distribution on the log-probability graph paper. After these three numbers are chosen, little effort

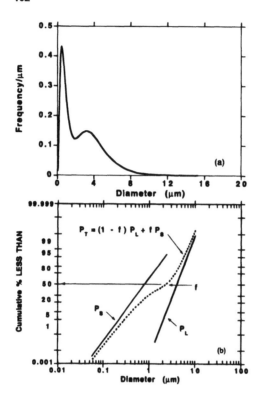

Figure 7 Characteristic flattened, S-shaped curve for the cumulative summation of two, overlapping log-normal distributions: a normalized frequency and b cumulative frequency of the combined distributions.

is required to calculate the second distribution (Equation 3) and adjust the combined cumulative distribution function (P_T) to fit the data. The process lends itself well to computer analysis, especially if a digitizing table can be used to read in points from the curve. When only a few curves are to be analyzed, the analog capabilities of the human eye can be used to estimate a best fit in a reasonably short time. With care, the results of this method can be within 5% of the best answer from a statistical fit obtained using a computer.

If the two log-normal distributions overlap, as in Fig. 4b, then the cumulative curve will have a flattened S shape (Fig. 7). The fraction of smalls, f, is still initially estimated from the inflection point, and the two distributions can be extracted as described above. If the geometric standard deviations are fairly close (within 20%) and the overlap significant, the combined cumulative distributions become indistinguishable from the cumulative distribution of a single log-normal distribution.

GENERATION OF AEROSOLS

Aerosols may be generated by condensation of vapors, dispersion of dry particles, or dispersion of liquids or of a suspension of solids in a liquid. This section gives an overview of currently available, commonly applied techniques used to produce aerosols for inhalation toxicology. The overview is based on material from references in the selected bibliography at the end of this chapter (specifically see Willeke and Baron, 1993; and Cohen, 1995).

Condensation   In the condensation process, a vapor condenses on nuclei or small clusters of atoms or molecules to form larger particles. In this process, polydisperse or monodisperse aerosols may be formed. Condensation aerosols are also formed through various chemical reactions and through combustion processes.

Polydisperse Aerosol Production.   Polydisperse condensation aerosols are frequently used in filter testing, both for inplace filters in buildings (as in laboratory hood systems), and in building ventilation where high-efficiency filtration of airborne material is required. Polydisperse and monodisperse aerosols produced by condensation are also used in evaluating personal respirators, producing new materials, and creating exposure atmospheres of fine and ultrafine particles (Wu et al., 1993; Pratsinis, 1988; Pratsinis et al., 1989). In one such generator designed to produce a polydisperse condensation aerosol of liquid droplets, a stream of oil is impinged on a heated block. The vapor condenses on naturally occurring nuclei in the dilution and cooling air. The condensation process in such a generator is the important step in determining the final size of the aerosol produced. Such generators (Fig. 8) are generally designed to produce particles in the 0.3-μm diameter range.

The condensation method is also useful for producing aerosols of metal, metal oxide, and organic compounds for inhalation studies or for basic aerosol research (Shaw and Lawman, 1994; Tu et al., 1981; Decker et al., 1982; McCarthy et al., 1982; Kanapilly et al., 1978; Japuntich, 1992; Kim and Kim, 1994; Ohshima et al., 1993). The aerosol material is either nebulized as a

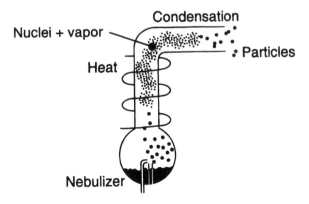

Figure 8   Evaporation condensation aerosol generator using a nebulizer.

liquid or contained in a Pyrex or quartz boat, which is placed inside a tube furnace. The furnace is heated at a constant temperature to evaporate the material which is then carried by nitrogen gas to the condensation chamber. The material cools and condenses into an aerosol when mixed with diluting air. The aerosol concentration is determined primarily by the evaporating temperature in the tube furnace. Sometimes, additional heat treatment at high temperature (up to 1700°C) is required to completely oxidize a metal oxide aerosol.

Monodisperse Aerosol Production   Monodisperse condensation aerosol generators use the same principle as those for polydisperse aerosols, except that greater care is taken to maintain a constant rate of vapor generation and to control the condensation process. The material is either vaporized directly from a reservoir or nebulized to an aerosol. It is then heated to a vapor and condensed onto nuclei produced during nebulization of the starting fluid. If the material is vaporized directly from a reservoir, a source of nuclei (e.g., NaCl), may have to be added. The vapor, nuclei, and growing particles must remain in the condensation region long enough to permit controlled condensation to occur. The greater the instability of flow in the condensation region, the greater the spread in size distribution. The mass and diameter of the particle of average mass can be estimated from the quotient of the vapor concentration, C, divided by the nuclei concentration, N, provided the density, ρ, of the condensed vapor is known.

Mass of the particle of average mass (g) = C/N

$$d_{\bar{m}} = \left[\frac{6}{\rho \pi} \frac{C}{N} \right]^{1/3} \qquad\qquad (7)$$

Chemical Reactions   Fuming sulfuric acid is an aerosol familiar to students of organic and inorganic chemistry. The sulfuric acid reacts with water vapor in air to form a mist. The reaction of titanium tetrachloride with water vapor in air is another chemical reaction that produces a visible aerosol. Titanium oxide and hydrocholoric acid are formed, resulting in a dense white cloud used to detect the direction of air flow in hoods and between rooms in a laboratory.

Simple reactions of vapors onto surfaces of particles have been successfully used in inhalation toxicology (Hemenway et al., 1990; Jakab and Hemenway, 1993) and in the production of specific nucleation aerosols (Anisimov et al., 1994; Foss and Davis, 1994). More complex systems have been developed using high-temperature condensation aerosol generators, as mentioned above, and even include the production of aerosols through the application of supercritical fluids (Tom 1991).

Combustion Generating aerosols relevant to combustion processes is one of the more difficult tasks for a toxicologist involved in modeling with animals the toxic stress that people experience during a fire. For a model, laboratory smoke must be designed to duplicate the smoke age and generation conditions expected to occur during a real fire. Techniques used to model combustion toxicology can be found in Kaplan et al. (1983). Precise control of the combustion process appears to be possible in some cases such as in spray pyrolysis (Ohshima et al., 1993; Pluym et al., 1993; Xiong and Kodas, 1993) or by combining a spray system with an inductively coupled plasma (Kagawa et al., 1993).

Inhalation studies have also been conducted for other types of combustion products, including automobile engine exhaust and cigarette smoke. Diesel engines have been mounted on test stands and computer-controlled to stimulate certain driving patterns (Mokler et al., 1984; Heinrich et al., 1985). Several kinds of smoke machines are available that produce both mainstream and sidestream cigarette smoke for inhalation studies (Teague et al., 1994; Guerin et al., 1979; Baumgartner and Coggins, 1980; Griffiths and Standafer, 1985).

Dry Dispersion

One of the most difficult challenges in inhalation toxicology research is the production of a consistent and reproducible exposure atmosphere when the bulk material is in powder form. The necessary techniques for delivery and dispersion of the powder and for discharge of the airborne particles are often unique for each material.

Delivering the Powder Powder delivery systems utilize either hoppers, screw feeders, rotating disks, compressed cylindrical packs in which powder is delivered by scraping off the top layer, or conveyor belts composed of chains, tubing, brushes, or troughs (Table 1). For further information on these devices, see Cohen (1995), Willeke and Baron (1993), and Hinds (1982) as well as specific discussions in papers by Higuchi and Steinhagen (1991), Rajathurai et al. (1990), O'Shaughnessy and Hemenway (1994), and Holländer et al. (1987). Gravity delivery systems, such as the National Bureau of Standards or the MDA (Sibata/MDA) aerosol generators, are composed of simple hoppers designed to drop their contents into grooves cut in a plate or between the teeth of a gear. The Thermal Systems Inc. (TSI) fluidized-bed system uses a chain to deliver a powder from a hopper to the multicomponent bed. Commercial screw feeders (Cheng et al., 1985), flexible-walled brush bristle augers (Milliman et al., 1981; Bernstein et al., 1984), and rotating grooved plates (the TSI, Small-Scale Powder Disperser, Minneapolis, MN) have been used to successfully deliver powders to where they can be dispersed into an airstream.

Table 1   Characteristics of Dry-Powder Generators

Generator	Delivery mechanism	Dispersion mechanism	Flow rate (L/min)	Mass concentration (mg/m³)	Test material
NBS	Gravity	Venturi	50–85	≥1500	Nonsticky powder
Wright Dust Feed	Rotating blade	Airstream	10–40	2–1100	Compactable powder
TSI model 3410	Rotating brush	Airstream	10–50	1–100	Nonsticky powder
MDA Micro Feed	Rotating disk	Venturi	30–50	10–300	Nonsticky powder
TSI model 3433	Rotating disk	Venturi	12–21	0.3–40	Nonsticky powder
Lovelace 4-in. FBG	Gravity	Fluidized bed	200	1–100	Nonsticky powder, fiber
TSI model 3400	Chain conveyor	Fluidized bed	5–15	10–100	Nonsticky powder
Jet-O-Mizer	Screw feed	Air mill	300–400	2–1000	Sticky powder
Battle Micronizer	Dual brush	Air mill	30–50	5–5000	Sticky powder
Microjet	Screw feed	Air mill	300–1000	2–1000	Fiber
Small Jet Powder Disperser (TSI)	Rotating plate	Venturi	12–21	0.3–40	Nonsticky powder

Sticky powders can be delivered at a constant rate to a preselected point by placing them in Tygon tubing that is slit down its length. The tubing opens when it is pulled around a sharp bend and closes when it has passed that point. This property also facilitates loading (DeFord et al., 1982).

In the field of toxicology and radiation biology, the aerosol generator most often used is the Wright dust feed mechanism (Fig. 9). It consists of a cup in which the material is packed and a scraper blade that moves through the packed material. The material removed from the pack is transported to a central tube by air flowing along the edge of the blade. This generator is very dependable if the material can be compressed into the cup. If the cup is not well packed, the air blowing along the edge of the scraper may cause large amounts of material to slough off and become airborne. This may clog the generator or cause the output to fluctuate between high and low concentrations of aerosol. These limitations have been overcome to some extent by changing the duration and interval between pulses of the scraper blade drive motor (O'Shaughnessy and Hemenway, 1994).

Dispersing the Powder   Powder dispersal must accomplish the tasks of aerosolization, dilution, and deagglomeration. Airstreams, venturi tubes, air jet mills, and fluidized beds have been used to disperse powders into fine aerosols (Cheng et al., 1989). The most common methods used in dispersing

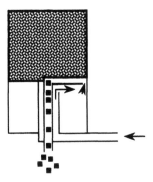

Figure 9   Basic air flow in a Wright dust feed mechanism.

a powder are to feed the dust into a high-velocity airstream or to blow air over the powder. The shear forces in the turbulent airstream disperse the powder and break up agglomerates. In the Venturi design, a high-velocity air jet blows across a nozzle or restriction in the pipe to produce suction that draws clumps of powders into the shear flow of air.

Fluid energy mills used for dispersion include the Trost jet mill (Bernstein et al., 1984), the Jet-O-Mizer (Cheng et al., 1985), and the Microjet mill (Lee et al., 1983). The fluid energy is delivered in high-velocity streams that circulate around a grinding and classifying chamber (Fig. 10), where turbulence and centrifugal forces deagglomerate particles. Fine particles carried by the fluid exit at the center of the chamber; coarse particles are recirculated for further size reduction. The Trost jet mill utilizes two opposing jets to effect a comminution to reduce particle sizes.

In a one- or two-component fluidized bed, the minimal air flow that causes the bed to fluidize is used so that only the smallest particles are released. Commercial and laboratory powder generators listed in Table 1 consist of a combination of these delivery and dispersion mechanisms. Many are material-specific, since delivery strongly depends on the bulk powder properties, including size, shape, compactness, and stickiness. For example, sticky powder tends to stay in a gravity feed tube or rotating disk, resulting in a reduced feed rate or complete stoppage. Compactness of the bulk material is essential for delivery by a Wright dust feeder, where the powder is packed under pressure to form a solid cylinder. Loose or uneven packing causes the pack to break up during the generation process. Most generators work best for nonsticky, dry powders. When sticky materials are dispersed as powder, they tend to form clumps that require more energy for dispersion, or they cannot be broken up, thus clogging the generator.

The kinetic energy of an air dispersion system is proportional to the square of the air velocity. The fluidized bed has the least kinetic energy and

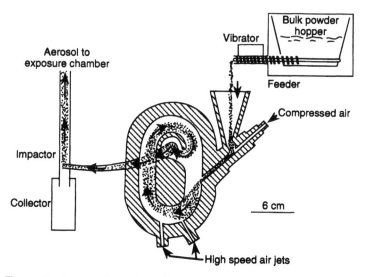

Figure 10   Aerosol dispersion with a turgulent jet fluid energy mill.

therefore cannot be used to generate sticky powders. The Venturi and fluid energy mills have the highest velocity, some operating at sonic speeds.

Discharging Particles   Airborne particles are generally discharged or brought into charge equilibrium (Boltzmann equilibrium) by passing them through a cloud of bipolar air ions. The time needed to reach equilibrium is independent of particle size or initial charge. Particles that have a high electrical charge require time in the presence of a bipolar cloud of air ions to reach equilibrium and be discharged. A guideline for the length of time, t, given an ion concentration of N (ions/cm^3), is that the product of N and t, $(N)(t)$, should at least equal 6×10^6 ions s/cm^3:

$$(N)(t) > 6 \times 10^6 \qquad\qquad (8)$$

In the laboratory, high bipolar concentrations of ions are produced with radioactive sources (such as krypton-85, nickel-63, and americium-241) (Mercer, 1964), or they can be produced by corona discharge in an air jet of sonic velocity (Whitby, 1961; Mercer, 1973; Willeke and Baron, 1993; Adachi et al., 1993).

Storing the Aerosol   An alternative for developing the technology for constant, reproducible generation of a powder aerosol is to temporarily store airborne material in an aerosol drum chamber. Such chambers have the capability of storing particles in the 0.5 to 20 micron range for several days with less than 10% loss (Asgharian and Moss, 1992; Gruel et al., 1987).

Wet Dispersion: Liquids

Wet dispersion of liquids is generally accomplished by filament or wave breakup into initial droplets. The final particle size of an evaporating solution is proportional to the cubic root of the mass fraction of the material in solution, $f^{1/3}$.

$$\text{Final Diameter} \approx f^{1/3} \qquad (9)$$

For example, a 0.1% solution of dye ($f = 0.001$) will produce a final dry particle that has 1/10 ($0.001^{1/3} = 0.1$) the diameter of the original droplet.

Filament Breakup   Filament breakup is generally accomplished using nebulizers, vibrating orifices, or spinning disks. A comparison of the output of devices using these different techniques can be found in a paper by Hewitt (1993). With the basic compressed air nebulizer (Fig. 8), a filament of fluid is sucked into the shear flow of a high-velocity air stream. The filament of liquid breaks apart into large droplets and secondary, small droplets. The latter droplets are formed as the large droplets pull apart and away from the original liquid stream.

The large droplets hit the sides of the nebulizer, and the secondary droplets escape as aerosol. One limitation of the basic compressed air nebulizer is that the fluid within the reservoir of the generator recycles. Each time liquid is drawn into the air stream, some evaporation takes place. Over time, the steady but slight concentration increase of the liquid in the reservoir causes the size distribution of the dried droplets to increase. The basic nebulizer is also sensitive to changes in the level of fluid in the reservoir: the output decreases as the fluid level drops. The output of a compressed-air nebulizer depends on the location of the impaction surfaces and baffles. Particles with diameters of less than 1 μm to several hundred microns can be produced through nebulization. Table 2 lists characteristics of some commercial nebulizers that are frequently used in inhalation studies. Further information on the output characteristics of currently available compressed air and ultrasonic nebulizers can be found in Willeke and Baron (1993).

For inhalation studies with exposures of more than one hour, the nebulizer reservoir must be modified to minimize the changes in aerosol concentration and particle size that result from the concentration of solution in the reservoir. DeFord et al. (1981) described a system in which an additional reservoir with a larger capacity is used to continuously supply fresh solution. In another system, a syringe pump meters fresh solution to a nebulizer (Liu and Lee, 1975).

Vibrating Orifice   A vibrating orifice produces filament breakup by passing the liquid through a piezoelectric crystal that imparts a frequency to

Table 2   Characteristics of Nebulizers

Nebulizer	Operating pressure (psi)	Flow rate (L/min)	Output concentration (μg/L)	MMAD (μm)	GSD
Laskin	20	84.0	4.8	0.7	2.1
In Tox	30	25.0	32.0	6.1	1.9
Solosphere	20	NA[a]	1.5	4.5	NA
Ohio	20	NA	0.5	4.5	NA
DeVilbiss	20	16.0	14.0	3.2	1.8
Hospitak	20	11.0	23.0	1.0	2.1
Collision	20	7.1	7.7	2.0	2.0
Reteck X-70/N	20	5.4	53.0	5.7	1.8
Lovelace	20	1.5	40.0	5.8	1.8

[a]NA: Data not available.

the liquid stream. The stream breaks apart with the same frequency patterns as it exits from the crystal. Large and secondary monodisperse particles are formed. After droplet evaporation, aerosol particle diameter is between 0.5 and 30 μm. The instrument is very useful for generating monodisperse particles to accurately calibrate aerosol instruments, but it is seldom used in inhalation studies because of its low output rate and the inability to generate droplets from solid suspensions.

Spinning Disk   Another way to form droplets is to maintain a thin layer of fluid on the upper surface of a top rotating at a constant speed. At the edge of the top, fluid is thrown off, and the characteristic bimodal distribution of bigs and smalls is formed. The inertia of the large particles is used to separate them from the small particles. The median sizes of the two distributions usually differ by a factor of 16. Large particles (diameters of 20–100 μm) can be produced in this manner as well as monodisperse aerosols of small droplets (Willeke and Baron, 1993; Melton et al., 1991).

Wave Breakup   Ultrasonic nebulizers are used to produce high concentrations of particles in the range of 5–10 μm in diameter. Ultrasonic energy from a piezoelectric crystal is focused just below the liquid surface. The resulting capillary waves break up, forming a dense aerosol cloud (Hinds, 1982; Willeke and Baron, 1993).

Electrospray   Electric fields in the range of 1000 to 6000 volts can be used to produce micron-sized droplets that, under the proper conditions of voltage and conductivity, are formed into a fan-shaped spray containing particles of only one size. Submicron diameter droplets can be produced by this method, but the fluid flow rate is low (Willeke and Baron, 1993; Grace and Marijnissen, 1994).

Wet Dispersion: Solids

Wet dispersion of solids is accomplished by generating a liquid droplet that contains a solid particle. The liquid evaporates, leaving a single or aggregated particle.

Droplet Lifetimes   When generating suspensions of solids in water, droplet lifetimes must be considered. At 20% relative humidity (RH), a droplet with a diameter of 40 μm will last only 1.3 s. At 50% RH, that same particle will take approximately 2 s to evaporate. At relative humidities near 100%, the droplet may grow rather than evaporate (Table 3).

Nebulization   Nebulization of monodisperse solid particles such as polystyrene latex spheres in a dilute water solution can be used for roughly calibrating the size of the initial liquid droplets escaping from the generator (Hinds, 1982). Single particles and double particles that are produced in generating a solution of monodisperse solid spheres are used in this technique which requires some means of readily detecting such particles in a representative sample of dried aerosol. The diameter of average volume, $d_{\bar{v}}$, can be calculated for the initial distribution of liquid droplets:

$$d_{\bar{v}} = d_p \left[\frac{2}{f} \frac{(\text{No. of doublets})}{(\text{No. of singlets})} \right]^{1/3} \tag{10}$$

Given the volume fraction, f, of spheres in solution, the droplet diameter of the average volume is directly proportional to the diameter of the solid spheres, d_p, and the cube root of the quotient of the number of doublets to the number of singlets counted in the output aerosol and inversely proportional to the cube root of the volume fraction, f.

CHARACTERIZING AIRBORNE PARTICLES

Characterizing test atmospheres includes defining aerosol concentration, particle size and shape, and chemical composition. Many methods and in-

Table 3   Water Droplet Lifetimes

Size (μm)	Lifetime (s) at 20% RH[a] (sec)
0.01	0.000002
0.1	0.00003
1.0	0.001
10.0	0.03
40.0	1.3

[a]At 50% relative humidity (RH), lifetime increases ~1.5 times; at 100% RH, lifetime increases 110–1000 times.

struments have been developed for sampling aerosols from pollution emissions, laboratory aerosols, and inhalation studies. Aerosols of test atmosphere must be characterized to quantitate toxicant concentration, stability, and particle size distribution during exposure. This information is used to establish exposure concentration and estimate dose delivered to the test animals. The discussion in this section is limited to determination of concentration and particle size distribution.

Aerosol Concentration

In most inhalation studies, the mass concentration and stability of the aerosol in the exposure chamber is determined even though mass may not provide the best correlation between inhaled dose and initial response in the lung. At minimum, the number concentration and size distribution should be measured in addition to the mass concentration. In special cases, the surface area concentration may need to be measured. If monodisperse aerosols are used in carefully controlled disposition and retention studies, particle number concentrations may have to be measured in inspiratory and expiratory air to determine deposition efficiency.

Mass Concentration   The time-integrated mass (or activity) concentration is measured most directly using filters, impingers, or impactors. The mass collected in the device, the average flow during the sampling period, and the duration of sampling must be known to calculate the concentration:

Time-integrated mass concentration   (11)

$$= \frac{(\text{mass collected})}{(\text{average flow})(\text{duration of sample})}$$

The pressure drop across the collection device may be great enough to affect the monitor if the flow monitor was not calibrated with the collection device in place. With membrane filters, for example, the collection efficiency and pressure drop across the filter increases with decreasing pore size. The increase in pressure drop affects the flow through the filter sample, and a correction must be made to obtain the actual flow. If a rotameter calibrated at atmospheric pressure is used to measure the flow rate through a sampling device with substantial pressure drop, for example, the actual flow rate through the device will not be the flow rate indicated on the rotameter:

$$Q_{actual} = Q_{indicated} \, (1 - \Delta P/P)^{1/2} \tag{10}$$

Care must be taken when collecting aerosol samples with filters to ensure that the weights of the filters remain stable. The detection limit of the filter samples is limited by the sensitivity of balances, weight stability of filters, sampling time, and flow rate (Hinds, 1982).

Real-Time Mass Monitors   Other methods of determining the mass concentrations of aerosols include beta-attenuation, piezobalance, and photometers. These three instruments are real-time, continuous monitors. Although a filter sample gives the time-averaged concentration, the stability of the aerosol concentration cannot be adequately defined because of the long sampling time (up to 8 h) required to collect enough material for gravimetric analysis. The use of a real-time aerosol mass monitor enables the operator to monitor the stability of aerosol concentration and detect problems related to aerosol generation and delivery. In addition, the unit can be used for periodic adjustment of aerosol concentration, determination of rise and fall times, and monitoring aerosol uniformity (Cheng et al., 1988).

Direct measurement of particle mass and concentration is accomplished with a piezoelectric balance. With this device, particles are precipiated onto an AT-cut quartz crystal that is oscillating at a stable frequency. The attachment of particles to the crystal surface results in a decrease in frequency that is directly proportional to the mass deposited. The response of the instrument is sensitive to the load on the vibrating surface and the quality of the attachment of the particles to the surface. Commercially available instruments (see Cohen, 1995; Willeke and Baron, 1993) are most suitable for measuring aerosol concentrations of less than 10 mg/m³. For higher aerosol concentrations, beta-attenuation instruments can be used to determine aerosol concentrations. Particles are collected on filters or impacted on Mylar film situated between the beta source (usually carbon-14) and the detector. The material collected prevents the beta-radiation from reaching the detector, and the system is calibrated for the mass buildup on the film.

Photometers are used as indirect measures of particle concentrations (Fig. 11) and are useful in inhalation studies as indicators of aerosol generator stability. In the basic photometer, the intensity of light-scattering is a function of particle size and number concentration as well as a function of the index of particle refraction. The photometer measures light intensity rather than directly measuring mass concentration. With careful calibration, a well-designed photometer provides instantaneous reading of mass concentration. Photometers that consist of incandescent light and photomultiplier detectors have poor detection efficiency and are not very sensitive. Recently developed instruments are based on a light-emitting diode (LED) that emits near-monochrome light with a wavelength between 0.8 and 1.0 μm. A silicon detector that has a response coinciding with the emission maximum of the LED is used to detect the scattered light. This type of device is lightweight and often battery-powered. The concentrations it can detect range from 0.01–200 mg/m³ which are suitable for most inhalation experiments. Both forward and backward light-scattering instruments are available. Newer photometers

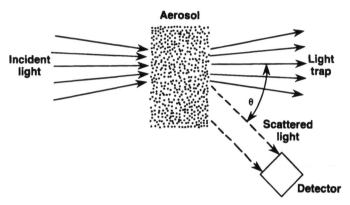

Figure 11 Indirect measurement of mass or count concentration by using forward light scatter.

that use forward or near-forward light-scattering units have a response curve that simulates the inhalable or respirable fraction of the aerosol (see discussions in Cohen, 1995).

Number Concentrations The measurement of number concentration may be very important in establishing a relationship between inhaled dose and observed response in the lung. The total number of particles inhaled and exhaled is an essential measurement in research on the deposition of inhaled particles in humans or animals. Number concentrations are obtained semiautomatically with nuclei counters, optical counters, and electrometers and by hand with microscopy.

Condensation nuclei counters, used to measure submicron-size particles, are based on light-scattering and condensation principles. Vapors such as water or alcohol condense on particles by supersaturation. In the earliest nuclei counters, adiabatic expansion was used to produce supersaturation of water vapor which condensed on particles in the chamber. A simple light source and detector were used to estimate the number of particles and nuclei originally present. An example of subsequent development in nuclei counter technology is the TSI model 3020 nuclei counter (Fig. 12) which is a continuous-flow device. Particles enter a chamber of saturated butanol vapor that condenses on the particles as the airstream passes through a condensation tube maintained at 10°C. The enlarged particles finally pass through an optical chamber. The impact of configuration, vapor composition, particle size, and particle number on the measured concentration are specific to the instrument being used (Ahn and Liu, 1990a, 1990b; Dreiling et al., 1986; Ensor et al., 1989; Zhang and Liu, 1990, 1991; Noone and Hansson, 1990; Niessner et al., 1990).

The single particle detector in a condensation nuclei counter is based on either a laser or incandescent light source. Light pulses from individual parti-

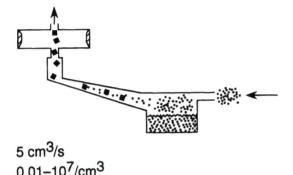

5 cm^3/s
0.01–10^7/cm^3

Figure 12 Continuous-flow nuclei counter.

cles are detected, amplified, and counted to determine number concentration. Maximum concentration is normally limited to about 100 particles per cubic centimeter, but because of coincidence errors, it can approach 10,000 in some instruments with adjustment to flow and correction of the reading for coincidence errors (see Willeke and Baron, 1993; Cohen, 1995; Julanov et al., 1984, 1986). Because scattered light intensity decreases with particle size, the detection limit is about 0.3 μm for optical counters that have an incandescent light source and 0.1 μm for laser optical counters.

Individual particles can be counted electrically by placing a constant charge on each particle and measuring the current produced when the particles are collected in an electrometer consisting of a filter contained in a Faraday cup. The relationship between average current and particle number concentration is determined (Hinds, 1982).

Counting particles by hand under an optical or electron microscope has been used to determine number concentrations of fibrous particles such as asbestos in filter, impinger, or electrostatic precipitator samples of air from working environments or exposure chambers. When fibrous and nonfibrous particles are counted in the microscope, an optimal number density on the substrate is approximately 10^7 particles per 47-mm filter (or $\approx$ 5800 particles per mm^2).

Particle Size and Shape

Characterizing the size and shape of particles composing an exposure atmosphere provides the key physical parameters necessary to make estimates of the collection efficiency and deposition pattern in the lungs of test animals. This information also provides insight into the degree of change in particle size occurring during aerosol generation and transport when samples from different locations in the exposure system are compared with the bulk material. Powder generators with high fluid energy reduce particle size. On the other hand, both the largest and smallest particles in the distribution can be selectively removed from the exposure atmosphere as they move through the delivery system and exposure chamber.

Microscopy must be used to measure the shape and physical size of particles collected on filters or on substrates. The projected area diameter (diameter of a sphere having the same projected area is that of the particle) of even irregularly shaped particles can be determined by using a Porton eyepiece graticule (Hinds, 1982), a Zeiss particle size analyzer, or an image analyzer with appropriate software. The width and length of fiber-shaped particles should be recorded in a table so that the correlation between length and diameter as well as the other 5 parameters of the bivariate lognormal distribution or 11 parameters of the bimodal bivariate lognormal distribution (Fig. 13) can be estimated (Cheng, 1986; Moss et al., 1994).

Dynamic size measurements, made using instruments with detection or collection capabilities based on the inertia or mobility of the particle, are frequently used in inhalation studies to determine the aerodynamic and mobility equivalent diameters.

Inertial Instruments   Aerosol sampling instruments such as the cascade impactor, cyclone, aerosol centrifuge, elutriator, and time-of-flight particle sizer can be classified as inertial sampling instruments. Such instruments use either the inertia or terminal settling velocity of particles to measure the aerodynamic equivalent diameter.

Cascade impactors are samplers designed to collect airborne material according to particle inertia. Those particles that cannot follow the flow streamline around a sharp turn make contact with the collection surface and

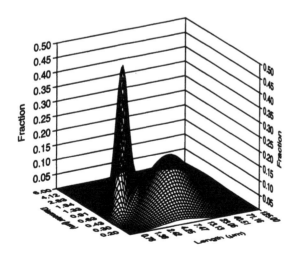

Figure 13   Bimodal, bivariate log-normal distribution.

are removed from the air stream. For a given flow rate through the cascade impactor, there is a particle diameter, the effective cutoff diameter, d_{50}, associated with each stage. Half the particles of this diameter deposit on the collection surface, and half pass to the subsequent stage. The raw data produced by a cascade impactor comprise the mass collected on each stage and the respective effective cutoff diameters. The cumulative particle size distribution function of the sampled aerosol is calculated from work sheets similar to the one shown in Table 4. A graph of this function on log-probability paper (Figs. 2 and 14) can be used to estimate the mass median aerodynamic diameter and geometric standard deviation.

The effective cutoff diameters in a well-designed cascade impactor generally can be predicted according to the theory developed by Marple (1970):

$$d_{50} = (C\,d_{ae})^{1/2} = \begin{cases} 40.7\ (nW^3/Q)^{1/2} \ldots \text{for round jets} \\ 69.0\ (W^2L/Q)^{1/2} \ldots \text{for rectangular jets} \end{cases} \tag{10}$$

where Q is flow rate (L/min), W(cm) is the nozzle diameter for a round-jet impactor or nozzle width for a rectangular jet impactor, L(cm) is slit length for a rectangular jet, and n is the number of jets in a stage. Flow rates lower than those used in the design of the impactor are recommended to minimize the problem of particle bounce. Details on the history, operation, data analysis, design, theory, and application of cascade impactors are included in a monograph by Lodge and Chan (1986) and in books on aerosol measurement (Willeke and Baron, 1993; Hinds, 1982; Cohen, 1995).

While cascade impactors are used to measure the time-averaged size distribution based on mass, time-of-flight particle sizers are used to measure the aerodynamic equivalent diameter in real time. Two similar units are the Aerosol Particle Sizer (APS) and the Aerosizer (Aerosol Particle Sizer, TST, Min-

Table 4 Cascade Impactor Data Reduction. Sample worksheet for the calculation of "cumulative mass fraction less than the maximum diameter d_j (bold numbers in column 7)" on each stage of an "n" stage cascade impactor containing an "xxx" mg sample.

Stage	Initial mass (mg)	Final mass (mg)	Net mass (mg)	Mass fraction (%)	d_{50} (μm)	Size range on stage (μm)	Cumulative mass fraction LESS THAN d_j (%)
1	—	—	—	—	d_1	d_1 to > d_1	100
2	—	—	—	—	d_2	d_2 to d_1	—
3	—	—	—	—	d_3	d_3 to d_2	—
—	—	—	—	—	—	—	—
n	—	—	—	—	d_n	d_n to d_{n-1}	—
n+1 (filter)	—	—	—	—	≈ 0	≈ 0 to d_n	—
		Total net mass = xxx					

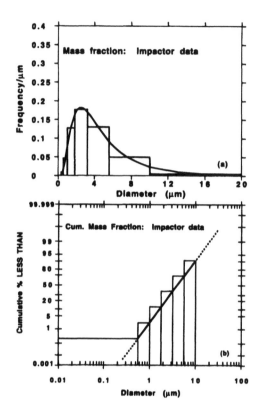

Figure 14 Analysis of cascade impactor data.

neapolis, MN; Aerosizer, Amherst Process Instruments, Inc., Hadley, MA). Time-of-flight particle sizers are based on the acceleration of particles through a nozzle. Because of their inertia, the particles of different size have different velocities downstream from the nozzle (Fig. 15). The magnitude of the lag between particle velocity and gas velocity depends on the aerodynamic diameter of the particles and is determined by the transit time of the particle as it passes through two laser beams. A calibration curve (relating the aerodynamic equivalent diameter to the transit time) restricted to a given set of operating conditions (ambient pressure, pressure drop across nozzle, and flow rate) is provided with each instrument. The frequency distribution by count of the aerodynamic diameter is estimated and transformed by calculation to the volume distribution, assuming spherical particles of uniform density.

The chemical composition of the aerosol as a function of particle size cannot be measured with a time-of-flight particle sizer since the particles are not collected during operation. However, such information can be obtained from the mass of material collected on each stage of the cascade impactor. If both an impactor and, for example, an APS are used to characterize a test

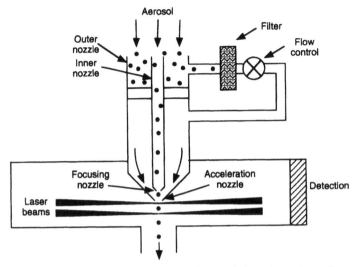

Figure 15   Time-of-flight particle sizer flow and detection schematic.

atmosphere, then size distributions obtained from the two instruments may not be the same. Several factors affect the response of the APS:

1   particle and gas density,
2   irregular particle shape, and
3   surface tension of liquid droplets (Baron, 1986; Chen et al., 1985; Willeke and Baron, 1993; Cohen, 1995; Cheng et al., 1993a,b; Marshall et al., 1991: Cheng et al., 1990; Rader et al., 1990; Lee et al., 1990).

Electrical Mobility Diameter Instruments   The inertial type of aerosol sampling instrument is used to measure aerodynamic diameters larger than about 0.5μm. For particles less than 0.5 μm in diameter, most inertial instruments other than low-pressure cascade impactors (Hering et al., 1978, 1979; Marple et al., 1981) are not effective in separating and measuring aerosol size. For the detection of such small particles, instruments are used that separate particles according to their ability to diffuse or to move in an electric field. The electrical mobility equivalent diameter and the diffusion equivalent diameter (Willeke and Baron, 1993) are used to predict collection efficiency in the respiratory tract, individual particle charging rate, and coagulation rate.

The electrical aerosol analyzer (EAA) and the related differential mobility analyzer (DMA) are based on principles of diffusion charging and electrical mobility (Fig. 16). The analyzers consist of two concentric cylinders. The outer cylinder is electrically grounded, and the central rod is maintained at a given voltage. Aerosol drawn into the EAA passes through a unipolar-ion diffusion charger. The charged particles are drawn to the central rod. Particles with mobility (Z) less than the cutoff mobility, as determined by the

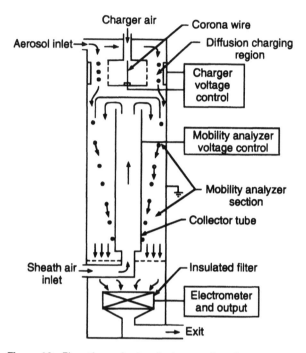

Figure 16 Flow-through electrical aerosol analyzer.

voltage on the system, exit the analyzer and are collected by a high-efficiency filter. An electrometer continuously monitors the current produced by the capture of the charged particles on the filter. The mobility distribution is obtained by varying the voltage on the center rod and measuring the current change in the electrometer (Horn, 1991; Reischl, 1991; Lehtimäki, 1987: Juozaitis et al., 1993; Adachi et al., 1990).

Screen diffusion batteries are used to separate particles according to their ability to diffuse to the internal surfaces of each collection cell (Fig. 17). The number of particles that penetrate each cell is usually measured with a condensation nuclei counter. Aerosol penetration through a stack of fine screens with uniform diameter and geometry is a function of the diffusion coefficient, defined by the fan model filtration theory (Cheng and Yeh, 1980). For submicron-size particles (diameter less than a few tenths of a micron) aerosol penetration increases with particle diameter. Data analysis requires numerical inversion techniques (Willeke and Baron, 1993; Cohen, 1995; Cooper and Wu, 1990; Yee, 1989; Wang, 1993; Ramamurth and Hopke, 1990).

Sampling Train No single instrument can measure aerosol size distribution of particles with diameters from 0.005 to 10 μm (Fig. 18). Yet some test atmospheres and many environmental atmospheres contain bimodal and

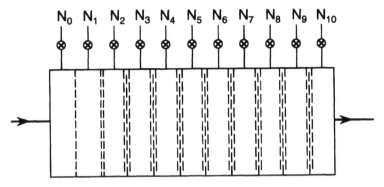

Figure 17 Diffusion battery.

trimodal distributions of particles. The toxicity of mixtures that have different size distributions is increasingly important. To obtain the full size distribution, sampling trains should consist of inertial and mobility types of instruments in series. Careful analysis of the data is required, since the impactor data are used to estimate mass distribution, and EAA and screen diffusion battery data are used to estimate number distribution (Cheng et al. 1986; Willeke and Baron, 1993; Cohen, 1995).

SUMMARY

The purpose of this overview is to provide the reader with background information about aerosols that will help in understanding current texts containing instructions for generating and measuring aerosols for use in inhalation toxicology exposure systems. A selected bibliography of these texts follows the reference list.

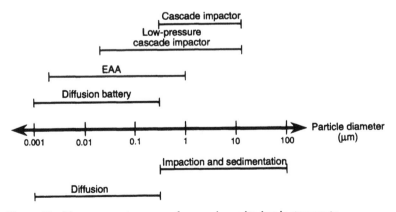

Figure 18 Measurement ranges of aerosol monitoring instruments.

ACKNOWLEDGMENT

This work was supported by the Chemical Industry Institute of Toxicology, Research Triangle Park, NC, and under contract DE-AC04-76EV01013 at the Lovelace Inhalation Toxicology Research Institute, Albuquerque, NM.

REFERENCES

Adachi, M, Okuyama, K, Kousaka, Y, Moon, SW, Seinfeld, JH: Facilitated aerosol sizing using the differential mobility analyzer. Aerosol Sci and Tech 12:225–239, 1990.

Adachi, M, Pui, DYH, Liu, BYH: Aerosol charge neutralization by a corona ionizer. Aerosol Sci and Tech 18:48–58, 1993.

Ahn, KH, Liu, BYH: Particle activation and droplet growth processes in condensation nucleus counter—I. Theoretical background. J Aerosol Sci 21:249–261, 1990a.

Ahn, H, Liu, BYH: Particle activation and droplet growth processes in condensation nucleus counter—II. Experimental study. J Aerosol Sci 21:263–275, 1990b.

Anisimov, MP, Hämeri, K, Kulmala, M: Construction and test of laminar flow diffusion chamber: Homogeneous nucleation of DBP and n-hexanol. J Aerosol Sci 25:23–32, 1994.

Asgharian, B, Moss, OR: Particle suspension in a rotating drum chamber when the influence of gravity and rotation are both significant. Aerosol Sci and Tech 17:268–277, 1992.

Asgharian, B, Yu, CP: Deposition of fibers in the rat lung. J Aerosol Sci 20:355–366, 1989.

Baron, PA: Calibration and use of the aerodynamic particle sizer (APA 3300). Aerosol Sci Technol 5:55–67, 1986.

Baumgartner, H, Coggins, CRE: Description of a continuous-smoking inhalation machine for exposing small animals to tobacco smoke. Beitr Tabakforsch 10:169–174, 1980.

Bernstein, DM, Moss, O, Fleissner, H, Bretz, R: A brush feed micronising jet mill powder aerosol generator for producing a wide range of concentrations of respirable particles. In: Aerosols, ed. BYH Liu, DYH Pui, HJ Fissan, pp. 721–724. Elsevier, New York, 1984.

Chen, BT, Cheng, YS, Yeh, HC: Performance of a TSI aerodynamic particle sizer. Aerosol Sci Technol 4:89–97, 1985.

Cheng, Y.S: Bivariate lognormal distribution for characterizing asbestos fiber aerosols. Aerosol Sci. Technol. 5(3): 359–368, 1986.

Cheng, YS, Barr, EB, Benson, JM, Damon, DG, Medinsky, MA, Hobbs, CH, Goehl, TJ: Evaluation of a real-time aerosol monitor (RAM-S) for inhalation studies. Fundam Appl Toxicol 10:321–328, 1988.

Cheng, YS, Barr, EB, Marshall, IA, Mitchell, JP: Calibration and performance of an API Aerosizer. J Aerosol Sci 24:501–514, 1993a.

Cheng, YS, Barr, EB, Yeh, HC: A venturi disperser as a dry powder generator for inhalation studies. Inhalat Tox 1:367–371, 1989.

Cheng, YS, Chen, BT, and Yeh, HC: Behaviour of isometric nonspherical aerosol particles in the aerodynamic particle sizer. J Aerosol Sci 21:701–710, 1990.

Cheng, YS, Chen, BT, Yeh, HC, Marshall, IA, Mitchell, JP, Griffiths, WD: Behavior of compact nonspherical particles in the TSI aerodynamic particle sizer model APS33B: Ultra-Stokesian drag forces. Aerosol Sci and Tech 19:255–269, 1993b.

Cheng, YS, Marshall, TC, Henderson, RF, Newton, GJ: Use of a jet mill dispersing dry powder for inhalation studies. Am Ind. Hyg. Assoc. J. 46:449–454, 1985.

Cheng, YS, Yeh, HC: Theory of a screen type of diffusion battery. J Aerosol Sci 11:313–320, 1980.

Cheng, YS, Yeh HC, Newton, GJ: Sampling in tandem with other instruments. In: Cascade Impactor Sampling and Analysis, ed. JP Lodge, TL Chan, pp. 129–150, American Industrial Hygiene Association, Akron, OH, 1986.

Cohen, BS (ed): Air Sampling Instruments for Evaluation of Atmospheric Contaminants, 8th Ed., American Conference of Governmental Industrial Hygienists, Cincinnati, 1995.

Cooper, DW, Wu, JJ: The inversion matrix and error estimation on data inversion: Application to diffusion battery measurements. J Aerosol Sci 21:217–226, 1990.

Decker, J, Moss, O, Kay, B: Controlled-delivery vapor generator for animal exposures. Am Ind Hyg Assoc J 43:400–402, 1982.

DeFord, HS, Clark, ML, Decker, JR: A laboratory-scale conveyor/metering device and combustor for generating smoke from solids at a uniform rate. Am Ind Hyg Assoc J 43:764–766, 1982.

DeFord, HS, Clark, ML, Moss, OR: A stabilized aerosol generator. Am Ind Hyg Assoc J 42:602–604, 1981.

Dreiling, V, Haller, P, Helsper, C, Kaminski, U, Pomp, A, Raes, F, Roth, C, Schier, J, Schuermann, G: Intercomparison of eleven condensation nucleus counters. J Aerosol Sci 17:565–570, 1986.

Ensor, DS, Viner, AS, Johnson, EW, Donovan, RP, Keady, PB, Weyrauch, KJ: Measurement of ultrafine aerosol particle size distributions at low concentrations by parallel arrays of a diffusion battery and a condensation nucleus counter in series. J Aerosol Sci 20:471–475, 1989.

Foss, WR, Davis, EJ: Formation of coated aerosol particles by simultaneous diffusion and chemical reaction. J Aerosol Sci 25:33–48, 1994.

Grace, JM Marijnissen, JCM: A review of liquid atomization by electrical means. J. Aerosol Sci 25:1005–1019, 1994.

Griffiths, RB, Standafer, S: Simultaneous mainstream-sidestream smoke exposure systems. II. The rat exposure system. Toxicology 35:13–24, 1985.

Gruel, RL, Reid, CR, Allemann, RT: The optimum rate of drum rotation for aerosol aging. J. Aerosol Sci 18:17–22, 1987.

Guerin, MR, Stokely, JR, Higgins, CE, Moneyhun, JH, Holmberg, RW: Inhalation bioassay chemistry—Walton horizontal smoking machine for inhalation exposure of rodents to cigarette smoke. JNCI 63:441–448, 1979.

Heinrich, U, Pott, P, Mohy, U, Stöber, W: Experimental methods for the detection of carcinogenicity and/or cocarcinogenicity of inhaled polycyclic-aromatic-hydrocarbon-containing emissions. In: Carcinogenesis, Vol. 8, pp. 131–146. Raven Press, New York, 1985.

Hemenway, DR, Jakab, GJ, Risby, TH, Sehnert, SS, Bowes, SM, Hmieleski, R: Nose-only inhalation system using the fluidized-bed generation system for coexposures to carbon black and formaldehyde. Inhalat Tox 2:69–90, 1990.

Hering, SV, Flagan, RC, Friedlander, SK: Design and evaluation of new low-pressure impactor 1. Environ Sci Technol 12:667–673, 1978.

Hering, SV, Friedlander, SK, Collins, JJ, Richards, LW: Design and evaluation of a new low pressure impactor 2. Environ Sci Technol 13:184–188, 1979.

Hewitt, AJ: Droplet size spectra produced by air-assisted atomizers. J Aerosol Sci 24:155–162, 1993.

Higuchi, MA, Steinhagen, WH: Modification and characterization of dry material feeder delivery of red and violet dye mixtures. Inhalat Tox 3:223–236, 1991.

Hinds, WC: Aerosol Technology. Properties, Behavior, and Measurement of Airborne Particles. John Wiley, New York, 1982.

Holländer, W, Holländer, M, Beyer, A, Koch, W: Development of a method for the high-production generation of fibrous material with aerodynamic diameters less than 3 microns. J Aerosol Sci 18:903–905, 1987.

Horn, HG: Optimization of the electrostatic classification for the generation of monodisperse aerosols for filter testing. J Aerosol Sci 22:S339–S342,1991.

Jakab, GJ, Hemenway, DR: Inhalation coexposure to carbon black and acrolein suppresses alveolar macrophage phagocytosis and TNF-α release and modulates peritoneal macrophage phagocytosis. Inhalt Tox 5:275–289, 1993.

Japuntich, DA, Stenhouse, JIT, Liu, BYH: An aerosol generator for high concentrations of 0.5–5 μm solid particles of practical monodispersity. Aerosol Sci and Tech 16:246–254, 1992.

Julanov, YuV, Lushinkov, AA, Nevskii, IA: Statistics of multiple counting in aerosol counters. J Aerosol Sci 15(1): 69–79, 1984.

Julanov, YuV, Lushnikov, AA, Nevskii, IA: Statistic of multiple counting in aerosol counters— II. J Aerosol Sci 17(1):87–93, 1986.

Juozaitis, A, Ulevicius, Girgzdys, A, Willeke, K: Differentiation of hydrophobic from hydrophilic submicrometer aerosol particles. Aerosol Sci and Tech 18:202–212, 1993.

Kagawa, M, Suzuki, M, Mizoguchi, Y, Hirai, T, Syono, Y: Gas-phase synthesis of ultrafine particles and thin films of Y-Al-O by the spray-ICP technique. J Aerosol Sci 24:349–355, 1993.

Kanapilly, GM, Tu, KW, Larsen, TB, Fogel, GR, Luna, RJ: Controlled production of ultrafine metallic aerosols by vaporization of an organic chelate of the metal. J Colloid Interface Sci 65:533–547, 1978.

Kaplan, HK, Grand, AF, Hartzell, GE: Combustion Toxicology, Principles and Test Methods. Technomic, Lancaster, PA, 1983.

Kim, YK, Kim, SS: Effects of mixed convection flow on aerosol generation in a condenser tube. Aerosol Sci and Tech 20:136–148, 1994.

Lee, KP, Kelly, DP, Kennedy, GL: Pulmonary response to inhaled Kelvlar synthetic fibers in rats. Toxicol Appl Pharmacol 71:242–253, 1983.

Lee, KW, Kim, JC, Han, DS: Effects of gas density and viscosity on response of aerodynamic particle sizer. Aerosol Sci and Tech 13:203–212, 1990.

Lehtimäki, M: New current measuring technique for electrical aerosol analyzers. J Aerosol Sci 18:401–407, 1987.

Liu, BYH, Lee, KW: An aerosol generator of high stability. Am Ind Hyg Assoc J 36:861–865, 1975.

Lodge, JP, Chan, TL: Cascade Impactor Sampling and Analysis. Am Ind Hyg Assoc, Akron, OH, 1986.

Marple, VA: A fundamental study of inertial impactors, PhD Thesis, University of Minnesota, Minneapolis, 1970.

Marple, VA, Liu, BYH, Kuhimey, GA: A uniform deposit impactor. J Aerosol Sci. 12:333 1981.

Marshall, IA, Mitchell, JP, Griffiths, WD: The behaviour of regular-shaped non-spherical particles in a TSI aerodynamic particle sizer. J Aerosol Sci 22:73–89, 1991.

McCarthy, JF, Yurek, GJ, Elliott, JF, Amdur, MO: Generation and characterization of submicron aerosols of zinc oxide. Am Ind Hyg Assoc J 48:880–886, 1982.

Melton, PM, Harrison, RM, Burnell, PKP: The evaluation of an improved spinning top aerosol generator and comparison with its predecessor. J Aerosol Sci 22:101–110, 1991.

Mercer, TT: Aerosol production and characterization: Some considerations for improving correlation of field and laboratory derived data. Heatlh Phys 10:873–877, 1964.

Mercer, TT: Aerosol Technology in Hazard Evaluation. Academic Press, New York, 1973.

Milliman, EM, Chang, DPY, Moss, OR: A dual flexible-brush dust-feed mechanism. Am Ind Hyg Assoc J 52:747–751, 1981.

Mokler, BV, Archibeque, FA, Beethe, RL, Kelley, CPJ, Lobez, JA, Mauderly, JL, Stafford, DL: Diesel exhaust exposure system for animal studies. Fundam Appl Toxicol 4:270–277, 1984.

Moss, OR, Wong, BA, Asgharian, B: Bimodal Bivariat Log-normal Distributions in the Application of Inhalation Toxicology Specific to the Measurement of Fiber and Particle Dosimetry. In: Toxic and Carcinogenic Effects of Solid Particles in the Respiratory Tract, ed. U Mohr, pp. 623–628. International Life Sciences Institute, Washington DC, 1994.

Niessner, R, Daeumer, B, Klockow, D: Investigation of surface properties of ultrafine particles by application of a multistep condensation nucleus counter. Aerosol Sci and Tech 12:953–963, 1990.

Noone, KJ, Hansson, HC: Calibration of the TSI 3760 condensation nucleus counter for nonstandard operating conditions. Aerosol Sci and Tech 13:478–485, 1990.

O'Shaughnessy, PT, Hemenway, DR: Computer automation of a dry dust generating system. Inhal Tox 6:95–113, 1994.

Ohshima, K, Tsuto, K, Okuyama, K, Tohge, N: Preparation of ZnO-TiO$_2$ composite fine particles using the ultrasonic spray pyrolysis method and their characteristics on ultraviolet cutoff. Aerosol Sci and Tech 19:468–477, 1993.

Pluym, TC, Powell, QH, Gurav, AS, Ward, TL, Kodas, TT, Wang, LM, Glicksman, HD: Solid silver particle production by spray pyrolysis. J Aerosol Sci 24:383–392, 1993.

Pratsinis, SE: Simultaneous nucleation, condensation and coagulation in aerosol reactors. J. Colloid Interface Sci. 124:416–427, 1988.

Pratsinis, SE, Landgrebe, JD, Mastrangelo, SVR: Design correlations for gas phase manufacture of ceramic powders. J Aerosol Sci. 20:1457–460, 1989.

Rader, DJ, Brockmann, JE, Ceman, DL, Lucero, DA: A method to employ the aerodynamic particle sizer factory calibration under different operating conditions. Aerosol Sci and Tech 13:514–521, 1990.

Rajathurai, AM, Roth, P, Fißan, H: A shock and expansion wave-driven powder disperser. Aerosol Sci and Tech 12:613–619, 1990.

Ramamurth, M, Hopke, PK: Simulation studies of reconstruction algorithms for the determination of optimum operating parameters and resolution of graded screen array systems (nonconventional diffusion batters). Aerosol Sci and Tech 12:700–710, 1990.

Reischl, GP: The relationship of input and output aerosol characteristics for an ideal differential mobility analyser particle standard. J Aerosol Sci 22:297–312, 1991.

Shaw, BD, Lawman, J: Analysis of constant-rate aerosol reactors. Aerosol Sci and Tech 20:363–374, 1994.

Teague, SV, Pinkerton, KE, Goldsmith, M, Gebremichael, A, Chang, S, Jenkins, RA, Moneyhun, JH: Sidestream cigarette smoke generation and exposure system for environmental tobacco smoke studies. Inhalat Tox 6:79–93, 1994.

Tom, JW, Debenedetti, PG: Particle formation with supercritical fluids—a review. J Aerosol Sci 22:555–584, 1991.

Tu, KW, Kanapilly, Gm, Mitchell, CE: Generation and characterization of condensation aerosols of benzo(a)pyrene. J Environ Sci 7:353–363, 1981.

Wang, HC: Thermal rebound of nanometer particles in a diffusion battery. Aerosol Sci and Tech 18;180–186, 1993.

Whitby, KT: Generator for producing high concentration of small ions. Rev Sci Instruments 32(12): 1351–1355, 1961.

Willeke, K, PA Baron (eds): Aerosol Measurement Principles, Techniques, and Applications. Van Nostrand Reinhold, New York, 1993.

Wu, MK, Windeler, RS, Steiner, CKR, Börs, T, Friedlander, SK: Controlled systhesis of nanosized particles by aerosol processes. Aerosol Sci and Tech 19:527–548, 1993.

Xiong, Y, Kodas, TT: Droplet evaporation and solute precipitation during spray pyrolysis. J Aerosol Sci 24:893–908, 1993.

Yee, E: On the interpretation of diffusion battery data. J Aerosol Sci 20:797–811, 1989.

Zhang, ZQ, Liu, BHY: Dependence of the performance of TSI 3020 condensation nucleus counter on pressure, flow rate, and temperature. Aerosol Sci and Tech 13:493–504, 1990.

Zhang, Z, Liu, BYH: Performance of TSI 3760 condensation nuclei counter at reduced pressures and flow rates. Aerosol Sci and Tech 15:228–238, 1991.

BIBLIOGRAPHY

Cohen, BS (ed): Air Sampling Instruments for Evaluation of Atmospheric Contaminants, 8th Ed. American Conference of Governmental Industrial Hygienists, Cincinnati, 1995.

Hinds, WC: Aerosol Technology. Properties, Behavior, and Measurement of Airborne Particles. John Wiley, New York, 1982.

Lodge, JP, Chan, TL: Cascade Impactor Sampling And Analysis. Am Ind Hyg Assoc, Akron, OH, 1986.

Mercer, TT: Aerosol Technology in Hazard Evaluation. Academic Press, New York, 1973.

Phalen, RJ: Inhalation Studies: Foundations and Techniques. CRC Press, Boca Raton, FL, 1984.

Raabe, OG: The Generation of Aerosols of Fine Particles. In: Fine Particles. Ed. Liu BYH. Academic Press, New York, 1976.

Willeke, K (ed): Generation of Aerosols. Ann Arbor Science, Ann Arbor, MI, 1980.

Willeke, K, PA Baron (eds): Aerosol Measurement Principles, Techniques, and Applications. Van Nostrand Reinhold, New York, NY, 1993.

Part Three

Basic Biology of the Respiratory Tract

Chapter Five

Morphology of the Respiratory Tract

Robert F. Phalen, Hsu-Chi Yeh and Shankar B. Prasad

INTRODUCTION

The respiratory tract is faced with a problem; it must bring large quantities of air into intimate contact with the blood and at the same time defend itself against the countless irritants, oxidants, allergens, carcinogens, pathogens, and other potentially harmful contaminants found in the atmosphere. That the average person inhales about 400 million liters of air in a lifetime while maintaining a healthy lung is a tribute to the architecture of the respiratory tract. (The volumetric rate of intake of air is about 5000 times greater than that of water or food).

The structure of the respiratory tract is being continually elucidated. Research is often specialized and directed at specific levels of organization such as the molecular, cellular, or tissue levels, or at specific regions, the nasal turbinates, the tracheobronchial tree, or the alveoli for instance. In some cases the specialized efforts have been integrated—the mucociliary system is an example—but in other cases the integration of existing morphologic information is far from complete. The current state of understanding of the structure of the mammalian lung has been summarized in several reviews. The 1990 international workshop on Respiratory Tract Dosimetry proceed-

ings (Guilmette & Boecker, 1991) covers anatomy along with particle deposition and clearance. The 1983 American Review of Respiratory Disease supplement entitled "Comparative Biology of the Lung" contains over 20 papers
on comparative mammalian lung anatomy and physiology. The series entitled
"Lung Biology in Health and Disease" of about 20 volumes edited by Claude
Lenfant and published by Marcel Dekker is another source of summary information on the respiratory system. The Handbook of Physiology, Section
3, The Respiratory System (Fishman and Fisher, 1985) also covers many aspects of our knowledge of respiratory-tract structure and function as does
Inhalation Toxicology: The Design and Interpretation of Inhalation Studies
and Their Use in Risk Assessment (Dungworth et al., 1988).

What follows here is an introduction to the gross and subgross structure
of the respiratory system as it relates to inhalation toxicology. The emphasis
is on the healthy adult human, with some mention of the events occurring
during growth and development, and some information on comparative
mammalian airway structure. Much of the information is updated from an
earlier chapter by the authors (Phalen and Prasad, 1989).

RESPIRATORY-TRACT REGIONS

The main anatomical structures of the respiratory tract include the

    1   nose, consisting of the nares, vestibule, and nasal cavity proper (with
the conchae or turbinates);
    2   nasopharynx;
    3   lips and oral cavity;
    4   oropharynx;
    5   laryngopharynx;
    6   larynx;
    7   trachea;
    8   bronchi;
    9   bronchioles;
  10   respiratory bronchioles;
  11   alveolar ducts;
  12   alveolar sacs; and
  13   alveoli.

These structures are commonly grouped into larger regions, or compartments, for the purpose of simplification and mathematical modeling. Several
compartmentalization schemes have been proposed, but three very similar
models have been particularly useful to inhalation toxicologists. The models
are those of Task Groups of the International Commission on Radiological
Protection (Morrow et al., 1966; Bair, 1991), the Ad Hoc Working Group to
Technical Committee 146-Air Quality of the International Standards Organization (ISO, 1983), the Air Sampling Procedures Committee of the American

Conference of Governmental Industrial Hygienists (Air Sampling Procedures Committee, ACGIH, 1985) and the Task Group of the National Council of Radiation Protection (Cuddihy and Yeh, 1988). These compartmental systems which are largely based on the pioneering morphometric work of Ewald Weibel (1963) are shown in Table 1.

Region 1—containing the airways of the head and neck—begins at the anterior nares and includes the respiratory airway down through the larynx. Particle deposition in this region includes those larger particles whose inertial properties cause impaction in the oral or nasal passages and particles smaller than about 0.5 μm in diameter that diffuse to airway walls. Two pathways, each having a half-time of 4 minutes, were used by the previous ICRP Task Group (Morrow et al., 1966) to describe the clearance of particles which deposit in the nasal airways. The first describes uptake of relatively soluble material into the blood; the second represents physical clearance by mucociliary transport to the throat for subsequent swallowing. Experimental data indicate that the anterior one-third of the nose, where 80% of 7-μm-diameter particles deposit, does not clear except by blowing, wiping, or other extrinsic means, and effective removal of insoluble particles may be slower (Morrow, 1977; Cuddihy and Yeh, 1988).

Region 2—the tracheobronchial region—begins below the larynx and includes the trachea and ciliated bronchial airways down to and including the terminal bronchioles. A relatively small fraction of all sizes of particles that pass through the airways of the head and neck will deposit in the tracheobronchial region. The mechanisms of inertial impaction at airway bifur-

Table 1   Compartmental Models of the Human Respiratory System as Developed by the ICRP Task Group on Lung Dynamics, the International Standards Organization, and the American Conference of Governmental Industrial Hygienists.

Region	Anatomic structures included	Task group region	ISO region	ACGIH region
1	Nose, mouth nasopharynx, oropharynx, laryngopharynx, larynx	Nasopharynx (NP)	Extrathoracic (E)	Head airways region (HAR)
2	Trachea, bronchi, bronchioles (to terminal bronchioles)	Tracheo-bronchial (TB)	Tracheo-bronchial (B)	Tracheo-bronchial region (TBR)
3	Respiratory bronchioles alveolar ducts, alveolar sacs, alveoli	Pulmonary (P)	Alveolar (A)	Gas exchange region (GER)

cations, sedimentation, and Brownian diffusion all cause deposition. Interception can be an important deposition mechanism for fibers. During mouth breathing, the benefits of the collection of larger particles in the nose are largely lost, and the larger particles generally tend to deposit in the tracheobronchial region with higher efficiency. An important characteristic of this region is that it is both ciliated and equipped with mucus-secreting elements, so that in the healthy individual the clearance of most of the deposited particles occurs within 24 hours by mucociliary action to the throat for swallowing. Relatively soluble material may quickly enter the bloodstream.

The rate of mucus movement is slowest in the smaller airways and increases toward the trachea. Particles depositing in the tracheobronchial tree are distributed somewhat differently depending on their size, with smaller particles tending to deposit more distally. Thus, one expects larger particles to clear more quickly. Clearance of material in this region cannot usually be described by a single exponential rate constant (Cuddihy and Yeh, 1988).

Region 3—the gas exchange region—includes the functional gas exchange sites of the lung. It includes respiratory bronchioles, alveolar ducts, alveolar sacs, and alveoli. For particles to reach and deposit in this region, they must penetrate the two more proximal regions on inspiration and by either settling, diffusion, or interception come into contact with deep lung surfaces. Since there is gas exchange between tidal residual air, a portion of each breath remains unexhaled, and the times available for deposition may be long for some particles. Clearance from this region is not completely understood, but several mechanisms are believed to exist, including

1   the dissolution of relatively soluble material with absorption into the systemic circulation,
2   direct passage of particles into the blood,
3   phagocytosis of particles by macrophages with subsequent translocation, and
4   transfer of particles to lymphatic channels, vessels, and nodes.

The three-region model does have some rather important drawbacks. For example, the detailed pattern with which particles deposit within a given region is usually not addressed. The assumption that deposition within a given region is uniform may eventually lead to improper estimation of risk. For example, bifurcations in the tracheobronchial region can be sites of high regional deposition. Also, there is not adequate separation of the region between the terminal bronchioles and the alveolar ducts and sacs. This junctional region contains respiratory bronchioles that are unique in structure in that they have both air conducting properties and gas exchange properties. Here, deposition of inhaled particles appears to be greater than in more distal regions, presumably because of lack of penetration of air beyond the respiratory bronchioles. Clearance from the respiratory bronchioles is not well understood. Since this portion of the gas exchange region is often the site of airway disease in humans, it should not be overlooked by the toxicologist.

GROSS ANATOMY
Nose, Nasopharynx, and Larynx

The mammalian nose and its immediately postnasal cavities comprise an elaborate organ system that provides for olfaction, detection of airborne irritants, collection of noxious gases and particles, humidification and temperature adjustment of inspired air, and drainage of fluids from the eyes, sinuses, and inner ears. The importance of these functions to maintaining good health also makes the nasopharyngeal region an important target for airborne agents. Despite this fact, the nasal region is often an overlooked region in toxicology. It must frequently deal with air pollutants in their raw unfiltered state at ambient concentrations, and failure of any of its critical functions can lead to serious, even life-threatening conditions.

In humans, the nose contains two channel-shaped nasal cavities that are separated by a cartilaginous and bony septum. The average adult male's nose has an air volume of about 17 cm^3. Each nasal cavity is entered through a naris (nostril) having a cross-sectional area of about 0.7 cm^2 (Landahl, 1950). The nasal cavity is supported by bone, cartilage, and connective tissue that provide sufficient rigidity to prevent total collapse during breathing. The anterior (nearest to the nares) one-third of the nasal cavity is covered with skin much like that on the face that does not have a continuous coating mucus. The posterior two-thirds of the cavity is covered with mucus that moves rearward driven by cilia at an average velocity of about 1.0 cm/min. to a point where it is swallowed. This mucus, produced by goblet cells and glands, is mixed with fluids, including tears, that drain into the nasal cavity from the eyes and sinus cavities of the facial bones. The anterior portion of the nasal cavity is partially covered with hair that traps large inhaled bodies and signals their presence via nerves at the base of the hair follicles.

Posteriorly, the nasal cavity narrows and turns sharply downward. This area—the nasopharynx—is a region of impaction of large particles that eluded previous capture. The nasopharynx, roughly tubular in shape, is joined by the oral pharynx (rear portion of the mouth) a few centimeters down its length. The pharynx then divides at the epiglottis to turn and enter either the larynx and trachea or continues downward to the esophagus. The pharynx is coated with mucus.

The mouth, entered through the variable-size opening between the lips, is divided into two regions—anterior vestibule or labial cavity (including the inner lips, cheeks, and teeth), and a posterior or buccal cavity which joins the oropharynx. The oral cavity is normally about 70–75 mm in length (anteroposterior dimension), 40–45 mm in horizontal dimension, and 20–25 mm in vertical dimension. The tongue has a volume of approximately 60–70 cm^3

(Task Group on Reference Man, 1975). The shape and cross-sectional area of the lip opening are highly variable.

The epiglottis is a muscular flap that moves to cover the entrance to the larynx and trachea during swallowing. Other muscular action also prevents swallowed material from entering the trachea because persons whose epiglottis has been surgically removed can still swallow without choking.

The larynx (or voice box) is a short cavity that has a slit like, variable-size narrowing in its central portion. The narrowing is caused by two pairs of folds in the walls of the larynx. The uppermost folds are called the false vocal cords, and the lower folds the vocal cords. The adult larynx is about 3.5–5 cm long and has a variable cross section that depends on the air flow rate passing through it (Stanescu et al., 1972). The larynx represents a major resistive element to air flow and also forms an inspiratory air jet that leads to particle impaction on the wall of the trachea (Schlesinger and Lippmann, 1976). The larynx is encased by muscle, bone, and cartilage and is lined by a mucus-covered membrane very similar to that found in the rear portion of the nasal cavities and pharynx. In the larynx, mucus is propelled upward for swallowing.

Tracheobronchial Tree

The distal larynx smoothly transitions into the trachea, a flexible tube that, in humans has about 20 roughly U-shaped cartilages set in its wall that prevent its collapse. The gap between the ends of the cartilaginous rings is covered with a flexible muscular sheet of tissue. Thus, in cross section the trachea tends to have a D or O shape, depending on the internal air pressure. During breathing, the trachea elongates on inspiration. The inner walls of the trachea are covered with mucus supplied by goblet cells and mucous glands.

In humans, the trachea divides into two main branches called major bronchi. The bronchi enter the right and left lungs and continue to divide for several generations (averaging about 16 in humans) before alveoli (air sacs) begin to appear in the bronchiolar walls as openings into the lumen. This appearance of alveoli marks the end of the tracheobronchial tree and the beginning of the gas exchange or alveolarized region.

Bronchi and bronchioles (Fig. 1) are roughly circular in cross section, and smooth muscle encircles bronchial airways. The U-shaped cartilages of the trachea are replaced in the bronchial walls by irregularly shaped cartilage plates situated outside the smooth muscle. Further down the tracheobronchial tree, where the tube diameters are about 1.0 mm or less, the cartilage disappears. These tubes are called bronchioles. Bronchioles have mucus-secreting goblet cells but do not have mucous glands in their walls. The outermost layer of the bronchi consists of a mixture of connective tissue and elastic fibers.

The inner lining of the bronchi is pseudostratified columnar epithelium

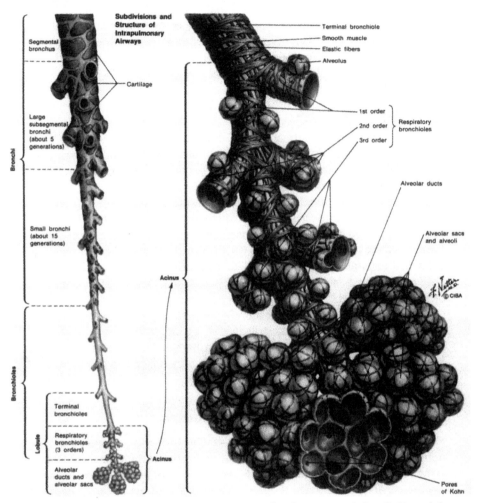

Figure 1 Bronchi, bronchioles, alveolar ducts, alveolar sacs, and alveoli. (Copyright 1979, 1980, CIBA pharmaceutical Company, Division of CIBA-GEIGY Corporation. Reprinted with permission from the CIBA Collection of Medical Illustration illustrated by Frank H. Netter, MD. All rights reserved.)

having ciliated cells, mucus-secreting goblet cells, and underlying mucus-secreting glands. Thus, the tracheobronchial tree possesses an active clearance mechanism because of the propulsion of mucus toward the pharynx. The bronchioles are lined with ciliated columnar epithelium that is not pseudostratified.

The symmetrical tracheobronchial tree model for the human, described by Weibel (1963), is widely used and contains information on airway lengths, diameters, and numbers. Similar, but asymmetric, human models have been published by Weibel (1963), Horsfield and Cumming (1968), and Yeh and Schum (1980). The fractal (self-similar) nature of the tracheobronchial tree has been recently described (West, 1990). Before B. Mandelbrot introduced the term fractal, K. Horsfield and others explored similar properties of branching systems, including the bronchial airways (Horsfield, 1976).

Respiratory Bronchioles

In humans the terminal bronchioles of the tracheobronchial tree, i.e., those with diameters of about 0.6 mm, branch to form the first-order respiratory bronchioles. These bronchioles continue to divide and branch to give a total of about two to five orders of tubes.

Respiratory bronchioles, as they branch, exhibit an increasing number of alveoli opening into their lumina. These alveoli are thin-walled, are surrounded by blood capillaries, and presumably participate in the gas exchange function of the lung. Because of this gas exchange function, these bronchioles are called respiratory. Within alveoli, ciliated cells are not found, and clearance of deposited debris is presumably by those mechanisms associated with deeper-sited alveoli.

Two major points must be made with respect to the respiratory bronchioles. First, they have been acknowledged as an important site for disease in humans. And second, these structures form part of the "silent zone" of the lung; a region in which respiratory disease is very difficult to detect by conventional pulmonary function testing.

Comparative Tracheobronchial Tree Structure

Substantial variation is recognized in tracheobronchial structure among different species of mammals. The major variations include tracheal length, symmetry of branchings, length-to-diameter ratios for airway tubes, presence or absence of a lobe branching from the mid-region of the trachea, shapes of the air flow dividers at branches, the number of generations in the bronchial tree, and the presence or absence of respiratory bronchioles.

When examining replica casts of the airways, two basic types of structure are seen in mammals (Fig. 2). Most mammals have a very asymmetric tracheobronchial branching structure characterized by long tapering airways, each having numerous small branches (typically making a 60° angle with the major airway). The other type of branching called regular dichotomous or symmetric is characterized by two daughter tubes at each branch that are nearly equal in diameter and branching angle. In reality, perfectly symmetrical bronchial trees have not been seen, but the human's structure is remarkably close. Highly asymmetric branching is seen in most other species including the dog, rabbit, guinea pig, rat, mouse, and others. In general, branching becomes more symmetric as one progresses deeper down the bronchial tree.

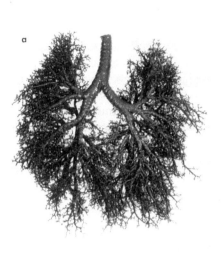
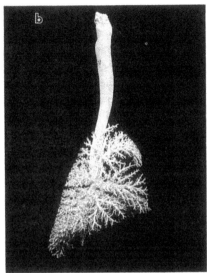

Figure 2 Replica silicone rubber casts of the tracheobronchial airways of the (a) human, and (b) dog. These casts demonstrate the more regular dichotomous and a more monopodial (asymmetric) branching system, respectively. Photographs supplied by the Inhalation Toxicology Research Institute.

Another striking variation is seen in the number of divisions of respiratory bronchioles. Animals that have an average of about 4 or more divisions include humans, monkeys, dogs, goats, and ferrets. The rabbits, guinea pigs, and golden hamster have about 1–2 orders of respiratory bronchioles. Respiratory bronchioles are essentially absent in laboratory rats, and probably in mice.

Parenchyma

Parenchyma is a term that relates to the primary functional tissue of an organ as distinguished from its supporting framework or secondary tissues. When applied to the lung, the parenchyma relates to the alveoli and does not include the trachea and bronchial tree which are often viewed as merely conductive airways for the purpose of delivering air to and from the gas exchange region. Major structural elements of the parenchyma of the lung include alveolar ducts, alveolar sacs, alveoli, alveolar capillaries, and the pulmonary lymphatics.

The alveolar duct is a tubular structure whose walls are completely covered with alveoli. It usually branches to either two other alveolar ducts or two blind-ended tubes called alveolar sacs. With respect to the total number of ducts and sacs, Weibel's (1963) figures of 7×10^6 ducts and 8.4×10^6 sacs are probably reliable. The dimensions of the pulmonary acinus, which consists of a terminal bronchiole and the structures supplied by it, have been described by Schreider and Raabe (1981).

Although often depicted as spherical, the alveoli more closely resemble incomplete polyhedra. The open face of the alveolus is exposed to the air in either a respiratory bronchiole—an alveolar duct, or an alveolar sac—the closed portions being surrounded by a network of fine blood capillaries. Thus, in the alveolus the atmosphere and the blood are brought into intimate contact where equilibration of CO_2 and O_2 can take place. In addition to the surrounding capillary net, alveoli are partially surrounded by elastic and nonelastic fibers that provide mechanical support. Alveoli, capillaries, and fibers are embedded in an interstitium or connective tissue. The average diameter of the adult's approximately 300 million alveoli is about 200–300 μm (Task Group on Reference Man, 1975).

CELLS AND TISSUES OF THE RESPIRATORY TRACT
Ciliated Mucosa

The tissue that lines the rear of the nose, larynx, trachea, bronchi, and bronchioles is ciliated mucosa (Fig. 3). As the name implies, such tissue is characterized by the presence of cells with numerous tiny hairlike projections (cilia) and by the presence of individual cells and glands that secrete the components that make up mucus (a sticky viscoelastic fluid). The cilia beat in a coordinated fashion (resulting in movement of the overlying mucus) toward the glottis where it is swallowed.

The ciliated cells of the human respiratory system have cell nuclei and are columnar in shape—about 10–15 μm in diameter and 20–40 μm in height. The ciliated cells are attached at their bases to a basement membrane, and new replacement cells appear to form beneath the mature cells and move upward to replace cells that are lost. At the top surface, protruding into the lumen of the airway, there are 15 to perhaps 100 or more filamentous cilia that are 5–15 μm long and about 0.2 μm in diameter. The cilia bend and then lash forward at rates up to several hundred cycles per minute. It is the coordinated beating of cilia on adjacent cells that propels the mucus.

Interspersed among the ciliated cells are columnar goblet cells similar in size to the ciliated cells but lacking cilia and having a narrow base (and thus a drinking-goblet shape). These cells, also attached to the basement membrane, manufacture mucus and, when filled, open at the top and discharge their contents onto the airway surface. Beneath the basement membrane there are glands consisting of clusters of mucus-secreting cells that secrete into a duct that leads to the epithelial surface. The action of the ciliated mucus-secreting tissues is responsible for sweeping surfaces of the airways free of particulate contamination. This function depends on the quality and quantity of mucus and the quantity and synchronization of cilia. Viral and

Mucus

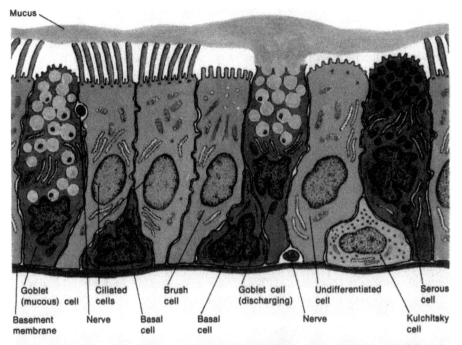

Goblet (mucous) cell	Ciliated cells	Brush cell	Goblet cell (discharging)	Undifferentiated cell	Serous cell
Basement membrane	Nerve	Basal cell	Basal cell	Nerve	Kulchitsky cell

Trachea and large bronchi. Ciliated and goblet cells predominant, with some serous cells and occasional brush cells, undifferentiated (intermediate) cells, and Clara cells. Numerous basal cells and occasional Kulchitsky cells present

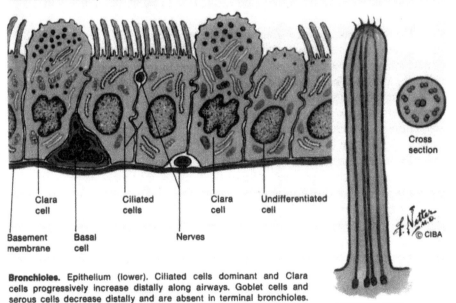

Clara cell	Ciliated cells	Clara cell	Undifferentiated cell
Basement membrane	Basal cell	Nerves	

Cross section

Bronchioles. Epithelium (lower). Ciliated cells dominant and Clara cells progressively increase distally along airways. Goblet cells and serous cells decrease distally and are absent in terminal bronchioles. Occasional undifferentiated and brush cells may be present. Basal cells and especially Kulchitsky cells are uncommon in distal airways

Magnified detail of cilium

Figure 3 Cells of bronchial walls. (Copyright 1979, 1980, CIBA Pharmaceutical Company, Division of CIBA-GEIGY Corporation. Reprinted with permission from The CIBA Collection of Medical Illustrations illustrated by Frank H. Netter, MD. All rights reserved.)

bacterial infections as well as various toxicants can lead to oversecretion or undersecretion of mucus and to loss or paralysis of cilia. During such states sneezing and coughing become the major clearance mechanisms that serve to clear the mucociliary epithelium. Often thought to be an annoying symptom, coughing can be a health-preserving mechanism for removing mucus, toxicants, and infectious organisms from the respiratory tract.

The Alveolus

As previously mentioned, the adult human's alveolus (Fig. 4): is a polyhedral structure, about 200–300 μm in diameter, having one face open to the airway. The walls of this structure are formed by very thin alveolar epithelial cells whose nuclei sometimes bulge into the alveolar airspace. In reality, there is more than one type of alveolar epithelial cell. At its thinnest portions, the type I (also called type A) alveolar epithelial cell is about 0.1 μm or slightly less in thickness (Nagaishi, 1972). These cells appear to have relatively smooth surfaces and lie on top of a basement membrane that is about 0.02–0.04 μm thick. Another basement membrane supports the blood capillary endothelial cells. These endothelial cells join to form the capillary wall and are quite similar in size and shape to the thin alveolar cell. The total thickness of the air-blood interface has been measured by Meessen (1960) and reported by Weibel (1964) and by Weibel and Gil (1977) to be between 0.36 and 2.5 μm.

A thicker, roughly cube-shaped cell, the type II (or type B) epithelial cell of the alveolus has a surface covered with small protrusions called microvilli. These microvilli greatly increase the surface area of this cell and imply, along with the presence of inclusions within the cell body, that this cell manufactures and secretes substances onto the surface of the alveolus. Biochemical and other evidence indicates that this cell is involved in the manufacture and secretion of surfactant, a surface-tension-lowering agent that reduces the tendency alveoli have for collapsing (Comroe et al., 1973; Pattle, 1965). Abnormalities in lung surfactant can be related to a variety of disease states. The type II cells, which are capable of mitotic division, may serve as precursors to the type I cell during lung growth and repair.

Other cells present in the alveolar region include the macrophage, alveolar brush cells (type III), and interstitial cells. The alveolar brush cell sits on the alveolar basement membrane and protrudes into the alveolar air space. It has large microvilli on its air-exposed side and has as yet largely unknown functions.

In some areas the basement membranes of the alveolus and capillaries are separated by a space called the interalveolar septum or interstitium. This interstitium contains both elastic and inelastic fibers and cells called fibroblasts. Fibroblasts are irregular-shaped cells that are involved in the formation of connective tissue. Rarely, nerve fibers have been seen in the intersti-

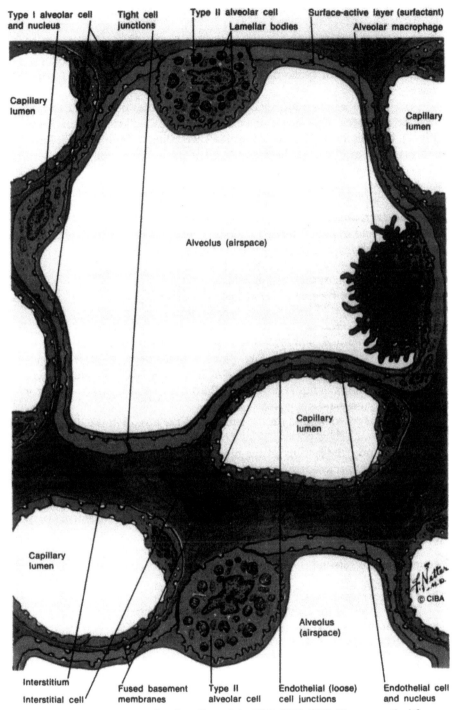

Type I alveolar cell and nucleus   Tight cell junctions   Type II alveolar cell   Surface-active layer (surfactant)
Lamellar bodies   Alveolar macrophage

Capillary lumen

Capillary lumen

Alveolus (airspace)

Capillary lumen

Capillary lumen

Alveolus (airspace)

Interstitium   Fused basement membranes   Type II alveolar cell   Endothelial (loose) cell junctions   Endothelial cell and nucleus
Interstitial cell

Figure 4 The pulmonary alveolus. (Copyright 1979, 1980, CIBA Pharmaceutical Company, Division of CIBA-GEIGY Corporation. Reprinted with permission from The CIBA Collection of Medical Illustrations illustrated by Frank H. Netter, MD. All rights reserved.)

tium (Nagaishi, 1972). In pathologic conditions such as edema (fluid accumulation) and infection, the interstitial space may become enlarged owing to the presence of excess fluid and cells such as blood leukocytes (white cells).

Alveolar walls are frequently observed to have pores that appear to connect the airspaces of adjacent alveoli. These pores, called pores of Kohn (Fig. 1), were discovered by Adriani in 1847 according to Miller (1947). Little information is available on their shapes, numbers, or dimensions.

The Macrophage

Alveolar macrophages are relatively large nucleated cells that possess the ability to move and to engulf foreign materials. Roughly similar to the familiar amoeba, macrophages can change shape presumably by

1 liquefaction of their cell membrane,
2 subsequent flowing of the cell contents, and
3 re-formation of their surface membrane.

Phagocytosis and pinocytosis are two terms used to describe the engulfment of substances in varying states by cells such as macrophages. Phagocytosis refers to the incorporation of solid materials; pinocytosis refers to the incorporation of liquid droplets. A third term—endocytosis—includes both phagocytosis and pinocytosis.

Macrophages are found on the surfaces of the alveoli in the deep lung, but they are not a fixed part of the alveolar epithelial wall. They are credited with maintaining the sterility of the lung by virtue of their ability to engulf and kill infectious microorganisms such as bacteria. Macrophages also engulf particles that deposit in the deep lung. It appears that these pulmonary alveolar macrophages undergo chemotaxis—movement in response to chemical stimuli. Chemotaxis may be positive (toward the stimulus) or negative (away from it).

The process of phagocytosis has been described as occurring in seven sequential steps (Stossel, 1976):

1 target recognition,
2 reception of the message to initiate phagocytosis,
3 transmission of the message to an effector,
4 attachment of the macrophage membrane to the target,
5 formation of the pseudopodia,
6 engulfment by the pseudopodia, and
7 fusion of the pseudopodia with the macrophage cell body.

Failure of any of these subprocesses could result in the inactivation of the function of the macrophage in providing for defense of the lung.

Macrophages have an amazing efficiency in engulfing particles. Within minutes of deposition of an inhaled particle, the pulmonary alveolar macrophage is seen to have begun ingestion. These cells also appear to be able to phagocytize even when packed nearly full of debris. On the other hand, certain dusts are clearly toxic to the macrophage and result in their death or debilitation. Macrophages appear to efficiently engulf a relatively narrow size range of particles (Fenn, 1921, 1923). Holma (1969) suggested a diameter of 1.5 μm for maximally efficient uptake by macrophages. He found that phagocytic uptake had an upper particle diameter limit of 8 μm, a size that usually does not penetrate to alveoli. Fibers are exceptions in that alveolar deposition occurs for particles whose lengths exceed the limits for phagocytosis. Physiologically realistic models that quantitatively describe particle clearance sequestration by macrophages are under active development and validation (Stöber et al., 1994). The macrophage also plays an important role in immunological responses (Dinarello, 1985).

Mucus-Secreting Glands

Mucus-secreting glands are present in the nose and tracheobronchial tree (Fig. 5). These glands are present in great numbers in large airways and become more sparse moving down to smaller airways, finally disappearing at the level of the bronchiole. Along with goblet cells, these glands produce the mucus that covers the ciliated portions of the respiratory tract. Since these glands lie beneath the mucous membrane, they are called submucosal glands. They have a branched tubuloacinar structure ("acinar" referring to the blind ends of the tubes that branch and form each gland). The tubes into which mucus is secreted join a collecting duct that becomes ciliated just before it enters the bronchial air space. These ciliated ducts appear as pinholes on the surfaces of bronchi, having a maximum surface concentration of about one opening per square millimeter in the trachea (Netter, 1979).

Two types of cell—mucous and serous—rest on a basement membrane and line the tubules. Serous cells are found lining the blind ends of the tubules, and mucous cells line the more proximal portions. The secretion of these cells—mucus—is primarily an acid glycoprotein, which is both viscous and elastic.

THE LYMPHATIC SYSTEM

The lymphatic system of lungs play an important supportive role in maintaining liquid homeostasis, respiratory defenses, and in the clearing of inhaled toxicants and particulate matter. The large flow of lymph from the interstitium of the lungs toward the blood also helps in removing the excess fluid from the tissue spaces.

The lymphatics of the lung are subdivided into a superficial and a deep

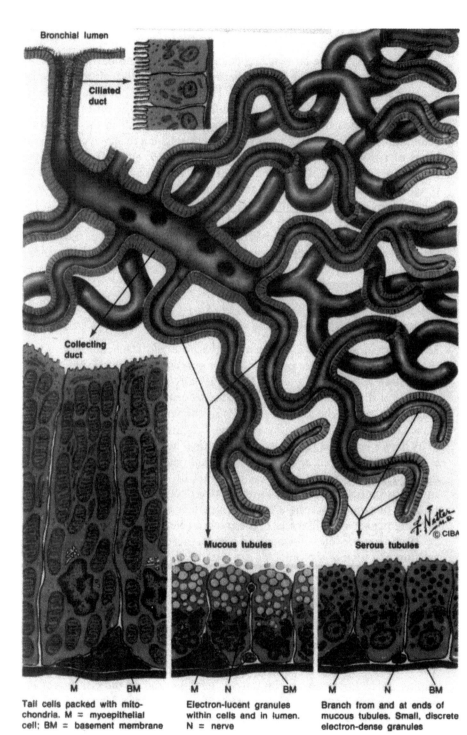

Bronchial lumen

Ciliated duct

Collecting duct

Mucous tubules

Serous tubules

| M | BM | M | N | BM | M | N | BM |

Tall cells packed with mitochondria. M = myoepithelial
cell; BM = basement membrane

Electron-lucent granules
within cells and in lumen.
N = nerve

Branch from and at ends of
mucous tubules. Small, discrete
electron-dense granules

Figure 5 Submucosal glands of bronchial walls. (Copyright 1979, 1980, CIBA Pharmaceutical Company, Division of CIBA-GEIGY Corporation. Reprinted with permission
from The CIBA Collection of Medical Illustrations illustrated by Frank H. Netter, MD.
All rights reserved.)

plexus depending on their location. The former is located in the connective tissue of visceral pleural layer and the latter in the peribronchovascular connective tissue. These are interlinked by vessels in the interlobular septa with a potential for bidirectional flow. In addition, the lymphatics line pulmonary arteries, veins, and bronchi and converge at the pulmonary hilus into the hilar lymph nodes. There are lymphatic vessels between the alveoli and lung capillaries. The pumping motion of the lungs (and in some regions valves) facilitate lymph flow.

The major groups of lymph nodes of the lung include:

1   bronchopulmonary lymph nodes around the divisions of lobar bronchi,
2   hilar nodes around the upper and lower bronchi,
3   paratracheal nodes on either side of trachea (more prominent on right side), and
4   the azygos node adjacent to the azygos vein at the junction of right upper lobar bronchus and the right main bronchus.

From these nodes the lymph drains into the thoracic duct on the left side and into the right lymphatic duct on the right side. These in turn drain into systemic venous circulation at the junctions of subclavian and internal jugular veins.

In addition, there are lymphatic vessels along the distribution of internal mammary arteries, intercostal arteries, and the anterior and posterior mediastinum, all of which receive drainage primarily from the chest wall. The lung's lymphatic system is in communication with the lower deep cervical nodes above and with the abdominal nodes below.

Commonly, the lower lobes drain into the hilar nodes while the upper lobes drain directly into paratracheal nodes, although this drainage pattern may show wide variations. One problem associated with this extensive lymphatic system is the spread of bronchogenic carcinoma so readily out of the lung to distant sites.

INNERVATION OF THE RESPIRATORY SYSTEM

The nervous system receives, generates, conveys, stores, and processes information. Portions of the nervous system, found in nearly every tissue of the body, play an important part in the voluntary and involuntary control and coordination of muscles, organs, glands and their subunits, tissues, and cells. In the respiratory system, nerves are responsible for

1   control of muscles for breathing, adjustment of the size of bronchial airways, and control of the cough, sneeze, and gag reflexes;
2   the initiation and control of protective breathing patterns;
3   the control of secretions;

4 adjustment of the distribution of blood flow; and
5 provision of sensory information on odor, irritancy, and the composition of lung tissue fluids and blood.

As for the body in general, much of the information that is carried by the nervous systems of the respiratory tract is not noticed at the conscious level.

Especially important are nerves that trigger the cough reflex; nerves that lead from pressure, stretch, and chemical receptors; and nerves involved in bronchial muscle constriction, protective breathing patterns, and mucous gland secretion. It is clear that the innervation of the respiratory tract is extensive and, in fact, present in nearly every region from the nose down to the alveoli. The interaction of inhaled airborne toxicants with this system is relatively poorly understood.

POSTNATAL LUNG DEVELOPMENT

The mammalian respiratory system goes through a period of significant differentiation and development after birth and the transition to air breathing. The sequence of development appears to be similar in all mammals examined thus far (mice, rats, cats, rabbits, dogs, and humans), but the timing of events varies considerably (Dunnill, 1962; Boyden and Tompsett, 1965; Emery, 1969; Crocker et al., 1970; Hislop and Reid, 1974; Burri et al., 1974; Thurlbeck, 1975; Boyden, 1977). Although most investigators conclude that the full number of tubular airways are present at birth, there is a significant increase in the number of alveoli throughout the period of early maturation after birth (Kerr et al., 1975; Thurlbeck, 1977; Reid, 1977; Jeffery and Reid, 1977; Burri and Weibel, 1977; Brody and Vaccaro, 1979).

In humans the number of alveoli present at birth, estimated at about 20–70 million, increases over about the first 5–10 years of life to the adult complement of about 200–500 million (Thurlbeck, 1975; Reid, 1977); the rate of increase in number decreases with age after birth. Great variability in the number of alveoli is seen at all ages, presumably due to genetic and environmental factors superimposed on differences in counting techniques as well as the difficulty in identifying immature forms of alveoli. The sites of alveolar development, as elucidated in human and laboratory animals studies, include the distal bronchioles (Thurlbeck, 1975, 1977), where saccular protrusions in the bronchiolar wall, which is initially covered by cuboidal epithelial lining cells, are eventually replaced by thin alveolar epithelium (Boyden and Tompsett, 1961). The process of formation of alveoli appears to proceed along these airways in a direction from the distal airways toward more proximal ones.

The tracheobronchial region is perhaps the best understood portion of the respiratory tract with respect to dimensional changes during growth and development. One reason for this is that particle dosimetry models indicate that newborns, infants, and children may receive larger inhalation doses than adults (Hofmann, 1982a,b; Phalen et al., 1985; Xu and Yu, 1986). Also, the tracheobronchial region has been relatively easy to measure quantitatively through the use of replica silicone rubber casts and dissection techniques.

SUMMARY

The anatomical characteristics of the respiratory tract are continually being elucidated via ongoing research. Although more than a dozen major anatomical structures, from the nose down to the alveoli, have been described. Inhalation toxicologists tend to use a 3-compartment model consisting of the airways of the head, the tracheobronchial airways, and the gas exchange airways. These three regions have characteristic lumen shapes, cell populations, particle clearance mechanisms, and particle deposition characteristics. Although mammalian species show some obvious variability in airway structure, such differences are relatively well understood.

Particulate material that has deposited on airway surfaces is cleared from the airways by several mechanisms; some such as coughing and mucociliary flow act rapidly, and others such as lymphatic drainage and particle dissolution may act very slowly. The nervous system of the respiratory tract has many functions that are toxicologically significant, including control of breathing depth and frequency, triggering of the cough reflex, and control of mucus production.

At birth, mammalian lungs undergo a transition to air breathing. This transition is followed by a period of lung development which includes the addition of alveoli. In humans, lung development is initially rapid, but 5–10 years may be required before the full number of alveoli is present. An area of considerable uncertainty which is important in inhalation toxicology is the observed individual-to-individual variability in lung anatomy, lung defenses, and rate of maturation.

REFERENCES

Air Sampling Procedures Committee, American Conference of Governmental Industrial Hygienists (ACGIH): Particle Size-selective Sampling in the Workplace. Cincinnati: American Conference of Governmental Industrial Hygienists, 1985.

Bair, WJ: Overview of ICRP respiratory tract model. Radiat Prot Dosim 38:147–152, 1991.

Boyden, EA: Development and growth of airways. In: Development of the Lung, edited by WA Hodson, pp. 3–35, New York: Marcel Dekker, 1977.

Boyden, EA, Tompsett DH: The postnatal growth of the lung in the dog. Acta Anat 47:185–215, 1961.

Boyden, EA, Tompsett DH: The changing patterns in the developing lungs of infants. Acta Anat 61:164–192, 1965.

Brody, JS, Vaccaro, C: Postnatal formation of alveoli: Interstitial events and physiologic consequences. Fed Proc 38:215–233, 1979.

Burri, PH, Weibel, ER: Ultrastructure and morphometry of the developing lung. In: Development of the Lung, edited by WA Hodson, pp. 215–268. New York: Marcel Dekker, 1977.

Burri, PH, Dbaly, J, Weibel, ER: The postnatal growth of the rat lung. Anat Rec 178:711–730, 1974.

Comroe, JH, Forster, RE, Dubois, AB, et al.: The Lung, 2nd Ed. Chicago: Year Book, 1973.

Crapo, JD, Barry, BE, Gehr, P, et al.: Cell numbers and cell characteristics of the normal human lung. Am Rev Respir Dis 126:332–337, 1982.

Crocker, TT, Teeter, A, Nielsen, B: Postnatal cellular proliferation in mouse and hamster lung. Cancer Res 30:357–361, 1970.

Cuddihy, RG, Yeh, HC: Respiratory tract clearance of particles and substances dissociated from particles. In Inhalation Toxicology: The Design and Interpretation of Inhalation Studies and Their Use in Risk Assessment, edited by D Dungworth, G Kimmerle, J Lewkowski, R McClellan, W. Stöber, pp 169–193. New York: Springer-Verlag, 1988.

Dinarello, CA: An update on human interleukin 1: From molecular biology to clinical relevance. J Clin Immunol 5:287–297, 1985.

Dungworth, D, Kimmerle, G, Lewkowski, J, McClellan, R, Stöber, W (eds): Inhalation Toxicology: The Design and Interpretation of Inhalation Studies and Their Use in Risk Assessment. New York: Springer-Verlag, 1988.

Dungworth, DL, Phalen, RF, Schwartz, LW, Tyler, WS: Morphological methods for evaluation of pulmonary toxicity in animals. Annu Rev Pharmacol Toxicol 16:381–401, 1976.

Dunnill, MS: Postnatal growth in the lung. Thorax 17:329–333, 1962

Emery, J (ed): The Anatomy of the Developing Lung. Suffolk, U.K.: Heinemann, 1969.

Emery, JL: The postnatal development of the human lung and its implications for lung pathology. Respiration 27 (Suppl):41–50, 1970.

Fenn, WO: The phagocytosis of solid particles. I. Quartz. J Gen Physiol 3:575–593, 1921.

Fenn, WO: The phagocytosis of solid particles. IV. Carbon and quartz in solutions of varying acidity. J Gen Physiol 5:311–325, 1923.

Fishman, AP, Fisher, AB (eds): The Respiratory System, Section 3, Handbook of Physiology. Bethesda, MD: American Physiological Society, 1985.

Guilmette, RA, Boecker, BB (eds): Respiratory Tract Dosimetry: Proceedings of a Workshop held in Albuquerque July 1–3, 1990. Ashford, Kent: Nuclear Technology Publishing, 1991 (Also, Radiat Prot Dosim 38, Nos 1–3, 1991).

Hislop, A, Reid, L: Development of the acinus in the human lung. Thorax 29:90–94, 1974.

Hofmann, W, Steinhausler, F, Pohl. E: Dose calculations for the respiratory tract from inhaled natural radioactive nuclides as a function of age. I. Compartmental deposition, retention, and resulting does. Health Phys. 37:517–532, 1979.

Hofmann, W: Mathematical model for the postnatal growth of the human lung. Respir Physiol 49:115–129, 1982a.

Hofmann, W: Dose calculations for the respiratory tract from inhaled natural radioactive nuclides as a function of age. II. Basal cell dose distributions and associated lung cancer risk. Health Phys 43:31–44, 1982b.

Holma, B: The acute effects of cigarette smoke on the initial course of lung clearance in rabbits. Arch Environ Health 18:171–173, 1969.

Horsfield, K: Some mathematical properties of branching trees with application to the respiratory system. Bull Math Biol 38:305–315, 1976.

Horsfield, K, Cumming, G: Morphology of the bronchial tree in man. J Appl Physiol 24:373–838, 1968.

ISO: Air Quality-Particle Size Fraction Definitions for Health Related Sampling, ISO/TR 7708–1983 (E). Geneva: International Standards Organization, 1983.

Jeffery, PK, Reid, LM: Ultrastructure of airway epithelium and submucosal gland during development. In: Development of the Lung, edited by WA Hodson, pp. 87–134. New York: Marcel Dekker, 1977.

Kerr, GR, Couture, J, Allen, JR: Growth and development of the fetal rhesus monkey. VI. Morphometric analysis of the developing lung. Growth 39:67–84, 1975.

Landahl, HD: On the removal of airborne droplets by the human respiratory tract. I. The lung. Bull Math Biophys 12:43–56, 1950.

Meessen, H: Die Pathomorphologie der Diffusion und Perfusion. Verhandl Dsch Ges Pathol 44–98, 1960.

Miller, WS: The Lung. Springfield, IL: Charles C. Thomas, 1947.

Morrow, PE: Clearance kinetics of inhaled particles. In: Respiratory Defense Mechanisms, Part II, edited by JD Brain, DF Proctor, and LM Reid, p. 491–543. New York: Marcel Dekker, 1977.

Morrow, PE, Bates, DV, Fish, BR, Hatch, TF, Mercer, TT: Deposition and retention models for internal dosimetry of the human respiratory tract (report of the International Commission on Radiological Protection: ICRP: Task Group on Lung Dynamics). Health Phys 12:173–207, 1966.

Nagaishi, C: Functional Anatomy and Histology of the Lung. Baltimore: University Park Press, 1972.

Netter, FH: The CIBA Collection of Medical Illustrations, Vol. 7: Respiratory System. Summit, NJ: CIBA Pharmaceutical Corporation, 1979.

Pattle, RE: Surface lining of lung alveoli. Physiol Rev 45:48–79, 1965.

Phalen, RF: Inhalation Studies: Foundations and Techniques. Boca Raton, FL: CRC Press, 1984.

Phalen, RF, Oldham, MJ, Beaucage, CB, Crocker, TT, Mortensen, JD: Postnatal enlargement of human tracheobronchial airways and implications for particle deposition. Anat Rec 212:368–380, 1985.

Phalen, RF, Prasad, SB: Morphology of the respiratory tract. In: Concepts in Inhalation Toxicology, edited by RO McClellan, RF Henderson, pp. 123–140, New York: Hemisphere, 1989.

Reid, L: The Lung: Its growth and remodeling in health and disease. Am J Roentgenol 129: 777–7888, 1977.

Schlesinger, RB, Lippmann, M: Particle deposition in the trachea: In vivo and in hollow casts. Thorax 31:678–684, 1976.

Schreider, JP, Raabe, OG: Structure of the human respiratory acinus. Am J Anat 162:221–232, 1981.

Stanescu, DC, Pattijn, J, Clement, J. Van de Woestijne, KP: Glottis opening and airway resistance. J Appl Physiol 32:460–466, 1972.

Stöber, W, Morrow, PE, Koch, W, Morawietz, G: Alveolar clearance and retention of inhaled insoluble particles in rats simulated by a model inferring macrophage particle load distributions. J Aerosol Sci 25:975–1002, 1994.

Stossel, TP: The mechanism of phagocytosis. J Reticuloendothel Soc 19:237–245, 1976.

Task Group on Reference Man: Report of the Task Group on Reference Man. International Commission of Radiological Protection No. 23, pp. 122–124. New York: Pergamon Press, 1975.

Thurlbeck, WM: Postnatal growth and development of the lung. Am Rev Respir Dis 11:803–844, 1975.

Thurlbeck, WM: Structure of the lungs. In: International Review of Physiology, Respiratory Physiology II, Vol. 14, edited by JG Widdicombe, pp. 1–34. Baltimore: University Park Press, 1977.

Weibel, ER: Morphometry of the Human Lung. New York: Academic Press, 1963.

Weibel, ER: Morphometrics of the lung. In: Respiration, VI, edited by WO Fenn, H Rahn, pp. 285–307. Washington, DC: American Physiological Society, 1964.

Weibel, ER, Gil, J: Structure-function relationships at the alveolar level. In: Bioengineering Aspects of the Lung, edited by JB West, pp. 1–81. New York: Marcel Dekker, 1977.

West, BJ: Physiology in fractal dimensions: Error tolerance. Ann Biomed Engr 18:135–149, 1990.

Xu, GB, Yu, CP: Effects of age on deposition of inhaled aerosols in the human lung. Aerosol Sci Technol 5:349–357, 1986.

Yeh, HC, Schum, GM: Models of human lung airways and their application to inhaled particle deposition. Bull Math Biol 42:461–480, 1980.

Chapter Six

Lung Biochemistry and Intermediary Metabolism for Concepts in Inhalation Toxicology

Aron B. Fisher

LUNG BIOCHEMISTRY AND INTERMEDIARY METABOLISM

The primary function of the lung is gas exchange, a function of immediate major importance for survival of the organism. Relevant to this primary role, the physiologist's model of the lung consists of conducting airways for transport of gas into and out of the lung and the alveolar membranes where gas exchange occurs by diffusion from the alveolus to the capillary. While this fundamental view of the lung is convenient for describing gas exchange, the cell biologist sees that the conducting airways and alveolar septae are comprised of metabolically active cells and that effectiveness of the lung for gas exchange depends on the integrity of its cellular components. These cells, like metabolically active cells of other organs, utilize complex biochemical reactions of intermediary metabolism to maintain their intracellular milieu, to generate chemical energy, and to carry out a broad array of specialized cell functions. These aspects of lung cell function comprise the topic of this chapter. The emphasis will be on oxygen uptake, ATP generation, glucose utilization, and lipid metabolism, although it must be recognized that the topic of "lung biochemistry," if inclusive, could comprise a full volume.

Like all organs, the lung is composed of a variety of cell types which serve varying roles but can function as an integrated unit. Unfortunately for those interested in the study of lung metabolism, the variety of different cell types is probably greater for this organ than for any other. A widely quoted early publication described 42 histologically different cell types in the mammalian lung (Sorokin, 1970) while more recently, nearly that many different cell types have been identified in the conducting airways alone. As techniques for molecular and immunologic identification improve, it is likely that an increasing number of distinct lung cell types will be described. What, then, is the lung biochemist to do? Ideally, one would like to know the metabolic characteristics of each individual cell type as well as its frequency in the lung in order to calculate its contribution to the metabolism of the whole organ. Clearly, this approach is not yet possible due to the relatively small number of cell types that have been isolated and characterized. The alternative is to regard the lung as a unit while recognizing the possible diverse influences of the different types of lung cell. This chapter will focus on the integrated metabolic characteristics of the lung and will describe the properties of individual lung cell types where appropriate. In depth reviews of these topics have been published previously (Fisher, 1984; Fisher, 1989; Fisher and Forman, 1985; Tierney and Young, 1985).

Oxygen Uptake by the Lung

Oxygen utilization by all organs of the body (ignoring the small component of O_2 derived by diffusion through the skin) ultimately relies on gas exchange in the alveoli of the lungs and represents the total oxygen uptake by the lung. This parameter has been well studied and is used as the basis for calculation of organismal basal metabolic rate or for calculation of cardiac output by the Fick principle. Within the lung, oxygen freely diffuses through the cells comprising the alveolar septae (epithelium, interstitium, endothelium) into the pulmonary capillaries for eventual distribution by the systemic blood flow. A small portion—generally less than 0.5%—of the total oxygen uptake by the lung is due to metabolic utilization by the lung cells themselves. Due to the availability of O_2 from the alveolar space, the alveolar cells do not rely on pulmonary perfusion for O_2 delivery. Thus, alveolar gas represents the major oxygen source for lung cells as it is the ultimate source of oxygen for all cells of the body.

The cells of the major conducting bronchi represent an exception to the generalization concerning the source of O_2 for lung cell metabolism. While the bronchial epithelium derives O_2 by diffusion from the airway lumen, oxygen consumed by the deeper tissues of the conducting airways is delivered via the systemic circulation through the bronchial arteries. Although difficult to define precisely, less than 15% of oxygen for lung cellular metabolism is derived normally through the bronchial circulation. Lung disease with cellular proliferation and hypertrophy of the bronchial arterial supply may lead to significant increases in this value.

Cellular Oxygen Utilization Because lung cellular oxygen utilization represents such a small fraction of total lung oxygen uptake, its measurement in the intact lung has been difficult. Calculation of the mass balance for oxygen (ignoring the bronchial circulation) requires the accurate measurement of the cardiac output and the oxygen contents of the alveolar gas and both the pulmonary arterial and venous blood. Our current techniques for measurement of the required parameters are not sufficient for use *in-vivo* where all of the parameters are constantly changing and small inaccuracies would produce large errors in calculated O_2 utilization. Consequently, there have been no reliable measurements of oxygen consumption by the normal lung *in-vivo*. Several investigators have utilized the isolated, perfused lung preparation where the required parameters can be more readily controlled and steady state conditions can be maintained. Reported mean values for oxygen consumption for isolated lungs of rabbit and dog are 39–48 μl/min/g dry weight (Koga, 1958; Weber and Visscher, 1969). Although use of the isolated perfused lung represents a possible approach, the relatively small differences among the several oxygen contents and the difficulty in minimizing diffusion of air and other leaks have rendered these types of experiments problematic.

More commonly, oxygen consumption by lung tissue has been measured using lung tissue slices. This preparation shares the limitations of tissue slices from any organ, including possible artifacts associated with injury due to the slicing procedure and the nonphysiologic routes for delivery of O_2 and removal of metabolic products. Further, this preparation is limited for study of possible physiological modulation of metabolism such as changes associated with lung ventilation.

Lung slices from rats have provided the greatest number of reported measurements of lung oxygen consumption. The range of mean values from various studies is 48–148 μl/min/g dry weight of lung with a mean of approximately 110 μl/min/g dry weight (Caldwell and Wittenberg, 1974; Stadie et al., 1945; Krebs, 1950; Barron et al., 1947; Edson and Leloir, 1936; Simon et al., 1947). The relatively wide range could reflect variation in slicing and incubation techniques. After a systematic study of factors affecting oxygen uptake measurement of rat lung slices, the measured value using optimal conditions was similar to the mean mentioned above (approximately 100 μl/min/g dry weight) (Levy and Harvey, 1974). Oxygen consumption by rat lung slices is similar to that of various rat organs such as intestine, pancreas, and resting skeletal muscle (Edson and Leloir, 1936; Engelbrecht and Maritz, 1974; Fisher and Forman, 1985; Krebs, 1950; Stadie et al., 1945; Caldwell and Wittenberg, 1974). However, it is somewhat less than that for the metabolically very active organs such as heart, kidney, brain, and liver—possibly related in part to the relatively high proportion of interstitial connective tissue in the lung. Within the lung itself, it is likely that individual lung cells types have a broad range of oxygen consumption with some cell types such as the granular pneumocyte (type II epithelial cell) and alveolar macrophage having oxygen consumption that is considerably greater than that of the average lung cell (Fisher et al., 1980). By morphometric analysis, the granular pneumocyte has a mitochondrial volume density that is approximately twice the combined value for membranous pneumocytes (type I epithelium) and endothelial cells (Massaro et al., 1975). This result presumably reflects the relative rates of O_2 consumption for these cell types.

The oxygen consumption of lung slices from various species, even when normalized to dry weight of the slice, varies inversely as a function of body size (Krebs, 1950; Massaro et al., 1975). Similar observations have been reported for slices from other organs and this effect has been attributed to both a greater proportion of interstitial nonmetabolizing connective tissue in larger species as well as intrinsic metabolic differences of the individual cells. Morphometric studies have provided evidence for intrinsic cellular metabolic differences by demonstrating that the volume density of mitochondria in granular pneumocytes from various species is directly proportional to lung tissue oxygen consumption (Massaro et al., 1975). The relation of oxygen consumption to body size is an approximate semilogarithmic function of body weight for individual species (Fisher and Forman, 1985). As an example, the oxygen consumption per unit weight is approximately twice as great for a rat lung slice as for a lung slice from the horse.

Pathways of Oxygen Consumption   As with most organs, the major pathway for cellular oxygen consumption by the lung is via the electron transport chain of mitochondria by which oxygen is chemically reduced to H_2O with the generation of ATP. Oxygen consumption by lung homogenate is decreased by more than 85% when cytochrome oxidase, the terminal electron acceptor of the mitochondrial electron transport chain, is inhibited (Freeman and Crapo, 1981; Rossouw and Engelbrecht, 1978). The approximate 10% of lung tissue oxygen consumption that occurs through other pathways can be accounted for by a variety of tissue oxidases such as cytochrome P450, lipo- and cyclo-oxygenases, prolyl- and lysyl oxidases, monoamine oxidase, and NADPH oxidase (Fisher and Forman, 1985). In addition, auto-oxidation of tissue components such as catechols, thiols, flavins, and mitochondrial ubi(semi)quinone can occur. These oxidase reactions and auto-oxidations may generate partially reduced oxygen species (that is, $O_2^{-\cdot}$ or H_2O_2) that are reactive and have been implicated in the pathogenesis of oxidant lung injury. From the standpoint of pulmonary toxicology, auto-oxidation of drugs and toxins, e.g. paraquat, could increase lung tissue oxygen consumption in the short term. The radicals produced provide a mechanism for pulmonary injury which can result in the eventual inhibition of cellular respiration.

Critical PO_2 of the Lung   The critical PO_2 for an organ can be defined as the ambient PO_2 at which oxygen consumption declines due to limited availability of O_2. As a reactant for isolated mitochondria, the critical PO_2 is less than 1 mmHg (Chance, 1965). For most tissues, the critical PO_2 is considerably greater in order to provide a sufficient oxygen diffusion gradient for maintenance of normal rates of mitochondrial oxygen utilization. On the other hand, the lung presents an ideal situation for maximizing gas exchange between its tissue and its source of O_2, the alveolar space. As required for the role of the lung in gas exchange, the surface area available for diffusion is large and the diffusion distances are extremely short so that the PO_2 on the alveolar and capillary sides of the lung cell are equal, or immeasurably different. Furthermore, cyclical alveolar ventilation maintains alveolar PO_2 at a relatively constant value. As a result, the critical PO_2 of the lung—measured from the effects of graded hypoxia on substrate metabolism and energy state—has been shown to approach the value obtained for isolated mitochondria (Fisher et al., 1976; Fisher and Dodia, 1981). Stated another way, the availability of O_2 for lung cell metabolism does not become rate-limiting until the alveolar PO_2 approaches 1 mmHg. These extremely low values for PO_2 if generalized would not be compatible with life and could occur only in abnormal segments of lung with absent ventilation and perfusion.

Energy Utilization of the Lung

In subserving its function in gas exchange, the lung can be looked upon as a "pump" that moves the respiratory gases to and from the lung air spaces. In that sense, the lungs are mechanically analogous to the heart. However, in contrast to the heart, the lung "pump" does not require metabolic energy from the intrinsic cells of the organ. Rather, the energy required for the pumping activity of lungs is supplied by the skeletal muscle of the chest wall and diaphragm while the lung tissue passively follows the changes in thoracic gas volume generated by the respiratory muscles. Consequently, the lungs *per se* do not have the great energy requirements that characterize the heart muscle or other tissues with a high physical work load. Rather, energy use by the lung is required for a wide range of physiologic functions representing the composite functions of individual lung cell types. These specialized energy-requiring functions include:

The ciliary beat of the tracheobronchial ciliated cells;
Secretion of mucus, surfactant, and other complex products by the tracheobronchial and alveolar epithelium;
Active transport of electrolytes, macromolecules, and vasoactive amines by epithelial and endothelial cells;
Constriction of airways and blood vessels by bronchial and vascular smooth muscle cells;

Synthesis and secretion of interstitial collagen and other connective tissue elements by interstitial fibroblasts;

Endocytosis of microorganisms and other particles by alveolar macrophages; and

Metabolic transformation of xenobiotics and other compounds by Clara cells.

This incomplete list of the energy requiring cellular reactions illustrates the need for metabolic energy generation by lung cells and provides a basis for understanding the pattern of lung functional impairment when ATP synthesis is inhibited.

Lung ATP Synthesis and Energy Stores   Analysis of rapidly frozen rat lung tissue has indicated an ATP content of approximately 10 μmol/g dry weight (Bassett and Fisher, 1976a; Bassett and Fisher, 1976b; Fisher and Dodia, 1981), comparable to values obtained for rat liver, kidney and brain (Williamson and Corkey, 1969). Extrapolation to the intact organ indicates a mean intracellular ATP concentration of approximately 5 mM. The greater fraction of the lung adenine nucleotide pool is present as ATP and the ratio of ATP to ADP is approximately 8. Therefore, the energy state of the lung reflects that of a metabolically active tissue. The lung content of phosphocreatine—a high energy compound that can generate ATP via creatine kinase—is only about 20% of lung ATP content on a molar basis (Buechler and Rhoades, 1980); it represents a relatively minor component of lung energy stores and may be associated predominantly with lung smooth muscle.

ATP generation by the rat lung, estimated from an analysis of metabolic flux using isolated perfused lung, is approximately 300 μmol/h/g lung dry weight (Bassett and Fisher, 1976a; Fisher and Forman, 1985). The basis for the estimate is calculation of ATP generation from glucose for both mitochondrial and non-mitochondrial (cytosolic) pathways. Under normal aerobic conditions, 85% of cellular ATP generation is via mitochondrial pathways and approximately 15% via substrate level phosphorylation associated with glycolysis. Mitochondrial inhibition (by ventilating the lung with high concentrations of CO for example) results in a doubling of flux through the glycolytic pathway (Pasteur effect), although total ATP production and lung energy stores are markedly decreased under these conditions (Bassett and Fisher, 1976a; Fisher and Forman, 1985). Ventilation of lungs with N_2 may be ineffective for producing cellular anoxia because of the low critical PO_2 of lungs (Fisher and Dodia, 1981) and the difficulty of excluding extraneous O_2. These results for ATP synthesis indicate that aerobic metabolism by lung mitochondria is required in order to maintain the normal energy status of the lung.

Lung Mitochondria   Mitochondria, the major site of cellular energy generation, are present in all lung cells although their relative cellular density varies. A survey in the alveolar region of the lung has indicated that volume density of mitochondria (that is, the fraction of cell volume represented by mitochondria) is greatest in the type II epithelial cell and alveolar macrophage, least in the type I epithelial cell, and intermediate in the capillary endothelium (Massaro et al., 1975). Furthermore, electronmicroscopic cytochemistry suggests significantly greater cytochrome oxidase activity per mitochondrion in type II vs. type I alveolar epithelial cells (Hirai et al., 1989). Presumably, this difference in cellular mitochondrial activity reflects differences in the energy requirement of these cells for their specialized functions. The relatively low level of mitochondrial activity in type I epithelium raises the possibility that this cell may rely to a greater extent on glycolytic pathways for ATP generation although this point has not been adequately evaluated.

Mitochondrial isolates have been prepared from lungs of a variety of species and have been evaluated for rates of oxygen utilization, substrate preferences, and metabolic coupling indicated by the respiratory control ratio (oxygen uptake in the presence and absence of ADP as a phosphate acceptor) (Reiss, 1966; Mustafa and Cross, 1974; Fisher et al., 1973; Fisher et al., 1975; Spear and Lumeng, 1978). Isolation methods can influence the metabolic parameters of the isolated organelles and could account for some of the differences that have been observed in various reports. Problems encountered in isolation of lung mitochondria include:

difficulty in homogenization of lungs due to the high content of connective tissues;

contamination of the preparation with blood-associated components related to the high pulmonary vascularity;

damage to membranes by fatty acids or lyso lipids due to the high lipid (surfactant) content of the lung;

and the great cellular heterogeneity of this organ which may result in a variable population of recovered mitochondria.

Isolated lung mitochondria in general show rates of oxygen uptake (50–100 nmol O_2/min/mg protein) and respiratory control ratios (4–8) that are only slightly lower than values obtained with many other preparations such as isolated liver mitochondria (Fisher et al., 1973; Fisher et al., 1976; Spear and Lumeng, 1978). The organelles contain the usual complement of cytochromes and other components of the respiratory chain and can oxidize the usual mitochondrial substrates including pyruvate, tricarboxylic acid cycle intermediates, fatty acids, glutamate, and glycerol 3-phosphate. Ketone bodies are relatively poor substrates for lung mitochondria although some species variation has been observed (Mustafa and Cross, 1974; Fisher et al., 1975). These studies have indicated that lung mitochondria have the characteristics necessary to carry out the complex oxidative reactions associated with the generation of ATP.

Substrate Utilization by the Lung

The cells of the lung require a carbon substrate source in order to generate the intermediary metabolites necessary for mitochondrial oxidation and production of ATP as well as to provide the building blocks for synthesis of complex carbohydrates and other macromolecules. Although a variety of interconvertible carbon sources can be oxidized by lung cells, glucose remains the most widely studied substrate and represents the predominant carbon source for intermediary metabolism under physiological conditions. Lactate and pyruvate—the 3-carbon metabolites of glucose—can enter the intracellular metabolic pool and could serve as an alternative carbon source when their blood concentrations are high. Indeed, the presence of high lactate of pyruvate (>0.5 mM) in the pulmonary perfusate results in a compensatory decrease in lung glucose utilization (Fisher and Dodia, 1984). Under usual physiologic circumstances, the lung shows a net release of these 3-carbon products into the circulation (O'Neil and Tierney, 1974; Bassett and Fisher, 1976a; Das, 1985). Glycerol after phosphorylation and fatty acids can also be oxidized but these substrates normally account for less than 10% of lung CO_2 production (Rhoades, 1974; Bassett et al., 1981; Fisher, 1989). These alternative carbon sources could assume greater physiological importance under conditions when availability of glucose for lung cells is limited.

Glucose Transport   Transport of glucose across the cell membrane represents the initial step for glucose utilization. Glucose transport has been studied in a variety of cell types from different organs, and several different mechanisms have been described. A preliminary characterization has been accomplished for the lung, but is very likely that transport mechanisms differ with lung cell type. For essentially all cells (including lung cells), glucose uptake for metabolic utilization occurs by facilitated diffusion in which a membrane carrier protein serves to equilibrate the transmembrane glucose concentrations. Studies showing uptake of several glucose analogues by isolated perfused lung are consistent with the presence of this mechanism (Kerr et al., 1981; Das and Steinberg, 1984). Accumulation of these analogues (e.g., 3-O-methylglucose) is not against a concentration gradient and is dependent on neither ATP nor Na^+.

On the other hand, Na^+-dependent accumulation of some glucose analogues (e.g., α-methylglucoside and 2-deoxyglucose) against a concentration gradient has been demonstrated in the isolated perfused rat lung and in isolated lung type II epithelial cells (Kerr et al., 1981; Das and Steinberg, 1984; Kerr et al., 1982; Kemp and Boyd, 1992). This finding is consistent with the presence of a Na^{++}-dependent active transport system such as has been described in other epithelia such as renal cortex and intestine. Interestingly and similar to its orientation in other epithelia, the Na^{++}-dependent glucose transport system has been localized to the apical surface of the type II epithelial cell; that is, the side facing the alveolus (Wangensteen and Bartlett, 1984; Basset et al., 1988). Glucose uptake from the alveolar space is significantly greater for the D-analogue providing evidence for specificity of the carrier mechanism.

Thus, the current model for glucose uptake by the lung epithelium is carrier mediated facilitated transport of glucose at the baso-lateral surface of the cell (i.e., the blood side) with Na^{++}-dependent active transport of glucose from the apical (alveolar) surface (Fig. 1). While the former serves the putative function of providing substrate for intermediary metabolism, the physiological significance of active transport from the alveolar surface is not yet apparent. This activity may be vestigial related to the embryonic derivation of the lung or could represent a mechanism to regulate the composition of the extra cellular alveolar fluid.

In muscle, adipose tissue, and some other organs, glucose transport into the cell can be regulated by the presence of insulin. Although initial studies suggested that the presence of insulin had little effect on lung glucose utilization, more recent results have indicated a possible role for the hormone. Insulin-specific receptors have been demonstrated in the lung of adult rat and fetal rabbit (Morishige et al., 1977; Neufeld et al., 1981). Depression of glucose utilization with reversal by insulin has been demonstrated in lungs

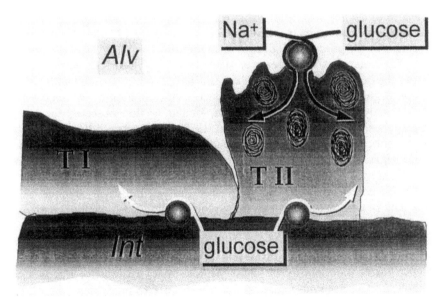

Figure 1   A scheme for glucose transport by lung epithelium. Glucose in the interstitial space, derived from the circulation, is transported into alveolar type I epithelial (TI) and alveolar type II epithelial (TII) cells by a facilitated transport system at the baso-lateral surface of the cells. Glucose in the alveolar space liquid is accumulated by a Na^{++}-dependent active transport system that is present on the apical membrane of the alveolar type II cell.

from rats with experimental diabetes (Fricke and Longmore, 1979; Das and Steinberg, 1984). Stimulation of glucose utilization in the presence of insulin has been reported for perfused rat lung (Stubbs et al., 1977; Kerr et al., 1979; Das and Steinberg, 1984); the modest effect of insulin with isolated lung systems may be related to a relatively slow turnover of the endogenously bound hormone. Isolated alveolar epithelial type II cells have a relatively higher density of insulin receptors and show stimulation of uptake of a glucose analogue by approximately 70% with insulin (Sugahara et al., 1984). Thus, this latter cell type may account for a major fraction of the insulin effect in the lung.

Glucose Utilization and Metabolic Products   Estimation of glucose utilization by the intact lung is beset by problems similar to those described above for measurement of O_2 utilization. With a normal resting blood glucose concentration of 5.6 mM, glucose delivery to the lungs of a normal rat (200g) is about 0.15 mmol/min. These lungs would be expected to metabolize 0.3 μmol/min (see below), representing only 0.2% of the glucose delivered through the pulmonary circulation. Consequently, availability of glucose is unlikely to be rate limiting for lung glucose utilization. In addition, the relatively small change in predicted glucose concentration between the pulmonary arterial and venous blood indicates that measurement of arteriovenous difference is not reliable for calculation of lung glucose utilization. Rather, glucose utilization has been evaluated through use of radiolabelled glucose and measurement of metabolic flux, that is, by the rate of radiolabel incorporation into the products of glucose metabolism.

The isolated rat lung perfused with ^{3}H- or ^{14}C-labeled glucose has provided the bulk of estimates for glucose utilization. When perfused with physiological glucose concentrations, the glucose consumption of rat lung is approximately 1 μmol/min/g dry weight, similar to that observed for other metabolically active organs (Kerr et al, 1979). The concentration of glucose that resulted in half-maximal glucose utilization by the isolated rat lung was approximately 3.5 mM; 10mM was required to saturate the glucose metabolizing pathways (Kerr et al, 1979; Maritz, 1993).

Isolated rat lungs metabolize glucose through the usual channels with the bulk of carbon flux through the glycolytic (Embden-Meyerhof) pathway and the tricarboxylic acid (Krebs) cycle (O'Neil and Tierney, 1974; Bassett and Fisher, 1976a; Bassett and Fisher, 1976b). The major end products of these pathways are lactate plus pyruvate and CO_2, respectively. The lung also has an active hexose monophosphate shunt (pentose cycle) pathway for glucose metabolism which accounts for approximately 25% of CO_2 production by the lung and results in the generation of NADPH (Bassett and Fisher, 1976c). The remainder of CO_2 production (75%) is from mitochrondrial sources.

The 3-carbon (lactate/pypruvate) products of glycolysis are permeable to the cell membrane and their production can be evaluated by analysis of the perfusate. Under normal conditions, approximately 50% of the carbon atoms derived through glucose catabolism can be recovered as lactate plus pyruvate; the ratio of lactate to pyruvate production, reflecting the redox state of the lung cells, is approximately 10 (Bassett and Fisher, 1976a; Fisher and Dodia, 1981; Fisher and Dodia, 1984; Fisher et al, 1976a). CO_2 production accounts for an additional 25% of glucose-carbon utilization. For the lung, the bulk of the CO_2 produced appears in the expired gas from lung ventilation. The remainder of the carbon atoms derived from glucose are recovered in lipids, polysaccharides, proteins, nucleic acids and other complex macromolecules representing the broad biosynthetic pathways of the various lung cell types (Yeager and Hicks, 1972; O'Neil and Tierney, 1974; Bassett and Fisher, 1976a). Relatively little of the glucose carbon is converted to glycogen in the adult lung, although lung glycogen stores are considerable in the fetus. A marked decrease in glycogen occurs in late gestation and during the neonatal period (Yeager and Hicks, 1972). Glycogen content of the adult lung is about 10 mg/g dry weight (Rhoades et al., 1978; Kerr et al., 1979) and is similar to other organs that do not specifically store glycogen.

Although all organs generate lactate plus pyruvate from the metabolism of glucose, the percent metabolized to the three carbon fragments by the lung is relatively high for an aerobic organ. For example, lactate production by the perfused rat heart represents less than 25% of glucose utilization, or one-half of the lung rate (Williamson, 1965). The high rate of lactate production by the lung may reflect its high blood flow to tissue ratio with consequent removal of the membrane-permeable lactate. Alternatively, high lactate production could represent a reliance on glycolytic pathways by lung cells such as the type I epithelial cell with its low mitochondrial volume density. The high lactate production does not represent cellular hypoxia since decreased alveolar oxygen has no effect on lung lactate production until extremely low levels of PO_2 are attained (see above: critical PO_2 of the lung).

The values for metabolic flux described above were attained during perfusion of lungs with glucose as the sole metabolizable substrate. When the concentration of lactate in the perfusate is increased, this substrate can enter the metabolic pool where it is metabolized via mitochondrial pathways (Fisher and Dodia, 1984). Under these conditions, the utilization of glucose and net production of lactate can decrease markedly. One suggested possibility for this effect is inhibition of the rate limiting enzyme phosphofructokinase by citrate generated in the mitochondria (Fisher and Dodia, 1984).

Modulation of Metabolic State   The profile for intermediary metabolism described above represents the balance of metabolic reactions necessary to maintain homeostasis. Key considerations are the energy and redox states of the cell and the availability of carbon intermediates and NADPH for biosynthetic reactions. The response of these parameters to distortion of normal metabolic processes can be evaluated as a measure of metabolic capacity of the organ.

Anoxia represents a commonly utilized metabolic challenge which reflects inhibition of the mitochondrial respiratory chain. Because of the difficulties of producing anoxia with the lung preparation (see above), mitochondrial inhibitors have been used to simulate the effects of anoxia. In the presence of high carbon monoxide—an inhibitor of the terminal cytochrome oxidase—lungs show a marked increase in glucose utilization and lactate production with essentially complete inhibition of CO_2 production via both mitochondrial metabolism and pentose shunt pathways (Bassett and Fisher, 1976a; Basset and Fisher, 1976c). The lactate to pyruvate ratio under these conditions increases about five-fold indicating an increased $NADH/NAD^+$ and marked reduction of the redox state (Fisher et al., 1976a). Estimated ATP production decreased by approximately 60% with a corresponding marked decrease in tissue ATP content and ATP/ADP (Fisher and Forman, 1985). These results confirm the aerobic dependence of lung intermediary metabolism and indicate that increased flux through glycolytic pathways with anoxia is insufficient to maintain normal lung energy balance. The consequence of anoxic energy depletion is a significant decrease in energy-linked lung functions such as phospholipid synthesis and amine uptake (Steinberg et al., 19875; Fisher et al., 1974; Fisher et al., 1985).

Anoxia represents a model for a metabolic perturbation that would decrease the rate of ATP generation. Use of an uncoupler of oxidative phosphorylation, such as dinitrophenol (DNP), represents the opposite perturbation, namely an apparent increase in energy utilization ("potential energy" is dissipated biochemically rather than through increased work). In the presence of this agent, glucose utilization was significantly stimulated with marked increases in production of lactate plus pyruvate as well as CO_2 generated through mitochondrial oxidations (Bassett and Fisher, 1976b). The estimated rate of ATP generation (although obviously dissipated by the presence of the uncoupler) was significantly increased. In contrast to the results with anoxia, the redox state (lactate/pyruvate) and ATP content of the lung were maintained near normal values.

The significance of these responses of lung to anoxia and uncoupling of oxidative phosphorylation is its appropriateness in terms of our present understanding of the pathways of metabolism. Thus, these metabolic pathways are physiologically regulated and designed to maintain normal energy and redox status. A summary of lung metabolism under control conditions and in response to metabolic perturbation is indicated in Table 1.

Metabolism of Lipids and Lung Surfactant

The lung tissue plays an active role in lipid metabolism, both for local needs as well as for whole body homeostasis. Some lipid-related metabolic functions include:

Table 1   Effect of Metabolic Perturbation on Intermediary Metabolism of Isolated Perfused Rat Lung

	Control	Anoxia*	DNP†
Glucose utilization, μmol/h/g dry weight	44.4	63.3	59.7
Lactate + pyruvate production, μmol/h/g dry weight	33.0	129	70.8
Lactate/pyruvate, μmol/μmol	10.0	53.0	13.0
CO_2 production from glucose, μmol/h/g dry weight	49.8	11.4	107.0
Pentose pathway, % of glucose utilization	11.9	1.0	5.2
ATP generation (estimated), μmol/h/g dry weight	425	166	1030‡
ATP content, μmol/g dry weight	11.0	4.9	9.3
ATP/ADP, μmol/μmol	8.5	2.7	5.0

*Lungs ventilated with 95% CO:5% CO_2.
†Lungs perfused with 0.5 mM dinitrophenol.
‡Represents "potential" ATP synthesis (see text).
Data from (Bassett and Fisher, 1976a; Bassett and Fisher, 1976b; Bassett and Fisher, 1976c).

synthesis and secretion of prostaglandins and other eicosanoids from arachidonate precursors,
uptake of various eicosanoids from the blood stream and intracellular degradation,
hydrolysis of lipoproteins via endothelial lipoprotein lipase,
oxidation of fatty acids via mitochondrial metabolism, and
synthesis of phosphatidylcholine and other complex lipids.

The last item is closely related to the question of lung surfactant metabolism and will be the major topic for the remainder of this chapter.

Lung Surfactant: Overview   Lung surfactant is a phospholipid-protein complex that lines the alveolar surface of the lung and acts to stabilize the lung alveoli during the respiratory cycle. It functions primarily by reducing the expected high surface tension at the air-liquid interface in the alveolus, thereby minimizing the physiologic effects of surface tension. This effect is most important during lung deflation (i.e., expiration) when elevated surface tension could result in alveolar collapse. The surfactant system develops relatively late in gestation, and its presence is essential for air breathing during the neonatal period. Deficiency of surfactant is manifested by the neonatal respiratory distress syndrome.

The life cycle of lung surfactant is complex (Fisher and Chander, 1985; Wright and Dobbs, 1991; Fisher and Dodia, 1994). Turnover is rapid with an estimated half-time in the rat of approximately 5–10 h. The surfactant components are synthesized in the endoplasmic reticulum of the type II alveolar epithelial cell, assembled in the characteristic storage/secretory organelle called the lamellar body, and exocytosed into the extracellular lung lining fluid (the alveolar subphase). The secreted material—tubular myelin—generates the monomolecular film of lung surfactant that inserts into the air-liquid interface. Excess and/or spent surfactant (possibly in the form of micelles or small vesicles) is endocytosed by type II cells and to a lesser degree by alveolar macrophages. Some fraction of the endocytosed phospholipid is resecreted while the bulk is degraded and components reutilized for surfactant synthesis. The fate of the protein components is still under investigation as are other aspects of the lung surfactant cycle. The major question—namely the physiological reason for the rapid turnover of lung surfactant—has not been answered satisfactorily but may be related to the necessity for renewal of spent or otherwise inactivated surfactant material.

Lung Surfactant Composition   The composition of lung surfactant is complex and has been difficult to define precisely because of difficulties in obtaining an uncontaminated sample. The standard approach for obtaining lung surfactant is by lung lavage through the airway using saline followed by differential centrifugation of the lavaged material to separate cells and other non-surfactant components. This procedure yields a surface-active fraction with a relatively reproducible composition, although variable contamination with material from the conducting airways or air spaces can occur. Lung surfactant (defined as the secreted product) can also be obtained by isolation of the secretory organelle—the lung lamellar body—from lung homogenates. These, of course, may be contaminated with extraneous material from the homogenate. Despite these constraints, the composition of these isolated organelles is remarkably similar to that for surfactant obtained by lung lavage (King and Clements, 1985).

Approximately 10–15% of surfactant by weight is protein. Three unique surfactant-specific proteins have been identified and are called SP-A, SP-B, and SP-C (Hawgood, 1989). SP-A is a glycoprotein with varying glycosylation state and a monomeric molecular mass of 26–35 kD; it is secreted as an octadecamer with a nominal molecular mass of the secreted form of approximately 650 kD. This protein accounts for nearly 50% of the total surfactant associated protein. SP-B, a dimer of approximately 18 kD, and SP-C, approximately 3.5 kD, are both hydrophobic proteins of lower abundance than SP-A. SP-B and SP-C appear to be important for the surface tension lowering properties of lung surfactant while the major role of SP-A may be related to lung surfactant metabolism. Thus, SP-A appears to be responsible in part for regulation of surfactant secretion, surfactant reuptake, the intracellular processing and metabolism of surfactant lipids. Other proteins such as serum albumin, secretory IgA, lysozyme, and SP-D (a lectin-like protein similar to serum conglutinin) (Lu et al., 1993) may co-isolate with the surfactant fraction from lung lavage although they are found in greater concentration in the soluble fraction after centrifugation and at present no surfactant-associated role has been ascribed to them. The three specific surfactant-associated proteins appear to be secreted with the surfactant by type II alveolar epithelial cells. However, SP-A and SP-B are also synthesized by Clara cells. Their physiological role in this latter cell type is not understood.

The remaining portion of lung surfactant (85–90%) consists of a mixture of lipids (van Golde, 1988). Unlike the protein components of lung surfactant, none of the lipid components is unique. However, the relative proportions of the different components is characteristic for this material. Neutral lipids, predominately cholesterol, account for less than 10% of the lipid fraction with phospholipid accounting for the remainder. Thus, 75–80% of surfactant by weight is due to phospholipids, and in particular phosphatidylcholine (PC) which represents by far the greatest component. The phospholipid composition is shown in Table 2 for surfactant (represented by lamellar bodies) in comparison to the whole lung.

Surfactant phospholipids represent a mixture with PC accounting for three-quarters of the total. In contrast to most tissues, where the fatty acid in position 2 of PC is generally an unsaturated fatty acid, surfactant PC is primarily disaturated, i.e., both fatty acids are generally palmitate. Thus, dipalmitoyl PC accounts for approximately two-thirds of the phospholipid and approximately 50% by weight of the total lung surfactant. Synthetic dipalmitoyl PC can reproduce the static surface tension lowering effect of lung surfactant, although additional protein and/or lipid components are necessary to facilitate spreading of the saturated phospholipid and to reproduce the dynamic behavior of the biological product (King and MacBeth, 1979; Suzuki, 1982). Phosphatidylglycerol (PG) is also a major phospholipid component of lung surfactant. PG is not common in most mammalian tissues and its relatively high concentration distinguishes surfactant from other membranes and organelles. In comparison to the whole lung (see Table 2), surfactant shows a marked increase in disaturated PC and PG with corresponding decreases in the proportion of unsaturated PC, phosphatidylethanolamine, and sphingomyelin.

Table 2 Phospholipid Composition of Rat Lung Tissue and Isolated Rat Lung Lamellar Bodies

	% of Phospholipid	
	Lung	**Lamellar Bodies**
PC	54.3	75.0
(% disaturated)	(32.6)	(87.7)
PG	3.9	11.7
PE	20.2	6.1
SM	12.6	0.7
Other PL	9.0	6.5

Abbreviations: PC—phosphatidylcholine PG—phosphatidylglycerol PE—phosphatidylethanolamine SM—sphingomyelin PL—phospholipid.
Data from Godinez et al. (1975) and Engle et al. (1976).

Lung Surfactant Phospholipid Synthesis   Published studies of lung surfactant phospholipid metabolism have emphasized synthesis and degradation of phosphatidylcholine, the predominant phospholipid component of lung surfactant. The two major pathways that have been described for PC synthesis in other tissues are the de novo pathway via cytidinediphosphate (CDP)-choline and the methylation pathway, involving the stepwise addition of three methyl groups to phosphatidylethanolamine. Both pathways have been demonstrated in the lung, although the methylation pathway accounts for a trivial fraction of total PC synthesis (Yost et al., 1986).

The de novo pathway via CDP-choline represents the major route for lung PC synthesis (van Golde, 1988; King and Clements, 1985). The substrates required for this pathway include glycerol 3-phosphate, free fatty acids, and choline; ATP, CTP, and co-enzyme A participate in the reactions. Fig. 2 shows the steps in the generation of PC. The initial steps require ATP for the generation of acyl Co-A and the phosphorylation of choline to form choline phosphate. PC is generated by the condensation of diacylglycerol (DAG) and CDP-choline. The metabolic machinery for production of these substrates will now be considered.

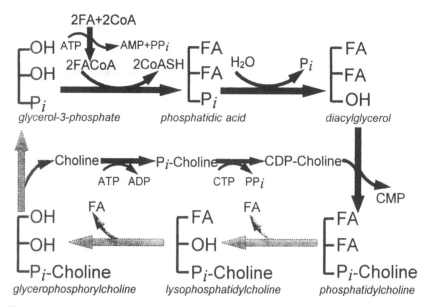

Figure 2   Metabolic scheme for synthesis of phosphatidylcholine (PC) via the *de novo* pathway (indicated by solid arrows) and its degradation via phospholipase A_2 (indicated by dashed arrows). The cycle of synthesis and degradation has been demonstrated in alveolar type II epithelial cells. PC synthesized by this pathway is generally monoenoic with a palmitate in position one. Dipalmitoyl PC is synthesized by reacylation of palmitoyl lyso PC with palmitoyl Co-A. Note that the cycle functions to conserve choline, the rate-limiting substrate for PC synthesis.

Diacylglycerol is a product of intermediary metabolism. Glycerol 3-phosphate, derived by reduction of dihydroxyacetone phosphate (DHAP) in the glycolytic pathway, or to a lesser extent, by phosphorylation of glycerol (Fisher and Chander, 1982), is acylated by fatty acyl Co-A to form phosphatidic acid. Alternatively, phosphatidic acid can be formed by acylation and reduction of DHAP (Fisher et al., 1976b; Mason, 1978); diacylglycerol is then generated from phosphatidic acid by dephosphorylation.

The enzymes involved in the sequential metabolism of choline include choline kinase for phosphorylation, choline phosphate cytidyltransferase (CPCT) for generation of CDP-choline and choline phosphotransferase (CPT) which catalyzes formation of PC from CDP-choline and diacylglycerol. Although each of these enzymes may be rate limiting under some circumstances, CPCT appears to be the usual rate limiting enzyme (Weinhold et al., 1984; Post et al., 1984). This enzyme is present in a cytosolic, largely inactive form that is activated by translocation to membranes in the presence of free fatty acids (Chander and Fisher, 1988).

The de novo pathway generates PC that reflects in part the fatty acid composition of the cell interior (Batenburg et al., 1978; Mason and Dobbs, 1980). Most studies have suggested that monoenoic PC, i.e., PC with an unsaturated fatty acid (such as oleate) at position 2 is the primary product of the de novo pathway. Remodeling of this unsaturated PC is necessary in order to generate the dipalmitoyl PC that is the major component of the lung surfactant. Remodeling proceeds through the action of a phospholipase A_2 to remove the unsaturated fatty acid in position 2 followed by reacylation with palmitoyl Co-A to generate dipalmitoyl PC. Specificity in this reaction is probably determined by the acyl transferase.

Since all other components for synthesis of lung surfactant dipalmitoyl PC are derived from intermediary metabolism, choline represents the rate limiting substrate. Choline is derived primarily through the diet; spontaneous choline deficiency occurs rarely but has been produced experimentally (Yost et al., 1986b). Choline deficiency in the rat results in compensatory increase in activity of the CDP-choline pathway that can maintain the lung PC pool. This appears to involve allosteric regulation of the CPCT enzyme although details have not been worked out (Yost et al., 1986b). Cells obtain choline from the blood (normal plasma concentration about 7 μM) and uptake occurs by specific transport systems. Type II lung epithelial cells transport choline by both Na^{++}-dependent high affinity (Km ~ 1μM) and Na^{++}-independent low affinity (Km ~ 40μM) membrane transport systems (Fisher et al., 1992b; Kleinzeller, et al., 1994). Intracellular free choline concentration in the type II alveolar epithelial cell appears to be at least 20 fold greater than the serum concentration, so that cell membrane transport of choline should not normally limit PC synthesis. Further, choline produced within the cell by phospholipase activity is retained and utilized for resynthesis of PC as discussed next.

Degradation of Lung Surfactant Phospholipids As discussed above, lung surfactant lipids are extensively recycled by endocytosis into type II cells. Since these cells constitute only 5 percent of the alveolar surface, they must possess specific mechanisms for efficient uptake of alveolar DPPC. Current evidence suggests that surfactant protein A binds DPPC and markedly increases the rate of uptake of this phospholipid by type II cells (Wright et al., 1987; Tsuzuki et al., 1993; Fisher et al., 1994). The presumed mechanism is through receptor mediated endocytosis of the surfactant protein A: DPPC complex, although the definitive demonstration of an SP-A cell membrane receptor has remained elusive. Type II cells also recognize and internalize DPPC through a receptor-independent process, apparently through cell membrane domains with increased DPPC affinity.

Once internalized, a fraction of DPPC can be routed to lamellar bodies for resecretion (Jacobs et al., 1985). Under normal circumstances, the major fraction of internalized DPPC in the adult lung is degraded by phospholipases, perhaps in lysosomes although the specific organellar compartment has not been precisely identified (Fisher and Dodia, 1994; Fisher et al., 1992a). The entire spectrum of phospholipases appears to be present in lung tissue. These include phospholipase D which generates choline plus phosphatidic acid, phospholipase C which generates choline phosphate plus diacylglycerol, and phospholipase A which generates lysoPC and free fatty acid. PLA_1 has specificity for the fatty acyl group in position 1 while PLA_2 attacks position 2. Each of these phospholipases may represent a family of enzymes which have not been fully characterized. For example, lung tissue contains both type I and type II secretory PLA_2 (14 kD) (Seilhamer et al., 1986; Nakano and Arita, 1990), cytosolic PLA_2 (85 kD) (Neagos et al., 1993) and a Ca^{2+}-independent PLA_2 that is active at acidic pH (Fisher et al., 1994; Wang et al., 1994). The latter has been localized to the lamellar body fraction of the lung, but is probably also present in lysosomes. Recent data have suggested that the activity of the acidic PLA_2 is modulated by surfactant protein A providing a mechanism whereby the surfactant protein can regulate phospholipid degradation (Fisher et al., 1994).

The precise contribution of each of these phospholipases to DDPC degradation in the lung type II cell has not yet been determined, although the Ca^{2+}-independent phospholipase A_2 appears to play a major role (Fisher et al., 1992). The lysoPC that is generated can be reacylated (as described above under PC synthesis) or further degraded by a combination of lysophospholipase and glycerolphosphorylcholine hydrolase to generate free choline and glycerophosphate (Chander et al., 1987; Fisher et al., 1987). The choline and other metabolic products liberated by these reactions reenter the intracellular metabolite pool and are reutilized. Thus, the type II cell shows a pattern of DPPC synthesis, secretion, reuptake, degradation, and reutilization that can be characterized as a lung surfactant cycle. By contrast, the alveolar macrophage can internalize and degrade dipalmitoyl PC, but components are not specifically reutilized and this represents a catabolic rather than a salvage pathway (Miles et al., 1988).

SUMMARY

Although the lung was once considered a passive sack functioning only for gas exchange, an increasing number of studies during the past 25 years have demonstrated the presence of active metabolic processes in this organ. This chapter has reviewed some special aspects of lung intermediary metabolism including the utilization of glucose, oxygen utilization, and energy production while lung phospholipid metabolism has been considered from the standpoint of lung surfactant. This chapter has not addressed the equally important topics of:

Protein synthesis and degradation;
Production of eicosanoids, cytokines, and other cellular factors;
Processing of circulating amines, lipoproteins, and other bioactive agents; and
Degradation and transformation of xenobiotics.

While no one of the metabolic pathways described for the lung is unique, the relationships among utilization of specific substrates, the interplay of the various metabolic pathways, and the individual roles of different lung cell types characterize this organ as metabolically complex. Clearly, the possibility of metabolic derangement by inhaled toxins represents fertile ground for characterization of toxic effects of these agents and could provide grist for the mill for future generations of inhalation toxicologists.

ACKNOWLEDGMENTS

I thank Chandra Dodia for helping to assemble the data, Abu Al-Mehdi for composing the figures, and Elaine Primerano and Susan Turbitt for excellent secretarial assistance.

REFERENCES

Barron, ESG, Miller, ZB, Bartlett, GR: Studies of biological oxidations. XXI The metabolism of lung as determined by study of slices and ground tissue. J Biol Chem, 171:791–800, 1947.

Basset, G, Saumon, G, Bouchonnet, F, Crone, C: Apical sodium-sugar transport in pulmonary epithelium in situ. Biochim Biophys Acta, 942:11–18, 1988.

Bassett, DJP, Fisher, AB: Metabolic response to carbon monoxide by isolated lungs. Am J Physiol, 230:658–663, 1976a.

Bassett, DJP, Fisher, AB: Stimulation of rat lung metabolism with 2,4-dinitrophenol and phenazine methosulfate. Am J Physiol, 321:898–902, 1976b.

Bassett, DJP, Fisher, AB: Pentose cycle activity of the isolated perfused rat lung. Am. J. Physiol, 231:1527–1532, 1976c.

Bassett, DJP, Hamosh, M, Hamosh, P, Rabinowitz, JL: Pathways of palmitate metabolism in the isolated rat lung. Exp Lung Res, 2:37–47, 1981.

Batenburg, JJ, Longmore, WJ, van Golde, LMG: The synthesis of phosphatidylcholine by adult rat lung alveolar type II epithelial cells in primary culture. Biochim Biophys Acta, 529:160–170, 1978.

Buechler, KF, Rhoades, RA: Fatty acid synthesis in the perfused rat lung. Biochim Biophys Acta, 619:186–195, 1980.

Caldwell, PRB, Wittenberg, BA: The oxygen dependency of mammalian tissue. Am J Med, 57:447–452, 1974.

Chance, B: Reaction of oxygen with the respiratory chain in cells and tissues. J Gen Physiol, 49:163–188, 1965.

Chander, A, Fisher, AB: Choline phosphate cytidyl transferase activity and phosphatidylcholine synthesis in rat granular pneumocytes are increased with exogenous fatty acids. Biochim Biophys Acta, 958:343–351, 1988.

Chander, A, Reicherter, J, Fisher, AB: Degradation of dipalmitoyl phosphatidylcholine by isolated rat granular pheumocytes and reutilization for surfactant synthesis. J Clin Invest, 79:1133–1138, 1987.

Chevalier, G, Collet, AJ: In vivo incorporation of choline-^{3}H, leucine-^{3}H and galactose-^{3}H in alveolar type II pneumocytes in relation to surfactant synthesis. Anat Rec, 174:289–310, 1972.

Das, DK, Steinberg, H: Effect of starvation and diabetes on glucose transport in the lung. Clin Physiol Biochem, 2:239–248, 1984.

Das, DK: Nutritional and hormonal control of glucose and fructose utilization by lung. Clin Physiol Biochem, 3:240–248, 1985.

Edson, NL, Leloir, LF: Ketogenesis-antiketogenesis. V. Metabolism of ketone bodies. Biochem, 30:2319–2332, 1936.

Engle, MJ, Sanders, RL, Longmore, WJ: Phospholipid composition and acyltransferase activity of lamellar bodies isolated from rat lung. Arch Biochem Biophys, 173:586–595, 1976.

Engelbrecht, FM, Marit, G: Influence of substrate composition on in vitro oxygen consumption of lung slices. A Afr Med J 48:1882–1884, 1974.

Fisher, AB: Intermediary metabolism of the lung. Envir Hlth Persp, 44:149–158, 1984.

Fisher, AB: Intermediary metabolism of the lung. In: Lung Cell Biology, edited by D Massaro. pp. 737–770. New York: Marcel Dekker, 1989.

Fisher, AB, Chander, A: Glycerol kinase activity and glycerol metabolism of rat granular pneumocytes in primary culture. Biochim Biophys Acta, 711:128–133, 1982.

Fisher, AB, Chander, A: Intracellular processing of surfactant lipids in the lung. Ann Rev Physiol, 47:789–802, 1985.

Fisher, AB, Dodia, C: The lung as a model for evaluation of critical intracellular PO_2 and P_{CO}. Am J Physiol (Endo Metab 4) 241:E47–E50, 1981.

Fisher, AB, Dodia, C: Lactate and regulation of lung glycolytic rate. Am J Physiol (Endo Metab 9) 246:E426-E429, 1984.

Fisher, AB, Dodia, C: Regulation of surfactant metabolism: degradation of internalized alveolar phosphatidylcholine. Prog Resp Res 27:74–83, 1994.

Fisher, AB, Forman, HJ: Oxygen utilization and toxicity in the lungs. In: Handbook of Physiology. Section 3: The Respiratory System. Vol 1:743–750, eds AP Fishman and AB Fisher, Bethesda, MD, Am Physiol Soc, 1985.

Fisher, AB, Scarpa, A, Lanoue, KF, Bassett, D, Williamson, JR: Respiration of rat lung mitochondria and the influence of Ca^{2+} on substrate utilization. Biochem, 12:1438–1445, 1973.

Fisher, AB, Steinberg, H, Bassett, D: Energy utilization by the lung. Am J Med, 57:437–446, 1974.

Fisher, AB, Huber, GA, Bassett, DJP: Oxidation of alpha-glycero-phosphate by mitochondria from lungs of rabbits, sheep, and pigeons. Comp Biochem Physiol, 50B:5–8, 1975.

Fisher, AB, Furia, L, Chance, B: Evaluation of redox state of the isolated perfused rat lung. Am J Physiol, 230:1198–1204, 1976a.

Fisher, AB, Huber, GA, Furia, L, Bassett, D, Rabinowitz, JL: Evidence for lipid synthesis by the dihydroxyacetone phosphate pathway in rabbit lung subcellular fractions. J Lab Clin Med, 87:1033–1040, 1976b.

Fisher, AB, Furia, L, Berman, H: Metabolism of rat granular pneumocytes isolated in primary culture. J Appl Physiol, 49:743–750, 1980.

Fisher, AB, Dodia, C, Chander, A: Energy dependence of lung phosphatidylcholine biosynthesis studied with CO hypoxia. Am J. Physiol (Endo Metab 12) 249:E89-E93, 1985.

Fisher, AB, Chander, A, Reicherter, J: Uptake and degradation of natural surfactant by isolated rat granular pneumocytes. Am J Physiol (Cell Physiol 22), 253:C792-C796, 1987.

Fisher, AB, Dodia, C, Chander, A, Jain, MK: A competitive inhibitor of phospholipase A_2 decreases surfactant phosphatidylcholine degradation by the rat lung. Biochem J, 288:407–411, 1992a.

Fisher, AB, Dodia, C, Chander, A, Kleinzeller, A: Transport of choline by plasma membrane vesicles from lung-derived epithelial cells. Am J Physiol 263 (Cell Physiol 32):C1250-C1257, 1992b.

Fisher, AB, Dodia, C, Chander, A: Inhibition of lung calcium-independent phospholipase A_2 by surfactant protein A. Am J Physiol 267 (Lung Cell Mol Physiol 11):L335–L341, 1994.

Freeman, BA, Crapo, JD: Hyperoxia increases oxygen radical production in rat lungs and lung mitochondria. J Biol Chem, 256:10986–10992, 1981.

Fricke, RF, Longmore, WJ: Effects of insulin and diabetes on 2-deoxy-D-glucose uptake by the isolated perfused rat lung. J Biol Chem, 254:5092–5098, 1979.

Godinez, RI, Sanders, RL, Longmore, WJ: Phosphatidylglycerol in rat lung. I. Identification as a metabolically active phospholipid in isolated perfused rat lung. Biochem, 14:830–834, 1975.

Hawgood, S: Pulmonary surfactant apoproteins: a review of protein and genomic structure. Am J Physiol, (Lung Cell Mol Physiol 1) 257:L13-L22, 1989.

Harai, KI, Ogawa, K, Wang, GY, Ueda, T: Varied cytochrome oxidase activities of the alveolar type I, type II, and type III cells in rat lungs: quantitative cytochemistry. J. Electron Microsc, 38:449–456, 1989.

Jacobs, HC, Ikegami, M, Jobe, AH, Berry, DD, Jones, S: Reutilization of surfactant phosphatidyl choline in adult rabbits. Biochim Biophys Acta, 837:77–84, 1985.

Kemp, PJ, Boyd, CAR: Pathways for glucose transport in type II pneumoctyes freshly isolated from adult guinea pig lung. Am J Physiol, 263 (Lung Cell Mol Physiol 7):L612-L616, 1992.

Kerr, JS, Fisher, AB, Kleinzeller, A: Transport of glucose analogues in rat lung. Am J Physiol (Endo Metab 4) 241:E191-E195, 1981.

Kerr, JS, Baker, NJ, Bassett, DJP, Fisher, AB: Effect of perfusate glucose concentration on rat lung glycolysis. Am J Physiol, (Endo Metab Gastrointest Physiol 5) 236:E229-E233, 1979.

Kerr, JS, Reicherter, J, Fisher, AB: 2-Deoxy-D-glucose uptake by rat granular pneumocytes in primary culture. Am J. Physiol, (Cell Physiol 12) 243:C14-C19, 1982.

King, RJ, MacBeth, MD: Physico-chemical properties of dipalmitoyl phosphatidylcholine after interaction with an apoprotein of pulmonary surfactant. Biochim Biophys Acta, 557:86–101, 1979.

King, RJ, Clements, JA: Lipid synthesis and surfactant turnover in the lungs. In: Handbook of Physiology, Section 3: The Respiratory System, Vol 1:309–336, eds. AP Fishman and AB Fisher, Bethesda, MD, Am Physiol Soc, 1985.

Kleinzeller, A, Dodia, C, Chander, A, Fisher, AB: The Na^+-independent systems of choline transport by plasma membrane vesicles of the A549 lung cell line. Am J Physiol 267 (Cell Physiol 36): C1279–C1287, 1994.

Koga, H: Studies on the function of isolated perfused mammalian lung. Kumamoto Med J, 11:1–11, 1958.

Krebs, HA: Body size and tissue respiration. Biochem Biophys Acta, 4:249–269, 1950.

Levy, SE, Harvey, E: Effect of tissue slicing on rat lung metabolism. J Appl Physiol, 37:239–240, 1974.

Lu, J, Wiedemann, H, Holmskov, U, Thiel, S, Timpl, R, Reid, KBM: Structural similarity between lung surfactant protein D and conglutinin. Two distinct, c-type lectins containing collagen-like sequences. Eur J Biochem, 215:793–799, 1993.

Maritz, FS: Carbohydrate metabolism of lung tissue of suckling and adult rats. Cell Biol Internatl, 17:773–780, 1993.

Mason, RG: Importance of the acyldihydroxyacetone phosphate pathway in the synthesis of phosphatidylglycerol and phosphatidylcholine in alveolar type II cells. J Biol Chem 253:3367–3370, 1978.

Mason, RJ, Dobbs, LG: Synthesis of phosphatidylcholine and phosphatidylglycerol by alveolar type II cells in primary culture. J Biol Chem, 255:5101–5107, 1980.

Massaro, GD, Gail, DB, Massaro, D: Lung oxygen consumption and mitochondria of alveolar epithelial and endothelial cells. J Appl Physiol, 38:588–592, 1975.

Miles, PR, Ma, JYC, Bowman, L: Degradation of pulmonary surfactant disaturated phosphatidylcholines by alveolar macrophages. J Appl Physiol, 64:2474–2481, 1988.

Morishige, WK, Uetake, CA, Greenwood, FC, Akaka, JA: Pulmonary insulin responsivity: in vitro effects of insulin binding to lung receptors in normal rats. Endocrin, 100:1710–1722, 1977.

Mustafa, MG, Cross, CE: Lung cell mitrochondria: rapid oxidation of glycerol-1-phosphate but slow oxidation of 3-hydroxybutyrate. Am Rev Respir Dis, 109:301–303, 1974.

Nakano, T, Arita, H: Enhanced expression of group II phospholipase A_2 gene in the tissues of endotoxin shock rats and its supression by glucorticoid. FEBS Lett, 273:23–26, 1990.

Neagos, GR, Feyssa, A, Peters-Golden, M: Phospholipase A_2 in alveolar type II epithelial cells: biochemical and immunologic characterization. Am J Physiol, (Lung Cell Mol Physiol 8) 264:L261-L268, 1993.

Neufeld, ND, Corbo, LP, Kaplan, SA: Plasma membrane insulin receptors in fetal rabbit lung. Ped Res 15:1058–1062, 1981.

O'Neil, JJ, Tierney, DF: Rat lung metabolism: Glucose utilization by isolated perfused lungs and tissue slices. Am J Physiol, 226:867–876, 1974.

Post, M, Batenburg, JJ, Smith, BT, van Golde, LMG: Pool sizes of precursors of phosphatidylcholine formation in adult rat type II cells. Biochim Biophys Acta, 795:552–557, 1984.

Reiss, OK: Studies of lung metabolism: I. Isolation and properties of subcellular fractions from rabbit lung. J Cell Biol, 30:45–57, 1966.

Rhoades, RA: Net uptake of glucose, glycerol, and fatty acids by the isolated perfused rat lung. Am J Physiol, 226:144–149, 1974.

Rhoades, RA, Shaw, ME, Eskew, ML, Wali, S: Lactate metabolism in perfused rat lung. Am J Physiol, (Endo Metab 6) 235:E619-E623, 1978.

Rossouw, DJ, Engelbrecht, FM: The effect of paraquat on the respiration of lung cell fractions. S Afr Med J, 54:1101–1104, 1978.

Seilhamer, JJ, Randall, RL, Yamanaha, M, Johnson, LK: Pancreatic phospholipase A_2: isolation of the human gene and cDNAs from porcine pancreas and human lung. DNA, 4:519–527, 1986.

Simon, FP, Potts, AM, Gerrard, RW: Metabolism of isolated lung tissue: Normal and in phosgene poisoning. J Biol Chem, 167:303–311, 1947.

Sorokin, SP: The Cells of the Lungs, Proc Biol Div Oak Ridge Natl Lab Atomic Energy Commission, Symp Series 21, pp. 3–43, 1970.

Spear, RK, Lumeng, L: A method for isolating lung mitochondria from rabbits, rats, and mice with improved respiratory characteristics. Anal Biochem, 90:211–219, 1978.

Stadie, WC, Riggs, BC, Haugaard, N: Oxygen poisoning. IV. The effect of high oxygen pressures upon the metabolism of liver, kidney, lung and muscle tissue. J Biol Chem, 160:209–216, 1945.

Steinberg, H, Basset, DJP, Fisher, AB: Depression of pulmonary 5-hydroxytryptamine uptake by metabolic inhibitors. Am J. Physiol, 288:1298–1303, 1975.

Stubbs, WA, Morgan, I, Lloyd, B, Alberti, KBMM: The effect of insulin on lung metabolism in the rat. Cllin Endocrinol, 7:181–184, 1977.

Sugahara, K, Freidenberg, GR, Mason, RJ: Insulin binding and effects on glucose and transepithelial transport by alveolar type II cells. Am J Physiol, (Cell Physiol 16) 247:C472-C477, 1984.

Suzuki, Y: Effect of protein, cholesterol, and phosphatidylglycerol on the surface activity of the lipid-protein complex reconstituted from pig pulmonary surfactant. J Lipid Res 23:62–69, 1982.

Tierney, DF, Young, SL: Glucose and intermediary metabolism of lungs. In: Handbook of Physiology, Section 3: The Respiratory System, Vol 1:255–275, eds AP Fishman and AB Fisher, Bethesda, MD, Am Physiol Soc, 1985.

Tsuzuki, A, Kuroki, Y, Akino, T: Pulmonary surfactant protein A-mediated uptake of phosphatidylcholine by alveolar type II cells. Am J Physiol, (Lung Cell Mol Physiol 9) 265:L193-L199, 1993.

van Golde, LMG: The pulmonary surfactant system: biochemical aspects and functional significance. Physiol Rev, 68:374–455, 1988.

Wang, R, Dodia, C, Jain, MK, Fisher, AB: Purification and characterization of a calcium-independent acidic phospholipase A_2 from rat lung. Biochem J 304:131–137, 1994.

Wangensteen, D, Bartlett, M: D- and L-glucose transport across the pulmonary epithelium. J Appl Physiol: Respirat Environ Exercise Physiol, 57:1722–1730, 1984.

Weber, KC, Visscher, MB: Metabolism of the isolated canine lung. Am J Physiol, 217:1044–1052, 1969.

Weinhold, PA, Rounsifer, ME, Williams, SE, Brubaker, PG, Feldman, DA: CTP: phosphoryl choline cytidyltransferase in rat lung. The effect of free fatty acids on the translocation of activity between microsomes and cytosol. J Biol Chem, 259:10315–10321, 1984.

Williamson, JR: Glycolytic control mechanisms. I. Inhibition of glycolysis by acetate and pyruvate in the isolated, perfused rat heart. J Biol Chem, 240:2308–2321, 1965.

Williamson, JR, Corkey, BE: Assays of intermediates of the citric acid cycle and related compounds by fluorometric enzyme methods. Methods in Enzymology, 13:434–513, 1969.

Wright, JR, Dobbs, LG: Regulation of pulmonary surfactant secretion and clearance. Ann Rev Physiol, 53:395–414, 1991.

Wright, JR, Wager, RE, Hawgood, S, Dobbs, L, Clements, JA: Surfactant apoprotein Mr = 26,000–36,000 enhances uptake of liposomes by type II cells. J Biol Chem, 262:2888–2894, 1987.

Yeager, H, Jr, Hicks, PS: Glucose metabolism in lung slices of late fetal, newborn, and adult rats. Proc Soc Exptl Biol Med, 141:1–3, 1972.

Yost, RW, Chander, A, Dodia, C, Fisher, AB: Stimulation of the methylation pathway for phosphatidylcholine synthesis in rat lung by choline deficiency. Biochim Biophys Acta, 875:122–125, 1986a.

Yost, RW, Chander, A, Dodia, C, Fisher, AB: Synthesis of phosphatidylcholine by rat lung during choline deficiency. J Appl Physiol, 61:2040–2044, 1986b.

Chapter Seven

Metabolic Characteristics of the Respiratory Tract

Alan R. Dahl

INTRODUCTION

The fate of inhaled material is often thought of as being the same as material administered partly intravenously and partly orally. Thus, a portion of the inhaled material deposited in the pulmonary region diffuses into the blood and is transported via the blood first to the heart and from there, throughout the body. This portion of the inhaled material then shares the same fate as intravenously administered material. On the other hand, inhaled material deposited in the nose or the lung conducting airways may be cleared by muco-ciliary clearance to the alimentary tract. From there, its fate is identical to that of an orally administered dose. Conceptualizing the fate of inhaled toxicants in terms of an oral/IV model is, therefore, valid to some degree. In its simplest expression, however, the model overlooks two important aspects of inhalation toxicology—complex tissue-dose relationships for inhaled materials, and respiratory tract metabolism.

Recognition of these aspects is necessary to explain numerous toxicological phenomena of inhalants. Examples of these phenomena include:

1 The observation that the uptake of inhaled vapors increases with the magnitude of the water/air (PC w/a) or blood/air (PC b/a) partition coefficient up to about 50, then reaches a plateau at about 75% uptake (Fig. 1); and

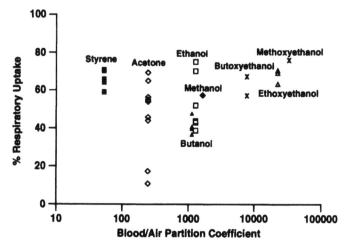

Figure 1 Experimentally observed relative respiratory uptake of some polar solvents in relation to their blood solubility (Johanson, 1991; Johanson and Filser, 1992).

2 The remarkable specificity of some inhalants for producing toxic effects in lung Clara cells or in the nasal olfactory epithelium (Table 1).

Understanding both the enzymology of the respiratory tract and the dosimetry of inhalants in the respiratory tract is necessary, ultimately, to explain these and many other phenomena.

ENZYME KINETICS

The rate at which an enzyme metabolizes its substrate depends on the properties of the enzyme (e.g., V_{max} and K_m, Fig. 2) and the substrate concentration.

Table 1   Some Clara Cell and Olfactory Tissue-Specific Toxicants

Toxicant	Activating P450 Enzymes	Location of Lesions	Species in Which Lesions Occur	Reference
Naphthalene	1A1, 2B1	Clara cells	Mice	Kanekal et al., 1991
Ipomeanol	4B1	Clara cells	Rodents, cattle	Chichester et al., 1991
3-Methylindole	4B1	Clara cells	Goat	Yost et al., 1989
2,6-dichlorocyanobenzene	?	Olfactory epithelium	Rats, mice	Bakke et al., 1988
Ferrocene	?	Olfactory epithelium	Rats, mice	Sun et al., 1991

Reaction rate* = $\dfrac{V_{max} \cdot c}{k_m + c}$ **(c = Concentration of substrate)**

For c >> k_m, rate $\approx V_{max}$

For c << k_m, rate $\approx (V_{max}/k_m) \cdot c$

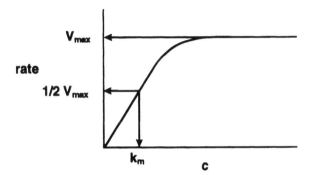

* For the reaction E + S $\rightleftharpoons$ ES $\longrightarrow$ EP $\longrightarrow$ E + P

Figure 2   Enzyme kinetics. E = free enzyme; S = free substrate; P = product; ES and EP = enzyme substrate complex and enzyme product complex, respectively.

For a given delivery rate of substrate (i.e., inhalant) to tissue, the substrate concentration in the tissue is a function of the overall (physical and enzymatic) clearance rate (Fig. 3), while enzyme properties—such as the turnover rate of the enzyme-substrate complex to product (V_{max}) and the affinity of the substrate for the enzyme ($1/K_m$)—are intrinsic to the enzyme.

At low substrate concentrations (c) the reaction rate is approximated by $V_{max}/K_m \times c$, while at high concentrations the rate is approximated by V_{max} (Fig. 2) and is, hence, independent of substrate concentration. The major enzyme activities in the respiratory tract can be divided with respect to maximum metabolic rate (V_{max}) into two categories:

1   High V_{max} metabolizers—exemplified by esterases and rhodanese (which metabolizes cyanide) and
2   Low V_{max} metabolizers—exemplified by cytochrome P450-dependent and flavin-containing monooxygenases (FAD-MO, Table 2).

The total tissue maximum metabolic rates of the enzymes (V_{max}) can be related to an upper limit on the concentration of inhalant in air that can be completely metabolized (Dahl, 1988).

The fraction of an inhalant that is metabolized in the respiratory tract is important for understanding the mechanism of toxicity of the inhalant.

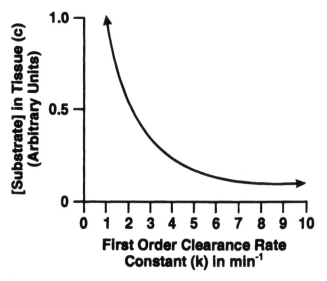

Figure 3 The steady state concentration of a substrate inhaled at a constant rate depends on the clearance rate. Calculated for inhaled rate of 1 arbitrary unit/min. Note: $t^{1}/_2 = 0.69/k$.

Table 2   Some Examples of Inhalants Activated or Detoxicated by Respiratory Tract Enzymes

Inhalant	Enzyme	Species	Tissue	V_{max}	V_{max}/K_m[a]	Reference
Benzyl cyanide	P450	Rat	Nasal respiratory	25[b]	20	Dahl and Waruszewski, 1989
CN^-	Rhodanese	Rat	Nasal respiratory	52,000[b]	10	Dahl and Waruszewski, 1989
BaP	P450	Rat	Nasal total	0.005[c]	1–3	Bond, 1983
Methyl acrylate	Esterase	Mouse	Nasal total	370[d]	120	Stott and McKenna, 1985
Formaldehyde	Aldehyde dehydrogenase		Nasal respiratory	0.9[d]	0.35	Casanova-Schmitz et al., 1984
N,N-Dimethylaniline	FAD-MO	Rat	Nasal respiratory	2.5[c]	[e]	McNulty et al., 1983

[a] All K_m units are μM.
[b] nmol/g tissue per min.
[c] nmol/mg microsomal protein per min.
[d] nmol/mg tissue protein per min.
[e] K_m not reported.

Mechanistic understanding, in turn, is necessary for accurate prediction of human response from animal data. In cases where the inhalant itself is the toxicant (e.g., cyanide), it is important to be able to estimate the fraction that will be detoxicated by respiratory tract metabolism. In cases where a metabolite of the inhalant is the toxicant (e.g., carboxylic esters, benzo(a)pyrene), it is necessary to know how much of the substance will be metabolized in the respiratory tract thus leading to local toxic responses. In either case, both the animal species-specific rate constant, V_{max}/K_m, and the substrate concentration, c, are needed to calculate the rate of metabolism. V_{max}/K_m, in principle, can be determined using *in vitro* methods, but estimation of substrate concentration *in vivo* is more difficult, requiring knowledge about deposition rates and clearance rates of the inhalant.

Some relationships between physical and enzymatic clearance of inhalants are illustrated in Fig. 4. Specifically shown are the relationship between the rate of clearance of an inhalant via diffusion into the perfusing blood (not the only mechanism for clearing unmetabolized inhalant: desorption of vapors [discussed below] and mucociliary clearance also contribute), and the ratio of free to enzyme-bound substrate for three classes of compounds with different ratios of diffusion/perfusion rate to metabolism rate. There are three points important in understanding respiratory tract metabolism of inhalants.

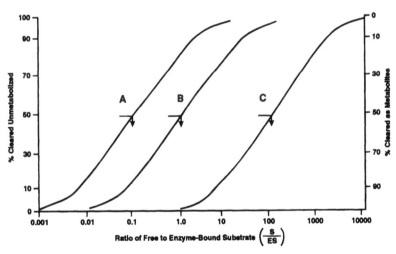

Figure 4 The relative importance of clearance of inhalants by metabolism compared to clearance by diffusion depends on both the relative rates of diffusion and intrinsic metabolic capacity, and on the ratio of free substrate (S) to enzyme-bound substrate (ES).

A Rate of diffusion and perfusion = 10× metabolism rate.
B Rate of diffusion and perfusion = metabolism rate.
C Rate of diffusion and perfusion = 1% of metabolism rate.

1   If inhaled concentrations are high enough to produce high tissue levels of inhalant relative to the enzyme levels present, then S >> ES, and clearance via diffusion/perfusion of the parent compound will exceed clearance via metabolism.

2   If inhaled concentrations are low enough to produce tissue concentrations of inhalant substantially less than the available enzyme, then S << E, and the ratio S/ES will depend on the affinity $(1/K_m)$ of the substrate for the enzyme. If for a given affinity the enzyme/substrate turnover rate (V_{max}) is high, metabolism may contribute substantially to keeping S << E, and higher air concentrations of inhaled substrate can be accommodated.

3   If metabolism is fast relative to clearance via diffusion/perfusion—either because of fast metabolism or particularly slow diffusion or a combination of both—then metabolism will contribute substantially to clearance even if S/ES is large as a result of low-affinity $(1/K_m)$, high-substrate concentration, or both. Inhalants in this last category include highly lipophilic inhalants, such as benzo(a)pyrene.

It has now been fairly well established that highly lipophilic organic compounds (always associated with particles because compounds with sufficiently high oil/water partition coefficients have very low vapor pressures [Gerde et al., 1994]) diffuse slowly through the epithelium of the lung conducting airways into the blood (Gerde et al., 1993; Dahl and Gerde, unpublished observations). Thus, because of slow diffusion, highly lipophilic compounds reach relatively high substrate concentrations in the mucosa. For such compounds, respiratory tract metabolism may contribute substantially to total metabolism—as well as to local toxicity—despite low rates of metabolism. The influence of local concentrations of inhalants in the respiratory tract mucosa on metabolism is discussed next.

DOSIMETRY OF INHALANTS IN THE RESPIRATORY TRACT

To understand the toxic effects of an inhalant and its metabolism, local tissue concentrations must be known. The concentration of an inhalant in the various sites of the respiratory tract depends on its rates of deposition and clearance. The differences in sites of initial deposition of inhaled gases (a vapor is a type of gas) compared to aerosolized particles are ascribable to differences in sizes of the components (Fig. 5). Thus, the deposition of small particles (Cheng et al., 1988, 1990) and the initial deposition of gases, which are very small particles, are dependent on a combination of diffusion and convection, while other mechanisms contribute substantially to the deposition of larger particles. Although the deposition of particles is covered in Chapter 8 in this volume by Schlesinger, current concepts and references regarding the uptake of inhaled vapors are discussed.

For purposes of modeling uptake, gases can be classified as stable or reactive as shown in Fig. 5. For stable gases (Dahl, 1990) having PC w/a less than about 50, uptake mainly takes place in the pulmonary region, and

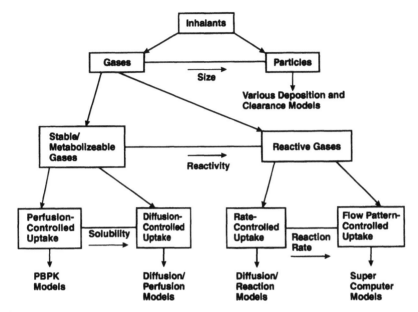

Figure 5 Models for inhalant uptake.

physiologically based pharmacokinetic models (e.g., Andersen, 1991) appropriately describe total uptake. For such gases, metabolism in the respiratory tract usually will not contribute significantly to total body uptake; nonetheless, metabolism may be involved in causing local effects within the respiratory tract. Most such gases are metabolized by P450 enzymes (Table 2) and belong to the class of inhalants described by curve A in Fig. 4.

For stable gases having PC w/a (or PC b/a) greater than about 50, uptake is mainly in the nasal cavity and conducting airways of the lung, and diffusion/perfusion models adequately describe uptake in the various sites of the respiratory tract (Gerde and Dahl, 1991; Johanson, 1991). For these gases, metabolism in the respiratory tract may contribute not only to local toxic effects, but also substantially to total uptake if enzymes having high values for V_{max}/K_m (such as carboxylesterases [Table 2]) are involved in their metabolism. Carboxylic esters, for example, are extensively metabolized in the nasal cavity (Morris, 1990) and belong to the class of inhalants described by curve B in Fig. 4.

Gases that have PC w/a or b/a over ~ 50 and are metabolized by cytochrome P450 will clear mainly by diffusion into the blood unless they are inhaled at very low concentrations and have high affinity for the metabolizing enzyme (curve A of Fig. 4). Nonetheless, even the small fraction that is metabolized may cause local toxicity as is the case for the potent rat nasal carcinogen, hexamethylphosphoramide (Dahl et al., 1982).

For stable gases, uptake in the respiratory tract via diffusion into the blood is dependent on PC w/a, and uptake during the inspiratory portion of the breathing cycle should approach 100% (Dahl et al., 1991) (Fig. 6). The observation that net vapor uptake approaches 75%, and not 100% (Fig. 1), is explained by the desorption of vapors from the nasal mucosa (Gerde and Dahl, 1991) and the lung conducting airways (Johanson, 1991) during the expiratory portion of the breathing cycle.

Highly reactive gases behave much like particles; they react quickly and irreversibly on first contact with airway walls and do not reentrain. Uptake

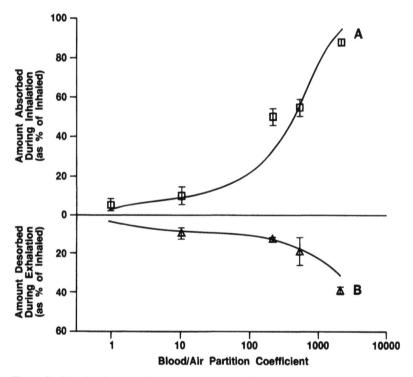

Figure 6 Uptake of vapors in the nasal airway of the Beagle dog during cyclic breathing as a function of the blood/air partition coefficient. Model simulation results (solid lines) are shown together with experimental data (□ and △). Curve A—Nasal absorption on inhalation. Curve B—Nasal desorption on exhalation. Experimental data are shown for the following vapors, given in the order of increasing PC:

    2,4-dimethylpentane
    propyl ether
    butanone
    dioxolane
    ethanol

Error bars represent standard errors of the means.

of such reactive gases (Dahl, 1990) is flow-dependent. For these gases, the relative exposure of different tissues in the respiratory tract is totally dependent on the delivery rate of the gas to those tissues via the inhaled air. Such gases are largely removed from the inhaled air in the anterior part of the nasal cavity; their deposition has been partially described by a super computer model (Kimbell et al., 1993). Member of this class of inhalants are not dependent upon local or systemic metabolism to effect uptake. Nonetheless, metabolism may contribute to detoxication as is the case for formaldehyde (Casanova-Schmitz et al., 1984).

Gases that are removed from inhaled air by reversible reaction in the respiratory tract mucosa or at a rate that permits some of the gas to reentrain in the air after initial contact—exemplified by ozone, chlorine, and nitrogen dioxide—can cause lesions throughout the respiratory tract. Although the concentration of such a gas in the distal portion of the lung is diminished by removal in the anterior airways, differences in sensitivity of the epithelium in different parts of the lung may actually lead to lesions more severe in some distal portions than in anterior portions. For example, ozone causes the severest lung lesions in airway generation ~ 18 where the liquid lining offers little protection for the epithelial cells, but ozone concentration in air is still substantial. Models have been developed that qualitatively predict such sensitivity (Miller et al., 1985).

RESPIRATORY-TRACT ENZYMES INVOLVED IN XENOBIOTIC METABOLISM
Monooxygenases

Cytochrome P450 and the FAD-MO are monooxygenases—enzymes that add one oxygen atom to their substrates—found in the respiratory tract and in many other organs. The substrate specificities and other characteristics of the cytochrome P450 enzymes have been reviewed (Guengerich, 1990), and the recommended nomenclature for this super family of proteins has recently been updated (Nebert and Nelson, 1991). Extensive research on respiratory tract P450 has localized many forms to specific sites (Table 3). Localization of enzyme activity often presents plausible explanations for the tissue-selective toxicity of inhalants in exposed animals. However, extrapolating toxicity data obtained in animals to expected effects in humans is complicated by the fact that the concentrations of cytochrome P450 present in the various portions of the respiratory tract vary greatly among animal species (Dahl and Hadley, 1991; Dahl and Lewis, 1993; Sabourin et al., 1988). For example, in a review of cytochrome P450-dependent metabolism in human lungs, it was reported that total human lung P450 is 2 pmol/mg of microsomal protein—only 5% of that found in rat lung and less than 1% of that in rabbit lung (Wheeler and Guenther, 1991).

Before leaving the subject of cytochrome P450, it is important to point

Table 3   Some P450 Isoforms Reported in the Rat, Rabbit, and Human Respiratory Tract[a]

| Isozyme | Nasal Tissue | | Lung | | |
	Rat	Rabbit	Rat	Rabbit	Human
1A1	Respiratory and olfactory	Absent	Bronchi, Clara cells, Type II cells	Bronchi, Endothelial cells	Bronchi, Lung
1A2	Olfactory	Olfactory	—	—	—
2A	Olfactory	—	Lung	—	—
2B1	Respiratory and olfactory	Respiratory and olfactory	High levels in Clara cells	Clara cells, Type II cells	—
2E1	Respiratory and olfactory	Olfactory	Lung	Lung	—
2F1	—[b]	—	—	—	Lung
2G1	Olfactory	Olfactory	Absent	Absent	—
3A	Respiratory and olfactory	—	Clara cells, Type II cells	—	—
4B1	—	Respiratory and olfactory	Lung	Clara cells, Type II cells	Lung

[a]From reviews by Dahl and Hadley, 1991 and Dahl and Lewis, 1993.
[b]Not listed in cited reviews.

out that the P450-dependent monooxygenases are necessary for life. P450 enzymes are associated with key reactions in endogenous steroid metabolism as well as with metabolism of xenobiotics. In the last case, the oxidation of lipophilic xenobiotics is often a necessary step for increasing their polarity and, hence, their excretion rates, thereby preventing harmful buildup of xenobiotics in lipidic tissues. Toxicologists tend to dwell on the rare cases where P450-dependent metabolism produces toxic products, but it is important to recall that this is unusual.

FAD-MO has been reported in the nasal cavity (McNulty et al., 1983; Sabourin et al., 1988) and in the lungs (Lawton et al., 1991; Sabourin et al., 1988) of a number of species. These enzymes catalyze oxidation of the hetero atom in sulfur- and nitrogen-containing compounds and sometimes compete with P450 for substrates; thus, in the rat nasal cavity, N,N-dimethylaniline is metabolized only by the FAD-MO, whereas dimethylamine is metabolized by P450 as well (McNulty et al., 1983).

To assess the role of respiratory tract monooxygenases to enhance or alleviate the toxicity of inhalants, it is important to remember that a given compound may be a substrate for more than one enzyme as is the case for dimethylamine. A given substrate may also be metabolized by respiratory tract enzymes to a number of different metabolites; these, in turn, may be subject to further metabolism. In order to assess the contribution of metabolites to toxicity, it is necessary to know their rates of formation and detoxification via further metabolism. The enzymes discussed below are often involved in further metabolism of products formed by the monooxygenases.

Dehydrogenases

The dehydrogenases include alcohol and aldehyde dehydrogenases, both of which are present in the respiratory tract (Bogdanffy et al., 1985; Morris and Cavanagh, 1987). The nasal cavity in particular has high levels of aldehyde dehydrogenase activity which may influence the toxicity of inhaled acetaldehyde and compounds that are metabolized to aldehydes. For example, carboxylesterase (discussed later) hydrolyzes vinyl acetate to acetic acid and transitory vinyl alcohol which promptly rearranges to acetaldehyde. Aldehydes are also produced by decomposition of the alpha hydroxy ethers and amines that are frequently the products of P450-catalyzed oxidation of ethers and amines.

Like cytochrome P450, aldehyde dehydrogenase is not uniformly distributed in the respiratory tract. In rats, the ciliated cells of the nasal respiratory mucosa have high levels of aldehyde dehydrogenase, whereas the cells of the olfactory mucosa, including Bowman's gland cells and basal cells, have less activity. This is an uncommon reversal of the usual order for enzyme distribution in the nasal cavity—for most enzymes, the olfactory mucosa has higher levels than the respiratory mucosa (Dahl and Hadley, 1991). Traveling down the respiratory tract, a more familiar pattern occurs in the distribution of aldehyde dehydrogenases—low concentrations of xenobiotic-metabolizing enzyme in the anterior tracheobronchial tree and higher concentrations in the lower airways—especially in the Clara cells. Interestingly, the lung Clara cells, which are nonciliated, have the highest levels of aldehyde dehydrogenase activity; whereas in the nasal cavity the highest levels are associated with ciliated cells (Bogdanffy et al., 1985).

Carboxylesterases

The carboxylesterases catalyze the hydrolysis of carboxylic esters to carboxylic acids and alcohols. Along with rhodanese (discussed later), carboxylesterase activity is among the highest in the nasal cavity (Table 2), and several forms have also been reported in the lung (Dahl and Lewis, 1993).

Extensive studies have compared nasal carboxylesterase activities in a number of species, including humans (Lewis et al., 1994a; Mattes and Mattes, 1992). Common lesions of the human nasal respiratory mucosa such as squamous metaplasia are associated with profoundly decreased carboxylesterase activity. Although such lesions undoubtedly reduce the effectiveness of the nose in warming and humidifying inhaled air, their presence may actually decrease the toxicity of inhaled esters because carboxylesterase activity is instrumental in the nasal toxicity of inhaled esters, including dibasic carboxylic esters (Bogdanffy et al., 1991) and vinyl acetate (Bogdanffy et al., 1994).

Transferases

Glutathione transferases (GSH-Ts) catalyze the addition of the tripeptide, glutathione, to a reactive moiety. Usually, but not always, GSH-T activity results in metabolites less toxic than the substrate molecule. The dimeric GSH-Ts constitute several classes; each class has a number of forms. GSH-T occurs in the nasal cavity (Dahl and Hadley, 1991), and at least six forms from three different classes occur in human lung (Dahl and Lewis, 1993).

UDP-glucuronyl transferase (UDP-G) transfers glucuronic acid to the hetero atom of alcohols, phenols, and other compounds. Activity by this enzyme greatly increases the solubility of its substrates and is virtually always a detoxification step. UDP-G activity has been demonstrated in the nasal cavity (Bond, 1983) and in the lung (Vaino and Hietanen, 1980).

Epoxide Hydrolase

Epoxide hydrolase (EH) catalyzes the hydrolysis of epoxides—such as butadiene monoxide—to vicinal diols. Such oxides are common products from P450 oxidation of unsaturated hydrocarbons and are often potent alkylating agents, binding to DNA and other macromolecules. Thus, EH activity almost always leads to detoxification. The enzyme occurs in fairly similar concentrations in mouse, rat, monkey, and human lung tissue (Fig. 7); high levels of activity also occur in nasal tissue (Dahl and Hadley, 1991). Differences in the relative amounts of EH activity versus P450 oxidase activity among animal species (Fig. 7) may contribute to differences in toxicity.

Rhodanese

Rhodanese metabolizes the cyanide anion to the far less toxic thiocyanate anion. The activity of this enzyme in rat, cow, and human nasal tissues is high (Dahl, 1989, Lewis et al., 1991, 1992) and probably alleviates the toxicity of both inhaled cyanide and cyanide released from other sources; e.g., the cyanide released from organonitriles after P450-catalyzed oxidation (Table 2).

INDUCTION AND INHIBITION OF RESPIRATORY-TRACT XENOBIOTIC-METABOLIZING ENZYMES

Some lung cytochrome P450 enzymes—notably P450 1A1, which is involved in activation of polycyclic aromatic hydrocarbons—are highly inducible by administration of their substrates (Philpot and Wolf, 1981). As an example, the metabolism of nitropyrene (NP) by isolated, perfused rat lungs is signifi-

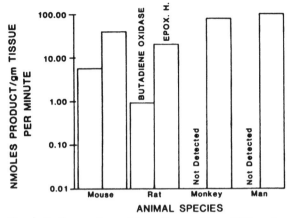

Figure 7   Comparison of butadiene oxidase (BO) and epoxide hydrolase (EH) activities in the lungs of mice, rats, monkeys, and man (from data reported by Schmidt and Loeser, 1986). ND = activity not detected.

cantly enhanced in rats pretreated with 3-methylcholanthrene (Bond and Mauderly, 1984). In contrast, phenobarbital—an inducer of some liver cytochrome P450 enzymes—is not an effective inducer of lung cytochrome P450-dependent NP-hydroxylase. Induction of lung enzymes in smokers is well documented and may have a major effect on the toxicity and fate of inhalants (Dahl and Lewis, 1993).

Nasal P450 is more difficult to induce than lung P450, although induction of rabbit nasal P450 2E1 by acetone and ethanol has been reported (Ding and Coon, 1990). In contrast, nasal P450 enzymes are easily inhibited at low concentrations by methylenedioxyphenyl (MDP) compounds such as isosafrole, a component of sassafras oil (Dahl and Brezinski, 1985). Thus, nasal cytochrome P450-dependent hexamethylphosphoramide N-demethylase is over 50% inhibited by 5 μM isosafrole. Inhibition of nasal carboxylesterase by bis(p-nitrophenyl)phospate (BNP) substantially alters the nasal uptake and toxicity of inhaled esters (Morris, 1990). Both MDP compounds and BNP irreversibly bind to the active sites of their respective enzymes, effectively lowering maximum metabolic capacity, V_{max} (Fig. 2). Nasal carboxylesterase is induced by cigarette smoke (Lewis et al., 1994b) and by pyridine vapor (Nikula et al., 1995). Nasal transferase enzymes are induced by treatment with phenobarbital in the rat (Bond, 1983).

SUMMARY

Both the nasal cavity and the lungs contain enzymes that metabolize xenobiotics. These enzymes can metabolize some inhaled xenobiotics to products that are less toxic than the inhalant; in other cases, the metabolites may be more toxic than the parent molecule. The presence of these enzymes in the respiratory tract indicates that the distribution of potentially toxic metabolites of inhaled material may be different from that after administration of the same material by other routes. In turn, the sites susceptible to toxic effects may be different for an inhaled toxicant compared to the same toxicant administered by noninhalation routes.

The fraction of an inhalant that is metabolized at a particular site within the respiratory tract depends not only upon the enzyme activity present, but also upon the concentration of the inhalant available for metabolism. Although complex tissue-dose relationships operate for inhalants, the remarkable sensitivity to toxicants of the olfactory tissues of the nose and the Clara cells of the lung is, in many instances, related to the high enzyme activities present in these tissues, not only to the dose of inhalant they receive. These tissues are sometimes uniquely susceptible to the toxic effects of compounds administered by noninhalation routes (Brittebo and Brandt, 1990).

Respiratory-tract enzymes may either be inhibited or induced by materials administered either systemically or by inhalation. Induction or inhibition may alter the toxicity of inhalants. Thus, diet, preadministration of other materials, or the coadministration of a second inhalant may affect the toxicity of inhalants.

ACKNOWLEDGMENT

Research supported by DOE/OHER under Contract No. DE-AC04-76EV01013.

REFERENCES

Andersen, ME: Physiological modelling of organic compounds. Am Occup Hyg 3:309–321, 1991.

Bakke, JE, Larsen, GL, Struble, C, Feil, VJ: Metabolism of 2,6-dichlorobenzonitrile, 2,6-dichlorothiobenzamide in rodents and goats. Xenobiotica 18:1063–1075, 1988.

Bogdanffy, MS, Randall, HW, Morgan, KT: Histochemical localization of aldehyde dehydrogenase in the respiratory tract of the Fischer-344 rat. Toxicol Appl Pharmacol 82:560–567, 1985.

Bogdanffy, MS, Kee, CR, Hinchman, CA, Trela, BA: Metabolism of dibasic esters by rat nasal mucosal carboxylesterase. Drug Metab Dispos 19:124–129, 1991.

Bogdanffy, MS, Dreef-van der Meulen, HC, Beems, RB, Feron, VJ, Cascieri, TC, Tyler, TR, Vinegar, MB, Rickard, RW: Chronic toxicity and oncogenicity inhalation study with vinyl acetate in the rat and mouse. Fundam Appl Toxicol 23:215–229, 1994.

Bond, JA: Some biotransformation enzymes responsible for polycyclic aromatic hydrocarbon metabolism in rat nasal turbinates; effects on enzyme activities of in vitro modifiers and intraperitoneal and inhalation exposure of rats to inducing agents. Center Res 43:4805–4811, 1983.

Bond, JA, Mauderly, JL: Metabolism and macromolecular covalent binding of ^{14}C-1-nitropyrene in isolated, perfused and ventilated rat lungs. Cancer Res 44:3924–3929, 1984.

Brittebo, EB, Brandt, I: Interactions of xenobiotics in the respiratory tract following noninhalation routes of exposure. Int J Epidemiol 19:S24-S31, 1990.

Casanova-Schmitz, M, David, RM, Heck, HD'A: Oxidation of formaldehyde and acetaldehyde by NAD$^+$-dependent dehydrogenases in rat nasal mucosal homogenates. Biochem Pharmacol 33:1137–1142, 1984.

Cheng, YS, Yamada, Y, Yeh, HC, Swift, DL: Diffusional deposition of ultrafine aerosols in a human nasal cast. J Aerosol Sci 19:741–751, 1988.

Cheng, YS, Hansen, GK, Su, YF, Yeh, HC, Morgan, KT: Deposition of ultrafine aerosols in rat nasal molds. Toxicol Appl Pharmacol 106:222–233, 1990.

Chichester, CH, Philpot, RM, Weir, AJ, Buckpitt, AR, Plopper, CG: Characterization of the cytochrome P-450 monooxygenase system in nonciliated bronchiolar epithelial (Clara) cells isolated from mouse lung. Am J Respir Cell Mol Biol 4:179–186, 1991.

Dahl, AR: The effect of cytochrome P-450-dependent metabolism and other enzyme activities on olfaction. In: Molecular Neurobiology of the Olfactory System, edited by FL Margolis, TV Getchell, pp. 51–70. Plenum Publishing Corporation, 1988.

Dahl, AR: The cyanide-metabolizing enzyme rhodanese in rat nasal respiratory and olfactory mucosa. Toxicol Lett 45:199–205, 1989.

Dahl, AR: Dose concepts for inhaled vapors and gases. Toxicol Appl Pharmacol 103:185–197, 1990.

Dahl, AR, Brezinski, DA: Inhibition of rabbit nasal and hepatic cytochrome P-450-dependent hexamethyl phosphoramide (HMPA) N-demethylase by methylenedioxyphenyl compounds. Biochem Pharmacol 34:631–636, 1985.

Dahl, AR, Hadley, WM: Nasal cavity enzymes involved in xenobiotic metabolism: Effects on the toxicity of inhalants. CRC Rev Toxicol 21:345–372, 1991.

Dahl, AR, Lewis, JL: Respiratory tract uptake of inhalants and metabolism of xenobiotics. Annu Rev Pharmacol Toxicol 32:383–407, 1993.

Dahl, AR, Waruszewski, BA: Metabolism of organonitriles to cyanide by rat nasal tissue enzymes. Xenobiotica 19(11):1201–1205, 1989.

Dahl, AR, Hadley, WM, Hahn, FF, Benson, JM, McClellan, RO: Cytochrome P-450-dependent monooxygenases in olfactory epithelium of dogs: Possible role in tumorigenicity: Science 216:57–59, 1982.

Ding, X, Coon, MJ: Induction of cytochrome P-450 isozyme 3a (P-450IIE1) in rabbit olfactory mucosa by ethanol and acetone. Drug Metab Dispos 18:742–745, 1990.

Gerde, P, Dahl, AR: A model for the uptake of inhaled vapors in the nose of the dog during cyclic breathing. Toxicol Appl Pharmacol 109:276–288, 1991.

Gerde, P, Muggenburg, BA, Henderson, RF: Disposition of polycyclic aromatic hydrocarbons in the respiratory tract of the beagle dog. III. Mechanisms of the dosimetry. Toxicol Appl Pharmacol 121:328–334, 1993.

Gerde, P, Muggenburg, BA, Henderson, RF, Dahl, AR: Particle-associated hydrocarbons and lung cancer; the correlation between cellular dosimetry and tumor distribution. In: Effects of Mineral Dusts on Cells, NATO/ASI Series, Vol. H85, edited by JMG Davis, MC Jaurand, pp. 337–346. Heidelberg: Springer-Verlag, 1994.

Guengerich, FP: Enzymatic oxidation of xenobiotic chemicals. Crit Rev Biochem Mol Biol 25(2):91–153, 1990.

Johanson, G: Modelling of respiratory exchange of polar solvents. Ann Occup Hyg 35:323–339, 1991.

Johanson, G, Filser, JG: Experimental data from closed chamber gas uptake studies in rodents suggest lower uptake rate of chemical than calculated from literature values on alveolar ventilation. Arch Toxicol 66:291–295, 1992.

Kanekal, S, Plopper, C, Morin, D, Buckpitt, A: Metabolism and cytotoxicity of naphthalene oxide in the isolated perfused mouse lung. J Pharmacol Exp Ther 256:391–401, 1991.

Kimbell, JS, Gross, EA, Joyner, DR, Godo, MN, and Morgan, KT: Application of computational fluid dynamics to regional dosimetry of inhaled chemicals in the upper respiratory tract of the rat. Toxicol Appl Pharmacol 121:253–263, 1993.

Lawton, MP, Kronbach, T, Johnson, EF, Philpot RM: Properties of expressed and native flavin-containing monooxygenases: Evidence of multiple forms in rabbit liver and lung. Mol Pharmacol 40:692–698, 1991.

Lewis, JL, Rhoades, CE, Gervasi, P-G, Griffith, WC, Dahl, AR: The cyanide-metabolizing enzyme rhodanese in human nasal respiratory mucosa. Toxicol Appl Pharmacol 108:114–120, 1991.

Lewis, JL, Rhoades, CE, Bice, DE, Harkema, JR, Hotchkiss, JA, Sylvester, DM, Dahl, AR: Interspecies comparison of cellular localization of the cyanide metabolizing enzyme rhodanese within olfactory mucosa. Anat Rec 232:620–627, 1992.

Lewis, JL, Nikula, KJ, Novak, R, Dahl, AR: Comparative localization of carboxylesterase in rat, dog, and human nasal tissue. Anat Rec 239:55–64, 1994a.

Lewis, JL, Nikula, KJ, Sachetti, LA: Induced xenobiotic-metabolizing enzymes localized to eosinophilic globules in olfactory epithelium of toxicant-exposed F344 rats. Inhal Toxicol 6 (Suppl):422–425, 1994b.

Mattes, PM, Mattes, WB: α-naphthyl butyrate carboxylesterase activity in human and rat nasal tissue. Toxicol Appl Pharmacol 114:71–76, 1992.

McNulty, MJ, Casanova-Schmitz, M, Heck, HD'A: Metabolism dimethylamine in the nasal mucosa of the Fischer 344 rat. Drug Metab Dispos 11:421–425, 1983.

Miller, FJ, Overton, Jr., JH, Jaskot, RH, Menzel, DB: A model of the regional uptake of gaseous pollutants in the lung. I. The sensitivity of the uptake of ozone in the human lung to lower respiratory tract secretions and exercise. Toxicol Appl Pharmacol 79:11–27, 1985.

Morris, JB, Cavanagh, DG: Metabolism and deposition of propanol and acetone vapors in the upper respiratory tract of the hamster. Fundam Appl Toxicol 9:34–40, 1987.

Morris, JB: First-pass metabolism of inspired ethyl acetate in the upper respiratory tracts of the F344 rat and Syrian hamster. Toxicol Appl Pharmacol 102:331–345, 1990.

Nebert, DW, Nelson, DR: P450 gene nomenclature based on evolution. Methods Enzymol 206:3–11, 1991.

Nikula, KJ, Novak, RF, Chang, IY, Dahl, AR, Kracko, DA, Lewis, JL: Induction of nasal carboxylesterase in F344 rats following inhalation exposure to pyridine. Drug Metab Dispos, 1995 (in press).

Philpot, RM, Wolf, CR: The properties and distribution of the enzymes of pulmonary cytochrome R-450-dependent monooxygenase systems. In: Biochemical Toxicology, edited by E Hodgson, JR Bend, RM Philpot, Vol. 3, pp. 51–76. New York: Elsevier/North Holland, 1981.

Sabourin, PJ, Tynes, RE, Philpot, RM, Winquist, S, Dahl, AR: Distribution of microsomal monooxygenases in the rabbit respiratory tract. Drug Metab Dispos 16:557–562, 1988.

Schmidt, U, Loeser, E: Epoxidation of 1,3-butadiene in liver and lung tissue of mouse, rat, monkey and man. In: Biological Reactive Intermediates. III. Advances in Experimental Medicine and Biology, Vol. 197, pp. 951–958. New York: Plenum, 1986.

Stott, WT, McKenna, MJ: Hydrolysis of several glycol ether acetates and acrylate esters by nasal mucosal carboxylesterase in vitro. Fundam Appl Toxicol 5:399–404, 1985.

Sun, JD, Dahl, AR, Gillett, NA, Barr, EB, Crews, ML, Eidson, AF, Burt, DG, Dieter, MP, Hobbs, CH: Two-week repeated inhalation exposure of F344/N rats and B6C3F₁ mice to ferrocene. Fundam Appl Toxicol 17:150–158, 1991.

Vaino, H, Hietanen, E: Role of extrahepatic metabolism in drug disposition and toxicity. In: Concepts in Drug Metabolism, edited by P Jenner, B Testa, part A, pp. 251–284. New York: Marcel Dekker, 1980.

Wheeler, CW, Guenther, TM: Cytochrome P450-dependent metabolism of xenobiotics in human lung. J Biochem Toxicol 6:163–169, 1991.

Yost, GS, Buckpitt, AR, Roth, RA, McLemore, TL: Contemporary issues in toxicology: Mechanisms of lung injury by systemically administered chemicals. Toxicol Appl Pharmacol 101:179–195, 1989.

Part Four

Disposition of
Inhaled Material

Chapter Eight

Deposition and Clearance of Inhaled Particles

Richard B. Schlesinger

INTRODUCTION

The biologic effects of inhaled particles are a function of their disposition. This, in turn, depends upon their patterns of deposition, i.e., the sites within which they initially come into contact with airway epithelial surfaces and the amounts removed from the inhaled air at these sites, and clearance, i.e., the rates and routes by which deposited particles are physically removed from the respiratory tract. For materials, such as irritants, which exert their action upon surface contact, the initial deposition is the predicator of toxic response. In many other cases, however, it is the net result of deposition and clearance—namely retention, i.e., the amount of particles remaining in the respiratory tract at specific times after exposure—which influences toxicity. This chapter provides an overview of the processes by which airborne particles are deposited within and cleared from the respiratory tract.

DEPOSITION OF INHALED PARTICLES
Deposition Mechanisms

There are five significant mechanisms by which particles may deposit in the respiratory tract. These are impaction, sedimentation, Brownian diffusion, electrostatic precipitation, and interception; they are depicted schematically in Fig. 1.

Impaction is the inertial deposition of a particle onto an airway surface. It occurs when the particle's momentum prevents it from changing course in an area where there is a rapid change in the direction of bulk airflow. Impaction is the main mechanism by which particles having diameters ≥ 0.5 μm deposit in the upper respiratory tract and at or near tracheobronchial tree branching points. The probability of impaction increases with increasing air velocity, rate of breathing, particle size, and density.

Sedimentation is deposition due to gravity. When the gravitational force on an airborne particle is balanced by the total of forces due to air buoyancy and air resistance, the particle will fall out of the air stream at a constant rate—the terminal settling velocity. The probability of sedimentation is proportional to the particle's residence time in the airway, particle size, and density and decreases with increasing breathing rate. Sedimentation is an important deposition mechanism for particles with diameters ≥ 0.5 μm which penetrate to airways where air velocity is relative low, e.g., mid to small bronchi and bronchioles.

Submicrometer-sized particles (especially ultrafines, which are those having diameters < 0.1 μm) acquire a random motion due to bombardment by surrounding air molecules. This motion may then result in particle contact with the airway wall. The displacement sustained by the particle is a function of a parameter known as the diffusion coefficient and is inversely related to particle size (specifically cross-sectional area) but is independent of particle density. The probability of deposition by diffusion increases with increasing particle residence time within the airway, and diffusion is a major deposition mechanism where bulk flow is low or absent, e.g., bronchioles and the pulmo-

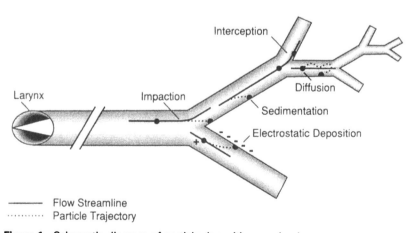

Figure 1   Schematic diagram of particle deposition mechanisms.

nary region (alveolated airways). However, extremely small ultrafine particles can show significant deposition in the upper respiratory tract, the trachea and larger bronchi. This likely occurs by turbulent diffusion.

Some freshly generated particles can be electrically charged and may exhibit enhanced deposition over that expected from size alone. This can be due to image charges induced on the surface of the airway by these particles, and/or to space-charge effects whereby repulsion of particles containing like charges results in increased migration towards the airway wall. The effect of charge on deposition is inversely proportional to particle size and airflow rate. Since most ambient particles become neutralized naturally due to the presence of air ions, electrostatic deposition is generally a minor contributor to overall particle collection by the respiratory tract. It may, however, be important in some laboratory studies.

Interception is a significant deposition mechanism for fibrous particles, which are those having length to diameter ratios > 3:1. While fibers are also subject to all of the same deposition mechanisms as are more spherical or compact particles, they have the additional possibility of deposition when an edge contacts, or intercepts, an airway wall. The probability of interception increases as airway diameter decreases, but it can also be fairly significant in both the upper respiratory tract and upper tracheobronchial tree. While interception probability increases with increasing fiber length, the aerodynamic behavior of a fiber and impaction/sedimentation probability is more influenced by fiber diameter.

Factors Controlling Deposition

The extent and loci of particle deposition depend upon various controlling factors (Table 1). These are characteristics of the inhaled particles, geometry of the respiratory tract, and breathing pattern.

Characteristics of Inhaled Particles   From the discussion above, it should be evident that the major particle characteristic which influences deposition is size. But it is important that this be expressed in the proper manner. The deposition probability for particles with geometric diameters $\gtrsim 0.5$ μm is governed largely by their equivalent aerodynamic diameter (D_{ae}), while the deposition probability for smaller particles is governed by actual physical diameter. It therefore follows that aerodynamic diameter is the most appropriate size parameter for describing particles subject to deposition by sedimentation and impaction, but not diffusion. Since particles are generally inhaled not singly, but as constituents of aerosols, the mass median aerodynamic diameter (MMAD) is an appropriate parameter to use for those aerosols in which most particles have actual diameters $\gtrsim 0.5$ μm, while the median size of aerosols containing particles with diameters less than this

Table 1   Some Factors That May Control or Affect Particle Deposition

Particle characteristics
  Geometric size
  Shape
  Density
  Hygroscopicity
  Electrical charge
Respiratory tract geometry
  Airway caliber
  Airway branching pattern
  Path length to terminal airways
Ventilation
  Mode of breathing—oral, nasal, oronasal
  Respiratory rate
  Tidal volume
  Flow rate and velocities
  Interlobular distribution of ventilation
  Length of respiratory pauses
Other factors
  Irritant exposure
  Respiratory tract disease
  Growth from newborn to adult
  Aging from maturity (?)
  Gender (?)

should be expressed in terms of a diffusion diameter, such as thermodynamic equivalent diameter, or by using actual geometric size.

The distribution of particle sizes within an aerosol, which is generally characterized as either monodisperse ($\sigma_g = 1.2$) or polydisperse ($\sigma_g > 1$), is also important in terms of ultimate deposition pattern. If the σ_g of a polydisperse aerosol is < 2, the total amount of deposition within the respiratory tract will probably not differ substantially from that for a monodisperse aerosol having the same median size (Diu and Yu, 1983). However, size distribution is critical in determining the spatial pattern of deposition, since the latter depends upon the sequential removal of particles within each region of the respiratory tract which, in turn, depends upon the actual particle sizes present within the aerosol. For example, when the deposition (in hamsters) of a monodisperse and a polydisperse aerosol having similar median aerodynamic sizes was compared (Thomas and Raabe, 1978), the latter was found to deposit to a greater extent in the upper respiratory tract due to the presence of a certain fraction of large particles that were effectively removed by impaction. Total respiratory tract deposition of the two aerosols (expressed as a percentage of the amount inhaled) was comparable.

A particle characteristic which may dynamically alter its size after inhalation is hygroscopicity. Hygroscopic particles will grow substantially while

they are still airborne within the respiratory tract and will deposit according
to their hydrated (rather than their initial dry) size.

Respiratory Tract Geometry   Respiratory tract structure affects particle
deposition in many ways. For example, airway diameter sets the displacement
required for a particle to contact a surface, while the cross-section determines
the air velocity and type of flow for a given inspiratory flow rate. Furthermore, flow characteristics depend upon branching angle and branching pattern. Differences in pathway lengths within different lung lobes may affect
regional deposition. For example, if particles subject to impaction or sedimentation are inhaled, those lobes with the shortest average path length between the trachea and terminal bronchioles may have the highest pulmonary
region concentration of deposition. On the other hand, differences in regional deposition become less obvious for ultrafine particles, which tend to
deposit more evenly in all lobes regardless of path length, but rather in proportion to relative ventilation.

The tracheobronchial airways and alveoli show a considerable degree of
size variability between different individuals. This is likely the primary factor
responsible for the large inter-individual differences in deposition which are
observed experimentally (Heyder et al., 1982).

Ventilation Pattern and Mode of Inhalation   The pattern of breathing
during particle exposure influences the sites and relative amounts of regional
deposition. For example, exercise or other enhanced activity may result in
increased respiratory rate and tidal volume and increased linear air velocities
within the conducting airways. This would tend to enhance impaction, but
decrease deposition due to sedimentation and diffusion. While total deposition within the respiratory tract may increase with exercise for particles >
0.2–0.5 μm in diameter (Harbison and Brain, 1983; Zeltner et al., 1991), a
shift in the deposition pattern towards the upper respiratory tract and central
bronchi and away from more distal conducting airways and the pulmonary
region can occur (Bennett et al., 1985; Morgan et al., 1984). Increased linear
velocities may also result in the development of turbulence, which tends to
enhance deposition of such particles. On the other hand, the deposition of
ultrafine particles within the upper respiratory tract and tracheobronchial
tree decreases as flow rate increases, and exercise may not increase total respiratory tract deposition of these particles even though it does result in greater
numbers inhaled (Hesseltine et al., 1986).

Tidal volume, i.e., the volume of air inhaled during a single breath,
affects regional deposition by determining the depth of penetration of inspired air. For a constant breathing frequency, an increase in tidal volume
would result in deeper penetration of inhaled particles, with a potential increase in deposition in the smaller conducting airways and pulmonary region.
Alterations in tidal volume may also dramatically affect total respiratory

tract deposition. For example, a doubling in tidal volume from 1.4 ml to 2.8 ml in the rat was predicted to increase the deposition of a 1 μm (median D_{ae}) aerosol by 7 times (Schum and Yeh, 1980). Finally, respiratory frequency and the duration of respiratory pauses influences sedimentation or diffusion deposition by affecting particle residence time in relatively still air.

A significant change that occurs in humans when activity level increases is a switch in the mode of breathing from nasal to oronasal (combined oral and nasal breathing). Since the nasal passages are more efficient than the oral in removing inhaled particles, even a partial bypassing of the nose could increase particle penetration into the lungs. Toxicological studies using aerosols may employ a variety of inhalation devices and protocols. When evaluating and comparing such studies, it is important to consider the effects of exposure technique upon subsequent deposition. Just as critical is assessment of the relationship between deposition in obligate nasal breathing experimental animals to humans breathing via the mouth.

Factors Modifying Deposition

Various factors may alter deposition patterns compared to those occurring in normal, healthy adult individuals—the group most commonly used in toxicologic assessments. As outlined in Table 1, these include previous or coexposure to airborne irritants, lung disease, and growth, all of which can affect deposition by changing its controlling parameters, namely ventilation pattern and/or airway geometry.

Irritant inhalation-induced bronchoconstriction would tend to increase impaction deposition in the upper bronchial tree. Likewise, deposition may be altered due to disease. Bronchial obstruction associated with various pulmonary diseases tends to increase total respiratory tract deposition via enhanced deposition within the upper respiratory tract and tracheobronchial tree (especially for particles > 1 μm), even though peripheral deposition may be reduced. The deposition of ultrafine particles is also increased in obstructive lung disease due to increased residence time and to flow perturbations resulting from reductions in airway lumen calibre. On the other hand, deposition may be entirely eliminated in portions of the lungs due to ventilation impairments (Thomson and Short, 1969; Thomson and Pavia, 1974; Lourenco et al., 1972).

Structural alterations in the lungs may affect deposition. For example, rodents with enzyme-induced emphysema showed a reduced deposition compared to normal controls (Damon et al., 1983; Hahn and Hobbs, 1979). This was likely due to an increase in alveolar size, resulting in greater distances to deposit on a surface and a concomitant reduction in pulmonary region deposition efficiency (Brain and Valberg, 1979). On the other hand, inhaled particles deposited more distally in rats with a fibrotic disease, i.e., coal or silica derived pneumoconiosis, than in normal animals (Heppelston, 1963).

One of the current concerns in inhalation toxicology involves differences in deposition between children and adults. A number of attempts have been made to estimate the influence upon deposition of anatomical and ventilatory changes during postnatal growth in humans (e.g., Hofmann, 1982; Crawford, 1982; Phalen et al., 1991). They indicate that the relative effectiveness of the major deposition mechanisms differs at various times during growth and that this, in turn, may alter regional deposition patterns. Taking into account anatomical differences and the greater ventilation per unit body weight in children, the deposition fractions for some particle sizes, especially those > 1 μm, within certain regions of the growing respiratory tract could be quite different, sometimes well above those found based upon studies with adults. Such differences would become even more significant when deposition is expressed on a per unit surface area basis. Since there are also regional differences in clearance rates, this infers that the dose to specific lung compartments from some inhaled particles may vary with age from newborn to adult Anatomical changes with aging post-maturity may also affect deposition for particles > 1 μm, increasing pulmonary region deposition in older adults compared to younger adults (Phalen et al., 1991). On the other hand, the deposition of ultrafine particles may not show dramatic differences between children and adults, nor with aging (Phalen et al., 1991; Swift et al., 1992).

Any differences in deposition between children and adults may be influenced by activity levels due to the manner by which breathing pattern changes. For example, increased ventilation with increasing activity in children occurs to a greater extent by increased respiratory frequency, while adults show greater increases in tidal volume. Since increased frequency is associated with decreased deposition of particles > 1 μm in diameter, the greater total respiratory tract deposition with increasing activity levels seen in adults is not seen in young children, and the latter may actually show somewhat of a decline (Becquemin et al., 1991).

Deposition within the Human Respiratory Tract

The deposition of particles within the human respiratory tract can be assessed with a number of techniques (Valberg, 1985). Unfortunately, the use of different experimental methods and assumptions results in considerable variations in reported values. Figs. 2A–D present experimentally determined values for spherical particle deposition within the human respiratory tract as a function of the median size of the inhaled aerosol. All values are expressed as deposition efficiency—the percentage deposition of the total amount inhaled.

Fig. 2A shows the pattern for overall respiratory tract deposition. Note the deposition minima over the 0.2–0.5 μm size range, with increasing deposition with increasing size for larger particles and with decreasing size for smaller

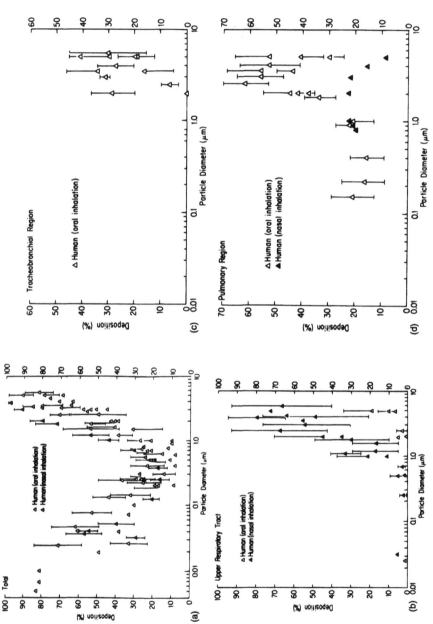

Figure 2 Particle deposition in the human respiratory tract. Deposition efficiency, i.e., the percentage deposition of the amount inhaled, is plotted as a function of particle size for: (a) total respiratory tract; (b) upper respiratory tract; (c) tracheobronchial tree; and (d) pulmonary region. Particle diameters are aerodynamic for those ≥0.5 μm and diffusion equivalent for those <0.5 μm. (Based upon data compiled by Schlesinger, 1985b, with additional data from Tu and Knudson, 1984; Wilson et al., 1985; Schiller et al., 1988; Heyder et al., 1986; Anderson et al., 1990; Becquemin et al., 1991).

198

ones. As previously discussed, particles with diameters $\gtrsim 0.5$ μm are subject to impaction and sedimentation, while the deposition of those $\lesssim 0.1$ μm is diffusion dominated. Particles with diameters between these values are minimally influenced by all three mechanisms and tend to have relatively prolonged suspension times in the inhaled air. They undergo minimal deposition after inhalation, and most are carried out of the respiratory tract in the exhaled air.

The effect of breathing mode upon deposition is evident from Fig. 2A. Inhalation via the nose results in greater total deposition than does oral inhalation for particles with diameters > 0.5 μm. This is due to enhanced collection in the upper respiratory tract with the former. On the other hand, there is little apparent difference in total deposition between nasal or oral breathing for those particles with diameters between 0.02–0.5 μm. For particles < 0.5 μm, an increase in total deposition with nose compared to mouth breathing also occurs, but the difference is smaller, amounting to ~ 5% for particles with diameters of 0.005 μm (Schiller et al., 1988).

The effects of hygroscopicity upon deposition deserves mention. If Fig. 2A is examined, it is evident that hygroscopic particles inhaled at 0.1–0.5 μm diameter would tend to show a decrease in total deposition if they grow to < 0.5 μm and will show a deposition increase only if their final hydrated diameter is > 1 μm. On the other hand, since particles > 5 μm may only minimally grow in one respiratory cycle, they may not show an increase in deposition at all compared to nonhygroscopic material (Ferron, 1988). Hygroscopic particles inhaled at 0.2–0.5 μm may show substantial changes in their deposition probability, particularly in the tracheobronchial and pulmonary regions.

Fig. 2B shows the pattern of deposition in the upper respiratory tract—the larynx and airways above it. Again, it is evident that nasal inhalation results in enhanced deposition compared to oral. The greater the deposition in the head, the less is the amount available for removal in the lungs. Thus, the extent of collection in the upper respiratory tract affects deposition in more distal regions.

Figure 2C depicts deposition in the tracheobronchial tree. There appears not to be as well a defined relationship between deposition and particle size as in other regions. Fractional tracheobronchial deposition is relatively constant over a wide particle size range.

Deposition in the pulmonary region is shown in Figure 2D. With oral inhalation, deposition increases with particle size after a minimum at ~ 0.5 μm. With nasal breathing, on the other hand, deposition tends to decrease with increasing particle size. The removal of particles in more proximal airways determines the shape of these pulmonary curves. For example, increased upper respiratory and tracheobronchial deposition would be associated with a reduction of pulmonary deposition. Thus, nasal breathing results in less pulmonary penetration of larger particles and a lesser fraction of deposition for entering aerosol than does oral inhalation. Thus, in the latter case, the peak for pulmonary deposition shifts upwards to a larger sized particle and is more pronounced. With nasal breathing, on the other hand, there is relatively constant pulmonary deposition over a wider particle size range.

The deposition of ultrafine particles is of great interest in inhalation studies since these particles present a large surface area for potential adsorption of other toxicants for delivery to the respiratory tract. There are a few inhalation studies in humans using ultrafine aerosols in the diameter range of 0.1–0.01 μm, and less for smaller sizes. The latter is partly due to technical difficulties in producing high quality monodisperse aerosols within this size range in sufficient quantity to allow evaluation of deposition. From Fig. 2A, it can be seen that total respiratory tract deposition increases as particle size decreases below 0.2 μm.

The regional deposition of ultrafine particles in humans has been examined using only mathematical and physical models (Cheng et al., 1991; 1988; 1993; NRC, 1991; Swift et al., 1992). These indicate that as particle size decreases below 0.2 μm, deposition within the upper respiratory tract and tracheobronchial tree increases substantially, while deposition within the pulmonary region is progressively reduced. Deposition efficiency in the nasal passages can be quite high, reaching over 80% for particles below about 0.002 μm, from a low of about 2% for particles in the 0.1–0.2 μm size range. Similar to larger particles, the deposition efficiency for ultrafine particles within the upper respiratory tract with oral breathing is somewhat less than that with nasal breathing. Estimates suggest that oral deposition is likely to be 70–90% of nasal deposition for comparable inspiratory flow rates.

The deposition efficiencies presented in Fig. 2 are for spherical or compact particles. Due to the potential toxicity of fibrous particle shapes, experimental fiber deposition data in humans is not available. However, studies in animals and use of mathematical and physical models provide some general indication of deposition patterns (Asgharian and Yu, 1988; 1989; Sussman et al., 1991; Hammad et al., 1982; Morgan et al., 1977). Long fibers (> 10 μm) tend to show enhanced deposition in the tracheobronchial tree, and reduced deposition in the pulmonary region, compared to shorter fibers. But fibers which are very long (e.g., > 50 μm) and thin (e.g., < 0.5 μm) can reach distal conducting airways, and significant amounts of such particles can deposit in the pulmonary region. But the deposition of fibers is much more complex than that for spherical particles. For example, the shape of the former is important, since straight fibers penetrate more distally than do curly ones.

Localized Patterns of Deposition

Particle deposition may not occur in a homogeneous manner along airway surfaces. Specific patterns of enhanced local deposition are important in determining the dose, which depends on the surface density of deposition. Nonuniformity implies that the initial dose delivered to specific sites may be greater than that occurring if a uniform density of surface deposit is assumed. This is important for inhaled particles which affect tissues on contact, e.g., irritants, and may be a factor in the site selectivity of certain diseases, e.g., bronchogenic carcinoma (Schlesinger and Lippmann, 1978).

In the upper respiratory tract, enhanced deposition occurs at areas characterized by constrictions, directional changes and high air velocities, e.g., the larynx, oropharyngeal bend, and nasal turbinates (Swift, 1981; Swift and Proctor, 1988). Likewise, the deposition of aerosols in the tracheobronchial tree is not homogeneous. In humans, air turbulence produced by the larynx results in enhanced localized deposition in the upper trachea and larger bronchi, while deposition is also greatly enhanced at bronchial bifurcations, especially along the carinal ridges, relative to the tubular airway segments (Schlesinger et al., 1982). This occurs for spherical particles > 0.5 μm diameter due to impaction, and for fibers due to both impaction and interception (Asgharian and Yu, 1989). However, enhanced deposition at bifurcations is also seen with submicrometer particles having diameters down to about 0.1 μm (Cohen et al., 1988). This is due to turbulent diffusion. As particle size decreases further, the effects of localized flow patterns upon particle behavior become less important, and more uniform deposition along airway surfaces occurs (Gradon and Orlicki, 1990). Thus, there may be a particle size below which enhanced deposition at bifurcations and other sites becomes insignificant. For example, no enhanced deposition of 0.04 μm particles was found at bifurcations in a cast of the human upper bronchial tree (Cohen et al., 1988).

The experimental conditions employed in the numerous tracheobronchial microdistribution studies varied widely, yet the relative enhancement distribution among the airways was found to be quite similar, suggesting that local patterns of deposition within the larger bronchi may be fairly insensitive to particle sizes > 0.1 μm and to air flow rates. It also appears that the proportional distribution of deposition in specific airways is relatively constant over a wide range of particle sizes and total lung deposition efficiencies (Schlesinger and Lippmann, 1978), as is the distribution of deposition in the various lobes of the lungs (Raabe et al., 1977).

There are a few data on localized deposition patterns for the pulmonary region. Fibers show nonuniform deposition in distal airways of animals, preferentially depositing on bifurcations of alveolar ducts near the bronchioalveolar junction (Brody and Roe, 1983; Warheit and Hartsky, 1990). While this has yet to be demonstrated in human lungs, the presence of early fiber-related lesions in similar regions suggests that it may occur in these as well (Brody and Yu, 1989).

Comparative Aspects of Deposition

Various animals are employed in experimental aerosol inhalation toxicology studies, with the ultimate goal being extrapolation to humans. To adequately apply the results to human risk assessment, however, it is essential to consider interspecies differences in total and regional deposition patterns. Since different species exposed to the same aerosol may not receive identical doses in comparable respiratory tract regions, and clearance processes are regionally distinct as will be discussed, the selection of a particular species may influence the estimated human lung (or systemic) dose as well as its relation to potential health effects.

Comparable deposition mechanisms operate for humans and animals, but the degree of similarity in deposition between different species may depend to some extent upon the deposition mechanism which predominates. For example, it has been suggested that interspecies particle deposition probabilities would be similar for sedimentation, but a function of body weight for diffusion (Stauffer, 1975).

Figures 3A–D present particle deposition profiles for a number of experimental animals. It is evident that there are few data on regional deposition of ultrafine particles. One study in the rat indicated that, at a normal inspiratory flow rate, nasal airway deposition efficiency ranged from 6% for 0.1 μm particles to 58% for 0.005 μm particles (Gerde et al., 1991). This is fairly comparable to values for humans indicated by studies in physical model systems, as discussed above.

In evaluating studies with aerosols, the amount of deposition expressed merely as a percentage of the total inhaled, i.e., deposition efficiency, may not be adequate information for relating results between species. For example, total respiratory tract deposition for the same size particle can be quite similar in humans and many experimental animals (Fig. 3A). It, therefore, follows that deposition efficiency is independent of body (or lung) size (McMahon et al., 1977; Brain and Mensah, 1983). However, different species exposed to identical particles at the same exposure concentration will not receive the same initial mass deposition. If the total amount of deposition is divided by body (or lung) weight, smaller animals would receive greater initial particle burdens per unit weight per unit exposure time than would larger ones. For example, the initial deposition of 1 μm particles in the rat will be 5–10 times that of humans, and in the dog 3 times that of humans, if deposition is calculated on a per unit lung or body weight basis (Phalen et al., 1977).

Humans differ from most other mammals used in inhalation toxicologic studies in various aspects of respiratory tract anatomy. But the implications of this to particle deposition have not been adequately appreciated One major interspecies difference is bronchial tree branching pattern (Schlesinger and McFadden, 1981). Humans show a relatively symmetrical dichotomous branching, while most quadrupeds have a highly asymmetric monopodial pattern. This can affect particle microdistribution patterns, in that the tendency for enhanced deposition noted at human airway bifurcations is reduced in monopodial branching systems (Schlesinger, 1980). Furthermore, branching pattern also influences the depth of particle penetration within the bronchial tree due to its effect upon airflow characteristics (Fang et al., 1993). However, the influence of interspecies anatomical differences upon deposition may depend upon particle type. For fibers, for example, the path length and number of branching divisions modulates deposition more than the branching angle or pattern. Thus, deposition of fibers in rodents may be quite relevant to that in humans (Pinkerton et al., 1986).

Interspecies anatomical differences in the upper respiratory tract can also influence deposition patterns. The greater complexity of the nasal passages in rodents compared to primates results in the bulk of impaction deposition occurring more anteriorly in the nasal passages of the former (Gooya and Patra, 1986; Schreider, 1986). Furthermore, rodents would tend to have consistently high deposition in the upper respiratory tract since they are obligate nasal breathers.

In the pulmonary region, alveolar size varies between species. This may be reflected by differences in the probability of deposition by diffusion and sedimentation due to differences in the distance between airborne particles and airway walls.

CLEARANCE OF DEPOSITED PARTICLES
Clearance Mechanisms

Particles which deposit upon airway surfaces may be cleared from the respiratory tract completely or may be translocated to other sites within this system. Clearance mechanisms are regionally distinct in terms of both specific routes (outlined in Table 2) and kinetics. Mechanisms are either absorptive, i.e., dissolution, or nonabsorptive, i.e., mechanical transport of intact particles, and these may occur simultaneously or with temporal variations. It should be mentioned that particle solubility in terms of clearance refers to

Table 2   Respiratory Tract Clearance Mechanisms

Upper respiratory tract
  Mucociliary transport
  Sneezing
  Nose wiping and blowing
  Dissolution (for soluble particles)
Tracheobronchial tree
  Mucociliary transport
  Endocytosis by macrophages/epithelial cells
  Coughing
  Dissolution (for soluble particles)
Pulmonary region
  Macrophages, epithelial cells
  Interstitial pathways
  Dissolution (for soluble and "insoluble" particles)

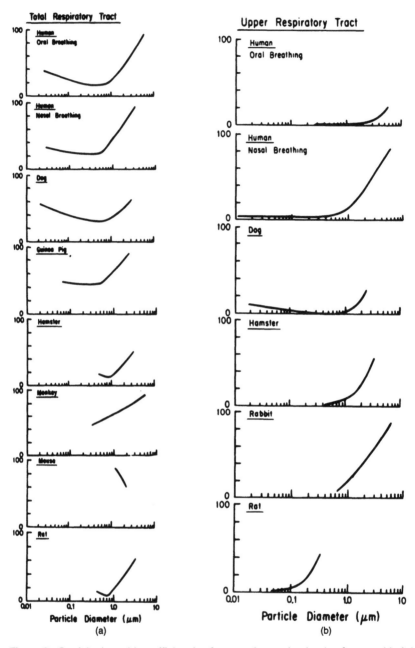

Figure 3   Particle deposition efficiencies for experimental animals often used in inhalation toxicological protocols plotted as a function of particle size for (a) total respiratory tract, (b) upper respiratory tract, (c) tracheobronchial tree, and (d) pulmonary re-

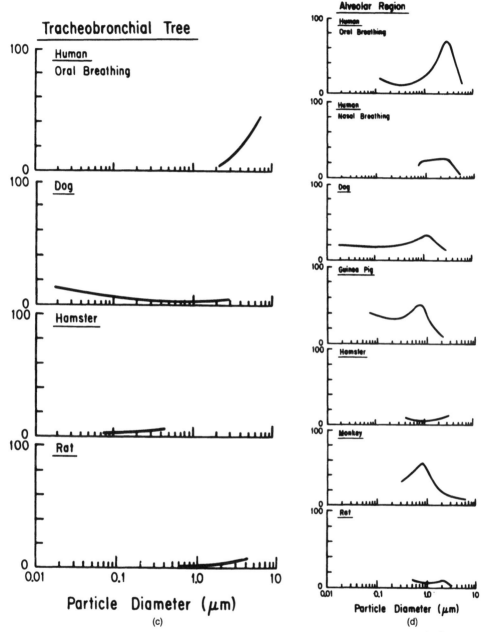

gion. Each curve represents an eye fit through mean values (or centers of ranges) of the data compiled by Schlesinger (1985b). Similar curves for humans are shown for comparison. Particle diameters are aerodynamic for those ≥0.5 μm and diffusion equivalent for those <0.5 μm.

205

solubility *in vivo* within respiratory tract fluids. Thus, an insoluble particle is considered to be one whose rate of clearance by dissolution is insignificant compared to its rate of clearance by mechanical processes. For the most part, all deposited particles clear by the same mechanisms whether they are fibers or compact spheres, with their ultimate fate a function of deposition site, physicochemical properties (including any toxicity), and deposited concentration.

Upper Respiratory Tract   Clearance of insoluble particles deposited in the nasal passages occurs via mucociliary transport, and the general flow of mucus is backwards—towards the nasopharynx. The epithelium in the most anterior portion of the nasal passages is not ciliated, and the mucus flow distal to this is forward, clearing deposited material to a site where removal is by sneezing, wiping, or blowing (extrinsic clearance). Soluble material deposited on the nasal epithelium will be accessible to underlying cells if it can diffuse them through the mucus prior to removal via mucociliary transport. Since there is a rich vasculature in the nose, uptake into the blood may occur rapidly.

Clearance of insoluble particles deposited in the oral passages is by swallowing into the gastrointestinal tract. Soluble particles are likely rapidly absorbed after deposition (Swift and Proctor, 1986).

Tracheobronchial Tree   Like the nasal passages, insoluble particles deposited on tracheobronchial tree surfaces are cleared primarily by mucociliary transport, with the net movement of fluid towards the oropharynx. Some insoluble particles may traverse the epithelium by endocytotic processes, entering the peribronchial region (Masse et al., 1974; Sorokin and Brain, 1975). Clearance may also occur following phagocytosis by airway macrophages, located on or beneath the mucus lining throughout the bronchial tree, which then move cephalad on the mucociliary blanket, or via macrophages which enter the airway lumen from the bronchial or bronchiolar mucosa (Robertson, 1980). Soluble particles may be absorbed through the mucus layer, into the blood, via intercellular pathways between epithelial cell tight junctions or by active or passive transcellular transport mechanisms.

The bronchial surfaces are not homogeneous; there are openings of daughter bronchi and normal islands of non-ciliated cells at bifurcation regions. In the latter, the usual progress of mucus movement is interrupted, and bifurcations may be sites of relatively retarded clearance. The efficiency with which non-ciliated obstacles are traversed is dependent upon the traction of the mucus layer.

Pulmonary Region   Clearance from the pulmonary region occurs via a number of mechanisms and pathways, but the relative importance of each is not always certain.

Nonabsorptive clearance processes, shown schematically in Fig. 4, are

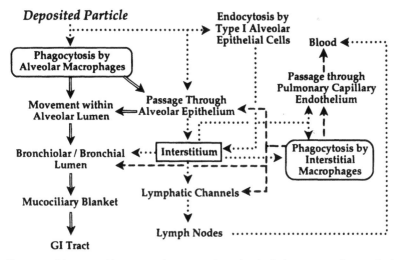

Figure 4 Diagram of known and suspected mechanical clearance pathways for insoluble particles depositing in the pulmonary region. (Dissolution is not included.)

mediated primarily via alveolar macrophages. These cells reside on the alveolar epithelium, and phagocytize and transport deposited material which they contact by random motion or, more likely, via directed migration under the influence of local chemotactic factors. Some deposited particles may be translocated to areas where macrophages congregate, due to pressure gradients or via capillary action within the alveolar surfactant lining (Schurch et al., 1990; Parra et al., 1986).

Alveolar macrophages normally comprise ~3–5% of the total alveolar cells in healthy (nonsmoking) humans and other mammals (Gehr, 1984), and represent the largest subpopulation of nonvascular macrophages in the respiratory tract (Lehnert, 1992). However, the actual cell count is influenced by particle loading. Low numbers of deposited particles may not result in an increase in cell number, but above some level macrophage numbers increase proportionally to particle number until a saturation point is reached (Adamson and Bowden, 1981; Brain, 1971). Since the magnitude of this increase is related more to the number of deposited particles than to total deposition by weight, equivalent masses of an identical deposited substance may not produce the same response if particle sizes differ. Thus, smaller particles tend to result in a greater elevation in cell number than larger ones.

Particle-laden macrophages may be cleared from the pulmonary region along a number of pathways. The primary route is cephalad transport via the mucociliary system after the cells reach the distal terminus of the mucus blanket. However, the manner by which macrophages actually attain this is not certain. The possibilities are:

Chance encounter;
Passive movement along the alveolar surface due to surface tension gradients between the alveoli and conducting airways;
Directed locomotion along a gradient produced by chemotactic factors released by macrophages ingesting deposited material (Sorokin and Brain, 1975; Kilburn, 1968); or
Passage through the alveolar epithelium and the interstitium (Brundelet, 1965; Green, 1973; Corry et al., 1984; Harmsen et al., 1985).

Some cells which follow interstitial clearance pathways are likely resident interstitial macrophages which have ingested free particles transported through the alveolar epithelium, probably via endocytosis by Type I pneumocytes (Brody et al., 1981; Bowden and Adamson, 1948). Such endocytosis is often seen for fibers that cannot be fully ingested by alveolar macrophages. Particle-laden interstitial macrophages can also migrate across the alveolar epithelium, becoming part of the alveolar macrophage cell population.

Macrophages which are not cleared via the bronchial tree may actively migrate within the interstitium to a nearby lymphatic channel or, along with uningested particles, be carried in the flow of interstitial fluid towards and into the lymphatic system (Harmsen et al., 1985). Passive entry into lymphatic vessels is fairly easy since the vessels have loosely connected endothelial cells with wide intercellular junctions (Lauweryns and Baert, 1974). Lymphatic endothelium may also actively engulf particles from the surrounding interstitium (Leak, 1980). Particles within the lymphatic system may be translocated to tracheobronchial lymph nodes, which often become reservoirs of retained material. Particles penetrating the nodes and subsequently reaching the postnodal lymphatic circulation may enter the blood.

Uningested particles or macrophages in the interstitium may traverse the alveolar-capillary endothelium, directly entering the blood (Raabe, 1982; Holt, 1981). Endocytosis by endothelial cells followed by exocytosis into the vessel lumen seems, however, to be restricted to particles <0.1 μm, and may increase with increasing lung burden (Lee et al., 1989; Oberdörster, 1988). Once in the systemic circulation, transmigrated macrophages, as well as free particles, can travel to extrapulmonary organs. Some species have pulmonary intravascular macrophages which can remove particles from circulating blood (Warner and Brain, 1990) and which may play some role in the clearance of material deposited in the alveoli.

Free particles and macrophages within the interstitium may travel to perivenous, peribronchiolar or subpleural sites, where they become trapped and increase the particle burden. The migration and grouping of particles and macrophages within the lungs can lead to the redistribution of initially diffuse deposits into focal aggregates (Heppleston, 1953). Some particles, notably fibers, can be found in the pleural space, often within macrophages which have migrated across the visceral pleura (Sebastien et al., 1977; Hagerstrand and Siefert, 1973). Resident pleural macrophages do occur, but any role in clearance is not known.

During clearance, particles can be redistributed within the alveolar macrophage population (Lehnert, 1992). One mechanism is by death of the macrophage and the release of free particles to the epithelium followed by uptake by other macrophages. Some of these newly freed particles may, however, translocate to other clearance routes.

Clearance by the absorptive mechanism involves dissolution in the alveolar fluid, followed by transport through the epithelium and into the interstitium, and diffusion into the lymph or blood. Some soluble particles translocated to and trapped in interstitial sites may be absorbed there. Although the factors affecting the dissolution of deposited particles are poorly understood, it is influenced by the particle's surface-to-volume ratio and other surface properties (Morrow, 1973; Mercer, 1967). Thus, materials generally considered to be relatively insoluble may have high dissolution rates and short dissolution half times if the particle size is small.

Some deposited particles may undergo dissolution after phagocytic uptake by macrophages. For example, metals may dissolve in the acidic milieu of the phagolysosomes (Lundborg et al., 1985). It is, however, not certain whether the dissolved material then emigrates from the macrophage. Finally, some particles can bind to epithelial cell or other cell components, delaying clearance from the lungs.

Fibrous particles deposited in the pulmonary region may be additionally subject to a process of disintegration, which involves the subdivision of a large particle into smaller segments. This can occur by leaching within the fibrous structure which then fractures, or by surface etching, resulting in a change in the external dimensions of the fiber. Some fiber types break up by length; others will disintegrate into smaller diameter particles (Lippmann, 1992).

Clearance Kinetics

Although deposited particles may be cleared completely from the respiratory tract, the actual time frame over which this occurs affects dose delivered to the respiratory tract, and to extrapulmonary organs. Particle-tissue contact and subsequent dose in the upper respiratory tract and tracheobronchial tree are often limited by the rapid clearance from these regions and are, thus, proportional to toxicant concentration and exposure duration. On the other hand, the dose from material deposited in the pulmonary region is highly dependent upon the characteristics of both the particle matrix and any substances associated with it.

Various experimental techniques have been used to assess clearance rates in both humans and experimental animals (Schlesinger, 1985a). Because of technical differences and the fact that measured rates are strongly influenced by the specific methodology, comparisons between studies are often difficult to perform. However, regional clearance rates, i.e., the fraction of the deposit which is cleared per unit time, are well defined functional characteristics of an individual human or experimental animal when repeated tests are performed under the same conditions. But, as with deposition, there is a substantial degree of inter-individual variability.

Upper Respiratory Tract   Mucus flow rates in the posterior nasal passages are highly nonuniform. Regional velocities in the healthy adult human may range from < 2 to > 20 mm/min (Proctor, 1980), with the fastest flow occurring in the midportion of the nasal passages. The median rate is about 5 mm/min. The overall result is a mean transport time for insoluble particles over the entire region of ~ 10–20 minutes (Stanley et al., 1985; Rutland and Cole, 1981).

Particles which deposit in the nonciliated anterior portion of the nasal passages may be cleared slowly (1–2 mm/hr) by mucus moved by traction due to more distal cilia (Hilding, 1963). Since this may take upwards of 12 hours, such deposits are usually more effectively removed by sneezing, wiping, or nose blowing, in which case clearance may occur in under 30 minutes (Morrow, 1977; Fry and Black, 1973).

Tracheobronchial Tree   Mucus transport in the tracheobronchial tree occurs at different rates in different local regions. The velocity of mucus movement is fastest in the trachea, and it becomes progressively slower in more distal airways. Measured rates in the human trachea range from 4–20 mm/minute, depending upon the experimental technique used. Anesthesia and/or invasive procedures affect transport, resulting in observed rates which are apparently slower than normal. In unanesthetized, healthy nonsmokers using noninvasive procedures, average tracheal mucus transport rates have been measured at 4.3–5.7 mm/minute (Leikauf et al., 1981, 1984; Yeates et al., 1975, 1981b; Foster et al., 1980). Furthermore, the rate of insoluble particle transport seems to be independent of the nature—shape, size, and composition—of the material being cleared (Man et al., 1980).

The mean mucus velocity in the human main bronchi has been experimentally found to be ~ 2.4 mm/minute (Foster et al., 1980). While rates of movement in smaller airways cannot be directly measured, those in human medium bronchi have been estimated at between 0.2–1.3 mm/minute and those in the most distal ciliated airways as low as 0.001 mm/minute (Yeates and Aspin, 1978; Morrow et al., 1967b).

The total duration of bronchial clearance, or some other time parameter, is often used as an index of mucociliary function. In healthy adult nonsmoking humans, 90% of insoluble particles depositing on the tracheobronchial tree will be cleared from 2.5 to 20 per hour after deposition, depending upon the individual subject and the size of the particles (Albert et al., 1973). The latter does not affect surface transport, but does affect the depth of particle penetration and deposition and the subsequent pathway length for clearance. Due to differences in regional transport rates, clearance times from different regions of the bronchial tree will differ. In most cases, however, removal of a tracheobronchial deposit will generally be 99% completed 48 hours after exposure (Bailey et al., 1985a).

Studies with both rodents and humans have indicated that a small fraction (~ 1%) of insoluble material may be retained for a prolonged period of time within the upper respiratory tract (nasal passages) or tracheobronchial tree (Patrick and Stirling, 1977; Gore and Patrick, 1982; Watson and Brain, 1979; Radford and Martell, 1977). The mechanism(s) underlying this long-term retention is unknown, but may involve endocytosis by epithelial cells with subsequent translocation into deeper tissue, or merely passive movement into this tissue. The retained particles may eventually be cleared to regional lymph nodes, but with a long half time that may be > 80 days (Patrick, 1989; Oghiso and Matsuoka, 1979).

Long-term tracheobronchial retention patterns are not uniform. There appears to be an enhancement at bifurcation regions (Cohen et al., 1988; Radford and Martell, 1977; Henshaw and Fews, 1984), perhaps the result of both greater deposition and ineffectual mucus clearance within these areas. Thus, doses calculated based upon uniform surface retention density may be misleading, especially if the material is, toxicologically, slow acting. Soluble material may also undergo long-term retention in ciliated airways due to binding to cells or macromolecules.

Pulmonary Region   Clearance kinetics in the pulmonary region are not definitively understood, although particles deposited there generally remain longer than those deposited in airways cleared by mucociliary transport. There are limited data on rates in humans, while within any species rates vary widely due to different properties of the particles used in the various studies. Furthermore, some of these studies employed high concentrations of insoluble particles, which may itself have interfered with normal clearance mechanisms, producing rates different from those which would occur at lower exposure levels. Prolonged exposure to high particle concentrations is associated with what is termed particle "overload." This is a nonspecific effect noted in experimental studies using many different kinds of insoluble particles, including fibers, and results in clearance slowing or stasis, with an associated inflammation and aggregation of macrophages in the lungs. (Muhle et al., 1990; Lehnert, 1990). While it is, however, likely to be of little relevance for most "real world" exposures to humans, it is of concern in interpreting some long-term experimental exposure data.

There are numerous pathways of pulmonary region clearance which may depend upon the nature of the particles being cleared. Thus, kinetic generalizations are hard to make, especially since the manner in which particle characteristics affect kinetics is not resolved. Nevertheless, pulmonary region clearance can be described as a multiphasic process. Each component is considered to represent removal by a different mechanism or pathway, characterized by increasing half times of clearance with time postexposure. For example, an initial fast phase, which has a half time of ~ 2–6 weeks, presumably represents rapid clearance via macrophages, while a phase of prolonged clearance, with a half time of months to years, represents removal by dissolution. This latter is extremely variable, but it likely dominates the long-term clearance of relatively insoluble particles (Kreyling et al., 1988). An intermediate phase with a half time on the order of months may represent a slower phase of macrophage clearance via interstitial pathways.

Clearance of inert, insoluble particles in healthy, nonsmoking humans has been generally observed to consist of two phases, the first having a half-time measured in days, and the second in hundreds of days. Table 3 presents half times for the longer second phase of clearance as reported in a number of studies. Although wide variations in clearance times reflect a dependence upon the nature of the deposited material, e.g., particle size, once dissolution is accounted for, mechanical removal to the gastrointestinal tract and/or lymphatic system appears to be independent of size, especially for particles < 5 μm (Snipes et al., 1983). Although not evident from Table 3, there is considerable intersubject variability in the clearance rates of identical particles, which appears to increase with time postexposure (Philipson et al., 1985; Bailey et al., 1985a). The large differences in clearance kinetics among different individuals suggest that equivalent chronic exposures to insoluble particles may result in large variations in respiratory tract burdens.

Although the kinetics of overall clearance from the pulmonary region have been assessed to some extent, much less is known concerning relative

Table 3 Long-term Particle Clearance from the Pulmonary Region in Nonsmoking Humans

Particle		Clearance half-time[a]	
Material	Size(μm)	(days)	Reference
Polystyrene latex	5.0	150–300	Booker et al., 1967
Polystyrene latex	5.0	144–340	Newton et al., 1978
Polystyrene latex	0.5	33–602	Jammett et al., 1978
Polystyrene latex	3.6	296	Bohning et al., 1982
Teflon	4	200–2500	Philipson et al., 1985
Aluminosilicate	1.2	330	Bailey et al., 1982
Aluminosilicate	3.9	420	Bailey et al., 1982
Iron oxide (Fe_2O_3)	0.8	62	Morrow et al., 1967a,b
Iron oxide (Fe_2O_3)	0.1	270	Waite & Ramsden, 1971
Iron oxide (Fe_3O_4)	2.8	70	Cohen et al., 1979

[a]Represents the half-time of clearance for the slowest phase observed

rates along specific pathways. The usual initial step in clearance, i.e., uptake of deposited particles by alveolar macrophages, is very rapid. Unless the particles are cytotoxic or very large, ingestion by macrophages occurs within 24 hours of a single inhalation (Naumann and Schlesinger, 1986; Lehnert and Morrow, 1985). But the actual rate of subsequent macrophage clearance is not certain. Perhaps 5% or less of their total number is translocated from the lungs each day (Lehnert and Morrow, 1985; Masse et al., 1974). The actual time for the clearance of particle-laden alveolar macrophages via their main route (the mucociliary system) depends upon the site of uptake relative to the distal terminus of the mucus blanket at the bronchiolar level. Furthermore, clearance pathways and subsequent kinetics may depend to some extent upon particle size. For example, ultrafine particles < 0.02 μm are less effectively phagocytosed than are larger ones (Oberdörster, 1993). But once ingestion occurs, alveolar macrophage-mediated kinetics are independent of the particle involved as long as solubility and cytotoxicity are low.

Free particles may penetrate into the interstitium (largely by Type I cell endocytosis) within a few hours following deposition (Ferin and Feldstein, 1978; Sorokin and Brain, 1975; Brody et al., 1981). This transepithelial passage seems to increase as particle loading increases, especially to a level above the saturation point for increasing macrophage number (Adamson and Bowden, 1981; Ferin, 1977). It may also be particle-size dependent since insoluble ultrafine particles < 0.05-μm diameter show increased access to and greater lymphatic uptake than larger ones (Oberdörster et al., 1992). Similarly, a depression of phagocytosis by toxic particles or the deposition of large numbers of smaller ultrafine particles may increase the number of free particles in the alveoli, enhancing removal by other routes. In any case, free particles and alveolar macrophages may reach the lymph nodes within a few days after deposition (Lehnert et al., 1988; Harmsen et al., 1985). However, the bulk of translocation to the lymphatic system is very slow, on the order of 0.02–0.003%/day (Snipes, 1989), and elimination from the lymph nodes is even slower, with half times measured in tens of years (Roy, 1989).

Soluble particles depositing in the pulmonary region are rapidly cleared via absorption through the epithelial surface into the blood, but there are few data on dissolution and transfer rates to blood in humans. Actual rates depend upon the size of the particle, i.e., molecular size, with smaller ones clearing faster than larger ones. Some solubilized material may be retained in lung tissue due to binding with cellular components, preventing it from passing into the circulation.

Factors Modifying Clearance

A number of host and environmental factors may modify normal clearance patterns, affecting the dose delivered by exposure to inhaled particles. As

Table 4   Some Factors That May Affect Particle Clearance

Factor	Upper respiratory tract and/ or tracheobronchial tree (mucociliary transport)	Pulmonary
Gender	Probably no effect	(?)[a]
Aging	Possible retardation	(?)
Exercise	Possible acceleration with heavy exercise	Possible acceleration
Irritant exposure	Acceleration or retardation depending on dose	Acceleration or retardation depending on dose
Lung disease		
Chronic bronchitis	Retardation	Retardation
Asthma	Retardation	(?)
Influenza	Retardation	Retardation

[a](?) = Effect has not been evaluated

outlined in Table 4, these include aging, gender, workload, disease and irritant inhalation. However, in many cases, the exact role of these factors is not resolved.

The evidence for aging-related effects on mucociliary function in healthy individuals is equivocal, with studies showing either no change or a slowing in clearance function with age after maturity (Goodman et al., 1978; Yeates et al., 1981a). However, it is difficult to determine whether any observed functional decrement was due to aging alone, or to long-term, low-level ambient pollutant exposure (Wanner, 1977).

There are no data to allow assessment of age-related changes in clearance from the pulmonary region. Although functional differences have been found between alveolar macrophages of mature and senescent mice (Esposito and Pennington, 1983), no age-related decline in macrophage function has been seen in humans (Gardner et al., 1981).

There is also insufficient data to assess changes in clearance in the growing lung. Nasal mucociliary clearance time in a group of children (average age 7 years) was found to be ~ 10 min (Passali and Ciampoli, 1985). This is within the range for adults. There is one report of bronchial clearance in children (12 years old), but this was performed in patients hospitalized for renal disease (Huhnerbein et al., 1984).

In terms of gender, no difference in nasal mucociliary clearance rate was observed between male and female children (Passali and Ciampoli, 1985), nor in tracheal transport rates in adults (Yeates et al., 1975). Slower bronchial clearance has been noted when male adults are compared to female adults, but this was attributed to differences in lung size (and resultant clearance pathway length) rather than to inherent gender-related differences in transport velocities (Gerrard et al., 1986).

The effect of increased physical activity upon mucociliary clearance is also unresolved; the available data indicate no change to a speeding with exercise (Wolff et al., 1977; Pavia, 1984). There are no data concerning changes in pulmonary region clearance with increased activity levels, but CO_2-stimulated hyperpnea (rapid, deep breathing) was found to have no effect on early pulmonary clearance and redistribution of particles (Valberg et al., 1985). Increased tidal volume breathing was noted to increase the rate of particle clearance from the pulmonary region. This was suggested to be due to distension-related evacuation of surfactant into proximal airways resulting in a facilitated movement of particle-laden macrophages or free particles because of the accelerated motion of the alveolar fluid film (John et al., 1994).

Various respiratory tract diseases are associated with clearance alterations. Nasal mucociliary clearance is prolonged in humans with chronic sinusitis, bronchiectasis, or rhinitis (Majima et al., 1983; Stanley et al., 1985) and in cystic fibrosis (Rutland and Cole, 1981). Bronchial mucus transport may be impaired in people with bronchial carcinoma (Matthys et al., 1983), chronic bronchitis (Vastag et al., 1986), asthma (Pavia et al., 1985), and in association with various acute infections (Lourenco et al., 1971; Camner et al., 1979; Puchelle et al., 1980). In certain of these cases, coughing may enhance mucus clearance, but it generally is only effective if excess secretions are present.

Rates of pulmonary-region particle clearance appear to be reduced in humans with chronic obstructive lung disease (Bohning et al., 1982) and in experimental animals with viral infections (Creasia et al., 1973). The viability and functional activity of macrophages was found to be impaired in human asthmatics (Godard et al., 1982). Studies with experimental animals have also found disease-related clearance changes. Hamsters with interstitial fibrosis showed an increased degree of pulmonary clearance (Tryka et al., 1985). Rats with emphysema showed no clearance difference from control (Damon et al., 1983), although the copresence of inflammation resulted in prolonged retention (Hahn and Hobbs, 1979). Inflammation may enhance particle and macrophage penetration through the alveolar epithelium into the interstitium by increasing the permeability of the epithelium and the lymphatic endothelium (Corry et al., 1984). Neutrophils, which are phagocytic cells present in alveoli during inflammation, may contribute to the clearance of particles via the mucociliary system (Bice et al., 1990).

Inhaled irritants, such as cigarette smoke, have been shown to have an effect upon mucociliary clearance function in both humans and experimental animals (Wolff, 1986). Single exposures to a particular material may increase or decrease the overall rate of tracheobronchial clearance, oftentimes depending upon the exposure concentration (Schlesinger, 1986). Alterations in clearance rate following single exposures to moderate concentrations of irritants are generally transient, lasting < 24 hours. However, repeated exposures may result in an increase in intra-individual variability of clearance rate and persistently retarded clearance. The effects of irritant exposure may be enhanced by exercise or by coexposure to other materials.

Acute and chronic exposures to inhaled irritants may also alter pulmonary region clearance (Cohen et al., 1979; Ferin and Leach, 1977; Schlesinger et al., 1986), which may be accelerated or depressed, depending upon the specific material and/or length of exposure. Alterations in alveolar macrophages likely underlie some of the observed changes since numerous irritants have been shown to impair the numbers and functional properties of these cells (Gardner, 1984).

Comparative Aspects of Clearance

As with deposition analyses, the inability to study the retention of certain materials in humans for direct risk assessment requires use of experimental animals. Since dosimetry depends upon clearance rates and routes, adequate toxicologic assessment necessitates that kinetics in these animals be related to those occurring in humans. The basic mechanisms of clearance from the respiratory tract appear to be similar in humans and most other mammals. However, regional clearance rates show substantial variation between species, even for similar particles deposited under comparable exposure conditions (Snipes, 1989). Dissolution rates and rates of transfer of dissolved substances into the blood may or may not be species independent, depending upon certain chemical properties of the deposited material (Griffith et al., 1983; Bailey et al., 1985b; Roy, 1989). For example, lipophilic compounds of comparable molecular weight are cleared from the lungs of various species at the same rate (dependent solely upon solute molecular weight and the lipid/water partition coefficient), but hydrophilic compounds show species differences. On the other hand, there are interspecies differences in rates of mechanical transport, e.g., macrophage-mediated clearance of insoluble particles from the pulmonary region (Bailey et al., 1985b); transport of particles from the pulmonary region to pulmonary lymph nodes (Snipes et al., 1983; Mueller et al., 1990); and mucociliary transport in conducting airways (Felicetti et al., 1981). This is likely to result in species-dependent rate constants for these clearance pathways. Thus, differences in regional (and perhaps total) clearance rates between some species are most likely due to these mechanical processes.

CONCLUSION

The toxic response from inhaled particles is dependent upon both the amount and pattern of deposition and the time frame of persistence in various sites. The deposition of particles on airway surfaces is the result of specific physical mechanisms that are influenced by particle characteristics, airflow patterns, and respiratory tract anatomy. Because regional deposition patterns determine the specific pathways and rates by which particles are ultimately cleared and redistributed, biological effects are often related more to the quantitative pattern of deposition at specific sites than they are to the total amount depositing in the respiratory tract. Clearance routes and kinetics are a function of the respiratory tract region and, in some cases, lung burden and the physico-chemical properties of the deposited material. The accurate interpretation of results from inhalation toxicological studies employing particles requires an appreciation of those factors which control and affect both their deposition and clearance.

REFERENCES

Adamson, IYR, Bowden, DH: Dose response of the pulmonary macrophagic system to various particulates and its relationship to transepithelial passage of free particles. Exp Lung Res 2:165–175, 1981.

Albert, RE, Lippmann, M, Peterson Jr, HT, Sanborn, K, Bohning, DE: Bronchial deposition and clearance of aerosols. Arch Intern Med 131:115–127, 1973.

Anderson, PJ, Wilson, JD, Hiller, FC: Respiratory tract deposition of ultrafine particles in subjects with obstructive or restrictive lung disease. Chest 97:1115–1120, 1990.

Asgharian, B, Yu, CP: Deposition of fibers in the rat lung. J Aerosol Sci 20:355–366, 1989.

Asgharian, B, Yu, CP: Deposition of inhaled fibrous particles in the human lung. J Aerosol Med 1:37–50, 1988.

Bailey, MR, Fry, FA, James, AC: Long-term retention of particles in the human respiratory tract. J Aerosol Sci 16:295–305, 1985a.

Bailey, MR, Fry, FA, James, AC: The long-term clearance kinetics of insoluble particles from the human lung. Ann Occup Hyg 26:273–290, 1982.

Bailey, MR, Hodgson, A, Smith, H: Respiratory tract retention of relatively insoluble particles in rodents. J Aerosol Sci 16:279–293, 1985b.

Becquemin, MH, Yu, CP, Roy, M, Bouchikhi, A: Total deposition of inhaled particles related to age: Comparison with age-dependent model calculations. Radiat Prot Dosim 38:23–28, 1991.

Bennett, WD, Messina, MS, Smaldone, GC: Effect of exercise on deposition and subsequent retention of inhaled particles. J Appl Physiol 59:1046–1054, 1985.

Bice, DE, Harmson, AG, Muggenburg, BA: Role of lung phagocytes in the clearance of particles by the mucociliary apparatus. Inhal Toxicol 2:151–160, 1990.

Bohning, DE, Atkins, HL, Cohn, SH: Long-term particle clearance in man: Normal and impaired. Ann Occup Hyg 26:259–271, 1982.

Booker, DV, Chamberlain, AC, Rundo, J, Muir, DCF, Thomson, ML: Elimination of 5 μm particles from the human lung. Nature 215:30–33, 1967.

Bowden, DH, Adamson, IYR: Pathways of cellular efflux and particulate clearance after carbon instillation to the lung. J Pathol 143:117–125, 1984.

Brain, JD, Mensah, GA: Comparative toxicology of the respiratory tract. Am Rev Respir Dis 128(Suppl):S87–S90, 1983.

Brain, JD, Valberg, PA: Deposition of aerosol in the respiratory tract. Am Rev Respir Dis 120:1325–1373, 1979.

Brain, JD: The effects of increased particles on the number of alveolar macrophages. In: Inhaled Particles III, V 1, edited by WH Walton, pp. 209–223. Old Woking, England: Unwin Bros, 1971.

Brody, AR, Hill, LH, Adkins Jr, B, O'Connor, RW: Chrysotile asbestos inhalation in rats: Deposition pattern and reaction of alveolar epithelium and pulmonary macrophages. Am Rev Respir Dis 123:670–679, 1981.

Brody, AR, Roe, MW: Deposition pattern of inorganic particles at the alveolar level in the lungs of rats and mice. Am Rev Respir Dis 128:724–729, 1983.

Brody, AR, Yu, CP: Particle deposition at the alveolar duct bifurcations. In: Extrapolation of Dosimetric Relationships for Inhaled Particles and Gases, edited by JD Crapo, FJ Miller, ED Smolko, JA Graham, A Wallace Hayes, pp. 91–99. San Diego: Acadmic Press, 1989.

Brundelet, PJ: Experimental study of the dust clearance mechanisms of the lung. Acta Pathol Microbiol Scand 175:1–141, 1965.

Camner, P, Mossberg, B, Philipson, K, Strandberg, K: Elimination of test particles from the human tracheobronchial tract by voluntary coughing. Scand J Respir Dis 60:56–62, 1979.

Cheng, Y-S, Su, Y-F, Yeh, H-C: Deposition of thoron progeny in human head airways. Aerosol Sci Technol 18:359–375, 1993.

Cheng, Y-S, Yeh, H-C, Swift, DL: Aerosol deposition in human nasal airway for particles 1 nm to 20 μm: A model study. Radiat Prot Dosim 38:41–47, 1991.

Cheng, YS, Yamada, Y, Yeh, HC, Swift, DL: Diffusional deposition of ultrafine aerosols in a human nasal cast. J Aerosol Sci 19:741–751, 1988.

Cohen, BS, Harley, NH, Schlesinger, RB, Lippmann, M: Nonuniform particle deposition on tracheobronchial airways: Implication for lung dosimetry. Ann Occup Hyg 32(Suppl 1):1045–1052, 1988.

Cohen, D, Arai, SF, Brain, JD: Smoking impairs long-term dust clearance from the lung. Science 204:514–517, 1979.

Corry, D, Kulkarni, P, Lipscomb, MF: The migration of bronchoalveolar macrophages into hilar lymph nodes. Am J Pathol 115:321–328, 1984.

Crawford, DJ: Identifying critical human subpopulations by age groups: Radioactivity and the lung. Phys Med Biol 27:539–552, 1982.

Creasia, DA, Nettesheim, P, Hammons, AS: Impairment of deep lung clearance by influenza virus infection. Arch Environ Health 26:197–201, 1973.

Damon, EG, Mokler, BV, Jones, RK: Influence of elastase-induced emphysema and the inhalation of an irritant aerosol on deposition and retention of an inhaled insoluble aerosol in Fischer-344 rats. Toxicol Appl Pharmacol 67:322–330, 1983.

Diu, CK, Yu, CP: Respiratory tract deposition of polydisperse aerosols in humans. Am Ind Hyg Assoc J 44:62–65, 1983.

Esposito, AL, Pennington, JE: Effects of aging on antibacterial mechanisms in experimental pneumonia. Am Rev Respir Dis 128:662–667, 1983.

Fang, CP, Wilson, JE, Spektor, DM, Lippmann, M: Effect of lung airway branching pattern and gas composition on particle deposition in bronchial airways. III. Experimental studies with radioactively tagged aerosol in human and canine lungs. Exp Lung Res 19:377–396, 1993.

Felicetti, SA, Wolff, RK, Muggenburg, BA: Comparison of tracheal mucous transport in rats, guinea pigs, rabbits, and dogs. J Appl Physiol 51:1612–1617, 1981.

Ferin, J, Feldstein, ML: Pulmonary clearance and hilar lymph node content in rats after particle exposure. Environ Res 16:342–352, 1978.

Ferin, J, Leach, LJ: The effects of selected air pollutants on clearance of titanic oxide particles from the lungs of rats. In: Inhaled Particles IV, Pt 1, edited by WH Walton, pp. 333–340. Oxford: Pergamon Press, 1977.

Ferin, J: Effects of particle content of lung on clearance pathways. In: Pulmonary Macrophage and Epithelial Cells, edited by CL Sanders, RP Schneider, GE Dagle, HA Ragan, pp. 414–423. Springfield, VA: NTIS, 1977.

Ferron, GA, Kreyling, WG, Haider, B: Influence of the growth of salt aerosol particles on the deposition in the lung. Ann Occup Hyg 32 (Suppl 1):947–955, 1988.

Foster, WM, Langenbach, E, Bergofsky, EH: Measurement of tracheal and bronchial mucus velocities in man: Relation to lung clearance. J Appl Physiol 48:965–971, 1980.

Fry, FA, Black, A: Regional particle deposition and clearance of particles in the human nose. J Aerosol Sci 4:113–124, 1973.

Gardner, DE: Alterations in macrophage functions by environmental chemicals. Environ Health Perspect 55:343–358, 1984.

Gardner, ND, Lim, STK, Lawton, JWM: Monocyte function in ageing humans. Mech Ageing Dev 16:233–239, 1981.

Gehr, P: Lung morphometry: In: Lung Modelling for Inhalation of Radioactive Materials, edited by H Smith, G Gerber, pp. 1–11. Luxembrough: Commission of the European Communities, 1984.

Gerde, P, Cheng, YS, Medinsky, MA: *In vivo* deposition of ultrafine aerosols in the nasal airway of the rat. Fundam Appl Toxicol 16:330–336, 1991.

Gerrard, CS, Gerrity, TR, Yeates, DB: The relationships of aerosol deposition, lung size, and the rate of mucociliary clearance. Arch Environ Health 41:11–15, 1986.

Godard, P, Chaintreuil, J, Damon, M, Coupe, M, Flandre, O, de Paulet, AC, Michel, FB: Functional assessment of alveolar macrophages: Comparison of cells from asthmatic and normal subjects. J Allergy Clin Immunol 70:88–93, 1982.

Goodman, RM, Yergin, BM, Landa, JF, Golinvaux, MH, Sackner, MA: Relationship of smoking history and pulmonary function tests to tracheal mucus velocity in nonsmokers, young smokers, ex-smokers and patients with chronic bronchitis. Am Rev Respir Dis 117:205–214, 1978.

Gooya, A, Patra, A: Deposition of particles in a baboon nose cast. Proceedings of the 34th Annual Conference on Engineering in Medicine and Biology 28:253, 1986.

Gore, DJ, Patrick, G: A quantitative study of the penetration of insoluble particles into the tissue of the conducting airways. Ann Occup Hyg 26:149–161, 1982.

Gradon, L, Orlicki, D: Deposition of inhaled aerosol particles in a generation of the tracheobronchial tree. J Aerosol Sci 21:3–19, 1990.

Green, GM: Alveolobronchiolar transport mechanisms. Arch Intern Med 131:109–114, 1973.

Griffith, WC, Cuddihy, RC, Boecker, BB, Guilmette, RA, Medinsky, MA, Mewhinney, JA: Comparison of solubility of aerosols in lungs of laboratory animals. Health Phys 45:233 (Abstr), 1983.

Hagerstrand, I, Siefert, B: Asbestos bodies and pleural plaques in human lungs at autopsy. Acta Pathol Microbiol Scand A81:457–460, 1973.

Hahn, FF, Hobbs, CH: The effect of enzyme-induced pulmonary emphysema in Syrian hamsters on the deposition and long-term retention of inhaled particles. Arch Environ Health 34:203–211, 1979.

Hammad, Y, Diem, J, Craighead, J, Weill, H: Deposition of inhaled man-made mineral fibres in the lungs of rats. Ann Occup Hyg 26:179–187, 1982.

Harbison, ML, Brain, JD: Effects of exercise on particle deposition in Syrian golden hamsters. Am Rev Respir Dis 128:904–908, 1983.

Harmsen, AG, Muggenburg, BA, Snipes, MB, Bice, DE: The role of macrophages in particle translocation from lungs to lymph nodes. Science 230:1277–1280, 1985.

Henshaw, DL, Fews, AP: The microdistribution of alpha emitting particles in the human lung. In: Lung Modelling for Inhalation of Radioactive Materials, edited by H Smith, G Gerber, pp. 199–208. Luxembourgh: Commission of the European Communities, 1984.

Heppelston, AG: Deposition and disposal of inhaled dust. Arch Environ Health 7:548–555, 1963.

Heppleston, AG: Pathological anatomy of simple pneumoconiosis in coal workers. J Pathol Bacteriol 66:235–246, 1953.

Hesseltine, GR, Wolff, RK, Mauderly, JL, Cheng, Y-S: Deposition of ultrafine aggregate particles in exercising rats. J Appl Toxicol 6:21–24, 1986.

Heyder, J, Gebhart, J, Rudolf, G, Schiller, CF, Stahlhofen, W: Deposition of particles in the human respiratory tract in the size range 0.005–15 μm. J Aerosol Sci 17:811–825, 1986.

Heyder, J, Gebhart, J, Stahlhofen, W, Stuck, B: Biological variability of particle deposition in the human respiratory tract during controlled and spontaneous mouth breathing. Ann Occup Hyg 26:137–147, 1982.

Hofmann, W: Mathematical model for the postnatal growth of the human lung. Respir Physiol 49:115–129, 1982.

Holt, PF: Transport of inhaled dust to extrapulmonary sites. J Pathol 133:123–129, 1981.

Huhnerbein, J, Otto, J, Thal, W: Untersuchungsergebnisse der mukoziliaren clearance bei lungengesunden kindern. Padiat Grenzgeb 23:437–443, 1984.

Jammet, H, Drutel, P, Parrot, R, Roy, M: Étude de l'epuration pulmonaire chez l'homme apres administration d'aerosols de particules radioactives. Radioprotection 13:143–166, 1978.

John, J, Wollmer, P, Dahlbäck, M, Luts, A, Jonson, B: Tidal volume and alveolar clearance of insoluble particles. J Appl Physiol 76:584–588, 1994.

Kilburn, KH: A hypothesis for pulmonary clearance and its implications. Am Rev Respir Dis 98:449–463, 1968.

Kreyling, W, Ferron, GA, Schumann, G: Particle transport from the lower respiratory tract. J Aerosol Med 1:351–370, 1988.

Lauweryns, JM, Baert, JH: The role of the pulmonary lymphatics in the defenses of the distal lung: Morphological and experimental studies of the transport mechanisms of intratracheally instilled particles. Ann NY Acad Sci 221:244–275, 1974.

Leak, LV: Lymphatic removal of fluids and particles in the mammalian lung. Environ Health Perspect 35:55–76, 1980.

Lee, KP, Trochimowicz, HJ, Reinhardt, CF: Transmigration of titanium dioxide (TiO_2) in rats after inhalation exposure. Exp Mol Pathol 42:331–343, 1989.

Lehnert, BE, Morrow, PE: Association of [59]iron oxide with alveolar macrophages during alveolar clearance. Exp Lung Res 9:1–16, 1985.

Lehnert, BE, Valdez, YE, Bomalaski, HS: Analyses of particles in the lung free cell, tracheobronchial lymph nodal, and pleural space compartments following their deposition in the lung is related to lung clearance mechanisms. Ann Occup Hyg 32(Suppl 1):125–140, 1988.

Lehnert, BE: Alveolar macrophages in a particle "overload" condition. J Aerosol Med 3(Suppl 1):S9–S30, 1990.

Lehnert, BE: Pulmonary and thoracic macrophage subpopulations and the clearance of particles from the lung. Environ Health Perspect 97:17–46, 1992.

Leikauf, G, Yeates, DB, Wales, KA, Albert, RE, Lippmann, M: Effects of sulfuric acid aerosol on respiratory mechanics and mucociliary particle clearance in healthy non-smoking adults. Am Ind Hyg Assoc J 42:273–282, 1981.

Leikauf, GD, Spektor, DM, Albert, RE, Lippmann, M: Dose-dependent effects of submicrometer sulfuric acid aerosol on particle clearance from ciliated human lung airways. Am Ind Hyg Assoc J 45:285–292, 1984.

Lippmann, M: Asbestos and other mineral fibers. In: Environmental Toxicants: Human Exposures and Their Health Effects, edited by M Lippmann, pp. 30–75. New York: Van Nostrand Reinhold, 1992.

Lourenco, RV, Loddenkemper, R, Cargan, RW: Patterns of distribution and clearance of aerosols in patients with bronchiectasis. Am Rev Respir Dis 106:857–866, 1972.

Lourenco, RV, Stanley, ED, Gatmaitan, B, Jackson, GG: Abnormal deposition and clearance of inhaled particles during upper respiratory and viral infections. J Clin Invest 50:62a, 1971.

Lundborg, M, Eklund, A, Lind, B, Camner, P: Dissolution of metals by human and rabbit alveolar macrophages. Br J Ind Med 42:642–645, 1985.

Majima, Y, Sakakura, Y, Matsubara, T, Murai, S, Miyoshi, Y: Mucociliary clearance in chronic sinusitis: Related human nasal clearance and in vitro bullfrog palate clearance. Biorheology 20:251–262, 1983.

Man, SFP, Lee, TK, Gibney, RTN, Logus, JW, Noujaim, AA: Canine tracheal mucus transport of particulate pollutants: Comparison of radiolabeled corn pollen, ragweed pollen, asbestos, silica, and talc to Dowex[R] anion exchange particles. Arch Environ Health 35:283–286, 1980.

Masse, R, Ducousso, R, Nolibe, D, Lafuma, J, Chretien, J: Passage transbronchique des particules metalliques. Rev Fr Mal Respir 1:123–129, 1974.

Matthys, H, Vastag, E, Kohler, D, Daikeler, G, Fischer, J: Mucociliary clearance in patients with chronic bronchitis and bronchial carcinoma. Respiration 44:329–337, 1983.

McMahon, TA, Brain, JD, LeMott, SR: Species difference in aerosol deposition. In: Inhaled Particles IV, Pt 1, edited by WH Walton, pp. 23–33. Oxford: Pergamon Press, 1977.

Mercer, TT: On the role of particle size in the dissolution of lung burdens. Health Phys 13:1211–1221, 1967.

Morgan, A, Evans, JC, Holmes, A: Deposition and clearance of inhaled fibrous minerals in the rat. Studies using radioactive tracer techniques. In: Inhaled Particles IV, Pt 1, edited by WH Walton, pp. 259–272. Oxford: Pergamon Press, 1977.

Morgan, WKC, Ahmad, D, Chamberlain, MJ, Clague, HW, Pearson, MG, Vinitski, S: The effect of exercise on the deposition of an inhaled aerosol. Respir Physiol 56:327–338, 1984.

Morrow, PE, Gibb, FR, Gazioglu, K: The clearance of dust from the lower respiratory tract of man: An experimental study. In: Inhaled Particles and Vapours II, edited by CN Davies, pp. 351–358. London: Pergamon Press, 1967a.

Morrow, PE, Gibb, FR, Gazioglu, K: A study of particulate clearance from the human lung. Am Rev Respir Dis 96:1209–1221, 1967b.

Morrow, PE, Gibb, FR, Johnson, L: Clearnace of insoluble dust from the lower respiratory tract. Health Phys 10:543–555, 1964.

Morrow, PE: Alveolar clearance of aerosols. Arch Intern Med 131:101–108, 1973.

Morrow, PE: Clearance kinetics of inhaled particles. In: Respiratory Defense Mechanisms, Pt II, edited by JD Brain, DF Proctor, LM Reid, pp. 491–543. NY: Marcel Dekker, 1977.

Mueller, H-L, Robinson, B, Muggenburg, BA, Gillett, NA, Guilmette, RA: Particle distribution in lung and lymph node tissues of rats and dogs and the migration of particle-containing alveolar cells in vitro. J Toxicol Environ Health 30:141–165, 1990.

Muhle, H, Creutzenberg, O, Bellmann, B, Heinrich, U, Mermelstein, R: Dust overloading of lungs: Investigations of various materials, species differences, and irreversibility of effects. J Aerosol Med 3 (Suppl. 1):S111–S128, 1990.

National Research Council: Comparative Dosimetry of Radon in Mines and Homes. Washington, DC: National Academy Press, 1991.

Naumann, BD, Schlesinger, RB: Assessment of early alveolar particle clearance and macrophage function following an actue inhalation of sulfuric acid mist. Exp Lung Res 11:13–33, 1986.

Newton, D, Fry, FA, Taylor, BT, Eagle, MC, Shorma, RC: Interlaboratory comparison of techniques for measuring lung burdens of low energy protein emitters. Health Phys 35:751–771, 1978.

Oberdörster, G, Ferin, J, Gelein, R, Soderholdm, SC, Finkelstein, J: Role of the alveolar macrophage in lung injury: Studies with ultrafine particles. Environ Health Perspect 97:193–199, 1992.

Oberdörster, G: Lung clearance of inhaled insoluble and soluble particles. J Aerosol Med 1:289–330, 1988.

Oberdörster, G: Lung dosimetry: Pulmonary clearance of inhaled particles. Aerosol Sci Technol 18:279–289, 1993.

Oghiso, Y, Matsuoka, O: Distribution of colloidal carbon in lymph nodes of mice injected by different routes, Jpn J Exp Med 49:223–234, 1979.

Parra, SC, Burnette, R, Price, HP, Takaro, T: Zonal distribution of alveolar macrophages, type II pneumocytes, and alveolar septal connective tissue gaps in adult human lungs. Am Rev Respir Dis 113:908–912, 1986.

Passali, D, Ciampoli, MB: Normal values of mucociliary transport time in young subjects. Int J Pediat Otorhinolaryngol 9:151–156, 1985.

Patrick, G, Stirling, C: The retention of particles in large airways of the respiratory tract. Proc Royal Soc London Ser V 198:455–462, 1977.

Patrick, G: Requirements for local dosimetry and risk estimation in inhomogeneously irradiated lung. In: Low Dose Radiation: Biological Bases of Risk Assessment, edited by KF baverstock, JW Stather, pp. 269–277. London: Taylor and Francis, 1989.

Pavia, D, Bateman, JRM, Sheahan, NF, Agnew, JE, Clarke, SW: Tracheobronchial mucociliary clearance in asthma: Impairment during remission. Thorax 40:171–175, 1985.

Pavia, D: Lung mucociliary clearance. In: Aerosols and the Lung, edited by SW Clarke, D Pavia, pp. 127–155. London: Butterworths, 1984.

Phalen, R, Kenoyer, J, Davis, J: Deposition and clearance of inhaled particles: Comparison of mammalian species. In: Proceedings of the Annual Conference on Environmental Toxicology, Vol 7, pp. 159–170, AMRL-TR-76-125, Springfield VA: NTIS, 1977.

Phalen, RF, Oldham, MJ, Schum, GM: Growth and ageing of the bronchial tree: Implications for particle deposition calculations. Radiat Prot Dosim 38:15–21, 1991.

Philipson, K, Falk, R, Camner, P: Long-term lung clearance in humans studied with teflon particles labeled with chromium-51. Exp Lung Res 9:31–42, 1985.

Pinkerton, KE, Plopper, CG, Mercer, RR, Roggli, VL, Patra, AL, Brody, AR, Crapo, JD: Airway branching patterns influence asbestos fiber location and the extent of tissue injury in the pulmonary parenchyma. Lab Invest 55:688–695, 1986.

Proctor, DF: The upper respiratory tract. In: Pulmonary Diseases and Disorders, edited by AP Fishman, pp. 209–223. NY: McGraw-Hill, 1980.

Puchelle, E, Zahm, JM, Girard, F, Bertrand, A, Polu, JM, Aug, F, Sadoul, P: Mucociliary transport in vivo and in vitro—relations to sputum properties in chronic bronchitis. Europ J Respir Dis 61:254–264, 1980.

Raabe, OG, Yeh, HC, Newton, GJ, Phalen, RJ, Velazquez, DJ: Deposition of inhaled monodisperse aerosols in small rodents. In: Inhaled Particles IV, Pt 2, edited by WH Walton, pp. 3–20. Oxford: Pergamon Press, 1977.

Raabe, OG: Deposition and clearance of inhaled aerosols. In: Mechanisms in Respiratory Toxicology, edited by H Witschi, P Nettesheim, pp. 27–76. Boca Raton FL: CRC Press, 1982.

Radford, EP, Martell, EA: Polonium-210: Lead 210 ratios as an index of residence times of insoluble particles from cigarette smoke in bronchial epithelium. In: Inhaled Particles and Vapours IV, Pt 2, edited by WH Walton, pp. 567–580. Oxford: Pergamon Press, 1977.

Robertson, B: Basic morphology of the pulmonary defense system. Eur J Respir Dis 61(Suppl 107):21–40, 1980.

Roy, M: Lung clearance modeling on the basis of physiological and biological parameters. Health Phys 57(Suppl 1):255–262, 1989.

Rutland, J, Cole, PJ: Nasal mucociliary clearance and ciliary beat frequence in cystic fibrosis compared with sinusitis and bronchiectasis. Thorax 36:654–658, 1981.

Schiller, CF, Gebhart, J, Heyder, J, Rudolf, G, Stahlhofen, W: Deposition of monodisperse insoluble aerosol particles in the 0.005 to 0.2 μm size range within the human respiratory tract. Ann Occup Hyg 32 (Suppl 1):41–49, 1988.

Schlesinger, RB, Driscoll, KE, Naumann, BD, Vollmuth, TA: Particle clearance from the lungs: Assessment of effects due to inhaled irritants. Ann Occup Hyg 32 (Suppl. 1):113–123, 1988.

Schlesinger, RB, Gurman, JL, Lippmann, M: Particle deposition within bronchial airways: Comparisons using constant and cyclic inspiratory flow. Ann Occup Hyg 26:47–64, 1982.

Schlesinger, RB, Lippmann, M: Selective particle deposition and bronchogenic carcinoma. Environ Res 15:424–431, 1978.

Schlesinger, RB, McFadden, L: Comparative morphometry of the upper bronchial tree in six mammalian species. Anat Rec 199:99–108, 1981.

Schlesinger, RB: Clearance from the respiratory tract. Fundam Appl Toxicol 5:435–450, 1985a.

Schlesinger, RB: Comparative desposition of inhaled aerosols in experimental animals and humans: A review. J Toxicol Environ Health 15:197–214, 1985b.

Schlesinger, RB: Particle deposition in model systems of human and experimental animal airways. In: Generation of Aerosols, edited by K Willeke, pp. 553–575. Ann Arbor: Ann Arbor Science, 1980.

Schlesinger, RB: The effects of inhaled acids on respiratory tract defense mechanisms. Environ Health Perspect 63:25–38, 1986.

Schreider, JP: Comparative anatomy and functions of the nasal passages. In: Toxicology of the Nasal Passages, edited by CS Barrow, pp. 1–25. NY: McGraw-Hill, 1986.

Schum, M, Yeh, HC: Theoretical evaluation of aerosol deposition in anatomical models of mammalian lung airways. Bull Math Biol 42:1–15, 1980.

Schurch, S, Gehr, IM, Hof, V, Geiser, M, Green, F: Surfactant displaces particles toward the epithelium in airways and alveoli. Respir Physiol 80:17–32, 1990.

Sebastien, P, Fondimare, A, Bignon, J, Monchaux, G, Desbordes, J, Bonnand, G: Topographic distribution of asbestos fibers in human lung in relation to occupational and nonoccupational exposure. In: Inhaled Particles IV, Pt 2, edited by WH Walton, pp. 435–446. Oxford: Pergamon Press, 1977.

Snipes, MB, Boecker, BB, McClellan, RO: Retention of monodisperse or polydisperse aluminosilicate particles inhaled by dogs, rats, and mice. Toxicol Appl Pharmacol 69:345–362, 1983.

Snipes, MB: Long-term retention and clearance of particles inhaled by mammalian species. Crit Rev Toxicol 20:175–211, 1989.

Sorokin, SP, Brain, JD: Pathways of clearance in mouse lungs exposed to iron oxide aerosols. Anat Rec 181:581–626, 1975.

Stanley, PJ, Wilson, R, Greenstone, MA, Mackay, IS, Cole, PJ: Abnormal nasal mucociliary clearance in patients with rhinitis and its relationship to concomitant chest disease. Br J Dis Ches 79:77–82, 1985.

Stauffer, D: Scaling theory for aerosol deposition in the lungs of different mammals. J Aerosol Sci 6:223–225, 1975.

Sussman, RG, Cohen, BS, Lippmann, M: Asbestos fiber deposition in a human tracheobronchial cast. I. Experimental. Inhal Toxicol 3:145–160, 1991.

Swift, DL, Montassier, N, Hopke, PK, Karpen-Hayes, K, Cheng, YS, Su, YF, Yeh, HC, Strong, JC: Inspiratory deposition of ultrafine particles in human nasal replicate cast. J. Aerosol Sci. 23:65–72, 1992.

Swift, DL, Montassier, N, Hopke, PK, Karpen-Hayes, K, Cheng, YS, Su, YF, Yeh, HC, Strong, JC: Inspiratory deposition of ultrafine particles in human nasal replicate cast. J Aerosol Sci 23:65–72, 1992.

Swift, DL, Proctor, DF: A dosimetric model for particles in the respiratory tract above the trachea. Ann Occup Hyg 32(Suppl 1):1035–1044, 1988.

Swift, DL: Aerosol deposition and clearance in the human upper airways. Ann Biomed Eng 9:593–604, 1981.

Thomas, RL, Raabe, OG: Regional deposition of ^{137}Cs-labelled monodisperse and polydisperse aluminosilicate aerosols in Syrian hamsters. Am Ind Hyg Assoc J 39:1009–1018, 1978.

Thomson, ML, Pavia, D: Particle penetration and clearance in the human lung. Arch Environ Health 29:214–219, 1974.

Thomson, ML, Short, MD: Mucociliary function in health, chronic obstructive airway disease and asbestosis. J Appl Physiol 26:535–539, 1969.

Tryka, AF, Sweeney, TD, Brain, JD, Godleski, JJ: Short-term regional clearance of an inhaled submicrometric aerosol in pulmonary fibrosis. Am Rev Respir Dis 132:606–611, 1985.

Tu, KW, Knutson, EO: Total deposition of ultrafine hydrophobic and hygroscopic aerosols in the human respiratory system. Aer Sci Technol 3:453–465, 1984.

Valberg, PA, Wolff, RK, Mauderly, JL: Redistribution of retained particles. Effect of hyperpnea. Am Rev Respir Dis 131:273–280, 1985.

Valberg, PA: Determination of retained lung dose. In: Toxicology of Inhaled Materials, edited by HP Witschi, JD Brain, pp. 57–91. Heidelberg: Springer-Verlag, 1985.

Vastag, E, Matthys, H, Zsamboki, G, Kohler, D, Daikeler, G: Mucociliary clearance in smokers. Eur J Respir Dis 68:107–113, 1986.

Waite, DA, Ramsden, D: The inhalation of insoluble iron oxide particles in the submicron range. Pt. I, Chromium 51 labelled aerosols, AEEW-R740, UK, Atomic Energy Authority, 1971.

Wanner, A: Clinical aspects of mucociliary transport. Am Rev Respir Dis 116:73–125, 1977.

Warheit, DB, Hartsky, MA: Species comparisons of alveolar deposition patterns of inhaled particles. Exp Lung Res 16:83–99, 1990.

Warner, AE, Brain, JD: The cell biology and pathogenic role of pulmonary intravascular macrophages. Am J Physiol 258:L1–L12, 1990.

Watson, MS, Brain, SD: Uptake of iron oxide aerosols by mouse airway epithelium. Lab Invest 40:450–459, 1979.

Wilson Jr, FT, Hiller, FC, Wilson, JG, Bone, RC: Quantitative deposition of ultrafine stable particles in the human respiratory tract. J Appl Physiol 58:223–229, 1985.

Wolff, RK, Dolovich, MB, Obminski, G, Newhouse, MT: Effects of exercise and eucapnic hyperventilation on bronchial clearance in man. J Appl Physiol 43:46–50, 1977.

Wolff, RK: Effects of airborne pollutants on mucociliary clearance. Environ Health Perspect 66:222–237, 1986.

Yeates, DB, Aspin, M, Levison, H, Jones, MT, Bryan, AC: Mucociliary tracheal transport rates in man. J Appl Physiol 39:487–495, 1975.

Yeates, DB, Aspin, M: A mathematical description of the airways of the human lungs. Respir Physiol 32:91–104, 1978.

Yeates, DB, Gerrity, TR, Garrard, CS: Particle deposition and clearance in the bronchial tree. Ann Biomed Eng 9:577–592, 1981a.

Yeates, DB, Pitt, BR, Spektor, DM, Karron, GA, Albert, RE: Coordination of mucociliary transport in human trachea and intrapulmonary airways. J Appl Physiol 51:1057–1064, 1981b.

Zeltner, TB, Sweeney, TD, Skornick, WA, Feldman, HA, Brain, JD: Retention and clearance of 0.9-μm particles inhaled by hamsters during rest or exercise. J Appl Physiol 70:1137–1145, 1991.

Chapter Nine

Pulmonary Retention of Particles and Fibers: Biokinetics and Effects of Exposure Concentrations

M. B. Snipes

INTRODUCTION

Particles and fibers are referred to individually in this chapter or are referred to collectively as "dust" when the intent is to generalize about materials deposited in the respiratory tract. Deposition of dusts in the pulmonary (P) region is the starting point for processes that sequester, redistribute, or clear them. Inhaled particles larger than about 10 μm have only a small probability for deposition in the P region because they deposit in the head airways or tracheobronchial region by impaction and gravitational effects. Therefore, discussions of particles deposited in the P region are generally restricted to particles having aerodynamic diameters less than about 10 μm. Unlike relatively symmetrical and compact particles, the physical attributes of fibers help to keep them suspended longer in the airstream during ventilation, thereby increasing their potential for deep penetration into the lung. Fibers 300 μm and longer have been recovered from samples of lung parenchyma of humans (Sebastien, 1991). The physical shape is also an important factor that influences retention and clearance patterns for fibers that deposit in the P region.

Particles and fibers that deposit in the P region are rapidly incorporated into phagocytic cells—primarily macrophages—or are trapped in the moist environment of lung tissue. Thereafter, the biokinetics of the deposited materials is linked to physical and chemical activities associated with cells and fluids in the lung. Materials are physically cleared from the P region by translocation in macrophages to the mucociliary transport system in the conducting airways of the tracheobronchial region, or to thoracic lymph nodes (TLNs) via pulmonary lymphatic channels. A competing clearance process—dissolution-absorption—transforms the deposited material into soluble constituents that can be absorbed into the circulatory system and either redistributed within the body or excreted into urine or feces. Physical transport within the lung is a relatively slow process, so that dissolution-absorption can be the dominant lung clearance process, especially for soluble and moderately soluble materials. Particles and fibers that are poorly soluble in biological fluids can be retained in the P region for months or years after deposition.

Inhalation exposures can be single events involving pulmonary deposition of a small mass of material, and small lung burdens of relatively nontoxic particles or fibers may not represent a threat to health. In fact, humans are constantly exposed to a variety of types of airborne particles and fibers, many of which deposit in the P region and remain there for very long times. On the other hand, inhalation exposures can involve deposition of toxic materials or substantial amounts of nontoxic materials in the respiratory tract. Repeated or chronic inhalation exposures can produce significant cumulative lung burdens that alter lung clearance patterns, disrupt other lung functions, and produce pathological effects because of physical and/or chemical damage to tissue. Even so-called nuisance dusts such as titanium dioxide (TiO_2), diesel exhaust particles, carbon black, or coal dust can produce adverse health effects if large amounts of the particles accumulate in the lungs of laboratory animals. The accumulation of excessive amounts of dust in the P region during repeated inhalation exposures, and the development of nonspecific adverse health effects related to the dust are conditions associated with "lung overload." This phenomenon has been well-documented for small laboratory animal species, primarily rats; its significance for humans is still being evaluated.

Chapter 8 discusses deposition and clearance for all regions of the respiratory tract. This chapter focusses on the biokinetics of inhaled particles and fibers after they deposit in the P region of the respiratory tract. Whereas deposition and clearance are important components of the biokinetics of inhaled materials, they are discussed in this chapter to the extent necessary to provide a more complete presentation of the biokinetics of inhaled materials. Additionally, inhalation exposures can be single events or repeated exposures to variable amounts of airborne materials. Therefore, information relevant to the effects of exposure conditions on the biokinetics of materials deposited in the P region and the concept of lung overload are included in this chapter.

THE PULMONARY REGION OF THE RESPIRATORY TRACT

The P region is the nonciliated, gas-exchange region of the respiratory tract, beginning just beyond the terminal bronchioles and ending with the pulmonary alveoli. This region includes respiratory bronchioles, alveolar ducts, alveolar sacs, atria, alveoli, interstitial tissues, alveolar capillaries, and pulmonary lymphatics. The basic features of the P region are common to all mammalian species, but substantial differences exist in gross and microscopic morphology. The state of understanding of species similarities and differences in anatomy of the P region has been reviewed by Phalen (1984), Warheit (1989), and Tyler and Julian (1991). To date, attempts to use anatomical differences among mammalian species to explain retention and clearance patterns for inhaled materials have been unsuccessful. However, it is important to continue these attempts in order to allow interpretations within and among the various species used in inhalation toxicology research.

Two types of migratory phagocytic cells—pulmonary alveolar macrophages (PAMs) and polymorphonuclear leukocytes (neutrophils)—are present in the P region and have important functions in the biokinetics of inhaled particles and fibers. Macrophages are the principal phagocytic cells in the lung and function in several capacities related to phagocytosis, isolation, digestion, clearance of foreign particles, fibers, microorganisms, and cellular debris. Macrophages that have phagocytized material release chemotactic factors, which results in recruitment of additional macrophages, as well as neutrophils, into the P region (Dauber and Daniele, 1980; Adamson and Bowden, 1982). Neutrophils are normally present in the P region only in small numbers, but numbers of neutrophils in the P region can be increased by recruitment from the circulatory system (Sorokin, 1977; Larsen et al., 1983; Lehnert et al., 1985; Sibille and Reynolds, 1990). The functions of neutrophils in the biokinetics of inhaled materials have not been clearly delineated. One role of neutrophils may be phagocytosis of foreign materials in the pulmonary airways to isolate them from contact with epithelial cells, thereby keeping the foreign materials from entering the interstitium. There are indications that neutrophils attain significant roles in the biokinetics of inhaled materials only if inhalation exposures result in pulmonary deposition of large amounts of materials that trigger an inflammatory response.

The lymphatic system receives a portion of materials that is physically cleared from the P region. While the basic components of the pulmonary lymphatic system appear to be common to all mammalian species, there are significant quantitative differences in the amount of tissue or lymphatic flow among species that probably influence effectiveness of the lymphatic systems in particle transport. Specifically, humans and large animal species have thick pleura with many interlobular septa that form an extensive network of lymphatic and collecting vessels throughout the connective tissue of the lung parenchyma. By comparison, the pleura is generally thin in small mammals, with a sparse distribution of lymphatic vessels within the pleura (Leak, 1977; Leak and Jamuar, 1983). These anatomical differences may allow a more efficient removal of materials from the P region via the lymphatic system in humans and other large mammalian species than in small mammals.

Two categories of lymphatic tissues are recognized in the lungs of mammals. One category is comprised of well-organized, encapsulated TLNs. The other category is bronchus-associated lymphoid tissue (BALT), which has been described morphologically, but its role in particle retention and clearance from the lung has not been fully elucidated. It is possible that particles are transported into the pulmonary interstitium and returned to bronchi through BALT, and placement of macrophages onto ciliated epithelium may be facilitated by pores or openings in the lymphoepithelium of BALT that allow what appears to be a one-way transport of phagocytized material onto the ciliated epithelium of bronchioles (Brundelet, 1965; Green, 1971, 1973; Tucker et al., 1973; Kilburn, 1974). In essence, BALT may represent the terminal component of an excretory pathway for particles, fibers, or effete phagocytic cells. Whether or not BALT contributes significantly to clearance of materials from the P region is still an unanswered issue that deserves additional attention. Also, the presence, type, and function of BALT appear to vary considerably among species (Sminia et al., 1989; Murray and Driscoll, 1992). To further complicate the issue, the amount and function of BALT in the lung may be influenced by environmental conditions. It appears probable that BALT could represent an important type of lung tissue with functions relevant to species differences in retention and physical clearance of materials deposited in the P region.

BIOKINETICS OF PARTICLES AND FIBERS DEPOSITED IN THE PULMONARY REGION
General Comments

Particles and fibers sequestered in the nonciliated airspaces beyond the terminal bronchioles are cleared from their retention sites by the two competing processes of translocation and dissolution-absorption. Translocation refers to the physical movement of materials within or out of the P region and is generally assumed to involve phagocytic cells, primarily macrophages. Translocation pathways and processes appear to be similar among mammalian species, but the relative amounts of material cleared via the pathways seem to be species-dependent.

Dissolution-absorption refers to a combination of processes that results in clearance of constituents of particles and fibers. Dissolution means the chemical processes in cells and body fluids that cause constituents of particles or fibers to dissociate from them or leach out. As a result, chemical constituents of the particles or fibers are available for transport into the lymphatic or circulatory systems, for metabolism, or for reactions with tissue constituents. Absorption is the phenomenon that results in a net movement of dissolved material into lymphatic vessels or into the circulatory system. The absorbed material is transported via the circulatory system to other body organs, or excreted into urine or feces. Fig. 1 summarizes the probable retention sites and transport mechanisms and pathways for particles, fibers, and their dissolved chemical constituents in the pulmonary region.

Early Events after Deposition

The fate of particles and fibers deposited in the P region depends on the sites of deposition and the physical/chemical characteristics of the particles and fibers. Physiological processes that determine the fate of materials that deposit in the lung are common to all mammalian species, but the biokinetics of the deposited materials vary considerably among species. The reasons for species differences are not understood and are the subject of ongoing experiments in numerous laboratories.

Some responses of the lung to inhaled materials are common among species. For example, most foreign materials that deposit in the P region are rapidly phagocytized by PAMs. Resident PAMs phagocytize small numbers of particles or fibers efficiently. Increased numbers of PAMs in alveoli occur in response to increased numbers of particles in the alveoli, whether the particles are instilled (Brain, 1971; Adamson and Bowden, 1978) or inhaled (Labelle and Brieger, 1960, 1961; Geiser et al., 1994). The increase in numbers of PAMs appears to be related more to the number of particles than to the total mass of particles. Therefore, equivalent masses of the same material may not produce the same macrophage response if the two masses are associated with particles having different physical sizes (Morrow, 1988). Additionally, large numbers of neutrophils appear when exposure levels are high, or when the exposure material produces an inflammatory response. The efficiency of the early phagocytic processes appears to be an important factor in the long-term fate of the deposited materials.

Macrophages can engulf a broad range of particle types and sizes, up to tens of μm in diameter. The rate and efficiency of phagocytosis appear to depend on properties of the deposited particles that include size, chemical composition, shape, and mass. For example, results of an *in vitro* study indicated that macrophages phagocytize fewer 6-μm polystyrene latex (PSL) microspheres than 3-μm or 1.5-μm microspheres, indicating that the 6-μm particles are not readily internalized by macrophages; also, carbon particles are more readily phagocytized than the PSL microspheres (Holma, 1967). Results of an *in vitro* study by Hahn et al. (1977) using spherical monodisperse aluminosilicate particles and $^{239}PuO_2$ particles demonstrated that macrophages phagocytize 2.2 μm particles more rapidly and in greater numbers than 0.3 μm particles; additionally, greater numbers of dense $^{239}PuO_2$ parti-

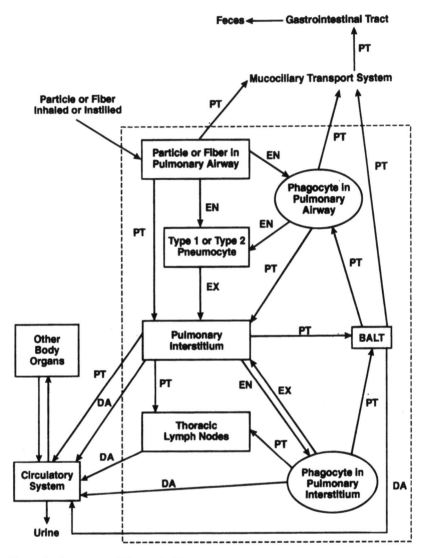

Figure 1 Summary of the probable retention sites and transport mechanisms and pathways for particles, fibers, and their dissolved chemical constituents in the pulmonary region of the respiratory tract. Arrows indicate physical transport (PT), endocytosis (EN), exocytosis (EX), and dissolution-absorption (DA). The dashed line indicates the boundary of the pulmonary region.

cles are phagocytized than fused aluminosilicate particles having the same geometric diameter. These investigators suggested that 2-μm particles are near the optimum size for phagocytosis by PAMs. Pratten and Lloyd (1986) demonstrated *in vitro* that phagocytosis of PSL microspheres by peritoneal macrophages increases steadily with size for microspheres in the size range of 0.1–1.1 μm. The cumulative results of these studies suggests that the size of the particles most efficiently phagocytized by macrophages may be in the range of 1.5–3 μm geometric diameter, and that efficiency of phagocytosis depends on the composition of the material.

Effective phagocytosis of fibers by macrophages is strongly influenced by fiber length. Bernstein et al. (1980) exposed rats to neutron-activated glass fibers by inhalation or intratracheal instillation. The fibers were either 1.5 μm diameter $\times$ 5 μm long, or 1.5 μm diameter $\times$ 60 μm long. Rats were sacrificed at times ranging from 18 weeks to 2 years after instillation, and samples of lung tissue and TLNs were examined microscopically. The longer fibers were not effectively phagocytized, and were not detected in TLNs. In contrast, the shorter fibers were phagocytized by macrophages, and some were cleared to TLNs. Similar results relative to short versus long fibers were noted by Wright and Kuschner (1977) and by Morgan et al. (1982). These results suggest that phagocytosis of fibers deposited in the lungs is requisite to movement of the fibers to TLNs and that long fibers are not effectively phagocytized and therefore have a low probability of being found in the lymph nodes.

Macrophages may function normally until they have accumulated amounts of material that render them immobile. The overloading of functional abilities of macrophages has been a subject of keen interest. Valberg et al. (1982) evaluated endocytosis of colloidal gold by macrophages as influenced by concentration of the gold colloid. Uptake of the gold colloid was linear until sufficient concentrations were used to saturate the ability of macrophages to phagocytize additional particles. Bowden (1987) noted that the recruitment of macrophages to the lung is closely associated with the burden of particles reaching the airways and alveoli and that the system can be overloaded by either too many particles or an inadequate cellular response. Morrow (1988) concluded that large numbers or masses of particles in the lung directly influence the functional abilities of macrophages and hypothesized that a critical volumetric loading of PAMs renders them unable to translocate from alveoli.

Oberdörster et al. (1988a, 1992) tested the hypothesis proposed by Morrow (1988). They intratracheally instilled mixtures of 3.3 and 10.3 μm plastic microspheres into the lungs of rats, then evaluated clearance kinetics numbers of microspheres present in macrophages over 202 days. Both sizes of microspheres were phagocytized by macrophages; one or more of the 3.3 μm microspheres was often seen in individual macrophages, but only single 10.3 μm microspheres, which represents about 60% of the volume of a rat PAM, could be phagocytized. The clearance rate of the 10.3 μm microspheres was 10 times less than the clearance rate of the 3.3 μm microspheres. Oberdöster et al. also demonstrated that ingestion of a single 10.3 μm microsphere by a macrophage severely compromises the mobility of the macrophage, which further supports the Morrow (1988) hypothesis. In a complementary study, Lehnert (1990) showed that a volumetric loading of macrophages by a number of 2.13 μm microspheres equivalent to about 60% of the macrophage volume severely reduces macrophage-mediated clearance and that the macrophage loading does not appear to be reversible.

There is overwhelming evidence that some particles can rapidly penetrate the pulmonary epithelium, become incorporated into Type 1 pneumocytes, or be transferred to the interstitium if the particles are not immediately phagocytized. Additionally, considerable evidence supports the hypothesis that increased interstitialization is related to both decreased size and increased numbers of particles deposited in the alveolar region of the lung (Adamson, 1990; Ferin et al., 1992). Several studies have indicated that both the ability to penetrate and the probability of penetrating the pulmonary epithelium increase with the amount of material deposited (Sorokin and Brain, 1975; Brody et al., 1981; Ferin and Feldstein, 1978). The transepithelial passage of free particles or fibers may be related to the number of PAMs available to ingest the deposited particles. With an increased particle load, a level exceeding the saturation point for increasing macrophage numbers may be reached (Lauweryns and Baert, 1977; Adamson and Bowden, 1978, 1981, 1982; Ferin, 1977), thereby increasing the probability of particle penetration into the interstitium.

Particles and fibers that move into the interstitium and remain there for long periods of time are predominantly found in subpleural and paraseptal positions, and in perivascular positions around pulmonary arterioles and venules associated with bronchioles (Adamson and Bowden, 1978; Bowden and Adamson, 1984, 1985). The peribronchiolar sites appear to be prominent, long-term retention sites for particles (Green, 1973) and fibers (Smith et al., 1987) retained in the lung.

To summarize, evidence available to date suggests that most particles or fibers deposited in the P region become phagocytized within a brief time after deposition. All sizes of respirable particles, but not all sizes of respirable fibers, can be completely phagocytized. Until deposited materials have been phagocytized, they are available for interaction with fixed cells in the lung and potential transfer to the pulmonary interstitium. The extent of transfer of particles or fibers into the pulmonary interstitium is influenced by the number of particles or fibers, or mass of deposited material.

EFFECTS OF SIZE ON BIOKINETICS OF PARTICLES AND FIBERS
Physical Transport Processes

The effect of particle size on physical clearance from the P region has been the subject of several studies using laboratory animals. Results of these studies demonstrate that physical clearance of relatively spherical particles in the size range of about 0.3 to 4 μm geometric diameter is not size selective (Cartwright and Skidmore, 1964; Morrow et al., 1967; Snipes et al., 1983; Morgan et al., 1986; Muhle and Bellmann, 1986; Kreyling, 1990). However, results from studies with larger particles yield different results. The larger particles can be phagocytized by PAM, but are not efficiently transported within the lung (Snipes and Clem, 1981; Snipes et al., 1984; Takahashi et al., 1987; 1992). Additionally, physical transport from the P region to TLNs decreases as the particle size increases, dropping dramatically for particle sizes larger than about 5 μm (Snipes and Clem, 1981; Snipes et al., 1984; Takahashi et al., 1992). These results for larger particles support the concept of macrophage overload discussed earlier.

In contrast to physical clearance of compact particles from the lung, physical clearance of fibers has been demonstrated to be size dependent. Several studies with rats have demonstrated preferential physical clearance of short fibers from the lung and slowed clearance of fibers 20 μm or longer (Timbrell and Skidmore, 1971; Leadbetter and Corn, 1972; Morgan et al., 1978; 1982; Morgan, 1980; Roggli and Brody, 1984; Roggli et al., 1987; Smith et al., 1987). In addition to size, shape may also influence the effectiveness of translocation mechanisms for fibers. For example, large fibers may be ingested by PAMs, but the mobility of the PAMs may be severely altered because of the length, shape, or mass of the fibers. Fibers such as asbestos (as well as some types of particles) may induce biochemical changes that adversely affect phagocytic cells attempting to transport them (Misra et al., 1978; Goodglick and Kane, 1986).

Results of studies evaluating fibers that had been removed from the lung in bronchoalveolar lavage fluid (BALF) support the concept of preferential clearance of short fibers from the lung. Kauffer et al. (1987) reported results from a study in which rats inhaled chrysotile asbestos. Some rats were sacrificed for evaluation immediately after exposure; others were sacrificed at times ranging to 90 days after the acute inhalation exposure. Samples of BALF and lung tissue were analyzed for their fiber content. Lengths of asbestos fibers measured in lung increased with time, again indicating faster clearance of shorter fibers. Interestingly, lavage fluid contained elevated numbers of short fibers. The mean diameter of fibers in lung tissue and BALF decreased as a function of time; this appeared to result from separation of the fibers into fibrils in the lung rather than from dissolution-absorption. Teschler et al. (1994) measured asbestos fibers in BALF and lung tissues of former asbestos workers and noted significant differences in sizes of fibers in these two types of samples. Asbestos fibers recovered from lung tissue were significantly longer than fibers recovered from BALF. However, fiber diameters were not significantly different in lung tissue and BALF. Morgan et al. (1994) reported similar results for rats intratracheally instilled with glass fibers. The longer fibers retained in the lung are apparently sequestered in a manner that makes them less available for removal from the lung using bronchopulmonary lavage procedures.

Physical transport of dust from the P region to the TLNs results in sequestration or trapping of the dust in what is generally perceived to be a dead-end compartment, and high concentrations of dust can accumulate in TLNs. Thomas (1968, 1972) demonstrated that translocation of radioactive particles from the P region to the TLNs after acute inhalation exposures can result in concentrations of particles in the lymph nodes that can be more than two orders of magnitude higher than concentrations of the particles retained in the lung. There are numerous indications in the open literature that excessive accumulations of dust in lymph nodes, especially radioactive particles, can damage the nodes and compromise their functional ability, apparently allowing dust to pass through the damaged lymph nodes, enter the circulatory system, and become trapped in the reticuloendothelial system of organs such as liver, spleen, and skeleton (Hourihane, 1965; Pooley, 1974; Gearhart et al., 1980; Lee et al., 1981a; Lefevre et al., 1982; Guilmette et al., 1987).

Results from several studies have indicated that macrophages can phagocytize foreign materials in the lung, then migrate directly into pulmonary blood vessels. In one study, Lee et al. (1981a) evaluated distribution and retention patterns of potassium octatitanate fibers (Fybex) in the lungs of rats, Syrian hamsters, and guinea pigs for as long as 24 months after subchronic inhalation exposures to the fibers. Most fibers < 5 μm long were phagocytized by PAMs; fibers ≥ 10 μm were phagocytized by foreign body giant cells. Fiber-containing macrophages were noted in the lumens of bronchial lymphatic or pulmonary blood vessels. Numerous fiber-containing cells were transported from the P region to the TLNs, where some of the macrophages penetrated into the blood or lymphatic circulation at the trabecular, hilar, and pericapsular regions of the lymph nodes. In addition, fiber-containing macrophages and free macrophages were noted in the lumens of postcapillary venules. Fibers found in lymph nodes were shorter than those in lung tissue, suggesting preferential transport of smaller fibers from the lungs to TLNs. In another study, Holt (1981) evaluated the lungs of guinea pigs exposed by inhalation to submicron-sized carbon particles for 100 hours or exposed for 24 hours to chrysotile asbestos fibers. After a few weeks, accumulations of dust-filled macrophages were found near pulmonary blood vessels, and fiber-containing macrophages were detected in the process of penetrating into blood vessels. Holt concluded that some inhaled dust is probably transported to extrapulmonary sites by macrophages in blood. Brody et al. (1981) also noted asbestos fibers in capillary endothelial cells of rats' lungs and lumens of capillaries 4 and 8 days after the rats were acutely exposed by inhalation to an aerosol of chrysotile asbestos.

Even though macrophages can penetrate blood vessels, it is probably a rare phenomenon because particles transported away from the lungs in macrophages appear to be almost exclusively transported to the mucociliary escalator or to TLNs. Macrophages are more likely to enter the lymphatic system than the circulatory system because lymphatic vessels have looser endothelial cell connections and wider intercellular junctions than capillaries (Lauweryns and Baert, 1977). Additionally, lymphatic vessels are open-ended tubes; macrophages may be able to enter the lymphatic vessels directly from interstitial spaces without the need to cross cellular barriers.

To summarize physical transport processes, particles and fibers can be phagocytized and physically removed from the lungs. Physical transfer to the mucociliary escalator results in elimination of the material from the body via the GI tract. Direct penetration of particles or fibers into lymphatic channels and the circulatory system has been reported, but appears to represent a very small portion of the lung burden of deposited material. Physical transport of particles or fibers from the P region to the TLNs represents a pathway for removing the material from the P region. However, particles or fibers physically transported to the TLNs remain there indefinitely, subject to dissolution-absorption processes, unless the TLNs are damaged to the extent that materials can pass through or around them to enter the circulatory system for redistribution to other body organs.

In Vivo Dissolution-Absorption

Factors that affect the dissolution of materials or leaching of their constituents in physiological fluids and absorption of their constituents are not fully understood. Solubility is influenced by the surface properties of particles, such as surface:volume ratios (Mercer, 1967; Morrow, 1973). The rates at which dissolution and absorption processes occur are also influenced by chemical composition of the material. Temperature history of materials is an important consideration for metal oxides. For example, the solubility of metal oxides is generally inversely related to the temperature of formation of the oxides; the higher temperatures also tend to result in compact particles having relatively small surface:volume ratios. It is sometimes possible to accurately predict dissolution-absorption characteristics of materials based on physical/chemical considerations. However, predictions for *in vivo* dissolution-absorption rates for most materials, especially if they contain multivalent cations or anions, should be confirmed experimentally.

Phagocytic cells, primarily macrophages, are involved with dissolution-absorption of particles retained in the respiratory tract. Some particles dissolve within the phagosomes of macrophages due to the acidic milieu in those organelles (Lundborg et al., 1984; 1985), but the dissolved material may remain with the phagosomes or other organelles in the macrophage rather than diffuse out of the macrophage to enter the circulatory system and either be excreted into urine or feces or deposited elsewhere in the body (Cuddihy, 1984). Examples of presumably soluble inorganic materials that exhibit delayed absorption are beryllium (Reeves and Vorwald, 1967) and americium (Mewhinney and Griffith, 1983). Species differences for absorption of Cd from the lung was reported by Oberdörster (1987), and Bailey et al. (1989) reported substantial differences among several mammalian species for absorption of Co from the lung. Delayed absorption has also been reported for organic materials deposited in the lungs. For example, covalent binding of benzo(a)pyrene or matabolites to cellular macromolecules results in an increased lung retention time for that compound after inhalation exposures of rats (Medinsky and Kampcik, 1985). Certain chemical dyes are also retained in the lungs (Medinsky et al., 1986), where they may dissolve and become adsorbed to lipids or react with other constituents of lung tissue. Understanding these phenomena and recognizing species similarities and differences are important for evaluating lung retention and clearance processes and interpreting results of inhalation studies.

Dissolution-absorption of fibers depends on fiber size and composition as demonstrated in several studies (Morgan et al., 1982; Le Bouffant et al., 1984; 1987; Johnson et al., 1984; Hammad, 1984; Hammad et al., 1988). To generalize, shorter fibers are subjected to more rapid dissolution-absorption than are longer fibers because they have a larger surface area available for chemical and biological attack. An additional factor is retention site. Morgan et al. (1982) attributed the dependency of dissolution on fiber length to the differences in pH encountered by the fibers. The shorter fibers, which are phagocytized by macrophages, are presumed to be exposed to a lower pH than long, nonphagocytized fibers in extracellular fluid. Examination of the surfaces of fibers recovered from lung generally reveals surface etching, which results in irregularities in outline, loss of electron density, and the appearance of pits along the edges of the fibers. The degree of etching increases with residence time in the lungs. The various types of man-made vitreous fibers (MMVFs) show different degrees of etching, with more etching of glass fibers and glass wool than of rock wool fibers (Johnson et al., 1984; Le Bouffant et al., 1984; 1987). The results of these studies indicate that physical and chemical attributes of the fibers are important factors in processes that dissolve or etch them.

The solubility of asbestiform minerals and MMVFs, both *in vivo* and *in vitro,* was reviewed by Morgan and Holmes (1986). These authors noted that MMVFs appear to be much more soluble *in vivo* than fibrous amphiboles. Also, glass fibers appear to be more soluble than rock wool fibers. However, the solubility of glass fibers varies considerably with chemical composition; thus it cannot be assumed that all glass fibers are more soluble than rock wool fibers. Some constituents of crystalline asbestiform minerals are leached preferentially which results in changes in surface properties of the fibers and presumably changes in their interactions with biological systems. With MMVFs, surface features change as a consequence of dissolution, but surface properties do not appear to change.

To summarize, dissolution-absorption of particles and fibers is clearly dependent on their chemical and physical attributes. While it is possible to predict rates of dissolution-absorption, it is prudent to experimentally determine this parameter in order to understand the importance of dissolution-absorption as a clearance process for the lungs, TLNs, and other body organs that might receive particles or fibers that enter the circulatory system from the lungs.

Movement of Free Particles and Fibers

Phagocytosis by macrophages and other cells in the lungs is the dominant method of trapping the particles or fibers in lung tissue, but some particles or fibers may directly enter the interstitium by passing through or between airway lining cells. Gross and Westrick (1954) hypothesized that very small particles can pass directly from lung tissue into the circulatory system, but did not indicate particle size limitations for this process. This hypothesis has been pursued in several studies, and evidence has been presented to support the idea that particles can pass directly from lung parenchyma into blood vessels (and lymphatic vessels) without being carried by mobile cells. For example, electron microscopic studies by Lauweryns and Baert (1974) demonstrated that carbon particles (25 nm diameter) are cleared from the lungs of rabbits via the lymphatics, while ferritin molecules (10 nm diameter) are absorbed by blood capillaries and lymphatics. Both types of particles were seen in the interstitium and the lymph vessels within 30 minutes after their intratracheal administration. The quantitative role of blood capillaries and lymphatics in alveolar clearance was evaluated by Meyer et al. (1969) who studied the removal of radioiodinated albumin from the alveolar spaces in dogs. These authors estimated that the efficiency of particle absorption via the lymphatics is about 37 times that of blood. In studies with $^{239}PuO_2$ (Smith et al., 1977) and $^{238}PuO_2$ (Stradling et al., 1978), aggregated particles of PuO_2 in suspensions were substantially disaggregated in water and presumably in lung fluids to 1 nm diameter sub-units. The small particles of PuO_2 were rapidly absorbed directly into the circulatory system and redistributed in the body or excreted into urine. These particles were small enough that they could form complexes that were readily transported across biological membranes.

Direct penetration into the circulatory system or lymphatic system in lungs may not be restricted to ultrafine particles or fibers. Huang et al. (1988) noted a direct relationship between the amount of fibers in extralung organs and the lung burdens of fibers in three human autopsy cases. The fibers apparently penetrated into the circulatory system and deposited in other internal organs. Oberdörster et al. (1988b) instilled neutron-activated amosite fibers into the lungs of mongrel dogs, then collected postnodal lymph to determine if fibers were penetrating the lymph node. Fibers less than about 9 μm long and about 0.5 μm diameter were found in postnodal lymph 1 day after instillation. The amount of fibers in postnodal lymph was small, but the study indicates that short asbestos fibers deposited in the lungs can penetrate directly into the lymphatic system, penetrate TLNs, and be transported to the circulatory system for deposition elsewhere in the body. Lee et al. (1988) suggested that the unique physical configuration of Kevlar fibrils appears to impair their penetration into lung tissue. These authors noted that the amount of transmigrated Kevlar particles in the TLNs is minimal compared to what is seen in studies with asbestiform fibers or TiO_2.

Particle and Fiber Fragmentation

Particles that are homogeneous do not appear to fragment unless they are physically damaged, for example by radiation (Fleischer and Raabe, 1977; Diel and Mewhinney, 1983). Fibers, on the other hand, can dissolve, and their constituents can leach out; some types can split into fibrils (Lee et al., 1981b; Roggli et al., 1987). Asbestos fibers can split longitudinally to create a larger number of fibers having the same length as the original fiber, but with smaller diameters. Timbrell et al. (1988) noted that a long residence time for chrysotile in human lungs results in almost complete dispersion of the chrysotile fibers into fibrils, which substantially increases the surface area of the retained fibers. In contrast to asbestos, glass fibers and MMVFs can fracture transversely, resulting in a larger number of shorter fibers with the same original diameter (Morgan and Holmes, 1984; Le Bouffant et al., 1984; 1987; Lee, 1985; Hesterberg et al., 1992; Kelly et al., 1993). Fragmentation *in vivo* results in pieces of the original fibers that no longer meet the definition of a fiber and provides an increased surface area for dissolution and leaching of constituents. Kevlar pulp fibers do not split longitudinally as asbestos, or transversely fracture as mineral fibers; they are strongly bonded and peel rather than split (Lee et al., 1988).

The physical configuration of fibers makes fragmentation plausible as a result of biological attack on the fibers by enzymes or by physical breakage at weak spots along the fiber. Churg et al. (1989) and Lippmann (1990) discussed breakup of fibers as a process that causes subdivision of the fibers into shorter segments which are more easily cleared from the lungs. Lippmann stated that breakup of MMVFs is virtually all by length, whereas splitting of MMVFs into smaller diameter fibrils, which is characteristic of asbestos and other durable fibers, is seldom seen. For MMVFs, the relative importance of breakage into shorter length segments, partial dissolution, leaching, and surface etching depends upon the size and composition of the fibers. Morgan et al. (1973) demonstrated that the morphology of chrysotile asbestos, as shown in electron micrographs, is not changed after all of the magnesium is removed from the fibers by leaching in 1 N HCl. However, partial dissolution or leaching may create pores or remove components of the fiber matrix and weaken the fibers without showing evidence visible in electron micrographs.

Redistribution of Particles and Fibers

Initial particle or fiber deposition patterns in the lung are different from long-term retention patterns. Experimental results from several inhalation studies indicate that particles deposited in lungs redistribute over time by mechanisms that are not understood. Redistribution of particles has been studied using PuO_2, which allows the use of autoradiography to precisely determine dispersion patterns for individual particles or clumps of particles. Studies have been conducted using dogs (Clarke and Bair, 1964; McShane et al., 1980), rats (Sanders and Dagle, 1974), Syrian hamsters (Diel et al., 1981), both rats and Syrian hamsters (Rhoads et al., 1982), and humans (Voelz et al., 1976). The migration of fiber-containing macrophages to the lung periphery was demonstrated in rats by Morgan et al. (1977) for acutely inhaled neutron-activated chrysotile asbestos fibers. Holt (1982) proposed that fibers phagocytized by PAMs are transported toward the lung periphery through alveolar walls; some of these cells aggregate in alveoli near larger bronchioles, then penetrate the bronchiolar wall and are cleared by mucociliary transport.

The consequences of particle and fiber redistribution in the lungs are uncertain. It can be argued that redistribution of individual particles or fibers in the lungs causes a changing pattern of cells at risk, and that concentrations of particles or fibers pose a greater hazard for the lungs. This would be especially relevant to radioactive materials. Also, clumping may reduce the rate of macrophage-mediated clearance and dissolution-absorption processes as a consequence of increasing effective size and reducing surface area of the material. A potential benefit of clumping is reduction in the amount of lung tissue exposed to the retained material. The magnitude of this potential benefit would depend on the physicochemical nature of the material and its potential for damage to lung tissue.

Movement and redistribution of materials retained in the lungs of animals appear to be similar for particles and fibers. These phenomenona may be common to all mammalian species, but this has not been clearly demonstrated. Over the course of several months or longer after deposition, materials retained in the lungs are preferentially found in peribronchial and perivascular connective tissues (Adamson and Bowden, 1978; Bowden and Adamson, 1984, 1985). Mechanisms responsible for redistribution of foreign materials deposited in the lungs have not been elucidated. It should be noted that even if the mechanisms are common to humans and other mammalian species, differences in rates and degree of redistribution and clumping can still result in species differences in retention patterns and biological responses to the materials.

Macrophage-mediated clearance rates may vary, depending on location within the lungs. Therefore, even in the absence of clumping, redistribution of materials retained in the lungs can result in changing patterns of clearance. Redistribution and variability in macrophage-mediated lung clearance from different alveolar regions within the same lung may contribute to part of the variation observed within and among species for lung retention and clearance of inhaled materials (Sorokin and Brain, 1975).

Summary Comments on the Biokinetics of Particles and Fibers

Table 1 summarizes and compares factors that influence the biokinetics of respirable-size particles and fibers after deposition in the P region. It is noteworthy that particles and fibers are similar with respect to most factors that influence lung clearance. Pulmonary alveolar macrophages can phagocytize any size particle that reaches the P region by inhalation, but large particles or numerous small particles can overload the PAMs and inhibit or stop macrophage-mediated clearance. Respirable fibers longer than about 20 μm cannot be completely phagocytized or transported by macrophages. The functional abilities of macrophages can therefore become overloaded by either type of dust. Both categories of dust can fragment, but fibers can also be segmented. These processes all increase the surface area of the dust, thereby enhancing clearance of dissolution-absorption. Both categories of dust redistribute within the P region over time.

Overall, the biokinetics of small amounts of particles and fibers deposited in the P region by inhalation appear to be very similar if the fibers are less than about 20 μm long. This simplified comparison is not valid if the particles or fibers elicit adverse responses in the lung because of their chemical composition, physical configuration, or if the amount of dust overwhelms normal macrophage functions.

EFFECTS OF EXPOSURE CONCENTRATIONS ON LUNG RETENTION AND CLEARANCE
General Comments on Lung Overload

Small amounts of poorly soluble dust that deposit in the pulmonary airspaces are efficiently phagocytized by PAMs. The phagocytized dust is subjected to macrophage-mediated clearance and dissolution-absorption. These competing processes clear the deposited dust with an effective half-time of

Table 1   Comparison of Factors that Influence the Biokinetics of Respirable Particles and Fibers Deposited in the Pulmonary Region

Factor	Particles	Fibers
Size-dependent phagocytosis	No	Yes
Size-dependent macrophage-mediated pulmonary clearance	Yes	Yes
Cause functional overload of macrophages	Yes	Yes
Fragment or split into sub-units	Yes	Yes
Disintegration of units and sub-units	Yes	Yes
Segmentation of units and sub-units	No	Yes
Cleared by dissolution-absorption	Yes	Yes
Redistribute in lung over time	Yes	Yes

about 2 months for most rodents and about 2 years for large mammalian species, including humans (Snipes, 1989). The PAMs can efficiently phagocytize dust during repeated exposures, and they appear to function normally in spite of the fact that the lung is accumulating dust. Under these exposure conditions, lung clearance proceeds at a normal rate and an equilibrium lung burden of dust is eventually reached when the pulmonary deposition rate equals the effective clearance rate.

Repeated inhalation of large amounts of dust results in a different scenario. An immediate response is recruitment of additional PAMs into the pulmonary airspaces (Brain, 1971; Adamson and Bowden, 1978; Brain, 1985). Unfortunately, continued inhalation of large amounts of dust overwhelms the functional abilities of macrophages. Exposure conditions that overwhelm PAMs produce an inflammatory response which includes the influx of large numbers of neutrophils into the pulmonary airspaces (Larsen et al., 1983; Lehnert et al., 1985). Lung clearance is also adversely affected, and the accumulation of dust in the lungs proceeds at an accelerated rate.

The concept of overloaded lung clearance has been discussed by numerous investigators who related altered lung clearance to the amounts of dust deposited in the lungs. For example, Labelle and Brieger (1961) suggested that a lung burden greater than 1.5 mg of dust in rats will significantly alter or block lung clearance mechanisms. In the first study that directly evaluated lung overload, Bolton et al. (1983) suggested that overloading of lung clearance occurs in rats when lung burdens of fibers (or particles) exceed about 1.5 mg/rat. It has been generally recognized for the past several decades that large lung burdens of dust can alter lung clearance and cause lung overload, and numerous recent studies have reported lung overload in rats that chronically inhaled relatively large amounts of diesel exhaust particles, carbon black, or TiO_2, as well as other types of poorly soluble dusts. It is almost certain that lung overload will result from repeated or chronic exposures of rats to large amounts of most types of relatively nontoxic, respirable dust.

The concept of lung overload was discussed in detail by Morrow (1986, 1988, 1992), who summarized the nonspecific adverse effects noted in lungs of rats exposed chronically to high concentrations of relatively nontoxic, poorly soluble particles. Pulmonary responses during overload include slowed or halted macrophage-mediated clearance, widespread accumulations of dust-laden PAMs within alveoli, enhanced appearance of dust deposits in the interstitium and TLNs, persistent inflammation, the eventual development of alveolitis and granulomatous lung disease, and the possible development of lung tumors. Several studies have produced lung tumors in rats that chronically inhaled large amounts of poorly soluble types of respirable particles (Mauderly et al., 1990; Mauderly, 1994). Noncytotoxic particles such as diesel exhaust particles, carbon black, and TiO_2 require relatively high weekly inhalation exposure rates to produce lung overload and lung tumors in rats; toxic particles such as quartz require lower exposure rates to cause these effects.

A limited number of studies relevant to lung overload have been conducted using fibers. Ferin and Leach (1976) demonstrated that accumulation
of fibers in the lungs of rats slows macrophage-mediated lung clearance.
These investigators exposed rats by inhalation to aerosols containing amosite
or chrysotile fibers at concentrations of 11 to 14 mg/m^3. Exposures ranged
from acute 1-hour exposures to repeated exposures 5.5 hours/day for 22 days.
After exposures to the fibers, the rats were exposed for 7 hours to an aerosol
containing about 15 mg TiO$_2$/m^3. Clearance of the TiO$_2$ was evaluated by
sacrificing rats at 1, 25, and 130 days after exposure to the TiO$_2$ and measuring lung burdens of Ti. Results of this study indicated that exposures for 1–3
hours to 10 mg fibers/m^3 or for more than 11 days to 1 mg fibers/m^3 suppresses lung clearance of TiO$_2$ particles. However, relationships between
amounts of fibers that accumulated in the lungs of the rats and the extent of
altered lung clearance were not determined.

Le Bouffant et al. (1984) produced lung burdens of about 9 mg JM 100
microfibers/g dry lung in rats using a 12-month chronic exposure 5 hours/
day, 5 days/week to about 5 mg fibers/m^3. This would be equivalent to lung
burdens of about 2 mg fibers/g wet tissue (Tillery and Lehnert, 1986). The
rats were housed in a clean air environment after the 12-month chronic exposure and lung clearance was evaluated. The lung clearance half-time for these
fibers was about 300 days, which is a slow lung clearance rate for poorly
soluble materials deposited in the lungs of rats. Macrophage-mediated lung
clearance was obviously slowed. Additionally, the concentrations of microfi-
bers in TLNs were quite large. The slowed macrophage-mediated lung clearance and high concentrations of dust in TLNs support the conclusion that
lung overload resulted from these chronic exposures.

Morrow (1986, 1988, 1992) reviewed early and contemporary work related to lung overload and concluded that the amount of dust associated with
overloading of rat lungs appears to be about 1 mg dust/g lung. Pritchard
(1989) summarized results from several chronic inhalation studies in which
lung clearance in rats was evaluated during exposures to a variety of poorly
soluble aerosols for long periods of time. Pritchard concluded that chronic
exposures of rats to a weekly average of 10 mg/m^3 for 1 yr will result in lung
overload. Two examples of exposures that would attain this weekly average
are a continuous exposure to 0.2 mg particles/m^3 or to 0.8 mg/m^3 8 hours/
day, 5 days/week. Muhle et al. (1990) concluded that retardation of alveolar
clearance is detectable when lung burdens are 0.5 mg of particles or more for
rats, or 1.0 mg or more for mice; specific lung burdens (mass of dust/g lung)
would be on the order of 0.4 mg/g lung for rats or 5 mg/g lung for mice.

The exposure conditions that cause lung overload in the rat lung have
not been conclusively determined, but two possibilities exist. One possibility
is that a critical lung burden of dust is reached at some point during repeated
inhalation exposures. Normal lung clearance occurs until the critical lung
burden of particles has accumulated and triggers the biological changes in

the lung that are associated with lung overload. This concept of lung overload prevailed until recent years. Alternatively, the amount of poorly soluble material deposited daily in the lungs is a dominant factor in lung overload. Muhle et al. (1988) proposed that overloading is primarily a function of the rate of pulmonary deposition. These researchers summarized the results from lung clearance tests in rats and Syrian hamsters exposed for several months to a variety of aerosols and concluded that lung clearance is retarded by chronic exposure to respirable particles at concentrations of 3 mg/m³ or higher. The smallest lung burdens associated with reduced rates of lung clearance in their studies was 0.8 mg of fly ash or 0.93 mg of glass fibers. Yu et al. (1989) described a similar concept, which they termed "critical deposition rate," which was based on mathematical analyses of lung clearance rates from several chronic inhalation studies. The critical deposition rate was defined as the "rate above which the overload condition will be present if the exposure time is sufficient." These authors included a calculation of the critical deposition rate, along with a table of theoretical times required to reach lung overload with several rates of particle deposition in the lungs. The higher the inhaled concentration of particles, the shorter the time required to reach the overload condition.

An alternative definition of critical deposition rate is the "rate above which macrophage-mediated lung clearance is overwhelmed." This may represent the exposure-response scenario that triggers lung overload, and could occur very soon after initiating a chronic inhalation study with rats if the concentration of respirable particles is sufficiently high. The critical deposition rate hypothesis for lung overload appears plausible because some of the biological indications of lung overload are seen within days or weeks of initiating chronic exposures that result in excessive lung burdens of dust. Additional experimental work should be done to determine which hypothesis best accounts for lung overload.

Henderson et al. (1992) attempted to evaluate the effect of exposure rate on pulmonary inflammatory responses in rat lungs. Rats were exposed to carbon black at concentrations of 3.5 mg/m³, 13 mg/m³, or 98 mg/m³; all rats received a cumulative exposure of 4700 mg•hour/m³ over a 12-week period. The rate of exposure to carbon black did not affect the accumulated lung burdens of carbon black, nor did the exposure rate affect the inflammatory responses that were quantified. The fact that Henderson et al. did not see a relationship between exposure rate and pulmonary responses might be because all three exposure concentrations produced pulmonary deposition rates of particles that exceeded the critical deposition rate.

To date, lung overload, with all of its adverse health effects, has been demonstrated repeatedly in rats, and the relevance of lung overload to other laboratory animal species and humans is still being investigated. Only a limited number of human studies have demonstrated a relationship between lung burdens of dust and altered pulmonary function. Lippmann and Timbrell

(1990) summarized human experience with dust loading of the lungs of workers who had long-term inhalation exposures to a variety of amphiboles. Timbrell conducted postmortem analyses of lungs from these individuals. A small percentage of heavily exposed workers had extensive fibrosis and severely altered lung clearance in which there was little or no clearance from parts of the lungs. Timbrell concluded that lung clearance of fibers is strongly dependent on the lung burden of fibers and the degree of pulmonary fibrosis. Interestingly, none of the individuals had lung tumors attributable to the accumulated burdens of fibers.

Coal miners represent another group likely to demonstrate one or more attributes of lung overload if the phenomenon applies to humans. Mauderly (1994) summarized information available from eight studies of coal miners in the United States and Europe in which lung burdens of dust were measured in a total of 1225 autopsied subjects. Substantial lung burdens of dust were measured in the lungs of these coal miners. The overall average specific lung burden was estimated by Mauderly (1994) to be in the range of 7 to 14 mg dust/g lung. No lung tumors were attributable to these specific lung burdens of coal dust in humans, but the burdens were sufficiently high to produce lung tumors if the human pulmonary response to accumulations of inhaled particles parallels the well-documented responses reported for rats.

Chemical Constituents of Inhaled Materials as Factors in Lung Overload

Very little information is available relevant to this subject. In general, any type of poorly soluble respirable dust can theoretically cause lung overload if certain exposure conditions are used, or if sufficiently large lung burdens of materials are reached during repeated or chronic exposures. The physical and chemical attributes of the inhaled materials may influence the rate with which the changes in lungs occur, and the relative amounts of inhaled material that are required to produce the overload condition.

Effects of Exposure Conditions on Clearance to Thoracic Lymph Nodes

The amount of dust deposited in the lungs has important consequences for the TLNs. The translocation rates of materials from the P region to lymph nodes are not constant for all types and amounts of materials deposited in the lungs. In an early study, Klosterkötter and Bünemann (1961) noted that increased lung burdens of particles causes increased translocation of the particles to TLNs in rats. Only small amounts of inert and fine-grained dusts of TiO_2 and gamma-aluminum oxide penetrated into the lymph passages, except when large amounts were deposited in the lungs. In contrast, over 50% of the amount of silica eliminated from the lungs was found in the mediastinal lymph nodes in some tests. Klosterkötter and Bünemann (1961) concluded that the lymphatic clearance route is used mainly for lung clearance of toxic substances (quartz, for example), while its role is negligible in the case of inert substances such as TiO_2. In a later report Klosterkötter and Gono (1971) noted that concurrent exposure to quartz and TiO_2 leads to a threefold increase in the quantity of TiO_2 observed in the lymph nodes of exposed rats as compared with the amounts observed after exposure to only TiO_2. This indicates that the accelerated lymphatic clearance is not specific for translocation of quartz. Kilpper et al. (1976) noted an increased clearance to regional lymph nodes in dogs for large lung burdens of insufflated tantalum.

Exposure concentrations clearly influence translocation of materials cleared from the lungs. Ferin (1977) concluded that low-exposure concentrations to insoluble particles favor a primary removal pathway for particle-laden macrophages via the mucociliary escalator in rats; however, with increasing exposure levels, translocation of free particles to the pulmonary lymphatic system increases. Ferin and Feldstein (1978) exposed rats to TiO_2 by inhalation or intratracheal instillation. Twenty-five days after exposure, the amounts of TiO_2 in the lymph nodes accounted for less than 1% of the total lung burden when lung burdens ranged from 0.1 and 1 mg. However, the TiO_2 content of the hilar lymph nodes increased substantially for lung burdens greater than 1 mg. With lung burdens in the range of 10 mg, about 4% of the TiO_2 was found in the lymph nodes. Fewer TiO_2 particles were found in lymph nodes after inhalation than after intratracheal instillation. The authors suggested that the carrier fluid might have facilitated particle penetration into the interstitium and that macrophages cannot contain all of the particles when lung burdens of particles are high, thereby increasing the probability that particles will enter the interstitium and translocate to lymph nodes. Vincent et al. (1987) reached a similar conclusion in a more recent study in which rats were chronically exposed to TiO_2 or quartz dust at concentrations ranging from 0.01 to 90 mg/m³. Exposure times ranged up to 222 days. Similar results for accumulations of particles in TLNs were observed for both types of mineral dusts. Lymph node burdens were negligible or nonexistent when exposure concentrations or lung burdens of the mineral dusts were low. However, for exposures to the higher concentrations of the dusts, or after lung burdens of the dusts were several mg, measurable amounts of the dust particles were noted in the lymph nodes. The limits of detection for TiO_2 and quartz in lymph nodes were estimated to be 12 μg and 10 μg, respectively. On the order of 1 mg or more of either type of dust was accumulated in lungs before detectable amounts were found in TLNs. An alternative interpretation of this result is that both types of dust accumulated in TLNs for all exposure conditions, but that the amounts could not be quantified for the lower-level exposures. Numerous studies with a variety of radioactive particles have demonstrated that particles are transported to TLNs under conditions where μg quantities of the particles were deposited in the lungs by intratracheal instillation or inhalation.

Chronic inhalation exposures to large amounts of dust clearly demonstrate that exposure conditions can markedly affect rates of particle translocation from lungs to TLNs. Vostal et al. (1981; 1982), Muhle et al. (1988), Strom et al. (1988), and Creutzenberg et al. (1990) reported elevated burdens of diesel exhaust particles in TLNs in rodents exposed for long periods of time to relatively high concentrations of diesel exhaust. Similar results have been obtained in studies with carbon black (Strom et al., 1989), for exposures to aerosols of TiO_2 and quartz (Vincent et al., 1987), and for photocopy toner (Bellmann et al., 1989).

Clearance of fibers from lungs to TLNs does not appear to be influenced by exposure conditions as much as is clearance of particles. Beattie and Knox (1961) evaluated the mineral content of lungs and TLNs from 50 workers in the asbestos industry. The workers had been exposed for times ranging from 3 months to 46 years, and the mean mineral content of their lung tissue ranged up to 10.1 mg/ g dried lung. The lung burdens appeared to increase about linearly at a rate of 0.25 mg/g dried lung per year. The highest concentration of mineral material within hilar lymph nodes was never greater than 6 times that in the lung tissue and was usually about a factor of 2–3 times that in the lung tissue. This ratio of TLN/lung burden is low relative to expectations for most poorly soluble particles (Thomas, 1968). Exposures to most types of respirable particles that produce lung burdens of this magnitude would result in lymph-node concentrations of at least an order of magnitude higher than that noted by Beattie and Knox (1961). This result would be expected for a moderately soluble type of material in the lungs, but not for poorly soluble fibers. A reasonable interpretation of this result is that translocation of fibers from the lungs to TLNs in these workers was biased towards small fibers which are more soluble than the larger fibers and therefore cleared from the TLNs relatively fast by dissolution-absorption.

Lung clearance rates for one material can be influenced by the presence of other materials. LaBelle and Brieger (1960; 1961) reported enhanced lung clearance of intratracheally injected UO_2 particles in rats when the UO_2 particles were injected together with other types of particles. The author's interpretation was that the mixtures of particles contained total numbers of particles sufficient to elicit an increase in the numbers of phagocytic cells in the lungs. The larger numbers of phagocytic cells enhanced lung clearance. Based on results of a similar study, Ferin et al. (1965) reported that lung clearance of SiO_2 particles is faster when macrophages are stimulated by an intratracheal injection of trypan blue, or by inhalation of TiO_2. Clearance is slower when the rats are pretreated with X-ray irradiation. The stimulus for recruitment of more macrophages into the lungs is offset by the radiation damage that reduces the ability of the lungs to recruit additional phagocytic cells. Adamson and Bowden (1982) reported an increased accumulation of carbon particles in hilar lymph nodes of mice under conditions where macrophage response to the burden of particles is reduced and interstitial burdens of the particles are elevated. They also attributed their results to an increased degree of direct penetration of the carbon particles into the interstitium, followed by transport to the lymph nodes.

Available evidence indicates that translocation from lungs to TLNs is independent of the type of particle or fiber unless the constituents are cytotoxic. All inhalation exposures to respirable particles or fibers result in the potential for interstitialization of some portion of the deposited dust. The larger the amount of deposited dust (or the larger the numbers of dust particles deposited in the lungs) the greater the potential for the dust to penetrate into the interstitium. Dust in the pulmonary interstitium then has the opportunity for translocation to TLNs. It has been well documented that the amount of dust that penetrates into the pulmonary interstitium and is subsequently transported to TLNs increases as the lung burden increases. This aspect of lung overload has not been fully explored and may have important implications for the lymphatic system and immune responses.

SUMMARY

This chapter summarizes contemporary scientific knowledge relative to the biokinetics of particles and fibers that deposit in the P region of mammalian species. With the exception of fibers longer than about 15–20 μm that cannot be completely phagocytized in the lungs, the fate of particles and fibers appears to be similar. Small fibers and all respirable particles appear to be readily endocytized by phagocytic cells. Particles and fibers can penetrate into the pulmonary interstitium to be retained there or translocated to TLNs, and amounts of both types of dust interstitialized and transported to TLNs appear to be proportional to the lung to the lung burdens of dust when lung burdens are relatively small. When exposure conditions overwhelm the ability of PAMs to rapidly phagocytize dust, the probability increases that dust can enter the pulmonary interstitium and be translocated to TLNs. Particles and fibers are subject to dissolution-absorption and size modification. Particles are subjected to dissolution and leaching, and some particles fragment; fibers are subjected to dissolution and leaching, and some fibers split and/or segment. Both types of dust cause altered lung clearance and lung overload in rats if exposure levels are sufficiently high.

The biokinetics of dusts deposited in the pulmonary region of the respiratory tract appear to be qualitatively similar among humans and other mammalian species. Inhalation exposure rates influence the biokinetics of inhaled dust, possibly to different degrees for different mammalian species. Unfortunately, available data do not allow thorough evaluations of the exposure-dependent biokinetics of inhaled dust for any species, and this remains an important issue. Another important issue yet to be resolved is the concept of lung overload as it relates to human inhalation exposures. Available evidence strongly suggests that human inhalation exposures to particles or fibers can produce substantial specific lung burdens of either type of dust and cause adverse health effects, but that the accumulations of dust do not cause the spectrum of nonspecific effects associated with lung overload as it has been applied to rats and other laboratory animal species.

ACKNOWLEDGMENTS

Portions of the research described in this chapter were sponsored by the U. S. Department of Energy, Office of Health and Environmental Research under Contract Number DE-AC04–76EV01013. The author is indebted to colleagues at the Inhalation Toxicology Research Institute and associates in the scientific community who produced the cumulative information used in this chapter.

REFERENCES

Adamson, IYR: Cellular responses and translocation of particles following deposition in the lung. J. Aerosol Med 3(Suppl 1):S31–S42, 1990.

Adamson, IYR, Bowden, DH: Adaptive responses of the pulmonary macrophagic system to carbon. II. Morphologic studies. Lab Invest 38:430–438, 1978.

Adamson, IYR, Bowden, DH: Dose response of the pulmonary macrophagic system to various particulates and its relationship to transepithelial passage of free particles. Exp Lung Res 2:165–175, 1981.

Adamson, IYR, Bowden, DH: Effects of irradiation on macrophagic response and transport of particles across the alveolar epithelium. Am J Pathol 106:40–46, 1982.

Bailey, MR, Kreyling, WG, Andre, S, Batchelor, A, Black, A, Collier, CG, Drosselmeyer, E, Ferron, GA, Foster, P, Haider, B, Hodgson, A, Masse, R, Metivier, H, Morgan, A, Müller, HL, Patrick, G, Pearman, I, Pickering, S, Ramsden, D, Stirling, C, Talbot, RJ: An interspecies comparison of the lung clearance of inhaled monodisperse cobalt oxide particles - Part I: Objectives and summary of results. J Aerosol Sci 20:169–188, 1989.

Beattie, J, Knox, JF: Studies of mineral content and particle size distribution in the lungs of asbestos textile workers. In: Inhaled Particles and Vapours, edited by CN Davies, pp. 419–433. Pergamon, Oxford, UK, 1961.

Bellmann, B, Muhle, H, Creutzenberg, O, Kilpper, R, Morrow, P, Mermelstein, R: Reversibility of clearance impairment after subchronic test toner inhalation. Exp Path 37:234–238, 1989.

Bernstein, DM, Drew, RT, Kuschner, M: Experimental approaches for exposure to sized glass fibers. Environ Health Perspec 34:47–57, 1980.

Bolton, RE, Vincent, JH, Jones, AD, Addison, J, Beckett, ST: An overload hypothesis for pulmonary clearance of UICC amosite fibres inhaled by rats. Brit J Ind Med 40:264–272, 1983.

Bowden, DH: Macrophages, dust, and pulmonary diseases. Exp Lung Res 12:89–107, 1987.

Bowden, DH, Adamson, IYR: Pathways of cellular efflux and particulate clearance after carbon instillation to the lung. J Pathol 143:117–125, 1984.

Bowden, DH, Adamson, IYR: Bronchiolar and alveolar lesions in the pathogenesis of crocidolite-induced pulmonary fibrosis in mice. J Pathol 147:257–267, 1985.

Brain, JD: The effects of increased particles on the number of alveolar macrophages. In: Inhaled Particles III, Vol. 1, edited by WH Walton. pp. 209–223. Old Woking, Surrey, England, Unwin Brothers Limited, 1971.

Brain, JD: Macrophages in the respiratory tract. In Handbook of Physiology, Section 3: The Respiratory System, edited by AP Fishman, AB Fisher, SR Geiger, pp. 447–471. American Physiological Society, Bethesda, MD. 1985.

Brody, AR, Hill, LH, Adkins, B, Jr., O'Connor, RW: Chrysotile asbestos inhalation in rats: Deposition pattern and reaction of alveolar epithelium and pulmonary macrophages. Am Rev Resp Dis 123:670–679, 1981.

Brundelet, PJ: Experimental study of the dust-clearance mechanism of the lung I. Histological study in rats of the intra-pulmonary bronchial route of elimination. Acta Pathol Microbiol Scand (Suppl 175):1–141, 1965.

Cartwright, J, Skidmore, JW: The size distribution of dust retained in the lungs of rats and dust collected by size-selecting samplers. Ann Occup Hyg 7:151–167, 1964.

Churg, A, Wright, JL, Gilks, B, DePaoli, L: Rapid short-term clearance of chrysotile compared with amosite asbestos in the guinea pig. Am Rev Resp Dis 139:885–890, 1989.

Clarke, WJ, Bair, WJ: Plutonium inhalation studies-VI: Pathologic effects of inhaled plutonium particles in dogs. Health Phys 10:391–398, 1964.

Creutzenberg, O, Bellmann, B, Heinrich, U, Fuhst, R, Koch, W, Muhle, H: Clearance and retention of inhaled diesel exhaust particles, carbon black, and titanium dioxide in rats at lung overload conditions. J Aerosol Sci 21(Suppl 1):S455-S458, 1990.

Cuddihy, RG: Mathematical models for predicting clearance of inhaled radioactive materials. In: Lung Modelling for Inhalation of Radioactive Materials, edited by H Smith, G Gerber, pp. 167–176. Proceedings of a meeting jointly organized by the Commission of European Communities and the National Radiological Protection Board, Eur9384EN, 1984.

Dauber, JH, Daniele, RP: Secretion of chemotaxins by guinea pig lung macrophages. I. The spectrum of inflammatory cell responses. Exp Lung Res 1:23–32, 1980.

Diel, JH, Mewhinney, JA: Fragmentation of inhaled $^{238}PuO_2$ particles in lung. Health Phys 44:135–143, 1983.

Diel, JD, Mewhinney, JA, Snipes, MB: Distribution of inhaled $^{238}PuO_2$ particles in Syrian hamster lungs. Radiat Res 88:299–312, 1981.

Ferin, J: Effect of particle content of lung on clearance pathways. In: Pulmonary Macrophage and Epithelial Cells, edited by CL Sanders, RP Schneider, GE Dagle, HA Ragan, pp. 414–423. Springfield, VA, ERDA, Technical Information Center, 1977.

Ferin, J, Feldstein, ML: Pulmonary clearance and hilar lymph node content in rats after particle exposure. Environ Res 16:342–352, 1978.

Ferin, J, Leach, LJ: The effect of amosite and chrysotile asbestos on the clearance of TiO_2 particles from the lung. Environ Res 12:250–254, 1976.

Ferin, J, Oberdörster, G, Penney, DP: Pulmonary retention of ultrafine and fine particles in rats. Am J Resp Cell Mol Biol 6:535–542, 1992.

Ferin, J, Urbankova, G, Vlckova, A: Pulmonary clearance and the function of macrophages. Arch Environ Health 10:790–795, 1965.

Fleischer, RL, Raabe, OG: Fragmentation of respirable PuO_2 particles in water by alpha decay - A mode of "dissolution." Health Phys 32:253–257, 1977.

Gearhart, JM, Diel, JH, McClellan, RO: Intrahepatic distribution of plutonium in beagles. Radiat. Res. 84:343–352, 1980.

Geiser, M, Baumann, M, Cruz-Orive, LM, Im Hof, V, Waber, U, Gehr, P: The effect of particle inhalation on macrophage number and phagocytic activity in the intrapulmonary conducting airways of hamsters. Am J Resp Cell Mol Biol 10:594–603, 1994.

Goodglick, LA, Kane, AB: Role of reactive oxygen metabolites in crocidolite asbestos toxicity to mouse macrophages. Cancer Res 46:5558–5566, 1986.

Green, GM: Alveolobronchiolar transport; observations and hypothesis of a pathway. Chest 59(suppl):1S, 1971.

Green, GM: Alveolobronchiolar transport mechanisms. Arch Intern Med 131:109–114, 1973.

Gross, P, Westrick, M: The permeability of lung parenchyma to particulate matter. Am J Pathol 30:195–213, 1954.

Guilmette, RA, Muggenburg, BA, Hahn, FF, Mewhinney, JA, Seiler, FA, Boecker, BB, McClellan, RO: Dosimetry of [239]Pu in dogs that inhaled monodisperse aerosols of PuO$_2$. Radiat. Res. 110:199–218, 1987.

Hahn, FF, Newton, GJ, Bryant, PL: In vitro phagocytosis of respirable-sized monodisperse particles by alveolar macrophages. In: Pulmonary Macrophage and Epithelial Cells, edited by CL Sanders, RP Schneider, GE Dagle, HA Ragan, pp. 424–436. U.S. Dept. of Commerce, Springfield, VA, ERDA Series 43, 1977.

Hammad, YY: Deposition and elimination of MMMF. In: Biological Effects of Man-Made Mineral Fibres, Vol. 2, pp. 126–142, World Health Organization, Copenhagen, 1984.

Hammad, Y, Simmons, W, Abdel-Kader, H, Reynolds, C, Weill, H: Effect of chemical composition on pulmonary clearance of man-made mineral fibres. Ann Occup Hyg 32(Suppl 1):769–779, 1988.

Henderson, RF, Barr, EB, Cheng, YS, Griffith, WC, Hahn, FF: The effect of exposure pattern on the accumulation of particles and the response of the lung to inhaled particles. Fund Appl Toxicol 19:367–374, 1992.

Hesterberg, TW, McConnell, EE, Miller, WC, Hamilton, R, Bunn, WB: Pulmonary toxicity of inhaled polypropylene fibers in rats. Fund Appl Toxicol 19:358–366, 1992.

Holma, B: Lung clearance of mono- and di-disperse aerosols determined by profile scanning and whole-body counting: A study on normal and SO$_2$ exposed rabbits. Acta Med Scand (Suppl. 474):1–103, 1967.

Holt, PF: Transport of inhaled dust to extrapulmonary sites. J Pathol 133:123–129, 1981.

Holt, PF: Translocation of asbestos dust through the bronchiolar wall. Environ Res 27:255–260, 1982.

Hourihane, DO'B: A biopsy series of mesotheliomata, and attempts to identify asbestos within some of the tumors. Ann. N. Y. Acad. Sci. 132:647–673, 1965.

Huang, J, Hisanaga, N, Sakai, K, Iwata, M, Ono, Y, Shibata, E, Takeuchi, Y: Asbestos fibers in human pulmonary and extrapulmonary tissues. Am J Ind Med 14:331–339, 1988.

Johnson, NF, Griffiths, DM, Hill, RJ: Size distribution following long-term inhalation of MMMF. In: Biological Effects of Man-Made Mineral Fibres, Vol. 2. pp. 102–125. World Health Organization, Copenhagen, 1984.

Kauffer, E, Vigneron, JC, Hesbert, A, Lemonnier, M: A study of the length and diameter of fibres, in lung and in broncho-alveolar lavage fluid, following exposure of rats to crysotile asbestos. Ann Occup Hyg 31:233–240, 1987.

Kelly, DP, Merriman, EA, Kennedy, GL, Jr., Lee, KP: Deposition, clearance, and shortening of Kevlar para-aramid fibrils in acute, subchronic, and chronic inhalation studies in rats. Fund Appl Toxicol 21:345–354, 1993.

Kilburn, KH: Clearance zones in the distal lung. Ann NY Acad Sci 221:276–281, 1974.

Kilpper, RW, Bianco, A, Gibb, FR, Landman, S, Morrow, PE: Uptake and retention of insufflated tantalum by lymph nodes. In: Radiation and the Lymphatic System, edited by JE Ballou, pp. 46–53. Washington, DC, Technical Information Center, ERDA, CONF-740930, 1976.

Klosterkötter, W, Bünemann, G: Animal experiments on the elimination of inhaled dust. In: Inhaled Particles and Vapours, edited by CN Davies, pp. 327–341. Oxford, UK, Pergamon Press, 1961.

Klosterkötter, W, Gono, F: Long-term storage, migration and elimination of dust in the lungs of animals, with special respect to the influence of polyvinyl-pyridine-n-oxide. In: Inhaled Particles III, Vol. 1, edited by WH Walton, pp. 273–281. Old Woking, Surrey, UK, Unwin, 1971.

Kreyling, WG: Interspecies comparison of lung clearance of "insoluble" particles. J Aerosol Med 3(Suppl 1):S93-S110, 1990.

LaBelle, CW, Brieger, H: The fate of inhaled particles in the early postexposure period. II. The role of pulmonary phagocytosis. Arch Environ Health 1:423–427, 1960.

LaBelle, CW, Brieger, H: Patterns and mechanisms in the elimination of dust from the lung, In Inhaled Particles and Vapours, edited by CN Davies, pp. 356–368. Oxford, UK, Pergamon, 1961.

Larsen, GL, Parrish, DA, and Henson, PM. Lung defense: The paradox of inflammation. Chest 83:1S-5S, 1983.

Lauweryns, JM, Baert, JH: The role of the pulmonary lymphatics in the defenses of the distal lung: Morphological and experimental studies of the transport mechanisms of intratracheally instilled particles. Ann NY Acad Sci 221:244–275, 1974.

Lauweryns, JM, Baert, JH: Alveolar clearance and the role of the pulmonary lymphatics, Am Rev Respir Dis 115:625–683, 1977.

Leadbetter, MR, Corn, M: Particle size distribution of rat lung residues after exposures to fiberglass dust clouds. Am Ind Hyg Assoc J 33:511–522, 1972.

Leak, LV: Pulmonary lymphatics and their role in the removal of interstitial fluids and particulate matter. In: Respiratory Defense Mechanisms, Part II, edited by JD Brain, DF Proctor, LM Reid, pp. 631–685. Marcel Dekker, Inc., NY, 1977.

Leak, LV, Jamuar, MP: Ultrastructure of pulmonary lymphatic vessels. Am Rev Resp Dis 128:S59-S65, 1983.

Le Bouffant, L, Daniel, H, Henin, JP, Martin, JC, Normand, C, Tichoux, G, Trolard, F: Experimental study on long-term effects of inhaled MMMF on the lungs of rats. Ann Occup Hyg 31:765–790, 1987.

Le Bouffant, L, Henin, JP, Martin, JC, Normand, C, Tichoux, G, Troland, F: Distribution of inhaled MMMF in the rat lung. In: Biological Effects of Man-Made Mineral Fibres, Vol. 2. pp. 143–168. World Health Organization, Copenhagen, 1984.

Lee, KP: Lung response to particulates with emphasis on asbestos and other fibrous dusts. CRC Crit Rev Toxicol 14:33–86, 1985.

Lee, KP, Barras, CE, Griffith, FD, Waritz, RS: Pulmonary response and transmigration of inorganic fibers by inhalation exposure. Am J Pathol 102:314–323, 1981a.

Lee, KP, Barras, CE, Griffith, FD, Waritz, RS, Lapin, CA: Comparative pulmonary responses to inhaled inorganic fibers with asbestos and fiberglass. Environ Res 24:167–191, 1981b.

Lee, KP, Kelly, DP, O'Neal, FO, Stadler, JC, Kennedy, GL, Jr.: Lung response to ultrafine Kevlar aramid synthetic fibrils following 2-year inhalation exposure in rats. Fund Appl Toxicol 11:1–20, 1988.

Lehnert, BE: Alveolar macrophages in a particle "overload" condition. J Aerosol Med 3(Suppl 1):S9-S30, 1990.

Lehnert, BE, Valdez, YE, and Bomalaski, SH. Lung and pleural "free-cell responses" to the intrapulmonary deposition of particles in the rat. J. Toxicol. Environ. Health 16:823–839, 1985.

LeFevre, ME, Green, FHY, Joel, DD, Laqueur, W: Frequency of black pigment in livers and spleens of coal workers: Correlation with pulmonary pathology and occupational information. Human Pathol. 13:1121–1126, 1982.

Lippmann, M: Man-made mineral fibers (MMVF): Human exposures and health risk assessment. Toxicol Ind Health 6:225–246, 1990.

Lippmann, M, Timbrell, V: Particle loading in the human lung - Human experience and implications for exposure limits. J Aerosol Sci 3(Suppl 1):S155–S1168, 1990.

Lundborg, M, Eklund, A, Lind, B, Camner, P: Dissolution of metals by human and rabbit alveolar macrophages. Brit J Ind Med 42:642–645, 1985.

Lundborg, M, Lind, B, Camner, P: Ability of rabbit alveolar macrophages to dissolve metals. Exp Lung Res 7:11–22, 1984.

Mauderly, JL: Contribution of inhalation bioassays to the assessment of human health risks from solid airborne particles. In: Toxic and Carcinogenic Effects of Solid Particles in the

Respiratory Tract, edited by U Mohr, DL Dungworth, JL Mauderly, G Oberdörster, pp. 355–365. Washington, DC, International Life Sciences Institute Press, 1994.

Mauderly, JL, Cheng, YS, Snipes, MB: Particle overload in toxicological studies: Friend or foe?. J Aerosol Med 3(Suppl 1):S169-S187, 1990.

McShane, JF, Dagle, GE, Park, JF: Pulmonary distribution of inhaled [239]PuO[2] in dogs. *In:* Pulmonary Toxicology of Respirable Particles, edited by CL Sanders, FT Cross, GE Dagle, JA Mahaffey, pp. 248–255. Springfield, VA, DOE Conf791002, National Technical Information Service, 1980.

Medinsky, MA, Kampcik, SJ: Pulmonary retention of [^{14}C]benzo[a]pyrene in rats as influenced by the amount instilled. Toxicology 35:327–336, 1985.

Medinsky, MA, Cheng, YS, Kampcik, SJ, Henderson, RF, Dutcher, JS: Disposition and metabolism of ^{14}C-solvent yellow and solvent green aerosols after inhalation. Fund Appl Toxicol 7:170–178, 1986.

Mercer, TT; On the role of particle size in the dissolution of lung burdens. Health Phys 13:1211–1221, 1967.

Mewhinney, JA, Griffith, WC: A tissue distribution models for assessment of human inhalation exposures to [241]AmO[2]. Health Phys 44 (Suppl 1):537–544, 1983.

Meyer, EC, Dominguez, EAM, Bensch, KG: Pulmonary lymphatic and blood absorption of albumin from alveoli: A quantitative comparison. Lab Invest 20:1–8, 1969.

Misra, V, Rahman, Q, Viswanathan, PN: Biochemical changes in guinea pig lungs due to amosite asbestos. Environ Res 16:55–61, 1978.

Morgan, A: Effect of length on the clearance of fibres from the lung and on body formation. In: Biological Effects of Mineral Fibres, Vol. 1, edited by JC Wagner, W Davis, pp. 329–335. IARC Scientific Publications No. 30, International Agency for Research on Cancer, 1980.

Morgan, A, Black, A, Moores, SR, Lambert, BE: Translocation of [239]Pu in Mice following inhalation of sized [239]PuO[2]. Health Phys 50:535–539, 1986.

Morgan, A, Evans, JC, Holmes, A: Deposition and clearance of inhaled fibrous minerals in the rat. Studies using radioactive tracer techniques. In: Inhaled Particles IV, Part 1, edited by WH Walton, B McGovern, pp. 259–274. Oxford, UK, Pergamon Press, 1977.

Morgan, A, Holmes, A: The deposition of MMMF in the respiratory tract of the rat, their subsequent clearance, solubility *in vivo* and protein coating. In: Biological Effects of Man-Made Mineral Fibres, Vol. 2, pp. 1–17, World Health Organization, Copenhagen, 1984.

Morgan, A, Holmes, A: The distribution and characteristics of asbestos fibers in the lungs of Finnish anthophyllite mine-workers. Environ Res 33:62–75, 1984.

Morgan, A, Holmes, A: Solubility of asbestos and man-made mineral fibers *in vitro* and *in vivo:* Its significance in lung disease. Environ Res 39:475–484, 1986.

Morgan, A, Holmes, A, Davison, W: Clearance of sized glass fibres from the rat lung and their solubility *in vivo.* Ann Occup Hyg 25:317–331, 1982.

Morgan, A, Lally, AE, Holmes, A: Some observations of trace metals in chrysotile asbestos. Ann Occup Hyg 16:231–240, 1973.

Morgan, A, Morris, KJ, Launder, KA, Hornby, SB, Collier, CG: Retention of glass fibers in the rat trachea following administration by intratracheal instillation. Inhal Toxicol 6:241–251, 1994.

Morgan, A, Talbot, RJ, Holmes, A: Significance of fibre length in the clearance of asbestos fibres from the lung. Brit J Ind Med 35:146–153, 1978.

Morrow, PE: Alveolar clearance of aerosols. Arch Intern Med 131:101–108, 1973.

Morrow, PE: The setting of particulate exposure levels for chronic inhalation toxicity studies. J Am Coll Toxicol 6:533–544, 1986.

Morrow, PE: Possible mechanisms to explain dust overloading of the lungs. Fund Appl Toxicol 10:369–384, 1988.

Morrow, PE: Dust overloading of the lungs: Update and appraisal. Toxicol Appl Pharm 113:1–12, 1992.

Morrow, PE, Gibb, FR, Gazioglu, K: The clearance of dust from the lower respiratory tract of man. An experimental study. In: Inhaled Particles and Vapours II, edited by CN Davies, pp. 351–359. Oxford, UK, Pergamon Press, 1967.

Muhle, H, Bellmann, B: Pulmonary clearance of inhaled particles in dependence of particle size. J Aerosol Sci 17:346–349, 1986.

Muhle, H, Bellmann, B, Heinrich, U: Overloading of lung clearance during chronic exposure of experimental animals to particles. Ann Occup Hyg 32(Suppl. 1):141–147, 1988.

Muhle, H, Creutzenberg, O, Bellmann, B, Heinrich, U, Mermelstein, R: Dust overloading of lungs: Investigations of various materials, species differences, and irreversibility of effects. J Aerosol Med 3(Suppl 1):S-111–S-128, 1990.

Murray, MJ and Driscoll, KE: Immunology of the respiratory system. In: Comparative Biology of the Normal Lung, Vol I, edited by RA Parent, pp. 725–746. Boca Raton, FL, CRC Press, 1992.

Oberdörster, G, Cox, C, Baggs, R: Long term lung clearance and cellular retention of cadmium in rats and monkeys. J Aerosol Sci 18:745–748, 1987.

Oberdörster, G, Ferin, J, Morrow, PE: Volumetric loading of alveolar macrophages (AM): A possible basis for diminished AM-mediated particle clearance. Exp Lung Res 18:87–104, 1992.

Oberdörster, G, Ferin, J, Morse, P, Corson, NM, Morrow, PE: Volumetric alveolar macrophage (AM) burden as a mechanism of impaired AM mediated particle clearance during chronic dust overloading of the lung. J Aerosol Med 1:207–208, 1988a.

Oberdörster, G, Morrow, PE, Spurny, K: Size dependent lymphatic short term clearance of amosite fibres in the lung. Ann Occup Hyg 32(Suppl. 1):149–156, 1988b.

Phalen, RF: Inhalation Studies: Foundations and Techniques. Boca Raton, FL: CRC Press, Inc., 1984.

Pooley, F: Locating fibers in the bowel wall. Env. Health Perspec. 9:235, 1974.

Pratten, MK, Lloyd, JB: Pinocytosis and phagocytosis: The effect of size of a particulate substrate on its mode of capture by rat peritoneal macrophages cultured in vitro. Biochem et Biophys Acta 881:307–313, 1986.

Pritchard, JN: Dust overloading - A case for lowering the TLV of nuisance dusts? J Aerosol Sci 20:1341–1344, 1989.

Reeves, AL, Vorwald, AJ: Beryllium carcinogenesis. II. Pulmonary deposition and clearance of inhaled beryllium sulfate in the rat. Cancer Res 27:446–451, 1967.

Rhoads, K, Mahaffey, JA, Sanders, CL: Distribution of inhaled $^{239}PuO_2$ in rat and hamster lung. Health Phys 42:645–656, 1982.

Roggli, VL, Brody, AR: Changes in numbers and dimensions of chrysotile asbestos fibers in lungs of rats following short-term exposure. Exp Lung Res 7:133–147, 1984.

Roggli, VL, George, MH, Brody, AR: Clearance and dimensional changes in crocidolite asbestos fibers isolated from lungs of rats following short-term exposure. Environ Res 42:94–105, 1987.

Sanders, CL, Dagle, GE: Studies of pulmonary carcinogenesis in rodents following inhalation of transuranic compounds. In: Experimental Lung Cancer, Carcinogenesis, and Bioassays, edited by E Karbe, JF Park, pp. 422–429. Berlin, Springer-Verlag, 1974.

Sebastien, P: Pulmonary deposition and clearance of airborne mineral fibers. In: Mineral Fibers and Health, edited by D Liddell, K Miller, pp. 229–248. Boca Raton, CRC Press, 1991.

Sibille, Y, Reynolds, HY: Macrophages and polymorphonuclear neutrophils in lung defense and injury. Amer. Rev. Resp. Dis. 141:471–501, 1990.

Sminia, T, van der Brugge-Gamelkoorn, GJ, Jeurissen, SHM: Structure and function of bronchus-associated lymphoid tissue (BALT). CRC Crit Rev Immun 9:119–150, 1989.

Smith, DM, Ortiz, LW, Archuleta, RF, Johnson, NF: Long-term health effects in hamsters and rats exposed chronically to man-made vitreous fibres. Ann Occup Hyg 31:731–754, 1987.

Smith, H, Stradling, GN, Loveless, BW, Ham, GJ: The *in vivo* solubility of plutonium-239 dioxide in the rat lung. Health Phys. 33:539–551, 1977.

Snipes, MB, Boecker, BB, McClellan, RO: Retention of monodisperse or polydisperse aluminosilicate particles inhaled by dogs, rats, and mice. Toxicol Appl Pharmacol 69:345–362, 1983.

Snipes, MB, Clem, MF: Retention of microspheres in the rat lung after intratracheal instillation. Environ Res 24:33–41, 1981.

Snipes, MB, Chavez, GT, Muggenburg, BA: Disposition of 3-, 7-, and 13-μm microspheres instilled into lungs of dogs. Environ Res 33:333–342, 1984.

Sorokin, SP: Phagocytes in the lungs: Incidence, general behavior, and phylogeny. In: Respiratory Defence Mechanisms, Part II, edited by JD Brain, DF Proctor, LM Reid, pp. 711–848. New York, Marcel Dekker, 1977.

Sorokin, SP, Brain, JD: Pathways of clearance in mouse lungs exposed to iron oxide aerosols, Anat Rec 181:581–626, 1975.

Stradling, GN, Ham, GJ, Smith, H, Cooper, J, Breadmore, SE: Factors affecting the mobility of plutonium-238 dioxide in the rat. Int J Radiat Biol 34:37–47, 1978.

Strom, KA, Chan, TL, Johnson, JT: Pulmonary retention of inhaled submicron particles in rats: Diesel exhaust exposures and lung retention model. Ann Occup Hyg 32(Suppl 1):645–657, 1988.

Strom, KA, Johnson, JT, Chan, TL: Retention and clearance of inhaled submicron carbon black particles. J Toxicol Environ Health 26:183–202, 1989.

Takahashi, S, Asaho, S, Kubota, Y, Sato, H, Matsuoka, O: Distribution of [198]Au and [133]Ba in thoracic and cervical lymph nodes of the rat following intratracheal instillation of [198]Au-colloid and [133]BaSO$_4$. J Radiat Res. 28:227–231, 1987.

Takahashi, S, Kubota, Y, Hatsuno, H: Effect of size on the movement of latex particles in the respiratory tract following local administration. Inhal Toxicol 4:113–123, 1992.

Teschler, H, Friedrichs, KH, Hoheisel, GB, Wick, G, Soltner, U, Thompson, AB, Konietzko, N, Costabel, U: Asbestos fibers in bronchoalveolar lavage and lung tissue of former asbestos workers. Am J Resp Crit Care Med 149:641–645, 1994.

Thomas, RG: Transport of relatively insoluble materials from lung to lymph nodes. Health Phys 14:111–117, 1968.

Thomas, RG: Tracheobronchial lymph node involvement following inhalation of alpha emitters. In: Radiobiology of Plutonium, edited by BJ Stover, WSS Jee, pp. 231–241. Salt Lake City, Utah, JW Press, 1972.

Tillery, SI, Lehnert, BE: Age-bodyweight relationships to lung growth in the F344 rat as indexed by lung weight measurements. Lab Animals 20:189–194, 1986.

Timbrell, V, Ashcroft, T, Goldstein, B, Heyworth, F, Meurman, LO, Rendall, REG, Reynolds, JA, Shilkin, KB, Whitaker, D: Relationships between retained amphibole fibres and fibrosis in human lung tissue specimens. Ann Occup Hyg 32(Suppl 1):323–340, 1988.

Timbrell, V, Skidmore, JW: The effect of shape on particle penetration and retention in animal lungs. In: Inhaled Particles III, Vol 1, edited by WH Walton, pp. 49–57. Old Woking, Surrey, England: Unwin Brothers Limited, 1971.

Tucker, AD, Wyatt, JH, Undery, D: Clearance of inhaled particles from alveoli by normal interstitial drainage pathways. J Appl Physiol 35:719–732, 1973.

Tyler, WS, Julian, MD: Gross and subgross anatomy of lungs, pleura, connective tissue septa, distal airways, and structural units. In: Comparative Biology of the Normal Lung, Vol I, edited by RA Parent, pp. 37–48. Boca Raton, CRC Press, 1991.

Valberg, PA, Chen, B-H, Brain, JD: Endocytosis of colloidal gold by pulmonary macrophages. Exp Cell Res 141:1–14, 1982.

Vincent, JH, Jones, AD, Johnston, AM, McMillian, C, Bolton, RE, Cowie, H: Accumulation of inhaled mineral dust in the lung and associated lymph nodes: Implications for exposure and dose in occupational lung disease. Ann Occup Hyg 31:375–393, 1987.

Voelz, G, Umbarger, J, McInroy, J, Healy, J: Considerations in the assessment of plutonium deposition in man. In: Diagnosis and Treatment of Incorporated Radionuclides, Symposium Proceedings, Vienna, 1975. pp. 163–175, International Atomic Energy Agency, Vienna, 1976.

Vostal, JJ, Chan, TL, Garg, BD, Lee, PS, Strom, KA: Lymphatic transport of inhaled diesel particles in the lungs of rats and guinea pigs exposed to diluted diesel exhaust. Environ Internat 5:339–347, 1981.

Vostal, JJ, Schreck, RM, Lee, PS, Chan, TL, Soderholm, SC: Deposition and clearance of diesel particles from the lung. In Toxicological Effects of Emissions from Diesel Engines, edited by J Lewtas, pp. 143–159. New York, Elsevier, 1982.

Warheit, DB: Interspecies comparisons of lung responses to inhaled particles and gases. CRC Crit Rev Toxicol 20:1–29, 1989.

Wright, GW, Kuschner, M: The influence of varying lengths of glass and asbestos fibres on tissue response in guinea pigs. In: Inhaled Particles IV, Part 2, edited by WH Walton, B McGovern, pp. 455–474. Oxford, UK, Pergamon Press, 1977.

Yu, CP, Chen, YK, Morrow, PE: An analysis of macrophage mobility kinetics at dust overloading of the lungs. Fund Appl Toxicol 13:452–459, 1989.

Chapter Ten

Regional Dosimetry of Inhaled Reactive Gases

Frederick J. Miller and Julia S. Kimbell

INTRODUCTION

The respiratory tract is the portal of entry for inhaled material and is particularly susceptible to injury from exposure to highly reactive gases. Experimentally studying the absorption of reactive gases such as formaldehyde, chlorine, phosgene, and ozone is technically challenging and has required the development of novel measurement methods (Casanova et al., 1989; Hatch and Aissa, 1987; Santrock et al., 1989). Experimental dosimetry data in laboratory animals are becoming more available. Thus, a data base is developing to evaluate the usefulness of theoretical dosimetry models in predicting absorption at specific sites in the respiratory tract and in making interspecies dosimetric extrapolations.

When translating an air concentration of a gas to some dose in the respiratory tract, species-specific anatomy, ventilatory patterns, and the physical and chemical properties of the gas combine to determine delivered dose. If the delivered dose is of sufficient magnitude (typically a short-term, high-level exposure or long-term exposure to lower levels), damage can occur at specific sites, and effects may be seen from the molecular to organ level. Various homeostatic systems such as mucociliary function, pulmonary mac-

257

rophages, antioxidant enzymes, growth factors, and mediators of inflammation operate to mitigate the damage. The relative dynamics between the degree of insult and repair processes determine whether recovery or progression
to a disease state occurs. Understanding interspecies differences in these relationships is critically dependent upon knowledge of dose. Dosimetry is
equally important for intraspecies comparisons of injury.

Reactive gases are difficult to work with in the laboratory, and accounting for all of the compound in the body once inhaled (conservation of mass,
or mass balance) requires detailed knowledge of the compound's disposition.
There are some experimental data available from studies in which the upper
respiratory tract airways are isolated (Morris and Smith, 1982; Morris and
Cavanagh, 1986; Morris et al., 1986; Morris and Blanchard, 1992). This
allows uptake to be determined for either the upper respiratory tract, total
respiratory tract, or lower respiratory tract by direct measurement or by
differencing methods. Techniques such as radiolabeling and isotope ratio
mass spectrometry can be used to study the uptake of the compound in
specific regions of the respiratory tract. Casanova-Schmitz et al. (1984)
measured dose in the nasal passages of F344 rats and rhesus monkeys for
inhaled ^{14}C-formaldehyde using DNA-protein cross-links as a dosimeter.
These studies were extended in the rat to two regions within the nasal
passages (Casanova et al., 1994), based on the locations of nasal tumors
that occurred in rats chronically exposed to high levels of formaldehyde.
Using isotope ratio mass spectrometry, Hatch et al. (1989) measured regional ozone dosimetry in F344 rats exposed to ^{18}O-zone. Histochemical
localization can be used to determine regional metabolic activity of inhaled
reactive gases (Keller et al., 1990; Bogdanffy et al., 1986, 1987).

Experimental dosimetry data alone cannot provide the necessary information for examining interspecies differences in dose. Since there are many
exposure scenarios for which it is not practical to obtain experimental data,
a major focus of this chapter will be on formulating and implementing dosimetry models of the regional absorption of inhaled reactive gases. For risk
assessment, the ultimate use of experimental or theoretical absorption data
is in converting concentration-response curves to dose-response curves for
intra- and interspecies comparisons of toxicological results.

Since the first edition of this book, much new anatomical information
has become available that is relevant to examining the regional absorption
of reactive gases. Here, we will highlight the nature of some of this information and indicate how such data can be incorporated into the formulation of theoretical dosimetry models. In view of equally rapid advances in
computer speed and storage capacity, we also provide a discussion of
computational approaches that can be used to implement the dosimetry
models.

FACTORS INFLUENCING ABSORPTION

The major factors affecting the absorption of reactive gases in the respiratory tracts of mammals are the structure of the respiratory tract; the route, depth, and rate of breathing; the physical and chemical properties of the gas; the physical and chemical processes governing gas transport in the lung; and the physical and chemical properties of the lining fluids and tissue of the airways and gas exchange units. For more in-depth, recent discussions of considerations useful in modeling respiratory-tract transport and absorption of reactive gases, the reader is referred to Miller et al. (1993b) and Miller (1994). Here, we briefly discuss only some of the important structural and physico-chemical factors that influence the absorption of reactive gases in the respiratory tract, noting major species differences when appropriate.

Respiratory Tract Structure

When examining toxicological effects or modeling dose, the respiratory tract is typically subdivided into three major regions: the upper respiratory tract (from the nose and mouth down to and including the larynx); the tracheo-bronchial tree (trachea to terminal bronchioles); and the pulmonary region (respiratory bronchioles to terminal alveolar sacs). The latter two regions comprise the lower respiratory tract (LRT). The anatomical features of each region are complex and differ significantly among laboratory animals and humans. Information on the anatomy, physiology, function, metabolism, types and numbers of cells, biochemistry of the airway lining fluid, and structure of the various respiratory tract regions can be found in a number of books, such as *The Nose* (Proctor & Anderson, 1982), *Concepts in Inhalation Toxicology* (McClellan & Henderson, 1989), *Toxicology of the Lung* (Gardner et al., 1988, 1993), and *Comparative Biology of the Normal Lung* (Parent, 1992). Given the extent of differences in respiratory tract structure among species, dosimetry models should ideally be anatomically based to facilitate interspecies dosimetric comparisons.

Upper Respiratory Tract   Within the URT, the anatomy varies among specific subregions. The shape and dimensions of these regions, patterns of mucous flow and airflow, and the composition of the mucus and types of cells lining the regions are all important determinants of absorption. An overview of these factors from a toxicological pathologist's viewpoint is available (Morgan, 1995).

There are major differences between rats and humans in the structure of the nose. Fig. 1 depicts serial cross sections of the nasal passages of one side of the F344 rat nose (Panel a) and of the human nose (Panel b). Note the tremendous difference between these species in the shape of the cross sections proceeding from the nostril towards more distal portions of the nose.

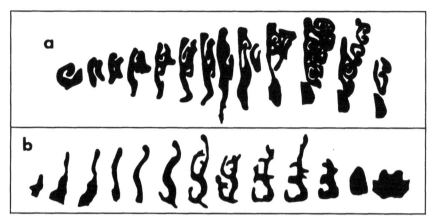

Figure 1 Aspects of upper respiratory tract structure important for dosimetry. (a) Cross sections of one side of the rat nasal passages, nostril at left (modified from Morgan et al., 1991). (b) Cross sections of one side of the human nasal passages, nostril at left (modified from Guilmette et al., 1989).

This complexity in airway structure affects bulk airflow patterns to such an extent that major airflow streams are created; hence, not all nasal surfaces receive the same exposure to an inhaled gas. The major inspiratory airflow streams in the rat nose, obtained from water-dye flow studies in clear acrylic models of the nose (Morgan et al., 1991), are shown in Fig. 2. Comparable airflow stream data are available for the rhesus monkey (Morgan et al., 1991) and the baboon (Patra et al., 1986). Limited data are available (Swift and Proctor, 1977; Hahn et al., 1993) on the nature of airflow in the URT of humans.

The nasal epithelium is lined by squamous, respiratory, or olfactory cells, depending upon location. In some regions, the absorbed gas must diffuse through the layer of mucus covering the epithelium before it can react with the underlying tissue. Hence, the pattern of mucous flow can also be an important factor that influences site-specific absorption of the gas. For the rat, the nature of these physiological factors is illustrated in Fig. 3.

Lower Respiratory Tract   Like the URT, the structure of the LRT varies both within and among species. The conducting airways form a complex branching structure that delivers oxygen efficiently and rapidly to the alveolar region, where an exchange with carbon dioxide occurs. Branching in the tracheobronchial (TB) tree of humans is often described as being symmetrically dichotomous (bipodal) in nature. In contrast, rodent TB airways are more monopodial in their branching pattern. For Long-Evans rats, Ménache et al. (1991) found greater intra- than interanimal variability in TB airway lengths and diameters. They also noted that intraanimal variability could be reduced by grouping airway generations after taking into account lobar position.

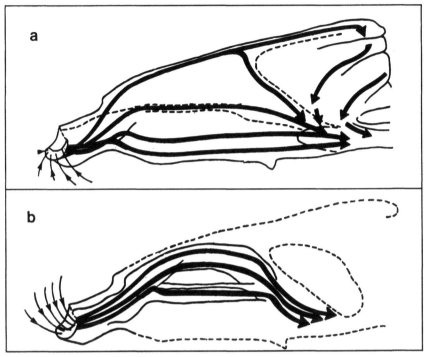

Figure 2 Diagrammatic representations of inspiratory airflow streams in the nose of the F344 rat. Panel (a) shows major medial streams, while panel (b) shows major lateral streams. Adapted from Morgan et al. (1991).

Significant variability is also present in the alveolar duct branching pattern within the pulmonary (P) region. A stick diagram for the rat alveolar duct system is shown in Fig. 4. The three-dimensional branching pattern in the rat is so complex that multiple branches can occur within the distance of a single alveolus (about 100 μm) (Crapo et al., 1990). Comparable morphometric data for the pulmonary region of humans is lacking.

The cellular composition of the alveolar septum is remarkably similar across most mammalian species (Crapo et al., 1982, 1983; Pinkerton et al., 1982; Stone et al., 1992). In contrast, there are significant differences among species in the number and type of cells in the TB airways (Plopper et al., 1983, 1989; Mariassy, 1992). Also, the mucous layer lining the TB airways varies considerably among species. For example, the thickness of the lining layer is 8.3 and 1.8 μm in human bronchi and bronchioles, respectively, compared with 3 and 2 μm for these same locations in the rat (Mercer et al., 1992). Implications for dosimetry modeling of the types of LRT data presented in this section have been described by Miller et al. (1993a).

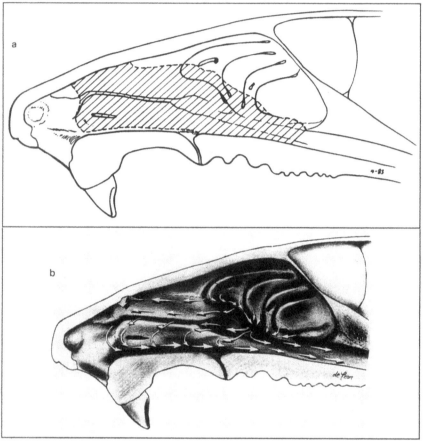

Figure 3 Site-specific types of cells and patterns of mucous flow in the rat upper respiratory tract. (a) Diagram of distribution of squamous and respiratory epithelium along the walls of the nasal passage. Zone left of cross-hatching = squamous; cross-hatching = respiratory (modified from Morgan et al., 1984). (b) Patterns of nasal mucus flow over the walls adjacent to the septum of the rat nasal passage (adapted with permission from Morgan et al., 1984).

Physical and Chemical Processes

Various physical and chemical processes are important for the gas-phase transport of reactive gases and their absorption in the liquid lining layers and tissues of the respiratory tract. The transport of a gas in the airstream is governed by the following processes: convection, molecular diffusion, turbulence, dispersion, and the loss or gain of gaseous species to and from the walls of the respiratory tract. Convection (bulk movement of air) in the gas phase is a major determinant of gas transport in the URT (Proetz, 1951) and in the first seven generations of the tracheobronchial tree (Wilson and Lin,

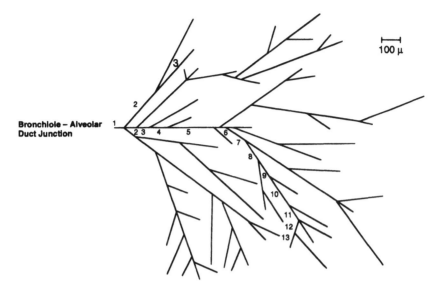

Figure 4 Stick drawing of the branching pattern in the alveolar duct system from an acinar reconstruction of a Sprague-Dawley rat lung. The bold line at the end of each path represents the terminal alveolar ducts (alveolar sacs). Although branching angles are not represented, the actual length of each duct and sac is proportional to the length depicted. Paths terminate after as few as 3 or as many as 13 generations. The 100-μm scale marker corresponds approximately to the diameter of an alveolus. Figure from Mercer and Crapo (1987).

1970). Molecular diffusion (Brownian motion) is the major process governing gas transport in the pulmonary (alveolar) region, with convection negligible during quiet breathing and only about 12% as important as diffusion even at flow rates 10–15 times those needed for normal respiration (Davidson and Fitz-Gerald, 1974). In contrast, Taylor diffusion (Taylor, 1953) is the main gas transport mechanism in the midportions of the TB region. The net effect of TB airway geometry, turbulence, and secondary flow patterns can be incorporated into an effective axial (longitudinal) diffusion coefficient (Scherer et al., 1975), thereby making modeling of gas transport in this region more tractable.

Whereas displacement and mixing of a gas in the airways is due to axial (i.e., longitudinal) transport, the absorption of the gas at specific sites is the result of gas transport in the radial direction. For a discussion of the relative importance of longitudinal transport and degree of reactivity of inhaled gases as well as implications for dosimetry modeling, the reader is referred to Ultman (1988).

Absorption of a gas by the liquid layers and tissues of the respiratory tract depends upon a number of factors such as airflow patterns, the thickness and biochemical composition of the liquid lining layer, tissue and blood compartments, and solubility and diffusion of the gas in these compartments. Chemical reactions between the gas and the constituents of the liquid lining layer, tissue, and blood are also important determinants of absorption. More detailed discussions of the physical and chemical processes influencing the absorption of reactive gases can be found in Miller et al. (1993b) and Miller (1994).

DOSIMETRY MODEL FORMULATION

Information on site-specific dosimetry is becoming increasingly important for interpreting the toxicologic effects of inhaled compounds and judging the potential for effects in humans. Knowledge of dose provides the critical link between exposure and response. Mathematical dosimetry models can be used to estimate exposure levels needed to produce the same dose at specific sites in different animals of the same or different species. Models provide an integration of our knowledge and can be used to test concepts, advance our understanding of toxicological events, and make more efficient use of resources. The reader is referred to the previous edition of this book (Table 2 of Chapter 8) for a summary of dosimetry models available at that time for modeling the absorption of reactive gases. Subsequently, other dosimetry models for reactive gases have been developed (Overton et al., 1987; Hanna et al., 1989; Mercer et al., 1991a; Hu et al., 1992; Casanova et al., 1991; Heck and Casanova, 1994). Here, we will focus more on modeling concepts and approaches.

The development of a quantitative model to predict dose in the respiratory tract, like the development of any model, is based on a series of steps: formulation, implementation, presentation and confirmation of results, and interpretation. During formulation of a dosimetry model, the hypothesis that the model represents is made explicit. Assumptions, known parameters upon which the model is based, and equations to be solved are also stated. The dosimetry model is ideally developed at the same level of detail as the data to which model output will be compared for hypothesis testing and confirming model accuracy. Sources of data for model structure are established, and approaches for implementation such as specific computational or experimental methods are determined. Confirmation of model predictions by comparison with a known scenario or specific experimental data provides credibility and error analysis. Results from a confirmed model are then interpreted in the context of the hypothesis tested and the assumptions made.

Hypothesis Definition

One of the first steps in developing a model is stating a testable hypothesis. A dosimetry model might be formulated to test the significance of anatomy, regional metabolism, epithelial type, mucus effects, airflow patterns, or morphometric parameters in determining dose to a specific respiratory tract region. The hypothesis determines whether the model will focus on the air phase, airway lining, or underlying tissues and blood and whether anatomy or morphology play important roles. The hypothesis also provides structure for model results and their interpretation.

The hypothesis guides the determination of how detailed or complex the model must be. The potential complexity of the model is illustrated schematically in Fig. 5. Depicted are the types of model compartments that may be included as well as some of the kinds of transport factors that are important for the various compartments. Knowledge of the structure and biology of the respiratory tract (URT, TB, or P) must be coupled with information on the physicochemical properties of the reactive gas in determining the level of model complexity needed to address the hypothesis. Another major determinant of level of detail in the structure of a model is the type of data to which model output will be compared for hypothesis testing and confirmation of model accuracy. Uptake of formaldehyde by the nasal passages of the F344 rat, as measured as being greater than 93% by Patterson et al. (1986), could be used in conjunction with a model in which the nasal passages are represented by one well-mixed compartment. The precise localization of nasal lesions induced in F344 rats after exposure to inhaled formaldehyde (Morgan et al., 1986a; Monticello et al., 1991) could be used along with a more detailed model to examine the site specificity of these responses.

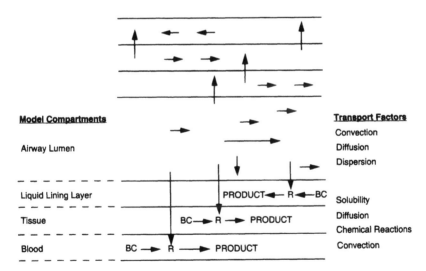

Figure 5 Diagram illustrating potential compartments and transport factors that may need to be included in the formulation of a dosimetry model for a reactive gas. The relative arrangement of the airway lumen, liquid lining layer, tissue, and blood compartments of the respiratory tract is shown. Since model compartments are symmetric about the airway lumen, only the lower half of the figure is fully labeled. The R represents chemical reactions between the absorbed gas and biochemical constituents (BC); "PRODUCT" represents the resulting products. Transport factors important in the various model compartments are noted. Modified from Overton (1990).

Analogously, for effects of ozone (O$_3$) in the lower respiratory tracts of animals and humans, single-path models (Miller et al., 1985; Overton et al., 1987) can be used to examine the similarity between species of the shape of tissue-dose curves. This allows predictions to be made for the likely site of maximal O$_3$-induced injury in man compared with known sites of maximal injury in animals (Dungworth et al., 1975; Schwartz et al., 1976; Barry et al., 1985). Chang et al. (1991) used a simplified model of cumulative ozone absorption in the proximal alveolar region to demonstrate that epithelial cell injury from low-level, subchronic exposure correlates with cumulative exposure. In a detailed analysis of ozone injury in the pulmonary region, Pinkerton et al. (1992) examined changes in alveolar wall and septal tip thickness as a function of distance from the bronchiolar-alveolar duct junction. In relating these observations of ozone dose, a detailed treatment of pulmonary acinar structure and gas transport, such as that of Mercer et al. (1991a), is needed.

Assumptions Underlying the Model

Another part of model formulation is the explicit statement of assumptions under which the model is constructed. A representation of the nasal passages as one well-mixed compartment reflects the assumption that factors governing uptake of the inhaled gas into the airway lining are more important than effects of local airflow patterns on uptake. Reaction rates measured *in vitro* are often assumed to be the same as those *in vivo*. As another example, a model of airflow in the respiratory tract in which the flow and uptake of an inhaled gas is modeled as steady-state (not dependent on time) assumes that uptake is virtually instantaneous, and that the cyclic nature of breathing is not important to the uptake process. For reactive gases, a pseudo-first-order chemical reaction scheme is typically assumed in the liquid lining layer, tissue, and blood compartments. This requires that the concentration of biochemical constituents available to react with the gas are orders of magnitude greater than the concentration of the gas in these compartments as is the case for ozone (Miller et al., 1985).

Assumptions can often be checked for reasonableness using available morphometric and physicochemical data or can be tested themselves as hypotheses. For example, the reasonableness of neglecting convection in a nonair-phase compartment can be examined by computing dimensionless variables that involve compartment parameters corresponding to thickness, characteristic length, and characteristic velocity, as well as other parameters such as the molecular diffusion coefficient and chemical rate constant (Miller et al., 1985). However, one way to avoid confusion in thinking about model structure is to be specific about the hypothesis being tested and then systematically subject assumptions to sensitivity analyses to determine how strongly model results are affected by changes in the assumptions.

In ozone dosimetry modeling, Miller and colleagues (1985) used sensitivity analyses to show that simulations were not sensitive to the dispersion or mass transfer coefficient but were highly dependent upon the assumed thickness of the mucous compartment. An important byproduct of sensitivity analyses is that they can be used to articulate research needs. For example, a sensitivity analysis of the thickness of mucus (Miller and Graham, 1988) led to experimental data (Mercer et al., 1992) that can be used to replace assumptions about mucous thickness.

A dosimetry model is ultimately a description of chemical transport, and the hypothesis determines which transport mechanisms are explicitly included. From this information, mathematical equations are developed to account for transport and absorption of the inhaled gas in the respiratory tract and its subsequent fate in the body. Depending on the statement of the hypothesis, chemical transport can be modeled to occur via respiratory airflow, mucus flow, absorption into the airway lining, reactions with airway lining and wall constituents, blood flow, or a combination of these factors.

Boundary and Initial Conditions

To solve the equations that describe chemical transport in a mathematical dosimetry model, information must be known about the quantities of chemical present at the edges of the regions represented in the model and at the starting time of a model run. The edges of model regions might be the entrances and exist of compartments representing respiratory tract regions or airway walls. The beginning of a breathing cycle is an example of the starting time of a model run. Chemical quantities must be explicitly given, or conditions that those quantities satisfy must be explicitly stated, at region edges for all times relevant to running the model. These specifications (or conditions) are called boundary conditions because they pertain to chemical quantities only at the boundaries of the model regions.

If chemical quantities change with time when running the model, then amounts of chemical present everywhere in the model at the starting time of a model run must also be specified. This specification is called an initial condition because it sets up the initial state of the model before time passes and the quantities modeled change. When using a one-dimensional equation of mass transport, some model parameters may be discontinuous at the boundaries between model compartments, depending upon the nature of the model description used to represent the structure of the respiratory tract (e.g., when model compartments correspond to specific tracheobronchial or pulmonary airway generations). In this case, solutions to the transport equation on either side of a compartment boundary can be matched by requiring the concentration and net mass transfer across the boundary to be continuous (Miller et al., 1985). Boundary and initial conditions, together with the equations of chemical transport throughout the model regions, form the core of a mathematical dosimetry model.

IMPLEMENTING THE DOSIMETRY MODEL
Types of Data Available

An explicit description of parameters taken to be fixed or known is necessary in dosimetry model formulation. The volumes and thicknesses of nasal airway, mucus, and tissue compartments in the URT and the volumetric flow rates of air, mucus, and blood have been reported for the nasal passages in some cases (Gross and Morgan, 1992; Morris et al., 1993; Morgan et al., 1986b) and used in modeling URT disposition at a macro level (Frederick et al., 1995; Morris et al., 1993). Nasal airway perimeters have been digitized to reconstruct accurate three-dimensional anatomical models to simulate regional URT uptake on a micro level (Kimbell et al., 1993, Kimbell, 1995; Scherer et al., 1995). Anatomical data analogous to those for the URT are also needed to implement LRT dosimetry models. Data on the lengths, diameters, and number of tracheobronchial and pulmonary airways as well as on the number of alveoli and their volumes and surface areas are available for various laboratory animal species (Kliment, 1972, 1973; Raabe et al., 1976; Yeh et al., 1979; Schreider and Hutchens, 1980). Similar data are also available for humans (Landahl, 1950; Weibel, 1963; Horsfield and Cumming, 1968; Raabe et al., 1976; Yeh and Schum, 1980).

Parent's (1992) *Comparative Biology of the Normal Lung* is replete with anatomical, biochemical, and functional data useful in gaseous dosimetry modeling. Sometimes parameters in a model are derived or calculated from known quantities. Airflow rates can be determined from breathing rates and minute volumes (Costa and Tepper, 1988; Lai, 1992), and allometric methods are sometimes used to scale these and other parameters from one species to another (Leith, 1984) or even between individuals of different sizes within a species (Takezawa et al., 1980).

Two sources of anatomical data suitable for three-dimensional reconstruction are computer-aided tomography (CAT) and magnetic resonance imaging (MRI) scans. A major advantage of these noninvasive methods is that they can be used to gather anatomical data *in vivo* or, in other cases, without destruction of tissue samples. At the present time, CAT scans evidently produce clearer images of hollow airways, while MRI may be superior for imaging other structures. Clarity and detail of images in both cases is dependent on the age and quality of the scanning and imaging equipment and on minimizing movement of the subject during scanning. CAT scans are not generally made *in vivo* at intervals of less than 1 or 2 mm to minimize the subject's exposure to radiation. This restriction limits the level of resolution of the anatomical data and becomes more of a limitation in smaller subjects. The resolution of CAT and MRI scans may be sufficient for determining interindividual differences in anatomical characteristics, however, and could provide data to customize a template model.

Another source of anatomical reconstruction data is specimen sectioning. With flexibility on preparation of the specimen prior to sectioning and section thickness, this method provides maximal resolution of detail on the structure of the airways, the cells comprising the airways, and even intracellular structure. In recent years, Mercer et al. at the Duke University Center for Extrapolation Modeling have used three-dimensional serial reconstruction methods in developing a wide spectrum of data on the structure of the pulmonary acinus of rats and humans and to a lesser degree on other laboratory animal species. This data base includes the alveolar duct-alveolar architecture of the rat lung (Mercer and Crapo, 1987), reconstruction of rat alveoli (Mercer et al., 1987), the spatial distribution of collagen and elastin fibers in rat and human lungs (Mercer and Crapo, 1990), the depth-distribution of nuclei and cytoplasm of various cell types in a number of different sized airways in rats and humans (Mercer et al., 1991b), and the structure of the alveolar septum in different species (Mercer et al., 1994). Allometric relationships of alveolar cell numbers and size across a wide range of mammalian lung weights have been determined by Stone et al. (1992). Two major disadvantages to specimen sectioning for reconstruction are that the specimen is destroyed in the process and that airways are reconstructed postmortem, with potentially significant differences from *in vivo* airway dimensions (Guilmette et al., 1989). Postmortem airway distortion can be addressed, however, by modification of the resulting reconstruction. In fact, postreconstruction alterations could accommodate interindividual differences in airway anatomy, so that one reconstruction may serve as a template for models with anatomical dimensions specific to an individual

Computational Approaches

There are two issues to consider when deciding what computational approach to take in implementing a mathematical dosimetry model: (1) how to solve the equations in the model and (2) how to present the solution once it is obtained. In some cases, an analytical, or closed-form solution can be found for the chemical transport equations and their boundary and initial conditions. Most of the time, however, numerical methods must be used. Numerical methods are also frequently used to present the solution as a summary of model output in a form that is easy to interpret and evaluate for accuracy.

To solve numerically the chemical transport equations subject to their boundary and initial conditions, the equations and conditions must be represented in discrete form. This means dividing up the model region (such as an anatomical structure) and possibly a time line into discrete pieces such as points, line or time segments, or smaller regions and approximating the equations by expressions using variables evaluated only at the designated locations and times. The amount of error introduced by these approximations is controlled by the way in which the model region is broken up (number, size, and shape of subregions) and by the ways in which mathematical operations such as differentiation and integration are represented. Taken together, the expressions approximating the original transport equations are called a discrete form of the equations.

There are a number of methods for deriving a discrete form of a set of equations, each having different merits and error capacities. Some standard methods include finite difference, finite volume, and finite element techniques. Each of these methods has as its basis the division of the model region into separate points or pieces, but each method represents a different interpretation of the approximation of the original equations in terms of this division. Some finite-difference techniques are very accurate when the chemical transport equations are complex, but they can be difficult to use when the model region has complex boundaries, making specification of boundary conditions in the discrete form difficult. Finite-element techniques, on the other hand, work well for regions of complex shape such as three-dimensional biological reconstructions and yet fall short on other aspects of the approximation of the original equations. Descriptions of these methods, sources for more information, and the circumstances under which the use of one method or another is recommended can be found in Press et al. (1986).

Once the equations are in discrete form, computational algorithms are then used to find a solution to the discrete expressions and their boundary and initial conditions. These algorithms may invert a matrix or iteratively search for a solution using Newton's method or penalty functions. The algorithms used depend on the form of the discrete expressions and on the tools available to perform the computations. Computational algorithms used in most mathematical dosimetry models are programmed to run on a personal computer, workstation, or supercomputer, depending on the size of the problem (i.e., number of model region divisions and equations being solved). A parallel machine or processor may need to be used for particularly large models.

The derivation of a discrete form of the equations, implementation of numerical algorithms to solve the discrete expressions, and even decomposition of model regions into small pieces can be done using commercially available, third-party software for many types of models. There are obvious time-saving advantages to using such software, but there is also a corresponding loss of direct control over the implementation of the model. Third-party software should be tested for accuracy by comparing model predictions with data from experimental systems or with a case for which there is a known mathematical solution.

The structure and complexity of a mathematical dosimetry model determine what numerical algorithms, computer programs, and computer platforms are required to run the model. A representation of the nasal passages as one well-mixed compartment produces relatively simple one- or two-dimensional equations that can be solved numerically using finite-differencing techniques programmed by hand and run on a personal computer. On the other hand, a representation of the nasal passages as a three-dimensional reconstruction in which the full Navier-Stokes equations of motion are to be solved numerically could require the use of several different algorithms to solve the equations, might be implemented through commercially available third-party software, and would be carried out on a supercomputer.

Recently, some interesting computational issues have come up with regard to presentation of model results. Increased accessibility to supercomputing and fast computer workstations in the last decade have facilitated the use of large-scale computing methods in some types of mathematical dosimetry models. These models can be quite complex, producing vast amounts of numerical output not suited for presentation in standard chart or curve and data-plotting formats. Yet model output must be presented in a form that investigators and peers can interpret to resolve hypotheses and to evaluate model accuracy. How best to represent these data and how to implement data processing into such forms are issues facing investigators using these types of models.

Techniques for processing such data (called visualization techniques) are rapidly evolving to address these issues. Visualization techniques include the use of three-dimensional graphics, color, texture, and shading and manipulation of such images in real time such as rotating, zooming, and "flying through" computer images. Early on, investigators often faced the tasks of devising algorithms to produce these graphic images and effects and writing computer programs to implement these algorithms in addition to development of the model itself. More recently, however, good visualization software packages have become commercially available on a variety of platforms. These packages provide ways to visualize many different types of data, making obsolete the need for an investigator to be a graphics programmer as well. Thus, the use of complex models and the subsequent processing of their output is becoming more accessible to a wider segment of the scientific community.

EXAMPLES OF DOSIMETRY MODELS
Modeling Formaldehyde Dosimetry in the Upper Respiratory Tract

To what level of inhaled formaldehyde can people safely be exposed over potentially long periods of time? To answer this and related questions, data from laboratory animal exposure-response experiments must be extrapolated to people to assess potential risks to human health. Site-specific responses in the upper respiratory tracts of the F344 rat and rhesus monkey have been reported, with overall patterns of response differing between these two species (Morgan and Monticello, 1990; Monticello et al., 1989). The site specificity of responses in both species and differences in response patterns between species imply that human responses to formaldehyde exposure could also be site-specific and could show an overall pattern different from either rats or primates. Thus, a dosimetry model is needed that can predict site-specific responses in species-specific patterns. The development of such a model (Kimbell et al., 1993) is described in this example.

Site specificity of responses to inhaled formaldehyde is attributed to a combination of clearance mechanisms, tissue susceptibility, and regional uptake patterns (Morgan and Monticello, 1990). The hypothesis tested by the dosimetry model in this example was that regional uptake patterns (as determined by inspiratory airflow patterns) are the key determinants of formaldehyde-induced lesion location in the respiratory tract. The model tested this hypothesis by including a detailed representation of the structure of the rat nose so that airflow patterns were faithfully recreated with all other factors thought to affect dosimetry excluded. As a chemical transport model, then, it included transport of inhaled formaldehyde in the nasal passages via inspiratory airflow and air-phase diffusion from regions of high to low concentration. Model-predicted uptake hot-spots along simulated airway walls were then compared with actual data on formaldehyde-induced lesion locations in the rat and primate. Good agreement would provide evidence in support of the hypothesis that airflow and anatomical structure are key determinants of formaldehyde dosimetry. If they are, these factors should be incorporated into a model for human formaldehyde dosimetry if accurate predictions are to be made.

To test the hypothesis that inspiratory airflow patterns are the major determinants of formaldehyde-induced lesion location, airflow and formaldehyde uptake during inspiration were modeled under the following assumptions:

1   all nasal airways walls were simulated to act equally as perfect sinks (instantaneous, unsaturably absorbing)
2   the presence of formaldehyde in the air did not affect breathing rates or airflow patterns;
3   all airway walls were static, i.e., no muscular movement, tissue swelling, or mucus movement was simulated;
4   no nasal hair was included;
5   temperature and humidity were assumed constant;
6   airflow was simulated as steady-state (constant speed) in the inspiratory direction; and
7   loss of formaldehyde from inspired air was simulated to occur only by unsaturable absorption by airway walls.

Anatomical structure, minute volume, viscosity of air, concentration of inhaled formaldehyde at the nostril, and the airphase diffusion coefficient of formaldehyde were taken to be fixed (known or given) quantities. The data to which model output could be compared consisted of localized formaldehyde-induced lesions observed on histologic sections through the nasal passages of male F344 rats (Morgan et al., 1986a). The level of detail in the model therefore must be sufficient to recreate these locations. The model was constructed from a complete set of histologic sections made serially at 50-μm intervals through the nasal passages of a male F344 rat. Nasal airway outlines on each section were traced on the digitizing tablet of a computer image analysis system, producing a series of files containing vertical and horizontal coordinates of each airway perimeter.

There are several ways to assemble such data into a three-dimensional reconstruction. The choice of method depends on the purpose of the model. In this example, the model was to be used for simulation of airflow and formaldehyde uptake, so a reconstruction appropriate for computational fluid dynamics (CFD) use was needed. The mathematical equations used in this model (see below) can be represented in discreet form using any of the methods described earlier. Since the finite element method works well with complex geometric shapes, a finite element model was constructed from the cross-sectional data. The data were processed into a three-dimensional model using the finite element CFD software package FIDAP (FDI, Evanston, IL). A review of several CFD packages, including FIDAP, is given by Lueptow (1988). Fig. 6 panel a shows several sections of the three-dimensional FIDAP mesh (Fig. 6, panels b,c) produced from the rat data.

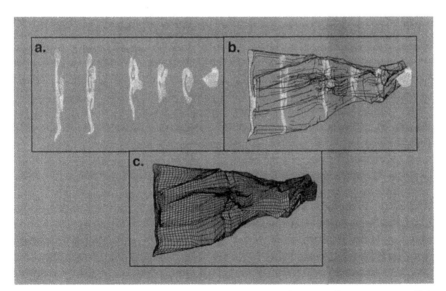

Figure 6 Sequential cross sections of the right side of the nasal passages are digitized, and a computational mesh is made of each cross section. The sections are aligned in three dimensions according to their anatomical spacing (a) and connected by lines in the longitudinal direction (b) to form a three-dimensional computational mesh (c). Nostril is to the right.

Simulation of nasal airflow and formaldehyde uptake involves solving a set of mathematical equations—several equations that solve for regional speed and direction of airflow (called the Navier-Stokes equations of motion), and one equation whose solution gives regional concentration of formaldehyde in nasal air passages (called a mass-transport equation). To solve the Navier-Stokes and mass-transport equations in the F344 rat dosimetry model, boundary conditions (e.g., airway walls, nostril) on airflow and formaldehyde concentration were needed. The speed of air flowing into the nose at the nostril was set by dividing the overall flow of air into the nostril (the minute volume is one of the flow rates used) by the surface area of the nostril. Airflow was set to zero at all airway walls in accordance with assumption 3 above that the walls were fixed and no mucus movement was modeled. Formaldehyde concentration was set to ambient air concentration levels at the nostril and zero at airway walls in accordance with assumption 1 above.

Steady-state inspiratory airflow was simulated in this model at various volumetric flow rates across the physiological range. The CFD software package FIDAP was used to solve the fluid flow and mass transport equations, and the model was run on a Cray Y-MP supercomputer. Visualizations of simulation results (Fig. 7) were created using either the postprocessing module of FIDAP or the visualization software package AVS (Advanced Visual Systems, Inc., Waltham, MA) on a Silicon Graphics workstation (Silicon Graphics, Inc., Palo Alto, CA). Airflow simulations recreated the major routes of flow and flow velocities observed in water-dye experiments with hollow acrylic molds of the F344 rat nasal passages (Morgan et al., 1991). Compare Figs. 2 and 7a. Simulated formaldehyde uptake along nasal passages walls varied and was generally highest in regions of largest bulk flow. Model predictions of airway wall uptake have been qualitatively confirmed (Morgan et al., 1992), and quantitative assessment of accuracy is in progress. Hot-spots of simulated formaldehyde uptake were predicted by the model in regions where formaldehyde-induced lesions occur in exposed rats.

These simulations of formaldehyde uptake provide evidence in support of the hypothesis that airflow patterns are the determining factor in the location of respiratory tract lesions induced in F344 rats exposed to inhaled formaldehyde. A similar model for the rhesus monkey is being constructed to test the hypothesis across species. Once confirmed, these models allow us to predict how much formaldehyde is associated in a given region (local dose) with specific respiratory tract lesions. In Fig. 8a, a cross-section through the nasal passages of a rat chronically exposed to 6 ppm inhaled formaldehyde shows the location of a lesion in the lateral meatus (Morgan et al., 1986a; Mery et al., 1994). Taking a cross section at the same level through the anatomical reconstruction of the rat dosimetry model, the predicted dose of formaldehyde to that same region can be determined as depicted in Fig. 8b. A human dosimetry model of airflow and formaldehyde uptake simulations in a three-dimensional anatomical reconstruction of the human nasal passages such as those described by Scherer et al., (1995) is then run for various inhaled concentrations of formaldehyde until the highest regional dose of formaldehyde to airway walls predicted in the human model attains the dose predicted in the lateral meatus of the rat. The model thus determines exposure scenarios under which humans would receive local doses of formaldehyde in their respiratory tracts at the same levels as those associated with lesions in laboratory animals.

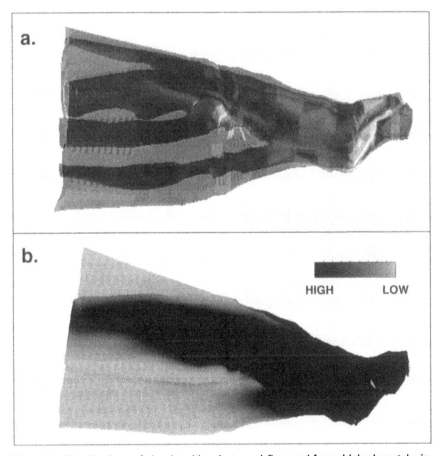

Figure 7 Visualizations of simulated inspiratory airflow and formaldehyde uptake in the F344 rat nasal passages, at a volumetric flow rate of 252 ml/min. Nostril is to the right. (a) A visualization software package (AVS, Advanced Visual Systems, Inc., Waltham, MA) was used to plot a speed isosurface (a plot of all points connected together where flow speed is the same) at 51.2 cm/s inside the airway, visualized through semitransparent airway walls. (b) Simulated formaldehyde uptake by F344 rat nasal airway walls was calculated and plotted using the software package FIDAP (FDI, Evanston, IL). Darker areas correspond to higher predicted formaldehyde uptake.

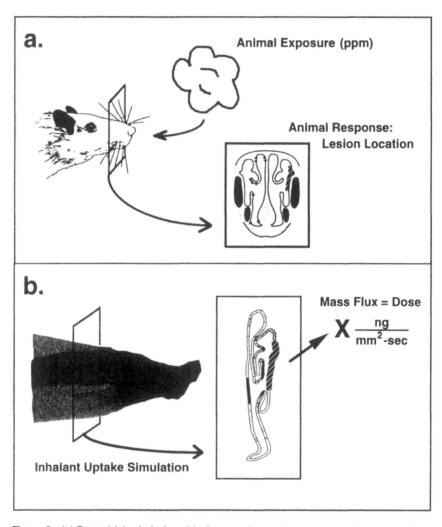

Figure 8 (a) Formaldehyde-induced lesions are located on cross section maps of the F344 rat nasal passages. (b) A cross section through a simulation of formaldehyde uptake by F344 rat nasal airway walls provides an estimate or prediction of dose to the lesion site.

This example describes a direct methodology for extrapolation of animal data to humans for health risk assessment by providing both a way to assign a dose to lesion locations in animals and a way to determine if similar doses occur in people. The formation of tissue lesions, however, involves more than delivery of formaldehyde to airway walls. Formaldehyde is metabolized within the tissue via metabolic pathways and forms DNA-protein cross-links (DPX; Casanova-Schmitz et al., 1984). Further work is needed to link regional formaldehyde dosimetry to airway walls with the disposition of form-aldehyde within nasal tissues via saturable and unsaturable metabolic path-ways and to determine how this information can be extrapolated to humans. The process of building the model is a systematic way to gain an understanding of these processes, since the model is a reflection of the current level of perception. As this research progresses, mathematical models and experimental analyses together can help synthesize our understanding of the dosimetry of inhaled gases.

Modeling Ozone Dosimetry in the Lower Respiratory Tract

Ozone is an ubiquitous air pollutant generated in the atmosphere by photo-chemical processes involving oxides of nitrogen and volatile organic chemicals. The chronic effects of this highly reactive gas are well documented in animal studies for high-exposure concentrations (U.S. EPA, 1986) and are being increasingly reported for near-ambient concentrations of O_3 (Chang et al., 1991; Hyde et al., 1989). A strong case can be made for homology between laboratory animals and humans on pulmonary physiological responses to O_3 from short-term exposures (Tepper et al., 1993). Hence, the question arises as to whether chronic effects occur in humans and, if so, at what levels of O_3 exposure.

While a qualitative answer to this question can be inferred (Miller et al., 1978), a quantitative evaluation of the magnitude of human risk involves extrapolating animal toxicological results of chronic O_3 exposure studies. Integral to such extrapolations is a dosimetry model for O_3 that can predict LRT dose-responses in various species. The dosimetry model needs to account for the heterogeneity in the magnitude of effects seen within various pulmonary acini with low-level exposures. Site specificity of acinar responses to inhaled O_3 is attributed to a combination of intraindividual variability in tracheobronchial path length and acinar volume in conjunction with the reactivity of O_3 with the LRT liquid lining layers (Miller et al., 1993b). The basis for a test of the hypothesis that these are the critical features and processes for O_3 absorption leading to the location and extent of O_3-induced lesions will be the focus of our example of LRT dosimetry modeling.

The hypothesis on O_3 dosimetry stated above was formulated from a body of work on the LRT dosimetry of O_3 spanning more than a decade (Miller et al., 1978, 1985; Overton et al., 1987, 1989; Overton and Graham, 1989; Mercer et al., 1991a). The type of morphometric or anatomical data available impart limitations on the nature and complexity of the mathematical dosimetry model formulation and on the types of hypotheses that can be tested. As improved anatomical data have become available, mathematical dosimetry models have been formulated to make use of the new data. The increase over time of the storage capacity and computational speed of computers is another factor that has contributed to the ability of investigators to incorporate more realistic representations of lung structure into dosimetry models. For a discussion of the evolution of our understanding of lung structure and resulting anatomical models for use in mathematically predicting the dosimetry of inhaled gases and particles, the reader is referred elsewhere (Crapo et al., 1990; Mercer and Crapo, 1992, 1993; Miller et al., 1993a,b).

To test the portion of the hypothesis relating to the reactivity of O_3 with the LRT liquid lining layer, the dosimetry model assumes a pseudo-first-order reaction scheme for reactions of O_3 with biochemical constituents of mucus in the TB and surfactant in the pulmonary regions. A synthesis of published data from a variety of sources on the concentrations of constituents (such as amino acids and unsaturated fatty acids) and on the stochiometry of reactions with O_3 (see Miller et al., 1985) is required. For a recent review of the comparative biochemistry of airway lining fluid, see Hatch (1992).

Any reactions of O_3 with constituents of mucus and surfactant serve to lessen the dose of free O_3 to underlying tissue. To implement the treatment of chemical reactions in the mucous and surfactant layers, the dosimetry model formulation must specify the thickness of the liquid lining layers in every TB airway and pulmonary generation. Initially, this was accomplished by assuming a linear decline in mucous thickness from the trachea to the terminal bronchioles and by assuming a constant thickness of surfactant of 1 μm in respiratory bronchioles and of 0.125 μm in alveolar generations (Miller, 1977). For some species such as rats and humans, recent work by Mercer et al. (1992) provides location-specific thickness data that can be used to refine assumptions about the relationship between mucous thickness and airway size.

Returning to the role of TB path length and acinar volume in determining absorption of O_3 and resultant injury, the development of evidence proceeds along a different track. While single-equivalent path-lung models have been typically derived as simplifications of thousands of measurements made on lung casts, they do not really reflect the complexity of branching patterns *in vivo*. To examine the extent to which O_3 dose might differ for various paths, Overton et al. (1989) compared the pattern of O_3 absorption in the shortest and longest TB paths of the rat using the data of Raabe et al. (1976). Their O_3 dosimetry model predicted a three-fold greater net dose in the first pulmonary generation of the shortest path compared to the longest one. Similar variability in the dose of inhaled gases to acini (ventilatory units) was shown by Mercer and Crapo (1989) using three-dimensional reconstruction data for a region of a rat lung. A ventilatory unit, which is defined to be the collection of all alveoli distal to a terminal bronchiole, includes about 1500 alveoli for the rat (Mercer and Crapo, 1987). These modeling results on dose variability are consistent with the findings of Boorman et al. (1980) and Schwartz et al. (1976) on differing degrees of injury in different centriacini of the same rat.

The above results are a reasonable basis for the hypothesis that site specificity of acinar responses to inhaled O_3 can be attributed to a combination of intraindividual variability and TB path length and acinar volume, in conjunction with the reactivity of O_3 with LRT liquid lining layers. To more fully test the hypothesis, the dosimetry model formulation needs to incorporate a physical model of lung structure having multiple individual TB paths and ventilatory units of different sizes attached at the end of the paths. Mercer et al. (1991a) developed such a model to study the inhomogeneity of ventilatory unit volume and its effect on uptake of a reactive gas. The representation of lung structure used by Mercer et al. (1991a) is given in panel a of Fig. 9. Using a finite-difference technique, the equation of gas transport was solved in all paths of the airway reconstruction and in a series of conical segments for each ventilatory unit.

The resulting distribution of ventilatory unit dose is shown in panel b of Fig. 9. A sizable fraction of ventilatory units receive a two-fold or greater dose of O_3 compared to the average dose across all ventilatory units. The larger ventilatory units that are ventilated by more proximal TB airways are the ones receiving the greatest dose (Mercer et al., 1991a). Previously published results showing the patchy lesion response following O_3 exposure have been obtained from random sections of parenchymal tissue, and thus the TB path leading to the site examined cannot be ascertained. However, studies to quantify the extent of O_3-induced injury in ventilatory units where the relevant TB path is known exactly can be conducted using a serial airway dissection technique (Plopper, 1990). Such studies are in progress, and preliminary results show good agreement between O_3 dosimetry model predictions and ventilatory unit injury when contrasting effects in units at the end of short versus long TB paths (Dr. Kent Pinkerton, UC Davis, personal communication). When these studies are completed, one will indeed be able to test the hypothesis that site specificity of acinar responses to inhaled O_3 is due to a combination of intraindividual variability in TB path length and ventilatory unit volume in conjunction with the reactivity of O_3 with LRT liquid lining layers.

On a final note, dosimetry models for reactive gases such as the model developed by Mercer et al. (1991a) also provide a means of examining the relationship between O_3-induced injury and distance within the ventilatory unit. Alveolar wall and septal tip thickness (as a function of distance from the bronchiolar-alveolar duct junction) were determined in rats exposed for 90 days to either filtered air or to 0.98 ppm O_3 (Pinkerton et al., 1992). Fig. 10 shows this relationship for changes from control in these parameters and also presents the predicted dose of O_3 using the model of Mercer et al. (1991a). Changes in septal tip thickness do not follow changes in predicted dose as closely as changes in alveolar wall thickness. This is probably due to the fact that the conical segments used in the dosimetry model formulation do not reflect the degree of microanatomy represented by septal tips. However, based on upon the similar curvilinear relationship for both predicted

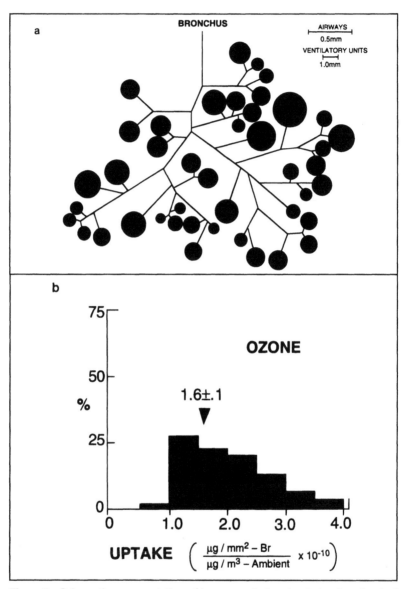

Figure 9 Schematic representation of lower respiratory tract structure for dosimetry model formulation (panel a) and the resulting predicted distribution of ozone uptake in ventilatory units (panel b). While branching angles are not maintained in the schematic, the length of each airway is proportional to the length of the line. Also, each ventilatory unit (symbolized by circles) is depicted so that the diameter of the circle is proportional to the volume-equivalent diameter of the actual ventilatory unit. The distribution of ventilatory unit uptake of ozone shown in panel b was obtained using an inspired O_3 concentration of $1\mu g/m^3$. Since the distribution uptake is independent of ozone concentration, the values of uptake per unit surface area shown can be adjusted to other exposure scenarios by multiplying them by the inspired O_3 concentration assumed to occur at the beginning of the bronchus. Based upon figures contained in Mercer et al. (1991a).

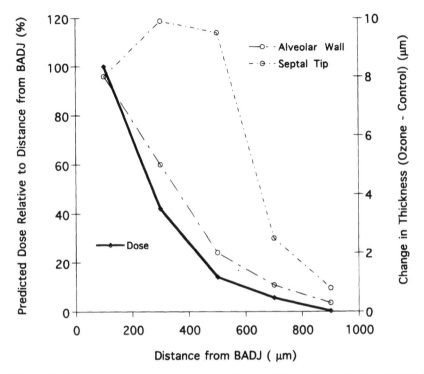

Figure 10 The predicted dose of ozone relative to distance from the bronchiolaralveolar duct junction (BADJ) and the corresponding changes from control in alveolar wall and septal tip thickness in rats exposed for 90 days to 0.98 ppm O_3. Predicted dose changes are related to the left-hand y-axis; changes in alveolar wall and septal tip thickness are related to the right-hand y-axis. The figure is based upon data contained in Table 1 of Pinkerton et al. (1992).

dose and changes in alveolar wall thickness relative to distance, one can infer that the O_3 dosimetry model can be used effectively in extrapolating animal toxicological results.

SUMMARY

The respiratory tract is particularly susceptible to injury from exposure to highly reactive gases. Obtaining experimental dosimetry data for reactive gases is particularly challenging, and there are many human exposure scenarios for which it is not possible to determine absorption experimentally. Hence, dosimetry models that can reliably predict absorption of reactive gases in the respiratory tract often must be developed. Defining a hypothesis, making explicit assumptions used to construct the model, and specifying boundary and initial conditions are critical steps in formulating a dosimetry model. This chapter focused on formulating and implementing dosimetry models of regional respiratory tract absorption of inhaled reactive gases. In so doing, some of the major factors influencing absorption were discussed. Also highlighted were recent advances in knowledge of respiratory tract structure and in computational approaches that can be used to implement dosimetry models. Aspects of formaldehyde and ozone dosimetry in the URT and LRT, respectively, were discussed as examples of modeling respiratory tract absorption of reactive gases. These examples illustrate some of the dynamics of this rapidly changing field.

REFERENCES

Barry, BE, Miller, FJ, Crap, JD: Effects of 012 and 025 ppm ozone on the proximal alveolar region of juvenile and adult rats. Lab Invest 53:692–704, 1985.

Bogdanffy, MS, Randall, HW, Morgan, KT: Histochemical localizaion of aldehyde dehydrogenase in the respiratory tract of the Fischer-344 rat. Toxicol Appl Parhmacol 82:560–567, 1986.

Bogdanffy, MS, Randall, HW, Morgan, KT: Biochemical quantitation and histochemical localization of carboxylesterase in the nasal passages of the Fischer-344 rat and B6C3F1 mouse. Toxicol Appl Pharmacol 88:183–194, 1987.

Boorman, GA, Schwarts, LW, Dungworth, DL: Pulmonary effects of prolongedited by ozone insult in rats. Lab Invest 43:108–115, 1980.

Casanova, M, Deyo, DF, Heck, Hd'A: Covalent binding of inhaledited by formaldehyde to DNA in the nasal mucosa of Fischer 344 rats: Analysis of formaldehyde and DNA by high-performance liquid chromatography and provisional pharmacokinetic interpretation. Fundam Appl Toxicol 12:397–417, 1989.

Casanova, M, Morgan, KT, Gross EA, Moss, OR Heck, Hd'A: DNA-protein cross-links and cell replication at specific sites in the nose of F344 rats exposedited by subchronically to formaldehyde. Fundam Appl Toxicol 23:525–536, 1994.

Casanova, M, Morgan, KT, teinhagen, WH, Everitt, JI, Popp, JA HeckHd'A: Covalent binding of inhaledited by formaldehyde to DNA in the respiratory tract of rhesus monkeys: pharmacokinetics, rat-to-monkey interspecies scaling, and extrapolation to man. Fundam. Appl Toxicol 17:409–428, 1991.

Casanova-Schmitz, M, Starr, TB, Heck, Hd'A: Differentiation . . . between metabolic incorporation and covalent binding in the labeling of macromolecules in the rat nasal mucosa and bone marrow by inhaledited by [^{14}C]- and [^{3}H]formaldehyde. Toxicol Appl Pharmacol 76:26–44, 1984.

Chang, L, Miller, FJ, Ultman, J, Huang, Y, Stockstill, BL, Grose, E, Graham, JA, Ospital, JJ, Crapo, JD: Alveolar epithelial cell injuries by subchronic exposure to low concentrations of ozone correlate with cumulative exposure. Toxicol Appl Pharmacol 109:219–234, 1991.

Costa, DL, Tepper, JS: Approaches to lung function assessment in small mammals In Toxicology of the Lung, edited by DE Gardner, JD Crapo, EJ Massaro, 147–174, New York, NY: Raven Press, 1988.

Crapo, JD, Barry, BE, Gehr, P, Bachofen, M, Weibel, ER: Cell number and cell characteristics of the normal human lung. Am. Rev. Respir Dis 126:332–337, 1982.

Crapo, JD, Chang, YL, Miller, FJ, Mercer, RR: Aspects of respiratory tract structure and function important for dosimetry modeling: interspecies comparisons. In Principles of Route-to-Route Extrapolation for Risk Assessment, edited by JR Gerrity, CJ Henry, 15–32. New York, NY: Elsevier, 1990.

Crapo, JD, Young SL, Fram, EK, Pinkerton, KE, Barry, BE, Crapo, RO Morphometric characteristics of cells in the alveolar region of mammalian lungs. Am. Rev. Respir Dis 128:542–546, 1983.

Davidson, MR, Fitz-Gerald, JM: Transport of 02 along a model pathway through the respiratory region of the lung. Bull Math Biol 36:275–303, 1974.

Dungworth, DL, Castleman, WL, Chow, CK, Mellick, PW, Mustafa, MG, Tarkington, B, Tyler, WS: Effect of ambient levels of ozone on monkeys Fedited by Proc 34:1670–1674, 1975.

Frederick, CB, Morris, JB, Kimbell, JS, Morgan, KT, Scherer, PW: Comparison of four biologically basedited by dosimetry models for the deposition of rapidly metabolizedited by vapors in the rodent nasal cavity. In Nasal Toxicity and Dosimetty of Inhaledited by Xenobiotics: Implications for Human Health, edited by FJ Miller, 135–157. Washington, DC: Taylor & Francis, 1995.

Gardner, DE, Crapo, JD, Massaro, EJ, editors: Toxicology of the Lung New York, NY: Raven Press, 1988.

Gardner, DE, Crapo JD, McClellan, RO, editors: Toxicology of the Lung New York, NY: Raven Press, 1993.

Gross, EA, Morgan, KT: Architecture of nasal passages and larynx. In Comparative Biology of the Normal Lung, edited by RA Parent, 7–25 Boca Raton, FL: CRC Press, 1992.

Guilmeffe, RA, Wicks, JD Wolff, RK: Morphometry of human nasal airways *in vivo* using magnetic resonance imaging. J Aerosol Med 2:365–377, 1989.

Hahn, I, Scherer, PW, Mozell, MM: Velocity profiles measuredited by for airflow through a large-scale model of the human nasal cavity. J Appl Physiol 75:2273–2287, 1993.

Hanna, LM, Frank, R, Scherer, PW: Absorption of soluble gases and vapors in the respiratory system. In Respiratory Physiology: An Analytical Approach, edited by HK Chang, M Paiva, 277–316. New York, NY: Marcel Dekker, Inc (Lenfant, C, edited by Lung biology in health and disease: v40), 1989.

Hatch, GE: Comparative biochemistry of airway lining fluid. In Comparative Biology of the Normal Lung, edited by RA Parent, 617–632, Boca Raton, FL: CRC Press, 1992.

Hatch, GE, Aissa, M: Determination of absorbedited by dose of ozone in animals and humans using stable isotope (oxygen-18) tracing Presentedited by at 80th annual meeting of the Air Pollution Control Association; June; New York, NY Pittsburgh, PA: Air Pollution Control Association, paper no 87-992, 1987.

Hatch, GE, Weister, JH, Overton, JH, Aissa, M: Respiratory tract dosimetry of 180-labeledited by ozone in rats: Implications for a rat-human extrapolation of ozone dose. In Atmospheric Ozone Research and its Policy Implications, edited by T Schneider, SD Lee, GTR Wolters, LD Grant, 553–560 New York, NY: Elsevier Science Publishers, 1989.

Heck, Hd'A, Casanova, M: Nasal dosimetry of formaldehyde: modeling site specificity and the effects of preexposure Inhal Toxicol 6:159–175, 1994.

Horsfield, K, Cumming, G: Morphology of the bronchial tree in man. J Appl Physiol 24:373–383, 1968.

Hu, SC, Ben-Jebria, A, Ultman, JS: Simulaton of ozone uptake distribution in the human airways by orthogonal colocation on finite elements. Comput Biomedited by Res 25:264–278, 1992.

Hyde, DM, Plopper, CG, Harkema, JR, St George, JA, Tyler, WS, Dungworth, DL: Ozoneinducedited by structural changes in monkey respiratory system. In Atmospheric Ozone Research and its Policy Implications, edited by T Schneider, SD Lee, GTR Wolters, LD Grant, 523–532 New York, NY: Elsevier Science Publishers, 1989.

Keller, D, Heck, Hd'A, Randall, HW, Morgan, KT: Histochemical localization of formaldehyde dehydrogenase. Toxicol Appl Pharmacol 106:311–326, 1990.

Kimbell, JS: Issues in modeling dosimetry in rats and primates. In Nasal Toxicity and Dosimetry of Inhaledited by Xenobiotics: Implications for Human Health, edited by FJ Miller, 73–83. Washington, DC: Taylor & Francis, 1995.

Kimbell, JS, Gross, EA, Joyner, DJ, Godo, MN, Morgan, KT: Application of computational fluid dynamics to regional dosimetry of inhaledited by chemicals in the upper respiratory tract of the rat Toxicol Appl Pharmacol 121:253–263, 1995.

Kliment, V: Geometry of guinea pig respiratory tract and application of Landahl's model of deposition of aerosol particles. J Hyg. Epidemiol Microbiol Immunol 16:107–114, 1972.

Kliment, V: Similarity and dimensional analysis, evaluation of aerosol deposition in the lungs of laboratory animals and man. Folia. Morphol (Warsz) 21:59–64, 1973.

Lai, Y-L: Comparative ventilation of the normal lung in Comparative Biology of the Normal Lung, Vol 1, edited by RA Parent, 217–240 Boca Raton, FL: CRC Press, 1992.

Landahl, HD: On the removal of airborne droplets by the human respiratory tract: I The lung Bull Math Biophys 12:43–56, 1950.

Leith, DE: Mass transport in mammalian lungs: comparative physiology In Fundamentals of Extrapolation Modeling of Inhaledited by Toxicants, edited by FJ Miller, DB Menzel, 71–91 Washington, DC: Hemisphere Publishing, 1984.

Lueptow, RM: Software for computational fluid flow and heat transfer analysis Comp Mech Eng 6:10–17, 1988.

Mariassy, AT: Epithelial cells of trachea and bronchia. In Comparative Biology of the Normal Lung, edited by RA Parent, 63–76, Boca Raton, FL: CRC. Press, 1992.

McClellan, RO, Henderson, RF, editors: Concepts in Inhalation Toxicology. New York, NY: Hemisphere Publishing Company, 1989.

Ménache, MG Patra AL, Miller, FJ: Airway structure variability in the Long-Evans rat lung Neurosci & Biobehav Rev 15:63–69, 1991.

Mercer, RR, Anjilvel, S, Miller, FJ, Crapo, JD: Inhomogeneity of ventilatory unit volume and its effects on reactive gas uptake. J Appl Physiol 70:2193–2205, 1991a.

Mercer, RR, Crapo, JD: Three-dimensional reconstruction of the rat acinus J. Appl Physiol 63(2): 785–794, 1987.

Mercer, RR, Crapo, JD: Anatomical modeling of microdosimetry of inhaledited by particles and gases in the lung in Extrapolation of Dosimetric Relationships for Inhaledited by Particles and Gases, edited by JD Crapo, ED Smolko, FJ Miller, JA Graham, AW Hayes, 69–78 San Diego, CA: Academic Press, 1989.

Mercer, RR, Crapo, JD: Spatial distribution of collagen and elastin fibers in the lungs J Appl Physiol 69(2): 756–765, 1990.

Mercer, RR Crapo, JD: Architecture of the acinus In Comparative Biology of the Normal Lung, edited by RA Parent, 109–119. Boca Raton, FL: CRC Press, 1992.

Mercer, RR, Crapo, JD: Three-dimensional analysis of lung structure and its application to pulmonary dosimetry models. In Toxicology of the Lung, 2nd ed, edited by DE Gardner, JD Crapo, RO McClellan, 155–186 New York, NY: Raven Press, 1993.

Mercer, RR, Laco, JM, Crapo, JD: Three-dimensional reconstruction of alveoli in the rat lung for pressure-volume relationships J Appl Physiol 62(4): 1480–1487, 1987.

Mercer, RR, Russell, ML, Crapo, JD: Radon dosimetry basedited by on the depth distribution of nuclei in human and rat lungs. Health Physics, 61(1): 117–130, 1991b.

Mercer, RR, Russell, ML, Crapo, JD: Mucous lining layers in human and rat airways Am Rev Respir Dis 145:355, 1992.

Mercer, RR, Russell, ML, Crapo, JD: Alveolar septal structure in different species J Appl Physiol 77:1060–1066, 1994.

Mery, S, Gross, EA, Joyner, DR, Godo, MN, Morgan, KT: Nasal diagrams: A tool for recording the distribution of nasal lesions in rats and mice. Toxicol Pathol 22:353–372, 1994.

Miller, FJ: A mathematical model of transport and removal of ozone in mammalian lungs PhD thesis, North Carolina State Univ, Raleigh, NC, 1977.

Miller, FJ: Dosimetry of inhaledited by gases In Respiratory Toxicology and Risk Assessment, edited by PG Jenkins, D Kayser, H Muhle, G Rosner, EM Smith, 111–144. Stuttgart, Germany: Wissenschaftliche Verlagsgesellschaft mbH, 1994.

Miller, FJ, Graham, JA: Research needs and advances in inhalation dosimetry identifiedited by through the use of mathematical dosimetry models of ozone Toxicol Lett 44:231–246, 1988.

Miller, FJ, Menzel, DB, Coffin, DL: Similarity between man and laboratory animals in regional pulmonary deposition of ozone. Environ Res 17:84–101, 1978.

Miller, FJ, Mercer, RR, Crapo, JD: Lower respiratory tract structure of laboratory animals and humans: Dosimetry implications. Aerosol Sci Technol 18:257–271, 1993a.

Miller, FJ, Overton, JH, Jr, Jaskot, RH, Menzel, DB: A model of the regional uptake of gaseous pollutants in the lung I The sensitivity of the uptake of ozone in the human lung to lower respiratory tract secretions and exercise Toxicol Appl Pharmacol 79:11–27, 1985.

Miller, FJ, Overton, JH, Kimbell, JS, Russell, ML: regional respiratory tract absorption of inhaledited by reactive gases. In Toxicology of the Lung, 2nd ed, edited by DE Gardner, JD Crapo, and RO McClellan, 485–525. New York, NY: Raven Press, 1993b.

Monticello, TM, Miller, FJ, Morgan, KT: Regional increases in rat nasal epithelial cell proliferation following acute and subchronic inhalation of formaldehyde Toxicol Appl. Pharmacol 111:409–421, 1991.

Monticello, TM, Morgan, KT, Everitt, JI, Popp, JA: Effects of formaldehyde gas on the respiratory tract of rhesus monkeys: Pathology and cell proliferation Am J Pathol 134:515–527, 1989.

Morgan, KT: Nasal dosimetry, lesion distribution, and the toxicologic pathologist: A brief review In Nasal Toxicity and Dosimetty of Inhaledited by Xenobiotics: Implications for Human Health, edited by FJ Miller, 41–57 Washington, DC: Taylor & Francis, 1995.

Morgan, KT, Asgharian, B, Ultman, JT, Kimbell, JS, Dasgupta, A: Comparison of wall mass flux simulations with flow tanks results and appliation to inhalation toxicology. Toxicologist 12:353, 1992.

Morgan, KT, Jiang, XZ, Patterson DL, Gross, EA: The nasal mucociliary apparatus. Am Rev Respir Dis 130:275–281 Morgan, KT, Jiang, XZ, Starr, TB, Kerns, WD 1986a. More precise localization of nasal tumors associated by with chronic exposure of F-344 rats to formaldehyde gas. Toxicol Appl. Pharmacol 82, 264–271, 1986a.

Morgan, KT, Kimbell, JS, Monticello, TM, Patra AL, Fleishman, A: Studies of inspiratory airflow patterns in the nasal passages of the F-344 rat and rhesus monkey using nasal molds: Relevance to formaldehyde toxicity. Toxicol Appl Pharmacol 110:223–240, 1991.

Morgan, KT, TM Monticello, TM: Airflow, gas deposition, and lesion distribution in the nasal passages. Environ Health Perspect 85:209–218, 1990.

Morgan, KT, Patterson, DL, Gross, EA: Responses of the nasal mucociliary apparatus of F-344 rats to formaldehyde gas. Toxicol Appl Pharmacol 82:1–23, 1986b.

Morris, JB, Blanchard, KT: Upper respiratory tract deposition of inspiredited by acetaldehyde. Toxicol Appl Pharmacol 114:140–146, 1992.

Morris, JB, Cavanagh, DG: Deposition of ethanol and acetone vapors in the upper respiratory tract of the rat. Fundam Appl Toxicol 6:78–88, 1986.

Morris, JB, Clay, RJ, Cavanagh, DG: Species differences in upper respiratory tract desposition of acetone and ethanol vapors. Fundam Appl Toxicol 7:671–680, 1986.

Morris, JB, Hasset, DD, Blanchard, KT: A physiologically-based pharmacokinetic model for nasal uptake and metabolism of nonreactive vapors. Toxicol Appl Pharmacol 123:120–129, 1993.

Morris, JB, Smith, FA: Regional deposition and absorption of inhaled hydrogen fluoride in the rat. Toxicol Appl Pharmacol 62:81–89, 1982.

Overton, JH: Respiratory tract physiological processes and the uptake of gases. In Principles of Route-to-Route Extrapolation for Risk Assessment, edited by TR Gerrity and CJ Henry, 71–91. New York, NY: Elsevier, 1990.

Overton, JH, Barnett, AE, RC Graham, RC: Significances of the variability of tracheobronchial airway paths and their airflow rates to dosimetry model predictions of the absorption of gases. In Extrapolation of Dosimetric Relationships for Inhaled Particles and Gasses, edited by JD Crapo, ED Smolko, FJ Miller, JA Graham, and AW Hayes, 273–291. New York, NY: Academic Press, 1989.

Overton, JH, Graham, RC: Predictions of ozone absorption in human lungs from newborn to adult. Health Phys 57:29–36, 1989.

Overton, JH, Graham, RC, Miller, FJ: A model of the regional uptake of gaseous pollutants in the lung II. The sensitivity of ozone uptake in laboratory animal lungs to anatomical and ventilatory parameters. Toxicol Appl Pharmacol 88:418–432, 1987.

Parent, RA, editor: Comparative Biology of the Normal Lung, Vol 1, Boca Raton, FL: CRC Press, 1992.

Patra, AL, Gooya, A, Morgan, KT: Airflow characteristics in a baboon nasal passage cast. J Appl. Physiol 61:1959–1966, 1986.

Patterson, DL, Gross, EA, Bogdanffy, MS, Morgan, KT: Retention of formaldehyde gas by the nasal passages of F-344 rats. Toxicologist 6:55, 1986.

Pinkerton, KE, Barry, BE, O'Neil, JJ, Raub, JA, Pratt, PC, Crapo, JD: Morphologic changes in the lung during the lifespan of Fischer 344 rats. Am J Anatomy 164:155–174, 1982.

Pinkerton, KE, Mercer, RR, Plopper, CG, Crapo, JD: Distribution of injury and microdosimetry of ozone in the ventilatory unit of the rat. J Appl Physiol 73:817–824, 1992.

Plopper, CG: Structural methods for studying bronchiolar epithelial cells. In Models of Lung Disease: Microscopy and Structural Methods, edited by J Gil, 537–559. New York, NY: Marcel Dekker, Inc., 1990.

Plopper, CG, Mariassy, AT, Loilini, LO: Effect of panting frequency on the plethysmographic determination of thoracic gas volume in chronic obstructive pulmonary disease. Am Rev Respir Dis 128:54–57, 1983.

Plopper, CG, Weir, A, St. George, J, Tyler, N, Mariassy, A, Wilson, D, Nishio, S, Cranz, D, Heidsiek, J, Hyde, D: Species differences in airway cell distribution and morphology. In Extrapolation of Dosimetric Relationships for Inhaled Particles and Gases, edited by JD Crapo, ED Smolko, FJ Miller, JA Graham, and AW Hayes, 19–34. San Diego, CA: Academic Press, 1989.

Press, WH, Flannery, BP, Teukolsky, SA, Vefterling, WT: Numerical Recipes. New York, NY: Cambridge University Press, 1986.

Proctor, DF, Anderson, I, editors: The Nose, Upper Airway Physiology and the Atmospheric Environment Amsterdam. Oxford. New York, NY: Elsevier Science Publishers, 1982.

Proetz, AW: Air currents in the upper respiratory tract and their clinical importance. Ann Otol Rhinol laryngol 60:439–467, 1951.

Raabe, OG, Yeh, HC, Schum, GM, Phalen, RF: Tracheobronchial geometry: human, dog, rat, hamster. Albuquerque, NM: Lovelace Foundation, LF-53, 1976.

Santrock, J, Hatchm GE, Slade, R, Hayesm JM: Incorporation and disappearance of oxygen-1 8 in lung tissue from mice allowed to breath 1 PPM $^{18}O_3^1$. Toxicol Appl Pharmacol 98:75–80, 1989.

Scherer, PW, Kehani, K, Mozell, MM: Nasal dosimetry modeling for humans. In Nasal Toxicity and Dosimetry of Inhaled Xenobiotics: Implications for Human Health, edited by FJ Miller, 85–97. Washington, DC: Taylor & Francis, 1995.

Scherer, PW, Shendalman, LH, Greene, NM, Bouhuys, A: Measurement of axial diffusivities in a model of the bronchial airways. J Appl Physiol 38:719–723, 1975.

Schreider, JP, Hutchens, JO: Morphology of the guinea pig respiratory tract. Anat Rec 196:313–321, 1980.

Schwartz, LW, Dungworth, DL, Mustafa, MG, Tarkington, BK, Tyler, WS: Pulmonary responses of rats to ambient levels of ozone: Effects of 7-day intermittent or continuous exposure. Lab Invest 34:565–578, 1976.

Stone, KC, Mercer, RR, Gehr, P, Stockstill, B, Crapo, JD: Allometric relationships of cell numbers and size in the mammalian lung. Am J Respir Cell Mol Biol 6:235–243, 1992.

Swift, DL, Proctor, DF: Access of air to the respiratory tract. In Respiratory Defense Mechanisms, Part I, edited by JD Brain, DR Proctor, LM Reid, 63–93. New York, NY: Marcel Dekker, Inc., 1977.

Takezawa, J, Miller, FJ, O'Neil, JJ: SIngle-breath diffusing capacity and lung volumes in small laboratory mammals. J Appl Physiol-Respirat Environ Exercise Physiol 48:1052–1059, 1980.

Taylor, GI: Dispersion of soluble matter in solvent flowing slowly through a tube. Proc Roy Soc A. 219:186–203, 1953.

Tepper, JS, Costa, DL, Lehmann, JR: Extrapolation of animal data to humans: Homology of pulmonary physiological responses with ozone exposure. In Toxicology of the Lung, 2nd ed, edited by DE Gardner, JD Crapo, RO McClellan, 217–251. New York, NY: Raven Press, 1993.

Ultman, JS: Transport and uptake of inhaledited by gases. In Air Pollution, the Automobile, and Public Health, edited by AY Watson, RR Bates, D Kennedy, 323–366. Washington, DC: National Academy Press, 1988.

US Environmental Protection Agency: Air quality criteria for ozone and other photochemical oxidants. Research Triangle Park, NC: Office of Health and Environmental Assessment, Environmental Criteria and Assessment Office; EPA report nos. EPA-600/8-84-O2OaF-eF 5v. Available from: NTIS, Springfield, VA; PB87-142949, 1986.

Weibel, ER: Morphometry of the Human Lung. New York, NY: Academic Press, 1963.

Wilson, TA, Lin, K: Convection and diffusion in the airways and the design of the bronchial tree. In Airway Dynamics Physiology and Pharmacology, edited by A Bouhuys, 5–19. Springfield, IL: Thomas, 1970.

Yeh, HC, Schum, GM: Models of human lung airways and their application to inhaled particle deposition. Bull Math Biol 42:461–480, 1980.

Yeh, HC, Schum, GM, Duggan, MT: Anatomic models of the tracheobronchial and pulmonary region of the rat. Anat Rec 195:483–492, 1979.

Chapter Eleven

Factors Modifying the Disposition of Inhaled Organic Compounds

James A. Bond and Michele A. Medinsky

INTRODUCTION

Polycyclic aromatic hydrocarbons (PAH), nitro-PAH, aromatic amines, heterocyclic amines, aliphatic compounds, aldehydes, ketones, and other classes of organic compounds have been detected and in many instances quantitatively measured in industrial settings, cigarette smoke and emissions, products, and byproducts associated with energy production and use. Volatile organic chemicals are also present in indoor air, household and industrial products, cigarette smoke, and gasoline vapors (Wallace et al., 1984). The disposition of organic chemicals is complicated by the fact that metabolites are more often toxic than the parent chemical. Additionally, these organic chemicals exist as components of chemical mixtures (either as individual components or adsorbed onto respirable particles) that can be either organic or inorganic. Little information is available on the potential human health risks associated with exposure to these classes of chemicals. Furthermore, the fate and target organs for these chemicals after inhalation in humans is largely unknown.

Since inhalation is a likely route of human exposure to many organic chemicals, determining the disposition of these chemicals after inhalation is important. The term disposition is applied to processes of distribution, metabolism, and elimination of chemicals from the body that together enable us to better understand the relationships between exposure concentration and dose to critical cellular macromolecules. Many factors (both chemical and biological) influence the disposition of inhaled chemicals. Disposition data are useful in predicting tissue concentrations of inhaled chemicals and metabolites and thus allow a more accurate assessment of their potential effects on human health after repeated or continuous exposure.

Inhaled chemicals can deposit in various regions of the respiratory tract. Factors that determine deposition include chemical reactivity and water solubility for gases and shape, physical size, and density for particles (Task Group on Lung Dynamics, 1966; Davies, 1985). The retention of organic compounds in the respiratory tract following deposition depends upon their reactivity and binding to respiratory tract tissue and their solubility in body fluids (Bond et al., 1986b; Ludden et al., 1976; Theodore et al., 1975; Mercer, 1973). These factors alone and in concert with one another can influence the biological fate of inhaled chemicals.

Because of the great diversity in chemical and physical properties of nonvolatile and volatile organics, examining the factors important for their uptake, distribution, and elimination is useful. Determinants can include those of a physiological nature, such as alveolar ventilation, cardiac output, blood flow to organs, and organ volumes. Chemical determinants include the partition coefficients that describe the distribution of the organic chemical between blood and air or between blood and tissues at equilibrium. Metabolic determinants such as the capacity for metabolism are extremely important in the disposition of most organics. The exposure history of an individual can modify metabolic determinants, resulting in inhibition or induction of metabolism.

The purpose of this chapter is to acquaint the reader with some of the more important factors that can influence the disposition of inhaled chemicals. The emphasis will be on factors that affect how the respiratory tract handles inhaled organic chemicals and how these factors may determine the ultimate fate of organic chemicals. Factors have been classified into two general categories—(1) physiochemical and (2) biological.

PHYSIOCHEMICAL FACTORS

Various physiochemical factors can influence the biological fate of inhaled chemicals. The following discussion will focus on three key factors: (1) particle size, (2) reactivity and (3) solubility.

Particle Size

The atmospheric concentration of a chemical does not, by itself, define the total dose or the specific sites of local doses of a chemical delivered to the respiratory tract. In general, it is the particle size of the chemical or the particle on which it may be associated that influences the specific sites of deposition for the chemical in the respiratory tract. The importance of particle size in influencing the extent and loci of deposited particles has been a topic of numerous reviews and will not be emphasized in this chapter. Particle size as a modifying factor is mentioned here to remind the reader that this important parameter can influence the biological fate of inhaled organic chemicals. Particles deposit at various sites within the respiratory tract by several mechanisms, many of which have been reviewed (Task Group on Lung Dynamics, 1966; Schlesinger, 1989; Lippmann and Schlesinger, 1984; Yeh and Schum, 1980; Yu, 1978; Yeh et al., 1976). Regional deposition determines the subsequent pathways for removal of a deposited chemical and, as such, is a major factor in determining the ultimate toxicological response to an inhaled chemical.

The aerodynamic diameter of an inhaled particle controls where a particle deposits, and the surface area, which is based on the physical diameter, controls the rate at which a chemical leaves the respiratory tract (Task Group on Lung Dynamics, 1966). These two diameters are related to one another by the following proportionality (Raabe, 1976):

$$\text{aerodynamic resistance diameter} = (\text{geometric diameter}) \times \sqrt{(\text{density})} \ (\text{slip correction factor})$$

Particle size of a chemical not only influences the sites of deposition, which in turn affect the mode by which a particle is cleared from the respiratory tract in a given region of the respiratory tract, but also influences the metabolic fate of the chemical. Different portions of the respiratory tract display different capabilities to metabolize chemicals (reviewed by Bond, 1933; Dahl, this book). Inhalation of large particles ($> 5\mu$m) results in the largest fraction of the inhaled chemical depositing in the nasal-pharyngeal region (Schlesinger, this book). Nasal tissue enzymes subsequently metabolize the chemical. On the other hand, Cheng et al. (1988) reported that inspiratory deposition efficiencies in a human nasal cast were 16% and 40% for 0.01-μm and 0.005-μm particles, respectively. Therefore, inhalation of submicrometer particles ($< 0.1 \ \mu$m) results in a large fractional deposition not only in the nasal-pharyngeal region but also in the deep lung, i.e., alveoli. Thus, metabolism of the inhaled chemical occurs at both these sites of the respiratory tract, and the metabolic fate of the chemical is a function of its metabolism at both sites of the respiratory tract.

In summary, perhaps the major influence of particle size on the biological fate of an inhaled chemical is to affect the site of deposition in the respiratory tract. Since different portions of the respiratory tract have different abilities to remove and metabolize inhaled chemicals, the eventual clearance and metabolism of a chemical is closely related to particle size.

Reactivity

The potential chemical reactivity of water soluble gases and vapors is quite variable, ranging from inert (e.g., methanol, ethanol, methoxyethanol) to highly reactive (e.g., ozone, formaldehyde, chlorine). Mucus, surfactant, lung tissue, and blood vary considerably in their biochemical composition. Thus, some knowledge of the nature of reactions of the gas or vapor with the constituents of these fluids, tissue, and blood is necessary to develop an understanding of how a given gas or vapor exerts its toxic effects in the respiratory tract.

Gases that are reactive, but of low water solubility, tend to deposit along the entire respiratory tract. Depending upon the concentration of inhaled gas, the duration of exposure, and the sensitivity of the epithelial cells in each of the three major regions, toxicity may be observed throughout the respiratory tract. However, the greater the reactivity and the lesser the solubility of the gas, sites of earliest and greatest toxicity tend to be in the more distal parts of the respiratory tract. Ozone is an example of a highly reactive gas with low water solubility. Studies in a number of species, such as rats, rabbits, and nonhuman primates, have consistently demonstrated effects of ozone at the level of the bronchoalveolar duct junction or respiratory bronchioles, depending upon the anatomy of the particular species (Stephens et al., 1973, 1974; Chang et al., 1992; P'an et al., 1972; Dungworth et al., 1975; Tyler et al., 1988). Dosimetry models (Miller et al., 1985; Overton et al., 1987; Mercer et al., 1991) have been developed for ozone that yield predictions of dose that are in relatively good agreement with available animal and human experimental dosimetry data. High reactivity does not necessarily ensure high respiratory tract uptake as evidenced by uptake experiments showing 50% removal of ozone in rodents (Wiester, 1987, 1988) compared to 75–95% in humans (Gerrity et al., 1988, 1989 1994; Wiester et al., 1994). However, high reactivity does result in minimal amounts if any of the gas or vapor penetrating to the blood without reacting in the mucous or surfactant layers or with lung tissue. Taken to the extreme, this may indicate that for gases or vapors with high reactivity and low water solubility, a cascade of reaction products may be responsible for cellular toxicity as has been postulated for ozone (Pryor, 1992).

Solubility

Vapors and Gases   An important factor that can influence the disposition of inhaled vapors and gases is the solubility of the chemical in the various biological matrices within the respiratory tract and body. For volatile organic chemicals, partitioning between different biological media and between media and air is a critical chemical factor in describing and predicting disposition and tissue dosimetry. Partition coefficients are measures of the affinity at equilibrium of the volatile organic chemical for one medium com-

Table 1 Partition Coefficients for Some Volatile Organic Chemicals

Chemical	Blood/Air	Fat/Air	Muscle/Air	Fat/Blood
Isoprene	3	72	2	24
Benzene	18	500	11	28
Styrene	40	2000	40	50
Methanol	1350	1500	1760	1.1

Source: Michele A. Medinsky, Critical Determinants in the Systematic Availability and Dosimetry of Volatile Organic Chemicals, in Principles of Route-to-Route Extrapolation for Risk Assessment, Gerrity/Henry ed., © 1990, p. 160. Reprinted by permission of Prentice-Hall, Inc., Englewood Cliffs, N.J.

pared with another. Partition coefficients for volatiles are often determined *in-vitro* using methods such as vial-equilibration (Gargas et al., 1989; Sato and Nakajima, 1979) and described in terms of blood:air, tissue:air, or tissue:blood ratios (Table 1).

In general, vapors and gases that are water soluble or have very large blood-air partition coefficients are removed from the inhaled air by the upper respiratory tract. For example, formaldehyde, a water-soluble vapor and rat nasal carcinogen, concentrates in rat nasal mucosa following inhalation (Heck et al., 1983). Dahl et al. (1991) determined that for highly water soluble solvents such as methanol or ethanol virtually all the vapor is removed from the airstream by tissues of the upper respiratory tract.

Gerde and Dahl (1991) developed a physiological model to simulate the uptake of inhaled vapors in the nasal airway during cyclic breathing. The model was used to predict the nasal absorption of a variety of vapors, and model predictions agreed reasonably well with the experimental data. Nasal uptake of vapors was highly dependent upon the blood:air partition coefficient. The model predicted that nasal absorption of vapors on inspiration increased from 1% for a compound with a blood:air partition coefficient of 1 to about 95% for a compound with a partition coefficient of 2000. The model also predicted that desorption of vapors from the nasal tissues on expiration increased from approximately 1% to approximately 30% over the same range of partition coefficients. In one breathing cycle, absorption occurred on inhalation and desorption occurred on exhalation. Nasal uptake over a complete breathing cycle was almost zero for low partition coefficient compounds and plateaued at approximately 65% for high partition coefficient compounds. This model demonstrated that the nasal tissues serve as a temporary storage compartment for chemicals on inhalation followed by desorption of vapors back into the air stream upon exhalation. Thus, the water solubility of volatile organic chemicals is a factor in determining the site for primary deposition of these volatiles in the respiratory tract.

Vapors that are not water soluble but are soluble in organic solvents will move down the respiratory tract and be absorbed into the blood in the alveolar region. The blood:air partition coefficient (P_b) is a critical determinant in

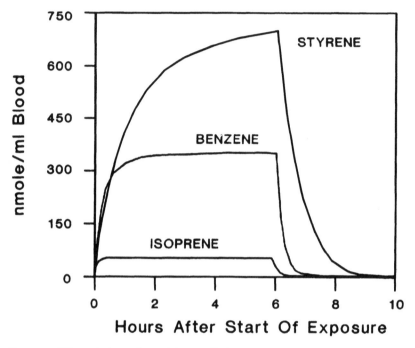

Figure 1   Effect of blood:air partition coefficient on the venous blood concentrations of three volatile organic chemicals both during and after 6-hour exposures of rats to 2000 mg/m³ of each chemical. Lines represent simulation results using a physiologically based pharmacokinetic model similar to that described by Ramsey and Andersen (1984). Blood:air partition coefficients in Table 1 were used in this model. (Taken from Michele A. Medinsky, Critical Determinants in the Systematic Availability and Dosimetry of Volatile Organic Chemicals, in Principles of Route-to-Route Extrapolation for Risk Assessment, Gerrity/Henry ed., © 1990, p. 161. Reprinted by permission of Prentice-Hall, Inc., Englewood Cliffs, N.J.)

the uptake of volatile organics into arterial blood. Assuming no clearance of the chemical from blood except by exhalation, the arterial blood concentration of the volatile organic at equilibrium is a product of the inhaled concentration and the blood-air partition coefficient. The net effect of this relationship is that chemicals with a high affinity or solubility in blood (i.e., large P_b) are absorbed to a greater extent than chemicals with a low solubility. For example, with increasing P_b, from 3 for isoprene to 40 for styrene (Table 1), the concentration of the volatile organic in the systemic circulation increases for a given exposure concentration(Fig. 1).

Partition coefficients are also a useful indicator of the tissues in the body that will achieve the highest concentration of the volatile following exposure. For example, for most volatile organic chemicals the fat-blood partition coefficients are large compared with partition coefficients for other tissues. Experimental studies have demonstrated that fat contains higher concentrations of volatile organics than other tissues (Ramsey and Andersen, 1984).

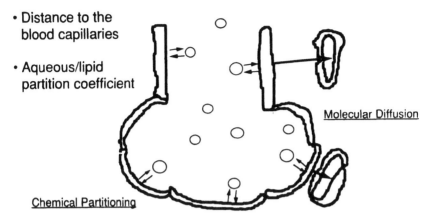

• Distance to the
blood capillaries

• Aqueous/lipid
partition coefficient

Molecular Diffusion

Chemical Partitioning

Figure 2 The rate of molecular diffusion and the lipid-aqueous partition coefficient are critical determinants for the dissolution of organic particulates into the blood from the surface of the respiratory tract tissues.

Particles Gerde et al. (1991b) developed a physiological model that identifies the lipid-aqueous partition coefficient as a critical determinant for the dissolution of lipophilic polycyclic organic particulates from the lungs to the blood (Fig. 2). In general, the larger the partition coefficient of a chemical particulate, the slower the rate of dissolution for the compound in the lung. Dissolution of chemicals in the tracheobronchial region into the blood will be slower than in the alveolar region because the distance for chemical diffusion to capillaries is considerably longer in the tracheobronchial region than the air-blood interface at the alveoli.

The model of Gerde et al. (1991b) was compared to experimental data on the disposition of PAH in the respiratory tract of beagle dogs. The experimental data supported the model predictions; highly lipophilic toxicants such as PAH were shown to be diffusion-limited during the clearance from the respiratory tract (Gerde et al., 1993a,b). This means that diffusion of the chemicals across the cell membranes and through the tissues of the respiratory tract was the rate-limiting event in the clearance of the chemicals, not the rate of blood perfusion to the respiratory tract. Volatile organic chemicals and organic particulates of lower lipohilicity are usually regarded as perfusion-limited during clearance from the lungs. Increased uptake into the blood with these organics is seen with increased blood flow to the lungs. The modeling studies of Gerde et al. (1993c) demonstrated that a perfusion-limited chemical is cleared within minutes after inhalation from all regions of the respiratory tract into the circulating blood. The clearance of diffusion-limited highly lipophilic PAHs into the blood from the alveolar region of the respiratory tract is slower (Fig. 3).

Regional differences in the respiratory tract clearance of the highly lipophilic PAH benzo(a)pyrene were also demonstrated (Gerde et al., 1993a,b). Even though transport of this chemical is diffusion-limited, clearance from

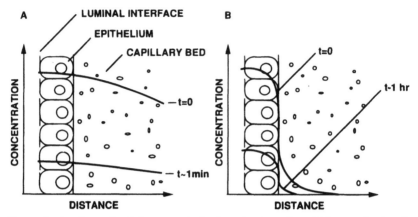

Figure 3 A schematic of the two principal mechanisms by which organic chemicals migrate from the respiratory tract tissues to the capillary blood. (A) Water-soluble to moderately lipophilic substances are blood flow-limited in their absorption. Molecules of these substances rapidly diffuse through the respiratory tract tissues to the capillaries and are cleared from the respiratory tract in minutes. As indicated in Panel A, the organic concentration (solid line) decreases in both the epithelium and capillary bed over a 1-minute period (difference between two lines). Likewise, the concentration gradient from the luminal surface to the capillary bed decreases only slightly indicating rapid transport between these two tissues. (B) Highly lipophilic substances are diffusion-limited in their absorption from the respiratory tract into the blood. These molecules slowly diffuse through the tissues, resulting in a steep concentration gradient from the luminal surface to the capillary bed. As a result the respiratory tract tissues can be exposed to these organics for a period of hours. (Taken from Gerde et, al 1993c, with permission)

the alveolar epithelium took only minutes, whereas clearance through the thicker epithelium of the conducting airways took hours. This phenomenon of slow upper airway clearance results from slower diffusion of highly lipophilic substances through the thicker air-blood barrier of the conducting airways compared to the thinner alveolar epithelium. The model predicted that a direct result of the slowed clearance is a high concentration of PAH in the bronchial walls and increased opportunity for metabolism to reactive forms (Gerde et al., 1993c). These simulations are important because they point to the bronchial epithelium as the preferential target of inhaled highly lipophilic toxicants. These studies focus on the importance of microdosimetry in contributing to local toxicity and suggest that bronchial cancer following inhalation exposures is more likely to be induced by highly lipophilic carcinogens such as the PAHs than by less lipophilic carcinogens.

The studies of Brown and Schanker (1983), who compared rates of pulmonary absorption of aerosolized and intratracheally instilled drugs in the rat (Fig. 4), support slower rates for dissolution from the upper respiratory tract compared with the lungs. These studies also demonstrated that mode of administration was an important factor in determining uptake into blood. Drugs were absorbed from the lungs about two times more rapidly when administered by inhalation than when administered by intratracheal instillation. The alveolar epithelium and the alveolar wall capillary boundry between the air and blood is significantly thinner than the corresponding boundries of the conducting airways. Chemicals that are inhaled as aerosols are more uniformly distributed and penetrate deeper to the alveoli than when administered by intratracheal instillation, where penetration is restricted primarily to the tracheobronchial region of the lung. Therefore, the faster clearance of chemicals after inhalation may be due to the fact that more chemical was deposited in the regions where the blood-tissue interface was thinnest, thus allowing for more rapid dissolution compared to instillation.

Schanker et al. (1986) also investigated the pulmonary absorption rates of lipid-insoluble and lipid-soluble drugs in a number of species, including dogs, rats, rabbits, mice, and guinea pigs following intratracheal instillation. The lipid-insoluble drugs represent several major drug types, including an organic anion, an organic cation, and a lipid-insoluble neutral molecule. Half-times for absorption of both lipid-insoluble drugs such as p-amino hippuric acid (PAHA), mannitol, and procainamide ethobromide (PAEB) and a lipid-soluble drug, procainamide, indicated that lipid-insoluble compounds tended to be absorbed relatively slowly among species (Table 2). Half-times for absorption ranged from 20–180 minutes. In contrast, the half-times for absorption for the lipid-soluble drug procainamide were similar across spe-

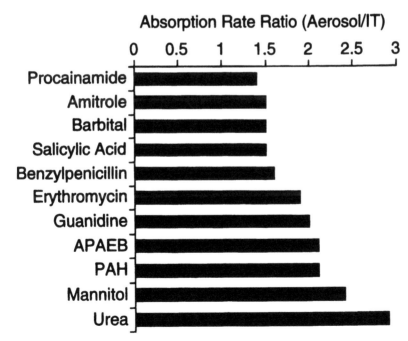

Figure 4 Pulmonary absorption rates after aerosol inhalation and intratracheal (IT) instillation in rats. APAEB = N-acetylprocainamide ethobromide; PAH = $p-$ aminohippuric acid. Data modified from Brown & Shanker, 1983 and Schanker et al., 1986.

Table 2   Pulmonary Absorption of Drugs in Several Animal Species[a]

	Halftime (min) for Absorption				
	Dog	Rat	Rabbit	Mouse	Guinea Pig
PAHA	56	45	103	21	25
Mannitol	69	65	139	23	27
PAEB	93	70	182	26	—
Procainamide	5.3	4.1	4.5	4.5	5.0

[a]Data from Brown and Schanker, 1983. PAHA = p-amino hippuric acid; PAEB = procainamide ethobromide.

cies and relatively rapid (~5 minutes). Based on data for procainamide, the investigators concluded absorption rates of lipid-soluble drugs in humans can probably be predicted with confidence directly from animal observations. In contrast, for the lipid-insoluble compounds that were tested, it is difficult to accurately predict human absorption rates from animal data.

The clearance from the respiratory tract of organic particles of environmental concern such as diesel exhaust, cigarette smoke, and other combustion products is also complicated by the fact that these particles rarely exist in the environment as pure particles. Most often they are absorbed onto the surface of submicrometer particles such as the carbonaceous core of diesel exhaust. Toxicity studies have suggested that instillation of larger amounts of PAH in combination with inert particles increases the development of lung tumors in rodents. The relevance of this experimental model for humans is not clear.

A model developed by Gerde et al. (1991a) was used to compare the clearance of PAH associated with large aggregates of inert dust with the clearance of PAH inhaled at environmental concentrations. When the PAHs are attached to an inert carrier dust before instillation, aggregation of particles is a highly significant factor in the retention of PAHs in the lung. The subsequent slow release of particle-associated PAHs results in a prolonged exposure to surrounding tissues from a limited number of intratracheal administrations. The increased incidence of tumors observed following these exposures led investigators to conclude that the increased retention of PAHs (due to their association with inert particles) is an important factor in PAH-induced pulmonary carcinogenesis. In contrast, during typical human exposures, large aggregates of inert dust are not likely to be inhaled. Individual particles are more likely to deposit on the surfaces of the respiratory tract without interference from other particles. Model simulations predicted that under these low-dose exposure conditions particle-associated PAHs were released rapidly from the particles. Thus, sustained exposure of target tissues to PAHs is the result of repeated exposures to the PAH and the rate limiting physical transport phenomena such as molecular diffusion and partitioning, as noted above. This distinction is important because in high-dose instillation experiments in animals, critical doses to cells, and thus tumors, should occur at sites at which the particles are retained. In contrast, the lower (but more frequent exposures common for humans) should instead lead to tumors at the sites at which the particles are initially deposited but not necessarily retained.

BIOLOGICAL FACTORS

Five major biological factors influencing the disposition of inhaled chemicals include: (1) physiological parameters, (2) capacity-limited metabolism, (3) interspecies differences in metabolism, (4) regional differences in metabolism, and (5) biochemical interactions. There are many other factors that could be discussed, including age, sex, hormonal status, disease state, and diurnal variation. While these factors are important and should not be ignored, they will not be emphasized.

Physiological Parameters

Respiration   An important factor that governs the rate at which inhaled chemicals can deposit in the respiratory tract is the rate at which an animal inhales the material. Clearly, breathing parameters are a factor that cannot be neglected if we are to understand the delivered dose of a chemical following inhalation. In a later chapter (Mauderly, this book), the effect of inhaled toxicants on various ventilatory parameters is discussed. In disposition studies involving inhalation as the route for exposure, the delivered dose of a chemical must be determined since this internal dose is what is available for distribution to tissues. A few examples will be discussed that emphasize the importance of measuring respiratory parameters in laboratory animals exposed to different toxicants and that demonstrate how alterations in breathing patterns can affect the subsequent dose inhaled.

Medinsky et al. (1985) studied the toxicokinetics of inhaled methyl bromide over a range of exposure concentrations (1.5–310 ppm). At each exposure concentration, respiratory parameters, including respiratory rate and tidal volume, were measured. Exposure to a high concentration of methyl bromide (310 ppm) caused alterations in the respiratory pattern of rats (Table 3). A significant depression in tidal volume was noted, resulting in a decreased minute volume. Rats exposed to this concentration of methyl bromide inhaled 25% less total volume of air than did rats exposed to lower concentrations of methyl bromide. Thus, at the two higher exposure concentrations, rats inhaled similar amounts of methyl bromide. The authors suggested that the altered respiratory parameters at the higher concentrations of methyl bromide resulted from the irritating properties of the chemical. A key

Table 3   Summary of Respiratory Parameters Measured in Rats During Exposure to Methyl Bromide[a,b]

Exposure concentration μmole/L (ppm)	Respiratory rate (breaths/min)	Tidal volume (mL)	Minute volume (mL/min)	Methyl bromide inhaled (μmole)
0.05 (1.5)	156±11	1.3±0.06	206±8	3.9±0.2
0.3 (9)	133±5	1.5±5	1.5±0.1	21.7±1.5
5.7 (170)	132±7	1.5±0.09	195±7	400±13
10.4 (310)	142±4	1.1±0.04[c]	153±18[c]	553±64

[a]Values are the mean ± SE.
[b]Data taken from Medinsky et al., 1985.
[c]Significantly different from each of the lower exposure concentrations at $p \leq 0.05$.

point from these studies is that the internal dose of methyl bromide would have been overestimated if measurements of respiratory parameters had not been included in the study design.

Landry et al. (1983) studied methyl chloride and methylene chloride, and also demonstrated altered breathing parameters with high concentrations of these chemicals. Exposure of rats to 1000 ppm methyl chloride resulted in a significant depression in both tidal volume and minute volume (Table 4). Exposure to 1500 ppm methylene chloride resulted in a significant depression of respiratory rate, tidal volume, and minute volume (Table 4). As was the case with methyl bromide, these depressed values resulted in a lower amount of chemical inhaled by the animal.

Studies by Chang et al. (1981) measured the effect of single or repeated formaldehyde exposure on minute volume of B6C3F$_1$ mice and F344 rats to determine if reflex apnea (a defense mechanism that limits the volume of air inhaled) was maintained on repeated exposure. As seen in Table 5, the formaldehyde concentration at which the minute volume decreased by 50% (RD_{50}) was significantly lower in mice than in rats. Naive mice had RD_{50} values of 4.9 ppm compared with RD_{50} values of 31.7 ppm for rats. The RD_{50} values within each species were similar in every group regardless of the pretreatment concentrations of formaldehyde. These data indicate that inhalation of formaldehyde was limited by the respiratory rate and suggest that reflex apnea is a defense mechanism against the inhalation of formaldehyde in both mice and rats. The difference in species sensitivity indicates that mice are more capable than rats of eliciting the observed reflex apnea. Rats, but not mice, develop nasal tumors on exposure to 15 ppm formaldehyde. Differences in inhaled dose may play a role in the differences in response.

In summary, these examples point to the importance of measuring respiratory parameters in animals exposed by inhalation to toxicants. Knowledge of the total amount of chemical inhaled is critical to an understanding of the relationships between exposure concentration and delivered dose. In the examples given here, the delivered dose would have been overestimated had respiratory measurements not been recorded.

Table 4   Summary of Respiratory Parameters Measured in Rats during Exposure to Methyl Chloride or Methylene Chloride[a,b]

Chemical	Exposure concentration (ppm)	Respiratory rate (breaths/min)	Tidal volume (mL)	Minute volume (mL/ min)
Methyl Chloride	50	122±13	1.5±0.1	179±18
	1000	135±34	1.1±0.3[c]	143±13[c]
Methylene Chloride	50	121±20	1.4±0.1	173±30
	1500	160±25[c]	0.8±0.1[c]	124±17[c]

[a]Values are mean ± SD; $n=3$.
[b]Data from Landry et al., 1983.
[c]Significantly different ($p \leq 0.05$) between mean values of rats exposed to low or high concentrations of methyl chloride or methylene chloride.

Organ Volumes and Clearance   The allometric relationships across species for certain physiological parameters, such as organ volumes or weights and processes like clearance and blood flow have been well studied (Medinsky, 1990). When extrapolating across species for tissue volumes including liver, kidney, blood, and heart, these organ volumes increase almost in direct proportion to increasing body weight. In contrast, for many physiological processes such as blood flow to organs, clearance, oxygen consumption, or glomerular filtration rate increase at a proportionally slower rate with increase in body weight. Thus, flow rates, for example, are slower in larger animals compared with smaller ones when expressed per unit of body weight.

The differences in the rates of increase of organ volumes and blood flow have significant impact on the disposition of volatile organics when extrapolating across species. For example, the terminal half-life ($t_{1/2}$) of a chemical tends to be shorter in smaller animals than larger ones. The terminal $t_{1/2}$ is proportional to the ratio of the volume of distribution to clearance. As size

Table 5   Effect of Formaldehyde (H_2CO) on RD_{50} and Minute Volume of Naive and Formaldehyde-pretreated Mice and Rats[a]

Pretreatment H_2CO concentration[b] (ppm)	B6C3F1 Mice		F-344 Rats	
	RD_{50}[c] (ppm)	% decrease in minute volume at RD_{50}	RD_{50} (ppm)	% decrease in minute volume at RD_{50}
Naive	4.9	46.8	31.7	45.1
2	5.9	51.2	29.5	50.1
6	2.2	44.4	28.6	42.4
15	3.6	54.1	22.7	37.5

[a]Data from Chang et al., 1981.
[b]Pretreatment lasted for 6 hours/day for 4 days
[c]Concentration of H_2CO at which breathing rate decreased by 50%.

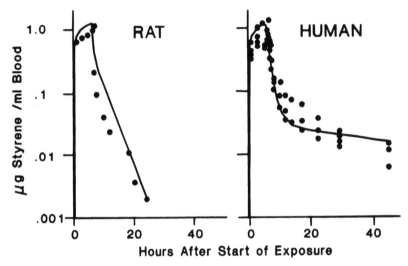

Figure 5 Concentration of styrene in the venous blood of rats and humans exposed to 80 ppm styrene for 6 hours. Data points represent individual measurements of styrene in blood. Lines are results of simulations using the physiologically based pharmacokinetic model of Ramsey and Andersen (1984). (Taken from Michele A. Medinsky, Critical Determinants in the Systematic Availability and Dosimetry of Volatile Organic Chemicals, in Principles of Route-to-Route Extrapolation for Risk Assessment, Gerrity/Henry ed., © 1990, p. 157. Reprinted by permission of Prentice-Hall, Inc., Englewood Cliffs, N.J.)

across species decreases, the volume of distribution (a measure of organ weight) decreases more than clearance (a measure of a flow). After an intravenous injection, the systemic availability (or blood concentration) of many organics is prolonged in larger animals such as man compared with smaller animals such as rodents (Kaye et al., 1986; Mordenti, 1985).

Species differences in systemic availability of volatile organics are illustrated in a study that measured blood concentrations of styrene in rats and humans (Ramsey and Andersen, 1984). Rats and humans were exposed to 80 ppm styrene for 6 hours, and blood samples taken both during and postexposure were analyzed for styrene. Blood concentrations of styrene were comparable in rats and humans at the end of the six hour exposure (Fig. 5). After the cessation of exposure, however, blood concentrations of styrene declined more rapidly in rats than humans. The more rapid decline in rats is due primarily to differences in clearance processes, such as blood flows and metabolism which tend to be much faster in rats than in humans when expressed per unit of body weight.

Capacity-limited Metabolism Another important parameter that can govern the disposition of inhaled chemicals is the capacity for metabolic transformation of the inhaled organic chemical. Capacity for metabolism can be exceeded if the exposure concentration is high. Depending upon the metabolic processes involved, the net result can be an alteration in chemical disposition. A few examples will be given to illustrate how increasing the concentration of a chemical in the exposure atmosphere can significantly affect its disposition. The basis for this alteration is most often attributed to saturation of one or more metabolic pathways.

Gerde et at. (1991b) developed a physiological model to describe the influence of BaP concentration on the retention and metabolism of this chemical by lung cells prior to translocation to other tissues. Their model indicates that the larger the dose, the larger the fraction that will be cleared unmetabolized from the lungs. This has significant implications, since studies performed in laboratory animals typically use very high exposure concentrations relative to the human exposures. Results from their model simulations suggest that enzyme systems are likely to be saturated during animal experiments in which very large doses are administered and only a minute fraction of the administered dose will be metabolized. In contrast, in human exposures, the doses are much lower, and a substantial fraction is probably metabolized. If the probability for malignant transformation from exposure to toxicants is assumed to be proportional to the cumulative amounts of metabolites produced in the lung, extrapolating from the results of animal experiments to the human exposure potentially could result in an underestimation of the risk for human exposure conditions.

McKenna et al. (1982) investigated the disposition of inhaled methylene chloride in rats. One of the pathways for methylene chloride metabolism involves formation of carbon monoxide. The rates of formation and disappearance of carboxyhemoglobin in rat blood during and after methylene chloride exposure were quantitated. There was no significant difference between apparent steady-state carboxyhemoglobin levels attained following either a 500- or 1500-ppm methylene chloride exposure (Fig. 6). The lack of a dose-related increase in blood carboxyhemoglobin level in steady state beyond the 500-ppm exposure concentration suggested that methylene chloride metabolism to carbon monoxide was saturated.

Another example of the effect of increasing exposure concentration on chemical disposition can be seen from work with hexane (Bus et al., 1982). Hexane is partly eliminated unchanged in expired air and partly metabolized. One of the metabolites is CO_2 which is also exhaled. Bus and coworkers demonstrated that the rate-limiting step in the pathway from n-hexane to CO_2 was capacity-limited (suggesting a saturable metabolism component). As a result, the percentage of the dose exhaled as CO_2 decreased with increasing dose, while the exhalation of unmetabolized n-hexane became proportionately more important at high doses (Fig. 7).

In a final example (Bond et al., 1986a), rats were exposed to graded concentrations of inhaled ^{14}C-butadiene, and pathways for excretion of butadiene, and metabolites were investigated. Both urine and exhaled air were the major routes of elimination of ^{14}C in rats exposed to all concentrations of butadiene (Table 6). As concentrations of butadiene increased, the ratio of ^{14}C eliminated in urine and exhaled air shifted. With increasing butadiene

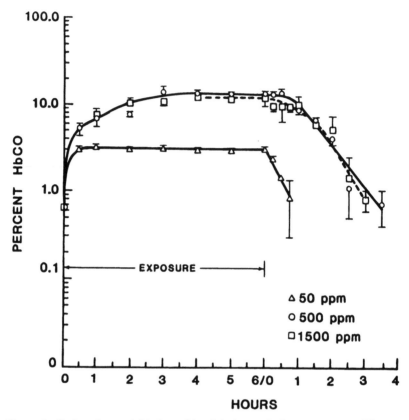

Figure 6 Carboxyhemoglobin in rat blood during and after exposure to different concentrations of methylene chloride. (Taken from McKenna et al., 1982, with permission.)

exposure concentration, the percentage of ^{14}C excreted in the urine decreased and that in exhaled CO_2 increased. The molar amount excreted in the urine was the same for the two higher exposure concentrations (Table 6), indicating saturation of the metabolic pathway(s) that produce urinary metabolites of butadiene.

Himmelstein et al. (1994, 1995) extended these studies to assess whether dose-dependent alterations in metabolic pathways would result in changes in blood concentrations of butadiene monoepoxide, a DNA-reactive metabolite of butadiene. The steady-state blood concentrations of butadiene were proportional to the exposure concentrations at 8000 and 1250 ppm butadiene. However, the steady-state concentrations of butadiene monoepoxide in the blood of male rats exposed to 8000 ppm butadiene (1.4 ± 0.07 μM) was similar to the concentration of butadiene monoepoxide in blood of rats exposed to 1250 ppm butadiene (1.3 ± 0.09 μM), clearly indicating saturation of butadiene monoepoxide formation. In this case, knowledge of metabolic saturation can provide clues with respect to the putative carcinogenic metab-

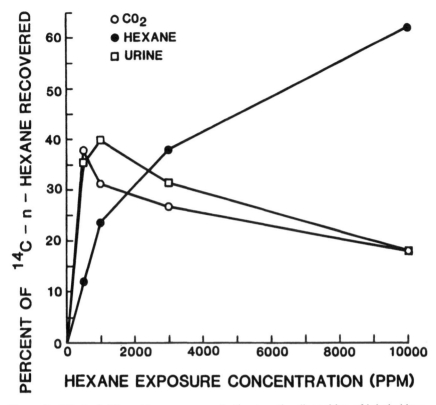

Figure 7 Effect of different hexane concentrations on the disposition of inhaled hexane in rats. (Modified from Bus et al., 1982)

olite of butadiene. Since significant tumor induction in male rats occurs only at 8000 ppm (Owen et al., 1987), butadiene monoepoxide blood levels are probably not predictive of a carcinogenic response.

In summary, the examples discussed above demonstrate that the concentration of an inhaled chemical can significantly alter chemical disposition if one or more of the metabolic pathways is capacity-limited. The shift in metabolism observed in many cases could have either a detrimental or beneficial effect. Incorporating multiple exposure concentrations into experimental protocols is essential for defining the range of concentrations over which metabolism of a chemical is linear.

Interspecies Differences in Metabolism

Not only can metabolism be a significant factor in chemical disposition, but it can also result in chemical species with increased toxicity. Although knowledge of metabolic pathways is important in predicting toxicity, information on metabolic pathways obtained in one animal species is not always pre-

Table 6   Distribution of ^{14}C in Urine and as Exhaled $^{14}CO_2$ in Rats Exposed to Different Concentrations of Butadiene at 65 hr after a 6-hr Exposure[a,b]

Exposure concentration (ppm)	Urine		CO_2	
70	8±0.4	(49)	2±0.5	(12)
1000	21±4	(31)	16±0.2	(24)
7100	19±2	(8)	124±4	(51)

[a]Values are the mean μmole ± SE. Numbers in parentheses are percentage of total ^{14}C absorbed and retained at the end of a 6-hr exposure.
[b]Data taken from Bond et al., 1986a.

dictive of pathways in a second species. The following example illustrates how the dose of an inhaled chemical is markedly different among species inhaling the same concentrations of the same chemical.

In studies by Himmelstein et al. (1994, 1995), male B6C3F1 mice and male Sprague-Dawley rats were exposed for 6 hours to 0, 62.5, 625, 1250, and 8000 (rats only) ppm butadiene. Rats were exposed to 8000 ppm butadiene because this concentration was used in the chronic bioassay for rats only (Owen et al., 1987). The concentrations of butadiene in blood and butadiene epoxides in blood, lung, and liver were measured. Both species formed butadiene monoepoxide from butadiene, although concentrations of butadiene monoepoxide were always higher in mice than rats for any given exposure concentration (Table 7). The most striking difference between rats and mice, however, was in the formation of butadiene diepoxide. Even when exposed to 8000 ppm butadiene, butadiene diepoxide could not be detected in the blood of rats. In contrast, mice had quantifiable concentrations of butadiene diepoxide in blood even at the lowest exposure concentration.

Similar species differences in butadiene epoxide levels were also noted in tissues taken from rats and mice exposed to butadiene. During exposure, the average concentrations of butadiene monoepoxide in mice were 4- to 8-fold higher in blood, 13- to 15-fold higher in lung, and 5–8 fold higher in liver compared to rats (Table 7 and Fig. 8). The concentration of butadiene diepoxide was greatest in the lungs of mice; butadiene diepoxide could not be detected in the livers of mice. Again, similar to what was observed for blood, butadiene diepoxide could not be detected in lungs or livers of rats (Fig. 8). The greater sensitivity of mice to butadiene-induced carcinogenicity can be explained in part by the higher levels of both butadiene monoepoxide and butadiene diepoxide in the blood, lungs, and livers of mice compared to rats.

Effect of Regional Metabolism on Uptake

Nasal uptake functions to protect the lungs from the potentially adverse effects of inspired vapors (Medinsky et al., 1993). Thus, pulmonary toxicity

Table 7 Concentrations of Butadiene Monoepoxide and Butadiene Diepoxide in Blood during 6 hour Nose-Only Inhalation of Butadiene[a]

Butadiene exposure concentration (ppm)	Concentration of analyte in blood (µM)[b]					
	Butadiene		Butadiene monoepoxide		Butadiene diepoxide	
	Mouse	Rat	Mouse	Rat	Mouse[c]	Rat
62.5	2.4±0.2 (24)	1.3±0.04 (15)	0.56±0.04 (19)	0.07±0.01 (15)	0.65±0.10 (6)	nd[d]
625	37±3 (26)	18±0.7 (21)	3.7±0.4 (23)	0.94±0.04 (21)	1.9±0.2 (6)	nd
1250	58±3 (33)	37±1.4 (20)	8.6±0.6 (25)	1.3±0.09 (22)	2.5±0.4 (11)	nd
8000	—[e]	251±8 (21)	—[e]	1.4±0.07 (21)	—[e]	nd

[a]Data taken from Himmelstein et al., 1994.
[b]Values are the mean ± SE (n) for blood samples analyzed during exposure (2–6 hours).
[c]Peak concentration measured at 6-hour exposure time.
[d]Not detected by gas chromatography-mass spectrometry; method detection limit for butadiene monoepoxide was 0.03 µM and 0.13 µM for butadiene diepoxide.
[e]Only rats were exposed to 8000 pppm butadiene.

can be strongly influenced by nasal function. As noted previously, highly soluble vapors may be efficiently absorbed in the nose. Due to the large xenobiotic-metabolizing capacity of the nasal mucosa (Dahl and Hadley, 1991), metabolism within nasal tissues could influence the systemic disposition of inspired vapors. Carboxylesterases are present in nasal tissues in large amounts. Studies by Morris and coworkers (Morris, 1990; Morris et al., 1991) have shown that administration of the carboxylesterase inhibitor bis-p-nitrophenyl phosphate (BNPP) decreases nasal uptake of the organic ester ethyl acetate without altering uptake of a nonmetabolized vapor (acetone) (Fig. 9). This observation provides strong evidence that inspired ethyl acetate

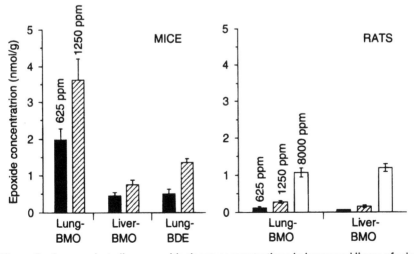

Figure 8 Average butadiene epoxide tissue concentrations in lungs and livers of mice and rats exposed for 6 hours to 625, 1250, and 8000 (rats only) ppm butadiene (modified from Himmelstein et al., 1995)

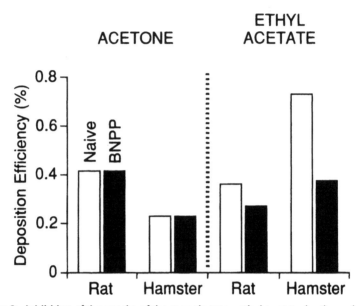

Figure 9 Inhibition of the uptake of the organic ester, ethyl acetate, by the carboxyles-terase inhibitor, bis-p-nitrophenyl phosphate (BNPP). The uptake of acetone which is not a substrate for the enzyme carboxylesterase is not influenced by BNPP. (Taken from Morris, 1990 with permission.)

is quantitatively metabolized in nasal tissues. As much as 65% of inspired ethyl acetate may be metabolized in the nasal cavity of the hamster (Morris, 1990).

If inspired ester vapors are extensively metabolized, then simultaneous inhalation of two esters should lead to competitive inhibition of metabolism and therefore to reduced deposition efficiencies. Studies by Morris et al. (1991) on dimethyl succinate (DMS) have confirmed this prediction. DMS is extensively metabolized by nasal tissues and is an effective inhibitor of nasal ethyl acetate metabolism *in-vitro.* Addition of DMS vapor to the inspired air resulted in a 12% reduction in nasal deposition of simultaneously inspired ethyl acetate. Moreover, DMS was found to be without effect on nasal deposition of isoamyl alcohol, a vapor that is not metabolized by carboxylesterase. These results not only provide further evidence that inspired ester vapors are metabolized in nasal tissues but also point to factors that might modify ester uptake in complex exposure scenarios.

Interactions

The behavior of chemicals in mixtures may differ greatly from that observed when compounds are tested as pure chemicals. For example, enzymatic metabolism of one component of the chemical mixture may be increased due to *induction* of the enzyme responsible for metabolism by a second component in that mixture. Alcohol is a good inducer of cytochrome P450, and a number of studies have shown that exposure to alcohol can increase the metabolism of a number of hydrocarbons. In contrast, if these chemicals are substrates for the same enzyme, each chemical could *inhibit* or reduce the metabolism of the other. If a metabolite of a chemical is the toxic species rather than the chemical itself, this interaction can result in decreased toxicity. Benzene-toluene mixtures are a good example of this type of interaction. Toluene inhibits the metabolism of benzene and reduces toxic effects such as bone marrow and blood cell damage. Yet another type of chemical interaction is the formation of a new product that is not initially present in the mixture. An excellent example of this interaction is the formation of 4-phenylglycolhexene in carpet latex by the interaction of styrene and butadiene. This new chemical is suspected as the agent responsible for chemical sensitivity of some individuals to new carpets.

Experimental methodologies now exist to study chemical interactions with a risk assessment orientation. These studies include tissue dosimetry studies which utilize both sensitive analytical techniques and biologically based mathematical modeling. They also include molecular dosimetry studies that focus both on specific hemoglobin adducts (Walker et al., 1992; Fennell et al., 1992) and DNA adducts using ^{32}P-postlabeling techniques (Bond et al., 1989, 1990) or mutagenesis studies using either conventional *in-vitro* tests employing the *hprt* mutation, or transgenic mice (Recio et al., 1992; Sisk et al., 1994).

Modulation of Butadiene Metabolism by Other Organics

Styrene-butadiene mixtures are often encountered in the workplace where both these monomers are used in the production of synthetic rubber. Unlike butadiene, styrene is highly soluble in tissues, especially lipid-rich tissues. Oxidation of styrene to styrene oxide occurs by cytochrome P450 2E1 (Guengerich et al., 1991), the same P450 isozyme responsible for oxidation of butadiene to butadiene monoepoxide (Csanady et al., 1992). The biochemical interaction between these two chemicals has been described using a physiological model that incorporates competitive inhibition of each chemical on the metabolism of the other (Bond et al., 1994). Model simulations predicted that coexposure to styrene would reduce the total amount of butadiene metabolized, but not proportional to the styrene exposure concentration (Fig. 10a). In contrast, simulations predicted that butadiene was not an effective inhibitor of styrene metabolism due to the low solubility of butadiene in tissues (Fig. 10b). The work of Laib and coworkers supports these predictions (Laib et al., 1992). Coexposure to styrene significantly decreased the rate of butadiene metabolism in Sprague-Dawley rats, but butadiene had no effect

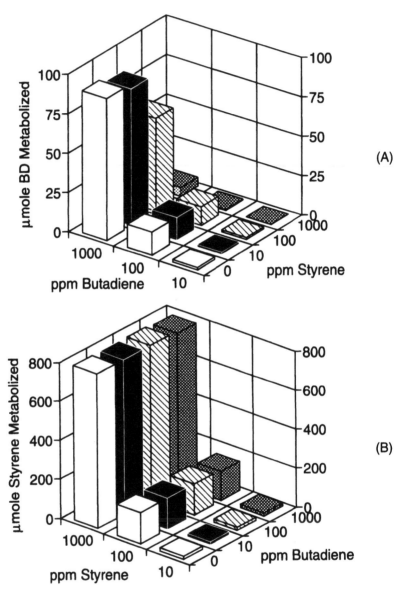

Figure 10 Physiological model predictions for the μmoles of butadiene (A) or styrene (B) metabolized by rats exposed to various combinations of butadiene and styrene assuming simple competitive inhibition. Each bar represents model predictions of the amount of chemical metabolized both during and after 8-hour simulated inhalation exposures to one of the 12 possible combinations of butadiene and styrene. For example, in simulating the effect of styrene on butadiene metabolism (A), as the inhaled concentration of styrene increases from 0 to 1000 ppm (right axis), the μmoles butadiene metabolized due to coexposure to 1000 ppm butadiene (left vertical axis) decreases. For 1000 ppm butadiene, the largest amount of butadiene metabolized occurs with no exposure to styrene (0 ppm styrene). The least butadiene is metabolized when 1000 ppm styrene is present. (Taken from Bond et al., 1994)

on the metabolism of styrene. Additionally, these authors determined that the inhibition constant for styrene in their system was much lower than its apparent Michaelis-Menten constant, suggesting that the interaction between butadiene and styrene might be more complex than the simple competitive inhibition description used in model simulations.

The effect of cytochrome P450 2E1 induction by ethanol on the metabolism of butadiene *in-vivo* was also simulated (Bond et al., 1994). Ethanol is a well known inducer of P450 2E1. The inductive effect of ethanol was assumed to give a twofold increase in the maximum capacity (V_{max}) for butadiene oxidation. Model simulations predicted that after 8 hours of exposure to 10, 100, or 1000 ppm butadiene, butadiene metabolism increased 113, 115, and 158%, respectively, suggesting that ethanol induction would not have a substantive effect on butadiene metabolism at the low concentrations relevant to human exposure. This marginal increase in the amount of butadiene metabolized at the lower concentrations predicted by the model is due to the fact that butadiene metabolism is probably blood-flow-limited at low inhaled concentrations of butadiene, not enzyme-capacity-limited. Thus, model simulations predicted sufficient enzyme at low butadiene concentrations to metabolize all the butadiene presented to the liver by the blood. Increasing the capacity for butadiene oxidation does not substantially increase the total butadiene metabolized (113%). At higher butadiene concentrations, however, when metabolism begins to become saturated, increasing enzyme capacity does increase the total amount of butadiene metabolized (158%).

From these simulations, two general points regarding chemical interactions were seen. The first relates to extrapolation from high to low doses for the saturable enzyme systems that biotransform most toxic organic chemicals. One cannot assume that inhibition effects demonstrated at high exposure concentrations will be proportional or even significant at lower exposure concentrations. Second, the effect of enzyme induction determined *in-vitro* may be diminished in the intact animal. In some cases, delivery of the chemical to the site of chemical interactions may be the rate-limiting factor rather than the quantity of the enzyme present.

Modulation of Benzene Metabolism Alters Toxicity

Enzymes implicated in the metabolic activation of benzene and its metabolites include the cytochrome P450 monooxygenases and myeloperoxidase. Other organic molecules such as toluene (which are also substrates for cytochrome P450) can inhibit the metabolism of benzene. Gad-El-Karim et al. (1984) investigated the genotoxicity of benzene in male and female mice treated with oral doses of benzene or combinations of benzene and toluene. Benzene exposure resulted in genetic damage to bone marrow cells of both male and female mice compared to controls. The genetic damage was reduced when both benzene and toluene were coadministered.

In an effort to correlate benzene toxicity and metabolism, other investigators administered radioactive benzene to mice with or without coexposure to toluene by subcutaneous injection (Andrews et al., 1977). Coconcentrations of benzene in various tissues (fat, liver, spleen, blood, or bone marrow) from mice given only benzene were very similar to those in which benzene was combined with toluene, suggesting that coadminisration of toluene did not alter the disposition of benzene itself. In contrast, the concentrations of benzene metabolites in tissues of mice given benzene alone were much higher than those found when benzene was coadministered with toluene. Coadminisration of toluene did not delay the appearance of benzene metabolites in tissues but markedly reduced the concentration of metabolites found in each tissue for all time periods. Taken together, these observations suggest that the metabolism of benzene is closely related to its genotoxicity. Thus it seems likely that toluene protects against benzene-induced genotoxicity by reducing the level of benzene metabolites in the bone marrow through suppression of benzene metabolism.

Some evidence suggests that suppression of benzene metabolism by toluene occurs in humans. Inoue et al. (1988) examined both the exposure concentration during a work shift and the benzene metabolite concentrations in the end shift urine of workers exposed to benzene, toluene, or a mixture of both chemicals. Urinary levels of the benzene metabolites phenol and hydroquinone were lower in the group exposed to both toluene and benzene compared with the group exposed to benzene alone. These studies suggest that coexposure to other chemicals is a potentially important modifying factor in the disposition of inhaled organic chemicals that can be metabolized.

CONCLUSIONS

In summary, metabolic capacity is probably the most important factor influencing the tissue dosimetry of organic chemicals because metabolites of organics are frequently more toxic than the parent chemical. Metabolism can be altered by the inhaled concentration, interspecies differences, regional differences, and other inducing or inhibiting chemicals. Perhaps the best examples of this were seen with butadiene where species differences in metabolism of butadiene defined the dose of inhaled butadiene in rats and mice and with benzene and where coexposure to another organic significantly affected both metabolism and toxicity.

There is a general lack of understanding of the various factors involved in production of the effective dose in the target tissue of an individual. In this chapter, many examples were given that demonstrated how different factors can be investigated. Data from these types of studies will increase our understanding of the mechanisms involved in xenobiotic-induced toxicity.

REFERENCES

Andrews, LS, Lee, EW, Witmer, CM, Kocsis, JJ, Snyder, R: Effects of toluene on the metabolism, disposition and hemopoietic toxicity of [³H]benzene. Biochem Pharmacol 26:293–300, 1977.

Bond, JA, Dahl, AR, Henderson, RF, Dutcher, JS, Mauderly, JL, Birnbaum, LS: Species differences in the disposition of inhaled butadiene. Toxicol Appl Pharmacol 84:617–627, 1986a.

Bond, JA, Sun, JD, Mitchell, CE, Dutcher, JS, Wolff, RK, McClellan, RO: Biological fate of inhaled organic compounds associated with particulate matter. In Aerosols. Edited by SD Lee, T Schreider, LD Grant, PJ Verkerk, Chelsea, Michigan:Lewis, pp. 579–592, 1986b.

Bond, JA, Gubin, JM, Johnson, NF: O⁶-Methylguanine-DNA methyltransferase activity in tissues and cells of the respiratory tract. Chem-Biol Interact, 71:255–263, 1989.

Bond, JA, Johnson, NF, Snipes, MB, Mauderly, JL: DNA adduct formation in rat alveolar type 11 cells—cells potentially at risk for inhaled diesel exhaust. Environ Molec Mutag, 915:64–69, 1990.

Bond, JA Metabolism of xenobiotics by the respiratory tract. In Toxicology of the Lung. Edited by DE Gardner, J Crapo, RO McClellan, 2nd edition, New York, NY: Raven Press, 187–215, 1993.

Bond, JA, Csanady, GA, Gargas, ML, Guengerich, FP, Leavens, T, Medinsky, MA, Recio, R: 1,3-Butadiene: linking metabolism, dosimetry, and mutation induction. Environ Hlth Perspect 102:87–94, 1994.

Brown, RA, Schanker, LS: Absorption of aerosolized drugs from the rat lung. Drug Metab Dispos 11:355–360, 1983.

Bus, JS, Deyo, D, Cox, M: Dose-dependent disposition of n-hexane in F-344 rats after inhalation exposure. Fundam Appl Toxicol 2:226–229, 1982.

Chang, JCF, Steinhagen, WH, Barrow, CS: Effect of single or repeated formaldehyde exposure on minute volume of B6C3F1 mice and F-344 rats. Toxicol Appl Pharmacol 61:451–459, 1981.

Chang, L-Y, Huang, Y, Stockstill, B, Graham, JA, Grose, EC, Ménache, MG, Miller, FJ, Costa, DL, Crapo, JD: Epithelial injury and interstitial fibrosis in the proximal alveolar regions of rats chronically exposed to a simulated pattern of urban ambient ozone. Toxicol Appl Pharmacol 115:241–252, 1992.

Cheng, YS, Yamada, Y, Yeh, HC, Swift, DL: Differential deposition of ultrafine aerosols in a human nasal cast. J Aerosol Sci 19:741–751, 1988.

Csanady, GA, Guengerich, FP, Bond, JA: Comparison of the biotransformation of 1,3-butadiene and its metabolite, butadiene monoepoxide, by hepatic pulmonary tissues from humans, rats and mice. Carcinogenesis 13:1143–1153, 1992.

Dahl, AR, Hadley, WM: Nasal cavity enzymes involved in xenobiotic metabolism: effects on the toxicity of inhalants. CRC Crit Rev Toxicol 21:345–372, 1991.

Dahl, AR, Snipes, B, Gerde, P: Sites for uptake of inhaled vapors in Beagle dogs. Toxicol Appl Pharmacol, 109:263–275, 1991.

Davies, CN: Absorption of gases in the respiratory tract. Ann Occup Hyg 29:1325, 1985.

Dungworth, DL, Castleman, WL, Chow, CK, Mellick, PW, Mustafa, MG, Tarkington, B, Tyler, WS: Effect of ambient levels of ozone on monkeys. Fed Proc 34:1679–1674, 1975.

Fennell, TR, Sumner, SCJ, Walker, VE: A model for the formation and removal of hemoglobin adducts. Cancer Epidemiol, Biomark Prevent 1:213–219, 1992.

Gad-El-Karim, MM, Harper, BL, Legator, MS: Modifications in the myeloclastogenic effect of benzene in mice with toluene, phenobarbital, 3-methylcholanthrene, Aroclor 1254 and SKF-525A. Mutat Res 135:225–243, 1984.

Gargas, ML, Burgess, RJ, Voisard, DE, Cason, GH, Andersen, ME: Partition coefficients of low-molecular-weight volatile chemicals in various liquids and tissues. Toxicol Appl Pharmacol 98:87–99, 1989.

Gerde, P, Muggenburg, BA, Hoover, MD, Henderson, RF: Disposition of polycyclic aromatic hydrocarbons in the respiratory tract of the beagle dog, I. The alveolar region. Toxicol Appl Pharmacol 121:313–318, 1993a.

Gerde, P, Muggenburg, BA, Sabourin, PJ, Hardema, JR, Hotchkiss, JA, Hoover, MD, Henderson, RF: Disposition of polycyclic aromatic hydrocarbons in the respiratory tract of the beagle dog. II. The conducting airways. Toxicol Appl Pharmacol, 121:319–327, 1993b.

Gerde, P, Muggenburg, BA, Henderson, RF: Disposition of polycyclic aromatic hydrocarbons in the respiratory tract of the beagle dog. III. Mechanisms of the dosimetry. Toxicol Appl Pharmacol 121:328–334, 1993c.

Gerde, P, Dahl, AR: A model for the uptake of inhaled vapors in the nose of the dog during cyclic breathing. Toxicol Appl Pharmacol 109:276–288, 1991.

Gerde, P, Medinsky, MA, Bond, JA: Particle-associated polycyclic aromatic hydrocarbons—a reappraisal of their possibile role in pulmonary carcinogenesis. Toxicol Appl Pharmacol 108:1–13, 1991a.

Gerde, P, Medinsky, MA, Bond, JA: The retention of polycyclic aromatic hydrocarbons in the bronchial airways and in the alveolar region—a theoretical comparison. Toxicol Appl Pharmacol 107:239–252, 1991b.

Gerrity, TR, Biscardi, F, Strong, A, Garlington, R: Ozone uptake in the conducting airways of humans. In press, 1994.

Gerrity, TR, McDonnell, WF, O'Neill, JJ: Experimental ozone dosimetry in humans. J Aerosol Med 2:129–139, 1989.

Gerrity, TR, Weaver, RA, Berntsen, J, House, DE, O'Neill, JJ: Extrathoracic and intrathoracic removal of O_3 in tidal-breathing humans. J Appl Physiol 65:393–400, 1988.

Guengerich, FP, Kim, D-H, Iwasaki, M: Role of human cytochrome P450 IIE1 in the oxidation of many low molecular weight cancer suspects. Chem Res Toxicol 4:168–179, 1991.

Heck, Hd'A, Chin, TY, Schmitz, MC: Distribution of [^{14}C] formaldehyde in rats after inhalation exposure. In *Formaldehyde Toxicity.* Edited by JE Gibson. Hemisphere Publishing Corporation, pp.26–37, Chapter 4, 1983.

Himmelstein, MW, Turner, MJ, Asgharian, B, Bond, JA: Comparison of blood concentrations of 1,3-butadiene and butadiene epoxides in mice and rats exposed to 1,3-butadiene by inhalation. Carcinogenesis 15:1479–1486, 1994.

Himmelstein, MW, Asgharian, B, Bond, JA: High concentrations of butadiene epoxides in livers and lungs of mice compared to rats exposed to 1,3-butadiene. Toxicol Appl Pharmacol, 132:281–288, 1995.

Inoue, O, Seiji, K, Watanabe, T, Kasahara, M, Nakatsuka, H, Yin, S, Li, G, Cai, S, Jin, C, Ikeda M: Mutual metabolic suppression between benzene and toluene in man. Int Arch Occup Environ Health 60:15–20, 1988.

Kaye, B, Cussans, NJ, Faulkner, JK, Stopher, DA, Reid JL: The metabolism and kinetics of doxazosin in man, mouse, rat and dog. Br J Clin Pharmacol 21:19S-25S, 1986.

Laib, RJ, Bucholski, M, Filser, JB, Csanady, GA: Pharmacokinetic interaction between 1,3-butadiene and styrene in Sprague-Dawley rats. Arch Toxicol 66:310–314, 1992.

Landry, TD, Ramsey, JC, McKenna, MJ: Pulmonary physiology and inhalation dosimetry in rats: Development of a method and two examples. Toxicol Appl Pharmacol 71:72–83, 1983.

Lippmann, M, Schlesinger, RB: Interspecies comparisons of particle deposition and mucociliary clearance in tracheobronchial airways. J Toxicol Environ Health 13:441–469, 1984.

Ludden, TM, Schanker, LS, Lanman, RC: Binding of organic compounds to rat liver and lung. *Drug Metab Dispos* 4:8–16, 1976.

McKenna, MJ, Zempel, JA, Braun, WH: The pharmacokinetics of inhaled methylene chloride rats. Toxicol Appl Pharmacol 65:1–10, 1982.

Medinsky, MA: Critical determinants in the systemic availability and dosimetry of volatile organic chemicals. In Principles of Route-to-Route Extrapolation for Risk Assessment. Edited by TR Gerrity, CJ Henry, Elsevier Science Publishing Co., Inc., p. 155–171, 1990.

Medinsky, MA, Kimbell, JS, Morris, JB, Gerde, P, Overton, JH: Advances in biologically based models for respiratory tract uptake of inhaled volatiles. Fund Appl Toxicol 20:265–272, 1993.

Medinsky, MA, Dutcher, JS, Bond, JA, Henderson, RF, Mauderly, JL, Snipes, MB, Mewhinney, JA, Cheng, YS, Birnbaum, LS: Uptake and excretion of [^{14}C]methyl bromide as influenced by exposure concentration. Toxicol Appl Pharmacol 78:215–225, 1985.

Mercer, RR, Anjilvel, S, Miller, FJ, Crapo, JD: Inhomogenity of ventilatory unit volume and its effects on reactive gas uptake. J Appl Physiol 70:2193–2205, 1991.

Mercer, TT: Aerosol Technology in Hazard Evaluation, Academic Press, New York, 1973.

Miller, FJ, Overton, JH, Jaskot, RH, Menzel, DB: A model of the regional uptake of gaseous pollutants in the lung. I. The sensitivity of the uptake of ozone in the human lung to lower respiratory tract secretions and exercise. Toxicol Appl Pharmacol 79:11–27, 1985.

Mordenti, J: Forecasting cephalosporin and monobactam antibiotic half-lines in humans from data collected in laboratory animals. Antimicrob Aqents and Chemother 27:887–891, 1985.

Morris, JB, Clay, RJ, Trela, BA, Bogdanffy, MS: Deposition of dibasic esters in the upper respiratory tract of the male and female Sprague-Dawley rat. Toxicol Appl Pharmacol 108:538–546, 1991.

Morris, JB: First-pass metabolism of inspired ethyl acetate in the upper respiratory tracts of the F344 rat and syrian hamster. Toxicol Appl Pharmacol 102:331–345, 1990.

Overton, JH, Graham, RC, Miller, FJ: A model of the regional uptake of gaseous pollutants in the lung. II. The sensitivity of ozone uptake in laboratory animal lungs to anatomical and ventilatory parameters. Toxicol Appl Pharmacol 88:418–432, 1987.

Owen, PE, Glaister, JR, Gaunt, IF, Pullinger, DH: Inhalation toxicity studies with 1,3-butadiene 3 two-year toxicity/carcinogenicity study in rats. Am Ind Hyg Assoc J 48:407, 1987.

P'an, A, Be'land, J, Jegier, Z: Ozone-induced arterial lesions. Arch Environ Hlth 24:229–232, 1972.

Pryor, WA: How far does ozone penetrate into the pulmonary air/tissue boundary before it reacts? Free Radic Biol Med 12:83–88, 1992.

Raabe, OG: Aerosol aerodynamic size conventions for inertial sampler calibration. Air Pollut Control Assoc J 26:856–860, 1976.

Ramsey, JC, Andersen, ME: A physiologically based description of the inhalation pharmacokinetics of styrene in rats and humans. Toxicol Appl Pharmacol 73:159–175, 1984.

Recio, L, Osterman-Golkar, S, Csanady, GA, Turner, MJ, Myhr, B, Moss, O, Bond, JA: Determination of mutagenicity in tissues of transgenic mice following exposure to 1,3-butadiene and N-ethyl-N-nitrosourea. Toxicol Appl Pharmacol 117, 58–64, 1992.

Sato, A, Nakajima, T: Partition coefficients of some aromatic hydrocarbons and ketones in water, blood, and oil. Br J Ind Med 36:231–234, 1979.

Schanker, LS, Mitchell, EW, Brown, RA: Pulmonary absorption of drugs in the dog: comparison with other species. Pharmacology 32:176–180, 1986.

Schlesinger, RB: Deposition and clearance of inhaled particles. In Concepts in inhalation toxicology. Edited by RO McClellan, RF Henderson, New York: Hemisphere Publishing Co., pp. 163–192, 1989.

Sisk, SC, Pluta, LJ, Bond, JA, Recio, L: Molecular analysis of lacI mutants from bone marrow of B6C3F1 transgenic mice following inhalation exposure to 1,3-butadiene. Carcinogenesis 15:471–477, 1994.

Stephens, RJ, Sloan, MF, Evans, MJ, Freeman, G: Alveolar type 1 cell response to exposure to 0.5 ppm O$_3$ for short periods. Exp Mol Pathol 20:11–23, 1974.

Stephens, RJ, Sloan, MF, Evans, MJ, Freeman, G: 1973. Early response of lung to low levels of ozone. Amer J Pathol 74:31–58.

Task Group on Lung Dynamics: Deposition and retention models for internal dosimetry of the human respiratory tract. Health Phys 12:173–207, 1966.

Theodore, J, Robin, ED, Gaudio, R, Acevedo, J: Transalveolar transport of large polar solutes (sucrose, insulin, and dextran). Am J Physiol 229:989–996, 1975.

Tyler, WS, Tyler, NK, Last, JA, Gillespie, MJ, Barstow, TJ: Comparison of daily and seasonal exposures of young monkeys to ozone. Toxicology 50:131–144, 1988.

Walker, VE, MacNeela, JP, Swenberg, JA, Turner, MJ, Jr., Fennell, TR: Molecular dosimetry of ethylene oxide: formation and persistence of N-(2-hydroxyethyl)valine in hemoglobin following repeated exposures of rats and mice. Cancer Res 52:4320–4327, 1992.

Wallace, LA, Pellizari, ED, Hartwell, T, Sparacino, C, Sheldon, L, Zelon, H: Personal exposures, indoor-outdoor relationships and breath levels of toxic air pollutants measured for 355 persons in New Jersey. Atmos Env 19:1651–1661, 1984.

Wiester, MJ, Stevens, MA, Ménache, MG, Gerrity, TR, McKee, JL: Ozone (O_3) uptake in healthy adult males during quiet breathing. In press, 1994.

Wiester, MJ, Tepper, JS, King, ME, Ménache, MG, Costa, DL: Comparative study of ozone (O_3) uptake in three strains of rats and in the guinea pig. Toxicol Appl Pharmacol 96:140–146, 1988.

Wiester, MJ, William, TB, King, ME, Ménache, MG, Miller, FJ: Ozone uptake in awake Sprague-Dawley rats. Toxicol Appl Pharmacol 89:429–437, 1987.

Yeh, HC, Phalen, RF, Raabe, OG: Factors influencing the deposition of inhaled particles. Environ Health Perspectives 15:147–156, 1976.

Yeh, HC, Schum, GM: Models of human lung airways and their application to inhaled particle deposition. Bulletin of Mathematical Biology 42:462–480, 1980.

Yu, CP: Exact analysis of aerosol deposition during steady breathing. Powder Technology 21:55–62, 1978.

Part Five

Evaluation of Responses of the Respiratory Tract to Inhaled Toxicants

Chapter Twelve

Carcinogenic Responses of the Lung to Inhaled Toxicants

Fletcher F. Hahn, D.V.M., Ph.D.

INTRODUCTION

Carcinogenic responses of the lung are important, because lung cancer is a leading cause of cancer deaths in the United States (Boring et al., 1994). Over 141,000 people died of lung cancer in 1990. The incidence rate is twice as high in men than women. In 1987, lung cancer death rate in women exceeded that of breast cancer and continues to increase. The high incidence in men and the increasing incidence in women have been attributed to cigarette smoking habits.

All known human lung carcinogens are taken into the body by the inhalation route. These carcinogens include cigarette smoke, radon, radon daughters, asbestos, arsenic compounds, chromium compounds, and mustard gas. In pulmonary carcinogenesis studies, the test compounds should be administered by the respiratory route and, if meaningful dose-response relationships are to be determined, the inhalation mode of exposure must be used. Carcinogenic compounds can reach the lung by other means, however, and these routes may be appropriate for some experimental purposes. For example, 4-(N-methyl-N-nitrosamino)-1-(3-pyridyl)-butanone (NNK) administered to rats by intravenous injection is metabolized by Clara cells in the lung, resulting in cytotoxicity in these cells and, ultimately, lung cancer (Belinsky et al., 1986).

Inhalation mode of carcinogens may result in neoplasms elsewhere in the body, other than the lung. A good example of this is inhalation exposure of workers in the rubber industry to vinyl chloride, resulting in liver cancers. Animals exposed to vinyl chloride by inhalation develop cancers of bone, skin, and lung (Viola et al., 1971). In this presentation of carcinogenic response of the lung, the inhalation mode of exposure and dose response for lung cancer will be emphasized.

GENERAL PRINCIPLES OF LUNG CARCINOGENESIS
Dosimetry

Deposition, retention, and metabolism of inhaled materials are the important factors in determining dose to the lung, as noted elsewhere in this book. The dose, or amount of inhaled material that reaches the target tissues, is not simply a function of air concentration to which the animal is exposed, but is dependent on factors inherent to the toxic material, e.g., solubility; particle size, shape, and density; bioavailability, etc., and the animal e.g., airway morphology, respiratory patterns, etc. Some inhaled chemicals, such as benzo(a)-pyrene (BaP), are toxic because they are metabolized to reactive compounds in the respiratory tract, while many others are detoxified when they are metabolized. Thus, the bioavailability of the chemicals for metabolism is an important factor. Inhaled chemicals may not be metabolized when deposited in locations where metabolic activities are weak or when complexed within particles or other chemicals.

Many factors, such as animal species, age at exposure, health status, and dose of inhaled material, can modify the dose to the lung by influencing the deposition, retention, and metabolism of inhaled materials. These factors have been discussed by others in this book.

Dose-Response Relationship

Two patterns of dose-response for lung cancer are usually observed with increasing dose:

    1 the incidence and multiplicity of neoplasms increases until toxic doses are achieved, reducing the incidence; and
    2 the time required for the appearance of neoplasms decreases.

Table 1 Dose-Response Relationships for Lung Tumors Induced by Inhalation of Toxic Vapors

Chemical	Species	Sex	Departure from linear dose response
Bis(chloromethyl)ether	Mouse	Male	None
	Rat	Male	None
1,2,Dibromo-3-chloropropane	Mouse	Female	Upward
	Mouse	Male	None
1,2-Dibromoethane	Mouse	Female	Upward
	Mouse	Male	None
Vinyl Chloride	Mouse	Both	Upward
	Mouse	Male	None

Source: Gold et al. (1984).

For example, high radiation doses to the lungs of rats from inhaled beta-emitting radionuclides can result in tumors as early as 15 months after exposure (Scott et al., 1987). Rats with higher doses to the lung died at earlier times from radiation pneumonitis. Rats with lower doses lived out a nearly normal life span, but had an increase in lung tumors. In rats that died within 18 months with high radiation doses, several had two lung tumors, each of a different morphologic type.

A difference in tumor phenotype with increased dose is another pattern that may be seen. High protracted radiation doses from beta-emitting radionuclides deposited in the lung of rats result primarily in squamous cell carcinomas and hemangiosarcomas of the lung, whereas lower doses result in adenocarcinomas of the lung (Hahn and Lundgren, 1992). The differences in tumor phenotypes are probably the result of different pathogenetic mechanisms or different responses of cells to radiation.

The shape of the dose-response curve is much discussed in carcinogenesis of the lung, as it is in other organ systems. A large database of standardized results of animal bioassays has been accumulated and a numerical index of carcinogenic potency developed for 770 chemicals (Gold et al., 1984). Not many inhalation carcinogenesis studies, however, have sufficient data to develop dose-response relationships. In studies of toxic vapors that have sufficient data, the results are mixed (Table 1). Male mice have linear dose-response relationships between exposure concentration and lung tumor incidence when exposed to bis(chloromethyl)ether, 1,2-dibromo-3-chloropropene, 1,2-dibromoethane, or vinyl chloride. In contrast, female mice have a curvilinear dose-response curve when exposed to the same compounds. The reason for this sex difference is unknown but may be the result of hormonal influences (Biancifiori, 1970).

Dose-response functions for inhaled alpha- and beta-emitting radioactive particles have been developed (ICRP 31, 1980). Most of the data are fit best by a curvilinear function (linear quadratic), although a linear dose-response model is usually difficult to reject statistically. For health protection purposes, a linear dose-response function is usually applied for extrapolation to low doses.

Dose protraction, either through repeated exposures or exposures over a prolonged period, influences the effectiveness of a carcinogen. This is an area in which little research has been done, but is important from the occupational health standpoint. Which is more important—the dose rate or the total dose? Table 2 shows that alpha-emitting radionuclides are increasingly effective in producing tumors if their dose is protracted.

Morphologic Progression of Lung Cancers

The progression of histologic and cytologic changes in lung tissues during the development of cancer has been described in man and in several animal species. These changes can be divided into the general categories of hyperplasia, benign neoplasia, and malignant neoplasia.

A series of early classic papers gave details of the progressive nature of the bronchial lesions in cigarette smokers (Auerbach et al., 1961). Hyperplasia, metaplasia, cellular atypia, and carcinoma in situ appeared to form a sequence of changes leading to invasive carcinoma. The incidence of these lesions was related to cigarette consumption and age at exposure.

Studies of Syrian hamsters instilled with a mixture of BaP and ferric oxide have shown a somewhat similar sequence of changes (Becci et al., 1978). Within 24 hours after instillation, an acute reaction occurs leading to sloughing of cilia and multiplication of mucus and goblet cells. With continued application of the carcinogen, the epithelium becomes hyperplastic, then metaplastic. These changes are reversible, and if the stimulus is removed, the epithelium can return to normal. If the application continues, nuclear atypia develops in the metaplastic areas leading to carcinoma in situ and finally invasive carcinoma.

Table 2 Influence of Dose Protraction on Effectiveness of Radioactive Compounds in Producing Lung Tumors

Radioactive compound	Type of radiation	Species	Exposure method	Effect of protraction
[144]Ce dioxide[a]	Beta	Mouse	Inhalation	None
[239]Pu dioxide[b]	Alpha	Mouse	Inhalation	Increases effectiveness
[210]Po[c]	Alpha	Syrian hamster	Intratracheal instillation	Increases effectiveness (at low doses)

[a]Hahn et al. (1980).
[b]Lundgren et al. (1987).
[c]Little et al. (1985).

The sequential morphologic alterations in bronchial cells of Syrian hamsters treated with nitrosamines have also been described during carcinogenesis (Reznik-Schuller, 1977). After 3 weeks of treatment with N-nitrosomorpholine, lamellar bodies, similar to those seen in Type II alveolar cells, develop in the nonciliated cells of the terminal bronchioles. These cells increase in number and eventually develop into invasive tumors.

Malignant tumors may arise from hyperplastic or metaplastic lesions as just described, or they may arise from benign tumors. The progression from a benign tumor to a malignant one has been described in lung tumors of mice (Kimura, 1971; Kauffman et al., 1979). Three morphologic patterns were described in the spontaneous and induced pulmonary tumors of Strain A mice: alveolar lining type; papillary type; and a transitional alveolar lining-papillary type (Kimura, 1971). Each type had different growth characteristics when transplanted into isologous hosts. The alveolar type required 200–400 days to form a tumor nodule, the transitional type required 60–200 days, and the papillary type required 30–60 days. When the papillary tumors were serially transplanted, less well-differentiated tumors developed in 14–30 days, and metastatic activity was exhibited, a sure sign of malignancy.

Multiple Stages of Carcinogenesis

The natural history of development of neoplasia in experimental systems may be separated conceptually into at least three different stages: initiation, promotion, and progression (Pitot, 1982). Initiation is the first stage of carcinogenesis and can be induced rapidly at any dose of the initiating agent. Initiation is irreversible and is related to direct effects on the DNA of the cell.

After initiation, promotion results in clonal expansion of initiated cells. The actions of promoting agents are reversible and only effective above certain threshold concentrations of the promoting agent. The promoting effects may be modulated by dietary, hormonal, or environmental factors. Many carcinogens can both initiate cells and promote those cells to the malignant state. Promoting agents cannot initiate cells, but may promote cells initiated by ambient environmental factors, thus giving the appearance of a weak carcinogen.

Progression converts the expanded clones into promotor-independent neoplastic populations allowing the tumor to grow, invade, and metastasize (Nowell, 1986).

Much of the experimental work on the multiple stages of carcinogenesis has been done using chemical compounds applied to the skin of mice. More recent studies have shown similar phenomenon in other organ systems, including the lung. Butylated hydroxytoluene (BHT) was found to significantly increase the number of pulmonary adenomas in mice injected with urethane, a compound that induces adenomas, in 14–24 weeks after administration (Witschi, 1986). BHT given after urethane not only increased the number of tumors, but accelerated their growth. BHT given before or simultaneously with the initiator, urethane, did not enhance tumor formation, a finding consistent with the action of promotors in the skin.

Genetic and Molecular Events in Rodent Lung Cancer

The characterization of genetic abnormalities in the lung cancer of laboratory animals has accelerated in the past few years (Reynolds and Anderson, 1991; You et al., 1991). Much of the work has focused on the tumor suppressor gene, *p53*, and the protooncogene, K-*ras*, because these two genes appear to play a major role in lung cancer in humans (Gazdar, 1992; Johnson and Kelley, 1993). Mutations of the *p53* gene are the most common genetic alteration identified in human cancers. In small cell lung cancers *p53* mutations are almost always present; however, in non-small cell lung cancer the mutation rate is less, about 65% in squamous cell carcinomas and 45% in adenocarcinomas. In the non-small cell cancers, the mutations do not have an obvious association with smoking history, gender, tumor stage, or survival, (Chiba et al., 1990). Normal *p53* negatively regulates cell growth and division, whereas mutated forms can stimulate cell division. The *p53* mutations may function to promote growth of cancer cells.

Alterations in the *p53* pathway are infrequent in lung cancers induced in rats and are rare in lung cancers of mice. *p53* protein in the nuclei of tumors has been identified immunohistochemically to document alterations in the *p53* pathway in rodents. Direct sequencing or single-strand conformational polymorphism (SSCP) has been used to localize these *p53* gene mutations with respect to specific exons. This approach has identified *p53* alterations in the lung tumors of rats exposed to x-irradiation (Belinsky et al., 1993), $^{239}PuO_2$ aerosols (Kelly and Hahn, 1993), and diesel exhaust or carbon black (Swafford et al., 1994) (Table 3). In each case, *p53* protein immunoreactivity was detected in a tumor with a squamous-cell phenotype but not in an adenocarcinoma. Direct sequencing of DNA from these tumors (exons 4,5–8,9) revealed few *p53* gene mutations in tumors with *p53* protein dysfunctions. These findings indicate that the increased *p53* immunoreactivity is not due to specific *p53* gene mutations but may indicate other dysfunctions within the *p53* regulatory pathway. In mice, direct sequencing or SSCP has been used to detect *p53* gene mutations at exons 5–8 in animals exposed to diethylnitrosamine, dimethyl benzanthracene (Goodrow et al., 1992), NNK (Devereux et al., 1993a), or methylene chloride (Hegi et al., 1993). In all cases very few (0–7%) *p53* gene mutations were found, indicating these mutations are rare events in murine lung carcinogenesis.

K-*ras* mutations in human lung cancer are found in nearly 30% of adenocarcinomas and 4% of squamous cell carcinomas, but not in the small cell carcinomas (Johnson and Kelley, 1993). The K-*ras* mutations are more frequent in the adenocarcinomas of smokers are more frequent than in non-

Table 3 Genetic Alterations in Lung Tumors of Rats Exposed to Pulmonary Carcinogens

Exposure	Genetic alterations		
	K-*ras* mutations	*p53* dysfunctions (IHC)[a]	Mutations (SSCP/direct sequence)
Unexposed (F344)[b]	2/8[c]	1/5	ND[d]
Beryllium metal[e]	2/12	0/24	0/12
[239]PuO$_2$[b]	33/71	2/28[f]	2/28
X-irradiation[g]	1/36	3/36	1/36
Diesel exhaust/carbon black[h]	3/50	6/12	0/12
N-nitrosobis(2-oxopropyl)amine[i]	6/24	ND	ND
Tetranitromethane[j]	19/19	ND	ND

[a]Immunohistochemistry
[b]Stegelmeier et al., 1991.
[c]Number with alteration/number examined.
[d]Not done.
[e]Nickell-Brady et al., 1994.
[f]Kelly and Hahn, 1993.
[g]Belinsky et al., 1993.
[h]Swafford et al., 1994.
[i]Ohgaki et al., 1993.
[j]Stowers et al., 1987.

smokers (30% vs. 7.5%); mutations of nonsmokers are G to C transversion rather than the more common G to T transversion seen in smokers.

The frequency of K-*ras* mutations in lung tumors of rats in dependent on the inducing agent (Table 3). In those tumors induced by [239]PuO$_2$ (Stegelmeier et al., 1991) or tetranitromethane (Stowers et al., 1987), the incidence is relatively high (46% and 100%, respectively). With other compounds the incidence is low, close to that in control rat lung tumors. In nearly all cases mutations are located in the 12th codon, but not correlated with the type of exposure. A notable example is in tumors induced by tetranitromethane in which all activating mutations are GC to AT transitions in the second base of the 12th codon (Stowers et al., 1987). In the rats exposed to [239]PuO$_2$, the mutations are not correlated with any one morphologic phenotype. On the other hand, the incidence of K-*ras* mutations is significantly higher in squamous cell carcinomas than in adenocarcinomas of rats exposed to N-nitrosobis(2-oxopropyl)amine (Ohgaki et al., 1993). K-*ras* mutations are also present in nonneoplastic proliferative lesions, indicating that they are early events in the carcinogenic process.

Many lung tumors of mice have K-*ras* mutations. The frequency of these mutations is dependent upon the strain of mouse and the nature of the initiating carcinogen. For example, the high susceptibility of the Strain A mouse to develop lung tumors has been directly linked to three "pulmonary susceptibility genes," one of which correlates with the K-*ras* gene (Malkinson, 1991; You et al., 1992). Activating K-*ras* mutations are found in a high percentage

Table 4   K-*ras* Mutations in NNK-induced Lung Tumors in Mice

Strain	Dose (mg/kg)	% Tumor incidence	K-*ras* mutation frequency	% K-*ras* mutation frequency	Reference
A/J	0	100	10/11	90	You et al., 1991
A/J	1050	100	10/11	90	Belinsky et all, 1989
C3H	0	25	3/7	43	Devereux et al., 1991
C3H	1050	97	11/11	100	Devereux et al., 1991
$C_{57}Bl/6$	0	4	1/6	17	Devereux et al., 1993a
$C_{57}Bl/6$	460	18	2/22	9	Devereux et al., 1993a

(>90%) of lung tumors in the Strain A mouse, regardless of how the tumors were induced. Other strains, with lower spontaneous incidence of lung tumors, do not have such a consistently high frequency of K-*ras* mutations, and one strain, $C_{57}Bl/6$, has a low frequency of K-*ras* mutations (Devereux et al., 1993a). This point is illustrated by the response of three strains of mice to NNK (Table 4).

The nature of the initiating carcinogen can affect the frequency of K-*ras* mutations in strains other than the A strain. For example, lung tumors induced in the CD-1 strain by 7,12-dimethylbenz[a]anthracene all contain K-*ras* mutations. In contrast, only 35% of lung tumors induced by N-nitrosodiethylamine contain K-*ras* mutations (Manam et al., 1992). In the $B6C3F_1$ strain, lung tumors induced by tetranitromethane all contain K-*ras* mutations, but in tumors induced by butadiene or methylene chloride, 66% and 20% respectively contain K-*ras* mutations (Stowers et al., 1987; Goodrow et al., 1990; Devereux et al., 1993b).

The spectra of K-*ras* mutations is dependent on the nature of the initiating carcinogen, but not on the strain of mouse (You et al., 1991; Devereux et al., 1991). Characteristic mutational spectra have been found in lung tumors induced by tetranitromethane (GGT > GAT codon 12), butadiene (GGC > CGC codon 13), and urethan (codon 61 mutations predominate) (Stowers et al., 1987; Goodrow et al., 1990; Ohmori et al., 1992). These characteristic mutational spectra are important as they may reveal mechanisms underlying tumor initiation by these carcinogens. It should be noted, however, that the dose of carcinogen may affect the mutational spectra (Chen et al., 1993).

Host Factors

Animal Species   A basic consideration in inhalation carcinogenesis studies is the selection of the test animal species. Rats, mice, and Syrian hamsters are commonly used. The spontaneous lung tumor incidence in these species varies widely (Tables 5a-c).

Table 5a Spontaneous Lung Tumor Incidence in Laboratory Rodents—Mice

Strain	Number in population	Sex	Length of observation	Lung tumor incidence (%) Benign	Lung tumor incidence (%) Malignant	Tumor types[a]	Reference
B6C3F₁	2328	Male	22–26 mo	12.1 ± 6.7	5.1 ± 4.3	Alveo.-bronch. adenoma/ca.	Haseman et al. (1984a)
B6C3F₁	2388	Female	22–26 mo	5.5 ± 3.6	2.0 ± 2.3	Alveo.-bronch. adenoma/ca.	Haseman et al. (1984a)
C₅₇Bl/lcrf	497	Male	Life span	6.0	0	Pap. adenoma or adenoma	Rowlatt et al. (1976)
C₅₇Bl/lcrf	293	Female	Life span	4.4	0.4	Pap. adenoma adenoma anapl. tum.	Rowlatt et al. (1976)
SENCAR	127	Male	Life span	6.2	14.9	Adenocarcinoma/ adenoma	Conti et al. (1985)
SENCAR	223	Female	Life span	8.5	5.8	Adenocarcinoma/ adenoma	Conti et al. (1985)
SWR/J	68	Male	Life span	3.0	4.4	Solitary nod. alveogenic	Rabstein et al. (1973)
SWR/J	243	Female	Life span	2.0	3.7	Solitary nod. alveogenic	Rabstein et al. (1973).
BALB/c (ctrl)	2397	Female	33 mo	(21)		Alveo.-bronch. neoplasms	Heath et al. (1982)
Swiss	101	Male	19 mo	4.0	6.0	Alveolar cell adenoma/ca.	Prejean et al. (1973)
Swiss	181	Female	19 mo	3.8	7.6	Alveolar cell adenoma/ca.	Prejean et al. (1973)

[a]Author designation.
Abbreviations: alveo.-bronch., alveolar-bronchiolar; anapl. tum., anaplastic tumors; ca., carcinoma; nod., nodule; pap., papillary.

Table 5b Spontaneous Lung Tumor Incidence in Laboratory Rodents—Rats

Strain	Number in population	Sex	Length of observation	Lung tumor incidence (%) Benign	Lung tumor incidence (%) Malignant	Tumor types[a]	Reference
F344	2305	Male	22–26 mo	1.5 ± 2.1	0.9 ± 1.6	Alveo.-bronch. adenoma/ca.	Haseman et al. (1984b)
F344	2354	Female	22–26 mo	0.8 ± 1.4	0.4 ± 0.9	Alveo.-bronch. adenoma/ca.	Haseman et al. (1984b)
F344	529	Male	Life span	0.6	2.3	Alveo.-bronch. adenoma/ca. squa. cell ca.	Solleveld et al. (1984)
F344	529	Female	Life span	0.2	1.3	Alveo. bronch. adenoma/ca. squa. cell ca.	Solleveld et al. (1984)
Osborne-Mendel	975	Male	26–29 mo	0.4	0.5	Alveo.-bronch. adenoma/ca. adenosqua. ca. mucoepeid. ca.	Goodman et al. (1980)
Osborne-Mendel	970	Female	26–29 mo	0.2	0.3	Alveo.-bronch. adenoma/ca.	Goodman et al. (1980)
Sprague-Dawley	179	Male	19 mo	0.5	1.1	Alveolar cell adenoma broncho. ca.	Prejean et al. (1973)
Sprague-Dawley	181	Female	19 mo	0	0.5	Bronchogenic carcinoma	Prejean et al. (1973)
WAG/RIJ (Wistar)	290	Female	Life span	0	0	—	Boorman and Hollander (1973)
WISTAR/CPB	197	Male	30 mo	0	0.5	Squamous cell carcinoma	Kroes et al. (1981)
WISTAR/CPB	182	Female	30 mo	0	0	—	
WISTAR/SPF Tox.	192	Male	30 mo	0	0.5	Adenocarcinoma	Kroes et al. (1981)
WISTAR/SPF Tox.	192	Female	30 mo	0	0	—	
Sprague-Dawley (Crl:CDB₂)	585	Male	24 mo	0.2	0.2	Adenoma adenocarcinoma	McMartin et al. (1992)
Sprague-Dawley (Crl:CDB₂)	585	Female	24 mo	0.2	0.2	Adenoma adenocarcinoma	MacMartin et al. (1992)
F344	740	Male	24 mo	—	—	—	Chandra and Frith (1992)
F344	740	Female	24 mo	—	0.3	Alveo.-bronch. ca.	Chandra and Frith (1992)

[a]Author designation.

Abbreviations: alveo.-bronch., alveolar-bronchiolar; adenosqua. ca., adenosquamous carcinoma; broncho. ca., bronchogenic carcinoma; ca., carcinoma; mucoepid. ca., mucoepidermoid carcinoma; squa. cell ca., squamous cell carcinoma.

Table 5c Spontaneous Lung Tumor Incidence in Laboratory Rodents—Syrian Hamsters

Strain	Number in population	Sex	Length of observation	Lung tumor incidence (%) Benign	Lung tumor incidence (%) Malignant	Tumor types[a]	Reference
Eppley Colony	170	Male	Life span	0.6	0	Bronchogenic adenoma	Pour et al. (1976)
Eppley Colony	150	Female	Life span	0.7	0	Bronchogenic adenoma	Pour et al. (1976)
Hannover Colony	80	Male	Life span	1.3	0	Adenoma	Pour et al. (1976)
Hannover Colony	115	Female	Life span	0.9	0	Adenoma	Pour et al. (1976)
Zucca Hamstery	1120	M/F	Life span	0	0	—	Dontenwill et al. (1973)

[a]Author designation.

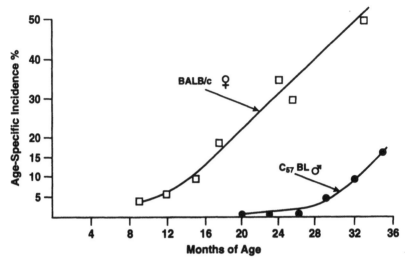

Figure 1 Age-specific incidence of lung tumors in a group of 495 unexposed male $C_{57}B1$ mice and 2397 unexposed female BALB/c mice. The crude incidences for the $C_{57}B1$ mice was 6.0% (number affected/ number examined × 100) and for the BALB/c mice, 21%. (Adapted from Rowlatt et al., 1976; Heath, et al., 1982.)

The best data on tumor incidence are in those strains, such as the F344 or Osborne-Mendel rats and the BALB/c or B6C3F₁ mice, that have historically been used in long-term or life-span carcinogenesis studies. Mice have the highest incidence of spontaneous lung tumors (>70% with Strain A mice), and Syrian hamsters have the lowest (1–2%). The incidence of lung tumors increases with age as shown in Fig. 1. The cumulative incidence of lung tumors was recorded in a group of $C_{57}B1$ mice (Heath et al., 1982) and BALB/c mice (Rowlatt et al., 1976). The incidence increased most rapidly after the median survival time, about 26 months.

Mice are valuable test animals because of the large number of genetically well-characterized inbred strains available. Slight genotypic differences can result in unique or unusual phenotypic differences that can be exploited in experimental studies. Lung tumor occurrence is one phenotypic difference that varies widely among mouse strains (Table 6). The high incidence in Strain A mice (20% by 6 months of age) has led to their use in a short-term carcinogenesis bioassay (Shimkin and Stoner, 1975). Recent evaluations, however, have shown a poor agreement in carcinogenicity test results from Strain A mice among different laboratories. The results also have poor agreement with results from genotoxicity tests and 2-year carcinogenicity tests (Maronpot et al., 1986). These factors must be recognized in evaluating inhalation carcinogenesis studies using mice.

The morphologic features of commonly occurring lung tumors in mice have recently been studied in detail (Dixon et al., 1991). Histologically, the

Table 6   Spontaneous Lung Tumor Incidence in Various Strains of Laboratory Mice

Mouse strain	Age (months)	Incidence (% of mice with lung tumors)
A	6	20
A	23	71
CBA	27	18
C3H	21	14
DBA	21	9
$C_{67}Bl$	22	7

Source: Data adapted from Shimkin and Stoner (1975) and Maronpot et al. (1986).

tumors can be classified as adenoma or carcinoma using standard diagnostic criteria. Adenomas are usually less than 4 mm in diameter and have a solid papillary or mixed pattern. Carcinomas are usually greater than 4 mm in diameter and have a papillary or mixed pattern. The tumors have been referred to by numerous synonyms (Table 5a), including alveolar-bronchiolar adenoma or carcinoma, alveolar-bronchiolar neoplasms, adenoma, papillary adenoma, adenocarcinoma, alveolar cell adenoma or carcinoma, and solitary alveogenic nodules. The plethora of names for ostensibly the same neoplastic process results from difficulty in clearly distinguishing benign from malignant lung tumors in mice, and the uncertainty as to the cellular origin of these tumors. The classification of tumors as benign or malignant, however, does have utility (McConnell et al., 1986). It is well recognized that lung tumors in mice vary in their differentiation. Well-differentiated tumors are generally benign, and poorly differentiated tumors are generally malignant. Some, but relatively few, do metastasize, irrefutable evidence of malignancy.

The cell of origin of the typical lung tumor in mice is not entirely clear; however, the preponderance of evidence is that lung tumors arise from alveolar Type II cells. Lung tumors in mice were first described as originating from alveolar lining cells (Grady and Stewart, 1940). Many subsequent morphological and ultrastructural studies since then provided additional evidence that both spontaneous and induced alveolar-bronchiolar neoplasms of mice came from alveolar lining cells, specifically the alveolar Type II cell (Sato and Kauffman, 1980; Shimkin and Stoner, 1975; Witschi and Haschek, 1983; Dixon et al., 1991).

Several investigators, however, have described morphologic evidence that many of the alveolar-bronchiolar neoplasms of mice originate from Clara cells in the bronchioles (Kauffman et al., 1979; Kauffman, 1981; Witschi and Haschek, 1983). A recent study used immunocytochemical localization of surfactant apoprotein and Clara cell antigen to identify the cells in chemically induced and naturally occurring alveolar-bronchiolar neoplasms in mice (Ward et al., 1985). This investigation provided evidence that a vast majority of mouse pulmonary tumors are alveolar Type II cell adenomas or carcinomas.

Rats are frequently used in inhalation carcinogenesis studies. They have a relatively low spontaneous incidence of lung tumors (Table 5b), yet develop lung cancers from a variety of carcinogenic compounds. In addition, the neoplastic alterations are closer morphologically to those seen in humans than the alterations seen in mice. The development of squamous cell carcinomas and adenocarcinomas in rats, although they do not mimic those in humans, are more similar to human lung tumors than are pulmonary adenomas of mice.

Adenomas and carcinomas in the bronchiolar-alveolar region have been reported as spontaneous lesions in the rat (Stula, 1975; Reznik, 1983). Histologically, the tumors have a papillary or tubular pattern and are found in the peripheral portions of the lung not associated with bronchi. Recent studies have shown that these tumors are composed of alveolar Type II cells.

An ultrastructural investigation of spontaneous and nitrosamine-induced tumors showed cells with eosinophilic lamellated inclusion bodies typical of Type II cells (Reznik-Schuller and Reznik, 1982). A recent combined immunocytochemical and ultrastructural study reached similar conclusions (Ohshima et al., 1985). They found that a vast majority of nitroso-methylurea-induced and spontaneous pulmonary neoplasms in F344 rats are alveolar Type II cell adenomas and carcinomas. In addition, they noted a progression of the lesions with tumors arising from foci of alveolar Type II cell hyperplasia.

Squamous cell carcinomas rarely occur as spontaneous tumors in rats but frequently as induced tumors, especially at high doses of chemical, radiological, or particulate carcinogens (Hahn and Lundgren, 1992). Adenosquamous carcinomas, tumors with a mixed morphology combining adenocarcinomatous and squamous patterns, are also seen in these situations. High lung burdens of relatively inert particles in the lung may also induce keratinizing squamous cysts or benign keratinizing tumors of the rats lung (Carlton, 1994). These squamous lesions and neoplasms, as well as other types of neoplasms, have been seen in rats exposed to high concentrations of insoluble isometric particles, such as diesel exhaust, carbon black, titanium dioxide, or petroleum coke (Morrow, 1994). The mechanism for the induction of these lesions is not certain, but is associated with impaired clearance of particles from the lung. The mere presence of inorganic particles in the lung, however, appears to be sufficient. For example, carbon black, which is not mutagenic, is equal to diesel exhaust in its ability to induce lung cancers in rats (Heinrich et al., 1994; Nikula et al., 1994). Diesel exhaust contains particles coated with mutagenic organics. The significance of these lung tumors in rats for the evaluation of risk in humans exposed to inert particles is not certain. At this point it is not scientifically justifiable to attempt risk calculations in humans based on these studies of particle-induced lesions in rats (Oberdörster, 1994). Syrian hamsters have several advantages for use in inhalation carcinogenesis studies. They are resistant to pulmonary infections and have a low spontaneous incidence of tumors (Table 5c). Unfortunately, they are also relatively insensitive to induction of lung cancer following inhalation of carcinogenic materials. The reason for this insensitivity is not clear because intratracheal instillation of carcinogenic polycyclic hydrocarbons and intravenous injection of carcinogenic nitrosamines readily causes lung cancers in Syrian hamsters. A good example of this paradox is the induction of lung cancers by radioactive materials. Inhaled plutonium, an alpha-emitting radionuclide, results in a low percentage of lung cancers in hamsters, less than 5% at the highest radiation doses (Lundgren et al., 1983). In contrast, intratracheally instilled ^{210}Polonium, an alpha-emitter, results in a high incidence of lung cancers within months and at radiation doses similar to or lower than those given by inhalation (Little et al., 1985).

The morphology of the naturally occurring lung tumors in Syrian hamsters is poorly documented, because so few of them occur (Table 5c). They are generally described as adenomas composed of goblet or uniform cylindrical cells similar to the normal cells of the bronchial epithelium. Because of this appearance, they have been called bronchogenic adenomas (Pour et al., 1976; Pour, 1983).

In spite of the large variation in spontaneous tumor incidence and response to inhaled carcinogens among animal species, long-term carcinogenicity tests in animals are predictive of human carcinogens. An analysis of the response of laboratory animals to known and suspected human carcinogens showed that 84% of the carcinogens caused cancer in animals (Wilbourn et al., 1986). The response may well have been higher had all the compounds been adequately tested experimentally. A key to this finding was the fact that at least two species, and many times more, were tested. This empirical approach of using at least two species in carcinogenicity testing will be required until the basic, underlying factors determining species differences are characterized (Hasemen et al., 1984b).

Respiratory Infections Respiratory infections are another prominent consideration in the design, conduct, and interpretation of inhalation carcinogenesis studies (Hamm, 1986). Chronic respiratory disease of rats and mice, caused by mycoplasma, was, until the past few years, a major compounding variable. Previously reported studies should be examined carefully because of this disease, which was ubiquitous in either rats or mice older than about 6 months of age. In the past few years, however, improved animal husbandry, animal quarters, and the commercial production of specific pathogen-free rodents have drastically reduced the incidence of this disease.

Two types of respiratory infections still remain, however, that can influence the results of an inhalation study; Sendai and sialodacryoadenitis virus infections. Both agents can cause lesions in the lower respiratory tract, thus directly affecting interpretation of lesions in the lung (Wojenski and Percy, 1986; Hall et al., 1986).

Infections resulting in a chronic respiratory disease can increase the number of lung tumors induced by a carcinogen. This point is illustrated by one study that compared the effects of a nitrosamine, N-nitrosoheptamethyleneimine, in specific pathogen-free rats or germ-free rats and rats with chronic respiratory disease induced by one of more microbial agents, including Sendai virus, rat corona virus, pneumonia virus of mice, or Mycoplasma pulmonis (Schreiber et al., 1972). Lung tumor incidence was markedly increased in the infected male rats. In addition, the number of tumors per rat in the infected animals was increased for both sexes. The reasons for increased neoplasia induction were not determined. Potential causes are enhanced promotion due to increased cell proliferation or oxidant stress from inflammatory cells, or enhanced progression due to suppression of immune mechanisms.

Studies of the influence of Sendai virus infection on the incidence of pulmonary adenomas induced by chemical carcinogens have yielded varied results. Sendai virus given several days before injection of the carcinogen, urethane, resulted in fewer lung tumors than seen in uninfected BALB/c mice (Nettesheim, 1974). Virus given weeks before or after urethane injection had no effect. Another study used Strain A mice with and without anti-Sendai antibody (Peck et al., 1983). The mice were injected with polycyclic hydrocarbon carcinogens. Sendai-exposed mice exposed to 7,12-dimethyl benzo-(a)anthracene had a higher incidence of lung tumors than controls. On the other hand, Sendai-infected mice exposed to 10-chloromethyl-9-chloroanthracene had a lower incidence of lung tumors than controls. The reasons for these contradictory results are unexplained. They may relate to the timing of virus infection in relation to the carcinogen administrations and immunologic responsiveness of the exposed mice.

AGENTS CAUSING TUMORS OF THE LUNG
Types of Agents

Lung tumors can be induced in laboratory rodents by a wide variety of vapors, dusts, and aerosols (Table 7). These can be divided into three general classes: organic chemicals; inorganic chemicals, including metallic, nonmetallic, and radioactive materials; and complex mixtures (Pepelko, 1984).

The inhaled organic chemicals that result in lung tumors are primarily in the vapor form. Several of these, such as bis(chloromethyl)ether and dimethyl sulfate, also cause nasal tumors. Others, such as vinyl chloride and butadiene, cause tumors in other organs, as well as the lung. Often, the neoplasms in other organs have a higher incidence and are more serious health risks than the lung tumors. Many studies of vinyl chloride in rats have shown a high incidence of Zymbal's gland carcinoma and hepatic hemangiosarcoma, but few lung tumors.

Organic chemicals in a particulate form have also been used to induce lung tumors by the route of inhalation. BaP caused a marginal increase in lung tumors in rats in inhaled with SO_2 (Laskin et al., 1970). On the other hand, Syrian hamsters developed lung tumors after inhalation of BaP, but the carcinogenesis was not enhanced by inhaled sulfate (Godleski et al., 1984). In most studies of inhaled BaP, however, Syrian hamsters developed tumors of the larynx and trachea, but not the lung.

Inhaled metallic inorganic compounds have caused lung tumors in rodents, but not tumors elsewhere in the body. Beryllium, cadmium, and nickel in several different chemical forms resulted in pulmonary carcinoma. The chemical form is important in determining the dose response, but the reasons for this have not been determined. For example, beryllium sulfate is more carcinogenic than beryllium oxide. This difference may relate to the solubility of these compounds in the lung and the actual dose of chemical to the target cells and tissues.

Asbestos is the classic nonmetallic inorganic fiber that causes lung cancers in rodents after inhalation. Mesotheliomas, tumors of the pleural lining of the lung and thoracic cavity, also result from inhaled asbestos. Many substitutes for the commercial uses of asbestos have been developed. In bioassays of these newly created fibers, asbestos is frequently used as a positive control. Two such new fibers, ceramic aluminosilicate and aramid, have been found to be carcinogenic in rats (Davis et al., 1984; Lee et al., 1988). Quartz or silica are generally not considered carcinogenic in exposed human populations, however, several studies in rats have demonstrated the induction of lung cancers (Dagle et al., 1985; Holland et al., 1985; Muhle et al., 1989).

Many studies of inhaled radionuclides have been conducted. For the purposes of inhalation hazards, radioactive materials can be divided into three general categories: particulate radioactive materials that emit alpha particles; particulate radioactive materials that emit beta particles; and radon gas, with its associated daughter radionuclides that emit alpha particles. When these radioactive materials in particulate form are inhaled and deposited in the lungs, a radiation dose is delivered primarily to the pulmonary parenchyma because many particles are delivered to the deep lung, and penetration of the emitted alpha or beta particles is low (micrometers to millimeters). Radon gas, on the other hand, delivers most of its radiation dose to the cells lining airways with little gas reaching the deep lung. All of these radiations can cause tumors in rats, mice, and dogs.

Complex mixtures include atmospheres that contain particles with several, if not hundreds, of organic and inorganic chemicals and various gases that may be toxic or carcinogenic. These atmospheres are usually generated

Table 7   Agents Causing Lung Tumors in Laboratory Animals Following Inhalation Exposure

I Organic chemicals
   A Vapors
      1 Benzene (Snyder et al., 1988; Farris et al., 1993)
      2 Bis(chloromethyl)ether (Leong et al., 1971; Kuschner et al., 1975; Dulak and Snyder, 1980)
      3 Dimethyl sulfate (Schlogel and Bannasch, 1970)
      4 Ethylene dibromide (NTP, 1982)
      5 Ethylene oxide (Huff et al., 1991)
      6 Methylene chloride (Mennear et al., 1985; Maltoni et al., 1985)
      7 Nitrobenzene (Cattley et al., 1994)
      8 Tetranitromethane (Bucher et al., 1991)
      9 3-Nitro-3-hexene (Anderson et al., 1966; Deichmann et al., 1963)
     10 Urethane (Otto and Platz, 1966; Leong et al., 1971)
     11 Vinyl chloride (Viola et al., 1971; Caputo et al., 1974; Maltoni and Lefemine, 1975)
     12 1, 3-Butadiene (Huff et al., 1985; Melnick et al., 1990)
     13 1, 2-Dibromo-3-chloropropane (Reznik et al., 1980)
     14 1,3-Dichloropropene (Lomax et al., 1989)
     15 1,2-Dibromoethane (Huff et al., 1991)
     16 1,2-Epoxybutane (Huff et al., 1991)
   B Particles
      1 Benzo(a)pyrene (Godleski et al., 1984; Laskin et al., 1970, 1976; Thyssen et al., 1981)
      2 Polymeric methylene diphenyl diisocyanate (Reuzel et al., 1994)
      3 Polyurethane dust (Laskin et al., 1972)
II Inorganic compounds
   A Metallic
      1 Antimony compounds (Groth et al., 1986)
      2 Beryllium compounds (Schepers et al., 1957; Schepers, 1961; Vorwald et al., 1966; Reeves et al., 1967; Wagner et al., 1969)
      3 Cadmium compounds (Takenaka et al., 1983; Glaser et al., 1990)
      4 Nickel compounds (Heuper, 1958; Sunderman, 1966; Ottolenghi et al., 1975)
      5 Titanium compounds (Lee et al., 1985, 1986)

directly from a source found in the environment, such as a cigarette or an automobile engine. Cigarette smoke has frequently been studied because of its importance to the human population. It is not, however, an impressive inducer of lung tumors in animals. Significant increases in lung tumors have been observed in long-term smoking studies with mice of the A (Essenberg, 1952), $C_{57}B1$ (Harris et al., 1974), and albino strains (Otto, 1963). Rats have developed a significant increase of lung tumors in only one study of cigarette smoke inhalation (Dalbey et al., 1980). Syrian hamsters have been used in many smoke inhalation studies and develop laryngeal cancer readily, but not lung tumors (Dontenwill et al., 1973). The reasons these rodent species do not develop lung cancer readily after inhalation of cigarette smoke is not

Table 7 *(Continued)*

B Nonmetallic
  1 Aramid fibers (Lee et al., 1988)
  2 Asbestos fibers (Davis et al., 1984, 1986; McConnell et al., 1984; Gross et al., 1967; Reeves et al., 1971; Wagner et al., 1974, 1977; Wehner et al. 1977)
  3 Carbon black (Heinrich, 1994; Nikula et al., 1994)
  4 Ceramic aluminosilicate fibers (Davis et al., 1984)
  5 Coal dust (Martin et al., 1977)
  6 Crystalline silica (Holland et al., 1985; Dagle et al., 1985; Muhle et al., 1989)
  7 Erionite (Johnson et al., 1984)
  8 Oil shale dust (Holland et al., 1985; Mauderly et al., in press)
  9 Talc (Hobbs et al., 1994)
C Radionuclides
  1 Alpha-emitting radionuclide particles (ICRP 31, 1980)
  2 Beta-emitting radionuclide particles (ICRP 31, 1980)
  3 Radon and radon daughters (NCRP 78, 1984)
III Complex mixtures
  A Artificial smog (Kotin and Falk, 1956; Kotin et al., 1958; Nettesheim et al., 1970)
  B Cigarette smoke (Henry and Kouri, 1986; Leuchtenberger and Leuchtenberger, 1974; Dalbey et al., 1980; Essenberg, 1952; Otto, 1963; Harris et al., 1974)
  C Coal tar aerosols (Mestitzova and Kossey, 1961; Horton et al., 1963; Tye and Stemmer, 1967)
  D Coal smoke (Liang et al., 1988)
  E Diesel engine exhaust (Campbell et al., 1981; Stöber, 1986; Mauderly et al., 1986; Iwai et al., 1986; Ishinishi et al., 1986; Brightwell et al., 1989)
  F Tar/pitch condensation aerosol (Heinrich et al., 1994)
  G Urban air (Ito et al., 1989)

clear, but may relate to the carcinogenic dose to the cells at risk. Animals are generally more susceptible to the acute toxic effects of cigarette smoke (due to nicotine and carbon monoxide) and cannot inhale as much smoke as humans without severe effects. In addition, many animals will avoid inhaling smoke if possible, thus reducing the dose to tissues.

Several large, long-term studies in the United States, Europe, and Japan have shown a carcinogenic effect of diesel exhaust in rats (Ishinishi et al., 1986). More recent studies have attempted to determine the role of the organic and inorganic portions of the diesel exhaust in production of these tumors (Heinrich et al., 1994; Nikula et al., 1994).

Examples of Dose-Response Relationships

Diesel Engine Exhaust The nature of diesel engine emissions, health effects in laboratory animals, and potential health risks from human exposures, have been studied at the Inhalation Toxicology Research Institute (Mauderly et al., 1986). Results indicate that high concentrations of diesel exhaust cause accumulation of diesel soot, pulmonary inflammation, interstitial fibrosis, and pulmonary neoplasia. The dose response for these tumors and the time of onset are of particular interest.

Male and female rats were randomized into four treatment groups of about 90 animals per group for long-term observation. Additional rats were used for sacrifice studies to determine diesel soot concentration in the lung. The rats were exposed 7 hours/day, 5 days/week, for up to 30 months to whole exhaust diluted to nominal soot concentrations of 0 (Control), 350 (Low), 3470 (Medium), and 7080 (High) $\mu g/m^3$ or to filtered air. Exhaust was generated by 1980 model 5.7L V8 engines. The exhaust also contained small amounts of carbon monoxide, nitric oxide, nitrogen dioxide, hydrocarbon vapor, and carbon dioxide. Exposures did not cause overt signs of toxicity, nor significantly alter the life span of either sex.

Two basic tumor types were found: bronchoalveolar tumors and squamous cell tumors, both arising from the alveolar parenchyma. Benign and malignant forms were found of each tumor. The tumors did not metastasize, were apparently slow growing, and did not kill the rats.

The bronchoalveolar adenomas (Fig. 2a) were generally small in size and composed of low cuboidal cells of uniform size arranged in a regular acinar or papillary pattern. The bronchoalveolar adenocarcinomas (Fig. 2b) often resembled the adenomas at their periphery, but had central areas with hyperchromatic cells of nonuniform size arranged in a solid, irregular pattern. Squamous cysts (Fig. 2c) had a wall of well-differentiated, uniform, squamous epithelial cells surrounding a central core of keratin. The larger cysts compressed adjacent alveoli, but there was no evidence of invasion. These were considered benign tumors. Two of these cysts had poorly differentiated cells in the cyst wall that invaded surrounding structures. These were classified as squamous cell carcinomas. These tumors were similar to those in other studies with rats exposed chronically to diesel exhaust (Mohr et al., 1986).

The percentage of rats with lung tumors is shown in Table 8. The lung tumor data were analyzed by estimating the prevalence of tumors as a function of time using logistic regression. The probability that a rat has a lung tumor i.e., estimated prevalence, is shown in Fig. 3 as a function of time for the four treatment groups. The estimated prevalences rise sharply near the end of exposure. One important feature of the progressive increase in lung tumor prevalence toward the end of life is the large fraction of rats with lung tumors after 2 years of exposure, a time when bioassays are frequently terminated. At 2 years of exposure, less than 20% of the lung tumors had been observed, and the difference between the high exposure group and the controls was not significantly different. This indicates that bioassays of potential inhaled carcinogens, especially if they are not highly potent, may take longer than the standard 2-year bioassay protocol.

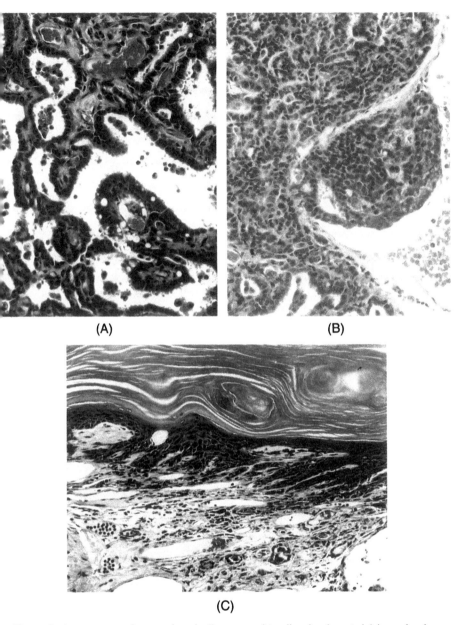

(A)   (B)

(C)

Figure 2 Lung tumors in rats chronically exposed to diesel exhaust: (a) bronchoalveolar adenoma showing papillary structure lined by uniform cuboidal cells, (b) bronchoalveolar adenocarcinoma showing solid growth of anaplastic cells and invasion of
blood vessel, and (c) squamous cyst showing keratin center surrounded by uniformly
flattened squamous cells.

Table 8   Rats with Lung Tumors after Chronic Exposure to Diesel Exhaust

	Percent with lung tumors			
Exposure atmosphere ($\mu g/m^3$)[a]	Adenomas	Adenocarcinomas or squamous cell carcinoma[b]	Squamous cysts	All tumors
7000	0.7	9.8[c]	5.6[c]	16.1[c]
3500	3.8	0.8	0	4.6
350	0	0.7	0	0.7
0	0	1.4	0	1.4

Source: Mauderly et al. (1986).
[a]Nominal diesel soot concentration.
[b]Only one squamous cell carcinoma (high level).
[c]Significantly higher than control at $p < 0.05$ by z-statistic.

Alpha-Emitting Radionuclides   A number of studies with inhaled radionuclides have been conducted at the Inhalation Toxicology Research Institute and by other organizations. One such study examined the carcinogenicity and dose response for mixed uranium and plutonium oxide fuels used in nuclear reactors (Mewhinney, 1986). Workers involved in fabrication of these mixed oxide fuels have a potential for accidental inhalation exposure. Two different types of powdered feed stock for these fuels were obtained from fuel fabrication glove boxes. Groups of 40 male and female rats were exposed

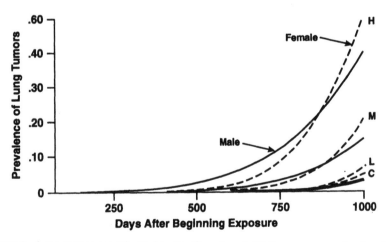

Figure 3   Lung tumor prevalence in rats after chronic lifetime exposure to diesel exhaust. Prevalence is related to time of exposure using a logistic regression model. Nominal exposure concentrations: H = 7080 $\mu g/m^3$, M = 3470 $\mu g/m^3$, L = 350 $\mu g/m^3$, and C = 0 $\mu g/m^3$.

briefly by inhalation to achieve one of three concentrations of radioactivity in the lung that was expected to produce significant incidences of lung tumors while having little or no effect on the median life span of the exposed rats. Groups of rats were designated for interval sacrifice to determine the retention of the radioactivity in the lung and calculate the radiation dose to the lung.

A number of morphologic tumor types were induced in the lung: bronchoalveolar adenoma and bronchoalveolar carcinoma, papillary adenocarcinoma; squamous cell carcinoma, adenosquamous carcinoma, and hemangiosarcoma (Hahn and Lundgren, 1992). All of these tumors originated in the alveolar parenchymal region of the lung. The tumor types seen after irradiation were generally similar to those seen in rats that were unexposed or exposed to chemical carcinogens. However, the squamous cell carcinomas in irradiated rats and those seen in rats chronically exposed to diesel exhaust were distinctly different. In irradiated rats, the squamous cell carcinomas were less well-differentiated and decidedly more locally invasive. Although cystic squamous tumors do occur after irradiation, the wall of these tumors was less well-differentiated (Fig. 4a). Local invasion was not unusual and invasion of blood vessels occurs (Fig. 4b). On occasion, one tumor had the morphologic features of both a squamous cell carcinoma and an adenocarcinoma. Whether these were two tumors that grew together or two phenotypically different clones within one tumor was undetermined.

The pathogenesis of the radiation-induced squamous tumors appears to be different than the pathogenesis of the diesel exhaust-induced tumors. Compared to the animals chronically exposed to diesel exhaust, very few particles were actually deposited in the lung with brief inhalation exposure to radioactive particles. Thus, no particle overload occurs in the case of the radiation-induced tumors. The one common feature in the pathogenesis of the two tumor types may be chronic injury to alveolar Type II cells.

The time of onset of lung tumors was dose dependent; higher radiation doses caused lung tumors as early as about 500 days after inhalation exposure (Fig. 5). Two years after exposure, 60% of the lung tumors in the highest exposure group were present.

A comparison of the lung tumor incidence with dose shows a dose response that is consistent with a linear relationship (Fig. 6). It also shows that the dose response for the two forms of mixed oxide fuels can be fit with a single linear function.

These two examples demonstrate some of the key features that can be shown with dose-response relationships, namely, time of onset of lung tumors, and models for extrapolation to the low dose region of the dose-response curve.

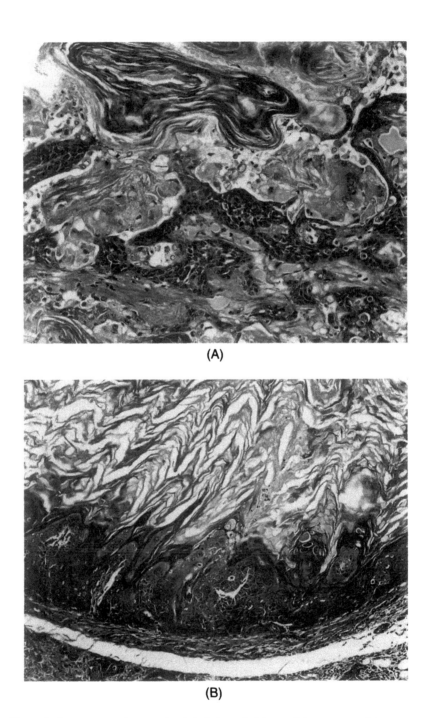

Figure 4 Lung tumors in rats after inhalation of alpha-emitting radionuclides (mixed oxide fuels). (a) Wall of cystic squamous cell carcinoma showing poorly differentiated squamous epithelium. (b) Invasion of blood vessel by squamous cell carcinoma.

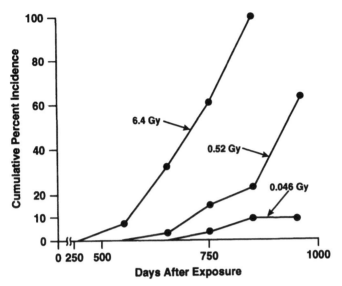

Figure 5 Cumulative incidence of lung tumors after inhalation of mixed-oxide fuels by F344 rats. Radiation doses calculated to lung to median life span.

SUMMARY

Studies in inhalation carcinogenesis hold a number of intriguing opportunities for future work. While it is generally true that inhalation bioassays can identify potential inhaled carcinogens, this is an empirical approach. Much is yet to be learned about the mechanisms of inhalation carcinogenesis in

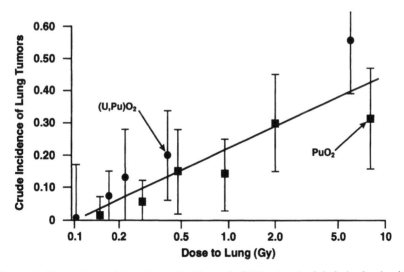

Figure 6 Comparison of lung tumor incidence in F344 rats who inhaled mixed-oxide fuels (average incidence ± SE).

man as it is experimentally induced in laboratory animals. Of particular importance is the pathogenesis of cigarette smoke-induced lung cancer because cigarette smoking is so pervasive in the society. Interactions between cigarette smoke and other carcinogens must be determined so that the true risks for lung cancer can be known.

ACKNOWLEDGMENTS

Portions of the research reported here were supported by the U.S. Department of Energy Office of Health and Environmental Research under Contract No. DE-AC04-76EV01013 in facilities fully accredited by the American Association for Accreditation of Laboratory Animal Care.

I acknowledge and thank the numerous members of the technical and professional staff of the Inhalation Toxicology research Institute who participated in these studies and reviewed the manuscript.

REFERENCES

Anderson, WAD, Deichmann, WE, MacDonald, WE, Lampe, KF, Dressler, I: Experimental lung cancers from inhalation of nitro-olefins, potential carcinogens in air pollution. IX Int Cancer Cong 9:122, 1966.

Auerbach, O, Stout, AP, Hammond, EC, Garfinkel, L: Changes in bronchial epithelium in relation to cigarette smoking and in relation to lung cancer. N Engl J Med 265:253–267, 1961.

Becci, PJ, McDowell, EM, Trump, BF: The respiratory epithelium. IV. Histogenesis of epidermoid metaplasia and carcinoma in situ in the hamster. J Natl Cancer Inst 61:577–586, 1978.

Belinsky, SA, White, CM, Boucheron, JA, Richardson, FC, Swenberg, JA, Anderson, MW: Accumulation and persistence of DNA adducts in respiratory tissue of rats following multiple administrations of the tobacco specific carcinogen 4-(N-methyl-N-nitrosamino)-1-(3-pyridyl)-1-butanone. Cancer Res 46:1280–1284, 1986.

Belinsky, SA, Devereux, TR, Maronpot, RR, Stoner, GD, Anderson, MW: Relationship between the formation of promutagenic adducts and the activation of the K-*ras* protooncogene in lung tumors from A/J mice treated with nitrosamines. Cancer Res 49:5305–5311, 1989.

Belinsky, SA, Mitchell, CE, Hahn, FF: K-*ras* and *p53* alterations in lung tumors induced in the F344 rat by x irradiation. In: Inhalation Toxicology Research Institute Annual Report 1993–1993. Edited by KJ Nikula, SA Belinsky, PL Bradley, ITRI-140, pp. 87–88. Springfield, VA: National Technical Information Service, 1993.

Biancifiori, C: Ovarian influence on pulmonary carcinogenesis by hydrazine sulfate in BALF/c/CB/Se mice. J Natl Cancer Inst 45:965–970, 1970.

Boorman, GA, Hollander, CF: Spontaneous lesions in the female WAG/Rij (Wistar) Rat. J Gerontol 28:152–159, 1973.

Boring, CC, Squires, TS, Tong, T, Montgomery, S: Cancer statistics, 1994. Cancer J Clin 44:7–20, 1994.

Brightwell, J, Fouillet, X, Cassano-Zoppi, A-L, Bernstein, D, Crawley, F, Duchosal, F, Gatz, R, Perczel, S, Pfeifer, H: Tumors of the respiratory tract in rats and hamsters following chronic inhalation of engine exhaust emissions. J Appl Toxicol 9(1):23–32, 1989.

Bucher, JR, Huff, JE, Jokinen, MP, Haseman, JK, Stedham, M, Cholakis, JM: Inhalation of tetranitromethane causes nasal passage irritation and pulmonary carcinogenesis in rodents. Cancer Letts 57:95–101, 1991.

Campbell, KI, George, EL, Washington, IS, Roberson, PK, Laurie, RD: Chronic Inhalation Oncogenicity Study of Diesel Exhaust in Sencar Mice. U.S. EPA 1981 Diesel Emissions Symposium, Raleigh, NC, 1981.

Caputo, A, Viola, PL, Bigotti, A: Oncogenicity of vinyl chloride at low concentrations in rats and rabbits. J Int Res Commun 2:1582, 1974.

Carlton, WW: Proliferative keratin cyst, a lesion in the lungs of rats following chronic exposure to para-aramid fibrils. Fund Appl Toxicol 23:304, 1994.

Cattley, RC, Everitt, JI, Gross, EA, Moss, OR, Hamm, Jr., TE, Popp, JA: Carcinogenicity and toxicity of inhaled nitrobenzene in B6C3F$_1$ mice and F344 and CD rats. Fundam Appl Toxicol 22:328–340, 1994.

Chandra, M, Frith, CF: Spontaneous neoplasms in aged control Fischer 344 rats. Cancer Letts 62:49–56, 1992.

Chen, B, Liu, L, Castonguary, A, Maronpot, RR, Anderson, MW, You, M: Dose-dependent *ras* mutation spectra in N-nitrosodiethylamine induced mouse liver tumors and 4-(methylnitrosamino)-1-(3-pyridyl)-1-butanone induced mouse lung tumors. Carcinogenesis 14:1603–1608, 1993.

Chiba, I, Takahashi, T, Nau, MM: Mutations in the *p53* gene are frequent in primary, resected non-small-cell lung cancer. Oncogene 5:1603–1610, 1990.

Conti, CJ, Clapp, N, Klein-Szanto, AJP, Nesnow, S, Slaga, TJ: Survival curves and incidence of neoplastic and non-neoplastic disease in SENCAR mice. Carcinogenesis 6:1649–1652, 1985.

Dagle, GE, Wehner, AP, Clark, ML, Buschbom, RL: Chronic inhalation exposure of rats to quartz. In: Silica, Silicosis, and Cancer—Controversy in Occupational Medicine: Cancer Research Monographs. Edited by DF Goldsmith, DM Winn, CM Shy, pp. 255–266. New York: Praeger, 1985.

Dalbey, WE, Nettesheim, P, Griesemer, R, Caton, JE, Guerin, MR: Chronic inhalation of cigarette smoke by F344 rats. J Natl Cancer Inst 64:383–390, 1980.

Davis, JMG, Addison, J, Bolton, RE, Donaldson, K, Jones, AD, Wright, A: The pathogenic effects of fibrous ceramic aluminosilicate glass administered to rats by inhalation or peritoneal injection. In: Biological Effects of Man-Made Mineral Fibers: Proceedings or a WHO/IARC Conference in Association with JEMRB and TIMA, pp. 303–322. Copenhagen: WHO, 1984.

Davis, JMG, Addison, J, Bolton, RE, Donaldson, K, Jones, AD, Smith, T: The pathogenicity of long versus short fibre samples of amosite asbestos administered to rats by inhalation and intraperitoneal injection. Br J Exp Pathol 67:415–430, 1986.

Deichmann, WB, MacDonald, WE, Anderson, WAD, Bernal, E: Adenocarcinoma in the lungs of mice exposed to vapors of 3-nitro-3-hexene. Toxicol Appl Pharmacol 5:445–456, 1963.

Devereux, TR, Anderson, MW, Belinsky, SA: Role of *ras* protooncogene activation in the formation of spontaneous and nitrosamine-induced lung tumors in the resistant C3H mouse. Carcinogenesis 12:299–303, 1991.

Devereux, TR, Belinsky, SA, Maronpot, RR, White, CM, Hegi, ME, Patel, AC, Foley, JF, Greenwell, A, Anderson, MW: Comparison of pulmonary O^6-methylguanine DNA adduct levels and Ki-*ras* activation in lung tumors from resistant and susceptible mouse strains. Mol Carcinog 8:177–185, 1993a.

Devereux, TR, Foley, JVF, Maronpot, RR, Kari, F, Anderson, MW: *ras* protooncogene activation in liver and lung tumors from B6C3F$_1$ mice exposed chronically to methylene chloride. Carcinogenesis 14:795–801, 1993b.

Dixon, D, Horton, J, Haseman, JK, Talley, F, Greenwell, A, Nettesheim, P, Hook, GE, Maronpot, RR: Histomorphology and ultrastructure of spontaneous pulmonary neoplasms in strain A mice. Exp Lung Res 17:131–155, 1991.

Dontenwill, W, Chevelier, HJ, Harke, HP, Lafrenz, U, Reckzeh, G, Schneider, B: Investigations of the effects of chronic cigarette-smoke inhalation in Syrian golden hamsters. J Natl Cancer Inst 51:1781–1832, 1973.

Dulak, NC, Snyder, CA: The relationship between the chemical reactivity and the inhalation carcinogenic potency of direct-activity chemical agents. In: Proceedings, 71st Annual Meeting of the American Association for Cancer Research, p. 106, 1980.

Essenberg, JM: Cigarette smoke and the incidence of primary neoplasms of the lung in the albino mouse. Science 116:561–562, 1952.

Farris, GM, Everitt, JI, Irons, RD, Popp, JA: Carcinogenicity of inhaled benzene in CBA mice. Fundam Appl Toxicol 20:503–507, 1993.

Gazdar, AF: Molecular markers for the diagnosis and prognosis of lung cancer. Cancer 69:1592–1599, 1992.

Glaser, U, Hochrainer, D, Otto, FJ, Oldiges, H: Carcinogenicity and toxicity of four cadmium compounds inhaled by rats. Toxicol Environ Chem 27:153–162, 1990.

Godleski, JJ, Melnicoff, MJ, Sadri, S, Garbeil, P: Effects of inhaled ammonium sulfate on benzo-(a)pyrene carcinogenesis. J Toxicol Environ Health 14:225–238, 1984.

Gold, LS, Sawyer, CB, Magaw, R, Backman, GM, de Veciana, M, Levinson, R, Hooper, NK: A carcinogenic potency database of the standardized results of animal bioassays. Environ Health Perspect 58:9–319, 1984.

Goodman, DG, Ward, JM, Squire, RA, Paxton, MB, Reichardt, WD, Chu, KC, Linhart, MS: Neoplastic and nonneoplastic lesions in aging Osborne-Mendel rats. Toxicol Appl Pharmacol 55:433–447, 1980.

Goodrow, T, Reynolds, S, Maronpot, R, Anderson, M: Activation of K-*ras* by codon 13 mutations in $C_{57}Bl/6 \times C3H\ F_1$ mouse tumors induced by exposure to 1,3-butadiene. Cancer Res 50:4818–4823, 1990.

Goodrow, TL, Storer, RD, Leander, KR, Prahalada, SR, van Zwieten, MJ, Bradley, MO: Murine *p53* intron sequences 5–8 and their use in polymerase chain reaction/direct sequencing analysis of *p53* mutations in CD-1 mouse liver and lung tumors. Mol Carcinog 5:9–15, 1992.

Grady, HG, Stewart, HL: Histogenesis of induced pulmonary tumors in strain A mice. Am J Pathol 16:417–432, 1940.

Gross, P, deTreville, RTP, Tolker, EB, Kaschak, M, Babyak, MA: Experimental asbestosis: The development of lung cancer in rats with pulmonary deposits of chysotile asbestos dust. Arch Environ Health 15:343–355, 1967.

Groth, DH, Stettler, LE, Burg, JR, Busey, W, Grant, GC, Wong, L: Carcinogenic effects of antimony trioxide and antimony ore concentrate in rats. J Toxicol Environ Health 18:607–626, 1986.

Hahn, FF, Lundgren, DL, McClellan, RO: Repeated inhalation exposure of mice to $^{144}CeO_2$. II. Biologic effects. Radiat Res 82:123–137, 1980.

Hahn, FF, Lundgren, DL: Pulmonary neoplasms in rats that inhaled cerium-144 dioxide. Toxicol Pathol 20:169–178, 1992.

Hall, WC, Lubet, RA, Henry, CJ, Collins, MJ: Sendai virus—disease processes and research complications. In: Complications of Viral and Mycoplasmal Infections in Rodents to Toxicology Research and Testing. Edited by TE Hamm, pp. 25–52. New York: McGraw-Hill, 1986.

Hamm, TE: Complications of Viral and Mycoplasmal Infection in Rodents to Toxicology Research and Testing. New York: McGraw-Hill, 1986.

Harris, RJC, Negroni, G, Ludgate, S, Pick, CR, Chesterman, FC, Maidment, BJ: The incidence of lung tumours in $C_{57}Bl$ mice exposed to cigarette smoke-air mixtures for prolonged periods. Int J Cancer 14:130–136, 1974.

Haseman, JK, Huff, J, Boorman, GA: Use of historical control data in carcinogenicity studies in rodents. Toxicol Pathol 12:126–135, 1984a.

Haseman, JK, Crawford, DD, Huff, JE, Boorman, GE, McConnell, EE: Results from 86 two-year carcinogenicity studies conducted by the National Toxicology Program. J Toxicol Environ Health 14:621–639, 1984b.

Heath, JE, Frith, CH, Wong, PM: A morphologic classification and incidence of alveolar-bronchiolar neoplasms in BALB/c female mice. Lab Anim Science 32:638–647, 1982.

Hegi, ME, Söderkvist, P, Foley, JF, Schoonhoven, R, Swenberg, JA, Kari, F, Maronpot, R, Anderson, MW, Wiseman, RW: Characterization of p53 mutations in methylene chloride-induced lung tumors from B6F3F₁ mice. Carcinogenesis 14:803–810, 1993.

Heinrich, U, Peters, L, Creutzenberg, O, Dasenbrock, C, Hoymann, H-G: Inhalation exposure of rats to tar/pitch condensation aerosol or carbon black alone or in combination with irritant gases. In: Toxic and Carcinogenic Effects of Solid Particles in the Respiratory Tract. Edited by U Mohr, DL Dungworth, JL Mauderly, G. Oberdörster, pp. 433–442. Washington, DC: ILSI Press, 1994.

Henry, CJ, Kouri, RE: Chronic inhalation studies in mice. II. Effects on long-term exposure to 2R1 cigarette smoke on (C_{57}B1/Cum × C3H/AnfCum)F mice. J Natl Cancer Inst 77(1):203–212, 1986.

Heuper, WC: Experimental studies in metal cancerigenesis. IX. Pulmonary lesions in guinea pigs and rats exposed to prolonged inhalation of powdered metallic nickel. AMA Arch Pathol 65:600–607, 1958.

Hobbs, CH, Abdo, KM, Hahn, FF, Gillett, NA, Eustis, SL, Jones, RK, Benson, JM, Barr, EB, Dieter, MP, Pickrell, JA, Mauderly, JL: Summary of the chronic inhalation toxicity of talc in F344/N rats and B6C3F₁ mice. In: Toxic and Carcinogenic Effects of Solid Particles in the Respiratory Tract. Edited by U Mohr, DL Dungworth, JL Mauderly, G Oberdörster, pp. 525–528. Washington, DC: ILSI Press, 1994.

Holland, LM, Wilson, JS, Tillery, MI, Smith, DM: Lung cancer in rats exposed to fibrogenic dusts. In: Silica, Silicosis, and Cancer—Controversy in Occupational Medicine: Cancer Research Monographs. Edited by DF Goldsmith, DM Winn, CM Shy, pp. 267–279. New York: Praeger, 1985.

Horton, AW, Tye, R, Stemmer, KL: Experimental carcinogenesis of the lung. Inhalation of gaseous formaldehyde or an aerosol of coal tar by C3H mice. J Natl Cancer Inst. 30:31–43, 1963.

Huff, JE, Melnick, RL, Solleveld, HA, Haseman, JK, Powers, M, Miller, RA: Multiple organ carcinogenicity of 1,3-butadiene in B6C3F₁ mice after 60 weeks of inhalation exposure. Science 227:548–549, 1985.

Huff, JE, Cirvello, J, Haseman, J, Bucher, J: Chemicals associated wity site-specific neoplasia in 1394 long-term carcinogenesis experiments in laboratory rodents. Environ Health Perspect 93:247–270, 1991.

ICRP 31: Report of ICRP Task Group on Biological Effects of Inhaled Radionuclides—International Commission on Radiological Protection. Annals of the ICRP 4, Nos. 1–2. International Commission on Radiation Protection. New York: Pergamon Press, 1980.

Ishinishi, N, Kuwabara, N, Nagase, S, Suzuki, T, Ishiwata, S, Kohno, T: Long-term inhalation studies on effects of exhaust from heavy and light duty diesel engines on F344 rats. In: Carcinogenic and Mutagenic Effects of Diesel Engine Exhaust. Edited by N Ishinishi, A Koizumi, RO McClellan, W Stöber, pp. 329–348. Amsterdam: Elsevier, 1986.

Ito, T, Ikemi, Y, Kitamura, H, Ogawa, T, Kanisawa, M: Production of bronchial papilloma with calcitonin-like immunoreactivity in rats exposed to urban ambient air. Exp Pathol 36:89–96, 1989.

Iwai, K, Udagawa, T, Yamagishi, M, Yamada, H: Long-term inhalation studies of diesel exhaust on F344 SPF rats. Incidence of lung cancer and lymphoma. In: Carcinogenic and Mutagenic Effects of Diesel Engine Exhaust. Edited by N Ishinishi, A Koizumi, RO McClellan, W Stöber, pp. 349–360. Amsterdam: Elsevier, 1986.

Johnson, NF, Edwards, RE, Munday, DE, Rowe, N, Wagner, JC: Pluripotential nature of mesotheliomata induced by inhalation of erionite in rats. J Exp Pathol 65:377–388, 1984.

Johnson, BE, Kelley, MJ: Overview of genetic and molecular events in the pathogenesis of lung cancer. Chest 103:1S-3S, 1993.

Kauffman, SL, Alexander, L, Sass, L: Histologic and ultrastructural features of the Clara cell adenoma of the mouse lung. Lab Invest 6:708, 1979.

Kauffman, SL: Histogenesis of the papillary Clara cell adenoma. Am J Pathol 103:174–180, 1981.

Kelly, G, Hahn, FF: *p53* alterations in [239]Pu-induced lung tumors in F344 rats. In: Ihalation Toxicology Research Institute Annual Report 1992–1993. Edited by KJ Nikula, SA Belinsky, PL Bradley, ITRI-140, pp. 83–84. Springfield, VA: National Technical Information Service, 1993.

Kimura, I: Progression of pulmonary tumor in mice. I. Histological studies of primary and transplanted pulmonary tumors. Acta Pathol Jpn 21:13–56, 1971.

Kotin, P, Falk, HL: The experimental induction of pulmonary tumors in strain-A mice after their exposure to an atmosphere of ozonized gasoline. Cancer 9:910–917, 1956.

Kotin, P, Falk, HL, McCammon, CJ: The experimental induction of pulmonary tumors and changes in the respiratory epithelium in $C_{57}B1$ mice following their exposure to an atmosphere of ozonized gasoline. Cancer 11:473–481, 1958.

Kroes, R, Bargis-Berkvens, JM, de Vries, T, van Nesselrooy, JHJ: Histopathological profile of a Wistar rat stock including a survey of the literature. J Gerontol 36:259–279, 1981.

Kuschner, M, Laskin, S, Drew, RT, Cappiello, V, Nelson, N: Inhalation carcinogenicity of alpha halo ethers. III. Lifetime and limited period inhalation studies with bis(chloromethyl)ether at 0.1 ppm. Arch Environ Health 30:73–77, 1975.

Laskin, S, Kuschner, M, Drew, RT: Studies in pulmonary carcinogenesis. In: Inhalation Carcinogenesis—A.E.C. Symposium Series. Edited by MG Hanna, P Nettesheim, JR Gilbert, pp. 321–351. Springfield, VA: U.S. Department of Commerce, 1970.

Laskin, S, Drew, RT, Cappiello, V, Kuschner, M: Inhalation studies with freshly generated polyurethane foam dust. In: Assessment of Airborne Particles. Edited by TT Mercer, PE Morrow, W Stöber, p. 382–404. Springfield, IL: Charles C Thomas, 1972.

Laskin, S, Kuschner, M, Sellakumar, A, Katz, GV: Combined carcinogen-irritant animal inhalation studies. In: Air Pollution and the Lung. Edited by EF Aharonson, A Ben-David, MA Klingberg, pp. 190–213. New York: Wiley, 1976.

Lee, KP, Trochimowicz, HJ, Reinhardt, CF: Pulmonary response of rats exposed to titanium dioxide (TiO_2) by inhalation for two years. Toxicol Appl Pharmacol 79:179–192, 1985.

Lee, KP, Kelly, DP, Schneider, PW, Trochimowicz, HJ: Inhalation toxicity study on rats exposed to titanium tetrachlorida atmospheric hydrolysis products for two years. Toxicol Appl Pharmacol 83:30–45, 1986.

Lee, KP, Kelly, DP, O'Neal, FO, Slader, JC, Kennedy, GL: Lung response to ultrafine Kevlar aramid synthetic fibrils following 2-year inhalation exposure in rats. Fundam Appl Toxicol 11:1–14, 1988.

Leong, BKJ, MacFarland, HN, Reese, WH: Induction of lung adenomas by chronic inhalation of bis(chloromethyl)ether. Arch Environ Health 22:663–666, 1971.

Leuchtenberger, C, Leuchtenberger, R: Differential response of Snell's and C57 black mice to chronic inhalation of cigarette smoke. Oncology 29:122–138, 1974.

Liang, CK, Quan, NY, Cao, SR, He, XZ, Ma, F: Natural inhalation exposure to coal smoke and wood smoke induces lung cancer in mice and rats. Biomed Environ Sci 1:42–50, 1988.

Little, JB, Kennedy, AR, McGanty, RB: Effect of dose rate on the induction of experimental lung cancer in hamsters by alpha radiation. Radiat Res 103:293–299, 1985.

Lomax, GL, Stott, WT, Johnson, KA, Calhoun, LL, Yano, BL, Quast, JF: The chronic toxicity and oncogenicity of inhaled technical-grade 1,3-dichloropropene in rats and mice. Fundam Appl Toxicol 12:418–431, 1989.

Lundgren, DL, Hahn, FF, Rebar, AH, McClellan, RO: Effects of the single or repeated inhalation exposure of Syrian hamsters to aerosols of [239]PuO_2. Int J Radiat Biol 43:1–18, 1983.

Lundgren, DL, Gillett, NA, Hahn, FF, Griffith, WC, McClellan, RO: Effects of protraction of the alpha dose to the lungs of mice by repeated inhalation exposure to aerosols of $^{239}PuO_2$. Radiat Res 111:201–224, 1987.

Malkinson, AM: Genetic studies on lung tumor susceptibility and histogenesis in mice. Environ Health Perspect 93:149–159, 1991.

Maltoni, C, Lefemine, G: Carcinogenicity bioassays of vinyl chloride: Current results. II. Carcinogenesis associated with vinyl chloride. Ann NY Acad Sci 246:195–218, 1975.

Maltoni, C, Cotti, G, Perino, G: Long-term carcinogenicity bioassays on methylene chloride administered by ingestion to Sprague-Dawley rats and Swiss mice and by inhalation to Sprague-Dawley rats. Ann NY Acad Sci 534:352–366, 1985.

Manam, S, Storer, RD, Prahalada, S, Leander, KR, Kraynak, AR, Hammermeister, CL, Joslyn, DJ, Ledwith, BJ, van Zwieten, MJ, Bradley, MO, Nichols, WW: Activation of the Ki-*ras* gene in spontaneous and chemically induced lung tumors in CD-1 mice. Mol Carcinog 6:68–75, 1992.

Maronpot, RR, Shimkin, MB, Witschi, HP, Smith, LH, Cline, JM: Strain A mouse pulmonary tumor test results for chemicals previously tested in the National Cancer Institute carcinogenicity tests. J Natl Cancer Inst 76:1101–1112, 1986.

Martin, JC, Daniel, H, LeBouffant, L: Short- and long-term experimental study of the toxicity of coal mine dust and some of its constituents. In: Inhaled Particles, Vol. IV. Edited by W Walton, pp. 361–370. Oxford: Pergamon Press, 1977.

Mauderly, JL, Jones, RK, McClellan, RO, Henderson, RF, Griffith, WC: Carcinogenicity of diesel exhaust inhaled chronically by rats. In: Carginogenic and Mutagenic Effects of Diesel Engine Exhaust. Edited by N Ishinishi, A Koizumi, RO McClellan, W Stöber, pp. 397–409. Amsterdam: Elsevier, 1986.

Mauderly, JL, Barr, EB, Eidson, AF, Harkema, JR, Henderson, RF, Pickrell, JA, Wolf, RK: Pneumoconiosis in rats exposed chronically to oil shale dust and diesel exhaust, alone or in combination. Ann Occup Hyg (in press).

McConnell, EE, Wagner, JC, Skidmore, JW, Moore, JA: A comparative study of the fibrogenic and carcinogenic effects of UICC Canadian chrysotile B asbestos and glass microfibre (JM 100). In: Biological Effects of Man-Made Mineral Fibers. Proceedings of a WHO/IARC Conference in Association with JEMRB and TIMA, pp. 234–252. Copenhagen: WHO, 1984.

McConnell, EE, Solleveld, HA, Swenberg, JA, Boorman, GA: Guidelines for combining neoplasms for evaluation of rodent carcinogenesis studies. J Natl Cancer Inst 76:283–289, 1986.

McMartin, DN, Sahota, PS, Gunson, DE, Hsu, HH, Spaet, RH: Neoplasms and related proliferative lesions in control Sprague-Dawley rats from carcinogenicity studies. Historical data and diagnostic considerations. Toxicol Pathol 20:212–225, 1992.

Melnick, RL, Huff, J, Chou, BJ, Miller, RA: Carcinogenicity of 1,3-butadiene in $C_{57}Bl/6 \times C3H$ F_1 mice at low exposure concentrations. Cancer Res 50:6592–6599, 1990.

Mennear, JH, McConnell, EE, Huff, JE, Renne, RA, Giddens, E: Inhalation toxicology and carcinogenesis studies of methylene chloride (dichloromethane) in F344/N rats and B6C3F$_1$ mice. Ann NY Acad Sci. 534:343–351, 1985.

Mestitzova, M, Kossey, P: Experimenteller Beitrag zum Problem der Genese der Lungenkrebses. Neoplasma 8:27–39, 1961.

Mewhinney, JA: Radiation Dose Estimates and Hazard Evaluations for Inhaled Airborne Radionuclides, NUREG/CR-4986. Washington, DC: U.S. Nuclear Regulatory Commission, 1986.

Mohr, U, Takenaka, S, Dungworth, DL: Morphologic effects of inhaled diesel engine exhaust on lungs of rats. In: Carcinogenic and Mutagenic Effects of Diesel Engine Exhaust. Edited by N Ishinishi, A Kiozumi, RO McClellan, W Stöber, pp. 459–470. Amsterdam: Elsevier, 1986.

Morrow, PE: Mechanisms and significance of particle overload. In: Toxic and Carcinogenic Effects of Solid Particles in the Respiratory Tract. Edited by U Mohr, DL Dungworth, JL Mauderly, G Oberdörster, pp. 17–25. Washington, DC: ILSI Press, 1994.

Muhle, H, Takenaka, S, Mohr, U, Dasenbrock, C, Mermelstein, R: Lung tumor induction upon long-term low-level inhalation of crystalline silica. Am J Ind Med 15:343–346, 1989.

NCRP 78: Evaluation of occupational and environmental exposures to radon and radon daughters in the United States—National Council on Radiation Protection and Measurements. National Council on Radiation Protection, Bethesda, MD, 1984.

NTP (National Toxicology Program): Carcinogenesis Bioassay of 1,2-Dibromoethane in F344 Rats and B6CF$_1$ Mice (Inhalation Study). NIH Publ. No. 82–1766, pp. 1–163, 1982.

Nettesheim, P, Hanna, MG, Doherty, DG, Newell, RF, Hellman, A: Effects of chronic exposure to artificial smog and chromium oxide dust on the incident of lung tumors in mice. In: Inhalation Carcinogenesis—A.E.C. Symposium Series, edited by MG Hanna, P Nettesheim, JR Gilbert, pp. 305–319. Springfield, VA: U.S. Department of Commerce, 1970.

Nettesheim, P: Multifactorial respiratory carcinogenesis. In: Experimental Lung Cancer. Edited by E Karbe, JF Park, pp. 157–160. New York: Springer-Verlag, 1974.

Nickell-Brady, C, Hahn, FF, Finch, GL, Belinsky, SA: Analysis of K-*ras*, *p53* and c-raf-1 mutations in beryllium-induced rat lung tumors. Carcinogenesis 15:257–262, 1994.

Nikula, KJ, Snipes, MB, Barr, EB, Griffith, WC, Henderson, RF, Mauderly, JL: Influence of particle-associated organic compounds on the carcinogenicity of diesel exhaust. In: Toxic and Carcinogenic Effects of Solid Particles in the Respiratory Tract. Edited by U Mohr, DL Dungworth, JL Mauderly, G Oberdörster, pp. 565–568. Washington, DC: ILSI Press, 1994.

Nowell, PC: Mechanisms of tumor progression. Cancer Res 46:2203–2207, 1986.

Oberdörster, G: Extrapolation of results from animal inhalation studies with particles to humans? In: Toxic and Carcinogenic Effects of Solid Particles in the Respiratory Tract. Edited by U Mohr, DL Dungworth, JL Mauderly, G. Oberdörster, pp. 335–358. Washington, DC: ILSI Press, 1994.

Ohgaki, H, Furukawa, F, Takahashi, M, Kleihues, P: K-*ras* mutations are frequent in pulmonary squamous cell carcinomas but not in adenocarcinomas of WBN/Kob rats induced by N-nitrosobis(2-oxopropyl)amine. Carcinogenesis 14:1471–1473, 1993.

Ohmori, H, Abe, T, Hirano, H, Murakami, T, Katoh, T, Gotoh, S, Kido, M, Kuroiwa, A, Nomura, T, Higashi, K: Comparison of Ki-*ras* gene mutation among simultaneously occurring multiple urethan-induced lung tumors in individual mice. Carcinogenesis 13:851–855, 1992.

Ohshima, M, Ward, JM, Singh, G, Katyl, SL: Immunocytochemical and morphological evidence for the origin of N-nitrosomethylurea-induced and naturally occurring primary lung tumors in F344/NCr rats. Cancer Res 45:2785–2792, 1985.

Otto, H: Experimentelle Untersuchungen an Mausen mit Passiver Zigarettenrauchbeatsung. Frankf Z Pathol 73:10–23, 1963.

Otto, H, Platz, D: Experimentelle Tumorinduktion mit Urethanaerosolen. Z Krebsforsch 68:284–292, 1966.

Ottolenghi, AD, Haseman, JK, Payne, WW, Falk, HL, MacFarland, HN: Inhalation Studies of nickel sulfide in pulmonary carcinogenesis of rats. J Natl Cancer Inst 54:1165–1172, 1975.

Peck, RM, Eaton, GJ, Peck, EB, Litwin, S: Influence of Sendai virus on carcinogenesis in strain A mice. Lab Anim Science 33:154–156, 1983.

Pepelki, W: Experimental respiratory carcinogenesis in small lab animals. Environ Res 33:144–188, 1984.

Pitot, HC: The national history of neoplastic development: The relation of experimental models to human cancer. Cancer 49:1206–1211, 1982.

Pour, P, Mohr, U, Cardesa, A, Althoff, J, Kmoch, N: Spontaneous tumors and common diseases in two colonies of Syrian hamsters. II. Respiratory tract and digestive system. J Natl Cancer Inst 56:937–948, 1976.

Pour, PM: Spontaneous respiratory tract tumors in Syrian hamsters. In: comparative Respiratory Tract Carcinogenesis. Edited by HM Reznik-Schuller, Vol. 1, pp. 131–170. Boca Raton, FL: CRC Press, 1983.

Prejean, JD, Peckham, JC, Casey, AE, Griswold, DP, Weisburger, EK, Weisburger, JH: Spontaneous tumors in Sprague-Dawley rats and Swiss mice. Cancer Res 33:2768–2773, 1973.

Rabstein, LS, Peters, RL, Spahn, GJ: Spontaneous tumors and pathologic lesions in SWR/J mice. J Natl Cancer Inst 50:751–758, 1973.

Reeves, AL, Deitch, D, Vorwald, AJ: Beryllium carcinogenesis. I. Inhalation exposure of rats to beryllium sulfate aerosol. Cancer Res 27:439–451, 1967.

Reeves, AL, Puro, HE, Smith, RG, Vorwald, AJ: Experimental asbestos carcinogenesis. Environ Res 4:496–511, 1971.

Reuzel, PGJ, Arts, JHE, Lomax, LG, Kuijpers, MHM, Kuper, CF, Gembardt, C, Feron, VJ, Löser, E: Chronic inhalation toxicity and carcinogenicity study of respirable polymeric methylene diphenyl diisocyanate (polymeric MDI) aerosol in rats. Fundam Appl Toxicol 22:195–210, 1994.

Reynolds, SH, Anderson, MW: Activation of proto-oncogenes in human and mouse lung tumors. Environ Health Perspect 93:145–148, 1991.

Reznik, G, Stinson, SF, Ward, JM: Lung tumors induced by chronic inhalation of 1,2-dibromo-3-chloropropane in B6C3F₁ mice. Cancer Lett 10:339–342, 1980.

Reznik, G: Spontaneous primary and secondary lung tumors in the rat. In: Comparative Respiratory Tract Carcinogenesis. Edited by H Reznik-Schuller, pp. 95–116. Boca Raton, FL: CRC Press, 1983.

Reznik-Schuller, H: Sequential morphologic alterations in the bronchial epithelium of Syrian golden hamsters during N-nitrosomorpholine-induced pulmonary tumorigenesis. Am J Pathol 89:59–66, 1977.

Reznik-Schuller, HM, reznik, G: Morphology of spontaneous and induced tumors in the bronchioloalveolar region of F344 rats. Anti-Cancer Res 2:53–58, 1982.

Rowlatt, C, Chesterman, FC, Sheriff, MU: Lifespan, age changes, and tumour incidence in an ageing C₅₇B1 mouse colony. Lab Anim 10:419–442, 1976.

Sato, T, and Kauffman, SL: A scanning electron microscopic study of type II and Clara cell adenoma of the mouse lung. Lab Invest 43:28–36, 1980.

Schepers, GWH, Durkan, TM, Delahant, AB, Creedon, FT: The biological action of inhaled beryllium sulfate. A preliminary chronic toxicity study on rats. AMA Arch Ind Health 15:32–58, 1957.

Schepers, GWH: Neoplasia experimentally induced by beryllium compounds. Prog Exp Tumor Res 2:203–244, 1961.

Schlogel, FA, Bannasch, P: Toxicity and cancerogenic properties of inhaled dimethyl sulfate. Arch Pharmacol 266:441, 1970.

Schreiber, H, Nettesheim, P, Lijinsky, W, Rickter, CB, Walberg, HE: Induction of lung cancer in germ free, specific pathogen free, and infected rats by N-nitrosoheptamethylenamine enhancement by respiratory infections. J Natl Cancer Inst 49:1107–1114, 1972.

Scott, BR, Hahn FF, Newton, GJ, Snipes, MB, Damon, EG, Mauderly, JL, Boecker, BB, Gray, DH: Experimental Studies of the Early Effects of Inhaled Beta-Emitting Radionuclides for Nuclear Accident Risk Assessment, NUREG/CR 5025. Washington, DC: U.S. Government Printing Office, 1987.

Shimkim, MB, Stoner, GD: Lung tumors in mice: Application to carcinogenesis bioassay. In: Advances in Cancer Research. Edited by G Klein, S Weinhouse, A Haddow, pp. 1–48. New York: Academic Press, 1975.

Snyder, CA, Sellakumar, AR, James, DJ, Albert, RE: The carcinogenicity of discontinuous inhaled benzene exposures in CD-1 and C₅₇B1/6 mice. Arch Toxicol 62:331–335, 1988.

Solleveld, HA Haseman, JK, McConnell, EE: Natural history of body weight gain, survival, and neoplasia in the F344 rat. J Natl Cancer Inst 72:929–940, 1984.

Stegelmeier, BL, Gillett, NA, Rebar, AH, Kelly, G: The molecular progression of plutonium-239-induced rat lung carcinogenesis. Ki-*ras* expression and activation. Mol Carcinog 4:43–51, 1991.

Stöber, W: Experimental induction of tumors in hamsters, mice and rats after long-term inhalation of filtered and unfiltered diesel engine exhaust. In: Carcinogenic and Mutagenic Effects of Diesel Engine Exhaust. Edited by N Ishinishi, A Koizumi, RO McClellan, W Stöber, pp. 421–439. Amsterdam: Elsevier, 1986.

Stowers, SJ, Glover, PL, Reynolds, SH, Boone, LR, Maronpot, RR, Anderson, MW: Activation of the K-*ras* protooncogene in lung tumors from rats and mice chronically exposed to tetranitromethane. Cancer Res 47:3212–3219, 1987.

Stula, EF: Naturally occurring pulmonary tumors of epithelial origin in Charles River-CD rats. Pharmacol Environ Pathol 3:3, 1975.

Sunderman, FW: Metastasizing pulmonary tumors in rats induced by the inhalation of nickel carbonyl. In: International Conference on Lung Tumors in Animals—Proc 3rd Quad Conf. Edited by L Severi, pp. 551–564. Perugia, Italy: Perugia Division of Cancer Research, 1966.

Swafford, DS, Nikula, KJ, Belinsky, SA: Genetic alterations in rat lung tumors induced by chronic exposure to diesel exhaust or carbon black. In: Proceedings of 12th Annual Meeting of Mountain West Chapter of the Society of Toxicology, Sept. 29–30, 1994, Albuquerque, NM, 1994.

Takenaka, S, Oldige, H, Konig, H, Hochrainer, D, Oberdörster, G: Carcinogenicity of cadmium chloride aerosols in W rats. J Natl Cancer Inst 70(2):367–373, 1983.

Thyssen, J, Althoff, J, Kimmerle, G, Mohr, U: Inhalation studies with benzo(a)pyrene in Syrian golden hamsters. J Natl Cancer Inst 66:575–577, 1981.

Tye, R, Stemmer, KL: Experimental carcinogenesis of the lung. II. Influence of phenols in the production of carcinoma. J Natl Cancer Inst. 39:175–186, 1967.

Viola, PL, Bigotti, A, Caputo, A: Oncogenic response of rat skin, lungs, and bones to vinyl chloride. Cancer Res 31:516–522, 1971.

Vorwald, AJ, Reeves, AL, Urban, ECJ: Experimental beryllium toxicology. In: Beryllium, Its Industrial Hygiene Aspects. Edited by HE Stokinger, pp. 201–234. New York: Academic Press, 1966.

Wagner, JC, Berry, G, Skidmore, JW, Timbrell, V: The effects of inhalation of asbestos in rats. Br J Cancer 29:252–269, 1974.

Wagner, JC, Berry, G, Cooke, TJ, Hill, RJ, Pooley, FD, Skidmore, JW: Animal experiments with talc. In: Inhaled Particles and Vapors. Edited by WC Watson, pp. 647–654. New York: Pergamon Press, 1977.

Wagner, WD, Groth, DH, Holtz, JL, Madden, GE, Stokinger, HE: Comparative chronic inhalation toxicity of beryllium ores, bertrandite and beryl, with production of pulmonary tumors by beryl. Toxicol Appl Pharmacol 15:10–29, 1969.

Ward, JM, Singh, G, Katyl, SL, Anderson, LM, Kovatch, RM: Immunocytochemical localization of the surfactant apoprotein and Clara cell antigen in chemically induced and naturally occurring pulmonary neoplasms of mice. Am J Pathol 118:493–499, 1985.

Wehner, AP, Zwicker, GM, Cannon, WC, Watson, CR, Carlton, WW: Inhalation of talc baby powder by hamsters. Food and Cosmet Toxicol 15:121–129, 1977.

Wilbourn, L, Haroun, L., Heseltine, E, Kaldor, J, Partensky, C, Vainio, H: Response of experimental animals to human carcinogens: An analysis based upon the IARC monographs programme. Carcinogenesis 7:1853–1863, 1986.

Witschi, H, Haschek, WM: Cells of origin of lung tumors in mice. J Natl Cancer Inst 70:991, 1983.

Witschi, HP: Separation of early diffuse alveolar cell proliferation from enhanced tumor development in mouse lung. Cancer Res 46:2675–2679, 1986.

Wojcinski, ZW, Percy, DH: Sialodacryoadenitis virus-associated lesions in the lower respiratory tract of rats. Vet Pathol 23:278–286, 1986.

You, M, Wang, Y, Lineen, A, Stoner, GD, You, L., Maronpot, RR, Anderson, MW: Activation of protooncogenes in mouse lung tumors. Exp Lung Res 17:389–400, 1991.

You, M, Wang, Y, Stoner, G, You, L, Maronpot, R, Reynolds, SH, Anderson, MW: Parental bias of Ki-*ras* oncogenes detected in lung tumors from mouse hybrids. Proc Natl Acad Sci USA 89:5804–5808, 1992.

Chapter Thirteen

Assessment of Pulmonary Function and the Effects of Inhaled Toxicants

Joe L. Mauderly

INTRODUCTION

Tests of pulmonary function are useful tools for evaluating the toxicology of inhaled materials. The purpose of this chapter is to introduce basic concepts of pulmonary function testing as it is applied in inhalation toxicology. The topics discussed include the functions measured, methods that have been applied to animals, and examples of typical results. This chapter focuses largely on respiratory function, although pulmonary function involves both respiratory and cardiovascular functions. Assessments of breathing patterns, lung volumes, and lung mechanical properties are discussed in greater detail than other assays, because they are the tests most frequently applied in inhalation toxicology.

There is a considerable body of literature describing the methodology, physiological principles, results, and interpretation of pulmonary function tests in man and experimental animals. Previous reviews of methods used for rodents were presented by O'Neil and Raub (1984), Costa (1985), Costa and Tepper (1988), and Costa et al. (1991). This chapter includes an extensive, although not exhaustive, review of the literature on pulmonary function tests of laboratory animals. The citations presented are representative and will point the reader toward the remaining literature. An attempt is made to cite classic references and the work of key contributors to the field, but no attempt is made to cite the work of all current investigators. The examples for which data are given are largely from this Institute. Discussion of equipment is limited to those devices with which the reader must be familiar to understand the information presented in this chapter.

An understanding of the fundamentals of respiratory and cardiovascular physiology is basic to understanding the measurement of pulmonary function. Fortunately, there are a number of good texts in this field, and the neophyte is encouraged to review one or more of these before attempting to distill more technique-specific information from journal articles. Several of the texts cited below are published as inexpensive paperbacks and are commonly available. For an excellent basic tutorial in respiratory physiology, the reader is referred to the recent fifth edition of the classic paperback text by West (1994). Other excellent texts on respiratory physiology and pathophysiology include (in alphabetical order) Burrows et al. (1975a), Cameron (1988), Chang (1988), Des Jardins (1993), Leff and Shumacker (1993), Matthews (1994), Mines (1993), Nunn (1993), Staub (1991a, 1991b), Swanson and Grodins (1990), Taylor (1989), and West (1977). Although many of these texts include information on assessment of respiratory function, the text by Cotes and Leathart (1993) focuses on this subject. Good examples of texts for cardiovascular physiology are Berne and Levy (1991), Kaplinsky (1993), and Smith and Kampine (1990). Additional older texts were listed in the previous edition of this book (Mauderly, 1989).

Individuals interested in pulmonary physiology should be aware of the series of handbooks of physiology published by the American Physiological Society, which includes volumes on respiration (Fenn and Rahn, 1964, 1965; Macklem and Mead, 1986a,b) and circulation (Hamilton, 1962, 1963, 1965). One of these (Macklem and Mead, 1986b) contains a chapter on tests of lung mechanical function (Anthonisen, 1986), which are the majority of pulmonary function tests commonly applied to laboratory animals. The reader should also be aware of the handbook published by the Federation of American Societies for Experimental Biology (Altman and Dittmer, 1983), which presents representative data and references for respiratory and circulatory physiological parameters for man and animals. Although these handbooks are several years old, they still serve as an excellent resource of information and theory that has not changed substantially to date.

ROLE OF PULMONARY FUNCTION TESTS

Pulmonary function tests are a nondestructive means of assessing the functional impact of alterations of lung structure. The tests provide information on the presence (whether or not function is impaired), nature (type of impairment), and extent (magnitude of impairment) of function loss. The tests are used in inhalation toxicology as an indicator of toxic response to characterize the pathogenesis of lung disease and to extrapolate the impact of toxicant-induced lung diseased from laboratory animals to man. Under given circumstances, pulmonary function tests might be the sole indicator of a response to an inhaled material (Silbaugh et al., 1981), or they might be used as one of many assays of response (McClellan, 1986). Sequential changes in pulmonary function measured serially provide information on the stage and progression of the underlying lung disease. Much is known about interrelationships among measured lung function, subjective feelings of respiratory illness, clinical lung disease, and occupational disability in man. Inhalation toxicological studies and studies using specific experimentally induced lung diseases in animals have shown that functional responses of man and animals to different types of lung injury are similar (Costa et al., 1991; Mauderly et al., 1983; Mauderly, 1984, 1988; Tepper et al., 1993). Based on this foundation, pulmonary function tests can be used to estimate the impact in man of responses to inhaled materials studied in laboratory animals (O'Neil and Raub, 1983).

It is important to understand that, although a functional change implies the presence of an underlying structural change, function tests themselves do not describe structural changes. As an example, lung compliance can be reduced by many structural changes, among which are fibrosis, edema, hemorrhage, smooth-muscle constriction, cell proliferation, and reduced lung volume. A finding of reduced compliance would not reveal the nature of the morphological cause. It is also important to understand that, although changes in function result from changes in structure, it is possible to find changes in one without demonstrable changes in the other. Transient airflow limitation from acute bronchoconstriction for example, can have a severe impact on lung performance but is very difficult to demonstrate by histopathology. Conversely, lung tumors characteristically have little effect on lung function unless they occupy a large portion of the lung tissue. Finally, it is important to remember that most pulmonary function tests evaluate the integrated function of the entire organ; focal lesions can exist without measurably affecting total organ function.

It is clear from the above that pulmonary function tests are not a substitute for histopathological evaluations. It should also be clear that the question of the relative sensitivity of pulmonary function tests and histopathology is irrelevant in a general sense. Either assay could be the more sensitive under a given circumstance. Although it has been argued that the ultimate criterion for the seriousness of lung disease is performance rather than pathology, the two approaches are complementary and are used to best advantage in concert. The examination of statistical correlates between pulmonary function data and data for morphological changes derived from semiquantitative scoring of lesions or from morphometry is a particularly powerful approach to understanding structure-function relationships (Mauderly et al., 1980, 1988).

As a general principle, qualitative changes in lung structure by light microscopy and quantitative changes in lung function by appropriate tests are detectable at about the same time in chronic, progressive lung diseases (Mauderly et al., 1988).

Although changes in pulmonary function are not pathognomonic for specific lesions, much about lung structure can be implied from functional changes. The underlying histopathology might be estimated with some accuracy in toxicological studies in which the nature of the toxicant, the general nature of the expected lung injury, and the exposure history are known. It is also possible to make preliminary estimates of the nature and extent of histopathology in human and animal lung diseases in which signs, symptoms, histories, and function data form familiar patterns. In such cases a battery of tests of different facets of lung function are usually applied, and results are expressed as classes of function disorders, e.g., obstructive or restrictive, that are "consistent with" classes of morphological changes, e.g., emphysema, bronchitis, and fibrosis.

Finally, it should be noted that a broad spectrum of pulmonary function tests has been developed, and there are many technical variations on each. In view of this, one must consider that the results are technique-dependent, and care must be exercised in comparing data for the parameter among studies. For example, the CO-diffusing capacity of a subject would differ depending on whether it was measured by the single-breath, steady-state end tidal, or equilibration methods. Even technical differences within the same method, such as differences in breath-holding time and CO concentration in the single-breath method, would produce different results in the same subject. Effort has been made to standardize some measures of human lung function, but there has been little attempt to standardize methods for animals.

A second important consideration is that the testing approach should be matched to the need. Some needs can only be satisfied by a battery of measurements requiring substantial equipment and expertise. In contrast, a simple measurement of the compliance of excised lungs might be a useful adjunct to some studies and might require nothing other than plastic tubing, ruler, syringe, stopcock, and a large-bore needle (Pickrell et al., 1976).

Symbols Used in Pulmonary Physiology

Publications in the field of pulmonary physiology are replete with symbols. Standardized symbols have recently been reaffirmed and adopted by the American Physiological Society (Macklem and Mead, 1986a,b). These symbols are nearly identical to those established in 1950 by the society (Pappenheimer et al., 1950) and recommended by the American College of Chest Physicians and American Thoracic Society in 1975 (Burrows et al., 1975b). These symbols form quite logical combinations when used properly, and they should be used for all publications in the field. Authors should not create new symbols unless no symbol has been previously established. Table 1 contains the most commonly used symbols relevant to the material covered in this chapter. The reader should also be aware of the useful glossary of respiratory and gas exchange terms adopted in 1973 by the International Union of Physiological Sciences (Bartels et al., 1973).

Apparatus Used to Measure Pulmonary Function

An understanding of a few items of apparatus commonly used in pulmonary function tests is requisite to a discussion of testing methods. In addition to items described in this section, respiratory function laboratories typically contain gas analyzers, differential pressure transducers, amplifiers, integrators, strip-chart recorders, and blood gas analyzers. The testing systems assembled in most laboratories incorporate a combination of laboratory-fabricated and commercially available equipment. There is only one current commercial supplier of integrated systems for performing tests in animals (Buxco Electronics, Sharon, CT). Although there are several commercial sources of human testing systems, attempts to adopt these systems for use with animals are rarely successful. The best approach to developing pulmonary function testing capability is to begin by contacting researchers currently working in the field. These individuals are typically quite willing to provide assistance.

Collection of Exhaled Gas   The commonest approach is to have the subject breathe through a nonrebreathing valve, attached by mask or by endotracheal or tracheostomy tube. A nonrebreathing valve consists of two one-way valves joined such that the subject inhales through one and exhales through the other. It is common to use manual or solenoid three-way values outside the inspiratory and expiratory ports of the nonrebreathing valve to initiate and terminate breathing of a test gas and to make timed collections of expirate. Small animals present a problem because of the need to minimize the "dead space" between the valve flaps and the mouth, since the tidal volume of the subject must be increased in compensation. The dead-space volume can be measured and subtracted from the tidal volume to approximate the "normal" tidal volume of the subject (Mauderly et al., 1979). Resistance to airflow can also alter the breathing pattern and is thus another concern. Valves suitable for any species can be constructed in the laboratory (Mauderly and Tesarek, 1975), and some are available commercially. For example, one supplier (Hans Rudolph, Kansas City, MO) manufactures nonrebreathing valves suitable for a wide range of species.

Spirometers are mechanical gas collection devices and are usually either of the Tissot (Fig. 1A) or wedge (Fig. 1B) type. Tissot spirometers consist of a metal chamber (dome or bell) sealed at the bottom by water and counterbalanced so that gas can be exhaled into the chamber with minimal resis-

Table 1   Symbols Commonly Used in Pulmonary Physiology

Breathing pattern	
f	Respiratory frequency (breaths/min)
$\dot{V}_T$	Tidal volume, the volume of each breath
$\dot{V}_E$	Minute volume (expired), the volume exhaled per minute
$\dot{V}_A$	Alveolar ventilation, V_E − dead-volume space ventilation
Lung volumes	
TLC	Total lung capacity, the lung volume at maximal inspiration
VC	Vital capacity, the maximum volume that can be exhaled in a single breath (TLC − RV)
FRC	Functional residual capacity, the lung volume at end of tidal expiration (TLC − IC)
RV	Residual volume, the lung volume after maximal expiration (TLC − VC)
IC	Inspiratory capacity, the maximum volume inhaled from FRC (TLC − FRC)
ERV	Expiratory reserve volume, the maximum volume exhaled from FRC (FRC − RV)
Lung mechanics	
P	Pressure
V	Volume
$\dot{V}$	Flow
Ppl	Pleural pressure, pressure at intrapleural space
Pes	Esophageal pressure, used as analog of Ppl
Paw	Airway pressure, pressure at airway opening (mouth)
Ptp	Transpulmonary pressure, Ppl—Paw or Pes—Paw
Cdyn	Dynamic lung compliance, compliance measured during breathing
Cqs	Quasistatic lung compliance, compliance measured during a slow inspiration or expiration (slope of pressure-volume plot)

tance. Gas law corrections for changes in temperature and water saturation are usually necessary for accurate calculation of the exhaled volume. Other versions of cylindrical spirometers have lightweight plastic chambers, and some have dry membrane or tubular rolling seals (rolling seal spirometer). Dry spirometers have the advantage that corrections for the additional humidification are not required. The rise and fall of the chamber (volume displaced) is measured by sight, using a fixed ruler-type scale, recorded by a pen connected to the chamber, or recorded electronically. Wedge spirometers function similarly but use collection chambers hinged to one side of the spirometer body. An older type of spirometer functioned as a bellows, but this type is now seldom used. Tissot-type spirometers as small as 1 L are available commercially. Smaller spirometers can be fabricated in the laboratory (e.g., Standaert et al., 1985), with special attention to minimizing resistance to airflow and movement of the collection chamber. It has become common to call systems using pneumotachographs (described below) for measuring flow

Table 1 (*Continued*)

Lung mechanics, continued	
Cst	Static lung compliance, compliance measured as Cqs, but with airflow coming to complete stop at discrete measurement steps
R_L	Total pulmonary resistance, Ptp difference divided by flow rate. Often mistakenly called airway resistance, but also contains small contribution from tissue forces
sRaw	Specific airway resistance, airway resistance multiplied by lung volume
sGaw	Specific airway conductance, the reciprocal of sRAW
Gas exchange	
F	Fractional concentration of a gas
P	Partial pressure of gas
A	Alveolar
a	Arterial
I	Inspired
E	Expired
	Examples: FIO_2 = Inspired O_2 concentration
	$PACO_2$ = Alveolar partial pressure of CO_2
	P_B = Barometric pressure
	$PA-aO_2$ = Alveolar-arterial PO_2 difference
Gas measurement conditions	
ATPD	Ambient temperature and pressure, dry
ATPS	Ambient temperature and pressure, saturated
BTPS	Body temperature and pressure, saturated
STPD	Standard temperature (273°K) and pressure (760 mm Hg), dry

and electronic integration of flow with time to calculate volume "spirometers"; however, these are seldom used for gas collection.

Pneumotachograph A penumotachograph consists of a resistance element fixed in a gas flow path, creating a pressure difference between the upstream and downstream sides of the resistance (Fig. 2). The pressure difference is proportional to flow rate as long as flow is laminar; thus, different-size pneumotachographs give linear responses within different ranges of flow. The flow path is typically round in cross section and sufficiently long to promote laminar flow. Sharp changes in direction or flow path diameter close to the pneumotachograph should be avoided as these promote turbulence. Although many types of resistances have been used for pneumotachographs, the resistance elements are most commonly wire screens, bundled capillary tubes, constrictions of the pneumotachograph wall (venturi-type), sections of honeycomb material, a rodlike structure in the center of the flow path, or

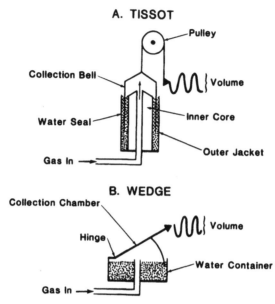

Figure 1 Two types of spirometers are illustrated in cross section. Both types use counterbalanced gas collection chambers sealed by water. The Tissot type (A) uses a cylindrical collection chamber suspended by a pulley, and the wedge type (B) uses a chamber hinged to one side of the water container. The movement of the collection chamber of both types is recorded mechanically or electronically and calibrated in units of volume.

a series of flat plates or vanes. A differential pressure transducer is used with an amplifier to measure the pressure difference, and pressure is calibrated in units of flow. In most applications, flow is integrated with time to produce a signal proportional to volume.

Pneumotachographs vary in their frequency response or the ranges of breathing frequency over which the relationship between actual and measured flows and volumes are identical. Good frequency response, i.e., having the capability for measuring very rapid events, can be important in measurements of small animals, and the study of the frequency responses of pneumotachographs (Finucane et al., 1972; Sinnett et al., 1981) and transducers (Jackson and Vinegar, 1979) was key to the development of sophisticated function tests for rodents. Although most types are satisfactory for measurements in the frequency range of spontaneous breathing, Venturi, bundled capillary, and Fleisch-type pneumotachographs are generally not adequate for evaluating forced expiratory events in animals. Pneumotachographs small enough for measuring tidal breathing of animals ranging from horses to hamsters are available commercially.

Pneumotachographs with very good frequency response characteristics can be easily made in the laboratory using wire screen (stainless-steel wire cloth) fixed across a section of tubing or simple a hole in a plethysmograph

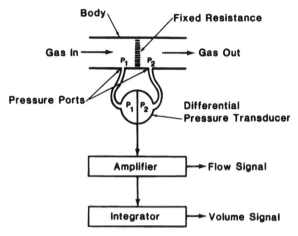

Figure 2 A pneumotachograph is illustrated in cross section. The tubular body has pressure ports on either side of a fixed resistance. Flow of gas through the pneumotachograph causes a difference in pressure between the two ports which is proportional to flow within a certain linear range. This pressure difference is measured by a differential pressure transducer. The amplified signal is calibrated in units of flow and is often integrated electronically with time to produce a volume signal.

wall. Wire screen pneumotachographs can be easily scaled for different species or applications using the following approach which has proved successful for species ranging from small rodents to horses (David Leith, personal communication, Harvard University, at Second Annual Conference of Respiratory Physiology of Small Animals, Davis, CA, May 1979). First, estimate the maximum flow to be measured. If data are lacking, use 10 vital capacities per second for peak forced expiratory flow, and the formula tidal volume × [6.28(respiratory frequency)/60] for peak tidal flow (peak flow of sine wave, Mauderly et al., 1979). Make the cross-sectional area of the pneumotachograph 6.45 cm² (1.0 in²)/L/sec expected peak flow. Add wire cloth in any combination of mesh sizes to achieve a pressure drop at peak flow equal to or below the maximum linear range of the transducer to be used. This approach will optimize frequency response. Experience has shown that such a pneumotachograph can be "desensitized" to the orientation of the screen meshes by separating the layers slightly with rings of cloth tape at their periphery.

Although not pneumotachographs per se, hot wire anemometers have been used to measure respiratory flows and with appropriate calibration can be accurate (Lundsgaard et al., 1979).

Plethysmograph A plethysmograph is a rigid-walled container into which the subject (at least the thorax) is placed and which allows measurement of respiratory flows and volumes without placing a measurement device in the breathing path itself. Plethysmography can be used for all species and

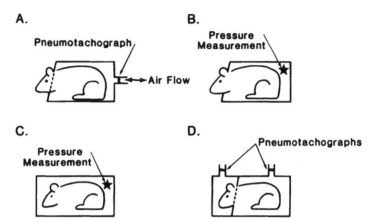

Figure 3   Four types of plethysmographs are illustrated. In the volume displacement (A) and constant volume (B) types, the animal breathes to the outside of the chamber. This is done either with the animal's nose or neck extending through a sealed port (as illustrated) or with the animal completely within the chamber but breathing to the outside through an airway attached to a mask or endotracheal catheter. In the volume displacement type (A), respiration is measured as the flow (pneumotachograph) or volume (spirometer) displaced from the chamber. In the constant-volume type (B), lung volume changes are measured as changes in pressure within the chamber. In the barometric type (C), the animal breathes within the chamber, and respiration is measured by pressure changes due to heating and humidification of inspired gas. There are several variations of plethysmographs in which separate flow or volume signals are obtained for the mouth and the thorax. One type is a "dual plethysmograph" in which the head and body are contained within two separate volume displacement plethysmographs (D). Flows and volumes at the mouth and thorax are measured using pneumotachographs, and the phase difference between the signals is used to calculate specific airway conductance.

is the most common approach to function testing of small animals. There are three fundamental types of plethysmographs (Fig. 3): volume displacement (often called "flow box"), constant volume ("pressure box"), and barometric ("Fenn box," for W. O. Fenn, an early user; Drorbaugh and Fenn, 1985). In the first two types, the subject breathes to the exterior, either via an airway or by having the head extend through a diaphragm in the wall. As the subject inhales, an equivalent volume is expelled from the volume displacement plethysmograph through a pneumotachograph for measuring flow or into a spirometer for measuring volume. The constant-volume plethysmograph is sealed, and an inspiration causes compression of the gas surrounding the body. This increase in pressure is proportional to the volume change; flow can be obtained by differentiation of the volume signal. In the barometric plethysmograph the subject breathes within the closed container, and volumes are detected as small pressure changes due to heating and humidification of the inspired gas. Differential pressure transducers are most often used as the source of signal for all of the above types, although microphones have also been used as transducers for barometric plethysmographs (Ellakkani et al., 1984; Travis et al., 1979).

Although not a technically different type, the "dual" plethysmograph constitutes a special case of either of the first two types above. The head of the subject is contained in one box and the body in another to measure differences in flows, volumes, or phases of the breath at the mouth versus the thorax due to flow restriction. Specific airway resistance, specific airway conductance, and other parameters reflecting airflow restriction can be derived using such a device, while avoiding the need for a transpulmonary pressure signal (Pennock et al., 1979, Silbaugh and Mauderly, 1984). Similar approaches use a combination of a plethysmograph and pneumotachograph (also contained within the plethysmograph) to measure airway resistance or specific airway conductance (Boyd et al., 1980; Agrawal, 1981).

There is considerably literature on the application of plethysmography, calibration methods, and sources of error. DuBois and Van de Woestijne (1969) and Leith and Mead (1974) are useful references for the volume displacement and constant volume types. Volume displacement plethysmographs are generally the most trouble-free and are the most useful for measurements of small animals if rapid events are to be measured (Sinnett et al., 1981). This type is least sensitive to leaks, differences in the ratio of animal to plethysmograph size, and frequency dependence (assuming an appropriate pneumotachograph-transducer combination is used). This author's bias is that this type would be the first choice unless conditions dictate otherwise.

Constant volume plethysmographs are the most sensitive for measuring small volume changes but are more sensitive to leaks, variations in body size, and frequency dependence. With care these problems can be overcome, and this type has been used successfully for accurate measurements of small animals (Costa et al., 1986; Vinegar et al., 1979). Particular care must be taken to avoid frequency dependence caused by artifacts from the heat of compression of gas (adiabatic vs. isothermal conditions) (Bargeton and Barres, 1969; Sinnett et al., 1981). Briefly, compression of gas surrounding the body during inspiration causes slight heating, and a resulting "extra" pressure rise that decays as the heat is transferred through the plethysmograph wall. Under adiabatic conditions, the measured volume change includes the entire heat artifact (no dissipation of heat at the time of measurement), whereas under isothermal conditions, heat is instantaneously dissipated, and there is no artifact. The plethysmograph can be calibrated assuming either condition, but errors occur when actual conditions are intermediate between the two extremes. Frequency dependence occurs because there is less time for heat to dissipate during rapid events. Addition of copper sponge to the chamber as a heat sink can extend the frequency range of "isothermal" operation (Leith, 1976; Vinegar et al., 1979; Watson et al., 1986).

In some situations, it is important to consider the frequency response characteristics of plethysmographs. While frequency response is seldom a problem when measuring events during tidal breathing, it can be very important when measuring more rapid events, such as forced exhalation or the fine details of flow during rapid breathing. Frequency response requirements

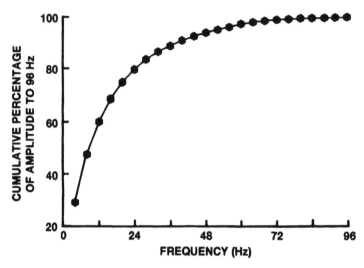

Figure 4   The mean frequency spectrum for forced exhalations of 22 male and 22 female 14 week-old F344 rats is shown. The flow signal was digitized, LaPlace transform was used to calculate amplitude at 4 Hz intervals from 4 to 96 Hz, and the amplitudes were plotted as cumulative percentages of the total amplitude of 96 Hz (Mauderly and Seiler, 1985).

are generally inversely related to the size of the species being measured, with the most stringent requirements associated with obtaining good data from rapid events in small species. Fig. 4 illustrates the mean frequency spectrum of the forced exhalations of adult F344 rats, measured using a well-tuned volume displacement plethysmograph (Mauderly and Seiler, 1985). Of the total amplitude, 90% was contributed by frequencies below approximately 40 Hz, and most of the meaningful detail of the flow-volume, or flow-time curve was contained in frequencies below 10 Hz. Disease-related changes in forced expiratory parameters were found to have little effect on the frequency response requirements of the measurement equipment (Mauderly et al., 1984). The requirements for larger species are less stringent. For humans, the highest frequency having an amplitude of 5% of the fundamental was found by Lemen et al. (1982) to be 5 Hz for normal lungs and 6.4 Hz for diseased lungs. Volume displacement plethysmographs with quite adequate frequency response characteristics can be constructed using simple wire screen pneumotachographs (Harkema et al., 1982).

Barometric plethysmographs allow measurements of the breathing of subjects contained within the chamber without anesthesia, restraint, or instrumentation (Lucey et al., 1982). This type, however, requires rigid calibration and operating conditions (Epstein and Epstein, 1978), is sensitive to movements of the subject and is generally less accurate than other types. These devices have been used to monitor respiration of small animals, but they are seldom used for pulmonary function tests. An exception is the use of barometric plethysmographs in conjunction with pneumotachographs to derive various indices of airflow restriction (Agrawal, 1981; Boyd et al., 1980; Ellakkani et al., 1984).

The use of nose-only exposure tubes as plethysmographs for measuring the volume of material inhaled by animals during inhalation exposures is a specialized application of plethysmography in inhalation toxicology. This useful procedure also permits the calculation of the fraction of the inhaled material that is deposited or absorbed and is integral to many pharmacokinetic studies (e.g., Medinsky et al., 1985). Although constant-volume plethysmographs could be used, volume displacement types using commercially available pneumotachographs are usually practical (Coggins et al., 1981; Mauderly, 1986). It is very important to remember that the volume measured is the volume of body expansion and equals the inhaled volume expanded because of warming and humidification. Unless the inhaled material is also at BTPS conditions, a conversion is required to calculate the actual volume of test atmosphere inhaled. The volume measured by plethysmograph is typically about 10% greater than the volume actually entering the nose. Simple gas law calculations are applied to make the correction.

Esophageal and Pleural Catheters   Transpulmonary pressure, the pressure difference between the pleural surface and the mouth, is the driving force for ventilating the lung. Therefore, this pressure is the relevant force term for many tests of lung mechanics. It is important to remember that the appropriate transpulmonary reference pressure is mouth (or airway opening) pressure. If the airway is connected to other equipment, such as a pneumotachograph or inhalation exposure chamber, referencing to room pressure will cause an artifact by including the pressure drop due to flow resistance of the external equipment.

Pleural pressure can be measured directly by inserting a catheter through the chest wall (Fig. 5). This technique has been widely used to measure lung mechanics in spontaneously breathing guinea pigs (Amdur and Mead, 1958) and is useful for conscious animals, such as guinea pigs, which tolerate esophageal catheters poorly. Its disadvantages are the requirement for surgical placement of the catheter under anesthesia, the frequency (even in experienced hands) of puncturing the lung surface, the limited time that the catheters can be maintained, and the pleural adhesions that often eventually form if the animal is maintained for weeks after the test. Functional pleural catheters have been maintained for up to 6 weeks at this Institute by filing with saline containing 100 IU heparin per milliliter, flushing every 3–4 days, and bandaging the chest.Pleural pressure is most commonly measured indirectly via the esophagus, using either an open, liquid-filled catheter or a balloon-tipped catheter (Fig. 5). This approach is suitable for anesthetized animals, is readily accepted by conscious dogs (Mauderly, 1974a), and has been used in conscious rats (Dorato et al., 1983) and guinea pigs (Skornik et al., 1981)

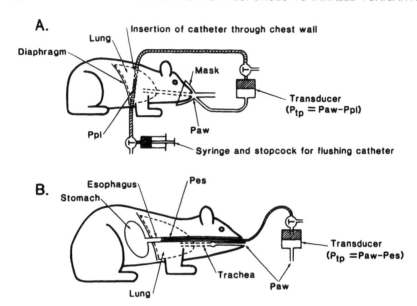

Figure 5 The most common methods for measuring transpulmonary pressure of animals are illustrated. Transpulmonary pressure (Ptp) is the difference between pressure at the pleural surface (Ppl) and the airway opening (Paw). Pressure at the pleural surface can be measured directly by a catheter surgically inserted through the chest wall (A). The catheter is filled with liquid and has small holes in the section within the pleural space. If an external airway is attached to the animal (as shown), the reference pressure (Paw) should be taken at the mouth opening to avoid including the pressure difference due to equipment flow resistance. A more common approach is to use esophageal pressure (Pes) as an analog of Ppl (B). Pes is measured with either an air-filled balloon catheter or a liquid-filled catheter (shown). If there is no external airway attached to the mouth, Paw is simply the ambient pressure. For liquid-filled pleural or esophageal catheters, it is important that the diaphragm of the transducer be at the same vertical height as the point of measurement to avoid hydrostatic pressure artifacts.

with close head restraint. There is considerable literature on the factors affecting measurement of pleural pressure and the validity of using esophageal pressure as an analog, the consensus being that the technique is reasonably valid (Beardsmore et al., 1980; Baydur et al., 1982). Esophageal pressures have been found to be similar to pleural pressure in man (Cherniack et al., 1955), dogs (Gillespie et al., 1973; D'Angelo, 1976), rabbits (Davidson et al., 1966), hamsters (Strope et al., 1980), and rats (Palecek, 1969).

Esophageal balloon catheters are made from thin latex balloons, purchased commercially or fabricated in the laboratory and fastened to a catheter with a rounded tip and multiple perforations in the catheter portion within the balloon. Laboratory-fabricated balloons are often made by dipping rod molds repeatedly into liquid latex, but they have also been made from condom rubber. It is important that the balloon wall be very compliant to avoid artifacts in the measured pressure (Lemen et al., 1974). For this reason, ordinary party balloons are usually unsuitable, and balloons having small enough diameters for rodents have inadequate compliance characteristics. The pressure-volume characteristics of a representative balloon catheter should be measured by collapsing the balloon, then injecting air while measuring pressure (a water manometer is suitable). Good balloons will have a "flat," highly compliant volume region before pressure rises with increasing volume, and they will remain in the highly compliant portion of their pressure-volume curve during respiration (Mead and Whittenberger, 1953; Maxted et al., 1977).

Balloon catheters are suitable for animals approximately the size of cats and larger. For use, the catheter is inserted to midthorax, the balloon is collapsed (readily felt if a glass syringe is used) and inflated to a volume representing the midpoint of its "flat" region, and then the catheter is attached to a transducer and adjusted to the depth producing the maximum tidal variation in pressure.

Liquid-filled catheters (without balloons) should have side holes near the tip to minimize the likelihood of plugging, and they can have either an open or closed tip. Stiffer catheters provide flatter response at higher frequencies (Maxted et al., 1977), but this is seldom a limitation. The end portion of an infant feeding tube makes a catheter satisfactory for rodents. The catheter-transducer system is filled with water and calibrated in the position in which it is to be used, i.e., with the tip in the same vertical relationship to the transducer as during animal measurements. The transducer should be adjusted so that its diaphragm is at the same vertical height as the catheter tip (Fig. 5). In another approach, a water reservoir flask is attached to a side arm of the catheter-transducer connection, and the water level is adjusted to the level of the esophagus (Koo et al., 1976). The catheter is filled with water before insertion and placed in midthorax, a fraction of a milliliter of water is injected to clear the tip, and then the depth is adjusted for maximum signal. An alternative method is to inset the catheter to the depth of the stomach (identifiable by the pressure trace) and withdraw it to the proper position. No attempt should be made to adjust the depth of either balloon or liquid-filled catheters to eliminate cardiogenic "artifacts" on the signal if the adjustment reduces the primary signal; electronic filtration can usually be used to remove the cardiac signal if necessary.

Measurement of Breathing during Inhalation Exposures   This is a specialized requirement encountered often in the field of inhalation toxicology. The most frequent need is to quantitate the amount of material inhaled, but these measurements are also used to detect changes in breathing pattern as responses, to document the normalcy of breathing during exposure, or to adjust artificial ventilation to a desired level. As is true for the assays of function described below, there are many approaches which can be used to obtain data. A range of approaches that was previously described in detail

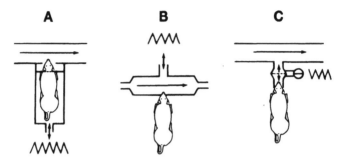

Figure 6 Three methods for measuring respiration during head-only or nose-only exposures with low-volume systems are shown. In A, an exposure tube is used as a plethysmograph. In B, a breathing signal is measured at a point outside the breathing path. In C, a pneumotachograph is interposed between the subject and the inhalant source.

(Mauderly, 1990) will be presented briefly here. Selection will depend on the purpose of the measurement, the type of exposure, and the equipment at hand.

Several approaches are illustrated in Figs. 6, 7, and 8 in which airflows are indicated by arrows and the point of measurement is shown by the saw-tooth figure indicating tidal breathing. Three approaches to measuring respiration during head-only or nose-only exposures are illustrated in Fig. 6. In Figure 6A, a plethysmograph is used as a head-only or nose-only exposure tube, by placing a seal at the nose or neck. Either a volume displacement or a constant volume plethysmograph can be used. With either type, the most frequent problem is the adequacy of the nose or neck seal. In Fig. 6B, a pressure or flow measurement device is attached to a low-volume inhalant flow path through which inhalant is metered through high-resistance inlet and outlet pathways at a constant flow rate. The animal's breathing causes pressure changes or volume displacement through the measurement port. In Fig. 6C, a pneumotachograph is interposed between the animal and the inhalant flow path.

Fig. 7 illustrates two approaches to measuring the respiration of animals breathing from an inhalant reservoir bag to an expirate collection bag. The total volume inhaled or exhaled can be determined by measuring the volumes contained in the collection bags. Minute volume is obtained by dividing collected volume by time and only f needs to be recorded to calculate V_T. In Fig. 7A, a thermistor or sensitive pressure transducer can be placed at the nose, or an impedance pneumograph or strain gauge can be placed across the thorax to obtain a breathing signal of uncalibrated amplitude. In Fig. 7B, both bags are placed in a box and a pneumotachograph or spirometer can be attached to the box to measure frequency and/or V_T.

Fig. 8 illustrates three approaches to measuring respiration during immersion exposures in the whole-body chambers. This is the most challenging exposure mode for breathing measurements, and no approached provides

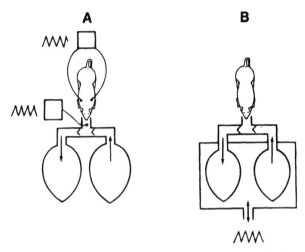

Figure 7   Two methods are shown for measurement of respiration during exposures using a bag-to-bag system. In A, the volume breathed is measured as the change in bag volume and an uncalibrated breathing signal is obtained at the airway or body to measure respiratory frequency. In B, both reservoir bags are contained in a box which is used as a plethysmograph to obtain breathing signals.

very accurate data for volume. In Fig. 8A, a low-volume exposure chamber forms a barometric plethysmograph, if the exposure atmosphere inlet and outlet are high-resistance pathways through which the inhalant is metered at a constant flow rate. Although this system will provide a breathing signal yielding good values for f, it is unlikely that V_T can be measured with acceptable accuracy. In Fig. 8B, telemetry is used to obtain a breathing signal. It is likely that the receiving antenna will have to be placed inside the chamber,

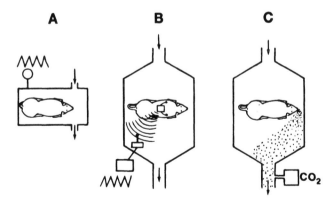

Figure 8   Three methods for measuring breathing in whole-body exposure systems are shown. In A, a barometric plethysmograph is used as a low-volume exposure chamber. In B, a breathing signal is obtained by telemetry. In C, the respired volume per unit of time is estimated from CO_2 output and an assumed expired CO_2 concentration.

and it is virtually impossible to derive accurate values for V_T. In Figure 8C, the concentration of CO_2 is measured at the chamber outlet and if chamber flow rate is known, the total CO_2 output of the animal (or the average for many animals) can be calculated. The minute volume of the animal(s) can be estimated using an assumed value for the mean CO_2 concentration in expired gas. If the approaches in Figures 8B and 8C are combined, the V_T can be estimated, but good accuracy should not be expected.

VENTILATION

The goal of respiration at the pulmonary level is to bring air and blood (in approximately equal volumes per unit of time) continuously into apposition across the thin alveolar-capillary membrane so that gas exchange can take place by simple diffusion. Ventilation involves movement of air into and out of the lung in a reciprocal (tidal) manner via the conducting airways in response to mouth-alveolar pressure differences (Ptp) caused by movements of the chest wall and diaphragm. Gas moves through the larger airways by bulk flow. With increasing cross-sectional areas in the deeper lung, bulk flow diminishes, and gas movement occurs by molecular diffusion. Physiological parameters of importance in ventilation include the rate at which air is moved, the portion reaching the gas exchange membrane, the work required to move the air, and the uniformity with which air is distributed to the gas exchange membrane. The corresponding ventilatory parameters of major concern in pulmonary function testing are the breathing pattern, lung volumes, lung mechanics, and gas distribution.

Breathing Patterns

The most frequently used indices of breathing pattern are respiratory frequency (f), tidal volume (V_T), and minute volume ($\dot{V}_E$), although inspiratory and expiratory flow rates and times are sometimes measured. The f is usually measured either by plethysmography or by a pneumotachograph at the mouth, but it can be counted manually. Tidal volume is usually measured using a plethysmograph or pneumotachograph, but it can also be measured by dividing the total volume expired during a time collection by the f. The respiration of rats has also been measured as pressure changes in a continuous flow of air past the nose (Leong et al., 1964). The $\dot{V}_E$ is calculated as the product of f and V_T or measured by a timed collection of expirate. By convention $\dot{V}_E$ is expressed in units BTPS.

The breathing pattern can have considerable interpretive value, but it must be remembered that it can be affected markedly by anesthetics and other drugs, excitement, activity, temperature, and the addition of deadspace volume or flow restriction to the airway. Conscious animals should be evaluated under uniform conditions of either rest or exercise. Anesthesia al-

Table 2 Effects of Radiation Pneumonitis-Fibrosis on Respiratory Function of Conscious Dogs after Inhaling ^{144}Ce in Insoluble Particles[a]

	Units	Preexposure baseline[b]	Moderate disease	Severe disease
Respiratory frequency	breaths/min	20	41	54
Tidal volume	ml	163	105	101
Minute volume	l/min	3.3	4.2	4.5
Dynamic lung compliance[c]	ml/cm H$_2$O	58	34	17
N$_2$ dilution constant[c]	B$_{1/2}$/FRC/V$_T$	2.1	2.6	3.0
CO-diffusing capacity	ml/min/mm Hg	8.7	2.8	2.2
Alveolar PO$_2$	mm Hg	80	82	88
Arterial PO$_2$	mm Hg	77	62	48
Alveolar-arterial PO$_2$	mm Hg	3	20	40
O$_2$ saturation of hemoglobin	%	92	84	64
Systemic arterial mean pressure	mm Hg	111	104	113
Pulmonary arterial mean pressure	mm Hg	17	30	60
Pulmonary wedge mean pressure	mm Hg	2	2	4
Pulmonary vascular resistance	dynes/sec/cm^5	750	1770	3480

[a]From the study reported by Mauderly et al. (1980). Beagle dogs were exposed once by inhalation to a mean of 47 μCi/kg of ^{144}Ce in a fused aluminosilicate particle matrix and measured serially as they developed radiation pneumonitis-fibrosis. The baseline values represent the means of 12 dogs, and the values for moderate and severe disease represent means of four dogs each (included within the 12 at baseline—the remaining four dogs were controls). Measurements were performed on conscious, spontaneously breathing dogs at an altitude of 1728 meters. Serial sacrifices revealed a progressive, patchy inflammation and fibrosis in the lungs and vascular lesions consisting of thickened vessel walls and patchy obliteration of capillaries.

[b]The data for vascular pressures and resistances were from time-matched controls rather than individual baselines.

[c]N$_2$ dilution constant is an indicator of intrapulmonary gas distribution uniformity (Mauderly, 1977). An increasing value indicates less uniform distribution.

ters breathing, but the breathing pattern of animals under consistent anesthetic conditions can still have interpretive value.

Minute volume is normally adjusted, within the limitations of the subject's ability to compensate, to produce an alveolar ventilation adequate for exchanging gas to meet the current metabolic need. Alveolar ventilation ($\dot{V}_A$) is the volume of gas reaching the alveoli, and during a single breath is the V_T minus the volume of the non-gas-exchanging dead space of the airway (both the anatomic and equipment deadspaces). The $\dot{V}_A$ is the key ventilatory parameter for gas exchange. In inhalation toxicology, it should be remembered that only inhaled material contained in the $\dot{V}_A$ can reach the deep lung. The combination of f and V_T used to produce a given $\dot{V}_A$ is variable and is usually adjusted to incur the least amount of respiratory work (Rohrer, 1925). With a stiff lung (restrictive lung disease), it is more efficient to achieve the required $\dot{V}_E$ by reducing V_T and increasing f; however, this requires an increase in $\dot{V}_E$ to compensate for the greater portion of V_T "wasted" ventilating deadspace. An example of this effect in dogs with radiation-induced pneumonitis and pulmonary fibrosis is given in Table 2. Conversely, with airflow restriction (obstructive lung disease) it is more efficient to take larger, slower breaths, thereby reducing flow rate and resistive work at the expense of additional stretching (or elastic) work.

The breathing patterns, particularly f and V_T have sometimes been used in toxicology as the sole indicators of acute responses to inhaled materials or of chronic lung disease. The largest data base on acute responses has been produced by Alarie and associates, who have used different plethysmographic methods to measure the f of mice during inhalation exposure as an indicator of the irritant potentials of a spectrum of materials (Alarie, 1973, 1981) and the f and V_T of guinea pigs to reflect acute or delayed airway constrictive events (Ellakkani et al., 1984; Karol et al., 1981; Matijak-Schaper et al., 1983; Schaper and Alarie, 1985). An example of chronic responses is the work of Travis and associates, who have used plethysmography to follow the course of radiation-induced lung disease in mice by measuring the increase in f and decrease in V_T (Travis et al., 1979, 1983). Such measurements are very useful toxicological tools; however, taken alone, they give very limited information about the functional status of the lung. The functional status of the lung cannot be described from these data unless the nature of the underlying physiological events or lung disease is already known or unless confirmatory studies using other tests are performed.

Lung Volumes

The classic physiological subdivisions of lung volume are illustrated in Fig. 9. Total lung capacity (TLC) is the total gas volume of the lung at maximum inspiration. This point is reached by voluntary effort in man and by inflation of animal lungs to a preselected Ptp, either by positive airway pressure (Harkema et al., 1982) or by negative pressure surrounding the body (Moorman

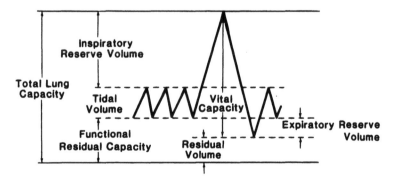

Figure 9 The classic physiological subdivisions of lung volume are illustrated as traces on the scale of maximum lung volume (total lung capacity). The vital capacity represents the range of lung volume within which the subject can voluntarily breathe. Functional residual capacity is the end-expiratory lung volume during tidal breathing. Residual volume is the volume that cannot be voluntarily expelled from the lung.

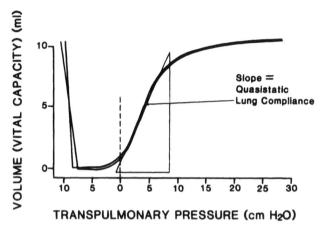

Figure 10 Duplicate tracings of the quasistatic pressure-volume curve of a rate lung measured *in vivo* are shown. The lung of the apneic rat was inflated to a transpulmonary pressure of 30 cm H_2O (defined as total lung capacity), then deflated slowly until flow stopped (residual volume). The quasistatic lung compliance is the slope of the curve of and is expressed as ml/cm H_2O. The average slope over the midvolume range is illustrated, but slopes of specific segments (or chords) of the curve are sometimes measured.

et al., 1975). Spontaneously breathing anesthetized animals are first hyperinflated to trigger the inflation reflex (Hering and Breuer, 1968) and cause brief apnea, or they are paralyzed by drugs and ventilated between measurements by respirators. The Ptp used to define the TLC of animals varies, but it is most commonly 25–30 cm H_2O and is based on the fact that the pressure-volume curve of the lung is nearly flattened within this pressure range (Fig. 10). Vital capacity (VC) is the maximum volume that can be exhaled starting at TLC. The VC of animals is often measured during a "quasistatic" (slow) deflation caused by applying a negative pressure to the airway after inflation to TLC while limiting expiratory flow rate by a valve in the external flow path. An expiratory flow rate of approximately one-third VC/second is a suitable balance between minimizing the time of apnea and preventing artifacts from flow-resistive forces (David E. Leith, personal communication, Harvard University, Second Annual Conference of Respiratory Physiology of Small Animals, Davis, CA, May 1979). The volume of gas in the lung at the end of this maximal exhalation is the residual volume (RV). The point of RV is easily recognized as both a cessation of flow and an abrupt deflection of Ptp.

The volume point at which V_T is placed within the TLC is called the functional residual capacity (FRC) and is described as the lung volume at the end of tidal exhalation. This volume is highly variable and in most subjects is based on functional (rather than on anatomical) factors. In subjects such as man and dog, which have stiff chest walls, FRC is typically the lung volume at which the counteracting forces of lung elastic recoil and outward expansion of the chest wall are in balance. The FRC of rodents with highly compliant chest walls is maintained above the relaxed volume during consciousness and is lower during anesthesia (Gillespie, 1983). Three primary methods are used to measure FRC of animals: the barometric method, open-circuit N_2 washout, and gas dilution. Descriptions of these methods and several references for each are given in the book by Clausen (1982).

The barometric method (DuBois et al., 1956) is most frequently used for anesthetized rodents being tested by plethysmography (Harkema et al., 1982 Costa et al., 1983) and is the most rapid method. The airway is blocked, either at end expiration or during apnea (apneic lung volume is considered equal to FRC), and the changes in airway pressure and lung volume are measured as the subject makes breathing efforts against the blocked airway. The FRC is calculated from the Boyle's law relationship of volume to pressure (sometimes called the "Boyle's law" method). The measured volume is sometimes called "thoracic gas volume," but it also includes the volume of the airways. This is the method most likely to measure all of the FRC in the presence of poor gas mixing, and it is sometimes used in concert with one of the other two methods in an attempt to measure the volume of poorly ventilated regions.

The open-circuit N_2 washout method (Darling et al., 1940) is suitable for spontaneously breathing subjects or subjects ventilated by a respirator. This technique has been used at this Institute for measuring the FRC of dogs (Mauderly, 1974a) and ponies (Mauderly, 1974b). The subject typically breathes air via a nonrebreathing valve having three-way valves at the inspiratory and expiratory ports. At the end of an expiration, the valves are turned so that the subject inhales pure O_2 and exhales into a gas collection device (usually a spirometer). Subsequent breaths sequentially dilute the N_2 originally in the lung and "wash it out" into the spirometer. If N_2 concentration is monitored at the mouth, a nitrogen washout curve with progressively decreasing end-tidal N_2 peaks can be recorded to evaluate gas mixing, as described in a later section. The washout continues until a low end-tidal N_2 is reached (typically 1–2%), but is not continued to zero N_2 because of the low concentration that persists for a long time due to prolonged washout of N_2 from blood. The volume of N_2, and thus the volume of air (FRC), originally contained in the lung and airways is calculated from the collected gas volume and N_2 concentration. This method might underestimate FRC in the presence of lung regions with very poor (or no) gas mixing.

The gas dilution method (McMichael, 1939) is based on dilution of an inert gas into the volume to be measured, can be used to measure TLC and RV as well as FRC, and is satisfactory for small animals (Koen et al., 1977; Takezawa et al., 1980; Costa et al., 1983). The subject's lung is placed at the volume to be measured, and a volume of test gas containing neon or helium is repeatedly injected and withdrawn to equilibrate the test gas with the unknown volume. The unknown volume is calculated from the original (known) volume and inert gas concentration of the test gas. Alternatively, the subject is switched at FRC to rebreathing from a bag containing the inert gas (Koen et al., 1977). As with the above approach, the FRC is calculated from a known volume and measured change in gas concentration. This technique can also underestimate the measured volume in the presence of poor gas mixing but is less likely to do so than the preceding method because of the large, slow lung movements typically induced during equilibration.

The RV is either calculated by subtraction of measured volumes or measured directly by gas dilution. For example, RV = FRC − ERV, or TLC − VC. The RV can be measured by gas dilution, beginning the dilution at RV, FRC, or TLC, depending on which other lung volumes are known.

Examples of changes in lung volumes of rats due to restrictive and obstructive lung diseases are presented in Tables 3 and 4, respectively. In restrictive disease, the lung is smaller and usually stiffer. In the example in Table 3, this effect resulted from interstitial fibrosis caused by intratracheal instillation of bleomycin followed by O_2 treatment. The TLC and VC of the fibrotic lungs were smaller than those of normal lungs, and the VC/TLC ratio was reduced. As will be noted later, lung compliance was also reduced, but there is no airflow obstruction. An example of similar lung volume changes in restrictive disease of rats caused by inhaled diesel exhaust can be found in Mauderly et al. (1988). In obstructive disease, the major impairment is a restriction of airflow; lung volumes may be altered, but they are not necessarily. In the example in Table 4, increased lung volumes and lung compliance accompanied a restriction of airflow resulting from emphysema caused by intratracheally instilled elastase. All of the lung volumes measured (TLC, VC, FRC, and RV) were increased, as were FRC/TLC and RV/TLC. It is useful to note that, although TLC and VC were affected differently by the restrictive and obstructive diseases, FRC, FRC/TLC, RV, and RV/TLC were increased, and VC/TLC was reduced by both. This indicates that both diseases reduced the volume range of voluntary breathing. Examples of similar lung volume changes in obstructive diseases of rats caused by inhaled O_3 and acrolein can be found in Costa et al. (1983) and Costa et al. (1986), respectively.

In interpreting changes in lung volumes of animals, it is very important to recognize the physical forces determining the values derived. A good example of this is the RV of rodents, which is set primarily by collapse of terminal airways by negative airway pressure during slow deflation. Tissue changes such as emphysema, which act to reduce the radial elastic tension on terminal airways, allow the airways to collapse earlier during deflation and raise the value of RV. This phenomenon is demonstrated by the data for emphysematous rats in Table 4. On the other hand, tissue changes acting to increase radial tension or stiffen the airway delay closure and decrease the RV. An example is the recent study of Harkema and Mauderly (1994) in which rats were tested after 20 months of chronic exposure to 0.12–1.0 ppm ozone. The only significant change in respiratory function was an increase in RV, which

Table 3 Effect of Pulmonary Fibrosis on Respiratory Function of Anesthetized Rats (Restrictive Lung Disease)[a]

	Units	Control (n = 12) X̄	SE	Fibrosis (n = 6) X̄	SE	p < 0.05[b]
Body weight	g	246	21	216	20	
Total lung capacity (TLC)	ml	14.2	0.8	10.8	0.9	*
Vital capacity (VC)	ml	11.6	0.7	8.1	0.8	*
VC/TLC	%	82	1	74	1	*
Functional residual capacity (FRC)	ml	3.8	0.2	4.3	0.2	
FRC/TLC	%	27	1	40	3	*
Residual volume (RV)	ml	2.5	0.2	2.8	0.2	
RV/TLC	%	18	1	26	1	*
Dynamic lung compliance (Cdyn)	ml/cm H_2O	0.39	0.03	0.30	0.05	
Cdyn/FRC	ml/cmH_2O/ml	0.106	0.025	0.069	0.011	
Quasistatic chord compliance	ml/cm H_2O	0.78	0.05	0.43	0.07	*
Slope III of single-breath N_2 washout	% N_2/ml	0.72	0.08	1.56	0.40	*
CO diffusing capacity (DLCO)	ml/min/mm Hg	0.22	0.01	0.13	0.02	*
DLCO/body weight	DLCO/kg	0.92	0.06	0.61	0.08	*
DLCO/alveolar volume	DLCO/ml	0.018	0.001	0.013	0.001	*
Forced vital capacity (FVC)	ml	11.3	0.7	8.1	0.8	*
Forced expired volume in 0.1 sec	% FVC	52	2	66	3	*
Peak expiratory flow rate (PEFR)	ml/sec	91	7	89	8	
PEFR/FVC	FVC/sec	8.1	0.3	11.4	1.0	*
Mean midexpiratory flow (MMEF)	ml/sec	52	6	50	5	
MMEF/FVC	FVC/sec	4.6	0.3	6.4	0.6	*
Expiratory flow at 50% FVC (EF_{50})	ml/sec	55	6	54	5	
EF_{50}/FVC	FVC/sec	4.8	0.4	6.9	0.6	*
Expiratory flow at 25% FVC (EF_{25})	ml/sec	33	3	27	3	
EFF_{25}/FVC	FVC/sec	2.9	0.2	3.4	0.3	
Expiratory flow at 10% FVC (EF_{10})	ml/sec	14	2	10	2	
EF_{10}/FVC	FVC/sec	1.3	0.1	1.2	0.1	

[a]Equal numbers of 12-week-old male and female F344 rats were instilled intratracheally with 1.6 IU of bleomycin/kg in 2 ml saline/kg and then exposed continuously to 80% O_2 for 72 hours. Control rats were instilled with saline only and not exposed to O_2. Tests were performed at 90 days after instillation by reported methods (Harkema et al., 1982). Treated rats sacrificed after function tests had patchy interstitial fibrosis, particularly in subpleural and central regions, and increased numbers of interstitial and alveolar inflammatory cells. Morphometry demonstrated that 25% of the lung parenchymal tissue of treated rats stained for collagen compared to 6% in controls. Measurements were performed at an altitude of 1728 m.

[b]Differences between group means tested by unpaired two-tailed t-test.

Table 4 Effect of Pulmonary Emphysema on Respiratory Function of Anesthetized Rats (Obstructive Lung Disease)[a]

	Units	Control (n = 12) X̄	Control (n = 12) SE	Emphysema (n = 16) X̄	Emphysema (n = 16) SE	$p < 0.05$[b]
Body weight	g	385	7	374	7	
Total lung capacity (TLC)	ml	15.8	0.4	19.6	0.3	*
Vital capacity (VC)	ml	14.1	0.3	17.0	0.3	*
VC/TLC	%	90	1	87	1	*
Functional residual capacity (FRC)	ml	2.9	0.1	4.7	0.2	*
FRC/TLC	%	19	1	24	1	*
Residual volume (RV)	ml	1.7	0.1	2.6	0.2	*
RV/TLC	%	10	1	13	1	*
Dynamic lung compliance (Cdyn)	ml/cm H_2O	0.50	0.03	0.88	0.06	*
Cdyn/FRC	ml/cmH_2O/ml	0.172	0.011	0.188	0.010	
Quasistatic chord compliance	ml/cm H_2O	0.99	0.03	1.13	0.03	*
Slope III of single-breath N_2 washout	% N_2/ml	0.48	0.03	0.54	0.02	
CO diffusing capacity (DLCO)	ml/min/mm Hg	0.30	0.01	0.29	0.01	
DLCO/body weight	DLCO/kg	0.79	0.02	0.78	0.02	*
DLCO/alveolar volume	DLCO/ml	0.019	0.001	0.015	0.001	
Forced vital capacity (FVC)	ml	14.8	0.3	17.6	0.3	*
Forced expired volume in 0.1 sec	% FVC	63	2	47	1	*
Peak expiratory flow rate (PEFR)	ml/sec	124	2	113	2	*
PEFR/FVC	FVC/sec	8.4	0.2	6.5	0.2	*
Mean midexpiratory flow (MMEF)	ml/sec	84	5	55	2	*
MMEF/FVC	FVC/sec	5.7	0.3	3.1	0.2	*
Expiratory flow at 50% FVC (EF_{50})	ml/sec	90	5	61	3	*
EF_{50}/FVC	FVC/sec	6.0	0.4	3.5	0.2	*
Expiratory flow at 25% FVC (EF_{25})	ml/sec	50	3	30	2	*
EF_{25}/FVC	FVC/sec	3.4	0.2	1.7	0.1	*
Expiratory flow at 10% FVC (EF_{10})	ml/sec	17	1	9	1	*
EF_{10}/FVC	FVC/sec	1.2	0.06	0.5	0.04	*

[a]Three-month-old male F344 rats were instilled with 0.5 IU porcine pancreatic elastase/g body weight in 1.0 ml saline; controls were instilled with saline alone. Respiratory function was measured 19 months later by reported methods (Harkema et al., 1982). Treated rats had mild pulmonary ephysema with dilation of terminal airspaces and disruption of alveolar septae. There was no accompanying inflammation. The mean diameter of terminal airspaces was 96 μm in controls and 162 μm in treated rats. Measurements were performed at an altitude of 1728 m.

[b]Differences between group means tested by unpaired two-tailed t-test.

accompanied a very subtle thickening of the terminal bronchiolar-alveolar duct junction. Another example of an interpretive issue is the fact that, because the TLC of animals is measured as the volume at a standard inflation pressure, TLC reflects lung elasticity as well as lung size. It may be speculated, but is frequently not known, how these volume changes might compare to those in humans having identical pathology, but performing the measurements voluntarily. Information on issues involved in comparing lung volume changes of humans and animals can also be found elsewhere (Mauderly, 1984; Costa, 1985; Costa et al., 1991).

Quasistatic and Static Lung Compliance

The elastic properties of lung tissue are most commonly evaluated by measuring its pressure-volume characteristics and expressing the results as compliance, the slope of the pressure-volume curve (volume change divided by pressure change). Among pulmonary function tests *in vivo,* static or quasistatic pressure-volume characteristics provide the most direct information on the intrinsic elastic properties of the lung parenchyma. The relationships of lung volume to Ptp are identical in intact and open-chested animals (Lai and Hildebrandt, 1978). If airway pressure is used instead of Ptp, the volume at a given pressure could be greater in open-chest preparations. Drazen (1984) reviewed the use of these tests in the differential diagnosis of pulmonary mechanical abnormalities, and Gibson and Pride (1976) reviewed their use in the clinical assessment of man.

If the lung is inflated to TLC and deflated slowly as described above for measuring vital capacity, and if lung volume is plotted versus Ptp during deflation, the result is a sigmoid pressure-volume curve as shown in Fig. 10. The curve is termed a "quasistatic" deflation pressure-volume curve if data are recorded during a slow continuous deflation, and then the slope is termed quasistatic lung compliance (Cqs). The maneuver is called "static" if deflation is stopped at selected volume or pressure points, and then the slope is termed static lung compliance (Cst). The pauses in deflation are intended to eliminate an artifactual pressure difference due to flow resistance and to allow tissue relaxation to take place. If low flow rates are used, quasistatic deflation is suitable for most purposes and requires shorter apnea.

The elasticity of lung tissue is similar among mammals, and pressure-volume curves of most mammals have similar shapes; thus, compliance varies among species because of differences in lung volume (Leith, 1976; Schroter, 1980). Even within species, compliance tends to increase with increasing lung volume; therefore, compliance is sometimes divided by volume as a normalization (often called "specific compliance," as in Mauderly, 1974a). A steeper slope represents greater elasticity, and a less steep slope represents a less elastic, stiffer lung. The curve is steepest (highest compliance) in the region just above zero Ptp; thus, the lung is easiest to inflate in the volume region of the normal V_T. The curve flattens at pressures above 20–25 cm H_2O, and this is why volumes at pressures ranging from 25 to 30 cm H_2O are accepted as TLC. The midportion of the curve is usually of greatest interest. Many indices of static and quasistatic pressure-volume relationships have been used, including volume at a given pressure, pressure at a given volume, the average slope of the midportion, the steepest slope (usually over some minimum curve segment), the slope of a chord of the curve between two stated volume or pressure points, the area under the curve, and curve shape constants. Lai and Diamond (1986) compared several methods for analyzing curves of rats.

Examples of opposite changes in Cqs of rates are given in Tables 3 and 4. The quasistatic chord compliance was calculated as the slope of the pressure-volume curve between 0 and 10 cm H_2O Ptp. The Cqs of the rats with fibrotic lungs was reduced (increased elastic recoil, or stiffer lung), and that of rats with emphysematous lungs was increased (reduced elastic recoil, or more flaccid lung).

Measurement of static or quasistatic lung pressure-volume characteristics is the most common pulmonary function test performed on excised lungs. This test is often a useful adjunct in an inhalation toxicological study and gives results similar to those obtained in vivo. The method can be very simple and inexpensive (Pickrell et al., 1976) or quite complex (Frazer et al., 1985). Lungs are excised, a cannula is tied into the trachea, and the cannula is attached to a device allowing inflation and measurement of pressure (a syringe and manometer in the simplest form). The lungs may be suspended in saline to minimize gravitational effects and are sometimes vacuum degassed before measurement; however, it has been suggested that degassing might induce artifacts (Gross, 1981). Measurements are sometimes performed using both air inflation and saline instillation to eliminate or measure the effects of surface tension (surfactant) (Sahebjami, 1986). In addition to compliance, indices such as the volume retained at a given inflation pressure (Morstatter et al., 1976) or the hysteresis of inflation-deflation limbs (Fedullo et al., 1980) have been used.

The static mechanical properties of the lung, chest wall, and total respiratory system can be evaluated separately in vivo if esophageal, airway, and transpulmonary pressures are recorded during static deflations (Rahn et al., 1946). This approach has been used little in toxicological studies but could be useful if the inhaled agent might affect chest wall structure or neuromuscular function.

Dynamic Lung Mechanics

After the breathing pattern itself, the mechanical properties of the lung during breathing ("dynamic") have been the functional parameters most commonly measured in toxicological studies. These assays address the work of breathing, which is composed of elastic, resistive, and inertial forces (Rodarte

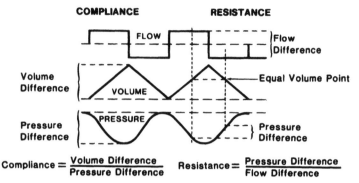

$$\text{Compliance} = \frac{\text{Volume Difference}}{\text{Pressure Difference}} \qquad \text{Resistance} = \frac{\text{Pressure Difference}}{\text{Flow Difference}}$$

Figure 11 The graphic method of Amdur and Mead (1958) for measurement of dynamic lung compliance and total pulmonary resistance from recordings of tidal flow, volume, and transpulmonary pressure is illustrated. The traces in the figure are idealized for simplicity; actual traces are more complex. In the illustration, a square-wave flow pattern results in a pyramidal volume trace. Compliance is measured from the volume and pressure differences between points of zero flow. Resistance is measured from the pressure and flow differences between points of equal inspiratory and expiratory volume. Volume points close to the maximum inspiratory and expiratory flows are typically selected.

and Rehder, 1986). The elastic work can be simply viewed as related to the force required to stretch the lung and chest wall during inspiration. Similarly, the resistive work is due to the force required to overcome airflow resistance, and inertial work is related to the force required to accelerate the lung, the chest wall, and the air column at the inflation of airflow. The factors most commonly measured are lung compliance (elastic work) and airflow resistance (resistive work), using Ptp as the measure of force.

The basics of dynamic lung mechanics are easily appreciated by examining the graphical method of Amdur and Mead (1958) for measuring dynamic lung compliance (Cdyn) and total pulmonary resistance (R_L)(Fig. 11). Total pulmonary resistance is the most accurate of several terms for the latter parameter, because certain tissue forces are included. This method separates the elastic and resistive components by measuring the former at points of zero flow and the latter at points of equal volume. Analog traces of Ptp, V_T and V recorded during tidal breathing are analyzed as illustrated in Fig. 11, which is simplified by assuming a square-wave flow pattern. The difference between end-inspiratory and end-expiratory volume is divided by the difference in Ptp at the same points, and the result is Cdyn. Vertical lines are extended trough the V and Ptp traces at equal volume points during inspiration and expiration. A volume point at approximately peak flow is usually selected and is often at 70–80% of V_T. The difference between these points in Ptp is divided by the difference in flow at the same points, and the result is R_L. Several breaths are typically measured, and the results are averaged.

Other methods have been used to calculate Cdyn and R_L from pressure, flow, and volume signals of animals. Most methods are derived from the theory and computational approach first described by Von Neergaard and Wirz (1927), which can be performed manually on analog traces and has also been incorporated into commercially available analyzers. Mauderly (1974a) used an analyzer (Model 794444, Honeywell) which derived values for Cdyn and R_L by manual closure of oscillographic pressure-flow and pressure-volume loops. Harkema et al. (1982) and Harkema and Mauderly (1994) used analyzers (Models 6 and LS-14, respectively, Buxco Electronics) which measured Cdyn and R_L using the Amdur-Mead (1958) method (Fig. 11) and requiring preselection of the fraction of V_T to be used for the equal volume point. Quan et al. (1990) used an analyzer (Model 8816A, Hewlett-Packard, Waltham, MA) which computed breath-by-breath values for Cdyn and R_L using the Von-Neergaard and Wirz (1927) algorithm directly. This algorithm, which can be programmed for computer analysis or performed manually, has the advantage of computing R_L early during inspiration and expiration without the need for selecting a single equal volume point.

The Cdyn reflects the integrated elastic work of breathing of the entire lung and, like Cqs, largely reflects the elasticity of lung tissue. However, Cdyn is measured during more rapid volume changes, which do not allow stress relaxation of the elastic tissues, and is also affected by other factors, such as the distribution of compliance within the lung and the frequency of breathing, which do not affect Cqs. For this reason, values for Cdyn are typically less than those for Cqs and are not as specific for tissue elasticity. The R_L almost entirely reflects flow resistance in the airway. However, since the bulk flow rate diminishes in smaller airways (gas moves progressively by diffusion) and flow becomes progressively less turbulent R_L largely represents resistive work in the larger airways. The R_L has been shown experimentally to be relatively insensitive to airflow obstruction in small airways (Brown et al., 1969). Reduced Cdyn in guinea pigs during inhalation exposures is often thought to reflect constriction of small airways; thus, the measurement of R_L and Cdyn gives some estimate of upper versus lower airway responses, respectively (Drazen, 1984). For example, Silbaugh and Mauderly (1984) observed a 72% decrease in the Cdyn of guinea pigs challenged with submicron histamine aerosol, whereas R_L was increased only 7%, suggesting a primarily lower-airway effect. Both Cdyn and R_L are somewhat dependent on lung volume; thus, they are sometimes normalized by volume and called specific compliance and specific airway resistance (sRAW), respectively. Conductance is the reciprocal of resistance, and specific airway conductance (sGAW) is sometimes measured to reflect resistance phenomena.

The frequency dependence of Cdyn has been used in man as a test for abnormalities of small airways (Woolcock et al., 1969). Regional differences in the elastic or flow-resistive properties of ventilating units cause differences in the speed with which the units respond to a driving pressure (commonly expressed as the "time constant," or the resistance × compliance of the unit). The ventilation of units with different time constants often becomes progressively out of phase with increasing f, and the effect is reflected by a decreased Cdyn. This effect can be examined in artificially ventilated animals, as done by D'Angelo (1976), who found Cdyn to be frequency-dependent in dogs, and by Diamond and O'Donnell (1977) and Yokoyama (1983), who did not find Cdyn to be frequency-dependent in rats.

Examples of the opposite effects of fibrosis and emphysema on Cdyn, measured by the above principle and calculated electronically, are presented in Tables 3 and 4, respectively. The Cdyn was reduced in fibrosis (Table 3) and increased in emphysema (Table 4), reflecting the increase and decrease, respectively, of lung elastic recoil. The lung volume at tidal breathing (FRC) was increased in both examples. Expressing compliance as specific compliance thus magnified the reduced Cdyn in fibrosis and reduced the magnitude of the increased Cdyn in emphysema. In both examples, the Cdyn of control rats was approximately one-half the magnitude of Cqs.

Another common method for measuring Cdyn and R_L involves the electronic subtraction of either the compliance or resistance component from hysteresis loops composed of combinations of Ptp, V, and V_T and displayed on an oscilloscope (Mead and Whittenberger, 1953). The angle of the flattened loop or the voltage required to remove hysteresis can be translated into Cdyn and R_L by appropriate equations. This method has been used for dogs at this institute (Mauderly, 1974a; Mauderly et al., 1980).

Several researchers have worked on methods that eliminate the need for measuring Ptp (and thus esophageal or pleural catheters) in tests of dynamic lung mechanics. The oscillation methods described below are one approach. Other methods have focused on the detection of changes in airflow resistance, largely related to studies of bronchoconstrictive responses. The general approach is to measure the difference in phase or magnitude between tidal signals obtained from expansion of the thorax and from airflow at the mouth. With increasing resistance to flow, there is an increasing difference between the two signals, due to slight decompression of intrapulmonary gas as the animal inhales. Two plethysmographs, either one inside the other (Silbaugh and Mauderly, 1984) or separate head and body chambers (Johanson and Pierce, 1971; Pennock et al., 1979), or a combination of a pneumotachograph at the nose within a plethysmograph (Agrawal, 1981; Boyd et al., 1980; Boyd and Mangos, 1981; Ellakkani et al., 1984), have been used to obtain the two signals. Some results have been expressed simply as ratios of the signal magnitudes (Silbaugh and Mauderly, 1984) or as area of X-Y loops representative of "flow-resistive work" (Ellakkani et al., 1984). Others have been expressed in units of airway resistance (Boyd et al., 1980; Boyd and Mangos, 1981), airway conductance (Johanson and Pierce, 1971), specific airway resistance (Pennock et al., 1979), or specific airway conductance (Agrawal, 1981).

All of these techniques have been shown to respond to airway constriction, and some can be used with conscious subjects. For accuracy of the measured parameter, however, they are technically demanding in terms of some combination of required frequency response, signal matching, signal-to-noise ratio, or temperature and humidity conditions. It should be noted that Watson et al. (1986) reported that the influence of lung volume on resistive work in guinea pigs was not completely removed by calculating sGAW.

Measurement of oscillation mechanics is another noninvasive method for evaluating dynamic mechanical behavior of the lung, as recently reviewed by Peslin and Fredberg (1986). An oscillating pressure signal is superimposed on tidal breathing, most commonly by attaching a pump or loudspeaker to the airway. An alternate approach is to apply the oscillating signal to the body surface. The resulting pressure, volume, and flow perturbations are measured and fit to mechanical or electrical models of the impedance of the respiratory system to calculate the inertance, elastance (compliance), and resistance of the system. The frequency of oscillation is set to exceed the frequency of breathing, and the oscillating pressure and volume changes are small. A single frequency might be used for routine measurement of resistance, e.g., Mink et al., 1984, used 4 Hz for dogs, or a spectrum of frequencies might be used to more completely characterize lung mechanical properties. For example, Jackson et al. (1984) used 4–64 Hz for dogs, Watson et al. (1986) used 4–40 Hz for guinea pigs, and Jackson and Watson 1982 used 20–90 Hz for rats. By using a spectrum of frequencies, the resistance, compliance, and inertance of central and peripheral airways can be differentiated (Jackson and Watson, 1982). Oscillation mechanics can be measured rapidly and noninvasively and have broad potential application in animals and man. However, measurement and data acquisition equipment can be complex, and detailed analyses are dependent on the accuracy with which the respiratory system is modeled. This, too, is an area of continuing development.

Murphy and Ulrich (1964) adapted a variation of the oscillation method of Mead (1960) for measuring total respiratory flow resistance of guinea pigs exposed to inhaled toxicants. The method is based on the fact that an oscillating frequency can be selected at which the ratio of respiratory flow to applied pressure is a function of flow resistance alone (resonant frequency). The guinea pigs were placed in a head-out, constant-volume plethysmograph, and a sinusoidal pressure signal at 5–8 Hz was applied to the body surface within the plethysmograph by a diaphragm pump. Diamond et al. (1975) also applied this technique to rabbits and cats in a pharmacologic study.

The "interrupter" method of measuring airway resistance is yet another noninvasive approach to evaluating lung mechanics during spontaneous breathing (Schlesinger et al., 1980). In this procedure, transpulmonary pressure is estimated as the difference between mouth pressure during a brief blockage of the airway (taken to represent alveolar pressure) and mouth pressure without blockage. The method is complex, since the blockage of airflow requires a rapidly responding trigger and valve device, and the appropriate value for pressure is extrapolated from recorded traces.

Forced Exhalation

Parameters measured during a maximal-effort (or forced) exhalation constitute a special case of assessing dynamic lung mechanics and are aimed at detecting airflow obstruction with greater sensitivity and descriptive value than measurements during tidal breathing. It is the function test most commonly performed in man and is frequently the only test applied in occupational and epidemiological studies. Because of its proven usefulness and ubiquitous use in man, considerable effort has been expended to adapt the procedure for animals. Much of the recent progress in plethysmography of small animals and in understanding the frequency response of function testing equipment resulted from this effort. Although forced exhalations of animals had been measured earlier (Leith, 1976), Diamond and O'Donnell (1977) first reported a positive-negative pressure plethysmographic system for incorporating forced exhalations into function tests of rats in pharmacological and toxicological studies.

The assay is actually a "stress test" of airflow performance under conditions rarely reached voluntarily except during cough. Human subjects perform the test by inhaling to TLC and exhaling as rapidly and deeply as possible to RV. The relationships among flow, volume, and time are measured during exhalation of the "forced vital capacity" (FVC) as illustrated in Fig. 12. The most common forced expiratory indices measured in man are the

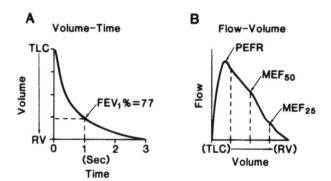

Figure 12 Volume-time (A) and flow-volume (B) curves representative of those recorded during a maximal (forced) exhalation from lung capacity (TLC) to residual volume (RV) are illustrated. In (A), the forced expired volume in 1.0 second is expressed as a percentage of the forced vital capacity (FEV$_1$%). In (B), the points of peak expiratory flow rate (PEFR) and maximal expiratory flow at 50% and 25% of force vital capacity (MEF$_{50}$ and MEF$_{25}$) are shown as three examples of the several measurements typically performed on such curves.

FVC and the forced expired volume in 1 sec (FEV_1), often expressed as a percentage of FVC ($FEV_1\%$) (Fig. 12A). The usual approach for animals is to plot the maximal expiratory flow-volume (MEFV) curve (Fig. 12B), a plot of expiratory flow versus volume or percent of FVC. The most commonly measured parameters are peak expiratory flow, flows at different lung volumes or percentages of FVC, and the mean flow over the midvolume range (usually from 75%–20% of FVC).

Forced exhalations of apneic animals are usually induced by negative airway pressures, as described above for measuring vital capacity and Cqs, except that the expiratory rate is not intentionally restricted. Best results are obtained by suddenly venting the airway to a vacuum reservoir via a low-impedance flow path and measuring flow and volume by plethysmography (airway pressure plethysmograph, or APP method). This method has been used in this institute for animals ranging from hamsters (Witschi et al., 1985) to dogs (Fig. 9) and by others for lungs of animals ranging from bats to whales (Leith, 1976). The volume of the negative-pressure reservoir should be sufficient to avoid a substantial change in driving pressure during the test. Volumes of 4.6 L for rats and 416 L for dogs are used at this Institute. Another method is the application of negative (for inflation) and then positive pressure at the body surface and measuring flow at the mouth by pneumotachography (body pressure pneumotachograph, or BPP method). This has been used for rats (Moorman et al., 1977), cats (Pepelko, 1982), monkeys (Moorman et al., 1975), and dogs (Johnson et al., 1982; Greene et al., 1984) using whole-body respirators in which pressure changes were induced by a hydraulically actuated diaphragm. A related method, used for human infants, is the application of pressure to the chest wall using either a body chamber connected to a positive-pressure reservoir (Taussig et al., 1982) or an inflatable thoracic cuff (Le Souef et al., 1986). Flow initiated from FRC is measured by a pneumotachograph attached to a face mask, thereby avoiding the need for intubation.

The APP method produces more satisfactory forced exhalations than the BPP method because of the more rapid application of the driving force. The relative peak flow rates can be used to compare the speed with which the driving pressures are applied by the two methods (the response time of the systems) because flow during early exhalation is effort-dependent (as described below). For comparison, we will assume that the largest practical endotracheal catheter was used in all cases. In a series of comparisons at this Institute, we measured mean peak flows in dogs of 7.8 FVC/sec by APP (e.g., Fig. 13), but only 2.5 FVC/sec by BPP (e.g., Johnson et al., 1982). An airway pressure of -200 cm H_2O and a solenoid valve was used for APP. For BPP, the chamber was pressurized to produce a Ptp of 30 cm H_2O, followed by sudden unplugging of the airway by jerking a string attached to a rubber stopper (a "cough" maneuver), rather than relying on the speed of the hydraulic diaphragm for deflation. Figures presented by Mink et al. (1984) indi-

RAT

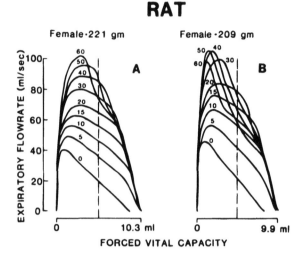

DOG

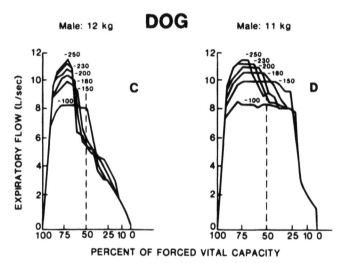

Figure 13 Families of flow-volume curves from two F344 rats (A, B) and two Beagle dogs (C, D) are shown. The curves were obtained by inflation of the lung to total lung capacity (defined as 30 cm H_2O transpulmonary pressure) and rapid deflation via a solenoid valve to a vacuum reservoir using different subambient reservoir pressures (given in cm H_2O). The curves for rats are actual tracings of signals recorded at high speed on tape, and forced vital capacities are given in milliliters. The zero-pressure rat curves were obtained by opening the airway to ambient pressure after lung inflation. The curves for dogs were reconstructed by connecting selected points on recorded curves, and volumes are expressed as percentages of forced vital capacity. The dashed lines intersect the curves at 50% of forced vital capacity as a reference point. These curves illustrate both the effect of driving pressure and individual differences in maximal flow-volume behavior of the lung.

cate a similar peak flow of 6.7–7.6 FVC/second for dogs measured by APP at an airway pressure of -238 cm H_2O.

Similar differences between these two methods can be observed in data from other species. Kosch et al. (1979) obtained peak flows of 5.4 FVC/second in rhesus monkeys and 8.0 FVC/second in bonnet monkeys by APP at an airway pressure of -185 cm H_2O, compared to 1.1 and 3.3 FVC/second obtained by BPP in cynomolgus monkeys by Moorman et al. (1975) and Knecht et al. (1985), respectively. At this institute, we obtained peak flows in F344 rats ranging from 8.1 (Table 3) to 9.4 FVC/second (Harkema et al., 1982) by APP at an airway pressure of -50 cm H_2O applied via a 1.8 mm 1.D. orotracheal catheter, compared to 5.3 FVC/second obtained by BPP in Wistar rats by Moorman et al. (1977) (size of tracheal cannula not specified). Diamond and O'Donnell (1977) obtained peak flows of 18.0 FVC/second by APP in rats at an airway pressure of -40 cm H_2O applied via a 2.3 mm 1.D. tracheal cannula placed surgically. Pepelko (1982) reported a mean peak flow of cats measured by BPP of 2.8 FVC/second, compared to a value of approximately 6 FVC/second shown for cats in a graph by Leith (1976). Although many variables complicate the comparison, the APP method appears preferable for obtaining the best forced exhalations in animals. The potential artifacts in measured flow induced by gas compression (BPP) or decompression (APP) in animals have not been studied but have been noted in man (Rodarte and Rehder, 1986). Despite the more rapid response of APP, BPP is likely to produce similar maximal flows at lower lung volumes (as described below) and therefore, it is likely to be adequate for detecting significant small airway disease.

To understand the interpretation of forced expiratory parameters, it is useful to examine families of MEFV curves produced by different levels of effort or, in the case of animals, different negative airway driving pressures. The families of MEFV curves for rats and dogs shown in Fig. 13 were selected because they illustrate individual differences in curve shape obtained by APP. The lungs of all the animals were all inflated to 30 cm H_2O Ptp (TLC), then rats were deflated at reservoir pressures from -0 (passive exhalation to open airway) to -60 cm H_2O and dogs from -100 to -250 cm H_2O. The peak flows increased with increasing driving pressure in all subjects except rat B, but flows at all forcing pressures stayed within an envelope on the descending limb of the curves which was not exceeded regardless of driving pressure.

Fig. 13 illustrates that the rapidity of the initial rise in flow and the peak flow achieved is dependent on driving pressure (called effort-dependent), but flows at middle and low lung volumes are not (called effort-independent). The same phenomena occur in man (Fry and Hyatt, 1960). It is interesting that in all four animals, flow at mid-FVC (dashed vertical lines) was maximal at intermediate driving pressures and reduced by increasing the driving force. This phenomenon has been observed in some human subjects and has been called negative effort dependence (Mead et al., 1967; Ingram and Schilder,

1966). These curves illustrate that the choice of driving pressure depends on the portion of the curve of most interest. Pressures of -50 and -200 cm H_2O have been used in studies of rats and dogs, respectively, at this Institute. The curves also illustrate that even if peak flows are not "maximal," as might occur with the BPP method, flows would probably still reach the limiting envelope at low lung volumes.

Much has been written about the several theories of the mechanisms of forced flow limitation, and they will not be discussed in detail here. For review, see the chapters by Hyatt, Rodarte and Rehder, as well as Wilson et al. in Macklem and Mead (1986a), and the papers by Fry and Hyatt (1960) and Mead et al. (1967). Suffice it to say that during forced exhalation, there are critical points ("choke points") in the airway that limit flow and that decreasing pressure downstream cannot increase flow above this limit. The location of the choke points is set by relationships among driving pressure, lung elastic recoil, compliance of the airway wall, density of the gas, and resistance to airflow upstream of the limiting point. The choke points first occur in larger airways and then move peripherally as lung volume decreases and upstream resistance increases. The location and movement of the choke points are what give MEFV curves their unique individual shapes, which are remarkably reproducible within subjects.

The practical implications of the above phenomena are that the rapidity with which forced flow rises and the peak flow is achieved are strongly influenced by driving pressure and the resistive characteristics of the tracheal catheter and external flow path. Reductions of peak flow and flow at high lung volumes are also indicative of large airway obstruction. Reductions of flows at progressively lower lung volumes are indicative of obstruction in progressively smaller airways. It is common to examine MEFV curves by calculating peak flow, the volume (or percent of FVC) exhaled in a given time (e.g., %FVC in 1.0 second in man and 0.1–0.2 seconds in rats are roughly equivalent), flows at specific fractions of FVC, and the mean flow between 75% and 25% of FVC. Because forced flows are dependent on lung volume, it is common to also examine flows in units of FVC/second to compare among species, sexes, ages, or subjects in which lung volume is altered by treatment (e.g., see Harkema et al., 1984).

Examples of altered forced expiratory values are presented in Tables 3 and 4. Fibrosis (Table 3) reduced lung volume but did not alter flows. The reduced elastic recoil (lower Cqs) actually helped maintain patency of small airways; thus, in the absence of airway disease, flows in FVC/second were increased above control values. Emphysema (Table 4) increased lung volume and reduced flows. Loss of lung elastic recoil (higher Cqs) gave less support to membranous airways, resulting in airway collapse when forcing pressure was applied. The obstructive nature of the disease is evident, even though there was no pathology in conducting airways.

Two other approaches to forced expiratory testing are worthy of note, although they are not common used. Because flow limitation is affected by gas density, the differences between flows with the subject breathing air versus breathing mixtures of helium and oxygen have been used as a method of detecting small airway disease. This test has been applied to dogs (Mink et al., 1984) and rats (Diamond and O'Donnell, 1977), and it yields results similar to those in man. The flow difference at a given volume point and the volume point at which flows with the two gases become identical ("volume of isoflow") are typically measured. Another approach is to perform "partial" forced exhalations by having the subject begin the exhalation at volumes below TLC. This test has been shown to have good sensitivity to some conditions in man (Bouhuys et al., 1969), and the %FVC expired in 0.1 second after inflation to 15 cm H_2O Ptp was found to have sensitivity equal to that using inflation to 30 cm H_2O in emphysematous rats in a preliminary trial at this institute (Mauderly, 1981). This suggests that forced expiratory tests might still be useful in situations in which inflations to maximal lung volumes were undesirable.

GAS DISTRIBUTION

Not all areas of the lung are ventilated or perfused equally during ordinary breathing. Optimum conditions for gas exchange occur when inhaled gas is distributed proportionately to the area of the alveolar-capillary membrane perfused by pulmonary arterial blood. Although reflex mechanisms actively help control matching of ventilation and perfusion, the distribution of gas within the lung is also affected by the mechanical properties of the airways and lung parenchyma. For this reason, the uniformity of intrapulmonary gas distribution (or intrapulmonary gas mixing) is often measured as an index of the uniformity of lung mechanical properties. Tests of gas distribution give information on the uniformity with which pathologic lesions are distributed within the lung.

Multiple-Breath Gas Washout

This method involves switching the subject from breathing air to breathing some other gas and evaluating the rapidity with which the new gas replaces, or washes out, the original gas in the lung. The subject can either breathe spontaneously, or can be ventilated using a respirator. One approach is to have the subject rebreathe from a bag or spirometer containing helium in O_2 and to measure successive changes in end-expiratory helium concentration (Briscoe, 1952). The commoner approach is to switch the subject from breathing air via a nonrebreathing valve to breathing O_2 and to measure the breath-by-breath decrease in end-expiratory N_2 concentration. The latter technique is called an open-circuit N_2 washout and is sometimes used to simultaneously measure FRC (as described earlier). This technique has been

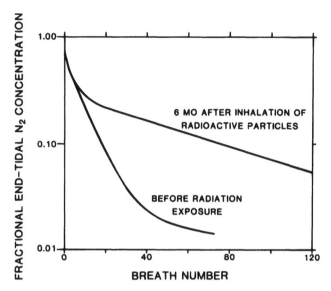

Figure 14 Shown are the effects of a patchy radiation-induced pneumonitis and fibrosis on the multiple-breath N_2 washout of a Beagle dog. The washout curves are reconstructed from successive end-tidal N_2 peaks after the dog was switched from breathing air to breathing O_2. The baseline curve was recorded before radiation exposure. The other curve was recorded 6 months after the dog inhaled 390 μCi [144]Ce in aluminosilicate particles which remained in the lung as an internal beta-gamma radiation source (Mauderly et al., 1980). The cumulative absorbed lung dose of radiation at the 6-month point was estimated to be 410 Gy. The N_2 washout rates of the two curves were similar during the first few breaths, but the slowing of washout after exposure is clearly evident in the remainder of the curves. The reduced slope of the postexposure washout curve reflects nonuniform gas mixing in the lung.

used for dogs (Mauderly et al., 1980) and rats (Costa et al., 1986) in inhalation toxicological studies. Multiple-breath washouts give a more representative view of gas distribution and mixing characteristics during normal breathing than the single-breath technique described below; however, variables such as the V_T/FRC ratio and changing FRC during the washout complicate the interpretation.

The breath-by-breath washout of gas from the lung follows a multiexponential decay function, as can be observed in a semilog plot of end-expiratory N_2 versus breath number. Nitrogen washout curves for a dog before and after induction of radiation pneumonitis-fibrosis by inhaled [144]Ce are shown in Fig. 14. The baseline curve shows the typical three components: (1) a small, rapidly emptying component due to washout of conducting airways; (2) a large component related to washout of most of the lung volume; and (3) a small component related to lung units with poor ventilation and to excretion of N_2 from the blood. The relative magnitudes and slopes of the individual components can be analyzed by curve-stripping techniques (Mauderly, 1977).

As seen in the figure, the slope (washout rate) of the second (or major) component is changed markedly by the presence of patchy lung fibrosis.

There are numerous methods for describing the efficiency of multiple-breath gas washout, ranging from simple to complex. The simplest indices are the washout time to reach a certain N_2 concentration, or the N_2 concentration after a certain washout time. Other simple approaches are to divide the volume of gas breathed by the FRC ("nitrogen clearance equivalent"—Luft et al., 1955) or to measure the ventilation required to reduce N_2 to a given concentration ("lung clearance index"—Bouhuys, 1963). Since the washout rate is dependent on the ratio of V_T to FRC, the slope of washout can be divided by this ratio as a first-order normalization ("nitrogen dilution constant"—Mauderly, 1977). An increase in the nitrogen dilution constant of dogs (caused by the same radiation-induced patchy fibrosis affecting the curves in Fig. 14) is shown in Table 2. Using another common approach, Costa et al. (1986) applied moment analysis of the log of end-expiratory N_2 concentration plotted against breath number to evaluate N_2 washout of rats.

Single-Breath Gas Washout

This method is based on the same principle of mixing a new gas with the gas in the lung as the multiple-breath technique, but the mixing occurs during a single, deep breath. The method and issues surrounding the test were summarized by Gold (1982). Oxygen is most commonly used to dilute N_2 in the lung as the tracer gas ("resident gas" technique); however, inert gases are also used as tracers ("bolus gas" technique). The air-breathing subject exhales to RV, inhales O_2 to TLC, and then exhales slowly to RV. This maneuver has been used to characterize lung disease in rats (Likens and Mauderly, 1982) and dogs (Mink et al., 1984) using positive and negative airway pressures as described above for measuring lung volumes. The test has been done using a plethysmograph respirator to characterize lung disease in monkeys (Knech et al., 1985).

The exhaled N_2 concentration is plotted versus the exhaled volume as the "single-breath N_2 washout curve," shown diagrammatically in Fig. 15. There is no N_2 in the first expirate (phase I) from the conducting airway dead space. After an initial rise in N_2 (phase II), the washout of curve of normal subjects flattens into a semiplateau that represents the washout of the majority of the alveolar bed (phase III). There is usually a final increase in N_2 (phase IV), but the point of inflection is often ill defined. The slope of phase III is used as an indication of the uniformity of intrapulmonary gas mixing and distribution, the more uniform the distribution, the flatter the plateau.

The lung volume at the onset of phase IV has been termed the closing volume. Several theories have been advanced concerning the mechanisms determining closing volume. It was first thought that the inflection of phase IV

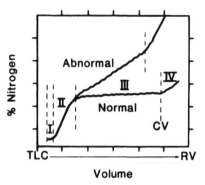

Figure 15 Idealized normal and abnormal single-breath N_2 washout curves are shown. Such curves are obtained by having the air-breathing subject exhale to residual volume, inhale O_2 to total lung capacity, then exhale again to residual volume while the volume and N_2 concentration of the expirate are measured. The normal curve is divided into four phases: phase I represents washout of O_2 from the large airways (anatomic dead space); phase II represents a mixture of dead space and alveolar gas; phase III represents the washout of the major portion of the alveolar bed; phase IV reflects emptying of lung units receiving less O_2 during inhalation. Cardiogenic oscillations on the trace are typical. The increase of the slope of phase III of the abnormal curve is taken as an indication of less uniform gas mixing (abnormal gas distribution), and the earlier onset of phase IV is thought to signal earlier closure of small airways.

signals the closure of small airways in dependent lung regions because of airway mechanical properties and vertical hydrostatic pressure gradients. At present, it is only agreed that closing volume represents a shift to the emptying of poorly ventilated units and is related in some way to the sequence of airway closure. The uncertainty about the mechanism has resulted in part from the observation of closing volume in small animals having little vertical pressure gradient (Likens and Mauderly, 1982). Regardless of the mechanism, lungs with nonuniform gas mixing produce a steeper slope of phase III and sometimes an earlier onset of phase IV (Fig. 15).

Examples of the influence of lung disease on the slope of phase III of rats are given in Tables 3 and 4. Patchy lung fibrosis (Table 3) markedly altered the regional gas distribution and doubled the slope of phase III. Emphysema distributed relatively uniformly throughout the lung (Table 4) caused little increase in the slope of phase III, even though the lung parenchyma was clearly abnormal. This illustrates that the slope of phase III reflects the distribution, rather than the presence, of lung lesions.

Closing volume has been increased in rats by emphysema (Likens and Mauderly, 1982) and in monkeys by acute inflammation (Knecht et al., 1985); however, experience at this Institute with rats who have a variety of different toxicant-induced lung diseases has suggested that the assay has limited interpretive value. At one time, closing volume was thought to be a likely candidate for early detection of small-airway disease in man. Although the test distinguishes between smoking and nonsmoking subjects, its validity as an early marker of small-airway disease is uncertain (Gold, 1982).

Radiolabeled Gas Distribution

Imaging of a radioactive gas with a gamma camera ("ventilation scan") is another method of evaluating the distribution of inhaled gas. The subject breathes the gas (usually $^{133}-Xe$), counts of gamma activity are measured by a radiation detector external to the thorax, and the counts are accumulated by computer to produce an image of the ventilated field. The wash-in of xenon can be analyzed as described above for multiple-breath gas washout, but the more common approach is to simply view an image of the lung. In contrast to the whole-lung integrative nature of most function tests, this assay depicts the anatomic location of the functional abnormality. The discriminatory ability of the test (usually approximately 1 cm) limits its use for small animals, but it works well with larger species such as dogs (King et al., 1985).

ALVEOLAR-CAPILLARY GAS EXCHANGE

The transfer of O_2 and CO_2 across the alveolar-capillary membrane is the ultimate result of all pulmonary function. This transfer occurs by passive diffusion of gases from regions of high to low partial pressures. The lack of any active mechanism for gas transfer makes mammalian life completely dependent on the ability to maintain alveolar ventilation and capillary perfusion, and on the nature of the diffusion pathway (membrane). Gas exchange functions are evaluated by measuring alveolar ventilation, diffusion of test gases, and transfer of O_2 and CO_2.

Measurement of Alveolar Gas

Measurements of alveolar ventilation are based on measuring the concentration of gases in alveolar air. Because sampling alveolar gas directly is impractical, end-expiratory gas is sampled as a reasonable surrogate. End-expiratory gas is sampled either as the last gas exhaled at the end of tidal breaths (called end-tidal gas) or as the last gas exhaled during a single, deep exhalation (usually called alveolar gas). End-tidal gas concentrations are usually measured and recorded by continuously sampling at the mouth with rapidly responding analyzers such as mass spectrometers. The end-tidal peak (or plateau during slow breathing) fractional concentration of the gas averaged over multiple breaths is multiplied by the barometric pressure minus the water vapor pressure at the subject's body temperature to calculate the alveolar partial pressure of the gas, such as PO_2 and PCO_2 (e.g., Mauderly, 1974a).

An alternate method of end-tidal gas sampling is to use an intermittent

Table 5   Effects of Radiation Pneumonitis-Fibrosis on Respiratory Function of a Conscious Dog after Inhaling ^{90}Y in Insoluble Particles[a]

		Weeks after inhaling 0.78 μCi/kg ^{90}Y			
	Units	4	8	12	16
Respiratory frequency	breaths/min	10	15	31	68
Tidal volume	ml	337	226	133	80
Minute volume ($\dot{V}_E$)	l/min	3.4	3.4	4.1	5.4
Alveolar ventilation	% of $\dot{V}_E$	65	43	39	40
Effective ventilation	% of $\dot{V}_E$	59	36	31	9
Alveolar dead-space ventilation	% of $\dot{V}_E$	6	7	8	31
Alveolar PO_2	mm Hg	81	82	86	100
Arterial PO_2	mm Hg	71	67	61	50
Alveolar-arterial PO_2	mm Hg	10	15	25	50
Arterial hemoglobin saturation with O_2	%	91	89	87	74
Arterial PCO_2	mm Hg	39	39	39	53
Alveolar PCO_2	mm Hg	35	32	32	12
Arterial-alveolar PO_2	mm Hg	4	7	7	41
Arterial pH		7.41	7.38	7.40	7.29
Ventilatory equivalent for O_2 at 3.0 mph	ml $\dot{V}_E$/ml O_2 uptake	49	54	68	98

[a]From Mauderly et al. (1973). This male Beagle was exposed only by inhalation to 780 μCi/kg of ^{90}Y in a fused aluminosilicate particle matrix and measured serially. The data represent single measurements of the conscious, spontaneously breathing dog at an altitude of 1728 meters. The values at 4 weeks were similar to preexposure values. The dog was euthanized in cardiorespiratory failure 2 weeks after the 16-week measurement. The lungs contained patchy inflammation and fibrosis, with alveolar epithelial hypertrophy and some large scars. The right ventricle was enlarged, and fluid had accumulated in the pleural cavity.

collection device to collect the last portion of tidal expirate into a reservoir, from which a continuous sample of end-tidal gas can be provided to a slowly responding analyzer. An example of such a device for dogs is shown diagrammatically in Mauderly (1972) and in the photograph in Fig. 2 of Mauderly (1974a). The "alveolar gas equation" (found in most physiology texts—see p. 166 in West, 1977 for example) is sometimes used to calculate alveolar PO_2 and PCO_2 measured in arterial blood. Although this is a useful approximation, it assumes no alveolar-arterial PCO_2 difference. There is often a measurable PCO_2 difference in normal subjects, and the difference can be substantial in subjects with lung disease (Table 5).

Oxygen and Carbon Dioxide Exchange

The O_2 uptake and CO_2 output are usually measured by collecting expirate over a given time, analyzing the fractional concentrations of mixed, expired O_2 and CO_2; measuring the volume of gas collected; and using equations (the so-called Haldane transformation) based on the differences between inspired and expired gas concentrations (Mauderly and Tesarek, 1975). The relevant equations are given in most physiology texts and are not presented here (see Wilmore and Costill, 1973). The Haldane transformation is based on the equality between inspired and expired N_2 concentration; however, studies have shown that the calculation is valid if these concentrations are only slightly different (Luft et al., 1973; Wilmore and Costill, 1973). An alternate approach to gas collection is the measurement of gas concentrations in air flowing continuously past the subject's airway (e.g., Withers, 1977). It is standard to express gas transfer values—such as O_2 uptake, CO_2 output, and CO diffusing capacity (discussed below)—corrected to standard temperature and pressure, dry (STPD) conditions.

Measurement of O_2 uptake and CO_2 output allows calculation of other useful indices. The respiratory exchange ratio (CO_2 output divided by O_2 uptake) indicates whether or not the subject was in a steady state of gas exchange at the time measurements were performed. A respiratory exchange ratio of 0.8 which matches the theoretical steady-state metabolic respiratory quotient is often considered normal; however, exchange ratios closer to 1.0 are more commonly obtained during measurement of conscious animals that are apparently relaxed (Mauderly, 1974a). Values substantially below 0.8 or above 1.0 indicate that the effects of hyperventilation or hypoventilation, either during or before the measurement, were being expressed. This is one method of estimating whether or not muscle activity or excitement was causing artifacts at the time of measurement. Dividing O_2 uptake (STPD) by minute volume (BTPS) yields the ventilatory equivalent for O_2 (sometimes called specific ventilation). This ratio of volume breathed per unit of O_2 consumed is a crude but useful index of the overall efficiency of ventilation. Values in the range of 20–40 ml/ml are typical for conscious laboratory animals at rest (Mauderly et al., 1979), and the value increases with lung disease. The ventilatory equivalent for O_2 during exercise was used at this Institute to detect early lung injury in dogs (Table 5) (Mauderly et al., 1973, 1980).

Alveolar Ventilation

Several useful indices of the apportionment of inspired gas among ventilated compartments (anatomic dead space, perfused alveoli, and alveolar dead space) can be calculated if inspired, expired, and alveolar gas concentrations, CO_2 output, and arterial PCO_2 are measured. The Bohr equation is used to calculate the physiological dead space/V_T ratio from the different between alveolar and mixed expired PCO_2 (in most texts; see pp. 164 in West, 1977 for example). Physiological dead space is the sum of anatomic dead space (including conducting airways and external, or equipment dead space) plus alveolar dead space (unperfused alveoli). Measuring this ratio over several breaths and multiplying it by V_T gives an approximate value for the actual volume of physiological dead space. Alveolar ventilation—that portion of the minute volume reaching the alveolar region—is calculated from the dilution of the CO_2 excreted from the lung. Alveolar ventilation equals CO_2 output divided by alveolar PCO_2 when correction factors for gas law conversions etc. are applied (Luft and Finkelstein, 1968; pp. 164 in West, 1979). If arterial PCO_2 is substituted for alveolar PCO_2, the calculated result is an estimate of the "effective" ventilation—the ventilation of perfused alveoli (Luft and Finkelstein, 1968). The difference between alveolar and effective ventilation is an estimate of alveolar dead space ventilation—the ventilation of unperfused alveoli. If the above values are expressed as fractions of V_T, and if external dead space is subtracted, the anatomic dead space volume can be estimated.

Blood Gases

The end result of alveolar-capillary gas exchange is the maintenance of normal partial pressures of O_2 and CO_2 in arterial blood which in turn control the transport of O_2 and CO_2 between the lung and the rest of the body tissues. Only small portions of these gases are transported as dissolved O_2 and CO_2; however, these portions exert the measured partial pressures and control both the loading of hemoglobin with O_2 (the primary O_2 delivery mode) and blood pH.

Arterial blood samples are analyzed for PO_2, PCO_2, and pH by electrodes in commercially available analyzers. The PCO_2 electrode is actually a pH meter in which the diffusion of CO_2 across a membrane into a buffer solution causes a change in the solution pH. The PO_2 is measured by a polarographic electrode in which O_2 passes through a membrane and causes an electrical current in an electrolyte in the presence of a suitable voltage. The O_2 content of blood can also be measured by chemical absorption such as the Van Slyke method (Van Slyke and Neill, 1924). Hemoglobin is usually measured spectrophotometrically. Total blood hemoglobin content and the degree of saturation of the hemoglobin with O_2 or CO_2 are typically measured by analyzers using the differential light absorption characteristics of the hemoglobin species (Silbaugh and Horvath, 1982). Another approach is the use of an oximeter, which measures in vivo O_2 saturation of blood noninvasively by light transmission (Douglas et al., 1979).

Although venous blood is often easier to obtain than arterial blood, it has limited interpretive value on its own. Arterial blood sampling is required for detailed evaluation of respiratory function. Arterial samples from small animals are usually obtained via surgically implanted catheters (Lucey et al., 1982; Silbaugh and Horvath, 1982). Samples from larger animals such as dogs can be obtained by percutaneous sampling with a syringe and needle (see photograph in Mauderly, 1974a). For valid interpretation, the subject should be in a steady state of respiration, either resting quietly or exercising at a constant level.

An example of the use of the above parameters to evaluate alterations of gas exchange in a dog is presented in Table 5. The dog was exposed by inhalation to particles containing a beat-gamma-emitting radionuclide and evaluated serially to study the pathogenesis of the resulting radiation pneumonitis (Mauderly et al., 1973). The progressive inflammation and fibrosis caused a stiffening of the lung and a shift in breathing pattern toward a higher f and smaller V_T. This shift wasted more of the V_T ventilating anatomic dead space and resulted in reduced percentages of alveolar and effective ventilation. The increased ventilatory equivalent for O_2 during mild exercise also reflected a reduced efficiency of gas exchange. Alveolar-capillary O_2 and CO_2 exchange were impaired progressively as reflected by the increasing alveolar-arterial PO_2 and PCO_2 differences. A substantial disruption of capillary perfusion was reflected by the increased percentage of alveolar dead space ventilation at the last measurement. At that time, respiratory failure was signaled by the inability to maintain reasonable arterial PO_2, the low O_2 saturation of hemoglobin, and the low arterial pH (uncompensated respiratory acidosis). It is useful to note in this example that the ventilatory equivalent for O_2 during exercise and the apportionment of ventilation and alveolar-arterial PO_2 difference at rest reflected the progressive gas exchange problem at a time when $\dot{V}_E$ and blood gases were normal.

Carbon Monoxide Diffusion Capacity

The diffusing capacity of the lung for CO (DLCO) has proved to be one of the most sensitive methods of detecting impairments of alveolar-capillary gas transfer at rest in both man and animals and is the commonest approach to evaluating gas exchange in toxicological studies. The DLCO is the rate of uptake of trace amounts of inhaled CO divided by the driving pressure for diffusion. Although the diffusing capacity for O_2 can be measured, the back pressure of PO_2 in the blood must be measured to determine the driving pressure difference. In contrast, the affinity of CO for hemoglobin eliminates (sufficiently for practical purposes) the back pressure of PCO from the blood and thus eliminates the need for blood samples. It should be noted that the diffusing capacity for O_2 can also be estimated from morphometric measurements; both functional and morphometric measurements of O_2-diffusing capacity were reviewed by Forster and Crandall (1976).

Although sensitive, DLCO is nonspecific, because it reflects the integrated effects of several different factors. The DLCO equals the sum of two terms: (1) membrane diffusion factors such as thickness and surface area and (2) the product of the volume of red blood cells in the pulmonary capillaries (capillary blood volume) and the combining rate of CO with hemoglobin. A straightforward review of the factors affecting DLCO and means of differential diagnosis was presented by Ayers et al. (1975). Comroe (1975) reviewed the usefulness of DLCO in man.

Three primary approaches (with multiple variations of each) have been adapted for measuring the DLCO of animals—the single-breath, the re-breathing, and the steady-state techniques. The most frequently used is the single-breath method, in which the subject's lung is inflated with a large single breath of gas containing CO and an inert gas such as Ne or He (Ogilvie et al., 1957). The dilution of the inert gas allows calculation of the gas volume in the lung, he breath is "held" for a measured length of time (usually 5–10 seconds) to allow calculation of uptake rate, and the CO concentration difference in the inhaled and alveolar (last gas exhaled) gas allows calculation of the fractional uptake of CO. Results are expressed in ml CO (STPD) taken up per minute per mm Hg alveolar-arterial (only alveolar is measured) PCO difference. Graham et al. (1980) gave a detailed theoretical analysis of this method.

Takezawa et al. (1980) studied in detail the influence of procedural variables and gave data for hamsters, rats, guinea pigs, and rabbits measured by the same technique. They found that inflation volume, lung volume at the beginning of inflation, repeated measurements, body position, and hyperventilation did not affect the outcome; CO concentration and breath-holding time did. This technique has been used in toxicological studies of monkeys (Knecht et al., 1985) and dogs (Young et al., 1963) as well as smaller species. The study of Knecht et al. (1985) is an example of the use of ^{18}CO to circumvent the problem of analyzing CO by mass spectrometry (CO and N_2 have the same mass).

An example of disease-induced changes in the single-breath DLCO of rats is presented in Table 3. The DLCO of rats with fibrotic lungs was reduced in the presence of lower body weight and smaller lung volume. The DLCOs per unit of body weight and lung volume were calculated to examine the influences of these changes, and the results show that these factors did not account for the reduced gas exchange efficiency. This effect likely resulted from the loss of surface area from fibrosis and the thickening of the diffusion path by fluid accumulation, cellular infiltration, and membrane thickening with fibrin. The DLCO of rats with mild emphysema (Table 4) was only slightly reduced, most likely because of loss of surface area. Studies at this institute and elsewhere have shown the effect of enzyme-induced emphysema on DLCO to be dose-related; thus, significant reductions occur with more severe disease.

The DLCO can also be measured by the uptake of CO during rebreathing of a fixed volume of test gas (Lewis et al., 1959). This method has been applied to hamsters (Snider et al., 1978), rats (Costa et al., 1986), and dogs (Mink et al., 1986). This approach has been thought to minimize potential artifacts due to nonuniform gas distribution in the lung. If the test is done using a mixture of CO, inert gas, and acetylene, it is possible to estimate capillary blood flow, capillary blood, and tissue volume and FRC as well as DLCO (Sackner et al., 1975).

The third method for measuring DLCO is the steady-state method in which the subject breathes from a continuously fresh source of test gas rather than rebreathing from a closed volume. One variation requires a blood sample for measuring PCO (Filley et al., 1954). This technique has been applied to rats (Turek et al., 1973) and dogs (Kilburn et al., 1963). Another variation does not require blood sampling but does require collection of an end-tidal gas sample (steady-state, end-tidal method) (Bates et al., 1955). An adjustable, automatic end-tidal sampler actuated by changes in Ptp was developed at this institute and used to measure the DLCO of dogs (Mauderly, 1972) and ponies (Mauderly, 1974b). It has been suggested (Woolf, 1964) that the fractional uptake of CO (part of the calculation of steady-state DLCO) has interpretive value by itself. A normal DLCO in the presence of a low fractional uptake is thought to reflect compensated abnormal gas exchange whereas a reduction of both parameters reflects uncompensated abnormal gas exchange.

An example of reduced steady-state end-tidal DLCO in dogs is presented in Table 2. The DLCO was reduced progressively in dogs with radiation pneumonitis-fibrosis and was correlated with developing focal inflammatory and fibrotic lesions (Mauderly et al., 1980). The effects of radiation-induced and nitrogen dioxide-induced lung injuries on DLCO were shown to be similar in man and dogs, and DLCO was shown to be one of the most sensitive tests to these injuries in both species (Mauderly et al., 1983).

PERFUSION

The perfusion of the alveolar capillary bed is just as critical to gas exchange as the ventilation of alveoli, although perfusion is less often evaluated in toxicological studies. Perfusion can be affected directly or indirectly by inhaled materials via reflexes, endothelial toxicity, the release of vasoactive mediators, or mechanical alterations of lung structure. The most frequently evaluated perfusion parameters are the flow of blood through the lung (cardiac output), the capillary blood volume, the distribution of blood, pulmonary vascular pressures, and the resistance to blood flow through the lung.

Pulmonary Blood Flow and Volume

Pulmonary blood flow is equal to the cardiac output, since the entire output from the right heart passes through the lung. Two methods are commonly used, both based on the flow-induced dilution of an injected indicator. The thermodilution method involves rapid injection of cool saline and integration of the blood temperature curve measured by a downstream thermistor. This can be done in rodents by injection via a catheter into the vena cava or right atrium and measurement of temperature in the carotid artery (Brackett et al., 1985). In larger animals (as well as in man), the procedure usually involves placement of a Swan-Ganz type catheter into the pulmonary artery which allows injection and measurement by a single catheter (Mauderly et al., 1980). The dye dilution method involves injection of a dye such as indocyanine green into the pulmonary artery, measuring the dye dilution curve in the aorta by a densitometer (Bisgard et al., 1975) .

Capillary blood flow can also be estimated by the rate of uptake of acetylene during rebreathing (Sackner et al., 1975). If the gas rebreathed also contains CO and an inert gas such as He, the procedure gives estimates for the combined pulmonary tissue and capillary blood volume, DLCO, and FRC. This technique was used by Abraham et al. (1980) to study the effects of inhaled nitrogen dioxide in conscious sheep.

Another method of estimating pulmonary capillary blood volume is the measurement of CO-diffusing capacity at two different levels of alveolar PO_2 (Roughton and Forster, 1957). This approach has been applied to rodents (Costa et al., 1984) and dogs (Young et al., 1963).

Distribution of Pulmonary Perfusion

Radioisotopic imaging using a gamma camera is the commonest method of evaluating the uniformity of intrapulmonary blood distribution. Small radiolabeled particles injected into the venous blood are retained for a short time in the pulmonary capillaries, and the distribution of gamma activity provides an image of blood distribution. An example is the use at this Institute of [99]m-technetium-labeled macroaggregated albumin particles to evaluate pulmonary thromboembolism in a dog (King et al., 1985). This technique generally has a resolution for lesions of approximately 1 cm; thus, the approach usually has insufficient sensitivity for use with smaller rodents.

The distribution of blood in limited portions of the lung can also be viewed by radiography or fluoroscopy after direct injection of radiopaque contrast media into a specific vessel. Good vascular images can be obtained; however, the area that can be practically viewed is small, and catheterization of specific, preselected areas can be difficult.

Pulmonary Vascular Resistance

The resistance to blood flow through the lung is commonly evaluated either indirectly by measuring pulmonary arterial pressure or by direct measurements of resistance. Measurement of vascular pressure is straightforward, using a fluid-filled catheter advanced from a peripheral vein (usually the jugular or femoral) into the pulmonary artery and is readily applied to both small (Silbaugh and Horvath, 1982) and large (Bisgard et al., 1975; Mauderly et al., 1980) animals.

Pulmonary vascular resistance (airflow resistance) is calculated by dividing the pressure drop (upstream minus downstream pressure) by the corresponding flow rate. Cardiac output is the flow term, and pulmonary arterial pressure is the upstream pressure term. The relevant downstream pressure is that in the pulmonary vein; however, this pressure is difficult to measure directly. Systemic arterial pressure—the pressure in the aorta or a more peripheral artery—is sometimes used as an estimate of downstream pressure (Bisgard et al., 1975). Pulmonary wedge pressure is preferable and is usually obtained using a Swan-Ganz-type catheter. This catheter has an inflatable balloon at the tip, which helps to "float" the catheter tip into the pulmonary artery and then into a small branch. The pressure measured at the tip of the catheter when it is wedged into a small branch (thereby precluding flow) is termed "wedge pressure" and is accepted as an estimate of the pressure on the downstream side of the capillary bed.

An example of increased pulmonary vascular resistance secondary to a lung disease is presented in Table 2. As described above, dogs inhaled [144]Ce in relatively insoluble particles and developed progressive radiation pneumonitis-fibrosis (Mauderly et al., 1980). Conscious dogs were catheterized with Swan-Ganz catheters via the femoral vein and simple catheters via the femoral artery. The catheters were advanced into the pulmonary artery and thoracic aorta, respectively. Cardiac output was measured by thermodilution, and pulmonary arterial and wedge pressures were measured via the Swan-Ganz catheter. Pulmonary vascular resistance was progressively increased by thickening of larger vessels and obliteration of portions of the capillary bed. Systemic arterial pressure and systemic arterial resistance were unchanged (not shown). Pulmonary arterial pressure and pulmonary vascular resistance were approximately doubled above those of controls at the moderate disease stage. At the severe, preterminal stage, pulmonary arterial pressure and pulmonary vascular resistance were increased 3.5- and 4.6-fold, respectively. Gas exchange was affected in parallel, with the dogs progressively losing the ability to compensate and maintain a satisfactory arterial PO_2 and O_2 saturation of hemoglobin. The cardiovascular and gas exchange effects were both considered to have contributed to the ultimate deaths of the dogs by cardiorespiratory collapse.

SUMMARY

Pulmonary function tests are useful tools for studies of the effects of inhaled toxicants. They provide a nondestructive means of evaluating the functional impact of transient or permanent alterations in respiratory system structure. These tests are not an alternative method for describing morphological changes; rather, physiological and morphological evaluations are best used together to study health effects. All pulmonary function tests used for man can be adapted for animals, and many are applied in toxicological studies. The tests range from simple measures of respiration to comprehensive evaluations of many facets of respiratory and cardiovascular performance. There are typically several techniques for measuring any given physiological function that allows the adaptation of tests to nearly any experimental situation.

Pulmonary function tests can provide information that enhances the extrapolation from inhalation toxicological studies of animals to the prediction of effects in man. Although measurement techniques may differ, results to date indicate that similar structural changes in man and animals result in similar functional changes. Properly interpreting results from animals and extrapolating them to man requires an understanding of the influences of measurement differences, differences in anatomy and physiology, and differences in sensitivity to inhaled toxicants. For this reason, pulmonary function testing usually requires professional input into selection of tests, design of measurement methods, and interpretation of results.

ACKNOWLEDGMENTS

The authors acknowledge the many members of the Inhalation Toxicology Research Institute's present and former scientific and technical staff who contributed to the studies from which data are presented. This work was supported by the Office of Health and Environmental Research, U.S. Department of Energy under contract DE-AC04-76EV01013. The research which results in the data used in this chapter was conducted in facilities fully accredited by the American Association for Accreditation of Laboratory Animal Care.

REFERENCES

Abraham, WM, Welkner, M, Oliver, W, Mingle, M, Januszkiewicz, AJ, Wanner, A, Sackner, MA: Cardiopulmonary effects of short-term nitrogen dioxide exposure in conscious sheep. Environ Res 22:61–72, 1980.

Agrawal, KP: Specific airway conductance in guinea pigs: Normal values and histamine-induced fall. Respir Physiol 43:23–30, 1981.

Alarie, Y: Sensory irritation of the upper airways by airborne chemicals. Toxicol Appl Pharmacol 24:279–297, 1973.

Alarie, Y: Toxicological evaluation of airborne chemical irritants and allergens using respiratory reflex reactions. In: Proceedings of the Inhalation Toxicology and Technology Symposium. Edited by BKJ Leong, pp. 207–231. Ann Arbor, MI: Ann Arbor Science Publishers, 1981.

Altman, PL, Dittmer, DS, eds.: Respiration and Circulation, Biological Handbooks. New York, NY: Elsevier Science, Inc. Bethesda MD: 1983.

Amdur, MO, Mead, J: Mechanics of respiration in unanesthetized guinea pigs. Am J Physiol 192:364–368, 1958.

Anthonisen, NR: Tests of mechanical function. In: Mechanics of Breathing, Handbook of Physiology Section 3: The Respiratory System. Edited by PT Macklem, J Mead, pp. 753–784. Bethesda, MD: American Physiological Society, 1986.

Ayers, LN, Ginsberg, ML, Fein, J, Wasserman, K: Diffusing capacity, specific diffusing capacity and interpretation of diffusion defects. West J Med 123:255–264, 1975.

Bargeton, D, Barres, G: Time characteristics and frequency response of body plethysmography.

In: Body Plethysmography, Progress in Respiration Research, Vol. 4. Edited by AB DuBois, KP van de Woestijne. Basel, Switzerland: Karger, pp. 2–23, 1969.

Bartels, H, Dejours, P, Kellog, RH, Mead, J: Glossary on respiration and gas exchange. J Appl Physiol 34:549–558, 1973.

Bates, DV, Boucot, NG, Dormer, AE: The pulmonary diffusing capacity in normal subjects. J Physiol 129:237–252, 1955.

Baydur, A, Behrakis, PK, Zin, WA, Jaeger, M, Milic-Emili, J: A simple method for assessing the validity of the esophageal balloon technique. Am Rev Respir Dis 126:788–791, 1982.

Beardsmore, CS, Helms, P, Stocks, J, Hatch, DJ, Silverman, MJ: Improved esophageal balloon technique for use in infants. J Appl Physiol Respir Environ Exercise Physiol 49:735–742, 1980.

Berne, RM, Levy, MN. Cardiovascular Physiology, 6th Edition. St. Louis, MO: Mosby Year Book, Inc., 1991.

Bisgard, GE, Orr, JA, Will, JA: Hypoxic pulmonary hypertension in the pony. Am J Vet Res 36:49–52, 1975.

Bouhuys, A: Pulmonary nitrogen clearance in relation to age in healthy males. J Appl Physiol 18:297–300, 1963.

Bouhuys, A, Hunt, V, Kim, B, Zapletal, A: Maximum expiratory flow rates in induced broncho-constriction in man. J Clin Invest 48:1159–1168, 1969.

Boyd, RL, Mangos, JA: Pulmonary mechanics of the normal ferret. J Appl Physiol Respir Environ Exercise Physiol 50:799–804, 1981.

Boyd, RL, Fisher, MJ, Jaeger, MJ: Noninvasive lung function tests in rats with progressive papain-induced emphysema. Respir Physiol 40:181–190, 1980.

Brackett, DJ, Schaefer, CF, Tompkins, P, Fagraeus, L, Peters, LJ, Wilson, MF: Evaluation of cardiac output, total peripheral resistance and plasma concentrations of vasopressin in the conscious, unrestrained rat during endotoxemia. Circ Shock 17:273–284, 1985.

Briscoe, WA: Further studies on the intrapulmonary mixing of helium in normal and emphyse-matous subjects. Clin Sci 11:45–58, 1952.

Brown, R, Woolcock, AJ, Vincent, NJ, Macklem, PT: Physiological effects of experimental air-way obstruction with beads. J Appl Physiol 27:328–335, 1969.

Burrows, B, Knudson, RJ, Kettel, LJ: Respiratory Insufficiency. Chicago: Year Book, 1975a.

Burrows, B, Huang, N, Hughes, R, Johnston, R, Kilburn, K, Kuhn, C, Miller, W, Mitchell, M, Snider, G: Pulmonary terms and symbols: A joint report of the ACCP-ATS Joint Commit-tee on Pulmonary Nomenclature. Chest 67:583–593, 1975b.

Cameron, JN: The Respiratory Physiology of Animals. New York, NY: Oxford University Press, 1988.

Chang, YA: Respiratory Physiology: A Quantitative Approach. New York, NY: Marcel Dekker, Inc., 1988.

Cherniack, RM, Farhi, LE, Armstrong, WB, Proctor, DF: A comparison of esophageal and intrapleural pressure in man. J Appl Physiol 8:203–211, 1955.

Clausen, JL, editor: Pulmonary Function Testing Guidelines and Controversies. Academic Press: New York, NY, 1982.

Coggins, CRE, Duchosal, F, Musy, C, Ventrone, R: Measurement of respiratory patterns in rodents using whole-body plethysmography and a pneumotachograph. Lab Anim 15:137–140, 1981.

Comroe, JH: Retrospectroscope-pulmonary diffusing capacity for carbon monoxide (DLCO). Am Rev Respir Dis 111:225–228, 1975.

Costa, DL, Tepper JS: Approaches to lung function assessment in small mammals. In: Toxicol of the Lung. Edited by DE Gardner, JD Crapo, EJ Massaro, pp. 147–174, Raven Press, NY, 1988.

Costa, DL: Interpretation of new techniques used in the determination of pulmonary function in rodents. Fundam Appl Toxicol 5:423–434, 1985.

Costa, DL, Kutzman, RS, Popenoe, EA, Drew, RT: A subchronic multidose ozone study in rats. In: Advances in Modern Environmental Toxicology. Edited by SD Lee, MG Mustafa, MA Muhiman, pp. 369–393. Princeton, NJ: Senate Press, 1983.

Costa, DL, Lehman, JR, Osheroff, MR, Bergofsky, EH: In vivo functional characterization of the pulmonary capillary bed in the rat after chronic hypoxia. Am Rev Respir Dis 129:A342, 1984.

Costa, DL, Kutzman, RS. Lehmann, JR, Drew, RT: Altered lung function and structure in the rat after subchronic exposure to acrolein. Am Rev Respir Dis 133:286–291, 1986.

Costa, DL, Tepper, JS, Raub, J: Interpretations and Limitations of Pulmonary Function Testing in Small Laboratory Animals. EPA Report No. EPA/600/D-91/112, NTIS, US Department of Commerce, 1991.

Cotes, JE; Leathart, GL: In Lung Function: Assessment & Application in Medicine, 5th Edition. Blackwell Scientific Publishers, Cambridge, MA, 1993.

Des Jardins, T: Cardiopulmonary Anatomy and Physiology: Essentials for Respiratory Care, 2nd edition. Albany, NY: Delmar Publishers, Inc., 1993.

D'Angelo, E: Effect of papain-induced emphysema on the distribution of pleural surface pressure. Respir Physiol 27:1–20, 1976.

Darling, RC, Cournand, A, Richards, DW: Studies on the intrapulmonary mixture of gases. III. An open circuit method for measuring residual air. J Clin Invest 19:609–618, 1940.

Davidson, JT, Wasserman, K, Lillington, GA, Schmidt, RW: Pulmonary function testing in the rabbit. J Appl Physiol 21:1094–1098, 1966.

Diamond, L., O'Donnell, M: Pulmonary mechanics in normal rats. J Appl Physiol Respir Environ Exercise Physiol 43:942–948, 1977.

Diamond, L, Adams, GK, Bleidt, B, Williams, B: Experimental study of a potential anti-asthmatic agent: SCH 15280. J Pharmacol Exp Ther 193:256–263, 1975.

Dorato, MA, Carlson, KH, Copple, DL: Pulmonary mechanics in conscious Fischer 344 rats: Multiple evaluations using nonsurgical techniques. Toxicol Appl Pharmacol 68:344–353, 1983.

Douglas, NJ, Brash, HM, Wraith, PK: Accuracy, sensitivity to carboxyhemoglobin and speed of response of the Hewlett-Packard 47201A ear oximeter. Am Rev Respir Dis 119:311–313, 1979.

Drazen, JM: Physiological basis and interpretation of indices of pulmonary mechanics. Environ Health Perspect 56:3–9, 1984.

Drorbaugh, JE, Fenn, WO: A barometric method for measuring ventilation in newborn infants. Pediatrics 16:81–87, 1955.

DuBois, AB, Van de Woestijne, KP: Body Plethysmography. Progress in Respiration Research, Vol. 4. Basel, Switzerland: Karger, 1969.

DuBois, AB, Botelho, SY, Bedell, GN, Marshall, R, Comroe, JH: A rapid plethysmographic method for measuring thoracic gas volume: A comparison with a nitrogen washout method for measuring functional residual capacity in normal subjects. J Clin Invest 35:322–326, 1956.

Ellakkani, MA, Alarie, YC, Weyel, DA, Mazumdar, S, Karol, MH: Pulmonary reactions to cotton dust: An animal model for byssinosis. Toxicol Appl Pharmacol 74:267–284, 1984.

Epstein, MAF, Epstein, RA: A theoretical analysis of the barometric method for measurement of tidal volume. Respir Physiol 32:105–120, 1978.

Fedullo, AJ, Jung-Legg, Y, Snider, GL. Karlinsky, JB: Hysteresis ratio: A measure of the mechanical efficiency of fibrotic and ephysematous hamster lung tissue. Am Rev Respir Dis 122:47–52, 1980.

Fenn, WO, Rahn, H: Respiration, Vol. 1. Handbook of Physiology, Section 3. Washington, DC: American Physiological Society, 1964.

Fenn, WO, Rahn, H: Respiration, Vol. 2. Handbook of Physiology, Section 3. Washington, DC: American Physiological Society, 1965.

Filley, GF, MacIntosh, DJ, Wright, GW: Carbon monoxide uptake and pulmonary diffusing capacity in normal subjects at rest and during exercise. J Clin Invest 33:530–539, 1954.

Finucane, KE, Egan, BA, Dawson, SV: Linearity and frequency response of pneumotachographs. J Appl Physiol 32:121–126, 1972.

Forster, RE, Crandall, ED: Pulmonary gas exchange. Annu Rev Physiol 38:69–93, 1976.

Frazer, DG, Weber, KC, Franz, GN: Evidence of sequential opening and closing of lung units during inflation-deflation of excised rat lungs. Respir Physiol 61:277–288, 1985.

Fry, DL, Hyatt, RE: Pulmonary mechanics. A unified analysis of the relationship between pressure, volume and gas flow in the lungs of normal and diseased human subjects. Am J Med 29:672–689, 1960.

Gibson, GJ: Pride, NB: Lung distensibility: The static pressure-volume curve of the lungs and its use in clinical assessment. Br J Dis Chest 70:143–187, 1976.

Gillespie, DJ, Lai, YL, Hyatt, RE: Comparison of eophageal and pleural pressures in the anesthetized dog. J Appl Physiol 35:709–713, 1973.

Gillespie, JR: Mechanisms that determine functional residual capacity in different mammalian species. Am Rev Respir Dis 128 (Suppl): S74–S77, 1983.

Gold, PM: Single breath nitrogen test: Closing volume and distribution of ventilation. In: Pulmonary Function Testing Guidelines and Controversies. Edited by JL Clausen, pp. 105–114. New York, NY: Academic Press, 1982.

Graham, BL, Dosman, JA, Cotton, DJ: A theoretical analysis of the single breath diffusing capacity for carbon monoxide. IEEE Trans Biomed Eng 27:221–227, 1980.

Greene, SA, Wolff, RK, Hahn, FF, Henderson, RF, Mauderly, JL, Lundgren, DL: Sulfur dioxide induced chronic bronchitis in beagle dogs. J Toxicol Environ Health 13:945–958, 1984.

Gross, NJ: Mechanical properties of mouse lungs: Effects of degassing on normal, hyperoxic and irradiated lungs. J Appl Physiol Respir Environ Exercise Physiol 51:391–398, 1981.

Hamilton, WF: Circulation, Vol. 1. Handbook of Physiology, Section 2. Washington, DC: American Physiological Society, 1962.

Hamilton, WF: Circulation, Vol. 2. Handbook of Physiology, Section 2. Washington, DC: American Physiological Society, 1963.

Hamilton, WF: Circulation, Vol. 3. Handbook of Physiology, Section 2. Washington, DC: American Physiological Society, 1965.

Harkema, JR, Mauderly, JL: Respiratory function alterations in rats following chronic ozone inhalation. Tropospheric Ozone: Critical Issues in the Regulatory Process: Orlando, FL, May 11–13, 1994.

Harkema, JR, Mauderly, JL, Hahn, FF: The effects of emphysema on oxygen toxicity in rats. Am Rev Respir Dis 126:1058–1065, 1982.

Harkema, JR, Mauderly, JL, Gregory, RE, Pickrell, JA: A comparison of starvation and elastase models of emphysema in rats. Am Rev Respir Dis 129:584–591, 1984.

Hering, E, Breuer, J: Die Selbststeuerrung der Athmung durch den Nervus Vagus. Sitzber Akad Wiss Wien 57:672–677, 1968.

Ingram, R, Schilder, D: Effect of gas compression on pulmonary pressure, flow and volume relationships. J Appl Physiol 21:1821–1826, 1966.

Jackson, AC, Watson, JW: Oscillatory mechanics of the respiratory system in normal rats. Respir Physiol 48:309–322, 1982.

Jackson, AC, Vinegar, A: A technique for measuring frequency response of pressure, volume and flow transducers. J Appl Physiol Respir Environ Exercise Physiol 47:462–467, 1979.

Jackson, AC, Watson, JW, Kotlikoff, MI: Respiratory system, lung, and chest wall impedances in anesthetized dogs. J Appl Physiol Respir Environ Exercise Physiol 57:34–39, 1984.

Johanson, WG, Pierce, AK: A noninvasive technique for measurement of airway conductance in small animals. J Appl Physiol 30:146–150, 1971.

Johnson WK, Mauderly, JL, Hahn, FF: Lung function and morphology of dogs after sublethal exposure to nitrogen dioxide. J Toxicol Environ Health 10:201–221, 1982.

Kaplinsky, E, Levy, MN: Cardiovascular Pathophysiology, St. Louis, MO: Mosby Year Book, Inc., 1993.

Karol, MH, Stadler, J, Underhill, D, Alarie, Y: Monitoring delayed-onset pulmonary hypersensitivity in guinea pigs. Toxicol Appl Pharmacol 61:277–285, 1981.

Kilburn, KH, Miller, HW, Burton, JE, Rhodes, R: Effects of altering ventilation on steady-state diffusing capacity for carbon monoxide. J Appl Physiol 18:89–96, 1963.

King, RR, Mauderly, JL, Hahn, FF, Wolff, RK, Muggenburg, BA, Pickrell, JA: Pulmonary function studies in a dog with pulmonary thromboembolism associated with Cushing's disease. J Am Anim Hosp Assoc 21:555–562, 1985.

Knecht, EA, Moorman, WJ, Clark, JC, Lynch, DW, Lewis, TR: Pulmonary effects of vanadium pentoxide inhalation in monkeys. Am Rev Respir Dis 132:1181–1185, 1985.

Koen, PA, Moskowitz, GD, Shaffer, TH: Instrumentation for measuring functional residual capacity in small animals. J Appl Physiol Respir Environ Exercise Physiol 43:755–758, 1977.

Koo, KW, Leith, DE, Sherter, CB, Snider, GL: Respiratory mechanics in normal hamsters. J Appl Physiol 409:936–942, 1976.

Kosch, PC, Gillespie, JR, Berry, JD: Flow-volume curves and total pulmonary resistance in normal bonnet and rhesus monkeys. J Appl Physiol Respir Environ Exercise Physiol 46:176–183, 1979.

Lai, YL, Diamond, L: Comparison of five methods of analyzing respiratory pressure-volume curves. Respir Physiol 66:147–155, 1986.

Le Souef, PN, Hughes, DM, Landau, LI: Effect of compression pressure on forced expiratory flow in infants. J Appl Physiol 61:1639–1646, 1986.

Leff, A.R., Schumacker PT: Respiratory Physiology: Basics and Applications, Philadelphia, PA: W B Saunders Co., 1993.

Leith, DE: Comparative mammalian respiratory mechanics. Physiologists 19:485–510, 1976.

Leith, DE, Mead, J: Principles of Body Plethysmography. Bethesda, MD: National Heart and Lung Institute, 1974.

Lemen, R, Benson, M, Jones, JG: Absolute pressure measurements with hand-dipped and manufactured esophageal balloons. J Appl Physiol 37:600–603, 1974.

Lemen, RJ, Gerdes, CB, Wegmann, MJ, Perrink, KJ: Frequency spectra of flow and volume events for forced vital capacity. Am Physiol Soc, pp 977984, 1982.

Leong, KJ, Dowd, GF, MacFarland, HN: A new technique for tidal volume measurement in unanesthetized small animals. Can J Physiol Pharmacol 42:189–198, 1964.

Lewis, B, Tai-hon, L, Hayfor-Nelson, E: The measurement of pulmonary diffusing capacity for carbon monoxide by a rebreathing method. J Clin Invest 38:2073–2086, 1959.

Likens SA, Mauderly, JL: Effect of elastase or histamine on single-breath N_2 washouts in the rat. J Appl Physiol 52:141–146, 1982.

Lucey, EC, Snider, GL, Javaheri, S: Pulmonary ventilation and blood gas values in emphysematous hamsters. Am Rev Respir Dis 125:299–303, 1982.

Luft, UC, Finkelstein, S: Hypoxia: A clinical-physiological approach. Aerospace Med 39:105–110, 1968.

Luft, UC, Roorbach, EH, MacQuigg, RE: Pulmonary nitrogen clearance as a criterion of ventilatory efficiency. Am J Tuberc 72:465–478, 1955.

Luft, UC, Myhre, LG, Loeppky, JA: Validity of Haldane calculation for estimating respiratory gas exchange. J Appl Physiol 34:864–865, 1973.

Lundsgaard, JS, Gronlund, J, Einer-Jensen, N: Evaluation of a constant temperature hot wire anemometer for respiratory gas flow measurements. Med Biol Eng Comput 17:211–215, 1979.

Macklem, PT, Mead, J: Mechanics of Breathing, Part 1, The Respiratory System, Handbook of Physiology, Section 3. Bethesda, MD: American Physiological Society, 1986a.

Macklem PT, Mead, J: Mechanics of Breathing, Part 2, The Respiratory System, Handbook of Physiology, Section 3. Bethesda, MD: American Physiological Society, 1986b.

Matijak-Schaper, M, Wong, KL, Alarie, Y: A Method to rapidly evaluate the acute pulmonary effects of aerosols in unanesthetized guinea pigs. Toxicol Appl Pharmacol 69:451–460, 1983.

Matthews, LR: Cardiopulmonary Anatomy and Physiology, Philadelphia, PA: J B Lippincott Co., 1994.

Mauderly, JL: Steady state carbon monoxide diffusing capacity of unanesthetized beagle dogs. Am J Vet Res 33:1485–1491, 1972.

Mauderly, JL, Pickrell, JA, Hobbs, CH, Benjamin SA, Hahn, FF, Jones, RK, Barnes JE: The effect of inhaled ^{90}Y fused clay aerosol on pulmonary function and related parameters of the beagle dog. Radiat Res 56:83–96, 1973.

Mauderly, JL: The influence of sex and age on the pulmonary function of the beagle dog. J Gerontol 29:282–289, 1974a.

Mauderly, JL: Evaluation of the grade pony as a pulmonary function model. Am J Vet Res 35:1025–1029, 1974b.

Mauderly, JL: A new technique for evaluating nitrogen wash-out efficiency. Am J Vet Res 38:69–74, 1977.

Mauderly, JL: Effects of elastase on maximal and partial flow-volume curves of rats. Fed Proc 40:385, 1981.

Mauderly, JL: Respiratory function responses of animals and humans to oxidant gases and to pulmonary emphysema. J Toxicol Environ Health 13:345–361, 1984.

Mauderly, JL: Respiration of F344 rats in nose-only inhalation exposure tubes. J Appl Toxicol 6:25–30, 1986.

Mauderly, JL: Comparisons of respiratory functional responses in laboratory animals and man. In: The Design and Interpretation of Inhalation Studies and Their Use in Risk Assessment. Edited by U Mohr, D Dungworth, G Kimmerle, J Lewkowski, RO McClellan, W Stober. New York: Springer-Verlag, Berlin, pp. 243–261, 1988.

Mauderly, JL: Effect of inhaled toxicants on pulmonary function. In: Concepts in Inhalation Toxicology, Chapter 13. Edited by RO McClellan and RF Henderson, Hemisphere, New York, NY: pp. 349–404, 1989.

Mauderly, JL: Measurement of respiration and respiratory responses during inhalation exposures. J Am College of Toxicol 9:397–405, 1990.

Mauderly, JL, Tesarek, JE: Nonrebreathing valve for respiratory measurements in unsedated small animals. J Appl Physiol 38:369–371, 1975.

Mauderly, JL, Tesarek, JE, Sifford, Linda J, Sifford, Lorna J: Respiratory measurements of unsedated small laboratory mammals using nonrebreathing valves. Lab Anim Sci 29:323–329, 1979.

Mauderly, JL, Muggenburg, BA, Hahn, FF, Boecker, BB: The effects of inhaled ^{144}cerium on cardiopulmonary function and histopathology of the dog. Radiat Res 84:307–324, 1980.

Mauderly, JL, Silbaugh, SA, Likens, SA, Harkema JR, Johnson, WK: Comparisons of the respiratory functional responses of animals and man. In: Proceedings of the 13th Annual Conference on Environmental Toxicology, Dayton, OH, 1982. USAF Report No. AFAMRL-TR-82-101, pp. 147–170, 1983.

Mauderly, JL, Seiler, FA, Buel, JA, Pino, MV: Effects of lung disease on the frequency content of forced exhalations of rats. In Inhalation Toxicology Research Institute Annual Report, 1983–1984, LMF-113, pp. 363–366, 1984.

Mauderly, JL, Gillett, NA, Henderson, RF, Jones, RK, McClellan, RO: Relationships of lung structural and functional changes to accumulation of diesel exhaust particles. Ann Occup Hyg 32:659–669, 1988.

Mauderly, JL, Seiler, FA: The frequency content of forced exhalations of normals rats. In Inhalation Toxicology Research Institute Annual Report, 1984–1985, LMF-114, pp. 404–407, 1985.

Maxted, KJ, Shaw, A, Macdonald, TH: Choosing a catheter system for measuring intra-esophageal pressure. Med Biol Eng Comput 15:398–401, 1977.

McClellan, RO: Health effects of diesel exhaust: A case study in risk assessment. Am Ind Hyg Assoc J 47:1–13, 1986.

McMichael, J: A rapid method of determining lung capacity. Clin Sci 4:167–173, 1939.

Mead, J: Control of respiratory frequency. J Appl Physiol 15:325–336, 1960.

Mead, J, Whittenberger, JL: Physical properties of human lungs measured during spontaneous respiration. J Appl Physiol 5:779–795, 1953.

Mead, J, Turner, J, Macklem, PT, Little, J: Significance of the relationship between lung recoil and maximum expiratory flow. J Appl Physiol 22:95–108, 1967.

Medinsky, MA, Dutcher, JS, Bond, JA, Henderson, RF, Mauderly, JL, Snipes, MB, Mewhinney, JA, Cheng, YS, Birnbaum, LS: Uptake and excretion of ^{14}C methyl bromide as influenced by exposure concentration. Toxicol Appl Pharmacol 78:215–225, 1985.

Mines, AH: Respiratory Physiology, 3rd Edition, New York, NY: Raven Press, Ltd., 3rd edition, 1993.

Mink, SN, Coalson, JJ, Whitley, L, Greville, H, Jadue, C: Pulmonary function tests in the detection of small airway obstruction in a canine model of bronchiolitis obliterans. Am Rev Respir Dis 130: 1125–1133, 1984.

Mink, SN, Gomez, A, Whitley, L, Coalson, JJ: Hemodynamics in dogs with pulmonary hypertension due to emphysema. Lung 164:41–54, 1986.

Moorman, WJ, Lewis, TR, Wagner, WD: Maximum expiratory flow-volume studies on monkeys exposed to bituminous coal dust. J Appl Physiol 39:444–448, 1975.

Moorman, WJ, Hornung, RW, Wagner, WD: Ventilatory functions in germfree and conventional rats exposed to coal dusts. Proc Soc Exp Biol Med 155:424–428, 1977.

Morstatter, CE, Glaser, RM, Jurrus, ER, Weiss, HS: Compliance and stability of excised mouse lungs. Comp Biochem Physiol 54A: 197–202, 1976.

Murphy, SD, Ulrich, CE: Multi-animal test system for measuring effects of gases and vapors on respiratory function of guinea pigs. Am Ind Hyg Assoc J 25:28–36, 1964.

Nunn, J.F.: Nunn's Applied Respiratory Physiology. Newton, MA: Butterworth-Heinemann, 4th edition, 1993.

Ogilvie, CM, Forster, RE, Blackemore, WS, Morton, SW: A standardized breath holding technique for the clinical measurement of the diffusing capacity for carbon monoxide. J Clin Invest 36:1–17, 1957.

O'Neil, JJ, Raub, JA: Can animal pulmonary function testing provide data for regulatory decision making? Environ Health Perspect 52:215–219, 1983.

O'Neil, JJ, Raub, JA: Pulmonary function testing in small laboratory animals. Environ Health Perspect 56:1–22, 1984.

Palecek, F: Measurement of ventilatory mechanics in the rat. J Appl Physiol 27:149–156, 1969.

Pappenheimer, JR, Comroe, JH, Cournand, A, Ferguson, JKW, Filley, GF, Fowler, WS, Gray, JS, Helmholtz, HF, Otis, AB, Rahn, H, Riley, RL: Standardization of definitions and symbols in respiratory physiology. Fed Proc 9:602–605, 1950.

Pennock, BE, Cox, CP, Rogers, RM, Cain, WA, Wells, JH: A noninvasive technique for measurement of changes in specific airway resistance. J Appl Physiol 46:399–406, 1979.

Pepelko, WE: EPA studies on the toxicological effects of inhaled diesel engine emissions. In: Toxicological Effects of Emissions from Diesel Engines, edited by J Lewtas, pp. 121–142. New York: Elsevier, 1982.

Peslin, R, Fredberg, JJ: Oscillation mechanics of the respiratory system. In: Mechanics of Breathing, Part 1: The Respiratory System, Handbook of Physiology, Section 3. Edited by PT Macklem J Mead. Bethesda, MD: American Physiological Society, pp. 145–177, 1986.

Pickrell, JA, Harris, DV, Benjamin, SA, Cuddihy, RG, Pfleger, RC, Mauderly, JL: Pulmonary collagen metabolism after lung injury from inhaled ^{90}Y in fused clay particles. Exp Mol Pathol 25:70–81, 1976.

Quan, SF, Lemen, RJ, Witten, ML, Sherrill, DL, Grad, R, Sobonya, RE, Ray, CG: Changes in lung mechanics and reactivity with age after viral bronchiolitis in beagle puppies. Am Physiol Soc:2034–2042, 1990.

Rahn, H, Otis, AB, Chadwick, LE, Fenn, WO: The pressure-volume diagram of the thorax and lung. Am J Physiol 146:161–178, 1946.

Rodarte, JR, Rehder, K: Dynamics of respiration. In: Mechanics of Breathing, Part 1: The Respiratory System, Handbook Of Physiology, Section 3. Edited by PT Macklem, J Mead. Bethesda, MD: American Physiological Society, pp. 131–144, 1986.

Rohrer, F: Physiologie der atembewegung. In: Handbuch der Normalen und Pathologischen Physiologie, Vol. 2. Edited by A Ellivoger, Berlin: Springer, Vol II, pp. 70–127, 1925.

Roughton, FJW, Forster, RE: Relative importance of diffusion and chemical reaction rates in determining the rate of exchange of gases in the human lung with special reference to true diffusing capacity of pulmonary membrane and volume of blood in the lung capillaries. J Appl Physiol 11:290, 1957.

Sackner, MA, Greeneltch, D, Heiman, MS, Epstein, S, Atkins, N: Diffusing capacity membrane diffusing capacity, capillary blood volume, pulmonary tissue volume and cardiac output measured by a rebreathing technique. Am Rev Respir Dis 111:157–165, 1975.

Sahebjami, H: Effects of postnatal starvation and refeeding on rat lungs during adulthood. Am Rev Respir Dis 133:769–772, 1986.

Schaper, M, Alarie, YC: The effects of aerosols of carbamylcholine, serotonin and propanolol on the ventilatory response to CO_2 in guinea pigs and comparison with the effects of histamine and sulfuric acid. Acta Pharmacol Toxicol 56:1–6, 1985.

Schlesinger, RB, Zeccardi, AV, Monahan, J: An interrupter method for measurement of airway resistance in small animals. J Appl Physiol Respir Environ Exercise Physiol 48:1092–1095, 1980.

Schroter, RC: Quantitative comparisons of mammalian lung pressure volume curves. Respir Physiol 42:101–107, 1980.

Silbaugh, SA, Horvath, SM: Effect of acute carbon monoxide exposure on cardiopulmonary function of the awake rat. Toxicol Appl Pharmacol 66:376–382, 1982.

Silbaugh, SA, Mauderly, JL: Noninvasive detection of airway constriction in awake guinea pigs. J Appl Physiol Respir Environ Exercise Physiol 56:1666–1669, 1984.

Silbaugh, SA, Mauderly, JL, Macken, CA: Effects of sulfuric acid and nitrogen dioxide on airway responsiveness of the guinea pig. Toxicol Environ Health 8:31–45, 1981.

Sinnett; EE, Jackson, AC, Leith, DE, Butler, JP: Fast integrated flow plethysmograph for small animals. J Appl Physiol Respir Environ Exercise Physiol 50:1104–1119. 1981.

Skornik, WA, Heimann, R, Jaeger, RJ: Pulmonary mechanics in guinea pigs: Repeated measurements using a nonsurgical computerized method. Toxicol Appl Pharmacol 59:314–323, 1981.

Smith, JJ, Kampine, JP: Circulatory Physiology—The Essentials. Baltimore, MD: Williams and Wilkins, 1990.

Snider, GL, Celli, BR, Goldstein, RH, O'Brien, JJ, Lucey, EC: Chronic interstitial pulmonary fibrosis produced in hamsters by endotracheal bleomycin. Am Rev Respir Dis 117:289–297. 1978.

Standaert, TA, LaFramboise, WA, Tuck, RE, Woodrum, DE: Serial determination of lung volume in small animals by nitrogen washout. J Appl Physiol 59:205–210, 1985.

Staub, NC: Basic Respiratory Physiology, New York, NY: Churchill Livingstone, Inc., 1991a.

Staub, NC: Comroe's Physiology of Respiration, St. Louis, MO: Mosby Year Book, 1991b.

Strope, GL, Cox, CL, Pimmel, RL, Clyde, WA: Dynamic respiratory mechanics in intact anesthetized hamsters. J Appl Physiol Respir Environ Exercise Physiol 49:197–203, 1980.

Swanson, GD, Grodins, FS, editors: Respiratory Control: A Modeling Perspective. New York, NY: Plenum Press, Ltd., 1990.

Takezawa, J, Miller, FJ, O'Neil, JJ: Single-breath diffusing capacity and lung volumes in small laboratory mammals. J Appl Physiol Respir Environ Exercise Physiol 48:1052–1059, 1980.

Taussig, LM, Landau, LI, Godfrey, S, Arad, I: Determinants of forced expiratory flows in newborn infants. J Appl Physiol Respir Environ Exercise Physiol 53:1220–1227, 1982.

Taylor, DC: Clinical Respiratory Physiology, Philadelphia, PA: W B Saunders CO., 1989.

Tepper JS, Costa DL, Lehmann JR: Extrapolation of animal data to humans: Homology of pulmonary physiological responses with ozone exposure. In: Toxicology of the Lung, 2nd edition. Edited by DE Gardner et al., Raven Press, Ltd., New York, NY: pp. 217–251, 1993.

Travis, EL, Vojnovic, B, Davies, EE, Hirst, DG: A plethysmographic method for measuring function in locally irradiated mouse lung. Br J Radiol 52:67–74, 1979.

Travis, EL, Parkins, CS, Down, JD, Fowler, JF, Thames, HD: Repair in mouse lung between multiple small doses of X rays. Radiat Res 94:326–339, 1983.

Turek, Z, Grandtner, M, Ringnalda, BEM, Kreuzer, F: Hypoxic pulmonary steady state diffusing capacity of CO and cardiac output in rats born in a simulated altitude of 3500 m. Pfluegers Arch 340:11–18, 1973.

Van Slyke, DD, Neill, JM: The determination of gases in blood and other solutions by vacuum extraction and manometric measurement. J Biol Chem 61:523–529, 1924.

Vinegar, A, Sinnett, EE, Leith, DE: Dynamic mechanisms determine functional residual capacity in mice, *Mus musculus*. J Appl Physiol Respir Environ Exercise Physiol 46:867–871, 1979.

Von Neergaard, K, Wirz, K: A method of measuring the lung elasticity in living humans, particularly in the presence of emphysema. Z Klin Med 105:35–50, 1927.

Watson, JW, Jackson, AC, Drazen, JM: Effect of lung volume on pulmonary mechanics in guinea pigs. J Appl Physiol 61:304–311, 1986.

West, JB: Pulmonary Pathophysiology—The Essentials. Baltimore, MD: Williams and Wilkins, 1977.

West, JB: Respiratory Physiology—The Essentials, 5th edition. Baltimore, MD: Williams and Wilkins, 1994.

Wilmore, JH, Costill, DL: Adequacy of the Haldane transformation in the computation of exercise VO_2 in man. J Appl Physiol 35:85–89, 1973.

Withers, PC: Measurement of VO_2, VCO_2 and evaporative water loss with a flow-through mask. J Appl Physiol Respir Environ Exercise Physiol 42:120–123, 1977.

Witschi, HP, Tryka, AF, Mauderly, JL, Haschek, WM, Satterfield, LC, Bowles, ND, Boyd, MR: Long-term effects of repeated exposure to 3-methylfuran in hamsters and mice. J Toxicol Environ Health 16:581–592, 1985.

Woolcock, AJ, Vincent, NJ, Macklem, PT: Frequency dependence of compliance as a test for obstruction in the small airways. J Clin Invest 48:1097–1106, 1969.

Woolf, CR: An assessment of the fractional carbon monoxide uptake and its relationship to pulmonary diffusing capacity. Dis Chest 46:181–189, 1964.

Yokoyama, E: Ventilatory functions of normal rats of different ages. Comp Biochem Physiol 75:77–80, 1983.

Young, RC, Nagano, H, Vaughan, TR, Staub, NC: Pulmonary capillary blood volume in dog: Effects of 5-hydroxytryptamine. J Appl Physiol 18:264–268, 1963.

Chapter Fourteen

Immunologic Responses of the Respiratory Tract to Inhaled Materials

David E. Bice

INTRODUCTION

The goal of this chapter is to help understand how interactions between the immune response and inhaled materials might cause pulmonary disease. To achieve this goal, it is first necessary to understand the mechanisms responsible for the induction of pulmonary immunity. Therefore, the cells and tissues responsible for pulmonary immune responses and their functions are described. Next, how inhaled materials might either suppress immunity or cause pulmonary hypersensitivity is presented. Finally, it is essential that animal models selected to study the effects of inhaled toxicants must have the same cellular and molecular responses in the induction and regulation of pulmonary immunity as humans. Otherwise, data from the animal model could not be used to estimate possible effects on human pulmonary immunity. Therefore, differences in pulmonary immunity among laboratory animal models and humans are also discussed.

PRIMARY IMMUNE RESPONSES AFTER
PULMONARY IMMUNIZATION

Many studies have evaluated the immune responses to antigens deposited in the lungs of experimental animals and humans. The results of these studies describe:

    1 the lymphoid tissues in which primary immune responses are produced after pulmonary immunization,
    2 the mechanisms responsible for the translocation of antigen from the lung to these tissues,
    3 the accumulation of immune cells in the lung, and
    4 the roles of immune cells that accumulate in the immunized lung in providing immune defense.

Although some studies used inhaled antigens, most have evaluated responses after instillation of antigen into the trachea of small laboratory species (Curtis and Kaltreider, 1989; Schnizlein et al., 1982; Bice and Schnizlein, 1980; Stein-Streilein and Hart, 1980; Stein-Streilein et al., 1979b; Curtis et al., 1990). In larger animals (e.g., dogs, nonhuman primates) and humans, antigen is usually instilled into selected airways of a single lung lobe by using a fiberoptic bronchoscope, while saline or other control materials are instilled into different lung lobes of the same animal (Bice et al., 1982a, 1982b; Mason et al., 1985, 1989). Both particulate and soluble antigens have been used (Bice et al., 1980a, 1982b; Mason et al., 1985, 1989; Weissman et al., 1992b).

Lung-Associated Lymph Nodes (LALN)

Production of Antigen-Specific T and B Lymphocytes Although lymphoid tissues and cells in the lung provide important immune defenses to antigen challenges, primary antibody and cellular immune responses to antigens instilled into the lungs are not produced in the lung, but in the LALN (Schnizlein et al., 1982; Bice et al., 1980b; Turner and Kaltreider, 1978; Stein-Streilein et al., 1979a; Bice and Schnizlein, 1980; Galvin et al., 1986; Hill and Burrell, 1979; Burrell and Hill, 1976). The LALNs function as effective filters to remove not only particulate materials but also antigens that clear from the lower respiratory tract via the lymphatics (Morrow, 1972; Snipes et al., 1983). Unless excessively high doses of antigen are used, the LALNs effectively remove antigen clearing from the lung. Lymphoid tissues that do not drain the lung are not normally involved in producing immunity after lung immunization. Only background numbers of antigen-specific antibody-forming cells (AFCs) are present in the spleen or lymph nodes that do not receive lymphatic drainage from the lung.

Translocation of Antigen to LALN   It is likely that most of the antigen deposited in the terminal airways and alveoli is cleared by phagocytosis by pulmonary alveolar macrophages (PAMs) and polymorphonuclear leukocytes (PMNs) with subsequent transport of the foreign material up the mucociliary escalator and out of the lung (Brain et al., 1978; Green et al., 1977; Morrow, 1972). However, some antigen is transported to the LALNs where a primary immune response is produced. How antigen leaves the lung and reaches these tissues is not completely understood.

Cell-free particles have been observed in the lymphatics, and antigen that enters the lung interstitium may pass directly to the LALNs via the lymphatics (Lauweryns and Baert, 1976). However, inflammation is produced in the lung after a primary lung immunization, and inflammation appears to be important in the induction of immunity in the LALNs (Bice and Muggenburg, 1985; Bice et al., 1980a, 1982a; Harmsen et al., 1985b, 1987; Hillam et al., 1985; Kaltreider et al., 1981). Soluble antigens and low doses of particulate antigens that do not induce pulmonary inflammation do not produce a primary immune response (Schatz et al., 1977). It is possible that an antigen dose that overwhelms normal phagocytic and clearance mechanisms in the lung is necessary for antigen to be cleared to the LALNs. PMNs that enter the lung during an inflammatory response can phagocytize particles in the lung and carry them to the lymph nodes that receive lymphatic drainage from the lung (Harmsen et al., 1987). Therefore, the production of lung inflammation and the accumulation of PMNs in the lung may enhance the transport of materials from the lung to the LALNs. In addition, PAMs also phagocytize particles in the alveoli and carry phagocytized particles to the LALNs (Corry et al., 1984; Harmsen et al., 1985b). It is not known if PAMs or PMNs also function as antigen-presenting cells in the LALNs, or if dendritic cells and/or lymph node macrophages are the cells responsible for antigen presentation in the LALNs.

Measurement of Immunity in the Blood after Lung Immunization

Large numbers of specific IgM, IgG, and IgA AFCs are found in the blood of dogs, cynomolgus monkeys, chimpanzees, and humans after immunization (Bice et al., 1980a, 1982a, 1982b; Mason et al., 1985; Mason and Bice, 1986; Weissman et al., 1994; Lue et al., 1988; Stevens et al., 1979; Thomson and Harris, 1977; Yarchoan et al., 1981). Antigen-specific AFCs in the blood after a primary immunization are produced by immune stimulation of the LALNs with a release of AFC into the efferent lymphatics (Bice et al., 1980a, 1982a; Brownstein et al., 1980; Kaltreider et al., 1981; Turner and Kaltreider, 1978). Antibody and antigen-specific T and B lymphocytes released from the LALNs would be distributed to all tissues of the body, including the lung. Memory lymphocytes would also be released into blood and would provide immune memory to the lung as well as to lymphoid tissues distant from the lung.

Immune Responses in Lung after Lung Immunization

Antigen-specific T lymphocytes are recruited from blood into the lung after a primary antigen exposure. The measurement of migration-inhibitory factor and the proliferation of lymphocytes after exposure of lung lymphocytes to antigen have been used to measure cell-mediated immunity in the lung (Bice and Schnizlein, 1980; Galvin et al., 1986; Ganguly and Waldman, 1977; Kaltreider et al., 1974; Lipscomb et al., 1981; Spencer et al., 1974; Waldman and Henney, 1971). Cell-mediated immunity to virus and other pathogens has also been measured in the lung (Spencer et al., 1974; Waldman et al., 1972).

In addition to antigen-specific T lymphocytes, large numbers of AFCs are found in lavage fluid from immunized lung lobes of dogs and nonhuman primates after localized pulmonary immunization. These AFCs are produced in the LALNs rather than in the lung (e.g., in bronchus-associated lymphoid tissues) (Bice and Muggenburg, 1985; Bice et al., 1980b, 1982a, 1982b; Brownstein et al., 1980; Hillam et al., 1985; Mason et al., 1985, 1989; Kaltreider et al., 1981; Van Der Brugge-Gamelkoorn et al., 1986). The AFCs produced in the LALNs are released into the blood, and large numbers of AFCs are recruited into immunized lung lobes. Although significantly lower numbers of AFCs are found in the saline-exposed control lung lobes, antigen-specific immune cells are present in lung lobes that were exposed only to saline or to nothing (Bice and Muggenburg, 1985; Bice et al., 1980a, 1982a, 1982b; Mason et al., 1985, 1989). Many lung AFCs mature to plasma cells and are found in both interstitial lung tissues and the alveoli (Bice et al., 1987a). The local production of antibody by AFCs recruited into the lung significantly increases the concentration of specific antibody in the alveoli (Harmsen et al., 1985a; Hill et al., 1983).

The recruitment of T lymphocytes and AFCs into the lung is not completely understood. Some data suggest that the recruitment of T lymphocytes into the lung is antigen specific (Lipscomb et al., 1982). However, the entry of AFCs into the lung is neither antigen- nor tissue-specific, and AFCs produced in lymphoid tissues that do not drain the lung enter an immunized lung lobe at the same rate as AFCs produced in the LALNs (Bice et al., 1982a; Hillam et al., 1985). Factors that appear to control recruitment of AFCs into the lung are inflammation and the activation status of the lymphocytes (North and Spitalny, 1974). Changes in adhesion molecules on the surface of lymphocytes and endothelial cells could be important. Although data are limited on the adhesion molecules on the surface of B lymphocytes, recent data suggest that most T lymphocytes which enter the lung are recently activated immune memory cells (CD45RO) (Marathias et al., 1991; Becker et al., 1990). In addition, inflammation causes phenotypic changes of endothelial cells that are important in the recruitment of lymphocytes. The expression of endothelial ligands for lymphocyte binding is either increased

(ICAM-1) or induced *de novo* (VCAM-1 and E-selectin) by inflammation (Shimizu et al., 1992). The interaction between adhesion molecules on the surface of recently activated lymphocytes and endothelial cells at sites of inflammation in the lung would result in recruitment. Recruitment by lymphocyte adhesion molecules and endothelial ligands would occur regardless of antigen specificity (Bice et al., 1982a) or the lymph node responsible for lymphocyte production (Hillam et al., 1985).

SECONDARY IMMUNE RESPONSES AFTER PULMONARY IMMUNIZATION
Immune Memory in Previously Immunized Lung Lobes

Memory lymphocytes for antibody- and cell-mediated immune responses are recruited and/or are produced in the lung after a primary exposure to antigen (Bice and Muggenburg, 1988; Mason et al., 1985, 1989; Hill and Burrell, 1979). The response of these cells in the lung to subsequent challenges with antigens, haptens, or infectious agents results in the amplification of T lymphocytes (Lipscomb et al., 1981; Rossman et al., 1988) and the localized production of high levels of antigen-specific antibody (Jones and Ada, 1987a, 1987b; Bice et al., 1993; Bice and Muggenburg, 1988; Mason et al., 1985). Pulmonary immune memory cells are present in the lung interstitium, but not in the alveoli (Mason et al., 1989).

Antigen Presentation in Secondary Immune Responses in the Lung

Alveolar macrophages, tissue macrophages, or dendritic cells could present antigen to immune memory cells in the lung. However, it is not clear what cell type or types are important in presenting antigens to immune cells in the lung.

Numerous studies have evaluated the function of PAMs as an antigen-presenting cell. Although the results of some studies suggest that PAMs will support antigen-stimulation of lymphocytes *in vitro* (Lipscomb et al., 1981; Schuyler and Todd, 1981; Weinberg and Unanue, 1981), other studies indicate that PAMs may actually suppress cellular (Demenkoff et al., 1980; Holt, 1978) and antibody immunity (Pennline and Herscowitz, 1981; Lawrence et al., 1982).

Lung dendritic cells can present antigen (Nicod et al., 1987; Wilkes and Weissler, 1994), and a number of studies have identified these cells in the lung (Holt and Schon Hegrad, 1987; Holt et al., 1985; Xia et al., 1991; Simecka et al., 1992). Cells responding to antigen challenge are in the interstitial tissues, rather than the alveoli, supporting the possibility that dendritic cells and/or tissue macrophages are the antigen-presenting cells (Mason et al.,

1989). It is possible that more than one cell type is important in antigen presentation in the lung. For example, some data support the possibility that dendritic cells and parenchymal macrophages may present antigen in T cell responses, while PAMs are more important in antibody responses (Wilkes and Weissler, 1994).

Long-Term Immunity in the Lung

After antigen challenge, antibody production continues in the lung for several years without any additional exposures to antigen (Bice et al., 1991, 1993). The cells responsible for this long-term antibody production are predominantly in the interstitial lung tissues of lung lobes exposed to antigen (Mason et al., 1989). Antibody is present in lavage fluid from control lung lobes and serum. However, no AFCs are found in blood or in lavage fluid or tissues from control lung lobes. In addition, at 2 years after the last exposure to antigen, no antibody production was observed in the LALNs or in any distant lymphoid tissues (Bice et al., 1993). Therefore, the antibody found in serum and in lavage fluid from control lung lobes must have been produced by AFCs in the immunized lung lobes. These observations support the possibility that antibody to antigens repeatedly deposited in the lung (e.g., allergens) is produced by lung tissues, while distant lymphoid tissues play little or no role in long-term antibody production.

The mechanisms responsible for the maintenance of long-term antibody production are not understood. Either long-lived AFCs (Holt et al., 1984; Miller, 1963) or the retention of antigen on the surface of antigen-presenting cells (Nicod et al., 1987; Wilkes and Weissler, 1994) could be responsible for maintaining antibody production in the lung for several years after the last exposure to antigen.

Fewer data are available concerning the maintenance of cellular immunity in the lung for extended times. However, some data support the possibility that T lymphocytes remain in the lung after antigen stimulation, and the exchange between these lymphocytes and the systemic immune system may be limited (Saltini et al., 1988).

IMMUNE CELLS AND TISSUES AT RISK TO INHALED MATERIALS

The identification of cells and tissues that are responsible for immunity to antigens deposited in the lung has also identified the immune cells and tissues at risk to inhaled materials. A primary toxicology concern is that inhaled toxicants might damage immune cells and lung tissues and suppress pulmonary immune responses. The suppression of pulmonary immunity would lead to increased pulmonary infectious diseases. However, it is also possible that pollutant exposures might elevate pulmonary immunity and increase sensitivity to inhaled antigenic materials. Listed below are changes that could be induced by inhaled pollutants which might suppress or stimulate pulmonary immune responses.

Changes in Antigen Clearance

Decreased Antigen Clearance   Damage to the pulmonary epithelial cells (e.g., Type I and/or Type II) and/or changes in numbers of epithelial cells (e.g., Type II cell hyperplasia) could decrease the lymphatic clearance of antigens from the lung to the LALNs (Nicod et al., 1987; Wilkes and Weissler, 1994). Changes in the numbers and functions of PAMs and PMNs that enter the lung could also decrease antigen clearance (Harmsen et al., 1985b, 1987).

Increased Antigen Clearance   It is also possible that exposure of the lung to toxicants can elevate immune responses to antigens deposited in the lung. Inflammation is produced by inhaled materials (Bice et al., 1985, 1987b; Schnizlein et al., 1982). Because PMNs and PAMs appear to be important in the transport of antigen to the LALNs, inflammatory responses in the lung could increase the amount of antigen that translocates from the lung to the LALNs.

Changes in Function of the LALNs

Some inhaled toxic particulates are cleared from the lung to the LALNs. Insoluble toxic particles accumulate in the LALNs and have a long half-life in these tissues (Snipes et al., 1983). The accumulation of toxic particles in the LALNs could alter the number of lymphocytes and/or the subpopulations of lymphocytes in the LALNs as well as the functions of these cells. The numbers and functions of antigen-handling cells in the LALNs could also be changed.

In addition to immune suppression, particles cleared to the LALNs increase the level of immunity in these lymph nodes after pulmonary immunization. Cellularity of the LALNs usually increases after exposures to particulate materials, and immune responses to antigens deposited in the lungs of these animals are frequently elevated (Bice et al., 1985, 1987b; Schnizlein et al., 1982).

Changes in the Recruitment of Immune Cells into the Lungs

Increasing the vascular permeability of the lung increases the recruitment of lymphoid cells and AFCs into the lung (Bice et al., 1982a). Therefore, the induction of inflammation and/or damage to lung tissues by inhaled materials could increase the recruitment of immune cells and antibody from the blood into the lung. These changes could increase immune sensitivity to inhaled antigens and allergens.

Changes in Function of Immune Cells in the Lung

The locations of AFCs and immune memory cells in the alveoli and interstitial tissues of the lung suggest that these cells are at risk to inhaled pollutants. A loss of immune memory or long-term antibody production in the lung after inhalation of toxicants could increase the incidence of recurrent pulmonary infections.

EFFECTS OF INHALED MATERIALS ON PULMONARY IMMUNE RESPONSES
Effects on Pulmonary Immune Defenses in Humans

Some data suggest that the inhalation of environmental pollutants suppresses pulmonary immune defenses. Individuals who smoke have an increased number of pulmonary infections (Finklea et al., 1969, 1971), and there is a higher rate of pulmonary infections in residents of highly polluted cities (Bouhuys et al., 1978; Green, 1974; Levy et al., 1977; Lloyd, 1978). Preliminary data show that cigarette smoking significantly suppresses immune responses to antigens deposited in the lung (Weissman et al., 1992a). Additional data are needed to determine if inhalation of relatively low levels of environmental pollutants can significantly suppress pulmonary immunity in humans.

Effects on LALNs Functions

Data suggest that probably all laboratory animals produce a primary immune response in the LALNs after pulmonary immunization (Bice and Shopp, 1988). Therefore, it is likely that any animal could be used to evaluate the effects of inhaled toxicants on immune responses in the LALNs. Studies have evaluated the effects of particulate materials, including pulmonary exposures to carbon (Miller and Zarkower, 1974; Zarkower, 1972), silica (Bice et al., 1987b; Miller and Zarkower, 1976), fly ash (Bice et al., 1987b), benzo-a-pyrene (Schnizlein et al., 1982), diesel exhaust (Bice et al., 1985), cigarette smoke (Sopori et al., 1989; Savage et al., 1991; Thomas et al., 1974a, 1974b), and radioactive particles (Bice et al., 1979; Galvin et al., 1989) on the development of immunity in the LALNs. These studies measured effects on immune responses in the LALNs as well as on immunity in distant lymphoid tissues. After either acute or chronic exposures, the immune functions of LALNs can be suppressed, and this damage can be permanent (Bice et al., 1987b). Damage of lung tissues by inhaled toxic gasses can also suppress the induction of immunity after lung immunization (Hillam et al., 1983; Schnizlein et al., 1980).

Whether immune responses in the LALNs of animals exposed to pollutants will be suppressed or elevated appears to be determined by several factors. For example, the dose and toxicity of the material inhaled are important factors. Frequently, suppression is observed only at high doses of toxic materials (Bice et al., 1987b; Miller and Zarkower, 1974, 1976). Exposures to lower doses or less toxic pollutants frequently elevates immunity to antigens in the lung (Bice et al., 1985, 1987b). The time of exposure in relation to when antigen is deposited in the lung also seems important. A single acute exposure can either suppress or elevate immunity to antigen deposited in the lung depending on when the animals are immunized in relation to the exposure (Hillam et al., 1983; Schnizlein et al., 1980, 1982).

Most of these exposures were at doses much higher than would be encountered in the environment. In addition, the immune system offers great redundancy, and even if the responses in the LALNs were suppressed, systemic immune responses would likely not be changed. Even if the LALNs were damaged, other lymphoid tissues could replace their filtration and immune functions to respond to antigens deposited in the lung.

Although immune responses in the LALNs can be altered by inhaled materials, the mechanisms responsible for the observed changes are poorly understood. In addition, it is not known if inhaled pollutants damage the immune response of the LALNs in humans.

Effects on Immune Cells in the Lung

The effects of inhaled materials on the recruitment of lymphocytes into the lung or on the immune functions of lymphoid cells in the lung have not been evaluated. Because AFCs and immune memory cells are in the alveoli and interstitial lung tissues, the immune functions of these cells could be damaged by the inhalation of toxic materials. Studies should be carried out to determine the effects of inhaled toxicants on the functions of lymphocytes in the lung.

Pulmonary inflammation caused by inhaled toxicants could increase the recruitment of lymphocytes into the lungs. Although an increased recruitment of lymphocytes into the lung may have no effect, such a change might increase the level of immunity to inhaled allergens. Studies should be carried out to determine if inhaled pollutants increase the immune recognition of commonly inhaled antigens and allergens (see below).

Although probably all species produce immune responses in the LALNs after lung immunization, only dogs and nonhuman primates have been shown to accumulate large numbers of lymphocytes in their alveoli and interstitial lung tissues. Rodents and rabbits can accumulate large numbers of lymphocytes along lung airways in bronchus-associated lymphoid tissues (BALT) (Bienenstock and Befus, 1984). However, humans most likely do not have BALT, and the use of animals with BALT may not provide data that can be extrapolated to humans (Pabst and Gehrke, 1990). The pulmonary immune responses that develop in dogs and nonhuman primates appear to be most like the responses in observed in humans after pulmonary exposure to antigenic materials (Bice, 1993; Bice and Shopp, 1988; Mason et al., 1989; Weissman et al., 1994). Of the rodents evaluated, only mice had increased AFCs in their lungs after antigen challenge. The numbers were small compared to dogs and nonhuman primates and were found only after high doses of antigen or after antigen challenge of immune mice (Curtis and Kaltreider, 1989; Kaltreider et al., 1983; McLeod et al., 1978).

INDUCTION OF PULMONARY HYPERSENSITIVITY BY INHALED MATERIALS

There are two ways in which inhaled materials could cause pulmonary hypersensitivity. First, it has been recognized for many years that some inhaled materials are antigens or haptens, and that the production of immune responses to these materials can cause serious lung disease in susceptible individuals. Second, it is possible that inhaled materials may alter pulmonary immunity to environmental allergens and increase the incidence of allergic lung diseases (e.g., allergic asthma). Both of these possibilities are discussed below.

Pulmonary Hypersensitivity Induced by Immunity to Inhaled Materials

Lung diseases induced by immunity to inhaled antigens or chemicals can be divided into three general categories. Examples of these diseases include (1) IgE-mediated asthma or rhinitis; (2) hypersensitivity pneumonitis (where cellular immunity or a combination of cellular and humoral immunity induce disease); and (3) chronic beryllium lung disease induced by cell-mediated immune responses.

Occupational Asthma, Rhinitis   Occupational asthma is characterized by variable airflow limitation and/or nonspecific bronchial hyperresponsiveness due to causes and conditions which are attributable to a particular occupational environment. Frequently, the etiological agents that cause occupational asthma are antigens or haptens, and asthma is produced by the induction of immunity. The production of IgE is frequently responsible for occupational asthma, although some etiological agents may cause asthma by non-IgE-dependent mechanisms. Most etiological agents that stimulate the production of IgE are high-molecular-weight proteins or polysaccharides (e.g., animal products, plant proteins, enzymes) (Botham et al., 1987; Flindt, 1969; Pepys et al., 1969), although some low-molecular-weight compounds can induce an IgE response as haptens (e.g., acid anhydrides, metals) (Bernstein and Brooks, 1993; Hunter et al., 1945; Zeiss and Patterson, 1993). Most low-molecular-weight compounds that induce occupational asthma apparently produce disease by non-IgE immune mechanisms that are not completely understood. Specific IgE antibodies have not been found or are found in only a few patients (Chan-Yeung and Malo, 1994). In experimental animals, airway responses are observed regardless of whether antibodies are present or not. Examples of low-molecular-weight agents include diisocyantes, western red cedar, and amines (Vandenplas et al., 1993; Hagmar et al., 1987; Chan-Yeung et al., 1973). The induction of pulmonary inflammation appears to play an important role in occupational asthma, and the pathology of a patient who died from isocyanate-induced asthma was the same as in individuals who have died from status asthmaticus (Fabbri et al., 1988). A few agents such as formaldehyde seem to cause asthma with no immunological mechanism (Chan-Yeung and Malo, 1994). Additional data are needed to fully understand the mechanisms responsible for occupational asthma caused by non-IgE and nonimmune responses.

Hypersensitivity Pneumonitis Hypersensitivity pneumonitis is characterized by a spectrum of lymphocytic and granulomatous, interstitial and alveolar-filling, pulmonary disorders. Hypersensitivity pneumonitis is associated with intense and often prolonged exposure to a wide range of inhaled antigens and possibly some haptens found in a variety of industries. The number of organic materials that can induce hypersensitivity pneumonitis continues to grow, with disease classification based on the material inhaled. It appears that chronic inhalation of high levels of nearly any organic dust can lead to hypersensitivity pneumonitis (Salvaggio, 1987).

Data indicating that hypersensitivity pneumonitis can be induced by inhaled chemicals are limited. However, it appears that immune responses to proteins altered by chemical binding or immune responses to hapten-protein conjugates may induce hypersensitivity pneumonitis. Toluene diisocyanate, trimellitic anhydride, diphenylmethane diisocyanate, and heated epoxy resin are examples of chemicals that may induce hypersensitivity pneumonitis (Musk et al., 1988; Salvaggio, 1987). Pulmonary opacities resulting from diisocyanate exposure have been reported (Blake et al., 1965), and diisocyanate exposure was suspected to be the cause of several cases of hypersensitivity pneumonitis (Charles et al., 1976). Additional reports implicate methylene diphenyl diisocyanate and hexamethylene diisocyanate as the cause of reactions similar to hypersensitivity pneumonitis (Baur et al., 1984; Fink and Schlueter, 1978; Malo et al., 1983; Malo and Zeiss, 1982; Zeiss et al., 1980). One report described the presence of isocyanate-specific IgG antibody in an individual with pulmonary disease similar to hypersensitivity pneumonitis (Le Quesne et al., 1976).

High levels of antigen-specific antibody are usually present and may also be involved in the induction of hypersensitivity pneumonitis. However, the roles of cellular and antibody-mediated immune responses in hypersensitivity pneumonitis are not completely clear (Salvaggio, 1987). Some data are available indicating that lymphocytes from immune animals can passively transfer sensitivity that results in lung lesions after pulmonary challenge (Bice et al., 1976; Schuyler et al., 1987, 1988). The results of these studies show that immune cells can transfer cellular immunity to the lung. However, the responses produced are transient and may not be representative of immune reactions in the lungs of the few individuals who develop hypersensitivity pneumonitis with lung damage.

Chronic Beryllium Lung Disease   Beryllium disease is a product of the 20th-century arrival of the high technologies of nuclear power, aerospace, and electronics. Beryllium was not produced commercially in the United States until the 1930s. However, numerous cases of disability and death from beryllium poisoning occurred among workers in the fluorescent lamp industry after the introduction of beryllium as a component of phosphors (Hardy, 1980). Although exposures to beryllium have been greatly reduced by improved exposure standards, the amount of beryllium produced and used has greatly increased, and new cases of chronic beryllium lung disease continue to be diagnosed.

Two different diseases are produced by inhaled beryllium. Acute beryllium lung disease (a severe chemical pneumonitis) is due directly to the toxicity from the inhalation of relatively high levels of inhaled beryllium. In contrast, chronic beryllium lung disease is caused by the induction of immunity to low levels of inhaled beryllium. Because beryllium ions are too small to be antigenic, it has been assumed that they must bind to a protein to form a hapten-carrier complex. Immune cells that have specificity for beryllium are recruited into and/or produced in the lung after beryllium exposures. The number of beryllium-specific lymphocytes are increased in the lung by a local amplification, probably in response to beryllium retained in the lung (Rossman et al., 1988; Newman et al., 1994a). Pulmonary immune responses to beryllium and the relatively long retention of beryllium may be important in the development of granulomatous lung disease. It is possible that the continual activation of immune cells in the lung by beryllium leads to a local secretion of lymphokines which could be responsible for the induction of lung granulomas.

One of the most difficult clinical problems of chronic beryllium disease is its differential diagnosis from sarcoidosis, a systemic granulomatous disease of unknown etiology (Sprince et al., 1976; Newman and Kreiss, 1992). The criteria for diagnosis of chronic beryllium disease adopted by the Beryllium Case Registry at the Massachusetts General Hospital require at least four of the following six features and must include at least one of the first two (Sprince and Kazemi, 1980):

1   epidemiologic evidence of significant beryllium exposure;
2   presence of beryllium in lung tissue, lymph nodes, or urine;
3   evidence of lower respiratory tract disease and a clinical course consistent with beryllium disease;
4   radiologic evidence of interstitial disease consistent with a fibronodular process;
5   evidence of a restrictive or obstructive ventilatory defect or diminished carbon monoxide diffusing capacity; and
6   pathologic changes consistent with beryllium disease on examination of lung tissue and/or lymph nodes.

More recently, the *in-vitro* lymphocyte stimulation assay to identify beryllium-sensitized lymphocytes in blood and cells obtained from the lung by bronchoalveolar lavage has been used (Rossman et al., 1988; Stokes and Rossman, 1991; Aronchick et al., 1987; Newman et al., 1994b; Mroz et al., 1991). Peripheral blood lymphocytes generally test positive when obtained from patients with beryllium disease, and negative in unexposed control subjects. Although there has been some problems in the ability to correctly identify patients with beryllium disease among laboratories, recent results show that lymphocyte stimulation using peripheral blood lymphocytes is an acceptable method to diagnosis beryllium sensitivity (Mroz et al., 1991). Other studies suggest that local proliferation of beryllium sensitized lymphocytes in the lungs of patients with chronic beryllium lung disease (or in the lungs of dogs that inhaled beryllium) provide a population of lymphocytes that can be lavaged from the lung and used in lymphocyte stimulation assays (Newman et al., 1994a; Haley et al., 1992; Rossman et al., 1988). These results suggest that lymphocyte stimulation assays using lung lavage cells may be more predictive of chronic beryllium lung disease than stimulation assays using lymphocytes from peripheral blood.

Factors Which Control Pulmonary Immunity to Hypersensitivity Antigens or Haptens

To induce pulmonary immunity and hypersensitivity, the chemicals or organic materials inhaled must be of a respirable size and deposit in the lower respiratory tract (Salvaggio, 1987). Particles greater than 10 μm generally do not induce pulmonary immunity or hypersensitivity lung disease. Low-molecular-weight chemicals apparently act as haptens, binding host proteins to form complete antigens that can induce hapten- or carrier-specific antibodies. Some highly reactive chemicals, (e.g., isocyanates or anhydrides) may induce the formation of neoantigens that stimulate antibody production (Butcher et al., 1982a; Zeiss et al., 1980). Therefore, immune responses to inhaled chemicals may be directed against a broad range of antigen specificities, from haptens to altered host protein neoantigens.

Genetic susceptibility appears to be an important factor in the development of pulmonary hypersensitivity (Demoly et al., 1993; Godfrey, 1993).

Atopy is a predisposing factor for the development of occupational asthma induced by high-molecular-weight sensitizers (e.g., enzymes, flour, animal products) (Newhouse et al., 1970). However, there is no relationship between atopy and the induction of lung disease by low-molecular-weight sensitizers such as toluene diisocyanate, plicatic acid, and trimellitic anhydrides. Other factors which may be important in the initiation of disease include the presence of preexisting bronchial hyperreactivity (Butcher et al., 1982b; Lam et al., 1979; O'Brien et al., 1979; Van Ert and Battigelli, 1975), altered adrenergic tone (Szentivanyi, 1968), recent or concurrent respiratory viral infections (Empey et al., 1976), or alterations in the integrity of tight junctions of basal membranes (McFadden, 1984). It is not known if environmental pollutants increase sensitivity to these inhaled antigens and the development of pulmonary hypersensitivity. However, smoking does not appear to be a predisposing factor (Chan-Yeung and Lam, 1986; Cotes and Steel, 1987). Many of the antigens that cause hypersensitivity pneumonitis are highly antigenic and may have adjuvant properties which may be important in the induction of disease (Bice et al., 1976). Frequently, chemicals or organic materials also induce inflammation that could increase the chance that an immune response will be produced.

Pulmonary Immunity versus Hypersensitivity   In the early studies of hypersensitivity, it was assumed that hypersensitivity lung disease would result any time a pulmonary immune response was induced to antigens that cause hypersensitivity. More recent data suggest that exposure to these antigens almost always results in immunity and that these immune responses can be maintained for long times without the development of clinical disease. A specific antibody is detected in the blood of most individuals exposed to antigens that can cause hypersensitivity pneumonitis, even though only about 3–4% of those exposed develop hypersensitivity pneumonitis. Therefore, the levels of antibody in blood do not correlate with disease (Pepys, 1986) and the induction of immunity to inhaled antigens may only indicate that an individual has been exposed. Not only do many individuals exposed to antigens that can cause hypersensitivity pneumonitis have elevated antibody levels with no symptoms, but also clinically normal individuals can have significantly elevated numbers of lymphocytes in their lungs for extended periods of time with no measurable pulmonary disease (Cormier et al., 1984, 1986, 1987).

The inhalation of allergens that cause atopic diseases (e.g., asthma) offer an additional example where immunity can be developed in the absence of disease. Most individuals produce antigen-specific antibody to inhaled allergens but the immune response produced does not cause atopic disease (Kemeny et al., 1989). Individuals who are genetically susceptible for atopic disease produce allergen-specific immune responses, including allergen-specific IgG subclasses and IgE, which cause asthma and rhinitis (Jimeno et al., 1992).

Although it is possible that chronic beryllium lung disease occurs only in individuals who can produce an immune response to beryllium, data from experimental animal studies suggest that most animals exposed to beryllium develop immunity and lung lesions (Haley et al., 1992). However, even though their histopathology is suggestive of chronic beryllium lung disease, these lesions resolve. It is not known if most humans exposed to beryllium produce an immune response to beryllium but only a few develop chronic beryllium lung disease. Published data suggest that only individuals who produce an immune response to beryllium develop chronic beryllium lung disease (Kreiss et al., 1993; Mroz et al., 1991).

Even though pulmonary hypersensitivity cannot be induced in the absence of an immune response, these examples suggest that factors other than the development of immunity must determine whether or not disease is produced. Disorders in the regulation of immunity (rather than the induction of immunity) may be responsible for the induction of hypersensitivity. Recent studies suggest that pulmonary immune responses to inhaled allergens which either cause asthma or result in no lung disease are controlled by subclasses of T helper lymphocytes (Cogan et al., 1994; Kay, 1991; Parronchi et al., 1991; Robinson et al., 1993). Normal individuals (those who are not asthmatic) theoretically respond to antigens by stimulation of Th1 helper cells. Asthmatics theoretically respond to inhaled antigens by a Th2 helper cell immune response. Th1 and Th2 immune responses can be differentiated by the production of cytokines by immune cells. T lymphocytes from normals (Th1) produce mRNA for IL-2 and IFN_γ, while T lymphocytes from asthmatics (Th2) produce mRNA for GM-CSF, IL-4, IL-5, and IL-10 (Cogan et al., 1994; Kay, 1991; Parronchi et al., 1991; Robinson et al., 1993). Immune responses in normals produced by Th1 helper T lymphocytes result in cell-mediated immunity and the production of IgG. Th2 helper T cells are responsible for the induction of IgE immunity important in allergic diseases. The possibility that immune responses in asthmatics and normals are controlled by Th1 and Th2 cells provides an important new insight to allergic disease.

As with asthma, chronic beryllium lung disease may develop only in individuals who are genetically susceptible. Recent data support this possibility and suggest that genetically susceptible individuals express the HLA-DPB1*0201-associated glutamic acid at residue 69, a position involved in the susceptibility to auto-immune disorders. Therefore, HLA-DP may have a role in conferring susceptibility, and residue 69 of HLA-DPB1 might be used in risk assessment for chronic beryllium lung disease (Richeldi et al., 1993).

Pulmonary Hypersensitivity to Environmental Allergens

Epidemiologic data suggest that the incidence of allergic lung disease (e.g., asthma or rhinitis induced by pollen inhalation) is increasing as is the number of deaths due to asthmatic responses (Evans et al., 1987; Gerstman et al., 1989; Manfreda et al., 1993). The factors responsible for this increasing incidence of allergic disease are unknown.

There is also a strong link between inhaled pollutants and airway hyperreactivity, and hospital admissions of asthmatics is significantly correlated with pollutant exposures (Seltzer et al., 1986; Moseholm et al., 1993; Pope, III, 1989; Pope, 1991). The inhalation of passive cigarette smoke by children increases the incidence of asthma. Children raised in the homes of smokers have a significantly higher incidence of asthma than children raised in the homes of nonsmokers (Ehrlich et al., 1992; Duff et al., 1993; Dekker et al., 1991; Martinez et al., 1992). Therefore, it appears likely that the inhalation of passive cigarette smoke by children triggers the development of asthma in susceptible individuals.

Although there is no clear link between pollutant inhalation and the induction of asthma to inhaled allergens, asthma is an urban disease, and the incidence of asthma is significantly lower in rural areas. It may be possible that the higher level of pollutants in urban areas is responsible for the increased incidence of asthma. It is also possible, however, that urban communities have higher levels of potent allergens (e.g., house dust mite, cockroach allergen) (Rizzo et al., 1993; Gelber et al., 1993; Sporik and Platts-Mills, 1992; Call et al., 1992; Duff and Platts-Mills, 1992). In contradiction to these findings, some data show no correlation between inhaled pollutants and asthma (von Mutius et al., 1994). Additional data are needed to determine if inhaled environmental pollutants other than cigarette smoke play any role in increasing the incidence of allergic sensitivity to inhaled allergens.

How inhaled cigarette smoke or environmental pollutants trigger the development of asthma in children is not known. It is possible that elevated inflammation in the lungs of children might increase the level of inhaled allergens that are cleared to the LALNs. Increasing the amount of allergens cleared to the LALNs could be responsible for the development of allergic immune responses in susceptible individuals. Experimental data show that a variety of inhaled materials frequently elevate immune responses in the LALNs after immunization of the lungs (Bice et al., 1985, 1987b; Hillam et al., 1983; Schnizlein et al., 1980, 1982).

In addition to elevated immune responses in the LALNs, inhaled pollutants might also increase the incidence of asthma by altering the number and function of immune cells in the lung. The immune response in the lung (e.g., local production of IgE or recognition of inhaled antigen by T lymphocytes) is probably much more important in the induction of pulmonary hypersensitivity disease than is systemic immunity. Elevated inflammation in the lung from inhalation of pollutants could increase the number of T and B lymphocytes recruited into the lung. There are no data on the possibility that inhaled pollutants may increase Th2 versus Th1 immune responses in the lung.

ANIMAL MODELS FOR HYPERSENSITIVITY

Although animal models have been devised to study the effects of inhaled materials on pulmonary immunity and hypersensitivity, differences in species have made the extrapolation of data from these studies to human disease difficult. Some of the problems with using animal models to study the effects of inhaled materials on pulmonary immunity are discussed below.

Animal Models Used to Evaluate Occupational Asthma

The guinea pig has frequently been used to evaluate asthmatic responses to chemicals (Karol, 1991; Griffiths-Johnson et al., 1993; Huang et al., 1993). Genetically hyperresponsive rats and dogs; ozone-exposed, hyperresponsive dogs and sheep; mice; and rabbit models of late allergic reactions have also been used to study bronchial hyperresponsiveness (Hirshman and Downes, 1985, 1986; Kepron et al., 1977; Tse et al., 1979; Woolcock, 1988). However, few animals spontaneously develop a response to inhaled antigens that is the same as asthma in humans. Nonhuman primates may be a good model because they do develop pulmonary hypersensitivity diseases. Dogs develop IgE-mediated skin disease to allergens (Kleinbeck et al., 1989; DeBoer et al., 1993; Hill and DeBoer, 1994), and injections of ragweed antigen with aluminum hydroxide as an adjuvant (starting within 24 hours of birth) may offer a valid model of asthma (Becker et al., 1989; Kepron et al., 1977). Therefore, the dog may also provide a good model to study the induction of occupational asthma.

Animal Models Used to Evaluate Hypersensitivity Pneumonitis

Several animal species have been used to evaluate pulmonary responses to antigens which can cause hypersensitivity pneumonitis (Keller et al., 1982; Richerson et al., 1972, 1978, 1991; Schuyler et al., 1992, 1994). Studies carried out with these animal models have provided important data concerning the induction of inflammation and immunity by either inhalation or instillation of antigens that can cause hypersensitivity pneumonitis. However, even though intense pulmonary immunity is produced, none of the experimental animals develop hypersensitivity pneumonitis with permanent lung damage as seen in humans. It is possible that when these experimental animals are exposed to these antigens, the response produced is the same as observed in approximately 97% of exposed humans who do not develop hypersensitivity pneumonitis. In fact, after multiple exposures of experimental animals, the accumulation of cells in the lung in response to antigen exposure can disappear (Butler et al., 1982; Richerson et al., 1981).

Animal Models Used to Evaluate Sensitivity to Beryllium

Rats and guinea pigs have been used to evaluate the induction of immunity to instilled or inhaled beryllium as well as the production of pulmonary lesions (Barna et al., 1981, 1984; Schepers et al., 1957). Different strains of guinea pigs who are either sensitive or nonsensitive for the induction of granulomas after instillation of beryllium have also been used. Unlike human chronic beryllium lung disease that is produced by low doses of beryllium, however, some studies with experimental animals used large doses of beryllium to induce immunity and/or lung lesions (Barna et al., 1981, 1984). In some animals, the histopathology of the lesions produced are not similar to the lesions observed in humans (Schepers et al., 1957; Haley, 1991; Haley et al., 1990).

Studies in dogs showed that this species develops pulmonary immunity to beryllium and lung lesions which mimics the responses observed in the human lung (Haley et al., 1989). As described above, there are significant differences in cellular responses in the lungs of various animals after lung immunization (Bice and Shopp, 1988). Because immune responses in the lungs of dogs appear to be the same as those observed in nonhuman primates, dogs may prove especially useful as a model for evaluating the effects of lung immunity on the development of pulmonary disease. However, a criticism of this model may be the same as that for other animal models used to study the induction of hypersensitivity lung diseases. The responses observed may represent the responses of normal individuals who do not develop chronic beryllium lung disease, rather than the responses of an individual that does develop chronic beryllium lung disease.

SUMMARY

The identification of tissues and cells responsible for the development of immune responses after lung immunization has helped understand the roles of pulmonary immunity in lung defense as well as in pulmonary hypersensitivity. These data also provide the background necessary to design studies on the effects of inhaled materials on pulmonary immunity and to determine how altered pulmonary immune responses could cause either suppress defenses in the lung or lead to hypersensitivity lung diseases. Because there are species differences in the development of pulmonary immune responses, the selection of the species is important in the design of pulmonary immunotoxicology studies.

ACKNOWLEDGMENT

The authors thank the personnel at the Inhalation Toxicology Research Institute for their suggestions and assistance in the preparation of this manuscript. This review is based in part on research performed under Depar ment of Energy Contract No. DE-AC04-76EV01013 and was conducted in facilities fully accredited by the American Association for Accreditation of Laboratory Animal Care.

REFERENCES

Aronchick, JM, Rossman, MD, Miller, WT: Chronic beryllium disease: diagnosis, radiographic findings, and correlation with pulmonary function tests. Radiology 163:677–682, 1987.

Barna, BP, Chiang, T, Pillarisetti, SG, Deodhar, SD: Immunologic studies of experimental beryllium lung disease in the guinea pig. Clin Immunol Immunopathol 20:402–411, 1981.

Barna, BP, Deodhar, SD, Chiang, T, Gautam, S, Edinger M: Experimental beryllium-induced lung disease. I. Differences in immunologic responses to beryllium compounds in strains 2 and 13 guinea pigs. Int Arch Allergy Appl Immunol 73:42–48, 1984.

Baur, X, Dewair, M, Rommelt, H: Acute airway obstruction followed by hypersensitivity pneumonitis in an isocyanate (MDI) worker. J Occup Med 26:285–287, 1984.

Becker, AB, Hershkovich, J, Simons, FER, Simons, KJ, Lilley, MK, Kepron, MW: Development of chronic airway hyperresponsiveness in ragweed-sensitized dogs. J Appl Physiol 66:2691–2697, 1989.

Becker, S, Harris, DT, Koren, HS; Characterization of normal human lung lymphocytes and interleukin-2-induced lung T cell lines. Am J Respir Cell Mol Biol 3:441–448, 1990.

Bernstein, IL, Brooks, SM: Metals. In: Asthma in the Workplace. Edited by IL Bernstein, M Chan-Yeung, JL Malo, DI Bernstein, pp. 459–479. New York: Marcel Dekker Inc., 1993.

Bice, DE: Pulmonary responses to antigen. Chest 103:95S–98S, 1993.

Bice, DE, Degen, MA, Harris, DL, Muggenburg, BA: Recruitment of antibody-forming cells in the lung after local immunization is nonspecific. Am Rev Respir Dis 1982: 635–639, 1982a.

Bice, DE, Gray, RG, Evans, MJ, Muggenburg, BA: Identification of plasma cells in lung alveoli and interstitial tissues after localized lung immunization. J Leuko Biol 41:1–7, 1987a.

Bice, DE, Hahn, FF, Benson, JM, Carpenter, RL, Hobbs, CH: Comparative lung immunotoxicity of inhaled quartz and coal combustion fly ash. Environ Res 43:374–389, 1987b.

Bice, DE, Harris, DL, Hill, JO, Muggenburg, BA, Wolff, RK: Immune responses after localized lung immunization in the dog. Am Rev Respir Dis 122:755–760, 1980a.

Bice, DE, Harris, DL, Muggenburg, BA: Regional immunologic responses following localized deposition of antigen in the lung. Exp Lung Res 1:33–41, 1980b.

Bice, DE, Harris, DL, Muggenburg, BA, Bowen, JA: The evaluation of lung immunity in chimpanzees. Am Rev Respir Dis 1982:358–359, 1982b.

Bice, DE, Harris, DL, Schnizlein, CT, Mauderly, JL: Methods to evaluate the effects of toxic materials deposited in the lung on immunity in lung-associated lymph nodes. Drug Chem Toxicol 2:35–47, 1979.

Bice, DE, Jones, SE, Muggenburg, BA: Long-term antibody production after lung immunization and challenge: role of lung and lymphoid tissues. Am J Respir Cell Mol Biol 8:662–667, 1993.

Bice, DE, Mauderly, JL, Jones, RK, McClellan, RO: Effects of inhaled diesel exhaust on immune responses after lung immunization. Fundam Appl Toxicol 5:1075–1086, 1985.

Bice, DE, Muggenburg, BA: Effect of age on antibody responses after lung immunization. Am Rev Respir Dis 132:661–665, 1985.

Bice, DE, Muggenburg, BA: Localized immune memory in the lung. Am Rev Respir Dis 138:565–571, 1988.

Bice, DE, Salvaggio, J, Hoffman, E: Passive transfer of experimental hypersensitivity pneumonitis with lymphoid cells in the rabbit. J Allergy Clin Immunol 58:250–262, 1976.

Bice, DE, Schnizlein, CT: Cellular immunity induced by lung immunization of Fischer 344 rats. Int Arch Allergy Appl Immunol 63:438–445, 1980.

Bice, DE, Shopp, GM: Antibody responses after lung immunization. Exp Lung Res 14:133–155, 1988.

Bice, DE, Weissman, DN, Muggenburg, BA: Long-term maintenance of localized antibody responses in the lung. Immunology 74:215–222, 1991.

Bienenstock, J, Befus, D: Gut and bronchus-associated lymphoid tissue. Anat Rec 170:437, 1984.

Blake, BL, Mackay, JB, Rainey, HB, Weston, WJ: Pulmonary opacities resulting from diisocyanate exposure. J Coll Rad Aust 9:45–48, 1965.

Botham, PA, Davies, GE, Teasdale, EL: Allergy to laboratory animals: a prospective study of its incidence and of the influence of atopy on its development. Br J Ind Med 44:627–632, 1987.

Bouhuys, A, Beck, GJ, Schoenberg, JB: Do present levels of air pollution outdoors affect respiratory health? Nature 276:466–471, 1978.

Brain, JD, Golde, DW, Green, GM, Massaro, DJ, Valberg, PA, Ward, PA, Werb, Z: Biologic potential of pulmonary macrophages. Am Rev Respir Dis 118:435–443, 1978.

Brownstein, DG, Rebar, AH, Bice, DE, Muggenburg, BA, Hill, JO: Immunology of the lower respiratory tract: Serial morphologic changes in the lungs and tracheobronchial lymph nodes of dogs after intrapulmonary immunization with sheep erythrocytes. Am J Pathol 98:499–514, 1980.

Burrell, R, Hill, JO: The effect of respiratory immunization on cell-mediated immune effector cells of the lung. Clin Exp Immunol 24:116–124, 1976.

Butcher, BT, Mapp, C, Reed, MA, O'Neill, CE, Salvaggio, JE: Evidence for carrier specificity of IgE antibodies detected by isocyanate-protein conjugates in sera of isocyanate sensitive individuals. J Allergy Clin Immunol 69(suppl): 123, 1982a.

Butcher, BT, O'Neil CE, Reed, MA, Salvaggio, JE, Weill, H: Development and loss of toluene diisocyanate reactivity: Immunologic, pharmacologic, and provocative challenge studies. Allergy Clin Immunol 70:231–235, 1982b.

Butler, JE, Swanson, PA, Richerson, HB, Ratajczak, HV, Richards, DW, Suelzer, MT: The local and systemic IgA and IgG antibody responses of rabbits to a soluble inhaled antigen. Am Rev Respir Dis 126:80–85, 1982.

Call, RS, Smith, TF, Morris, E, Chapman, MD, Platts-Mills, TA: Risk factors for asthma in inner city children (see comments) J Pediatr 121:862–866, 1992.

Chan-Yeung, M, Barton, GM, MacLean, L, Grzybowski, S: Occupational asthma and rhinitis due to western red cedar (Thuja plicata). Am Rev Respir Dis 108:1094–1102, 1973.

Chan-Yeung, M, Lam, S: Occupational asthma. Am Rev Respir Dis 133:686–703, 1986.

Chan-Yeung, M, Malo, JL: Aetiological agents in occupational asthma. Eur Respir J 7:346–371, 1994.

Charles, J, Berstein, A, Jones, B, Jones, DJ, Edwards, JH, Seal, RME, Seaton, A: Hypersensitivity pneumonitis after exposure to isocyanates. Thorax 31:127–136, 1976.

Cogan, E, Schandene, L, Crusiaux, A, Cochaux, P, Velu, T, Goldman, M: Brief report: Clonal proliferation of type 2 helper T cells in a man with the hypereosinophilic syndrome. N Engl J Med 330:535–538, 1994.

Cormier, Y, Belanger, J, Beaudoin, J, Laviolette, M, Beaudoin, R, Hebert, J: Abnormal bronchoalveolar lavage in asymptomatic dairy farmers. Am Rev Respir Dis 130:1046–1049, 1984.

Cormier, Y, Belanger, J, Laviolette, M: Persistent bronchoalveolar lymphocytosis in asymptomatic farmers. Am Rev Respir Dis 133:843–847, 1986.

Cormier, Y, Belanger, J, Laviolette, M: Prognostic significance of bronchoalveolar lymphocytosis in farmer's lung. Am Rev Respir Dis 135:692–695, 1987.

Corry, D, Kulkarni, P, Lipscomb, M: The migration of bronchoalveolar macrophages into hilar lymph nodes. Am J Pathol 115:321–328, 1984.

Cotes, JE, Steel, J: Work related lung disorders. In: Occupational asthma. Edited by Anonymous, pp. 345–372. Oxford: Blackwell Scientific Publications, 1987.

Curtis, JL, Kaltreider, HB: Characterization of bronchoalveolar lymphocytes during a specific antibody-forming cell response in the lungs of mice. Am Rev Respir Dis 139:393–400, 1989.

Curtis, JL, Warnock, ML, Arraj, SM, Kaltreider, HB: Histologic analysis of an immune response in the lung parenchyma of mice. Angiopathy accompanies inflammatory cell influx. Am J Pathol 137:689–699, 1990.

DeBoer, DJ, Ewing, KM, Schultz, KT: Production and characterization of mouse monoclonal antibodies directed against canine IgE and IgG. Vet Immunol Immunopathol 37:183–199, 1993.

Dekker, C, Dales, R, Bartlett, S, Brunekreef, B, Zwanenburg, H: Childhood asthma and the indoor environment. Chest 100:922–926, 1991.

Demenkoff, JH, Ansfield, MJ, Kaltreider, HB, Adam E.: Alveolar macrophage suppression of canine bronchoalveolar lymphocytes: The role of prostaglandin E2 in the inhibition of mitogen-responses. J Immunol 124:1365–1370, 1980.

Demoly, P, Bousquet, J, Godard, P, Michel, FB: The gene or genes of allergic asthma? Presse Med 22:817–821, 1993.

Duff, AL, Platts-Mills, TA: Allergens and asthma. Pediatr Clin North Am 39:1277–1291, 1992.

Duff, AL, Pomeranz, ES, Gelber, LE, Price, CW, Farris, H, Hayden, FG, Platts-Mills, TA, Heymann, PW: Risk factors for acute wheezing in infants and children: Viruses, passive smoke, and IgE antibodies to inhalant allergens. Pediatrics 92:535–530, 1993.

Ehrlich, R, Kattan, M, Godbold, J, Saltzberg, DS, Grimm, KT, Landrigan, PJ, Lilienfeld, DE: Childhood asthma and passive smoking. Am Rev Respir Dis 145:594–599, 1992.

Empey, DW, Laitinen, LA, Jacobs, L, Gold, WM, Nadel, JA: Mechanisms of bronchial hyperreactivity in normal subjects after respiratory tract infection. Am Rev Respir Dis 113:131–139, 1976.

Evans, R, Mullally, DI, Wilson, RW, Gergen, PJ, Rosenberg, HM, Grauman, JS, Chevarley, FM, Feinleib, M: National trends in the morbidity and mortality of asthma in the US Prevalence, hospitalization and death from asthma over two decades: 1965–1984. Chest 91:65S–74S, 1987.

Fabbri, LM, Danieli, D, Crescioli, S: Fatal asthma in a subject sensitized to toluene diisocyanate. Am Rev Respir Dis 137:1494–1498, 1988.

Fink, JN, Schlueter, DP: Bathtub refinisher's lung: An unusual response to toluene diisocyanate. Am Rev Respir Dis 118:955–959, 1978.

Finklea, JF, Hasselbald, V, Riggan, WB, Nelson, WC, Hammer EI, Newill, VA: Cigarette smoking and hemagglutination inhibition response to influenza after natural disease and immunization. Am Rev Respir Dis 104:368–376, 1971.

Finklea, JF, Sandifer, SH, Peck, FB, Manos, JP: A clinical and serologic comparison of standard and purified bivalent inactivated influenza vaccines. J Infec Dis 120:708–719, 1969.

Flindt, ML: Pulmonary disease due to inhalation of derivatives of Bacillus subtilis containing proteolytic enzyme. Lancet 1:1177–1181, 1969.

Galvin, JB, Bice, DE, Guilmette, RA, Muggenburg, BA, Haley, PJ: Pulmonary immune response of dogs after exposure to ^{239}PuO$_2$. Int J Radiat Biol 55:285–296, 1989.

Galvin, JB, Bice, DE, Muggenburg, BA: Comparison of cell-mediated and humoral immunity in the dog lung after localized lung immunization. J Leukoc Biol 39:359–370, 1986.

Ganguly, R, Waldman, RH: Respiratory tract cell-mediated immunity. Bull Eur Physiopathol Respir 13:95–102, 1977.

Gelber, LE, Seltzer, LH, Bouzoukis, JK, Pollart, SM, Chapman, MD, Platts-Mills, TA: Sensitization and exposure to indoor allergens as risk factors for asthma among patients presenting to hospital. Am Rev Respir Dis 147:573–578, 1993.

Gerstman, BB, Bosco, LA, Tomita, DK, Gross, TP, Shaw, MM: Prevalence and treatment of asthma in the Michigan medicaid patient population younger than 45 years, 1980–1986. J Allergy Clin Immunol 83:1032–1039, 1989.

Godfrey, S: Airway inflammation, bronchial reactivity and asthma. Agents Actions Suppl 40:109–143, 1993.

Green, GM: Air pollution, host immune defenses, and asthma; A review. In: Clinical Implications of Air Pollution Research. Edited by AJ Findel, WC Duel. pp. 147–163. Acton, MA.: Publishing Sciences Group Inc., 1974.

Green, GM, Jakab, GJ, Low, RB, Davis, GS: Defense mechanisms of the respiratory membrane. Am Rev Respir Dis 115:479–514, 1977.

Griffiths-Johnson, D, Jin, R, Karol, MH: Role of purified IgG1 in pulmonary hypersensitivity responses of the guinea pig. J Toxicol Environ Health 40:117–127, 1993.

Hagmar, L, Nielsen, J, Skerfving, S: Clinical features and epidemiology of occupational obstructive respiratory disease caused by small molecular weight organic chemicals. Epidemiology of allergic diseases. Monogr Allergy 21:42–58, 1987.

Haley, PJ: Mechanisms of granulomatous lung disease from inhaled beryllium: the role of antigenicity in granuloma formation. Toxicol Pathol 19:514–525, 1991.

Haley, PJ, Finch, GL, Hoover, MD, Cuddihy, RG: The acute toxicity of inhaled beryllium metal in rats. Fund Appl Tox 15:767–778, 1990.

Haley, PJ, Finch, GL, Hoover, MD, Mewhinney, JA, Bice, DE, Muggenburg, BA: Beryllium-induced lung disease in the dog following two exposures to BeO. Environ Res 59:400–415, 1992.

Haley, PJ, Finch, GL, Mewhinney, JA, Harmsen, AG, Hahn, FF, Hoover, MD, Bice, DE: A canine model of beryllium-induced granulomatous lung disease. Lab Invest 61:219–227, 1989.

Hardy, H: Beryllium disease: A clinical perspective. Environ Res 21:1–9, 1980.

Harmsen, AG, Bice, DE, Muggenburg, BA: The effect of local antibody responses on in vivo and in vitro phagocytosis by pulmonary alveolar macrophages. J Leuko Biol 37:483–492, 1985a.

Harmsen, AG, Mason, MJ, Muggenburg, BA, Gillett, NA, Jarpe, MA, Bice, DE: Migration of neutrophils from lung to tracheobronchial lymph node. J Leukoc Biol 41:95–103, 1987.

Harmsen, AG, Muggenburg, BA, Snipes, MB, Bice, DE: The role of macrophages in particle translocation from lungs to lymph nodes. Science 230:1277–1280, 1985b.

Hill, JO, Burrell, R: Cell-mediated immunity to soluble and particulate inhaled antigens. Clin Exp Immunol 38:332–341, 1979.

Hill, JO, Bice, DE, Harris, DL, Muggenburg, BA: Evaluation of the pulmonary immune response by analysis of bronchoalveolar fluids obtained by serial lung lavage. Int Arch Allergy Appl Immunol 71:173–177, 1983.

Hill, PB, DeBoer, DJ: Quantification of serum total IgE concentration in dogs by use of an enzyme-linked immunosorbent assay containing monoclonal murine anti-canine IgE. Am J Vet Res 55:944–948, 1994.

Hillam, RP, Bice, DE, Hahn, FF, Schnizlein, CT: Effects of acute nitrogen dioxide exposure on cellular immunity after lung immunization. Environ Res 31:201–211, 1983.

Hillam, RP, Bice, DE, Muggenburg, BA: Lung localization of antibody-forming cells stimulated in distant peripheral lymph nodes. Immunology 55:257–261, 1985.

Hirshman, CA, Downes, H: Antigen sensitization and methacholine responses in dogs with hyperreactive airways. J Appl Physiol 58:485–491, 1985.

Hirshman, CA, Downes, H: Airway responses to methacholine and histamine in basenji greyhounds and other purebred dogs. Respir Physiol 63:339–346, 1986.

Holt, PG: Inhibitory activity of unstimulated alveolar macrophages on T-lymphocyte blastogenic response. Am Rev Respir Dis 118:791–793, 1978.

Holt, PG, Degebrodt, A, O'Leary, C, Krska, K, Plozza, T: T cell activation by antigen-presenting cells from lung tissue digests: suppression by endogenous macrophages. Clin Exp Immunol 62:586–593, 1985.

Holt, PG, Schon Hegrad, MA: Localization of T cells, macrophages and dendritic cells in rat respiratory tract tissue: implications for immune function studies. Immunology 62:349–356, 1987.

Holt, PG, Sedgwick, JD, O'Leary, C, Krska, K, Leivers, S: Long-lived IgE- and IgG-secreting cells in rodents manifesting persistent antibody responses. Cell Immunol 89:281–289, 1984.

Huang, J, Aoyama, K, Ueda, A: Experimental study on respiratory sensitivity to inhaled toluene diisocyanate. Arch Toxicol 67:373–378, 1993.

Hunter, D, Milton, R, Perry, KMA: Asthma caused by the complex salts of platinum. Br J Ind Med 2:92–98, 1945.

Jimeno, L, Lombardero, M, Carreira, J, Moscoso Del Prado, J: Presence of IgG4 on the membrane of human basophils. Histamine release is induced by monoclonal antibodies directed against the Fab but not the Fc region of the IgG4 molecule. Clin Exp Allergy 22:1007–1014, 1992.

Jones, PD, Ada, GL: Persistence of influenza virus-specific antibody-secreting cells and B-cell memory after primary murine influenza virus infection. Cell Immunol 109:53–64, 1987a.

Jones, PD, Ada, GL: Influenza-specific antibody-secreting cells and B cell memory in the murine lung after immunization with wild-type, cold-adapted variant and inactivated influenza viruses. Vaccine 5:244–248, 1987b.

Kaltreider, HB, Barth, E, Pellegrini, C: The effect of splenectomy on the appearance of specific antibody-forming cells in the lungs of dogs after intravenous immunization with sheep erythrocytes. Exp Lung Res 2:231–238, 1981.

Kaltreider, HB, Byrd, PK, Daugherty, TW, Shalaby, MR: The mechanism of appearance of specific antibody-forming cells in lungs of inbred mice after intratracheal immunization with sheep erythrocytes. Am Rev Respir Dis 127:316–321, 1983.

Kaltreider, HB, Kyselka, L, Salmon, SE: Immunology of the lower respiratory tract. J Clin Invest 54:263–270, 1974.

Karol, MH: Comparison of clinical and experimental data from an animal model of pulmonary immunologic sensitivity. Ann Allergy 66:485–489, 1991.

Kay, AB: T lymphocytes and their products in atopic allergy and asthma. Int Arch Allergy Appl Immunol 94:189–193, 1991.

Keller, RH, Calvanico, NJ, Stevens, JO: Hypersensitivity pneumonitis in nonhuman primates. I. Studies on the relationship of immunoregulation and disease activity. J Immunol 128:116–122, 1982.

Kemeny, DM, Urbanek, R, Ewan, P, McHugh, S, Richards, D, Patel, S, Lessof, MH: The subclass of IgG antibody in allergic disease: II. The IgG subclass of antibodies produced following natural exposure to dust mite and grass pollen in atopic and non-atopic individuals. Clin Exp Allergy 19:545–549, 1989.

Kepron, W, James, JM, Kirk, B, Sehon, AH, Tse, KS: A canine model for reaginic hypersensitivity and allergic bronchoconstriction. J Allergy Clin Immunol 59:64–69, 1977.

Kleinbeck, ML, Hites, MJ, Loker, JL, Halliwell, RE, Lee KW: Enzyme-linked immunosorbent assay for measurement of allergen-specific IgE antibodies in canine serum. Am J Vet Res 50:1831–1839, 1989.

Kreiss, K, Mroz, MM, Zhen, B, Martyny, JW, Newman, LS: Epidemiology of beryllium sensitization and disease in nuclear workers. Am Rev Respir Dis 148:985–991, 1993.

Lam, S, Wong, R, Yeung, M: Nonspecific bronchial reactivity in occupational asthma. J Allergy Clin Immunol 63:28–34, 1979.

Lauweryns, JM, Baert, JH: Alveolar clearance and the role of the pulmonary lymphatics. In: Lung disease state of the art, edited by JF Murray, pp. 185–243. New York: American Lung Association, 1976.

Lawrence, EC, Theodore, BJ, Martin, RR: Modulation of pokeweed-mitogen-induced immunoglobulin secretion by human bronchoalveolar cells. Am Rev Respir Dis 126:248–252, 1982.

Le Quesne, PM, Axford, AT, McKerrow, CB, Jones, AP: Neurological complications after a single severe exposure to toluene di-isocyanate. Br J Ind Med 33:72–78, 1976.

Levy, D, Gent, M, Newhouse, MT: Relationship between acute respiratory illness and air pollution levels in an industrial city. Am Rev Respir Dis 116:167–173, 1977.

Lipscomb, MF, Lyons, CR, O'Hara, RM, Stein-Streilein, J: The antigen induced selective recruitment of specific T lymphocytes to the lung. J Immunol 128:111–115, 1982.

Lipscomb, MF, Toews, GB, Lyons, CR, Uhr, JW: Antigen presentation by guinea pig alveolar macrophages. J Immunol 126:286–291, 1981.

Lloyd, OL: Respiratory-cancer clustering associated with localized industrial air pollution. Lancet 1:318–320, 1978.

Lue, C, Tarkowski, AJ, Mestecky, J: Systemic immunization with pneumococcal polysaccharide vaccine induces a predominant IgA2 response of peripheral blood lymphocytes and increases of both serum and secretory anti-pneumococcal antibodies. J Immunol 140:3793–3800, 1988.

Malo, JL, Zeiss CR: Occupational hypersensitivity pneumonitis after exposure to diphenylmethane diisocyanate. Am Rev Respir Dis 125:113–116, 1982.

Malo, JL, Ouimet, G, Cartier, A, Levitz, D, Zeiss CR: Combined alveolitis and asthma due to hexamethylene diisocyanate (HDI), with demonstration of crossed respiratory and immunologic reactivities to diphenylmethane diisocyanate (MDI). J Allergy Clin Immunol 72:413–419, 1983.

Manfreda, J, Becker, AB, Wang, PZ, Roos, LL, Anthonisen, NR: Trends in physician-diagnosed asthma prevalence in Manitoba between 1980 and 1990. Chest 103:151–157, 1993.

Marathias, KP, Preffer, FI, Pinto, C, Kradin, RL: Most human pulmonary infiltrating lymphocytes display the surface immune phenotype and functional responses of sensitized T cells. Am J Respir Cell Mol Biol 19:470–476, 1991.

Martinez, FD, Cline, M, Burrows, B: Increased incidence of asthma in children of smoking mothers. Pediatrics 89:21–26, 1992.

Mason, MJ, Bice, DE: Enumeration of lymphocyte subpopulations in bronchoalveolar lavage of monkeys using an immunoenzymatic staining technique. Vet Immunol Immunopathol 13:347–355, 1986.

Mason, MJ, Bice, DE, Muggenburg, BA: Local pulmonary immune responsiveness after multiple antigenic exposures in the cynomolgus monkey. Am Rev Respir Dis 132:657–660, 1985.

Mason, MJ, Gillett, NA, Bice, DE: Comparisons of systemic and local immune responses after multiple pulmonary antigen exposures. Region Immunol 2:149–157, 1989.

McFadden, ER: Pathogenesis of asthma. J Allergy Clin Immunol 73:413–424, 1984.

McLeod, E, Caldwell, JL, Kaltreider, HB: Pulmonary immune responses in inbred mice. Appearance of antibody-forming cells in C57BL/6 mice after intrapulmonary or systemic immunization with sheep erythrocytes. Am Rev Respir Dis 118:561–571, 1978.

Miller, JJ: An autoradiographic study of plasma cell and lymphocyte survival in rat popliteal lymph nodes. J Immunol 92:673–681, 1963.

Miller, SD, Zarkower, A: Effects of carbon dust inhalation on the cell-mediated immune response in mice. Infec Immun 9:534–539, 1974.

Miller, SD, Zarkower, A: Silica-induced alterations of murine lymphocyte immunocompetence and suppression of B lymphocyte immunocompetence: A possible mechanism. J Reticuloendothel Soc 19:47–61, 1976.

Morrow, PE: Lymphatic drainage of the lung in dust clearance. Ann N Y Acad Sci 200:46–65, 1972.

Moseholm, L, Taudorf, E, Frosig, A: Pulmonary function changes in asthmatics associated with low-level SO_2 and NO_2 air pollution, weather, and medicine intake. Allergy 48:334–344, 1993.

Mroz, MM, Kreiss, K, Lezotte, DC, Campbell, PA, Newman, LS: Reexamination of the blood lymphocyte transformation test in the diagnosis of chronic beryllium disease. J Allergy Clin Immunol 88:54–60, 1991.

Musk, AW, Peters, JM, Wegman, DH: Isocyanates and respiratory disease: Current status. Am J Indus Med 13:331–349, 1988.

Newhouse, ML, Tagg, B, Pockock, SJ, McEwan, AC: An epidemiological study of workers producing enzyme washing powders. Lancet 1:689–693, 1970.

Newman, LS, Kreiss, K: Nonoccupational beryllium disease masquerading as sarcoidosis: identification by blood lymphocyte proliferative response to beryllium. Am Rev Respir Dis 145:1212–1214, 1992.

Newman, LS, Bobka, C, Schumacher, B, Daniloff, E, Zhen, B, Mroz, MM, King, TE, Jr.: Compartmentalized immune response reflects clinical severity of beryllium disease. Am J Respir Crit Care Med 150:135–142, 1994a.

Newman, LS, Buschman, DL, Newell, JD, Jr., Lynch, DA: Beryllium disease: assessment with CT. Radiology 190:835–840, 1994b.

Nicod, LP, Lipscomb, MF, Weissler, JC, Lyons, CR, Albertson, J, Toews, GB: Mononuclear cells in human lung parenchyma. Characterization of a potent accessory cell not obtained by bronchoalveolar lavage. Am Rev Respir Dis 136:818–823, 1987.

North, RJ, Spitalny, G: Inflammatory lymphocyte in cell-mediated antibacterial immunity: Factors governing the accumulation of mediator T cells in peritoneal exudates. Infec Immun 10:489–498, 1974.

O'Brien, IM, Newman-Taylor, AJ, Burge, PS, Harries, JG, Fawcett, IW, Pepys, J: Toluene diisocyanate-induced asthma II. Inhalation challenge tests and bronchial reactivity studies. Clin Allergy 9:7–15, 1979.

Pabst, R, Gehrke, I: Is the bronchus-associated lymphoid tissue (BALT) an integral structure of the lung in normal mammals, including humans? Am J Respir Cell Mol Biol 19:131–135, 1990.

Parronchi, P, Macchia, D, Piccini, MP, et al.: Allergen- and bacterial antigen-specific T-cell clones established from atopic donors show a different profile of cytokine production. Proc Natl Acad Sci 88:4538–4542, 1991.

Pennline, KJ, Herscowitz, HB: Dual role for alveolar macropages in humoral and cell mediated immune response: Evidence for suppressor and enhancing functions. J Reticuloendothel Soc 30:205–216, 1981.

Pepys, J: Occupational allergic lung disease caused by organic agents. J Allergy Clin Immunol 78:1058–1062, 1986.

Pepys, J, Longbottom, JL, Hargreave, FE, Faux, J: Allergic reactions of the lungs to enzymes of Bacillus subtilis. Lancet 1:1181–1184, 1969.

Pope, C: Respiratory hospital admissions associated with PM10 pollution in Utah, Salt Lake, and Cache Valleys. Arch Environ Health 46:90–97, 1991.

Pope, CA, III: Respiratory disease associated with community air pollution and a steel mill, Utah Valley. Am J Pub Health 79:623–628, 1989.

Richeldi, L, Sorrentino, R, Saltini, C: HLA-DPB1 glutamate 69: A genetic marker of beryllium disease (see comments). Science 262:242–244, 1993.

Richerson, HB: Acute experimental hypersensitivity pneumonitis in the guinea pig. J Lab Clin Med 79:745–757, 1972.

Richerson, HB, Coon, JD, Lubaroff, D: Experimental hypersensitivity pneumonitis in the rat: Histopathology and T-cell subset compartmentalization. Am J Respir Cell Mol Biol 19:451–463, 1991.

Richerson, HB, Richards, DW, Swanson, PA, Butler, JE, Suelzer, MT: Antigen-specific desensitization in a rabbit model of acute hypersensitivity pneumonitis. J Clin Immunol 68:226–234, 1981.

Richerson, HB, Seidenfeld, JJ, Ratajczak, HV, Richards, DW: Chronic experimental interstitial pneumonitis in the rabbit. Am Rev Respir Dis 117:5–13, 1978.

Rizzo, MC, Arruda, LK, Chapman, MD, Fernandez-Caldas, E, Baggio, D, Platts-Mills, TA, Naspitz, CK: IgG and IgE antibody responses to dust mite allergens among children with asthma in Brazil. Ann Allergy 71:152–158, 1993.

Robinson, DS, Hamid, Q, Jacobson, M, Ying, S, Kay, AB, Durham, SR: Evidence for Th2-type

T helper cell control of allergic disease in vivo. Springer Semin Immunopathol 15:17–27, 1993.

Rossman, MD, Kern, JA, Elias, JA, Cullen, MR, Epstein, PE, Preuss, OP, Markham, TN, Daniele, RP: Proliferative response of bronchoalveolar lymphocytes to beryllium. A test for chronic beryllium disease. Ann Intern Med 108:687–693, 1988.

Saltini, C, Hemler, ME, Crystal, RG: T lymphocytes compartmentalized on the epithelial surface of the lower respiratory tract express the very late activation antigen complex VLA-1. Clin Immunol Immunopathol 46:221–233, 1988.

Salvaggio, JE: Hypersensitivity pneumonitis. J Allergy Clin Immunol 79:558–571, 1987.

Savage, SM, Donaldson, LA, Cherian, S, Chilukuri, R, White, VA, Sopori, ML: Effects of cigarette smoke on the immune response. II. Chronic exposure to cigarette smoke inhibits surface immunoglobulin-mediated responses in B cells. Toxicol Appl Pharmacol 111:523–529, 1991.

Schatz, M, Patterson, R, Fink, J: Immunopathogenesis of hypersensitivity pneumonia. J Allergy Clin Immunol 60:27–37, 1977.

Schepers, GWH, Durkan, TM, Delahant, AB, Creedon, FT: The biological action of inhaled beryllium sulfate. A preliminary chronic toxicity study on rats. Arch Ind Health 15:32–58, 1957.

Schnizlein, CT, Bice, DE, Mitchell, CE, Hahn, FF: Effects on rat lung immunity by acute lung exposure to benzo(a)pyrene. Arch Environ Health 37:201–206, 1982.

Schnizlein, CT, Bice, DE, Rebar, AH, Wolff, RK, Beethe, RL: Effect of lung damage by acute exposure to nitrogen dioxide on lung immunity in the rat. Environ Res 23:362–370, 1980.

Schuyler, M, Cook, C, Listrom, M, Fengolio Preiser, C: Blast cells transfer experimental hypersensitivity pneumonitis in guinea pigs. Am Rev Respir Dis 137:1449–1455, 1988.

Schuyler, MR, Gott, K, Edwards, B: Adoptive transfer of experimental hypersensitivity pneumonitis: CD4+ cells are memory and naive cells. J Lab Clin Med 123:378–386, 1994.

Schuyler, MR, Gott, K, Shopp, G, Merlin, T: CD5-negative cell mediated experimental hypersensitivity pneumonitis. Am Rev Respir Dis 145:1185–1190, 1992.

Schuyler, M, Subramanyan, S, Hassan, M: Experimental hypersensitivity pneumonitis: Transfer with cultured cells. J Lab Clin Med 109:623–630, 1987.

Schuyler, MR, Todd, LS: Accessory cell function of rabbit alveolar macrophages. Am Rev Respir Dis 123:53–57, 1981.

Seltzer, J, Bigby, BG, Stulbarg, M, Holtzman, MJ, Nadel, JA, Ueki, IF, Leikauf, GD, Goetzl, EJ, Boushey, HA: O3-induced change in bronchial reactivity to methacholine and airway inflammation in humans. J Appl Physiol 60:1321–1326, 1986.

Shimizu, Y, Newman, W, Tanaka, Y, Shaw, S: Lymphocyte interactions with endothelial cells. Immunol Today 13:106–112, 1992.

Simecka, JW, Thorp, RB, Cassell, GH: Dendritic cells are present in the alveolar region of lungs from specific pathogen-free rats. Reg Immunol 4:18–24, 1992.

Snipes, MB, Boecker, BB, McClellan, RO: Retention of monodisperse or polydisperse aluminosilicate particles inhaled by dogs, rats and mice. Tox Appl Pharmacol 69:345–362, 1983.

Sopori, ML, Cherian, S, Chilukuri, R, Shopp, GM: Cigarette smoke causes inhibition of the immune response to intratracheally administered antigens. Toxicol Appl Pharmacol 97:489–499, 1989.

Spencer, JC, Waldman RH, Johnson, JE: Local and systemic cell-mediated immunity after immunization of guinea pigs with live or killed m. tuberculosis by various routes. J Immunol 112:1322–1328, 1974.

Sporik, R, Platts-Mills, TA: Epidemiology of dust-mite-related disease. Exp Appl Acarol 16:141–151, 1992.

Sprince, NL, Kazemi, H: U.S. beryllium case registry through 1977. Environ Res 21:44–47, 1980.

Sprince, NL, Kazemi, H, Hardy, HL: Current (1975) problem of differentiating between beryllium disease and sarcoidosis. Ann N Y Acad Sci 278:654–664, 1976.

Stein-Streilein, J, Gross, GN, Hart, DA: Comparison of intratracheal and intravenous inocula-tion of sheep erythrocytes in the induction of local and systemic immune responses. Infec Immun 24:145–150, 1979a.

Stein-Streilein, J, Gross, GN, Hart, DA: Comparison of intratracheal and intravenous inocula-tion of sheep erythrocytes in the induction of local and systemic immune responses. Infect Immun 24:145–150, 1979b.

Stein-Streilein, J, Hart, DA: Effect of route of immunization on development of antibody-forming cells in hilar lymph nodes. Infect Immun 30:391–396, 1980.

Stevens, RH, Macy, E, Morrow, C, Saxon, A: Characterization of a circulating subpopulation of spontaneous antitetanus toxoid antibody producing B cells following in-vivo booster immunization. J Immunol 122:2498–2503, 1979.

Stokes, RF, Rossman, MD: Blood cell proliferation response to beryllium: Analysis by receiver-operating characteristics (see comments). J Occup Med 33:23–28, 1991.

Szentivanyi, A: The beta adrenergic theory of the atopic abnormality in bronchial asthma. J Allergy Clin Immunol 42:203–232, 1968.

Thomas, WR, Holt, PG, Keast, D: Development of alterations in the primary immune response of mice by exposure to fresh cigarette smoke. Int Arch Allergy 46:481–486, 1974a.

Thomas, WR, Holt, PG, Keast, D: Recovery of immune system after cigarette smoking. Nature 248:358–359, 1974b.

Thomson, PD, Harris, NS: Detection of plaque-forming cells in the peripheral blood of actively immunized humans. J Immunol 118:1480–1482, 1977.

Tse, CST, Chen, SE, Bernstein, IL: Induction of murine reaginic antibodies by toluene diisocya-nate. An animal model of immediate hypersensitivity reactions to isocyanates. Am Rev Respir Dis 120:829–835, 1979.

Turner, FN, Kaltreider, HB: Immunology of the lower respiratory tract. III. Concentrations of antigen and of antibody-forming cells in pulmonary and systemic lymphoid tissues of dogs after intrapulmonary or intravenous administration of sheep erythrocytes. Clin Exp Immu-nol 33:128–135, 1978.

Van der Brugge-Gamelkoorn, GJ, Claassen, E, Sminia, T: Anti-TNP-forming cells in bronchus-associated lymphoid tissue (BALT) and paratracheal lymph node (PTLN) of the rat after intratracheal priming and boosting with TNP-KLH. Immunology 57:405–409, 1986.

Van Ert, M, Battigelli, MC: Mechanism of respiratory injury by TDI (toluene diisocyanate). Ann Allergy 35:142–147, 1975.

Vandenplas, O, Malo, JL, Saetta, M, Mapp, CE, Fabbri, L: Occupational asthma and extrinsic alveolitis due to isocyanates: Current status and perspectives. Br J Ind Med 50:213–228, 1993.

von Mutius, E, Martinez, FD, Fritzsch, C, Nicolai, T, Roell, G, Thiemann, HH: Prevalance of asthma and atopy in two areas of West and East Germany. Am J Respir Crit Care Med 149:358–364, 1994.

Waldman, RH, Gadol, N, Jurgensen, PF, Olsen, GN, Johnson, JE: Secretory and systemic cell-mediated and humoral immune response in humans and guinea pigs to the inactivated influenza virus vaccine. Adv Exp Med Biol 31:87–95, 1972.

Waldman, RH, Henney, CSA: Cell-mediated immunity and antibody responses in the respira-tory tract after local and systemic immunization. J Exp Med 134:482–494, 1971.

Weinberg, DS, Unanue, ER: Antigen-presenting function of alveolar macrophages: Uptake and presentation of Listeria monocytogenes. J Immunol 126:794–799, 1981.

Weissman, DN, Bice, DE, Crowell, RE, Romero, JL, Chilukuri, R, Koster, FT, Schuyler, MR: Human lung immunization. Impairment of humoral responses in smokers. Am Rev Respir Dis 145:A639, 1992a.

Weissman, DN, Bice, DE, Muggenburg, BA, Haley, PJ, Shopp, GM, Schuyler, MR: Primary immunization in the canine lung. Soluble antigen induces a localized response. Am Rev Respir Dis 145:6–12, 1992b.

Weissman, DN, Bice, DE, Crowell, RE, Schuyler, MR: Intrapulmonary antigen deposition in the human lung: Local responses: Am J Respir Cell Mol Biol (in press) 1994.

Wilkes, DS, Weissler, JC: Alloantigen-induced immunoglobulin production in human lung: differential effects of accessory cell populations on IgG synthesis. Am J Respir Cell Mol Biol 10:339–346, 1994.

Woolcock, AJ: Asthma—what are the important experiments? Am Rev Respir Dis 138:730–744, 1988.

Xia, W, Schneeberger, EE, McCarthy, K, Kradin, RL: Accessory cells of the lung. II. Ia+ pulmonary dendritic cells display cell surface antigen heterogeneity. Am J Respir Cell Mol Biol 19:276–283, 1991.

Yarchoan, R, Murphy, BR, Strober, W, Clements, ML, Nelson, DL: *In-vitro* production of anti-influenza virus antibody after intranasal inoculation with cold-adapted influenza virus. J Immunol 126:1958–1963, 1981.

Zarkower, A: Alterations in antibody response induced by chronic inhalation of SO₂ and carbon. Arch Environ Health 25:45–50, 1972.

Zeiss, CR, Kanellakes, TM, Bellone, JD, Levitz, D, Pruzansky, JJ, Patterson, R: Immunoglobulin E-mediated asthma and hypersensitivity pneumonitis with precipitating anti-hapten antibodies due to diphenylmethane diisocyanate (MDI) exposure. J Allergy Clin Immunol 65:346–352, 1980.

Zeiss, CR, Patterson, R: Acid anhydrides. In: Asthma in the Workplace. Edited by IL Bernstein, M Chan-Yeung, JL Malo, DI Bernstein. pp. 439–457. New York: Marcel Dekker Inc., 1993.

Chapter Fifteen

Biological Markers in the Respiratory Tract

Rogene F. Henderson

In the field of environmental health research, the use of biological markers (or biomarkers) to link exposures to specific chemicals and potential resultant adverse health effects has shown promise. Such biomarkers have been described in terms of markers of exposure, markers of effects, and markers of susceptibility (National Research Council, 1987). The respiratory tract, as a portal of entry for inhaled toxicants, has good potential for containing markers that document an exposure to a specific chemical. Biological markers of effects induced in the respiratory tract by exposure to chemicals in most cases will not be chemical-specific because the lung responds in a similar fashion to many chemicals. But biological markers of early effects that are predictive of long-term health outcome would be useful for intervention therapy. Biomarkers of susceptibility to respiratory tract disease represent an exciting area of current research. In the following chapter, there is first a discussion of what biomarkers are and how they might be useful. This is followed by a discussion of the biomarkers in the respiratory tract that show the most promise for helping to clarify the extent to which chemical exposures contribute to human adverse health effects. Finally, research strategies which are needed to improve the usefulness of biomarkers are described.

WHAT ARE BIOMARKERS?

A biological marker or biomarker has been defined as "an exogenous substance or its metabolite or the product of an interaction between a xenobiotic agent and some target molecule or cell that is measured in a compartment within an organism" (National Research Council, 1987). Biomarkers are always found within an organism and help to define the links between exposure, internal dose, initial biological effects, and induced adverse health effects (Fig. 1). Epidemiology studies have usually relied on external indicators of exposure (area or personal monitors, questionnaires, and histories) to find associations between past exposures and subsequent disease development. New molecular techniques have provided the tools for assessing the dose received by an individual by measuring the chemical adducts formed with macromolecules in the body. The same molecular techniques also provide information on the initial changes induced by the biologically effective dose—changes that may eventually lead to disease. The biomarkers to the far left in Fig. 1 are in large part the product of what are commonly called classical pharmacokinetic studies or in the case of toxicants toxicokinetic studies. On the other hand, the biomarkers on the far right are clinical signs and symptoms which are within the purview of standard clinical medicine. It is really the biomarkers shown in the middle of Fig. 1 that represent the new and exciting areas of cellular and molecular research which provide the hope of being able to link biologically effective doses to early events predictive of health outcome. In this chapter, the major, but not exclusive, emphasis will be on those biomarkers illustrated in the middle section of Fig. 1.

Biomarkers of Exposure   On the left side of Fig. 1 are illustrated various types of biomarkers of exposure. The external exposure atmosphere will result in some fraction of the inhaled toxicants being deposited in the body (internal dose). A portion of that internal dose (or total body burden) will be excreted either as parent compound or as metabolites without causing adverse health effects. A portion will end up in tissues that are not sensitive to the toxic effects of the chemical or its metabolites and will also cause no toxic effects. But some fraction will reach the target tissue and have the potential for causing adverse effects. Of the fraction of the total body burden that reaches the target tissue, only a small fraction may be biologically effective in causing those biological changes that eventually lead to adverse health effects. The types of markers named thus far (biomarkers of internal dose, dose-to-target tissue or biologically effective dose) are often referred to as biomarkers of exposure. The most useful biomarkers of exposure are those that are chemical-specific and are quantitatively relatable to the degree of the prior or ongoing exposure. Ideally, one would also want the biomarker of the extent of exposure to be predictive of health outcome, but is rare that sufficient information is available to make such predictions.

BIOLOGICAL MARKERS

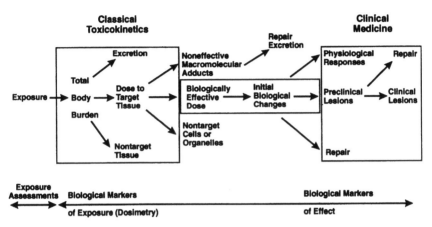

Figure 1 Biological Markers. Chemicals that enter the body can take many paths as illustrated in this figure. A portion of the chemical or its metabolites may lead to biological changes which eventually lead to disease. Biological markers are used to quantitate the different pathways illustrated above (a) to regulate exposures and to prevent disease and (b) to allow interventions at early critical points to prevent progression toward disease.

Biomarkers of Effects On the righthand side of Fig. 1 are illustrated potential types of biomarkers of effect. Many types of responses may be made to toxicants. Some of the induced effects may be merely physiological responses that are not deleterious. Other responses may be deleterious but are quickly repaired. But some responses represent the earliest indicators of a change that if persistent, can lead to an adverse health effect. The persistence and amplification of such a response leads to a clinical disease state. The most useful biomarker of an effect is one that is definitive for a specific adverse health effect, is quantitatively predictable for health outcome, and occurs early enough in the process that its detection allows intervention in the disease process.

While the biomarkers to the left in Fig. 1 are referred to as biomarkers of exposure and the markers to the right as biomarkers of effects, in practice the two areas overlap. DNA adducts may be markers of exposure, but if they occur at specific sites known to induce mutations leading to cancer, the adducts may also be biomarkers of effect.

Biomarkers of Susceptibility Indicators of individual or population differences which influence the response to environmental agents are called biomarkers of susceptibility. These might include such characteristics as an enhanced metabolic capacity for converting a chemical to its reactive, more toxic metabolite or an enhanced capacity to detoxicate reactive metabolites, or differences in number of receptor sites that are critical for a specific response. Such markers can be quite valuable in providing information that will allow protection of susceptible populations. The use of such markers is, however, fraught with legal and ethical problems because identification of persons with enhanced susceptibility to adverse health effects from exposure to chemicals could lead to discrimination against those persons in obtaining certain jobs and insurance (Wulfsberg et al., 1994). In this chapter we will limit discussion to potential markers of susceptibility in the respiratory tract, but we will not attempt to discuss the ethical or legal implications.

Uses of Biomarkers   As mentioned in the beginning of this chapter, a major potential use for biological markers is to link environmental exposures causally and quantitatively to health effects in individuals or populations. If the whole chain of events illustrated in Fig. 1 can be defined in a quantitative fashion for a single toxicant, the exposure to such a toxicant could be regulated to prevent adverse health effects with a reasonable degree of certainty. This would avoid over or underestimation of the risk from such a toxicant. In practice, this use of biomarkers is still in its infancy because of insufficient data to fill in all the information illustrated in Fig. 1. Strategies for improving our ability to use biomarkers to quantitate the ability of environmental agents to produce adverse health effects are discussed at the end of this chapter.

Other practical uses for biomarkers are to detect and quantitate prior or ongoing exposures to specific chemicals, to detect early stages of a disease to allow intervention, to determine the effectiveness of intervention strategies, to detect cells at risk from an inhaled toxicant, and to detect and protect sensitive populations.

BIOMARKERS OF EXPOSURE

Parent Compound and Metabolites   Because the respiratory tract is a portal of entry for inhaled toxicants, it is a good potential source of biomarkers of exposure (Table 1). The biomarkers commonly determined in toxicokinetic studies (such as parent compound or metabolites in exhaled breath, blood, or urine) have been reviewed recently (Henderson and Belinsky, 1993) and will not be covered in detail in this chapter. Biomarkers of exposure to a volatile organic compound (VOC) such as benzene will serve as an example. Benzene has been quantitated in the exhaled breath of gasoline station attendants immediately following the filling of a gas tank and quantities found to correlate with external monitoring of benzene exposure (Wallace, 1987). Urinary metabolites of benzene (such as muconic acid, phenylmercapturic acid, and hydroquinone) have been suggested as valid markers of benzene exposure (Sabourin et al., 1989). These metabolite are taking the place of

Table 1   Biomarkers of Exposure

Marker	Source	Example References
Parent compound or metabolites	Excreta, tissue	Henderson and Belinsky, 1993
DNA adducts		
Total (by ^{32}P-post labeling)	Target tissues	Bond et al., 1988
		Gupta et al., 1989
		Lee et al., 1993
		Phillips et al., 1988; 1990
Chemical-specific	Target tissues	Hecht et al., 1994
		Weston and Bowman, 1991
		Izzotti et al., 1991; 1992
Mutations (*hprt*) in lung cells	Tissue	Driscoll et al., 1994
Protein adducts		
Blood protein	Hemoglobin	Osterman-Golkar et al., 1976
		Hecht et al., 1994
	Albumin	Bechtold et al., 1992
Other proteins		
Nitric oxide-heme	BALF[a]	Maples et al., 1991
S-sulfonates	NALF[b]	Bechtold et al., 1993
Cholesterol-ozone products	Lung tissue	Pryor et al., 1992

[a]BALF = bronchoalveolar lavage fluid
[b]NALF = nasal lavage fluid

urinary phenol for biomonitoring of benzene exposure because phenol has too high a background in urine to be useful for monitoring low-level benzene exposures (Bechtold et al., 1991).

DNA Adducts   The total amount of DNA adducts formed at various sites along the respiratory tract has been used as a measure of dose to target tissues in animals exposed to complex mixtures such as diesel exhaust or cigarette smoke. The studies use the ^{32}P-post-labeling technique that separates adducted DNA bases from normal bases based on the greater bulk of the adducted bases. Bond et al. (1988) used ^{32}P-post-labeling techniques to determine the distribution of DNA adducts along the respiratory tract of rats exposed to high levels of diesel exhaust. The goal was to determine if the level of adducts (a measure of dose) corresponded with the number of tumors (a measure of an effect) at the various sites along the respiratory tract. The highest level of adducts was found in the peripheral lung, the site where tumors were also found.

To obtain information on cells potentially at risk from inhaled diesel exhaust, the same investigators isolated alveolar type II cells from control rats and from rats exposed chronically to either diesel soot or carbon black (Bond et al., 1990). There was a fourfold increase in the DNA adducts detected in the type II cells from exposed rats, but there was no difference in the amount of adducts in cells from the rats exposed to diesel exhaust and those exposed to carbon black—a carbonaceous particle essentially free of organic compounds. The bulk of the increased adducts in rats exposed to either type of particle seemed to be an increase in the adducts that were already present in the control, nonexposed rats. While this result was unexpected at the time, the data were predictive of what was found in rats chronically exposed to diesel exhaust and carbon black in terms of lung tumor formation (Nikula et al., 1994). Diesel exhaust soot and carbon black induced similar numbers and types of lung tumors in the exposed rats.

Using similar techniques, Bond et al. (1989) found that a 22-day exposure to cigarette smoke increased the total amount of DNA adducts in the lungs of rats examined the day after the exposures ended. The level of DNA adducts had returned to normal by 3 weeks after the exposure, indicating that the adducts were repaired fairly rapidly. Lee et al. (1993) used the ^{32}P-post-labeling technique to determine the concentration of aged, side-stream smoke required to increase DNA adducts in rats exposed for 90 days. They found an increase in the DNA adducts in rats exposed to 10 mg/m^3 of the smoke, but not in rats exposed to 0.1 or 1.0 mg/m^3 smoke. Gupta et al. (1989) used a ^{32}P-post-labeling technique to determine that DNA adducts increased in nasal and lung tissue of rats exposed to cigarette smoke, and that the concentration of the adducts in the DNA increased with longer exposure time. In both studies (Lee et al., 1993; Gupta et al., 1989), the DNA adduct level decreased with time after exposure, indicating DNA repair was occurring. In studies on human lung tissue, Phillips et al. (1988) reported that the level of DNA adducts (as determined by the ^{32}P-post-labeling technique) was higher in smokers versus nonsmokers, and that the relationship between the level of adducts and the daily or lifetime cigarette consumption was linear. In agreement with the animal studies, people who had given up smoking for at least 5 years had adduct levels similar to nonsmokers. Thus, measurement of nonspecific DNA adducts in respiratory tract tissues from animals and humans has proved useful in determining target tissues for inhaled mixtures and rates of repair for the adducted DNA.

When one moves from measurements in tissues to the more readily available peripheral blood lymphocytes, however, this type of biomarker does not appear useful. Human bronchial tissue from patients undergoing pulmonary surgery was examined for DNA adducts using ^{32}P-post-labeling techniques (Phillips et al., 1990). The adduct levels detected in tissue from nonsmokers and former smokers were the same, but the level in tissue from smokers was significantly higher. A comparison was made between the concentration of adducts in DNA from the bronchial tissue and that in peripheral-blood lymphocytes; no correlation was observed. This finding suggests that peripheral-blood lymphocytes are not appropriate surrogates for target tissue to estimate dosimetry from at least some inhaled carcinogens.

The above studies were based on unidentified DNA adducts that, because of their bulk, separate from normal DNA bases in the ^{32}P-post-labeling procedure. The use of the ^{32}P-post-labeling method for analysis of total DNA adducts formed during an exposure to a complex mixture is not quantitative because one cannot run the necessary standards to determine recovery of each potential adduct in the assay. If the agent of interest is a single chemical, or if a single chemical can be identified as a surrogate for tracking the dosimetry of a complex mixture, then analytical chemistry techniques can be used to develop quantitative techniques for analysis of specific DNA adducts. Using sophisticated gas chromatography/mass spectrometry techniques, Hecht et al. (1994) found that the levels of DNA adducts formed from two tobacco-specific nitrosamines in cigarette smoke, 4-(methylnitrosamino)-1-(3-pyridyl)-1-butanone (NNK) and N'-nitrosonornicotine (NNN), were higher in the lungs of smokers versus nonsmokers. Weston and Bowman (1991) describe techniques using immunoaffinity chromatography, hydrolysis, high performance liquid chromatography, and synchronous fluorescence spectroscopy to quantitate benzo(a)pyrene-induced adducts in human lung tissue with a detection limit of 1–40 adducts in 10^8 nucleotides.

Izzotti et al. (1991) used the synchronous fluorescence spectrophotometric techniques described by Vahakangas et al. (1985) to quantitate the amount of benzo(a)pyrene diolepoxide (BPDE)-derived DNA adducts in alveolar macrophages from smokers and nonsmokers. No adduct was found in samples from nonsmokers or former smokers, but 85% of samples from current smokers contained the adduct. The amount of adduct in macrophages from current smokers correlated with the number of cigarettes currently smoked, but not with the lifetime cumulative total of cigarettes smoked. This would be expected based on the known turnover of alveolar macrophages. Izzotti et al. (1992) later used the same techniques in a chemoprevention study. Rats were exposed to cigarette smoke, and the amount of BPDE-DNA adducts was monitored in both lung and heart tissue. There was a high correlation between the levels of adducts in the two organs, with higher levels achieved in the heart than in the lung. Administration of the chemopreventative agent—N-acetyl-L-cysteine (NAC)—prevented the formation of the adducts in both the lung and heart. In parallel studies in which the effect of NAC on the histopathology observed in smoke-exposed rats was addressed, the NAC prevented histopathologic alterations in the airways and the alveoli and inhibited the induction of micronuclei in alveolar macrophages (Bagnasco et al., 1992). Thus, the BPDE-DNA adducts appear to be valuable both as markers of exposure and predictors of effect.

Mutations Mutations leading to development of cancer are discussed in the next section on Biomarkers of Effects. However, some mutations that do not directly lead to cancer can be used as a measure of dose to target tissue of an agent capable of inducing altered DNA. An example in the respiratory tract is the recent study of Driscoll et al. (1994) in which lung epithelial cells were isolated from rats exposed to varying amounts of silica instilled intratracheally in the lung. The isolated epithelial cells had a dose-related increase in *hprt* mutations indicating the silica and the inflammatory milieux induced by the silica caused alterations in the DNA. Such mutations in tissues provide markers of tissues at risk for alterations in DNA which may lead to cancer.

Protein Adducts   Reactive compounds or metabolites will bind with tissue proteins as well as with DNA. Protein adducts formed in respiratory tract tissue have not, however, generally been used as biomarkers of exposure. This is most likely because such adducts have not been viewed as either readily accessible or on the pathway to disease induction.

An exception is the use of protein interaction products to detect exposures to nitrogen dioxide (NO_2) and sulfur dioxide (SO_2). The nitric oxide (NO)/heme protein complex can be used as a biologic marker of exposure to NO_2 (Maples et al., 1991). The complex forms in cells that are free in the respiratory tract lumen (mainly macrophages) upon exposure to NO_2, and can be assayed by electron spin resonance spectroscopy in cell pellets obtained by bronchoalveolar lavage (BAL). The marker is limited in its usefulness for exposures below approximately 1 ppm NO_2, however, because of the relatively high background of NO/heme complexes in respiratory tract cells due to endogenous NO.

A protein marker of SO_2 exposure can be found in the nose, the site of deposition of this highly water-soluble gas (Bechtold et al., 1993). It is known that SO_2 interacts with S-S bonds in proteins to form S-sulfonates (Keller and Menzel, 1985; Gunnison and Palmes, 1978). The concentration of S-sulfonates in nasal proteins obtained by nasal lavage can be assayed by cyanolysis to release sulfite, followed by ion chromatography to quantitate the sulfite (Bechtold et al., 1993).

By far the most common use of protein adducts as markers of exposure is the formation of blood protein adducts with reactive compounds from any route of exposure. These adducts are not unique to compounds that enter the body through the respiratory tract; therefore, only a few examples will be given.

Adducts formed with hemoglobin are advantageous for monitoring exposures to VOC because the turnover time for such adducts often is the same as for the red blood cell (Osterman-Golkar et al., 1976). Thus, the hemoglobin adducts can integrate the exposures of several months. Hecht et al. (1994) found hemoglobin adducts to the tobacco-specific nitrosamines, NNN and NNK, to be higher in the blood of a subset of smokers than in other smokers and in nonsmokers. The investigators suggested that the hemoglobin adducts were useful for indicating those individuals that were most able to convert NNN and NNK to reactive metabolites and therefore potentially at higher risk for cancer development. (See section on Markers of Susceptibility.)

While hemoglobin is the blood protein used most frequently to assay for markers of exposure, the protein albumin also has advantages. Albumin is more readily accessible to reactive compounds in the blood than is hemoglobin because interaction with hemoglobin requires that the reactive compound cross the red blood cell membrane. Thus, in some cases, one can detect albumin adduct at lower exposure levels than hemoglobin adducts. This was the case in workers occupationally exposed to 4 ppm benzene in which benzene-derived phenylcysteine adducts could be detected in blood albumin but not hemoglobin (Bechtold et al., 1992). While the life span of albumin in blood is only about 3 weeks, in situations where people are being exposed on a continuing basis to a VOC, the albumin adducts may provide more sensitive markers of individual exposures than the hemoglobin adducts.

Lipid Products While samples from fat depots are often analyzed for evidence of stored lipophilic compounds, the respiratory tract does not contain a large fat depot. However, there is at least one example of a lipid reaction product used as a marker of exposure in the respiratory tract. Pryor et al. (1992) reported the detection of cholesterol ozonation products in lung tissue extracts from rats exposed to 1.3 ppm ozone for 12 hours. The approach has the advantage that the target lipid, cholesterol, is found in all cell membranes. The usefulness of such an assay must be determined in feasibility studies of the assay for tissues from rats or humans exposed to lower ozone levels.

BIOMARKERS OF EFFECTS

Cancer Of all the health effects for which biomarkers might be useful in delineating a causal agent, perhaps the disease most affected by new techniques in molecular biology is cancer. The ability to detect, identify, and quantitate alterations in DNA provides an opportunity not only to quantitate the biologically effective dosimetry, but also to find the critical events leading to mutations which allow the development of cancer. In the sections above, DNA adducts were discussed as biomarkers of exposure. In this section, the mutations, activated oncogenes, and inactivated suppressor genes that result from such altered DNA will be discussed as biomarkers of effects (Table 2). In addition, the products from such altered genes that are indicative of early changes leading to cancer will be considered.

A more detailed review of oncogenes in human lung cancer has recently been published (Kratzke et al., 1993). Mutations in the $3p$, retinoblastoma (RB), $p53$, *myc,* and *ras* genes have all been associated with lung cancer. DNA loss on one allele of the $3p$ chromosome is in > 90% of all small cell lung cancer and in 50% of nonsmall cell lung cancers (Yokota et al., 1987; Kok et

Table 2   Biomarkers of Effects

Marker	Indication	Example Reference
Cancer		
Deletions on 3p gene	Small lung cell cancer	Yokota et al., 1987 Kok et al., 1987
Aberrant expression of RB gene	Small lung cell cancer	Harbour et al., 1988 Reissmann et al., 1990
Overexpression of myc oncogene	Lung tumors	Wong et al., 1986 Johnson et al., 1987, 1988 Brennan et al., 1991
Activated K-ras oncogene; overexpression of $p21$	Cell transformation	Brandt-Rauf, 1991 Harris, 1991
Mutated $p53$	Loss of tumor suppressor activity	Chiba et al., 1990 Vahakangas et al., 1992
Overexpression of p185^{neu}	Decreased survival of adenocarcinoma patients	Kern et al., 1990
Hypermethylation of DNA	Decrease in gene transcriptions	Sakai et al., 1991
Hypomethylation of DNA	Increase in gene transcriptions	Wilson et al., 1987; Liteplo et al., 1987
Elevated membrane glycoprotein, CD44	Squamous metaplasia, nonsmall cell lung tumors	Penno et al., 1994
Proliferating cell nuclear antigen	Proliferating cells	Lippman et al., 1990b Lee et al. 1990
Chromogranin A, leu 7, neuro-specific enolase and dopa decarboxylase	Small cell cancer of lung	Mulshine et al., 1992
Clara 1 kd protein, surfactant-associated peptides, carcino-embryonic antigen, activated K-ras oncogene	Adenomatous lung tumors	Mulshine et al., 1992

al., 1987). Almost all small cell lung cancers have loss of or an aberrant expression of the RB gene, while only 15% of nonsmall cell lung cancers have loss of RB function (Harbour et al., 1988; Reissmann et al., 1990).

Mutations in the tumor suppressor $p53$ gene are frequent (70–100%) in small cell lung cancer (Vahakangas et al., 1992; Chiba and Minna, 1990), while approximately 40% of nonsmall cell lung cancers have such mutations (Nigro et al., 1989; Takahashi et al., 1989; Iggo et al., 1990; Takahashi et al., 1991; Chiba et al., 1990). In lung tumors from miners exposed to radon and cigarette smoke, no K-ras mutations were found, but 9 of 19 tumors had $p53$ mutations (Vahakangas et al., 1992). The mutational spectra differed from that associated with tobacco smoking alone in that no G:C to T:A transver-

Table 2   Biomarkers of Effects (continued)

Marker	Indication	Example Reference
Noncancer lung disease:		
Elevated inflammatory cells in bronchoalveolar lavage fluid (BALF)	Inflammation	Baughman, 1992 Henderson, 1988
Elevated protein, lactate dehydrogenase activity in BALF	Cytotoxicity damage to alveolar/capillary barrier	Henderson, 1988
Elevated alkaline phosphatase in BALF	Increase in surfactant secretions	Henderson et al., 1995
Increase in release of tumor necrosis factor (TNF-α), interleukein-1 (IL-1), Fn from pulmonary macrophages	Fibrotic processes	Kunkel and Strieter, 1990 Driscoll and Maurer, 1991
Element urinary hydroxyproline, hydroxylysine	Connective tissue breakdown (not lung specific)	Yanagisawa et al., 1986
Elevated urinary desmosine, isodesmosine	Emphysematous process	Stone et al., 1991
Elevated urinary or plasma elastin peptide fragments	Emphysematous process, chronic obstructive lung disease	Rosenbloom, 1991
Elevated BALF IL-8 or release of IL-8 from pulmonary macrophages	Fibrotic processes	Lynch et al., 1992 Donnelly et al., 1993

sions were observed in the mutations. Approximately half of both the small cell cancers and the squamous cell cancers had the $p53$ mutations.

One of the *myc* oncogenes is overexpressed in approximately 10–40% of observed small cell lung tumors (Wong et al., 1986; Johnson et al., 1987, 1988; Brennan et al., 1991).

Activation of the K-*ras* oncogene on codon 12 is associated with 40% of human adenocarcinomas with a G:C to T:A transversion. Polycyclic aromatic hydrocarbons (PAHs) have been shown to cause specific mutations that lead to activation of the *ras* oncogene and expression of its $p21$ protein product in many lung cancers (Brandt-Rauf, 1991; Harris, 1991; Bos, 1989). Studies in strain A mice indicate that the most commonly detected mutation in *ras* genes in benzo(a)pyrene-induced lung tumors are G to T transversions. These findings suggest that PAH-induced mutations may be responsible for activation of the K-*ras* gene in some human adenocarcinomas.

The degree of methylation of genes can also affect their function with a decrease in methylation (hypomethylation) leading to an increase in gene transcription (Wilson et al., 1987; Liteplo and Kerbel, 1987) and hypermethylation leading to decreased gene transcription (Sakai et al., 1991). Hypermethylation has been found in regions of chromosome $11p$, a region thought to contain tumor suppressor genes, in small lung cell cancer and lymphoma (Bustros et al., 1988; Makos et al., 1992).

Other markers of mutational changes can be found in the mRNAs and protein products of the mutations. Enhanced H- or K-*ras* mRNA has been found in four of four human lung cancers (Slamon et al., 1984). Overexpression of $p21$ (fourfold to tenfold) was found in 82% of squamous cell lesions (Kurzrock et al., 1986). Thus, *ras* mutations appear to be more frequent in adenocarcinomas, and *ras* gene overexpression appears to be more common in squamous cell carcinomas.

The $p21$ protein product of the *ras* gene has potential as a marker because it is obtainable from body fluids (Brandt-Rauf, 1991). Serum from patients with nonsmall cell lung cancer have been reported to have frequent elevations of normal $p21$ and infrequent expression of mutated forms of $p21$. In four patients with nonsmall cell lung cancer, the serum level of $p21$ was followed during therapy (Perera et al., 1990). A lack of response to the therapy corresponded with a lack of decrease in serum $p21$.

Growth factors are known to mediate nonsmall cell lung cancer. In a report by Garver et al. (1993), the angiogenic growth factor pleiotrophin (PTN) and homolog midkine (MK) were assessed in normal and neoplastic lung tissue. Normal lung tissue consistently expressed PTN, but only two of 17 normal tissues expressed MK. In contrast, all 20 lung cancers expressed MK, but only one expressed PTN. More research is required to determine the significance of this reciprocal expression of growth factors in normal versus neoplastic lung tissue.

Elevated expression of $p185^{neu}$, the protein product of the HER2/*neu* protooncogene, has been found to be associated with diminished survival in patients with adenocarcinomas (Kern et al., 1990). The protein has the characteristics of a tyrosine kinase growth factor receptor and is thought to play a role in human ovarian and breast cancer.

In studies designed to detect and distinguish between the early stages of asbestos-induced mesothelioma and bronchogenic carcinoma, Pluygers et al. (1991; 1992) reported that four markers were useful in distinguishing between the two types of tumors and asbestosis. In the first study (Pluygers, 1991), tissue polypeptide antigen (TPA), carcino-embryonic antigen (CEA), hyaluronic acid, and ferritin were found to discriminate between the late stage of the two tumors and asbestosis. In the second study (Pluygers, 1992), these markers were used to screen asbestosis workers who were free of clinical disease. Five of the 19 workers showed patterns similar to those found in patients with tumors. Chemoprevention studies are beginning with the therapy aimed at antagonizing free radical carcinogenesis in persons with elevated TPA or ferritin and at inhibiting chemical carcinogenesis in persons having elevated CEA levels.

Recently, it has been reported that the membrane glycoprotein (CD44) is a novel biomarker for nonsmall cell lung tumors and squamous metaplasia of the lung. CD44 transcription and translation were consistently high in nonsmall cell tumors and rare in small cell tumors (Penno et al., 1994). In a cultured small cell cancer line, the insertion of a *ras* gene to induce characteristics of a nonsmall lung cancer also induced a fortyfold increase in CD44 expression. While the CD44 in normal lungs was confined to the surface of bronchial basal cells and alveolar macrophages, areas of squamous metaplasia showed high levels of CD44. Additional studies will be required to validate CD44 as a marker for identifying early forms of lung cancer.

Cancer Prevention   Much effort has been expended on developing biomarkers for chemoprevention trials to aid in early assessment of chemotherapy for cancer (Lippman et al., 1990a; Lee et al. 1992; Benner et al., 1992; Mulshine et al., 1992). Cancer incidence is the endpoint of concern in such studies, but does not allow a rapid determination of the efficacy of treatments. Biomarkers of intermediate stages in the development of cancer would be of tremendous value in speeding up the evaluation of cancer prevention trials. As reviewed by Lippman et al. (1990a), intermediate endpoints can be divided into three classes: genetic markers, proliferation markers, and differentiation markers. Selection criteria for such markers include:

1   differential expression in normal and high-risk tissue;
2   ability to be analyzed in small tissue samples;
3   correlation of the amount or pattern with the stage in carcinogenesis; and
4   preclinical or early clinical data supporting modulation of the marker by the study agent (Lee et al., 1992).

Genetic Biomarkers   A genetic biomarker which has been used in some chemoprevention trials for respiratory tract cancer is micronuclei frequency in exfoliated cells. However, Lippman et al. (1990a) report no correlation between micronuclei frequency and histological stage (hyperplasia, metaplasia, or dysplasia) in their bronchial premalignancy studies. In comparisons of ploidy in cells from seven lung tumors with cells in the normal tissue near the tumors, in all but one normal lung sample, the distribution of chromosome number per cell resembled that in the corresponding tumor specimen (Sohn et al., 1989). This suggests that many cytogenetic abnormalities found in tumor cells accumulated prior to malignant transformation.

Markers of Cell Proliferation   In a chemoprevention trial of retinoids in long-term smokers with bronchial metaplasia, three markers of proliferation were used: (1) proliferating cell nuclear antigen (PCNA), (2) Ki-67, and (3) DNA polymerase-α. The markers were analyzed by using antibodies on cells obtained from bronchial brushings and biopsies. The investigators observed good correlation between the PCNA expression and histological progression with the highest expression in the high-risk metaplasia group (Lippman et al., 1990b; Lee et al., 1990).

Markers of Differentiation   Because bronchial epithelia differentiate via squamous pathways in carcinogenesis, markers of squamous cell differentiation such as transglutaminase type I, involucrin, epidermal growth factor receptor (EGFR), cytokeratin, and the high-molecular-weight keratin K1 have been used in chemoprevention trials (Lippman et al., 1990a; Mulshine et al., 1992). Recently, there has been renewed interest in blood group antigens as markers of differentiation (Lee et al., 1992; Mulshine et al., 1992). ABH blood group antigens which are normally found on red blood cells and a variety of epithelial cells are markers of differentiation that are lost during carcinogenesis. In a study in which the expression of EGFR was found to be a favorable prognostic factor for surgically resected nonsmall lung cancer (Lee et al., 1989), it was noted that the antibody to EGFR cross-reacted with the blood group antigen A epitope. Subsequent studies showed that in patients with ABO blood type, lung tumor cell expression of blood group antigen A was a favorable prognostic factor in resected nonsmall cell lung cancer cases (Lee et al., 1991). Differentiation markers for other lung tumors are summarized in a review by Mulshine et al. (1992). Markers of developing neuroendocrine tumors (small cell) of the lung include chromogranin A, leu 7, neuro-specific enolase and dopa decarboxylase. Markers of adenomatous tumors include the Clara 1-kilodalton protein, surfactant-associated peptides, CEA, and K-*ras* oncogene.

Noncancer Lung Disease   As with biomarkers of carcinogenesis, those that are the greatest interest in noncancer chronic lung disease represent an early stage in the disease process when intervention to prevent further development of the disease is possible (Table 2). Markers of inflammation, developing pulmonary fibrosis, and developing emphysema will be discussed. The major emphasis will be on chronic disease, although biomarkers of acute pulmonary inflammation will be covered because it is often the first stage of a disease process leading to chronic disease. Infectious lung disease will not be discussed. Atopy and cystic fibrosis (CF) will be discussed in the next section on Biomarkers of Susceptibility.

Inflammation   In inhalation toxicology studies, an *in-vivo* screen for pulmonary inflammation has been developed to determine what level of exposure (concentration × time) to an airborne toxicant is required to induce pulmonary inflammation and toxicity (Henderson, 1991; Warheit et al., 1991; Henderson, 1988; Beck et al., 1982; Moores et al., 1980). The assay involves BAL of exposed animals with physiological saline (or other buffered fluids) to sample the epithelial lining fluid (ELF). The bronchoalveolar lavage fluid (BALF) is then analyzed for biomarkers which allow quantitation of toxicity and the inflammatory response. Total and differential cell counts indicate the degree of the cellular inflammatory response. Neutrophil counts are, of course, sensitive indicators of inflammation. In larger animals, the lymphocyte counts provide information on whether an immune response is involved. Eosinophil counts are used to indicate an allergic response. The acellular fraction of BALF is analyzed for lactate dehydrogenase activity—a measure of cell damage or lysis. Total protein or albumin assays are measures of damage to the alveolar/capillary barrier which allows transudation of serum proteins. Lysosomal enzyme activity in BALF measures macrophage phagocytic activity and lysis of macrophages (Henderson, 1989; Henderson et al., 1991). Alkaline phosphatase (AP) activity, because it is associated with surfactant secreted from type II cells, is a measure of type II cell secretory activity (Henderson et al., 1995). Recently, there has been a report that small amounts of a heat-stable form of AP (HSAP) is present in alveolar type I cells, and that this HSAP is elevated in the serum of smokers (Maslow et al., 1983; Nouwen et al., 1990). Additional work is required to validate HSAP as a marker of type I cell injury. A marker for Clara cell damage has been reported to be elevated levels of γ-glutamyltranspeptidase in BALF (Day et al., 1990). Arachidonate metabolites and various oxygen radical species produced by inflammatory cells are also sometimes used as measures of inflammation.

In humans, BALF has been analyzed mainly for its cellular content because of the difficulty of normalizing the soluble components to the volume of ELF sampled. (Animal studies can report values in comparison to control animals similarly treated or as the total removed from a known portion of the lung.) An example of the use of cellular biomarkers in human BALF is the early work of Hunninghake et al. (1981). In this work, it was shown that BALF cells correctly reflected the tissue inflammatory cell populations. The alveolitis associated with idiopathic pulmonary fibrosis (IPF) was characterized by high neutrophil counts, while the alveolitis associated with sarcoidosis was characterized by high lymphocyte counts. A summary of the use of BALF analysis in human clinical studies has recently been published (Baughman, 1992) and describes the use of BALF analysis to stage the progress of diseases such as sarcoidosis and adult respiratory distress syndrome (ARDS), to distinguish between IPF and sarcoidosis, and to help diagnose and manage such diseases as cryptogenic fibrosing alveolitis, hypersensitivity pneumonitis, drug-induced lung disease, and occupational lung disease.

Pulmonary Fibrosis Most of the early markers of developing pulmonary fibrosis have been found in the ELF, which as described above, can be sampled by BAL. The development of pulmonary fibrosis is usually a sequelae to an inflammatory response in the lung. The forces that determine whether an inflammation resolves or persists to become a chronic fibrotic condition are not fully understood. However, cytokines, which are intracellular messengers, are major players in the process.

Early mediators of pulmonary inflammation are tumor necrosis factor-α (TNF-α) and interleukin-1 (IL-1)—factors released from the resident macrophages in the alveoli (Kunkel and Strieter, 1990). These cytokines promote adherence of circulating inflammatory cells to the endothelium at the site nearest the pulmonary region involved by induction of intercellular adhesion molecule-1 or endothelial leukocyte adhesion molecule-1. Chemotactic factors such as leukotriene B$_4$, fibronectin (Fn) and platelet-activating factor have short half-lives and signal leukocytes (nonspecific as to type) to enter the lung. A longer acting chemotactic factor specific for neutrophils is interleukin-8 (IL-8), which can be synthesized by circulating monocytes, alveolar macrophages, pulmonary endothelial cells, fibroblasts, epithelial cells, and even activated neutrophils themselves (Strieter et al., 1992). A second chemotactic cytokine produced by fibroblasts in response to the early acting cytokines is monocyte chemotactic protein-1 (MCP-1), a chemotactic factor specific for monocytes (Kunkel et al., 1991). Studies by Rolfe et al. (1992a) have shown that endotoxin (a pulmonary inflammagen) is not a direct stimulus for either fibroblasts or pulmonary epithelial type II cells to produce IL-8 or MCP-1. However, IL-1 and TNF-α produced by endotoxin-treated macrophages were potent inducers of IL-8 production by alveolar epithelial cells and MCP-1 by fibroblasts. Thus, IL-1 and TNF-α are early acting cytokines in the inflammatory response, while IL-8 and MCP-1 are induced in response to IL-1 and TNF-α.

With this brief and greatly simplified examination of a few of the cytokines involved in inflammation in the lung, let us examine how these cytokines have been used as biomarkers of lung disease. Elevated expression of mRNA for IL-8 in macrophages obtained by BAL from humans was associated with patients having IPF but not patients with sarcoidosis (a granulomatous lung disease) or patients without lung disease (Lynch et al., 1992). Increased TNF-α was found in association with infiltrating interstitial monocytes and multinucleated giant cells in the alveolar spaces in transbronchial biopsies of a patient with hard metal pneumoconiosis (Rolfe et al., 1992b).

BALF levels of IL-8 were measured in 29 patients at risk for developing ARDS (Donnelly et al. 1993). The mean BALF IL-8 concentration was significantly higher for patients who subsequently progressed to ARDS. Immunocytochemistry suggested that the alveolar macrophage was an important source of IL-8 at this early stage of ARDS. High concentrations of IL-8 are also found in BALF from patients with CF; therapy leads to a reduction of IL-8 in BALF.

In animal studies, pulmonary fibrosis was induced by intratracheal instillation of α-quartz in rats, and control rats received either saline (vehicle) or titanium oxide, a nonfibrogenic particle. Alveolar macrophages obtained from the exposed rats were tested for their release of cytokines. Elevated release of Fn, IL-1 and TNF-α was found in macrophages isolated from the α-quartz-exposed rats but not in macrophages from rats exposed to the nonfibrogenic titanium oxide. The authors observed that elevated release of Fn correlated with developing fibrosis that elevated release of IL-1 was associated with developing granulomas and that elevated TNF release paralleled the degree of neutrophilic influx (Driscoll and Maurer, 1991). (Rats have not been shown to produce IL-8 in pulmonary inflammatory responses.) The association between Fn and developing fibrosis is logical because Fn is a chemoattractant for fibroblasts—an adhesive protein for the attachment of fibroblasts to the connective tissue matrix and also a competence factor for initiation of fibroblast cell replication.

The role of TNF-α in developing α-quartz-induced fibrosis in mice was shown by Piguet et al. (1990). The mRNA for TNF-α was increased in the lung tissues of α-quartz-instilled mice, while the mRNAs for Fn and IL-1 were not. Treating the mice with antibody to mouse TNF-α decreased the pulmonary collagen deposition induced by the quartz, and recombinant TNF-α restored the fibrotic process.

Growth factors are also important in the fibrotic process. Nagaoka et al. (1990) found an elevated secretion of platelet-derived growth factor (PDGF) A and B in pulmonary alveolar macrophages (PAM) from patients with IPF. Studies by Rom et al. (1987; 1991a; 1991b) also indicated that PAM from patients with IPF or asbestosis spontaneously released excessive amounts of PDGF and insulin-like growth factor. In these same studies, the investigators observed an increase in secretion of Fn in PAM from patients with occupational, dust-induced interstitial lung disease.

Oxidative stress has been proposed as one etiology for pulmonary fibrosis, and inflammatory cells in the alveoli are a source of oxygen radicals. An excess of oxygen radicals was produced by inflammatory cells obtained by BAL from patients with IPF, (Strausz et al., 1990) pneumoconiosis, or progressive fibrosis (Wallert et al., 1990). Maier et al. (1991) have proposed to use the ratio of oxidized to normal methionine in BALF proteins as a marker for developing fibrosis. The ratio was fourfold higher in BALF from patients with IPF than in healthy controls.

Emphysema Emphysema is a chronic disorder of the lower respiratory tract characterized by destruction of the walls of the alveoli (Snider et al., 1985). The condition is caused by an imbalance in proteolytic/antiproteolytic activities in the alveoli in favor of proteolytic activity, resulting in loss of elastin and the alveolar walls. An inherited deficiency in α-1-antitrypsin activity is associated with an increased incidence of emphysema (Crystal, 1991). Markers of this genetic susceptibility to emphysema are discussed in the next section on Biomarkers of Susceptibility. The major risk factor for emphysema is, however, not genetic, but a lifestyle that includes cigarette smoking

(Ogushi et al., 1991; Advisory Committee to the Surgeon General [Smoking and Health], 1984). Biomarkers that would predict progress toward emphysema in smokers would be useful. One hypothesis with respect to the mechanism of emphysema development in smokers with normal serum α-1-antitrypsin activity is that α-1-antitrypsin in the lower respiratory tract of smokers is less able to inhibit neutrophil elastase (neutrophils are present in higher numbers in smokers' lungs than in lungs of nonsmokers) because cigarette smoke oxidizes the protease, rendering it less competent as an inhibitor of proteolytic activity (Gadek et al., 1979). This hypothesis has been supported by analysis of the ability of α-1-antitrypsin from smokers' lungs to associate with neutrophil elastase (Ogushi et al., 1991). The association constant for smoker lower respiratory tract α-1-antitrypsin was significantly lower than that for nonsmokers or for the serum protease from either smokers or nonsmokers. Thus, the quality of the α-1-antitrypsin from the lower respiratory tract may be a biomarker for early events that can lead to emphysema.

Potential markers of developing emphysema are urinary desmosine and isodesmosine—cross-linking amino acids found only in elastin. Degradation of elastin during early emphysematous changes should result in elevated levels of these amino acids in urine. In animal studies in which elastase has been instilled in the lung to artificially produce emphysema, urinary desmosine and isodesmosine are elevated (Goldstein and Starcher, 1978; Janoff et al., 1983). Human urine from patients with apparent emphysema, however, does not have elevated levels of the amino acids (Davies et al., 1983; Harel et al., 1980). An improved analytical method developed by Stone et al. (1991) may resolve the usefulness of the elastin amino acids as early markers of developing emphysema.

An alternative set of markers for emphysema is elastin-derived peptide fragments in plasma or urine (Rosenbloom, 1991). Antibodies have been raised to elastin fragments obtained by treatment of purified human lung elastin with neutrophil elastase (Kucich et al., 1985). These antibodies were successfully used to distinguish between patients with chronic obstructive lung disease and a nonsmoking control group. The use of immunologic assays of elastin-derived peptides shows promise for detection of developing emphysema, but requires further validation.

BIOMARKERS OF SUSCEPTIBILITY

As mentioned earlier in this chapter, biomarkers of susceptibility to disease (Table 3) represent an exciting new area of research. Knowledge of individual susceptibilities to chemical-induced disease is both a blessing and a curse. Knowledge of susceptibility helps individuals protect themselves against disease and helps officials protect the public health. Knowledge of the mechanism of susceptibility can also be important in designing therapy for a disease.

Table 3   Biomarkers of Susceptibility

Marker	Susceptibility to	Example Reference
α-1-antitrypsin deficiency	Emphysema	Crystal, 1991
Abnormal cystic fibrosis gene	Cystic fibrosis	Collins, 1992
Elevated IgE responses, atopy	Allergies	Bice, this book
Glutathione-S-transferase μ gene deficiency	Lung cancer	Shields et al., 1993
CYP1A1 inducibility	Lung cancer	Bartsch et al., 1991; Kouri et al., 1982
Elevated CYPIID6	Lung cancer	Amos et al., 1992
Restriction fragment length polymorphism in CYPIIE1 gene	Lung cancer	Uematsu et al., 1991
Restriction fragment length polymorphisms in murine K-*ras* gene, mutated K-*ras* gene	Lung cancer	Ryan et al., 1987 You et al., 1989

Biomarkers of Susceptibility to Noncancer Disease of the Respiratory Tract   In the respiratory tract, there are at least two major noncancer diseases for which the genetic basis of susceptibility is known—CF and emphysema.

Cystic fibrosis is the most common potentially lethal autosomal recessive disease of Caucasians (Collins, 1992). Biological markers for this disease include the abnormal CF gene and the abnormal protein produced by that gene. The mutant gene has a 3-base pair deletion in exon 10 that results in the loss of a single amino acid—phenylalanine—at codon 508 (designated the ΔF508 mutation). The ability to detect the mutant form allows identification of carriers of the disease and genetic counseling of couples at risk for having children with the disease. Knowledge of the genetic basis of the disease is also generating new approaches for therapy for the condition, including gene therapy.

Alpha-1-antitrypsin deficiency is associated with a predilection for emphysema (Crystal, 1991). The deficiency in a normal antiprotease results in proteolytic destruction of alveolar walls, leading to emphysema. The α-1-antitrypsin deficiency is inherited in an autosomal codominate fashion. The following variants exist: MM represents the common, normal activity genotype; ZZ—the severely deficient genotype with approximately 15% of normal serum activity; SS—the second most common deficient class; and MZ—heterozygotes who have an intermediate level of α-1-antitrypsin activity in their blood. It is only the ZZ phenotype that results in early-onset lung disease. MZ phenotypes are carriers of the condition. The gene encoding the α-1-

antitrypsin protein is located on chromosome 14q32.1 (Cox et al., 1982). The regulatory elements for expression of the protein as well as the amino acid sequences of the protein variants are known (Carrell et al., 1982). This information is being used not only as biological markers of the disease but to aid in development of therapy for the condition.

Atopy refers to a condition associated with allergic tendencies. Individuals with atopy can be recognized by a high level of specific sensitizing antibody (IgE mediated) in response to an allergen and a prolonged maintenance of a high IgE response without further exposure to the allergen. Biological markers of this condition can be obtained by skin testing. For further information, see the chapter on "Immunologic Responses of the Respiratory Tract" in this book.

Biomarkers of Susceptibility to Cancer in the Respiratory Tract   Recent studies have examined the relationship between genetic polymorphisms of major metabolic enzymes and DNA adducts in the lung or lung cancer incidence. Shields et al. (1993), using a modified ^{32}P-post-labeling technique with an internal standard of BPDE-DNA adducts for quantitation, determined the relationship between the level of PAH-DNA adducts in human lung and genetic polymorphisms that might influence cancer incidence. They found no correlation between the level of adducts and cytochrome P-4501A1 gene (CYP1A1), the protein product of which mainly contributes to toxication reactions. On the other hand, there was a positive association with null phenotypes for the glutathione-S-transferase μ gene (GSTμ), the product of which leads mainly to detoxication reactions. Thus, it appears that the lack of a functional GSTμ gene may be a susceptibility marker in humans.

Bartsch et al. (1991) conducted a study to determine if individual variations in activity of pulmonary metabolic enzymes may affect cancer risk from cigarette smoking. Mean activities of CYP1A1 were significantly higher in nontumorous lung tissue from lung cancer patients who had smoked within 30 days of surgery than in cancer-free subjects with a similar smoking history, while GST activities were lower. In recent smokers, CYP1A1 activity in lung parenchyma was positively correlated with the level of tobacco smoke-derived DNA adducts. The authors concluded that CYP1A1 inducibility was associated with lung cancer risk.

Using more accessible cells, blood lymphocytes, Kouri et al. (1982) found a positive correlation between CYP1A1 inducibility in lymphocytes and cigarette-induced bronchogenic carcinoma. Cosma et al. (1993) tested blood lymphocytes from 68 persons of different racial backgrounds and found that CYP1A1 inducibility was lower in African-Americans than in persons of European or Asian descent. Petersen et al. (1991) examined a three-generation family of 15 individuals and found that the high-CYP1A1-inducibility phenotype segregates concordantly with an infrequent polymorphic site located 450 bases downstream from the CYP1A1 gene. Hayashi et al. (1991) also have identified a DNA polymorphism of the CYP1A1 gene (called *Msp*I) in individuals with a genetically high risk for lung cancer. The same investigators found that this polymorphism resulted in a substitution of an isoleucine for a valine at residue 462 of the heme-binding region of the CYP1A1 enzyme.

Two other CYP gene polymorphisms have been reported to be associated with lung cancer incidence. Case-control studies that assess the effects of CYPIID6 on lung cancer risk have shown a mildly decreased risk among poor metabolizers (Amos et al., 1992). A restriction fragment length polymorphism was detected in the human CYPIIE1 gene using the restriction endonuclease *Dra*I; the polymorphism was associated with host susceptibility to lung cancer (Uematsu et al., 1991).

In an animal model, Ryan et al. (1987) found that an Eco-R1 restriction fragment length polymorphism in the K-*ras* gene was associated with susceptibility to urethane-induced pulmonary adenomas in inbred mice. A susceptibility to both spontaneously induced or carcinogen-induced pulmonary adenomas was found in mice having 0.55-Kb K-*ras* alleles, whereas the resistant mice had a 0.70 Kb allele. In a later study in strain A mice, a strain known to be susceptible to spontaneous and chemically induced lung tumors, You et al. (1989) found K-*ras* protooncogene activation associated with the lung tumors. The point mutations were associated with codons 12 and 61. Thus, in these mice strains inducibility of the activated K-*ras* oncogene is a marker of susceptibility to lung tumors.

STRATEGIES FOR THE USE OF BIOMARKERS

Biomarkers can be used to answer "yes or no" type questions rather easily. That is, if one only wants to know if a teenage child has been drinking ethanol, a quick sniff of the exhaled breath may be sufficient to answer the question. However, if regulations are to be based on biomarkers, a much more quantitative approach must be used. The association between a quantitative measure of ethanol in the blood and a degree of inebriation must be established. In a like manner, the use of biological markers in setting regulations must be based on sound quantitative data that will allow estimation of prior or ongoing exposure levels or prediction of health outcome. The use of biomarkers to assess exposure has been successful; the use of biomarkers to predict health outcome is an area still in its infancy. Below are described strategies for use of biomarkers to assess exposures and to predict health outcome.

Strategies for the Use of Biomarkers to Quantitate Prior Exposure The amount of any biomarker of a chemical exposure at a specific site in the body will depend on the amount of the exposure (concentration × time), the kinetics of formation and removal of the marker, as well as physiological parameters such as breathing rate, blood flow, and partition coefficients for the chemical or its metabolites into the tissue of interest. If this type of information is available, mathematical models known as physiologically based pharmacokinetic (PBPK) models can be made to predict tissue dose or subcellular dose that will result from a given exposure regimen. Animal models can be used to determine the kinetics of formation and removal of biological markers of exposure, and PBPK models can be developed based on those kinetics and measured partition coefficients, respiration rates, and blood flows. The animal PBPK model can then be adjusted for known differences in human parameters and a human PBPK model developed. Kinetics of formation and removal of metabolites can be studied in human tissue samples to determine if there are major differences from the data obtained in animals. Finally, limited studies in humans can be used to determine the validity of the human PBPK model. Such an approach has been successfully used to develop a PBPK model for styrene in humans (Ramsey and Andersen, 1984).

Strategies for the Use of Biomarkers to Predict Health Outcome There have been as yet few validations of the use of molecular biomarkers to predict health outcome. There has been a great deal of optimism that such a use can be made of biomarkers, but much work must be done to use molecular markers to predict health outcome. If biomarkers are to be widely used to assess exposures, it is essential that the research needed to link the markers to expected health outcome be done. People who are assessed for biomarkers of exposure will quite naturally want to know what a positive result predicts for their health. At the present time, except in the rare cases, we cannot answer that question.

The following strategy used by Groopman et al. (1994) to validate the link between specific molecular markers of aflatoxin (AFB) B_1 exposure and the development of liver cancer serves to illustrate the approach that is required to validate biomarkers as predictors of health outcome. First, animal models were used to determine the quantitative relationship between administered dose and biomarkers of internal dose (urinary metabolites) and biologically effective dose (AFB liver DNA adducts). The log of the amount of one biomarker (AFB-N^7-guanine) in 24-hour urine samples was found to be linearly related to the log of AFB liver DNA adduct concentrations (pmol/mg DNA) at 24 hours. Then the animal model was used to determine if chemoprotective agents (1,2-thiole-3-thione, oltipraz), which are known to reduce the incidence of AFB-induced liver cancer, would reduce the amount of the biomarkers. Total AFB metabolites in urine did not reflect the reduced health risk in the rats receiving both AFB and the chemoprotective agent, but liver AFB-N^7-guanine DNA adducts, urinary AFB-N^7-guanine and AFB-albumin adducts did. Based on the animal model work, the investigators then characterized the same biomarkers in high-risk human populations in China as well as Gambia, West Africa. Dose was estimated by analysis of food samples for AFB. As in the animal model, urinary $AFB-N^7$-guanine was found to correspond with dietary AFB intake. Current prospective epidemiological studies in high-risk populations are underway to determine if the urinary $AFB-N^7$-guanine amount is predictive for risk of liver cancer. The biomarkers will also be used in chemoprotective trials in humans. It is this type of extensive research involving both basic research in animals and epidemiological research in high-risk populations that is required to make biomarkers useful as valid predictors of adverse health outcome.

SUMMARY

There are numerous biomarkers of exposure, effect, and susceptibility associated with the respiratory tract. Within the broad range of biomarkers described in this chapter, there are varying degrees of validation of the significance of the markers. For the markers of susceptibility, there is a plethora of information that provides dependable prediction of susceptibility to specific diseases. For the biomarkers of exposure, some progress has been made in relating the biomarkers in tissues to exposures to specific chemicals; much more needs to be done to define quantitative links between the level of the markers in tissues and the degree of prior exposure. Very little has been done to define quantitative links between the tissue levels of biomarkers of exposure and the predicted health outcome. Perhaps the greatest research challenge is to validate the significance of biological markers of effects. Research directed at defining the mechanisms of disease induction should provide the information needed to improve our ability to use biological markers as reliable, validated predictors of the role of chemical exposures in inducing adverse health effects in the respiratory tract.

ACKNOWLEDGMENT

I thank the many members of the staff of the Inhalation Toxicology Research Institute (ITRI) who helped in the preparation of this chapter. The ITRI facility is owned by the U.S. DOE/OHER and operated by the Lovelace Biomedical Environmental Research Institute under Contract No. DE-AC04-76EV01013.

REFERENCES

Amos, CI, Caporaso, NE, Weston A: Host factors in lung cancer risk: A review of interdisciplinary studies. Cancer Epidemiol Biomarkers Prev 1:505–13, 1992.

Bagnasco, M, Bennicelli, C, Camoirano, A, Balansky, RM, De Flora, S: Metabolic alterations produced by cigarette smoke in rat lung and liver, and their modulation by oral N-acetylcysteine. Mutagenesis 7:295–301, 1992.

Bartsch, H, Petruzzelli, S, De Flora, S, Hietanen, E, Camus, A-M Castegnaro, M, Geneste, O Camoirano, A, Saracci, R, Giuntini, C: Carcinogen metabolism and DNA adducts in human lung tissues as affected by tobacco smoking or metabolic phenotype: A case-control study on lung cancer patients. Mutat Res 250:103–14, 1991.

Baughman, RP (ed): Bronchoalveolar Lavage. St. Louis: Mosby Year Book, 1992.

Bechtold, WE, Lucier, G, Birnbaum, LS, Yin, S-N, Li, G-L, Henderson, RF: Muconic acid determinations in urine as a biological exposure index for workers occupationally exposed to benzene. Am Ind Hyg Assoc J 52:473–78, 1991.

Bechtold, WE, Willis, JK, Sun, JD, Griffith, WC, Reddy, TV: Biological markers of exposure to benzene: S-phenylcysteine in albumin. Carcinogenesis 12(7):1217–20, 1992.

Bechtold, WE, Waide, JJ, Sandström, T, Stjernberg, N, McBride, D, Koenig, J, Chang, IY, Henderson RF: Biological markers of exposure to SO₂: S-sulfonates in nasal lavage. J Exp Anal Environ Epidemiol 3:371–82, 1993.

Beck, BD, Brain, JD, Bohannon DE: An in vivo hamster bioassay to assess the toxicity of particles for the lungs. Toxicol Appl Pharmacol 66:9–29, 1982.

Benner, SE, Hong, WK, Lippman, SM, Lee, JS, Hittelman, WM: Intermediate biomarkers in upper aerodigestive tract and lung chemoprevention trials. J Cell Biochem Suppl 16G:33–38, 1992.

Bond, JA, Wolff, RK, Harkema, JR, Mauderly, JL, Henderson, RF, Griffith, WC, McClellan, RO: Distribution of DNA adducts in the respiratory tract of rats exposed to diesel exhaust. Toxicol Appl Pharmacol 96:336–46, 1988.

Bond, JA, Chen, BT, Griffith, WC, Mauderly, JL: Inhaled cigarette smoke induces the formation of DNA adducts in lungs of rats. Toxicol Appl Pharmacol 99:161–72, 1989.

Bond, JA, Johnson, NF, Snipes, MB, Mauderly, JL: DNA adduct formation in rat alveolar type II cells: Cells potentially at risk for inhaled diesel exhaust. Environ Mol Mutagen 16:64–69, 1990.

Bos, JL: ras oncogenes in human cancer: A review. Cancer Res 49:4682–89, 1989.

Brandt-Rauf, PW: Advances in cancer biomarkers as applied to chemical exposures: The ras oncogene and p21 protein and pulmonary carcinogenesis. J Occup Med 33(9):951–55, 1991.

Brennan, J, O'Connor, T, Makuch, RW, Simmons, AW, Russel, E, Linnoila, RI, Phelps, RM, Gazdar, AF, Ihde, DC, Johnson, BE: myc family DNA amplification in 107 tumors and tumor cell lines from patient with small cell lung cancer treated with different combination chemotherapy regimens. Cancer Res 51:1708–12, 1991.

Bustros, AD, Nelkin, BD, Silverman, A, Ehrlich, G, Poiesz, B, Baylin, SB: The short arm of chromosome 11 is a "hot spot" for hypermethylation in human neoplasia. Proc Natl Acad Sci USA, 85:5693–97, 1988.

Carrell, RW, Jeppsson, JO, Laurell, C-B, Brennan, SO, Owen, MC, Vaughan, L, Boswell, DR: Structure and variation of human α₁-antitrypsin. Nature 298:329–33, 1982.

Chiba, I, Takahashi, T, Nau, MM, D'Amico, D, Curiel, DT, Mitsudomi, T, Buchhagen, DL, Carbone, D, Piantadosi, S, Koga, H, Reissmann, PT, Slamon, DJ, Holmes, EC, Minna, JD: Mutations in the p53 gene are frequent in primary, resected nonsmall cell lung cancer. Oncogene 5:1603–10, 1990.

Chiba, I, Minna, JD: Mutations in the p53 genes are frequent in primary, resected nonsmall cell lung cancer. Oncogene 5:1603–10, 1990.

Collins, FS: Cystic fibrosis: Molecular biology and therapeutic implications. Science 256:774–79, 1992.

Cosma, G, Crofts, F, Currie, D, Wirgin, I, Toniolo, P, Garte, SJ: Racial differences in restriction fragment length polymorphisms and messenger RNA inducibility of the human CYP1A1 gene. Cancer Epidemiol Biomarkers Prev. 2:53–57, 1993.

Cox, DW, Markovic, VD, Teshima, IE: Genes for immunoglobulin heavy chains and for α₁-antitrypsin are localized to specific regions of chromosome 14q. Nature 297:428–30, 1982.

Crystal, RG: Alpha 1-antitrypsin deficiency: Pathogenesis and treatment. Hosp Pract (Off Ed) 26(2):81–94, 1991.

Davies, SF, Offord, KP, Brown, MG, Campe, H, Niewoehner, D: Urine desmosine is unrelated to cigarette smoking or to spirometric function. Am Rev Respir Dis 128:473–75, 1983.

Day, BJ, Carlson, GP, De Nicola, DB: γ-glutamyltranspeptidase in rat bronchoalveolar lavage fluid as a probe of 4-ipomeanol and α-napthylthiourea-induced pneumotoxicity. J Pharmacol Meth 24:1–8, 1990.

Donnelly, SC, Strieter, RM, Kunkel, SL, Walz, A, Robertson, CR, Carter, DC, Grant, IS, Pollok, AJ, Haslett C: Interleukin-8 and development of adult respiratory distress syndrome in at-risk patient groups. Lancet 341:643–47, 1993.

Driscoll, KE, Maurer, JK: Cytokine and growth factor release by alveolar macrophages: Potential biomarkers of pulmonary toxicity. Toxicol Pathol 4(1):398–405, 1991.

Driscoll, KE, Carter, JM, Howard, BW, Hassenbein, DG: Mutagenesis in rat lung epithelial cells after in vivo celica exposure or ex vivo exposure to inflammatory cells. Am J Resp Crit Care Med 149:A553, 1994.

Gadek, JE, Fells, GA, Crystal, RG: Cigarette smoking induces functional antiprotease deficiency in the lower resiratory tract of humans. Science 206:1315–16, 1979.

Garver, RI, Jr, Chan, CS, Milner, PG: Reciprocal expression of pleiotrophin and midkine in normal versus malignant lung tissues. Am J Respir Cell Mol Biol 9:463–66, 1993.

Goldstein, RA, Starcher, BC: Urinary excretion of elastin peptides containing desmosine after intratracheal injection of elastase in hamsters. J Clin Invest 61:1286–90, 1978.

Groopman, JD, Wogan, GN, Roebuck, BD, Kensler, TW: Molecular biomarkers for aflatoxins and their application to human cancer prevention. Cancer Res Suppl 54:1907s-11s, 1994.

Gunnison, AF, Palmes, ED: Species variability in plasma S-sulfonate levels during and following sulfite administration. Chem Biol Interact 21:315–29, 1978.

Gupta, RC, Sopori, ML, Gairola, CG: Formation of cigarette smoke-induced DNA adducts in the rat lung and nasal mucosa. Cancer Res 49:1916–20, 1989.

Harbour, JW, Lai, S-L, Whang-Peng, J, Gazdar, AF, Minna, JD, Kaye, FJ: Abnormalities in structure and expression of the retinoblastoma gene in SCLC. Science 241:353–47, 1988.

Harel, S, Janoff, A, Yu, SY, Hurewitz, A, Bergofsky, EH: Desmosine radioimmunoassay for measuring elastin degradation in vivo. Am Rev Respir Dis 122:769–73, 1980.

Harris, CC: Chemical and physical carcinogenesis: Advances and perspectives for the 1990s. Cancer Res Suppl 51:5023s-44s, 1991.

Hayashi, S, Watanabe, J, Nakachi, K, Kawajiri, K: Genetic linkage of lung cancer-associated MspI polymorphisms with amino acid replacement in the heme binding region of the human cytochrome P450IA1 gene. J Biochem 110:407–11, 1991.

Hecht, SS, Carmella, SG, Foiles, PG, Murphy, SE: Biomarkers for human uptake and metabolic activation of tobacco-specific nitrosamines. Cancer Res Suppl 54:1912s-17s, 1994.

Henderson, RF: Use of bronchoalveolar lavage to detect lung damage. In Toxicology of the Lung, ed. DE Gardner, JD Crapo, and EJ Massaro, 239–68. New York: Raven Press, 1988.

Henderson, RF: Bronchoalveolar lavage: A tool for assessing the health status of the lung. In Concepts in Inhalation of Toxicology, ed. RO McClellan and RF Henderson, 415–44. New York: Hemisphere Publishing, 1989.

Henderson, RF: Analysis of respiratory tract lining fluids to detect injury. In Pulmonary Pharmacology and Toxicology, Vol. II, ed. MA Hollinger, 2–18. Boca Raton: CRC Press, 1991.

Henderson, RF, Harkema, JR, Hotchkiss, JA, Boehme, DS: Effect of blood leucocyte depletion on the inflammatory response of the lung to quartz. Toxicol Appl Pharmacol 109:127–36, 1991.

Henderson, RF, Belinsky, SA: Biological markers of respiartory tract exposure. In Toxicology of the Lung, ed. DE Gardner, JD Crapo, and RO McClellan, 253–82. New York: Raven Press, 1993.

Hendreson, RF, Scott, GG, Waide, JJ: Comparison of alkaline phosphatase (AP) in alveolar type II cells with AP in bronchoalveolar lavage fluid in normal and injured lungs. Toxicol Appl Pharmacol (submitted), 1995.

Hunninghake, GW, Kawanami, O, Ferrans, VJ, Young, RC, Jr, Roberts, WC, Crystal, RG: Characterization of the inflammatory and immune effector cells in the lung parenchyma of patients with interstitial lung disease. Am Rev Respir Dis 123:407–12, 1981.

Iggo, R, Gatter, K, Bartek, J, Lane, D, Harris AL: Increased expression of mutant forms of p53 oncogene in primary lung cancer. Lancet 355:675–79, 1990.

Izzotti, A, Rossi, GA, Bagnasco, M, De Flora, S: Benzo(a)pyrene diolepoxide-DNA adducts in alveolar macrophages of smokers. Carcinogenesis 12(7):1281–85, 1991.

Izzotti, A, Balansky, RM, Coscia, N, Scatolini, L, D'Agostini, F, De Flora, S: Chemoprevention of smoke-related DNA adduct formation in rat lung and heart. Carcinogenesis 13(11):2187–90, 1992.

Janoff, A, Chanana, AD, Joel, DD, Susskind, H, Laurent, P, Yu, SY, Dearing R: Evaluation of the urinary desmosine radioimmunoassay as a monitor of lung injury after endobronchial elastase installation in sheep. Am Rev Respir Dis 128:545–51, 1983.

Johnson, BE, Ihde, DC, Makuch, RW, Gazdar, A, Carney, D, Oie, H, Russell, E, Nau, M, Minna, J: myc family oncogene amplification in tumor cell lines established from small cell lung cancer patients and its relationship to clinical staus and course. J Clin Invest 79:1629–34, 1987.

Johnson, BW, Makuch, RW, Simmons, AD, Gazdar, AF, Burch, D, Cashell, AW: myc family DNA amplification in small cell lung cancer patients' tumors and corresponding cell lines. Cancer Res 48:5163–66, 1988.

Keller, DA, Menzel, DB: Picomole analysis of glutathione, glutathione disulfide, glutathione S-sulfonate and cysteine S-sulfonate by high performance liquid chromatography. Anal Biochem 51:418–23, 1985.

Kern, JA, Schwartz, DA, Nordberg, JE, Weiner, DB, Greene, MI, Torney, L, Robinson, RA: p185neu expression in human lung adenocarcinomas predicts shortened survival. Cancer Res 50:5184–91, 1990.

Kok, K, Osinga, J, Carritt, B, Davis, MB, van der Hout, AH, van der Veen, AY, Landsvater, RM, de Leij, LF, Berendsen, HH, Postmus, PE, Poppema, S, Buys, CHC: Deletion of a DNA sequence at the chromosomal region 3p21 in all major types of lung cancer. Nature 330:578–81, 1987.

Kouri, RW, McKinney, CD, Slomiany, DJ, Snodgrass, DR, Wray, NP, McLemore, TL: Positive correlation between high aryl hydrocarbon hydroxylase activity and primary lung cancer as analyzed in cryopreserved lymphocytes. Cancer Res 42:5030–37, 1982.

Kratzke, RA, Shimizu, E, Kaye, FJ: Oncogenes in human lung cancer. In Oncogenes and Tumor Suppressor Genes in Human Malignancies, ed. CB Benz, and ET Liu, 61–85. Boston: Kluwer Academic Publishers, 1993.

Kucich, U, Christner, P, Lippmann, M, Kimbel, P, Williams, G, Rosenbloom, J, Weinbaum, G: Utilization of a peroxidase antiperoxidase complex in an enzyme linked immunosorbent assay of elastin-derived peptides in human plasma. Am Rev Respir Dis 131:709–713, 1985.

Kunkel, SL, Strieter, RM: Cytokine networking in lung inflammation. Hosp Pract 25(10):63–76, 1990.

Kunkel, SL, Standiford, T, Kasahara, K, Strieter, RM: Stimulus specific induction of monocyte chemotactic protein-1 (MCP-1) gene expression. Adv Exp Med Biol 305:65–71, 1991.

Kurzrock, R, Gallick, GE, Gutterman, JU: Differential expression of p21 ras gene products among histological subtypes of fresh primary human lung tumors. Cancer Res 46:1530–34, 1986.

Lee, JS, Ro, JY, Sahin, A, Hittelman, W, Brown, BW, Mountain, CF, Hong, WK: Expression of epidermal growth factor receptor (EGFR): A favorable prognostic factor for surgically resected nonsmall cell lung cancer (NSCLC). Proc Am Soc Clin Oncol 8:226, 1989.

Lee, JS, Ro, JY, Sahin, A, Hong, WK, Hittelman, WN: Quantitation of proliferating cell fraction (PCF) in nonsmall cell lung cancer (NSCLC) using immunostaining for proliferating cell nuclear antigen (PCNA). Proc Am Assoc Cancer Res 31:22, 1990.

Lee, JS, Ro, JY, Sahin, AA, Hong, WK, Brown, BW, Mountain, CF, Hittelman, WN: Expression of blood-group antigen A: A favorable prognostic factor in nonsmall-cell lung cancer: N Engl J Med 324:1084–90, 1991.

Lee, JS, Lippman, SM, Hong, WK, Ro, JY, Kim, SY, Lotan, R, Hittelman, WN: Determination of biomarkers for intermediate end points in chemoprevention trails. Cancer Res 52:2707s–10s, 1992.

Lee, CK, Brown, BG, Reed, EA, Coggins, CRE, Doolittle, DJ, Hayes, AW: Ninety-day inhalation study in rats, using aged and diluted sidestream smoke from a reference cigarette: DNA adducts and alveolar macrophage cytogenetics. Environ Appl Toxicol 20:393–401, 1993.

Lippman, SM, Lee, JS, Lotan, R, Hittelman, W, Wargovich, MJ, Hong, WK: Biomarkers as intermediate end points in chemoprevention trials. J Natl Cancer Inst 82:555–60, 1990a.

Lippman, SM, Lee, JS, Peters, E, Ro, J, Wargovich, M, Morice, R, Hittelman, W, Hong, W: Expression of proliferation cell nuclear antigen (PCNA) correlates with histologic stage of bronchial carcinogenesis. Proc Am Assoc Cancer Res 31–168, 1990b.

Liteplo, RG, Kerbel, RS: Reduced levels of DNA 5-methylcytosine in metastatic variants of the human melanoma cell line MeWo. Cancer Res 47:2264–67, 1987.

Lynch, JP, III, Standiford, TJ, Rolfe, MW, Kunkel, SL, Strieter, RM: Neutrophilic alveolitis, in idiopathic pulmonary fibrosis. Am Rev Respir Dis 145:1433–39, 1992.

Makos, M, Nelkin, BD, Lerman, MI, Alatif, F, Zbar, B, Baylin SB: Distinct hypermethylation patterns occur at altered chromosome loci in human lung and colon cancer. Proc Natl Acad Sci USA 89:1929–33, 1992.

Maier, K, Leuschel, L, Costabel, U: Increased levels of oxidized methionine residues in bronchoalveolar lavage fluid proteins from patients with idiopathic pulmonary fibrosis. Am Rev Respir Dis 143:271–74, 1991.

Maples, KR, Sandström, T, Su, Y-F, Henderson, RF: The nitric oxide/heme protein complex as a biologic marker of exposure to nitrogen dioxide in humans, rats, and in vitro models. Am J Respir Cell Mol Biol 4:538–43, 1991.

Maslow, WC, Muensch, HA, Azama, F, Schneider, SS: Sensitive fluorometry of heat-stable alkaline phosphatase (regan enzyme) activity in serum from smokers and nonsmokers. Clin Chem 29(2):260–63, 1983.

Moores, SR, Sykes, SE, Morgan, A, Evans, N, Evans, JC, Holmes A: The short-term cellular and biochemical response of the lung to toxic dusts: An in vivo cytotoxicity test. In In Vitro Effect of Mineral Dusts, ed. RC Brown, IP Gormley, M Chamberlain, et al., 297–303. New York: Academic Press, 1980.

Mulshine, JL, Linnoila, RI, Treston, AM, Scott, FM, Quinn, K, Avis, I, Shaw, GL, Jensen, SM, Brown, P, Birrer, MJ, Cuttitta, F: Candidate biomarkers for application as intermediate end points of lung carcinogenesis. J Cell Biochem Suppl 16G:183–6, 1992.

Nagaoka, I, Trapnell, BC, Crystal, RG: Upregulation of platelet-derived growth factor-a and -b gene expression in alveolar macrophages of individuals with idiopathic pulmonary fibrosis. J Clin Invest 85:2023–27, 1990.

National Research Council, Committee on Biological Markers. 1987. Biological markers in environmental health research. Environ Health Perspect 74:3–9, 1987.

Nigro, JM, Baker, SJ, Preisinger, AC, Jessup, JM, Hosletter, R, Cleary, K, Bigner, SH, Davidson, N, Baylin, S, Devilee, P, Glover, T, Collins, FS, Weston, A, Modali, R, Harris, CC, Vogelstein, B: Mutations in the p53 gene occur in diverse human tumor types. Nature 342:705–08, 1989.

Nikula, KJ, Snipes, MB, Barr, EB, Griffith, WC, Henderson, RF, Mauderly, JL: Comparative pulmonary toxicities and carcinogenicities of chronically inhaled diesel exhaust and carbon black in F344 rats. Fund Appl Toxicol (in press), 1995.

Nouwen, EJ, Buyssens, N, De Broe, ME: Heat-stable alkaline phosphatase as a marker for human and monkey type-I pneumocytes. Cell Tissue Res 260:321–35, 1990.

Ogushi, F, Hubbard, RC, Vogelmeier, C, Fells, GA, Crystal, RG: Cigarette smoking is associated with a reduction in the association rate constant of lung α_1-antitrypsin for neutrophil elastase. J Clin Invest 87:1060–65, 1991.

Osterman-Golkar, S, Ehrenberg, D, Seberbäck, D, Hällström I: Evaluation of genetic risks of alkylating agents. II. Hemoglobin as a dose monitor. Mutat Res 34:1–10, 1976.

Penno, MB, August, JT, Baylin, SB, Mabry, M, Linnoila, RI, Lee, VS, Croteau, D, Yang, XL, Rosada, C: Expression of CD44 in Human Lung Tumors. Cancer Res 54:1381–87, 1994.

Perera, F, Fischman, HK, Hemminki, K, Brandt-Ruaf, P, Niman, HL, Smith, S, Toporoff, E, O'Dowd, K, Tang, MX, Tsai, WY, Stoopler, M: Protein binding sister chromatid exchange and expression of oncogene proteins in patients treated with cisplatinum (cis DDP)-based chemotherapy. Arch Toxicol 64:401–06, 1990.

Petersen, DD, McKinney, CE, Ikeya, K, Smith, HH, Bale, AE, Mcbride, OW, Nebert, DW: Human CYP1A1 Gene: Cosegregation of the enzyme inducibility phenotype and an RFLP. Am J Hum Genet 48:720–25, 1991.

Phillips, DH, Hewer, A, Margin, CN, Garner, RC, King, MM: Correlation of DNA-adduct levels in human lung with cigarette smoking. Nature 336:790–92, 1988.

Phillips, DH, Schoket, B, Hewer, A, Bailey, E, Kostic, S, Vincze, I: Influence of cigarette smoking on the levels of DNA adducts in human bronchial epithelium and white blood cells. Int J Cancer 46:569–75, 1990.

Piguet, PF, Collart, MA, Grau, GE, Sappino, AP, Vassalli, P: Requirement of tumor necrosis factor for development of silica-induced pulmonary fibrosis. 344:245–47, 1990.

Pluygers, E, Baldewyns, P, Minette, P, Beauduin, M, Gourdin, P, Robinet, P: Biomarker assessments in asbestos-exposed workers as indicators for selective prevention of mesothelioma or bronchogenic carcinoma: Rationale and practical implementations. Eur J Cancer Prev 1:57–68, 1991.

Pluygers, E, Baldewyns, P, Minette, P, Beauduin, M, Gourdin, P, Robinet, P: Biomarker assessments in asbestos-exposed workers as indicators for selective prevention of mesothelioma or bronchogenic carcinoma: Rationale and practical implementations. Eur J Cancer Prev 1:129–38, 1992.

Pryor, WA, Wang, K, Bermúdez, E: Cholesterol ozonation products as biomarkers for ozone exposure in rats. Biochem. Biophys Res Commun 188(2):618–23, 1992.

Ramsey, JC, Andersen ME: A physiologically based description of the inhalation pharmacokinetics of styrene in rats and humans. Toxicol Appl Pharmacol 73:159–75, 1984.

Reissmann, PT, Koga, H, Takahashi, R, Benedict, WF, Figlin, R, Holmes, EC, Slamon, DJ, and the Lung Cancer Study Group: Inactivation of the retinoblastoma gene in nonsmall cell lung cancer. Proc Am Assoc Cancer Res 31:318, 1990.

Rolfe, MW, Kunkel SL, Standiford, TJ, Orringer, MB, Phan, SH, Evanoff, HL, Burdick, MD, Strieter, RM: Expression and regulation of human pulmonary fibroblast-derived monocyte chemotactic peptide-1. Am J Physiol 263:L536–45, 1992a.

Rolfe, MW, Paine, R, Davenport, RB, Strieter, RM: Hard metal pneumoconiosis and the association of tumor necrosis factor-alpha. Am Rev Respir Dis 146:1600–02, 1992b.

Rom, WN, Bitterman, PB, Rennard, SI, Cantin, A, Crystal, RG: Characterization of the lower respiratory tract inflammation of nonsmoking individuals with interstitial lung disease associated with chronic inhalation of inorganic dusts. Am Rev Respir Dis 136:1429–34, 1987.

Rom, WN, Pääkkö, P: Activated alveolar macrophages express the insulin-like growth factor-I receptor. Am J Respir Cell Mol Biol 4:432–39, 1991a.

Rom, WN, Travis, WD, Brody, AR: Cellular and molecular basis of the asbestos-related diseases. Am Rev Respir Dis 143:408–422, 1991b.

Rosenbloom, J: Biochemical/immunologic markers of emphysema. Ann NY Acad Sci 624(Supp): 7–12, 1991.

Ryan, J, Barker, PE, Nesbitt, MN, Ruddle, FH: K-ras2 as a genetic marker for lung tumor susceptibility in inbred mice. J Natl Cancer Inst 79:1351–57, 1987.

Sabourin, PJ, Bechtold, WE, Griffith, WC, Birnbaum, LS, Lucier, G, Henderson, RF: Effect of exposure concentration, exposure rate, and route of administration on metabolism of benzene by F344 rats and B6C3F$_1$ mice. Toxicol Appl Pharmacol 99:421–44, 1989.

Sakai, T, Toguchida, J, Ohtani, N, Yandell, DW, Rapaport, JM, Dryja, TP: Allele-specific hypermethylation of the retinoblastoma tumor-suppressor gene. Am J Hum Genet 48:880–88, 1991.

Shields, PG, Bowman, ED, Harrington, AM, Doan, VT, Weston, A: Polycyclic aromatic hydrocarbon-DNA adducts in human lung and cancer susceptibility genes. Cancer Res 53:3486–92, 1993.

Slamon, DJ, DeKernion, JB, Verma, IM, Cline, MJ: Expression of cellular oncogenes in human malignancies. Sciences 224:256–62, 1984.

Smoking and Health; Report of the Advisory Committee to the Surgeon General of the Public Health Service. U.S. Dept of Health, Education, and Welfare. 14–57. Washington, D.C., 1984.

Snider, GL, Kleinerman, J, Thurlbeck, WM, Bengali, ZH: The definition of emphysema. Report of a National Heart, Lung, and Blood Institute, Division of Lung Diseases Workshop. Am Rev Respir Dis 132:182–85, 1985.

Sohn, HY, Cheong, N, Wang, Z-W, Hong, WK, Hittelman, WN: Detection of aneuploidy in normal lung tissue adjacent to lung tumor by premature chromosome condensation. Cancer Genet Cytogenet 41:250, 1989.

Stone, PJ, Bryan-Rhadfi, J, Lucey, EC, Ciccolella, DE, Crombie, G, Faris, B, Snider, GL, Franzblau, C: Measurement of urinary desmosine by isotope dilution and high performance liquid chromatography. Am Rev Respir Dis 144:284–90, 1991.

Strieter, RM, Kasahara, K, Allen, RM, Standiford, TJ, Rolfe, MW, Becker, FS, Chensue, SW, Kunkel, SL: Cytokine-induced neutrophil-derived interleukin-8. Am J Pathol 141:397–407, 1992.

Strausz, JJ, Müller-Quernheim, J, Steppling, H, Ferlinz, R: Oxygen radical production by alveolar inflammatory cells in idiopathic pulmonary fibrosis. Am Rev Respir Dis 141:124–28, 1990.

Takahashi, T, Nau, MM, Chiba, I, Birrer, M, Rosenberg, R, Vinocur, M, Levitt, M, Pass, H, Gazdar, A, Minna, J: p53: A frequent target for genetic abnormalities in lung cancer. Science 246:491–94, 1989.

Takahashi, T, Takahashi, T, Suzuki, H, Hida, T, Sekido, Y, Ariyoshi, Y, Ueda, R: The p53 gene is very frequently mutated in small cell lung cancer with a distinct nucleotide substitution pattern. Oncogene 6:1775–78, 1991.

Uematsu, F, Kikuchi, H, Motomiya, M, Abe, T, Sagami, I, Ohmachi, T, Wakui, A, Kanamaru, R, Watanabe, M: Association between restriction fragment length polymorphism of the human cytochrome P450IIE1 gene and susceptibility to lung. Jpn J Cancer Res 82:254–56, 1991.

Vahakangas, KH, Haugen, A, Harris, CC: An applied synchronous fluorescence spectrophotometric assay to study benzo(a)pyrene dioepoxide-DNA adducts. Carcinogenesis 6:1109–16, 1985.

Vahakangas, KH, Samet, JH, Metcalf, RA, Welsh, JA, Bennet, WP, Lune, DP, Harris, CC: Mutations of p53 and *ras* genes in radon-associated lung cancer from uranium miners. Lancet 339:576–80, 1992.

Wallace, LA: The total exposure assessment methodology (TEAM) study: Summary and analysis. Vol. I Report EPA 660/6-87/002a. Washington, DC: U.S. Environmental Protection Agency, Office of Acid Deposition, Environmental Monitoring and Quality Assurance, 1987.

Wallert B, Lassalle, P, Fortin, F, Aerts, C, Bart, F, Fournier, E, Voisin, C: Superoxide anion generation by alveolar inflammatory cells in simple pneumoconiosis and in progressive massive fibrosis of nonsmoking coal workers. Am Rev Respir Dis 141:129–33, 1990.

Warheit, DB, Carakostus, MC, Hartsky, MA, Hansen, JF: Development of a short-term inhalation bioassay to assess pulmonary toxicity of inhaled particles: Comparisons of pulmonary iron and silica. Toxicol Appl Pharmacol 107:350–68, 1991.

Weston, A, Bowman, ED: Fluorescence detection of benzo(a)pyrene-DNA adducts in human lung. Carcinogenesis 12(8):1445–49, 1991.

Wilson, VL, Smith, RA, Longoria, J, Liotta, MA, Harper, CM: Chemical carcinogen-induced decreases in genomic 5-methyldeoxycytidine content of normal human bronchial epithelial cells. Proc Natl Acad Sci USA 84:3298–3301, 1987.

Wong, A, Rupert, J, Eggleston, J, Hamilton, SR, Baylin, SB, Volgelstein, B: Gene amplification of c-*myc* and N-*myc* in small cell carcinoma of the lung. Science 223:461–64, 1986.

Wulfsberg, EA, Hoffmann, DE, Cohen, MM: α_1-antitrypsin deficiency: Impact of genetic discovery on medicine and society. JAMA 271(3):217–22, 1994.

Yanagisawa, Y, Nishimura, H, Matsuki, H, Osaka, F, Kasuga, H: Personal exposure and health effect relationship for NO_2 with urinary hydroxyproline to creatinine ratio as indicator. Arch Environ Health 41:41–48, 1986.

Yokota, J, Wada, M, Shimosata, Y, Terada, M, Sugimura, T: Loss of heterozygosity on chromosomes 3, 13, and 17 in small cell carcinoma and on chromosome 3 in adenocarcinoma of the lung. Proc Natl Acad Sci USA 84:9252–56, 1987.

You, M, Candrian, U, Maronpot, R, Stoner, G, Anderson, M: Activation of the Ki-*ras* protooncogene in spontaneoulsy occurring and chemically induced lung tumors of the strain A mouse. Proc Natl Acad Sci USA 86:3070–74, 1989.

Chapter Sixteen

Role of Cytokines in Pulmonary Inflammation and Fibrosis

Kevin E. Driscoll

INTRODUCTION

The response of the lung to chemical exposure depends, in part, on the orchestration of a number of molecular and cellular events established by an elaborate signaling network (Kunkel and Strieter, 1990). In this respect, while different etiologic agents cause pulmonary toxicity, a number of stereotypic mechanisms are involved in mediating the ensuing inflammation and fibrosis. Inflammatory cell recruitment, cell proliferation and extracellular matrix synthesis are regulated by lipid and protein mediators released both proximal and distal to the site of injury(s). Cytokines are one class of molecules regulating these processes and thus are important factors influencing the response of the lung to exogenous agents. In this chapter general concepts of cytokine biology as it relates to inflammation and wound healing are reviewed. The roles of selected cytokines in pulmonary responses to various pneumotoxins are characterized; and the potential use of altered cytokine expression as *in vitro* or *in vivo* biomarkers of lung toxicity is discussed.

CYTOKINES

Cytokines are small (8–30 kDa), typically glycosylated proteins which regulate cellular differentiation, proliferation, and activation (Balkwill and Burke, 1989). Cytokines stimulate both immune and nonimmune cells via specific cell membrane receptors. Because these receptors are present on all nucleated cells, cytokines can play a key roll in cell:cell communication throughout the body. Initially, several cytokines were identified as secretory products of lymphocytes and monocytes and hence referred to as "lymphokines" or "monokines" to denote nonantibody protein mediators produced by immunocompetent cells. However, it has become clear that cells not traditionally considered to be immune effector cells (i.e., fibroblasts, epithelial cells, endothelial cells) are important sources of these mediators (Thorton et al., 1990; Standiford et al., 1990; Driscoll et al., 1993b). Thus, the term "cytokine" has been adopted to describe a diverse group of protein molecules, elaborated by a variety of cell types which serve as communication signals in normal physiologic and pathophysiologic processes. At present, there is no standard nomenclature for cytokines and many are named based on an originally defined biological activity (i.e., tumor necrosis factor, transforming growth factor). This can be a source of confusion since many cytokines exhibit pleiotropic and overlapping biological activities (Larrick and Kunkel, 1988).

While this chapter focuses on the role of cytokines in toxic responses of the lung, the activities of cytokines are not restricted to pathophysiologic responses. Normal tissue homeostasis is maintained by close coordination of processes such as cell differentiation, turnover, and secretion which are coordinated in part through cytokine mediated cell:cell communication. In this respect, cells may secrete cytokines to modify their own activity (autocrine effects) or the activity of other cells (paracrine effects). Cytokines influence such physiologic processes as the circadian rhythm of body temperature, changes in appetite, and patterns of sleep (Dinarello, 1989). Altered production of cytokines in response to injury or infection can result in significant physiologic changes (Sherry and Cerami, 1988; Dinarello, 1989; Balkwill and Burke, 1989) with the nature of the biological effect and site of action influenced by the magnitude of cytokine production (Kunkel et al., 1995). Low concentrations of cytokines are often involved in maintaining tissue homeostasis, after tissue injury or infection, increased production of cytokines contributes to the initiation and resolution of local inflammation and tissue repair. Further increases in cytokine levels and increased tissue permeability can produce systemic effects manifested as alterations in the physiology of the host. For example, increased systemic levels of TNFα and IL-1 are believed to be responsible for signs and symptoms of disease such as: fever, altered heart rate and blood pressure, cachexia, loss of appetite and activation of the acute phase protein response. In some situations, exaggerated expression of cytokines can be associated with such life-threatening diseases as bacterial sepsis or multiple organ failure. Thus, the concentration-dependent bioactivities of cytokines can transform host physiology from normal homeostasis to severe life threatening pathophysiology.

The role of cytokines in tissue injury, infection, and neoplasia is multifaceted and complex. Still, in recent years we have begun to more fully understand how cytokines influence the respiratory tract's response to inhaled or otherwise administered pneumotoxic agents. While it should be recognized that cytokines can exhibit a broad spectrum of activities, many can be viewed as playing key roles in specific aspects of response. With respect to pulmonary inflammation and fibrosis, cytokines play critical roles in the initiation of response; the recruitment of inflammatory and immune cells; stimulating tissue repair through increased cell proliferation and extracellular matrix synthesis; and facilitating the resolution of response.

Initiating Cytokines

"Initiating" cytokines are early mediators of response to injury; these molecules are induced and expressed rapidly after exposure to noxious agents or the recognition by host defenses of infection or neoplasia. Prototype cytokines in this category include interleukin-1 (IL-1) and tumor necrosis factor-α (TNFα) (Kunkel et al., 1995). Transforming growth factor β (TGFβ) may also act as an early mediator of cell recruitment and activation although its most important role is likely in wound repair. Secretion of TNFα and/or IL-1 initiates a cascade of events through cytokine networks which influence the recruitment and activation of inflammatory cells (Standiford et al., 1990; Driscoll et al., 1994). These cytokines can not only stimulate leukocytes to produce cytokines, but also induce cytokine expression by nonimmune cells such as fibroblasts, smooth muscle cells, epithelial cells, and endothelial cells (Standiford et al., 1990; Driscoll et al., 1993b; Strieter et al., 1989a,b). The ability of TNFα and IL-1 to rapidly amplify a response is due, in part, to the presence of specific receptors for these cytokines on all somatic cells. While TNFα and IL-1 are important "initiating cytokines," this does not mean their expression or effects are short lived. Increased expression of TNFα and IL-1, and consequently their effects, may be long term, particularly if the initiating stimulus persists or exposure to the stimulus is intermittent. Thus, TNFα and IL-1 may contribute to both acute and chronic manifestations of lung toxicity (Driscoll et al. 1990a). For example, persistent TNFα and IL-1 expression is seen in the pulmonary response to mineral dust particles such as crystalline silica, where increased expression is maintained because the eliciting agent is not readily cleared from the lungs. Ultimately, IL-1 and TNFα contribute to the development of silica-induced chronic inflammatory lesions such pulmonary granulomas and tissue fibrosis (Driscoll, 1995; Piguet et al., 1990a,b).

Table 1   Selected Cytokines Exhibiting Chemotactic Activity and Their Cellular Targets

Cytokine	Cell Type(s) Responding
Transforming Growth Factor β	monocyte, lymphocyte, fibroblast
Platelet-Derived Growth Factor	fibroblast
Interleukin-1	T lymphocyte
Chemokines	
Interleukin-8	neutrophil, lymphocyte, basophil
Macrophage Inflammatory protein-2	neutrophil
Monocyte Chemotactic peptide-1	monocyte, basophil
Macrophage Inflammatory Protein-1α,β	monocyte, neutrophil, lymphocyte
RANTES	monocyte, lymphocyte

Recruitment Cytokines

A high degree of redundancy exists in the signals which contribute to leukocyte recruitment, attesting to the critical role of this response in host survival. A diverse group of lipid and protein mediators have been identified which possess chemotactic activity for leukocytes. However, many of these chemotactic factors, such as the leukotrienes (e.g., LTB$_4$), platelet activating factor, the bacterial tripeptide, f-met-leu-phe (fMLP), and the cleavage fragment of the fifth complement component, C5$_a$, lack specificity for the recruitment of individual populations of leukocytes. Yet, inflammatory responses to various pneumotoxins are frequently characterized by accumulation of rather specific subpopulations of leukocytes. The mechanisms for selective recruitment of cell subpopulations have been at least partially clarified with the discovery of cytokines exhibiting cell-specific chemotactic activity. Table 1 lists selected cytokines and their cell targets of chemotactic activity. Since the mid 1980s significant attention has been focused on the chemokines, a supergene family of chemotactic cytokines (Miller and Krangel, 1992). As discussed below, chemokines possess a high degree of cellular specificity for recruitment of certain types of leukocytes (Baggiolini et al., 1989; Matsushima et al., 1988; Matsushima and Oppenheim, 1989). Increasing evidence suggests that members of the chemokine family play important roles in the respiratory tract's response to a variety of agents (Kunkel and Strieta, 1990; Driscoll, 1994).

Repair and Resolution Cytokines

Table 2 lists selected cytokines which can contribute to the repair and regeneration of damaged tissue. These cytokines are critical in the orchestration of tissue remodeling, neovascularization, and, in some instances, fibrosis. Several of the these cytokines, including platelet derived growth factor

Table 2 Cytokines Which Can Contribute to Tissue Repair Processes by Stimulating Cell Proliferation and Extracellular Matrix Synthesis

Cytokine	BioActivities Related to Tissue Repair
Platelet Derived Growth Factor (PDGF)	Stimulate proliferation of: • fibroblasts • smooth muscle cells • chondrocytes
Fibroblast Growth Factor (acidic, basic FGF)	Stimulate proliferation of: • endothelial cells • fibroblasts • chondrocytes
Insulin-Like Growth Factor I & II (IGF-I, -II)	Stimulate proliferation of: • fibroblasts • multiple other cell types
Transforming Growth Factor α (TGFα)	Stimulate proliferation of: • epithelial cells • fibroblasts • keratinocytes • lymphocytes
Transforming Growth Factor β (TGFβ)	Stimulate matrix synthesis Stimulate synthesis of protease inhibitors Inhibit synthesis of proteases Stimulate proliferation of fibroblasts

(PDGF); transforming growth factor α and β (TGF) and; the insulin-like growth factors, also known as the somatomedins, are potent stimulators of cell growth (Bonner et al., 1991; Madtes et al., 1988; Moses et al., 1990; Rom et al., 1988). In addition (although strictly speaking not a cytokine) fibronectin is included here since it can stimulate chemotaxis and proliferation of fibroblasts and is thought to contribute to the development of fibrosis. These growth factors are instrumental in reestablishing normal tissue structure after injury. In addition, cytokines such as TGFβ stimulate synthesis of extracellular matrix proteins critical in tissue repair and scar formation (Ignotz et al., 1986; Roberts and Sporn, 1990).

Cytokines can influence the resolution of response by down-regulating expression of mediators involved in response initiation and cell recruitment. As listed in Table 3 cytokines which may facilitate resolution of inflammatory responses include TGFβ, IL-1 receptor antagonist (IRAP), IL-4, IL-10, IL-6 and the IL-6-like cytokines, leukemia inhibitory factor (LIF), oncostatin M and IL-11 (Wahl, 1992; Hart et al., 1989; Fiorentino et al., 1991). For example, both IL-4 and IL-10 are effective in down-regulating expression of

Table 3 Cytokines Which Contribute to Resolution of Inflammatory Processes and Their Related Bioactivities

Transforming Growth Factor β
- down-regulate activated immune cells
- wound repair

Interleukin-4
- modulate monocyte-derived cytokines

Interleukin-10
- suppresses TNFα, GM-CSF and IL-1 secretion
- suppresses T cell IL-2 production
- suppresses Class II MHC expression (monocytes)
- decrease monocyte NO production

Interleukin-6–like Cytokines (IL-6; IL-11, Oncostatin M; LIF)
- stimulate acute phase response
- inhibit production of chemokines by structural cells (Oncostatin M; LIF)

Interleukin-1 Receptor Antagonist
- blocks activity of IL-1

the initiating cytokines, TNFα and IL-1, by macrophages (Hart et al., 1989; Fiorentino et al., 1991). Additionally, administration of many of the cytokines in Table 3 has been shown to attenuate the inflammatory response. For example, treatment of animals with TGFβ, IL-6, IRAP or LIF prior to exposure to endotoxin greatly attenuates the subsequent pulmonary inflammatory response; an effect at least in part mediated by attenuating TNFα and/or IL-1 production (Ulich et al., 1991a,b; Ulich et al., 1994). It is likely the balance between initiating/recruitment signals and repair/resolution signals is a key factor in determining the nature and progression of responses to pneumotoxic agents.

ROLE OF CYTOKINES IN PULMONARY INFLAMMATION AND FIBROSIS

In the lung, as in other tissues, cytokines are primary effectors of host defense. They mediate the cell:cell communication critical to the recruitment and activation of inflammatory and immune effector cells, a response contributing to the removal and inactivation of the inciting agent(s) or the debridement of damaged tissue. Cytokines regulate cell replication and secretion of extracellular matrix proteins, orchestrating tissue repair. Importantly, these inflammatory and repair processes, while critical to host survival, also can contribute to derangements in tissue structure and function. For example, inflammatory cells are well known for their ability to release mediators such as reactive oxygen species and proteases which can damage lung tissue as well as inactivate invading microorganisms. Thus, these cells may inflict tissue injury above and beyond that resulting from the eliciting agent. With respect to pulmonary fibrosis this response involves many of the same biologic processes and mediators underlying tissue homeostasis and normal tissue repair. However, excessive deposition of connective tissue (which is not an uncommon consequence of exposure to toxic agents) can impair lung function. Thus, regarding the lungs response to exogenous materials, the induction of cytokine expression can represent a double-edged sword.

There are a variety of mechanisms by which chemical exposures can alter cytokine production within the lung. Some materials possess an ability to directly stimulate cytokine release and this property is responsible to a large extent for their adverse effects. For example, bacterial endotoxin, an important component of cotton dust as well as other organic dusts, when complexed with LPS binding protein interacts with specific cell membrane receptors (i.e., CD14) and stimulates production of cytokines, such as IL-1 and TNFα (Wright et al., 1990). The ability of endotoxin to directly activate cytokine expression is responsible to a significant degree for its potent inflammatory activity (Ryan and Karol, 1991; Xing et al., 1994). In some instances, release of cytokines occurs secondarily to other effects, for example, as a result of tissue injury, activation of the complement cascade, or production of oxidative stress. Indirect mechanisms likely contribute to the elaboration of cytokines following exposure to excessive levels of relatively innocuous materials such as titanium dioxide (Driscoll et al., 1990a). Still other materials such as crystalline silica, asbestos, or bleomycin may influence cytokine release via both direct and indirect mechanisms (DuBois et al., 1989; Driscoll et al., 1990b). The inherent cytotoxic and cytokine activating properties of crystalline silica and asbestos combined with their persistence in the lung likely contributes to their high degree for pulmonary toxicity.

The induction and/or suppression of cytokine production is an important factor in a variety of respiratory tract responses to chemical exposures. In the subsequent pages the biology of selected cytokines will be reviewed and the evidence supporting their role in lung toxicity discussed.

Tumor Necrosis Factor α

TNFα was first identified in 1975 by Carswell and colleagues (1975) as a protein present in the sera of Bacillus-Calmette-Guerin (BCG) and endotoxin-treated animals which mediated hemorrhagic necrosis of tumors *in vivo* and was cytostatic or cytotoxic for tumor cells *in vitro*. Subsequently, several groups reported the isolation and purification of a protein which exhibited tumor necrotizing activity and, in the early 1980s, the human and murine TNFα genes were cloned and expressed (Pennica et al., 1984; Marmenout et al., 1985; Shirai et al., 1985). TNFα has also been known as "cachectin" a protein shown to contribute to cachexia, a wasting syndrome associated with

Table 4   Inflammatory and Immune Processes Stimulated by TNFα

T and B lymphocyte proliferation and activation
Acute phase protein response
Arachidonic acid metabolism (i.e., prostaglandin E2, prostacyclin)
Inflammatory cell oxidative burst and degranulation
Expression of adhesion molecules (e.g., ELAM, ICAM, VCAM)
Chemotactic protein (chemokine) expression (e.g., IL-8, MCP-1, MIP-2)
Cytokine expression (e.g., IL-1, TNF, IL-6, GM-CSF)

infectious disease. Shortly after its isolation, however, cachectin was demonstrated by N-terminal sequencing to be identical to TNFα (Beutler et al., 1985).

The human TNFα gene is located on the short arm of chromosome 6 and codes for a 233 amino acid, 26 kd precursor protein which is processed to a mature 157 amino acid, 17 kd protein (Pennica et al., 1984; Marmenout et al., 1985; Shirai et al., 1985). TNFα can be produced by a variety of cells including macophages, monocytes, polymorphonuclear leukocytes, lymphocytes, smooth muscle cells, and mast cells (Young et al., 1987; Dubravec et al., 1990; Warner et al., 1989). Although TNFα is a secreted protein, it lacks the typical 20-30 amino acid hydrophobic signal sequence characteristic of secretory proteins. TNFα exists as oligomers of the 17 kd protein ranging in molecular weight from 34–140 kd. It is, however, the TNFα trimer which binds with highest avidity to the TNFα receptor and expresses the greatest bioactivity (Smith and Baglioni, 1987). As indicated in Table 4, TNFα exhibits a variety of activities. TNFα can up-regulate expression of at least three different adhesion proteins on endothelial cells: intercellular adhesion molecule 1 (ICAM-1), endothelial leukocyte adhesion molecule-1 (ELAM-1), and vascular cell adhesion molecule-1 (VCAM-1) which are important to leukocyte recruitment (Pober et al., 1986; Osborn et al., 1989; Wallis et al., 1985). These adhesion molecules are ligands for complementary molecules (called integrins) on leukocytes (Kishimoto et al., 1989). Besides adhesion, inflammatory cell recruitment involves migration in response to local chemotactic gradients. In this respect, TNFα can stimulate production of cytokines (called chemokines) which possess potent chemotactic activity for inflammatory and immunocompetent cells (Strieter et al., 1989a,b; Driscoll, 1994). As discussed later, TNFα-induced chemokine expression appears to be an important mechanism of pulmonary inflammation in response to particle inhalation.

TNFα can also influence other inflammatory processes, including: expression of cytokines such as TNFα, IL-1, IL-6, PDGF and TGFβ; stimulation of arachidonic acid metabolism; and activation of leukocytes to release reactive oxygen and nitrogen species, hydrolytic enzymes, lysozyme, and lac-

Table 5   Dyscoordinate Expression of TNFα and IL-1 by Human Alveolar Macrophages and Peripheral Blood Monocytes in Response to Endotoxin

Cell Type	TNFα (Units/ml)†	IL-1 (Units/ml)†
Monocytes:		
Unstimulated	1±1	0
Endotoxin (10μg/ml)	60±84	78±61¥
Alveolar Macrophages:		
Unstimulated	30±12	0
Endotoxin (10μg/ml)	596±367¥	8±12

Source: Rich et al., 1989. Cells were cultured with or without endotoxin for 16 hours and the levels of cytokine in the cell supernatants determined.
†all values = X±SD n=7
¥significantly greater than respective monocyte or alveolar macrophage value.

toferrin (Klebanoff et al., 1986; Das et al., 1990; Hoffman et al., 1987; Richter et al., 1989). TNFα can also regulate cell growth. It enhances IL-2-induced T-lymphocyte proliferative (Lee et al., 1987) and stimulates B cell proliferation and immunoglobulin secretion (Jelinek et al., 1987). In addition, TNFα exerts a growth-inhibitory effect on some transformed cell lines while stimulating proliferation of normal fibroblasts from a variety of tissues (Sugarman et al., 1985). This latter effect appears to be mediated indirectly through TNFα-induced expression of epidermal growth factor (EGF) receptors (Palombella et al., 1987) and/or stimulation of PDGF release (Hajjar et al., 1987).

Many of the bioactivities described above for TNFα are shared with IL-1. However, several studies have indicated that of these two cytokines, TNFα may be of particular importance in the lung. As illustrated in Table 5, Rich et al. (1989) demonstrated human macrophages stimulated with endotoxin produce significantly more TNFα than IL-1, whereas peripheral blood monocytes exhibit an opposite profile of cytokine production. Similarly, there is a preferential release of TNFα versus IL-1 from rat and human alveolar macrophages exposed *in vitro* to mineral dust particles (Driscoll et al., 1990b; Gosset et al., 1991). The reduced ability of alveolar macrophages to release IL-1 appears to be due to ineffective processing of the IL-1 precursor protein by these cells (Wewers and Herzuk, 1989). While not very proficient in secreting IL-1, it is noteworthy that alveolar macrophages are effective producers of the IL-1 receptor antagonist (IRAP) which is an IL-1-like molecule that binds to (but does not activate) the IL-1 receptor (Jansen et al., 1991). Overall, these findings suggest TNFα is a primary initiator of lung responses and, in particular, responses to materials which interact with alveolar macrophages.

Pneumotoxic Agents Alter TNFα Expression

An extensive and growing data base indicates many particulate and non-particulate inflammagens effectively activate TNFα production within the respiratory tract. As summarized in Table 6, a variety of agents activate human and/or rat alveolar macrophages to release TNFα *in vitro* and this activity corresponds with the *in vivo* inflammatory activity of these materials. For example, endotoxin and inflammatory mineral dusts such as α-quartz, crocidolite, and chrysotile are effective agonists of alveolar macrophage TNFα release. Relatively innocuous dusts such as titanium dioxide, aluminum oxide, and latex beads, however, are markedly less effective in stimulating this response. These *in-vitro* observations appear to be consistent across species having been demonstrated for both rat and human alveolar macrophages.

In contrast to TNFα, the data on *in vitro* alveolar macrophage IL-1 release are less clear. For example, a few studies report increased IL-1-like bioactivity in conditioned media from α-quartz exposed rat alveolar macrophages (Oghiso and Kubota, 1986; Kang et al., 1992). However, mineral dust preparations first treated to remove endotoxin did not increase IL-1 release by human or rat alveolar macrophages, although stimulation of TNFα production was detected (Driscoll et al., 1990b; Gossett et al., 1991). Endotoxin present on mineral dust preparations may be a confounding factor in studies demonstrating *in vitro* stimulation of alveolar macrophage IL-1 release. Over-

Table 6 Alveolar Macrophage TNFα Production after *In Vitro* Exposure to Various Materials

Agent	Source of macrophages	TNFα response*	Reference
endotoxin	rat	+++++	Driscoll et al., 1990b
	human	+++++	Bachwich et al., 1986
α-quartz	rat	+++	Dubois et al., 1989; Driscoll et al., 1990b
	human	+++	Gosset et al., 1991; Zhang et al., 1993
crocidolite	rat	+++	Driscoll et al., 1990b
	human	+++	Zhang et al. 1993
chrysotile	rat	+++	Dubois et al., 1989
	human	+++	Zhang et al., 1993
coal dust	human	++++	Gosset et al., 1991
hydrogen peroxide	rat	+++	Simeonova and Luster, 1994
titanium dioxide	rat	−	Driscoll et al., 1990b
	human	+/−	Gosset et al., 1991
aluminum oxide	rat	−	Driscoll et al., 1990b
latex beads	rat	−	Dubois et al., 1989

*Relative response: (−) no response; (+++++) maximal response

all, these observations suggest that TNFα is a key alveolar macrophage-derived initiator of response.

The relevance of the *in vitro* observations on macrophage-derived TNFα are supported by *in vivo* studies demonstrating increased production of macrophage-derived cytokines in dust exposed animals and humans. For example in rats, intratracheal instillation of crystalline silica (e.g., α-quartz) results in pulmonary inflammation progressing to granulomatous inflammation and fibrosis similar to human silicosis. Instillation of α-quartz or titanium dioxide particles into rat lungs markedly increases release of TNFα by alveolar macrophages. Similarly, steady-state TNFα mRNA in rat lungs and immunoreactive TNFα protein in rat alveolar macrophages increase following prolonged inhalation of α-quartz or ultrafine titanium dioxide particles (Driscoll et al., 1994). Similarly, studies of occupationally exposed human populations also demonstrate activation of macrophage TNFα production after dust exposure. For example, alveolar macrophages obtained from coal workers release increased TNFα (Lassalle et al., 1990) with the magnitude of TNFα production being greatest in those individuals with the most severe lung disease. Additionally, macrophage TNFα production is elevated in asbestosis patients as well as individuals with the inflammatory lung disease, idiopathic pulmonary fibrosis (Zhang et al., 1993). Importantly, these *in vivo* observations on TNFα production are not unique to mineral dust particles. Studies on ozone inhalation demonstrate acute exposure of rats to this oxidant increases macrophage TNFα release. Furthermore, inhalation of cotton dust particles is associated with activation of rat alveolar macrophage TNFα production—a response attributed, at least in part, to the presence of endotoxin in this material (Ryan and Karol, 1991). Overall, these *in vivo* and *in vitro* observations indicate TNFα is a key mediator in pulmonary responses to a variety of materials.

While the above studies clearly document the upregulation of TNFα production by several inflammatory materials, it is important to recognize that pneumotoxic agents may also exert adverse effects by impairing macrophage cytokine production. For example, subchronic exposure of rats to cigarette smoke results in markedly decreased capacity of alveolar macrophage to produce TNFα as well as decreased macrophage cytotoxic activity against tumor cells (Flick et al., 1985). Studies on cigarette-smoking human populations support these observations in animals (Yamaguchi et al., 1993). Macrophages from smokers were found to produce significantly less TNFα in response to endotoxin than macrophages from nonsmokers. Similarly, alveolar macrophages from smokers have a decreased ability to release IL-1 (Yamaguchi et al., 1989). In this respect, the ability of cigarette smoking to alter macrophage TNFα and IL-1 production may be responsible, at least partly, for the impaired humoral immune responses and T cell function seen in smokers.

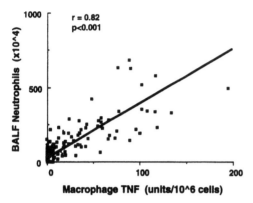

Figure 1 The relationship between mineral dust-elicited neutrophilic inflammation in the lung and activation of alveolar macrophage TNFα production. Groups of rats were intratracheally instilled with 1, 5, 10, or 100 mg of α-quartz or titanium dioxide particles. Animals were sacrificed 7, 14, and 28 days after exposure, bronchoalveolar lavage performed and the number of neutrophils in lavage fluid determined and *ex vivo* macrophage TNFα production characterized. Shown above is the relationship between lung neutrophils and macrophage TNFα. There was a significant positive correlation between activation of macrophage TNFα release and neutrophilic inflammation in the lung (p<0.001; r=0.82). A similar significant correlation with dust-induced neutrophil recruitment was not detected for macrophage IL-1 or IL-6 production. (*Adapted from:* Driscoll et al., 1990).

TNFα and Inflammatory Cell Recruitment

There is compelling evidence that a key action of TNFα is in the recruitment of inflammatory cells. As shown in Fig. 1 for α-quartz and titanium dioxide exposed rats, a significant positive correlation exists between *in vivo* activation of macrophage TNFα release and the recruitment of neutrophils to the lung. A similar correlation has been demonstrated in rats exposed to ultrafine particles (Driscoll and Maurer, 1991). Importantly, a significant relationship with inflammatory cell recruitment was not detected for other macrophage-derived cytokines such as IL-1 and IL-6 (Driscoll et al., 1990c). Similar to these findings in rats, a positive correlation has been shown between bronchoalveolar lavage fluid TNFα levels and neutrophilic inflammation in the lungs of asbestosis and IPF patients (Zhang et al., 1993). Further evidence that TNFα contributes to particle-elicited pulmonary inflammation comes from studies in which animals are passively immunized with antibody to TNFα. As shown in Fig. 2 pretreatment of rats with a monoclonal antibody to TNFα significantly attenuated the pulmonary recruitment of neutrophils in response to intratracheally instilled titanium dioxide (Driscoll et al., 1994). A similar effect of passive immunization against TNFα was demonstrated for the pulmonary inflammatory response of mice to inhaled α-quartz (Driscoll 1995). Collectively, these studies demonstrate a key role of macrophage-derived TNFα in inflammatory cell recruitment to the lung.

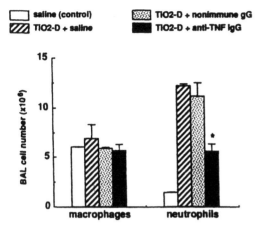

Figure 2 Effect of passive immunization with anti-TNF IgG on titanium dioxide-induced neutrophil recruitment. Rats were pretreated by intratracheal instillation with saline, anti-TNF IgG, or nonimmune IgG. Twenty-four hours after the pretreatments, animals were instilled with 10 mg/kg titanium dioxide—a group of animals without pretreatment but instilled with saline was included as a control. Twenty-four hours after instillation, the animals were euthanised, the lungs lavaged, and the number and type of bronchoalveolar lavage cells determined. (*) denotes a significant difference from the saline and nonimmune IgG pretreated groups; p<0.05. In saline pretreated animals, titanium dioxide increased the number of BAL neutrophils but not macrophages 24 post instillation. Pretreatment with anti-TNF IgG—but not nonimmune IgG—significantly attenuated the neutrophil response. (*From:* Driscoll et al., 1994)

TNFα and Pulmonary Fibrosis

TNFα has been shown to be critical in the pathogenesis of fibrosis in several tissues including lung. Direct subcutaneous administration of TNFα to mice via osmotic minipumps produces a local accumulation of neutrophils followed by formation of a tissue mass consisting of fibroblasts, collagen, and capillaries (Piguet et al., 1990a). Specific to the lung, passive immunization of mice with an anti-TNFα antibody prevented silica-induced increases in lung collagen deposition (Piguet et al., 1990b). This effect of anti-TNFα is not unique to silica-induced fibrosis since similar studies have shown anti-TNFα IgG attenuates bleomycin-induced pulmonary fibrosis in mice (Piguet et al., 1989). Overall, these findings indicate a causal relationship between increased TNFα and fibrosis in these murine models of pneumotoxicity.

Macrophage Inflammatory Proteins

Macrophage inflammatory proteins are potent mediators of immune and inflammatory responses. Murine macrophage inflammatory protein 1 (mMIP-1) was first identified in 1988 as a secretory product of the mouse macrophage cell line, RAW 264.7 (Davatelis et al., 1988; Wolpe et al., 1988). mMIP-1 was initially characterized as an acidic heparin binding protein which migrated as a doublet of ~8 kDa on SDS-PAGE, however, the doublet was subsequently resolved into two distinct but structurally related proteins, mMIP-1α and mMIP-1β. cDNA's for mMIP-1α and β predict proteins of 92 amino acids which show ~60% homology in amino acid sequence and exhibit a high degree of homology (~75%) with the human cytokines LD78 (also known as human MIP-1α) and Act-2 (also known as human MIP-1β) (Driscoll, 1994). MIP-1α is chemotactic for neutrophils, monocytes, and lymphocytes; MIP-1β has been shown to be chemotactic for monocytes (Miller and Krangel, 1992; Driscoll et al., 1995). The primary sources of MIP-1α and β are hematopoietic cells including macrophages, monocytes, and T and B lymphocytes (Miller and Krangel, 1992; Driscoll, 1994).

Murine macrophage inflammatory protein 2 (mMIP-2) was isolated from the conditioned media of LPS stimulated RAW 264.7 cells and is a basic heparin binding protein with a molecular weight of ~6 kDa (Wolpe et al., 1989). The mMIP-2 cDNA predicts a protein of 100 amino acids, the first 30 of which have features of a typical signal sequence (Tekamp-Olson, 1990). Two human homologues of mMIP-2 have been cloned and termed human MIP-2α and β (hMIP-2α and β). hMIP-2α and β share >70% amino acid sequence homology with murine form of MIP-2. MIP-2 is a potent neutrophil chemotactic factor *in vitro* and *in vivo* but is not chemotactic for monocytes or macrophages (Driscoll, 1994). The cell sources of MIP-2 differ from that of MIP-1 in that the former can be secreted by both hematopoietic cells and a broad range of nonhematopoietic cells which include fibroblasts, epithelial cells, and endothelial cells (Driscoll, 1994).

MIP-1 and MIP-2 are members of a supergene family of chemotactic cytokines called chemokines; the key characteristics of chemokine family members are presented in Table 7. Briefly, chemokines possess 4 positionally conserved cysteine residues, are heparin binding, and typically range in molecular weight from ~8–10 kDa. cDNAs for chemokine family members are recognized by their conserved single open reading frames, typical signal sequences in the 5' region, and AT rich sequences in the 3' untranslated regions. Members of the chemokine family can be divided into two branches based on the spacing of the N-terminal cysteines and to some degree their spectrum of target cells (Miller and Krangel, 1992). C-X-C branch (also known as the α branch) chemokines are characterized by having their first two cysteines separated by an intervening amino acid. In the C-C or β branch chemokines, the first pair of cysteines are in juxtaposition. Chemokines possess disulfide bonds between the first and third as well as the second and fourth cysteines which contribute to their highly stable biologic activity.

Table 8 lists members of chemokine cytokine family identified in humans and their likely murine homologs. In general, the C-X-C chemokines exhibit chemotactic activity for neutrophilic leukocytes, while the C-C chemokines are chemotactic for monocytes and in some instances subpopulations of lym-

Table 7   Characteristics of the Chemokine Chemotactic Cytokine Family

- heparin binding proteins which are ~6-10 kDa in size
- possess 4 positionally conserved cysteines
- two branches defined by spacing of cysteines, genomic organization, and function
- derived from multiple cell sources, including immune and nonimmune cells
- function in inflammatory, immune, and tissue repair processes

phocytes. Additional bioactivities, however, are being identified for the chemokines. For example, MIP-1α has been shown to influence the proliferation of hematopoietic precursor cells, and MIP-2 has been shown stimulate proliferation of alveolar epithelial cells (Driscoll et al., 1995).

MIP Chemokines and Pulmonary Inflammation

The bioactivities of MIPs, the multiplicity of their sources as well as the numerous factors which stimulate MIP production, suggest these cytokines contribute to the network of cell:cell signalling which orchestrates host responses to injury and infection. The importance of MIPs in respiratory tract defenses has been demonstrated in studies using animal models of silicosis, pulmonary sepsis, and oxidant-induced lung injury (Driscoll et al., 1993a; Huang et al., 1992; Xing et al., 1994). As shown in Fig. 3, expression of MIP-1α and MIP-2 mRNA is markedly increased in rats exposed to α-quartz, and upregulation of these chemokines precedes the influx of inflammatory cells into the respiratory airways. Long-term inhalation of cristobalite or titanium

Table 8   Members of the Chemokine Chemotactic Cytokine Family†

| α(C-X-C) Branch | | β (C-C) Branch | |
Human	Murine	Human	Murine
IL-8 (NAP-1, 3-10C)*	?	MCP-1/MCAF	JE
gro-MGSA (gro α);	KC, MIP-2,	LD78 (MIP-1α, PAT	MIP-1α (TY5), C10
gro β (MIP-2α);	CINC	464, GOS-27)	
gro γ (MIP-2β)		Act-2 (MIP-1β) hH400,	MIP-1β (H400)
		HC27, G-28	
Platelet factor 4	PF4 (rat)	I-309	TCA.3 (P500)
Platelet basic protein	?	RANTES	RANTES
γIP-10	CRG-2		
ENA-78	?		
?	mig		

†*Modified from:* Miller and Krangel, 1993; Driscoll, 1994.
*Alternate designations are presented in parenthesis.

dioxide particles also increases MIP-2 gene expression in rat lungs concurrent with lung inflammation (Driscoll et al., 1994). Collectively, these findings support a role for these chemokines in mineral dust-induced inflammation.

The contribution of MIPs, however, to lung inflammation is not unique to mineral dust effects. Bacterial endotoxin can produce acute lung injury in sepsis and gram-negative pneumonia—two conditions characterized by the activation of lung macrophages and a marked neutrophilic inflammatory response. *In vitro,* endotoxin is a potent inducer of MIP expression by several cell types, including macrophages, endothelial cells, and epithelial cells (Driscoll et al., 1993). *In vivo,* within 30 minutes of intratracheal instillation of endotoxin TNFα and MIP-2 mRNA are markedly increased in rat lung (Xing et al., 1994). Analysis of endotoxin-instilled lungs by *in situ* hybridization localizes MIP-2 expression to alveolar macrophages and infiltrating neutrophils. Importantly, in these studies neutrophil recruitment was not observed until 6 hours after endotoxin treatment, subsequent to induction of MIP-2. These findings indicate that in pulmonary sepsis chemokines, (which include MIP-2) are likely important mediators of inflammatory cell infiltration.

Inhalation of near ambient levels of the oxidizing air pollutant ozone can result in pulmonary inflammation characterized by increased numbers of neutrophils as well as transient alterations in pulmonary function (Driscoll et al., 1988; Koren et al., 1989). Exposure of mice to varying concentrations of ozone results in a marked induction of MIP-2 mRNA—a response associated with neutrophil recruitment. In light of the effects of ozone on MIP-2 expression it is noteworthy that oxidant stress *in vitro* (i.e., hydrogen peroxide exposure) can stimulate MIP-2 gene expression in alveolar macrophages (Driscoll et al., 1993a). In this respect, analysis of the 5' flanking region of mMIP-2 and related genes (i.e., hMIP-2α, β, gro α) demonstrates a consensus NFkB binding sequence is an operative cis regulatory element for LPS, IL-1, and TNFα responsiveness of MIP-2 and related genes (Widmer et al., 1993), NFkB is a trans-acting factor which can be activated by oxidant stresses, including hydrogen peroxide (Schreck et al., 1991). Thus, it is possible that oxidant stress represents a general mechanism for upregulating MIP-2 chemokine expression, and chemical agents which produce oxidative stress may elicit pulmonary inflammation, in part, via upregulation of MIP-2.

Chemokines and TNFα Mediated Cytokine Networking

An important concept in cytokine mediated inflammatory events is that they result from a network of cytokine and cell:cell interactions. Thus, initiating (or early response) cytokines produced by one cell are thought to stimulate a cascade of events ultimately resulting in cell recruitment. The putative role of TNFα as an initiating cytokine in the lung and the associated cell:cell

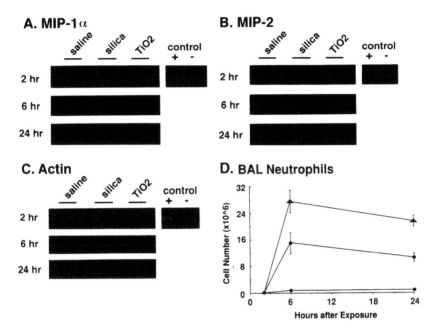

Figure 3 Time course of expression of the chemokines macrophage inflammatory protein 1α (MIP-1α) and macrophage inflammatory protein 2 (MIP-2) and neutrophilic inflammation in rat lung after acute exposure to α-quartz or titanium dioxide. Rats were intratracheally instilled with 2 mg of α quartz or titanium dioxide and euthanised 2, 6, and 24 hours after exposures. mRNA for MIP-1α, MIP-2 and GAPDH in lung tissue was determined using RT-PCR techniques (see Driscoll et al., 1993b for details). Number of neutrophils in the bronchoalveolar lavage fluid were also determined. Dust exposure increased steady-state levels of mRNA for MIP-1α and MIP-2, but not the housekeeping gene GAPDH. The increase in chemokine expression preceded the detection of increased neutrophils in the lung. (*From:* Driscoll et al., 1993b).

and cytokine interactions are illustrated in Fig. 4. Briefly, inflammagens are thought to activate lung macrophages to release TNFα which then acts via autocrine or paracrine pathways to stimulate production of MIPs and other chemokines which are the proximate mediators of inflammatory cell recruitment. In the context of lung effects, the existence of such a network is supported by several lines of evidence. First, many inflammagens can directly activate lung macrophages to release TNFα (see Table 6). Second, while TNFα is not directly chemotactic for inflammatory cells, it has been shown in a variety of *in vitro* systems to be a potent inducer of chemokine expression (including MIPs) by several cell types—lung macrophages, epithelial cells and fibroblasts (Strieter et al., 1989a,b; Standiford et al., 1990; Driscoll et al., 1993b). Further, passive immunization of rats with anti-TNFα antibody attenuates mineral dust-induced neutrophil recruitment and decreases lung MIP-2 expression (Driscoll et al., 1994).

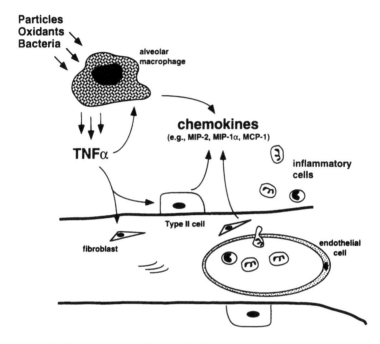

Figure 4 Cytokine networking and inflammatory cell recruitment to the lung. There is increasing evidence that the recruitment of inflammatory cells to the respiratory tract after exposure to particles and other agents involves a network of cytokine and cell:cell interactions. The cytokine TNFα appears to play a key role in this network. Activation of lung macrophages to release TNFα stimulates via autocrine and paracrine pathways lung cells, including macrophages, epithelial cells, fibroblasts, and endothelial cells to release potent chemotactic cytokines called chemokines. These chemokines then act as the proximate mediators of inflammatory cell recruitment. Also involved in this network is the upregulation on vascular endothelium and peripheral blood leukocytes of molecules mediating cell adhesion. (*Modified from:* Driscoll, 1994a)

Transforming Growth Factor β (TGFβ)

The transforming growth factors were discovered as mediators secreted from virally transformed cells which promoted anchorage-independent growth of nonneoplastic cells in culture. Two proteins isolated and shown to express this transforming activity are known as transforming growth factor α and β TGFα and TGFβ are distinct proteins which interact with independent receptor systems and are expressed by a variety of normal as well as neoplastic tissues. Regarding TGFβ, several isoforms have been identified and are recognized as members of a supergene family of proteins present in both mammalian and nonmammalian species (Roberts and Sporn, 1992). The TGFβ family includes: 5 different isoforms of TGFβ, the activins and inhibins, bone morphogenic protein, and the Vgl protein of amphibians (Roberts and Sporn, 1992). Members of the TGFβ family share significant homologies in nucleic acid and amino acid sequence and possess conserved cysteine residues which are key to bioactivity. They all appear to play a role in development, cell replication, and extracellular matrix protein synthesis.

Of the various TGFβ molecules, the TGFβ1 and TGFβ2 isoforms are the most abundant. TGFβ1 is a 25 kDa homodimer comprised of polypeptides derived from the 112 carboxy-terminal amino acids of a 390 amino acid precursor (Derynk et al., 1985). TGFβ1 has been isolated from a variety of tissues including bone, platelets, kidney, and lung. TGFβ2 is a 24 kDa homodimeric protein whose monomers are derived from a 412 amino acid precursor (Seyedin et al., 1985). TGFβ2 is present at relatively high levels in bone and has previosly been called cartilage-inducing factor (Seyedin et al., 1985). A heterodimeric TGFβ protein has also been identified which is comprised of TGFβ1 and β2 chains (Wahl et al., 1989). TGFβ1 and 2 share approximately 70% homology in amino acid sequence and possess a 20-23 amino acid hydrophobic secretory signal sequence. The TGFβ molecules are typically secreted in a biologically inactive or latent form and can be activated by a number of mechanisms including proteolytic cleavage by plasmin or cathepsin D as well as exposure to extremes of pH (Roberts and Sporn, 1990).

Functionally, TGFβ has the potential to play a role in all aspects of inflammation and wound healing. TGFβ is one of the most potent chemoattractants for monocytes with activity demonstrated at $\sim 10^{-12}$M. TGFβ also exhibits chemotactic activity for neutrophils and fibroblasts (Wahl et al., 1987; Wahl, 1992). In this respect, the release of TGFβ from the α granules of platelets or the activation of latent TGFβ bound to extracellular matrix proteins by locally released proteases may be an important early signal for cell recruitment. TGFβ can stimulate production of itself as well as other inflammatory cytokines, including IL-1, TNFα, and PDGF (Assoian et al., 1984; Roberts and Sporn, 1990). In this respect, TGFβ stimulation of fibroblast proliferation appears to occur, at least in part, by inducing PDGF which then acts as an autocrine growth factor (Majack et al., 1990). TGFβ has been shown to specifically inhibit IL-1-induced T lymphocyte proliferation; an effect which appears secondary to upregulation of TGFβ receptors on these cells (Wahl, 1992).

While TGFβ can function either as an agonist or antagonist of inflammation and cell proliferation, its stimulatory effect on extracellular matrix synthesis has been consistently demonstrated. TGFβ can markedly increase fibroblast expression of mRNAs for types I, II and V collagen and fibronectin (Dean et al., 1988; Ignotz and Massague, 1986). In addition to stimulating synthesis of various extracellular matrix proteins, TGFβ promotes matrix deposition by inhibiting production of connective tissue proteases and stimulating synthesis of protease inhibitors (Edwards et al., 1987). Thus, while TGFβ has the ability to promote inflammatory cell recruitment and activation, its predominant function may be on tissue repair by mediating the accumulation of monocytes and fibroblasts at sites of tissue injury and stimulating synthesis of connective tissue proteins. TGFβ also has the ability to down-regulate macrophages and lymphocytes once these cells differentiate into activated cells. This suppressive activity of TGFβ likely contributes to resolution of the inflammatory response as repair progresses (Wahl et al., 1989; Wahl 1992).

TGFβ and Fibrosis in the Lung

TGFβ clearly exhibits bioactivities which would suggest this cytokine can contribute to the pathogenesis of pulmonary fibrosis. Specifically, TGFβ can directly affect expression of extracellular matrix protein genes, inhibit synthesis of proteases and stimulate production of protease inhibitors, and indirectly stimulate fibroblast proliferation. Thus, it is not surprising that increased TGFβ expression has been observed in both human and animal lungs under conditions of developing fibrosis (Table 9). In human disease, elevated TGFβ expression has been demonstrated in idiopathic pulmonary fibrosis (IPF) and lung fibrosis associated with systemic sclerosis. In animal models of pulmonary fibrosis induction of TGFβ has been observed following exposure to silica, asbestos, cadmium, radiation, and bleomycin.

Bleomycin is an anti-neoplastic agent with the occasional side effect of causing lung inflammation and fibrosis and has been used extensively in animals as a model of interstitial lung disease. Exposure of mice, rats, or hamsters to high doses of bleomycin results in a mixed inflammatory infiltrate in the lung, endothelial cell and alveolar epithelial cell injury and a later (1–2 weeks) occurring lung fibrosis. Within a week after treatment, an increase in steady-state TGFβ mRNA can be detected in the lungs of bleomycin-treated animals. This increase in TGFβ precedes increases in the synthesis of extracellular matrix proteins, suggesting a role for this cytokine in the observed fibrosis. The association between TGFβ expression and increased collagen has also been demonstrated in IPF patients. Evaluation of IPF lung tissue with combined immunocytochemistry for TGFβ and *in situ* hybridization for collagen gene expression demonstrates a co-localization of these two activities. As with the rodent studies, this association in humans supports a role for TGFβ in fibrosis.

The key role of TGFβ in pulmonary inflammation and excess collagen deposition was directly shown in studies by Giri et al. (1993). These investigators treated mice with bleomycin followed by antibodies to TGFβ and characterized pulmonary inflammation, tissue injury, and lung collagen. As shown in Fig. 5, treatment of mice with antiTGFβ antibodies significantly decreased bleomycin-induced pulmonary fibrosis characterized by hydroxyproline content of lung tissue. These investigators also reported a significant attenuation of bleomycin elicited inflammatory cell recruitment. These findings provide direct evidence that TGFβ contributes to bleomycin-induced

Table 9 Human Fibrotic Lung Diseases and Animal Models of Pulmonary Fibrosis Associated with Increased Lung Expression of Transforming Growth Factor β

Disease/Animal Model	Reference
Human Interstitial Lung Disease:	
Idiopathic Pulmonary Fibrosis	Broekelmann et al., 1991
Systemic Sclerosis	Corrin et al., 1994
Animal Models of Pulmonary Fibrosis:	
Bleomycin (rat, mouse, hamster)	Phan and Kunkel, 1992; Hoyt and Lazo, 1988; Raghow et al., 1989
Silicosis (rat)	Williams et al., 1993
Asbestosis (rat)	Perdue and Brody, 1994
Cadmium (rat)	Ohno et al., 1993
Radiation (mouse)	Finkelstein et al., 1994

fibrosis. In this respect, the association of increased TGFβ expression with several animal models of pulmonary fibrosis as well as interstitial lung disease in humans indicates that excessive or persistent expression of TGFβ may be a common factor in the pathogenesis of fibrotic changes in the lung.

Fibronectin

A variety of studies have demonstrated fibronectins' ability to modulate cell function and, as discussed below, increased expression of fibronectin has been consistently associated with the development of fibrosis in humans and animals. The term fibronectin refers to a group of large molecular weight dimeric glycoproteins, comprised of ~230–250 kDa polypeptides crosslinked by two disulfide bonds (Rouslahti, 1988). Fibronectins exist in both soluble and insoluble forms with soluble fibronectin (i.e., plasma fibronectin) produced by hepatocytes and insoluble fibronectin (i.e., cellular fibronectin) being synthesized by a variety of cell types. There are 20 different isoforms of fibronectin which derive, in part, from differential splicing of the mRNA transcribed from a single gene (Paul et al., 1986; Dean, 1989). Three alternatively spliced regions of the fibronectin gene have been identified: EIIIA, EIIIB, and the variable or V region (Dean, 1989). EIIIA and EIIIB are absent from plasma fibronectin and, since their presence can be detected immunochemically, they can be used to distinguish between cellular and plasma fibronectin in tissue (Patel et al., 1987).

Within the lung, fibronectin is normally associated with the alveolar and capillary basal laminae as well as in larger blood vessels and in the basement membranes of airways (Torikata et al., 1985). The fibronectin in normal adult

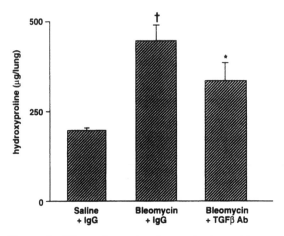

Figure 5 Effect of passive immunization with an antibody against TGFβ$_2$ on bleomycin-induced pulmonary fibrosis in mice. Mice were intratracheally instilled with saline or 0.125 units bleomycin followed by tail-vein injections of either nonimmune IgG or an anti-TGFβ IgG. Injections were given immediately after bleomycin (250 μg IgG) as well as on days 5 and 9 after bleomycin. Lung fibrosis was characterized on day 14 after bleomycin by determining levels of lung hydroxyproline—a biochemical marker for lung collagen. The results demonstrate that while bleomycin treatment causes fibrosis in both groups, this effect was significantly attenuated in the animals treated with antibody against TGFβ. (†) denotes significant difference from saline group; (*) denotes significant difference from saline and bleomycin + nonimmune IgG group. (*From:* Giri et al., 1993).

lung is largely plasma fibronectin. Local sources may, however, contribute significantly to lung fibronectin during development and in periods of tissue repair (Torikata et al., 1985; Broekelmann et al., 1988). In support of this, the amount of EIIIA- or EIIIB-containing fibronectin subunits (i.e., cellular fibronectin) are low in the normal adult lung, but are high in the developing lung and increased at sites of wound healing and fibrosis (Schwartzbauer, 1991; Torikata et al., 1985; Broekelmann et al., 1988). Potential sources of fibronectin in the lung include alveolar macrophages, fibroblasts, endothelial cells, smooth muscle cells, mesothelial cells, and epithelial cells (Rennard et al., 1981a; Rouslahti, 1988).

Functionally, fibronectin contributes to a variety of processes, including cell adhesion and migration, hemostasis and thrombosis, and cell differentiation and proliferation (Dean, 1989; Rouslahti, 1988). The diverse functions of fibronectin result from the presence of specific binding domains for interaction with other extracellular matrix proteins such as collagen and heparin, cell surface receptors such as the integrins, and coagulation pathway proteins such as fibrin. Fibronectin has been shown to be an acute phase protein whose production can be upregulated in hepatocytes by a variety of agents (Pick-Kober et al., 1986). Further, fibronectin can act as an opsonin and (due to its ability to bind simultaneously to cells and extracellular matrix) can influence the organization of newly deposited extracellular matrix protein.

Fibronectin and Pulmonary Fibrosis

The role of fibronectin as a mediator and potential biomarker of pulmonary fibrosis has attracted considerable attention. Fibronectin is a potent chemoattractant for fibroblasts (Rennard, et al., 1981a,b) and can facilitate the recruitment of these cells to sites of tissue injury. Once fibroblasts have been recruited, their subsequent proliferation and extracellular matrix synthesis is regulated by the presence of growth factors, including TGFβ discussed above. Studies by Bitterman and co-workers (1983) have demonstrated that fibronectin can also function as a competence growth factor and replace PDGF in stimulating fibroblast growth. Thus, fibronectin accumulating at sites of tissue injury can influence the recruitment and proliferation of fibroblasts.

As discussed above, during development and tissue injury, levels of cellular fibronectin (i.e., fibronectin containing EIIIA and/or EIIIB regions) increase. One source of fibronectin within the lung is the alveolar macrophage—a cell type whose numbers increase markedly at sites to tissue injury and developing fibrosis. Several studies have characterized alveolar macrophage fibronectin production in patients with fibrotic lung disease as well as animal models of pulmonary fibrosis (Table 10). These studies have consistently demonstrated increased alveolar macrophage fibronectin production in a variety of lung diseases, including IPF, sarcoidosis, fibrosis associated with asbestos, coal dust, or silica exposure as well as fibrosis associated with scleroderma or arthritis. Immunocytochemical studies have shown increased fibronectin deposition at sites of developing fibrotic lesions in coal workers, silicotics, and asbestosis patients (Wagner et al., 1982). In addition, studies by McDonald et al. (Limper et al., 1991; Broekelmann et al., 1988) have provided evidence that fibronectin is active early in the pathogenesis of pulmonary fibrosis in humans. These investigators characterized mRNA expression for several growth factors, including PDGF, TGFβ, and fibronectin as well as collagen type I in lung tissues from individuals with interstitial lung disease. In early lesions (characterized primarily by increased macrophages) the macrophages were expressing increased levels of fibronectin and TGFβ mRNA, while message for PDGF or collagen was not detectable. In lesions considered to represent later stages of disease, being associated with fibrosis, increased expression of fibronectin and TGFβ protein as well as collagen type I and PDGF mRNA were detected. The early and persistent upregulation of fibronectin seen in this study supports its role in the pathogenesis of the fibrotic lesions. Studies by Rom and colleagues (1991) have provided additional support for an important role of fibronectin in the pathogenesis of interstitial lung disease and its potential use as a marker of lung disease activity. These investigators demonstrated a positive correlation between elevated levels of alveolar macrophage fibronectin production and increasing severity of lung function impairment in patients with various forms of pneumoconiosis.

Table 10   Human Fibrotic Lung Diseases and Animal Models of Pulmonary Fibrosis Associated with Increased Production of Fibronectin by Alveolar Macrophages

Disease/Animal Model	Reference
Human Interstitial Lung Diseases:	
Silicosis	Rom et al., 1987; 1991
Coal Worker Pneumoconiosis	Rom et al., 1987; 1991
Asbestosis	Begin et al., 1986; Rom et al., 1987; 1991
Idiopathic Pulmonary Fibrosis	Rennard et al., 1981; Ozaki et al., 1990
Sarcoidosis	Rennard et al., 1981;
Rheumatoid Arthritis	Perez et al., 1989
Scleroderma	Kinsella et al., 1989
Animal Models of Pulmonary Fibrosis:	
Rat	
Silica	Davies and Erdogdu et al., 1989; Driscoll et al. 1990c; 1991
Titanium dioxide	Driscoll et al., 1990c
Asbestos	Driscoll et al., 1990d
Cadmium chloride	Driscoll et al., 1992
Sheep	
Silica	Begin et al., 1989
Asbestos	Begin et al., 1986
Monkey	
Paraquat	Schoenberger et al., 1984

Studies in animals have also demonstrated an association between increased macrophage fibronectin synthesis and the pathogenesis of fibrotic lung disease after exposure to silica, asbestos, titanium dioxide, bleomycin, and cadmium (see Table 10). In several of these studies an increase in macrophage fibronectin production was clearly shown to precede the development of histologically or radiographically detectable fibrosis. For example, in rats intratracheally instilled with α-quartz or titanium dioxide particles, a persistent increase in alveolar macrophage fibronectin production was observed within the first month after exposure, while histohemical evidence of increased collagen was not detectable in some groups until 3 months after exposure. Similar prospective observations have been reported for silica exposed sheep as well as cadmium exposed rats (Begin et al., 1989; Driscoll et al., 1992). Thus, like the human data, the animal studies demonstrate a relationship between macrophage-derived fibronectin and the development of pulmonary fibrosis.

It is noteworthy that unlike TNFα, direct *in vitro* exposure of alveolar macrophages to pneumotoxic agents does increase fibronectin production (Driscoll et al., 1990c). This observation suggests the lung environment is key in influencing the upregulation of macrophage fibronectin synthesis. In this respect, a potent inducer of fibronectin production is TGFβ which, as described above, is also a key effector of increased collagen synthesis and decreased collagen degradation. In this respect, it may be that the strong association between macrophage fibronectin and pulmonary fibrosis reflects not only the role of fibronectin in the disease process, but also the fact that increased fibronectin is a marker for a TGFβ-rich environment which is conducive to extracellular matrix deposition.

CYTOKINES AS BIOMARKERS OF LUNG TOXICITY

Studies on the biology of cytokines in the lung and their contribution to toxicant-induced alterations in lung structure and function are important for understanding basic mechanisms of disease. Better mechanistic understanding can lead to the development improved therapeutic interventions for diseases in the lung and other tissues. In addition, understanding the contribution of cytokines to responses to pneumotoxic agents can guide the development and validation of *in vitro* and *in vivo* approaches for assessing the toxicity of materials, the development of models for improved assessments of health risks, and the identification of biomarkers for monitoring exposure or lung disease activity.

As discussed earlier, there is increasing evidence certain cytokines play key roles in processes underlying responses to pneumotoxic agents. As these mediators are identified and their contribution to response understood, determining the effect of chemical exposure on their expression in *in vitro* or *in vivo* models should provide useful information to assess potential adverse health effects. For example, there now exists a significant data base demonstrating TNFα plays a key role in initiating lung inflammation and can contribute to the pathogenesis of pulmonary fibrosis. In addition, *in vitro* studies have demonstrated an association between a material's ability to directly activate macrophage TNFα release and its inflammatory activity *in vivo*. In this respect, our mechanistic understanding of TNFα's contribution to pulmonary inflammation and fibrosis supports the use of *in vitro* macrophage TNFα production as a screening tool to evaluate a materials potential *in vivo* inflammatory activity.

An example of the use of cytokines as biomarkers of *in vivo* effects is demonstrated through the research on macrophage fibronectin. The association between the development of pulmonary fibrosis and increased alveolar macrophage fibronectin release has been clearly demonstrated. In addition, when appropriate experimental designs have been used, increased macrophage fibronectin production has been shown to precede the development histologically definable fibrosis. In this respect, characterizing endpoints such as fibronectin in acute or subchronic inhalation safety studies may be useful to facilitate extrapolation from short- to long-term effects. Stated another way, including measurements of mechanistically based endpoints such as macrophage fibronectin, increases our sensitivity for detecting adverse responses as well as confirming the absence of any ongoing disease processes. In the end, given the key role cytokines play in both normal physiologic as well as pathophysiologic processes, understanding their contribution to respiratory tract response to chemical exposure will provide opportunities to identify and validate biomarkers for use in the safety assessment process.

CONCLUSION

Cytokines are small proteins which mediate the cell:cell communication needed to regulate cell differentiation, proliferation, and cell secretory activity in development and tissue homeostasis. Cytokines also contribute to host defense, mediating the cascade of events underlying the recruitment and activation of inflammatory and immune effector cells as well as the repair of damaged tissue. While cytokines play a key role in normal physiology, their altered expression can contribute to the pathophysiology of disease.

Certain cytokines have been shown to play critical roles in initiation of response, cell recruitment, tissue repair and resolution aspects of inflammation and fibrosis. In the lung, TNFα is a key initiator of response and this cytokine appears to play a critical role in mediating inflammatory cell recruitment after exposure to a variety of pneumotoxic agents. TNFα acts through a cytokine network involving TNFα-induced elaboration of other cytokines by both immune and nonimmune cells. Studies have demonstrated that members of the chemokine cytokine family which are induced by TNFα may act as the proximate mediators of inflammatory cell recruitment. Importantly, the chemokines show a high degree of specificity in the cells they attract and, thus, provide a mechanism by which certain diseases can be associated with specific populations or subpopulations of leukocytes. Repair of lung tissue is influenced by a variety of cytokines including TGFβ, IGF, PDGF and fibronectin. With respect to stimulating connective tissue synthesis, TGFβ appears to be an essential factor in both normal tissue repair and the pathogenesis of fibrosis. Fibronectin has the ability to influence several processes which contribute to the development of wound healing, including fibroblast chemotasis and proliferation. In this respect there is a growing database associating increased macrophage fibronectin production with the pathogenesis of fibrosis, providing support for the use of macrophage fibronectin production as a biomarker of interstitial lung disease.

REFERENCES

Assoian, RK, Fleurdelys, BE, Stevenson, HC, Miller, PJ, Madtes, DK, Raines, EW, Ross, R, Sporn, MB: Expression and secretion of type beta transforming growth factor by activated human macrophages. Proc Natl Acad Sci USA 84:6020–6024, 1987.

Bachwich, PR, Lynch, JP, Larrick, J, Spengler, M, Kunkel, SL: Tumor necrosis factor production by human sarcoid alveolar macrophages. Am J Path 125:421–425, 1986.

Baggiolini, M, Walz, A, Kunkel, SL: Neutrophil activating peptide-1/interleukin-8, a novel cytokine that activates neutrophils, J Clin Invest 84: 1045–49, 1989.

Balkwill, FR, Burke, F. The cytokine network, Immunol Today 10:299–310, 1989.

Begin, R, Martel, M, Desmarais, Y, Drapeau, G, Boileau, R, Rola-Pleszczynski, M, Masse, S: Fibronectin and procollagen 3 levels in bronchoalveolar lavage fluid of asbestos exposed human subjects and sheep. Chest 89:237–243, 1986.

Begin, R, Dufresne, A, Cantin, A, Possmayer, F, Sebastien, P: Quartz exposure, retention, and early silicosis in sheep. Exp Lung Res 15:409–428, 1989.

Beutler, B, Greenwald D, Hulmes JD, Chang M, Pan Y-CE, Mathison J, Ulvetich R, Cerami A: Identity of tumor necrosis factor and the macrophage secreted factor cachectin. Nature. 316:552–554, 1985.

Bitterman, PB, Rennard, SI, Adelberg, S, Crystal, RG: Role of fibronectin as a growth factor for fibroblasts. J Cell Biol 97:1925–1932, 1983.

Bonner, JC, Osornio-Vargas AR, Badgett, A, Brody, AR: Differential proliferation of rat lung fibroblasts induced by the platelet-derived growth factor (PDGF)—AA, AB and BB isoforms secreted by rat alveolar macrophages. Am J Resp Cell Mol Biol 5:539–47, 1991.

Broekelmann, TJ, Limper, AH, Colby, TV, McDonald, JA: Transforming growth factor $\beta1$ is present at sites of extracellular matrix gene expression in human pulmonary fibrosis. Proc Natl Acad Sci USA 88:6642–6, 1988.

Carswell, EA, Old, LJ, Kassel, RL, Green, S, Fiore, N, Williamson, B. An endotoxin-induced serum factor that causes necrosis of tumors. Proc Nat Acad Sci USA. 72:3666–70, 1975.

Corrin, B, Butcher, D, McAnulty, BJ, Dubois, RM, Laurent, GJ, Harrison, NK: Immunohistochemical localization of transforming growth factor-beta in the lungs of patients with systemic sclerosis, cryptogenic fibrosing alveolitis and other lung disorders. Histopath 24:145–50, 1994.

Das UN, Padma M, Sangeetha Sagar P, Ramesh G, Koratkar R. Stimulation of free radical generation in human leukocytes by various agents including tumor necrosis factor is a calmodulin dependent process. Biochem Biophys Res Comm 167:1030–1036 1990.

Davatelis, G, Tekamp-Olson, P, Wolpe, SD, Hermsen, K, Luedke, C, Gallegos, C, Coit, D, Merryweather, J, Cerami, A: Cloning and characterization of a cDNA for Murine Macrophage Inflammatory Protein (MIP), A Novel Monokine with Inflammatory and Chemokinetic Properties. J Exp Med 167:1939–1944, 1988.

Davis, R, Erdogdu, G: Secretion of fibronectin by mineral dust-derived alveolar macrophages and activated peritoneal macrophages. Exp Lung Res 15:285–297, 1989.

Dean, DC, Newby, RF, Bourgeois, S: Regulation of bironectin biosynthesis by dexamethasome, transforming growth factor β, and cAMP in human cell lines. J Cell Biol 106:2159–2170, 1988.

Dean, DC: Expression of the fibronectin gene. Am J Resp Cell Ml Biol. 1:5–10, 1989.

Derynck, R, Jarrett, JA, Chen, EY, Eaton, DH, Bell, JR, Assoian, RK, Roberts, AB, Sporn, MB, Goeddel, DV: Human transforming growth factor-beta cDNA sequence and expression in tumor lines. Nature 316:701–705, 1985.

Dinarello CA: Interleukin-1 and its biologically related cytokines. Adv Immunol 44:153–167, 1989.

Driscoll, KE, Schlesinger, RB: Alveolar macrophage-stimulated neutrophil and monocyte migration: effects of in vitro ozone exposure. Toxicol Appl Pharmacol 93:312–328, 1988.

Driscoll, KE, Lindenschmidt, RC, Maurer, JK, Higgins, JM, Ridder, G: Pulmonary response to silica or titanium dioxide:inflammatory cells, alveolar macrophage-derived cytokines and histopathology. Am J Resp Cell Mol Biol 2:381–390, 1990a.

Driscoll, KE, Higgins, JM, Leytart, MJ, Crosby, L. L: Differential effects of mineral dusts on the in vitro activation of alveolar macrophage eicosanoid and cytokine release. Toxic in vitro, 4:284–288, 1990b.

Driscoll, KE, Maurer, JK, Lindenschmidt, RC, Romberger, D, Rennard, SI, Crosby, L: Respiratory tract response to dust: Relationships between dust burden, lung injury, alveolar macrophage fibronectin release, and the development of pulmonary fibrosis. Toxicol Appl Pharm 106:88–101, 1990c.

Driscoll, KE, Higgins, J, Romberger, D, Rennard, SI, Crosby, L: Alveolar macrophage tumor necrosis factor and fibronectin release in asbestos or cadmium chloride-induced pulmonary injury and fibrosis. Am Rev Respir Dis 143:A416, 1990d.

Driscoll, KE, Maurer, JK: Cytokine and Growth Factor Release by Alveolar Macrophages: Potential Biomarkers of Pulmonary Toxicity. Tox Pathology, 19:398–405, 1991.

Driscoll, KE, Maurer, JK, Poynter, J, Higgins, J, Asquith, T, Miller, NS: Stimulation of rat alveolar macrophage fibronectin release in a cadmium chloride model of lung injury and fibrosis. Toxicol Appl Pharm, 116:30–37, 1992.

Driscoll, KE, Simpson, L, Carter, J, Hassenbein, D, Leikauf, GD: Ozone inhalation stimulates expression of a neutrophil chemotactic protein. Macrophage Inflammatory Protein 2. Toxicol Appl Pharmacol 127:306–309, 1993a

Driscoll KE, Hassenbein DG, Carter J, Poynter J, Asquith TN, Grant RA, Whitten J, Purdon MP, Takigiku R: Macrophage inflammatory proteins 1and 2: Expression by rat alveolar macrophages, fibroblasts, and epithelial cells in rat lung after mineral dust exposure, Am J Respir Cell Mol Biol 8:311–328, 1993b.

Driscoll, KE, Maurer, JK, Hassenbein, D, Carter, J, Janssen, YMW, Mossman, BT, Osier, M, Oberdorster, G: Contribution of macrophage-derived cytokines and cytokine networks to mineral dust-induced lung inflammation. In: Toxic and carcinogenic effects of solid particles in the respiratory tract. (Edited by D Dungworth, U Mohr, J Mauderly, G Oberdörster). ILSI Press, Washington, pp. 177–190, 1994.

Driscoll, KE: Macrophage Inflammatory Proteins: Biology and Role in Pulmonary Inflammation. Exp Lung Res 20:473–489, 1994.

Driscoll KE, Hassenbein DG, Howard BW, Isfort RJ, Cody D, Tindal MH, Suchanek M, Carter JM: Cloning, expression, and functional characterization of rat MIP-2. J Leukocyte Biol, 1995 (in press).

Driscoll, KE: The role of interleukin-1 and tumor necrosis factor in the lung's response to silica. In: Silica and silica-induced lung diseases, Current Concepts. (Edited by V Castranova, V Vallyathan, W Wallace) CRC Press, 1995 (in press).

DuBois, CM, Bissonnette, E, Rola-Pleszczynski, M: Asbestos fibers and silica particles stimulate rat alveolar macrophages to release tumor necrosis factor. Am Rev Respir Dis 139:1257–1264, 1989

Dubravec DB, Spriggs DR, Mannick JA, Rodrick ML: Circulating human peripheral blood granulocytes synthesize and secrete tumor necrosis factor. Proc Natl Acad Sci USA. 87:6758–63 1990.

Edwards, DR, Murphy, G, Reynolds, JJ, Whitman, SE, Docherty, AJP, Angel, P, Health, JK: Transforming growth factor beta modulates the expression of collagenase and metalloproteinase inhibitor. EMBO J 6:1899–1904, 1987.

Finkelstein, JN, Johnston, CJ, Baggs, R, Rubin, P: Early alterations in extracellular matrix and transforming growth factor beta gene expression in mouse lung indicative of late radiation fibrosis. Int J Radiat Oncol Biol Phys 28:621–631, 1994.

Fiorentino, DF, Zlotnik, A, Mosmann, TR, Howard, M, O'Gawa, A: IL-10 inhibits cytokine production by activated macrophages. J Immunol 147:3815–3822, 1991.

Flick, DA, Gonzalez-Rothi, J, Harris, JO, Gifford, GE: Rat lung macrophage tumor cytotoxin production: impairment by chronic in-vivo cigarette smoke. Cancer Res 45:5225–5229, 1985.

Gery, I, Davies, P, Derr, I, Krett, N, Barranger, JA: Relationship between production and release of lymphocyte-activating factor (interleukin-1) by murine macrophages. Effects of various agents. Cell Immunol 64:293–303 1981.

Giri, SN, Hyde, DM, Hollinger, MA: Effect of antibody to transforming growth factor b on bleomycin induced accumulation of lung collagen in mice. Thorax 48:959–966, 1993.

Godelaine, D, Beaufay, H: Comparative study of the effect of chrysotile, quartz and rutile on the release of lymphocyte-activating factor (interleukin-1) by murine peritoneal macrophages *in-vitro.* IARC Sci Publ. 90:149–155, 1989.

Gosset, P, Lassalle, P, Vanhee, D, Wallaert, B, Aerts, C, Voisin, C, Tonnet A-B: Production of tumor necrosis factor- and interleukin-6 by human alveolar macrophages exposed *in-vitro* to coal mine dust. Am J Resp Cell Mol Biol 5:431–436, 1991.

Hajjar KA, Hajjar DP, Silverstein RL, Nachman RL: Tumor necrosis factor-mediated release of platelet-derived growth factor from cultured endothelial cells. J Exp Med 166:235–245, 1987.

Hart, PH, Vitti, GF, Burgess, DR, Whitty, GA, Piccoli, DS, Hamilton, JA: Potential antiinflammatory effect of interleukin 4: suppression of human monocyte tumor necrosis factor alpha, interleukin-1 and prostaglandin E2. Proc Natl Acd Sci USA 86:3803–3807, 1989.

Hoffman M, Weinbert JB: Tumor Necrosis Factor-α induces increased hydrogen peroxide production and Fc receptor expression, but not increased Ia antigen expression by peritoneal macrophages. J Leukocyte Biol 42:704–7, 1987.

Hoyt, DG, Lazo, JS: Alterations in pulmonary mRNA encoding procollagens, fibronectin, and transforming growth factor- precede bleomycin-induced pulmonary fibrosis. J Pharmaco Exp Ther 246:765–71, 1988.

Huang, S, Paulauskis, JD, Godleski, JJ, Kobzik, L: Expression of Macrophage Inflammatory Protein-2 and KC mRNA in Pulmonary Inflammation, Am J Pathol, 141:981–988, 1992.

Ignotz, RA, Massague: J. Transforming growth factor-β stimulates the expression of fibronectin and collagen and their incorporation into the extracellular matrix. J Cell Biol 261:4337–4342, 1986.

Jansen RW, Hance KR, Arend WP: Production of IL-1 receptor antagonist by human *in-vitro*-derived macrophages: effects of lipopolysaccharide and granulocyte-macrophage colony-stimulating factor. J Immunol 147:4218–4226 1991.

Jelinek DF, Lipsky PE: Enhancement of human B cell proliferation and differentiation by tumor necrosis factor-α and interleukin. J Immunol. 139:2970–2976, 1987.

Kang, JH, Lewis DM, Castranova V, Rjanasakul Y, Banks DE, Ma JYC, Ma JKH: Inhibitory Action of tetrandrine on macrophage production of interleukin-1 (IL-1)-like activity and thymocyte proliferation. Exp Lung Res 18:715–729, 1992.

Kinsella, MB, Smith, EA, Miller, KS, LeRoy, EC, Silver, RM: Spontaneous production of fibronectin by alveolar macrophages in patients with scleroderma. Arth Rheum 32:577–583, 1989.

Kishimoto TK, Larson RS, Corbi AL, Dustin ML, Staunton DE, Springer TA: The leukocyte integrins. Adv Immunol 46:149–182, 1989.

Klebanoff SJ, Vadas MA, Harlan JM, Sparks LH, Gamble JR, Agosti JM, Waltersdorph AM: Stimulation of neutrophils by tumor necrosis factor. J Immunol 136:4220–4225, 1986.

Koren, HS, Devlin, RB, Graham, DE, Mann, R, McGee, MP, Horstman, DH, Kozumbo, WJ, Becker, S, House, DE, McDonnell, WF, Bromberg, PA: Ozone-induced inflammation in the lower airways of human subjects. Am Rev Resp Dis 139:407–415, 1989.

Kunkel SL, Strieter RM: Cytokine networks in lung inflammation, Hospital Practice 25: 63–75, 1990.

Kunkel SL, Driscoll KE, Ward PA, Nicoloff B, Strieter R: Mediators of inflammation. In: Pathology of Environmental and Occupational Disease. (Edited by J Craighead), Mosby, New York, pp. 385–396, 1995.

Larrick JW Kunkel SL: The role of tumor necrosis factor and interleukin-1 in the immunoinflammatory response, Pharm Res 5: 129–39, 1988.

Lassalle, P, Gosset, P, Aerts, C, Fourier E, Lafitte, JJ, Gegreef, JM, Wallaert, B, Tonnel, AB, Voisin, C: Abnormal secretion of interleukin-1 and tumor necrosis factor a by alveolar macrophages comparison between simple pneumoconiosis and progressiver massive fibrosis. Exp Lung Res 16:73–80, 1990.

Lee JC, Truneh A, Smith MF, Tsang KY: Induction of interleukin-2 receptor (TAC) by tumor necrosis factor in YT cells. J Immunol 139:1935–1938, 1987.

Limper, AH, Broekelmann, TJ, Colby, TV, Malizia, G, McDonald, JA: Analysis of local mRNA expression for extracellular matrix growth proteins and growth factors using *in-situ* hybridization in fibroproliferative lung disorders. Chest 99:55S–56S, 1991.

Madtes, DK, Raines, EW, Sakariassen, KS, Assoian, RK, Sporn, MB, Bell, GI, Ross, R: Induction of transforming growth factor a by activate human alveolar macrophages. Cell 53:255–63, 1988.

Majack, RA, Majesky, MW, Goodman, LV: Role pf PDGF-A expression in the control of vascular smooth muscle cell growth by transforming growth factor-β. J Cell Biol 111:239–247, 1990.

Marmenout A, Fransen L, Tavernier J, Van der Heyden J, Tizard R, Kawashima E, Shaw A, Johnson MJ, Semon D, Muller R, Ruysschaert MR, Van Vliet A, Fiers W: Molecular cloning and expression of human tumor necrosis factor and comparison with mouse tumor necrosis factor. Eur J Biochem 152:515–522, 1985.

Matsushima K, Oppenheim JJ: Interleukin-8 and MCAF. Novel inflammatory cytokines inducible by IL-1 and TNF. Cytokine 1:2–13, 1989.

Matsushima K, Morishita K, Yoshimura T, Lavu S, Obayashi Y, Lew W, Appella E, Kung HF, Leonard EJ, Oppenheim JJ: Molecular cloning of a human monocyte-derived neutrophil chemotactic factor (MDNCF) and the induction of MDNCF mRNA by interleukin-1 and tumor necrosis factor, J Exp Med 167:1883–1893, 1988.

Miller, MD, Krangel, MS: Biology and Biochemistry of the Chemokines: A Family of Chemotactic and Inflammatory Cytokines. Crit Rev Immunol 12:27–46, 1992.

Moses HL, Yang EY, Pietenpol, JA: TGF-b stimulation and inhibition of cell proliferation: new mechanistic insights. Cell 63:245–47, 1990.

Oghiso Y, Kubota Y: Interleukin 1-like thymocyte and fibroblast activating factors from rat alveolar macrophages exposed to silica and asbestos particles. Jpn J Vet Sci. 48:461–471, 1986.

Ohno, I, Driscoll, KE, Hassenbein, D, Miura, S, Takanashi, S, Gauldie, J, Jordana, M: Sequential expression of cytokines in lung tissues from rats exposed to admium chloride. Am Rev Resp Dis 147:A734, 1993

Osborn L, Hession C, Tizard R, Vassallo C, Luhowskyl S, Ghi-Rosso G, Lobb R: Direct expression cloning of vascular cell adhesion molecule 1, a cytokine-induced endothelial protein that binds to lymphocytes. Cell. 59:1203–1211, 1989.

Ozaki, T, Moriguchi, H, Nakamura, Y, Kamei, T, Yasuoka, S, Ogura, T: Regulatory effect of prostaglandin E2 on fibronectin release from human alveolar macrophages. Am Rev Resp Dis 141:965–969, 1990.

Palombella VJ, Yamashiro DJ, Maxfield FR, Decker SJ, Vilcek J: Tumor necrosis factor increases the number of epidermal growth factor receptors on human fibroblasts. J Biol Chem. 262:1950–54, 1987.

Patel, RS, Odermatt, JE, Schwarzbaurer, Hynes, RO: Organization of the fibronectin gene provides evidence for exon shuffling during evolution. EMBO J 6:2565–2572, 1987.

Paul, JI, Schwarzbauer, JE, Tamkun, JW, Hynes, RO: Cell-type specific fibronectin subunits generated by alternate splicing. J Biol Chem 261:12258–65, 1986.

Pennica D, Nedwin GE, Hayflick JS, Seeburg PH, Derynck R, Palladino MA, Kohr WJ, Aggarwal BB, Goeddel DV: Human tumor necrosis factor: precursor structure, expression and homology to lymphotoxin. Nature 312:724–728, 1984.

Perdue, TD, Brody, AR: Distribution of transforming growth factor-beta-1, fibronectin, and smooth muscle actin in asbestos-induced pulmonary fibrosis in rats. J Histochem Cytochem 42:1061–1070, 1994.

Perez, T, Farre, JM, Gosset, P, Wallaert, B, Duquesnoy, B, Voisin, C, Delcombre, B, Tonnel, AB: Subclinical alveolar inflammation in rheumatoid arthritis: superoxide anion, neutrophil chemotactic activity and fibronectin generation by alveolar macrophages. Eur Respir J 2:7–13. 1989.

Phan, SH, Kunkel, SL: Lung cytokine production in bleomycin-induced pulmonary fibrosis. Exp Lung Res. 18:29–43, 1992.

Pick-Kober, KH, Munker, D, Gressner, AM: Fibronectin is synthesized as an acute phase reactant in rat hepatocytes. J Clin Chem Clin Biochem 24:521–528, 1986.

Piguet PF, Collart MA, Grau GE, Kapanci Y, Vassalli P: Tumor necrosis factor/cachectin plays a key role in bleomycin-induced pneumopathy and fibrosis. J Exp Med. 170:655–663, 1989.

Piguet, PF, Grau, GE, Vassalli, P: Subcutaneous perfusion of tumor necrosis factor induces local proliferation of fibroblasts, capillaries, and epidermal cells, or massive tissue necrosis. Am J Pathol 136, 1990–1998, 1990a.

Piguet, PF, Collart, MA, Grau, GE, Sappino, AP, Vassalli, P: Requirement for tumor necrosis factor for development of silica-induced pulmonary fibrosis. Nature, 344, 245–247, 1990b.

Pober, JS, Gimbrone, MA Jr, Lapierre, LA, Mendrick, DL, Fiers, W, Rothlein, R, Springer, TA: Overlapping patterns of activation of human endothelial cells by interleukin-1, tumor necrosis factor and immune interferon. J Immunol 137:1893–1896, 1986.

Raghow, R, Irish, P, Kang, AH: Coordinate regulation of transforming growth factor gene expression and cell proliferation in hamster lungs undergoing bleomycin-induced pulmonary fibrosis. J Clin Invest 84:1836–42, 1989.

Rennard, SI, Hunninghake, GW, Bitterman, PB, Crystal, RG: Production of fibronectin by human alveolar macrophages: a mechanism for the recruitment of fibroblasts to sites of tissue injury in interstitial lung disease. Proc Natl Acad Sci USA 78:7147–51, 1981a.

Rennard, SI, Crystal, RG: Fibronectin in human bronchopulmonary lavage fluid. Elevation in patients with interstitial lung disease. J Clin Invest 69:113–122, 1981b.

Rich, EA, Panuska, JR, Wallis, RS, Wolf, CB, Leonard, ML, Ellner, JJ: Dyscoordinate expression of tumor necrosis factor-alpha by human blood monocytes and alveolar macrophages. Am Rev Resp Dis 139:1010–1016, 1989.

Richter, J, Andersson, T, Olsson, I: Effect of tumor necrosis factor and granulocyte/macrophage colony-stimulating factor on neutrophil degranulation. J Immunol 142:3199–3205, 1989.

Roberts, AB, Sporn, MB: The transforming growth factor βs. In: Peptides Growth Factors and Their Receptors. Edited by MB Sporn and AB Roberts. Springer-Verlag, Heidelberg, 419–472, 1990

Rom, WN, Bassett, P, Fells, GA, Nukiwa, T, Trapnell, BC, Crystal, RG: Alveolar macrophages release an insulin-like growth factor I-type molecule. J Clin Invest 82:1685–93, 1988.

Rom, WN, Bitterman, PB, Rennard, SI, Cantin, A, Crystal, RG: Characterization of the lower respiratory tract inflammation of nonsmoking individuals with interstitial lung disease associated with chronic inhalation of inorganic dusts. Am Rev Resp Dis 136:1429–34, 1987.

Rom, WN: Relationship of inflammatory cell cytokines to disease severity in individuals with occupational inorganic dust exposure. Am J Ind Med 19:15–27, 1991.

Ruoslahti, E: Fibronectin and its receptors. Ann Rev Biochem 57:375–413, 1988.

Ryan, LK, Karol, MH: Release of tumor necrosis factor in guinea pigs upon acute inhalation of cotton dust. Am J Resp Cell Mol Biol 5:93–98, 1991.

Schoenberger, CI, Rennard, SI, Bitterman, PB, Fikuda, Y, Ferrans, VJ, Crystal, RG: Paraquat-induced pulmonary fibrosis. Am Rev Resp Dis 129:168–174, 1984.

Schmidt JA, Oliver CN, Lepe-Zuniga JL, Green I, Gery I: Silica-stimulated monocytes release fibroblast proliferation factors identical to interleukin 1: A potential role for interleukin 1 in the pathogenesis of silicosis. J Clin Invest 73:1462–1472, 1984.

Schreck, R, Rieber, P, Baeuerle, PA: Reactive Oxygen Intermediates as Apparently Widely Used Messengers In The Activation of the NF-κB Transcription Factor and HIV-1 EMBO J 10:2847–2858, 1991.

Schwarzbauer, JE: Alternate splicing of fibronectin: three variants, three functions. Bioessays 13:527–533, 1991.

Seyedin, SM, Thomas, TC, Thompson, Rosen, DM, Piez, KA: Purification and characterization of two cartilage-inducing factors from bovine demineralized bone. Proc Natl Acad Sci USA 82:2267–2271, 1985.

Sherry B, Cerami A: Cachectin/tumor necrosis factor exerts endocrine, paracrine, and autocrine control of inflammatory responses, J Cell Biol 107:1269–1277, 1988.

Shirai T, Yamaguchi H, Ito H, Todd CW, Wallace RB: Cloning and expression in Escherichia coli of the gene for human tumour necrosis factor. Nature. 313:803–806, 1985.

Simeonova P, Lusten M: Iron and reactive oxygen species in asbestos-induced TNFα response from alveolar microphage. Am J Resp Cell Mol Biol, 1995 (in press).

Smith RA, Baglioni C: The active form of tumor necrosis factor is a trimer. J Biol Chem 262:6951–6954, 1987.

Standiford TJ, Kunkel SL, Basha MA, Chensue SW, Lynch JP, Toews GB, Strieter RM: Interleukin-8 expression by pulmonary epithelial cells: A model for cytokine networks in the lung, J Clin Invest 86:1945–1953, 1990.

Strieter RM, Phan SH, Showell HJ, Remick DG, Lynch JP, Genord M, Kunkel SL: Monokine-induced neutrophil chemotactic factor gene expression in human fibroblasts, J Biol Chem 264:10621–626, 1989a.

Strieter RM, Kunkel SL, Showell, HJ, Remick DG, Phan SH, Ward PA, Marks RM: Endothelial cell gene expression of a neutrophil chemotactic factor by TNF, LPS and IL-1, Science 243:1467–1469, 1989b.

Sugarman BJ, Aggarwal BB, Hass PE, Figari IS, Palladino MA Jr., Shepard HM: Recombinant human tumor necrosis factor-: Effects on proliferation of normal and transformed cells in vitro. Science. 230:943–45, 1985.

Tekamp-Olson, P, Gallegos, C, Bauer, D, McClain, J, Sherry, B, Fabre, M, van Deventer, S, Cerami, A: Cloning and Characterization of cDNAs for Murine Macrophage Inflammatory Protein 2 and its Human Homologues J Exp Med 272:911–919, 1990.

Thornton AJ, Strieter, RM, Lindley I, Baggiolini M, Kunkel SL: Cytokine-induced gene expression of a neutrophil chemotactic factor/interleukin-8 in tumor hepatocytes, J Immunol 144: 2609–2614, 1990.

Torikata, C, Villiger, B, Kuhn, C, McDonald, JA: Ultrastructural distribution of fibronectin in normal and fibrotic human lung. Lab Invest 52:399–408, 1985.

Ulich, TR, Yin, S, Guo, K, Yi, ES, Remick, D, Castillo, JD: Intratracheal injection of endotoxin and cytokines. II. Interleukin-6 and transforming growth factor beta inhibit acute inflammation. Am J Pathol 138:1097–1101, 1991a

Ulich, TR, Yin, S, Guo, K, Yi, ES, Castillo, JD, Eisenberg, SP: The intratracheal administration of endotoxin and cytokines. III. The IL-1 receptor antagonist inhibits endotoxin- and IL-induced acute inflammation. AM J Pathol 138:521–524, 1991b.

Ulich, TR, Fann, M, Patterson, PH, Williams, JH, Samal, B, Castillo, JD, Yin, S, Guo, K, Remick, D: Intratracheal instillation of LPS and cytokines. V. LPS induces expression of LIF and LIF inhibits accute inflammation. Am J Phys: Lung Cell Mol Physiol 11:L442–446, 1994.

Wagner, JC, Burns, J, Munday, DE, McGee, J: Presence of fibronectin in pneumoconiotic lesions. Thorax 37:54–56, 1982.

Wahl SM, Hunt DA, Wakefield LM, McCartney-Francis N, Wahl L, Roberts AB, Sporn MB: Transforming growth factor-beta induces monocyte chemotaxis and growth factor production, Proc Natl Acad Sci (USA) 84:5788–5792, 1987.

Wahl, SM, McCartney-Francis, N, Mergenhagen, SE: Inflammatory and immunomodulatory roles of TGFβ. Immuno Today 10:258–261, 1989.

Wahl, SM: Transforming growth factor β (TGFβ) in inflammation: a cause and a cure. J Clin Immunol 12:1–14, 1992.

Wallis, WJ, Beatty, PG, Ochs, HD, Harlan, JM: Human monocyte adherence to cultured vascular endothelium: monoclonal antibody-defined mechanisms. J Immunol 135:2323–2330, 1985.

Warner, SJC, Liby, P: Human vascular smooth muscle cells. Target for and source of tumor necrosis factor. J Immunol. 142:100–105 1989.

Wewers MD, Herzuk DJ: Alveolar macrophages differ from blood monocytes in human IL-1 release. J Immunol 143:1635–1641, 1989.

Widmer, U, Mangoue, KR, Cerami, A, Sherry, B: Genomic Cloning of Promoter Analysis of Macrophage Inflammatory Protein (MIP)-2, MIP-1, and MIP-1, Members of the Chemokine Superfamily of Proinflammatory Cytokines. J Immunol 150:4996–5012, 1993.

Williams, OA, Flanders, KC, Saffiotti, U: Immunohistochemical localization of transforming growth factor-b1 in rats with experimental silicosis, alveolar type II hyperplasia, and lung cancer. Am J Path 142:1831–1839, 1993.

Wolpe, SD, Davatelis, G, Sherry, B, Beutler, B, Hesse, DG, Nguyen, HT, Moldawer, LL, Nathan, CF, Lowry, SF, Cerami, A: Macrophages Secrete a Novel Heparin-Binding Protein with Inflammatory and Neutrophil Chemokinetic Properties, J Exp Med, 167:570–581, 1988.

Wolpe, SD, Sherry, B, Juers, D, Davatelis, G, Yurt, RW, Cerami, A: Identification and Characterization of Macrophage Inflammatory Protein 2. Proc Natl Acad Sci, 86:612–616, 1989.

Wright, SD, Ramos, RA, Tobia, PS, Ulevitch, RJ, Mathison, JC: CD14, a receptor for complexes of lipopolysaccharide (LPS) and LPS binding protein. Science 249:1431–33, 1990.

Yamaguchi, E, Akihide, I, Furuya, K, Miyamoto, H, Shosaku, A, Yoshikazu, K: Release of tumor necrosis factor a from human alveolar macrophages is decreased in smokers. Chest 103:479–483, 1993.

Young JD, Liu C, Butler G, Cohn ZA, Galli SJ: Identification, purification, and characterization of a mast cell-associated cytolytic factor related to tumor necrosis factor. Proc Natl Acad Sci USA 84:9175–9181, 1987.

Xing, Z, Jordana, M, Kirpalani, H, Driscoll, KE, Schall, TJ, Gauldie, J: Cytokine Expression by Neutrophils and Macrophages In-Vivo: Endotoxin Induces Tumor Necrosis Factor, Macrophage Inflammatory Protein-2, Interleukin-16 and Interleukin-6, but not RANTES or Transforming Growth Factor β_1 mRNAs Expression in Acute Lung Inflammation Am J Resp Cel Mol Biol 10:148–153, 1994.

Zhang Y, Lee TC, Guillemin B, Yu M-C, Rom WN: Enhanced IL-1β and tumor necrosis factor-α release and messenger RNA expression in macrophages from idiopathic pulmonary fibrosis or after asbestos-exposure. Am Rev Resp Dis, J Immunol 150:4188–4196, 1993.

Chapter Seventeen

Applications of Behavioral Measures to Inhalation Toxicology

Bernard Weiss and Alice Rahill

INTRODUCTION

Inhalation is the primary medium of exposure to toxic materials in the workplace and probably in the home as well. It surely predominates in the communal environment. California may have expended more resources on the control of air pollution than on any of the other health risks confronting its inhabitants, including earthquakes. Simply because it represents the exposure interface, the lung is the natural focus of inhalation risk assessment. Its morphological and functional properties dominate the choice of endpoints. But the lung is not invariably the ultimate target tissue for inhaled chemicals; it may serve merely as the exposure gatekeeper so to speak. For many agents, it is simply the first phase in a path to the brain for instance.

Behavioral toxicology is a discipline equipped, not only to quantify and amplify adverse nervous system effects posed by inhaled toxicants, but also to illuminate actions occurring at the pulmonary level. In this chapter, we discuss how behavioral measures are used to describe various aspects of inhalation toxicity broadly defined. We discuss the application of behavioral measures to questions stemming from direct toxicant effects on the lung, to the assessment of neurotoxicants that enter the body through inhalation, and to a new set of issues provoked by syndromes such as Multiple Chemical Sensitivity.

MEASURES OF AVERSIVE AND IRRITANT PROPERTIES

The Jogger's dilemma exemplifies one aspect of how behavioral measures may be slighted when functional assessment is confined only to pulmonary endpoints. Consider the jogger who runs in Pasadena on a high smog day to improve her cardiovascular fitness. At the same time, breathing at a high rate and probably largely through her mouth, she bathes her respiratory system in a blend of airborne contaminants that is possibly toxic and certainly unpleasant. One way to ascertain the impact of exercise under these conditions is to apply pulmonary function tests. Such tests provide important measures of pulmonary toxicity. But other kinds of measures may also serve as indices of how air pollution influences function. High school cross-country runners, on elevated oxidant days in southern California, performed below expected levels (Wayne et al, 1967). The authors noted that their results might be "more related to lack of maximal effort due to increasing discomfort than directly to physiologic capability." This is a behavioral interpretation. It is amenable to experimental manipulation and testing. Ozone research offers some useful examples.

OZONE

Reduced effort in animals can be measured by procedures that are analogous to the types of effort demanded from humans. Motor activity is common in psychobiology as a nonspecific index of the response to drugs and toxicants. Spontaneous motor activity demonstrated a connection with exercise in early ozone studies. For example, Murphy et al (1964) noted reduced activity in mice exposed at 0.3 ppm ozone, and Konigsberg and Bachman (1970) observed such an effect in rats exposed to 0.2 ppm.

Tepper et al (1982) studied 8 rats that were housed in small enclosures attached to running wheels located in large Rochester exposure chambers (Leach et al, 1959). Rodents are known to run an individually consistent amount (mostly at night) and to remain quiescent between bouts of running. The apparatus included electrical contacts on the wheels that transmitted digital signals to a computer. This arrangement allowed not only the measurement of total revolutions, but also the time per revolution (speed of running) and burst or episode length (number of revolutions before a pause, number of pauses, and pause duration). When exposed for 6 hours to ozone concentrations between 0.12 and 1.0 ppm, the rats showed a statistically significant decline in total revolutions beginning at the 0.12 ppm concentration.

The decline at 0.12 ppm became especially evident during the last 2 hours of exposure. Activity rebounded during the 3-hour post-exposure period for all concentrations.

Most human behavior, however, is not spontaneous in the sense described above, but is undertaken to achieve specific consequences. Operant conditioning paradigms are especially useful models of purposeful human effort. These paradigms (designated as varieties of schedule-controlled operant behavior) are defined by the conditions under which specified behaviors are followed by specified consequences (Weiss and Cory-Slechta, 1994). For example, the specified behavior could be pressing a lever attached to one wall of an experimental enclosure, the specified consequence could be the delivery of a small pellet of food, and the schedule of reinforcement could stipulate that the first response following a designated period of time since the previous pellet of food (or other consistent event) would trigger the delivery of another pellet. This particular contingency is known as a fixed-interval (FI) schedule of reinforcement. Because animals tend to respond at substantial rates during the interval although no responses are required (except for the first following the end of the interval), FI response rates can be taken as a reflection of the inclination (rather than the capacity) to respond. Adoption of a FI schedule also offers the experimenter a stable baseline (studied in many situations) against which to measure disruptions and provides the flexibility to explore them from several aspects.

An effortful response, such as running in the wheel to obtain food, may parallel many human settings (ranging from occupational to recreational) which require physical activity to effect a certain outcome. Rotating the wheel became the designated motor response (or operant) in an experiment (Tepper et al, 1983; Tepper & Weiss, 1986) in which three rats were trained to respond under a FI 5-minute schedule. On this schedule, a 45 mg food pellet was delivered for the first complete rotation occurring at least 5 minutes following the previous pellet delivery. FI schedules tend to produce a characteristic pattern of behavior which consists of a pause during the early part of the interval, succeeded by progressive increases in response rate until the end of the interval and the response that triggers food pellet delivery. Ozone exposures lasted 6 hours, with exposure concentrations ranging from 0.08 to 0.5 ppm. The total number of revolutions was reduced as a function of concentration and fell by 23% at 0.12 ppm. Two features of the findings were noteworthy. First, cumulative records indicated that the reduction in responding did not arise from lowered rates during bursts of responses as much as it did from long pauses without any responding and second, the effects of ozone became magnified during the latter parts of the session.

Premack and Schaeffer (1962) and more recently Iversen (1994) identified the reinforcing properties of running when they demonstrated that rats with restricted access to a wheel will press a lever to gain such access. The contingencies are similar to human efforts to engage in jogging. Tepper and

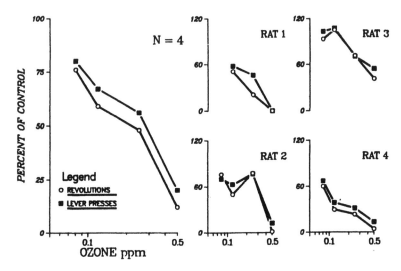

Figure 1   Results of an experiment in which rats worked to gain access to a running wheel. The rats were trained to press a lever mounted on the inner wall of the wheel to release a brake. Five lever presses (a fixed-ratio 5 schedule or reinforcement) released the wheel for 15 seconds, but the wheel was not locked again until a pause of 3.5 seconds or longer. Rat 1 remained on fixed ratio 1. Access to the running wheel was allowed only during the final hour of the 6-hour exposure period. Rats 2 and 4 showed decreases in lever presses at 0.08 ppm which exceeded 1 SD from control performance. (*From:* Tepper and Weiss, 1986.)

Weiss (1986) took advantage of such a situation to test whether the reinforcing properties of running would be modified by ozone. They trained 4 rats to make five lever presses (FR 5) to release the brake on a running wheel for a 15-second period. Ozone exposures ranged from 0.08 to 0.5 ppm. Three of the 4 rats reduced lever pressing by more than 1 standard deviation at 0.08 ppm (Fig. 1). These data confirm the hypothesis that the reinforcing potency of exercise is reduced during exposure to ozone.

The above experiments can be interpreted as reflections of ozone's aversive effects during exercise. Several experimental studies indicate that the adverse effects in humans of ozone exposure are also magnified by exercise (e.g., Follinsbee et al., 1978; McDonnell et al. 1983; & Schelegle & Adams, 1986), a finding consistent with the fact that exercise raises the delivered dose. However, even sedentary behavior can be modified by ozone exposure. Weiss et al. (1981) demonstrated that ozone modified performance on a well-studied (FI 5) schedule of food reinforcement requiring only lever pressing.

Tepper & Weiss (1986) also studied sedentary performances under ozone, using a more intricate operant schedule known as stochastic reinforcement of waiting (SRW) (Weiss, 1970) that specifies increasing probabilities of reinforcement as the time between successive responses lengthens. The advantage of the SRW schedule is that the frequency of reinforcement remains constant provided the subject responds on the average of at least once every t seconds, where t is the interresponse time corresponding to a reinforcement probability of 1.0. Consequently, even large variations in response rate do not affect the distribution of reinforcements which appear with a mean interval of t seconds. With $t = 120$ seconds and ozone concentrations between 0.12 and 1.0 ppm, 4 rats demonstrated a substantial lowering of response rate at the 0.5 ppm level, confirming previous data from Weiss et al., (1981).

All of the results described above can be interpreted as showing that the lowered rate of both effortful and sedentary responding arose from a reduced inclination to respond, not a lack of physiologic capacity. Even an act as simple as lapping water from a drinking tube, at least over a short period, declines in frequency during exposure to ozone levels from 0.2 to 0.8 ppm (Umezu et al, 1987). Assuming parallels with the findings from human studies, the decrement in rat performance can be viewed as analogous to human subjective complaints of throat irritation and painful breathing, with the reduction of activity seen as a means of attenuating such effects.

Alternate explanations, however, must be considered. Perhaps the animals were responding less because they were physiologically impaired due to trigeminal and vagal sensory irritation. This hypothesis can be tested with yet another behavioral paradigm common to pharmacology and toxicology—that of escape from aversive stimuli. Wood (1979) devised a situation in which mice could escape from a noxious inhalant (ammonia) by performing the prescribed response of inserting their snouts into a cone to interrupt a light beam which then terminated the ammonia infusion. The flow lasted for 60 seconds if the mouse did not respond. Tepper and Wood (1985) applied the same technique to ozone and found a clear effect of concentration. Significant aversive effects appeared at 0.5 ppm, and 50% of the deliveries were terminated at 1.7 ppm. The declines in running-wheel performance at levels as low as 0.08 ppm occurred progressively during experimental sessions. Sensory irritation would be expected to act immediately.

Perhaps the diminished responding over time was due to other varieties of physiological impairment. In a detailed multivariate analysis of wheel running, Tepper et al (1982) found that the number and duration of the pauses between episodes of running offered the best correlations with ozone concentration. The measures which would be expected to demonstrate a limited capacity to perform (running speed and burst length) changed little at low concentrations.

One unique contribution of behavioral science is the methodology by which subjective complaints can be converted into quantified behavioral measures. Frequent symptoms reported by humans exposed to ozone include nasal irritation, fatigue, generalized depression, increased lethargy or sleep, headaches, substernal pressure, generalized aches, and accelerated cardiac action (Kleinfeld & Giel, 1956; Langerwerff, 1963). The hypothesis that activity reduction is a means for modulating aversive subjective effects seems supported by the above series of experiments demonstrating that low concentrations of ozone (corresponding to current regulatory limits) reduce tendencies to engage in effortful activity. The wheel-running data support the hypothesis that avoidance of the noxious consequences of exercise during exposure to ozone is the source of reduced performance. Even the reduction of sedentary responses (such as lever presses) probably includes some degree of avoidance.

The question that offers a challenge to risk assessment is whether the reduced willingness to perform deserves to be interpreted as toxicity (Weiss, 1989). A reduced inclination to exercise (demonstrated in the laboratory with behavioral techniques) seems to model the observations by Wayne et al. (1967) of human runners in Los Angeles. Similar observations with human subjects in controlled experimental settings were made by Follinsbee et al. (1977), Adams and Schelegle (1983), and Schelegle and Adams (1986). A natural experiment was reported by Spektor et al. (1991) who measured ventilatory function twice daily on 46 healthy children during a 4-week period of summer camp in 1988, during which they also measured 1-hour ozone concentrations (highest was 150 ppb), 12-hour concentrations of acidic aerosols, and humidity. They concluded that ozone exposures in ambient air produce greater lung function deficits than does pure ozone administered in controlled chamber studies because of three factors: "1) longer exposures, 2) potentiation by other factors in the ambient exposure, and 3) the persistence of effects."

Because toxicity is often equated with overt tissue damage, it is not surprising that positive behavioral findings have encouraged a search for the mechanistic correlates. For example, Warren et al (1986) noted increases in total protein recovered from the lungs of rats exposed to ozone concentrations as low as 0.12 ppm. Although such data do not explain the behavioral findings, they do offer another source of support for their validity.

Behavioral measures with human subjects often take the form of neuropsychological tests. Because we will discuss such tests relative to solvent exposure later in this chapter, it seemed reasonable to include the results of neurobehavioral assessments based on exposure to ozone. To our surprise, we could not locate such research within the past 20 years despite earlier speculations that ozone exposure might affect cognitive ability (Glick, 1940; Peters, 1939; Ritchie, 1939). The only study found to address this question (Hore & Gibson, 1968) reported no differences in a multiple choice, group-administered intelligence test given to college students concurrent with 70 minutes of ozone in concentrations of 0.2 to 0.3 ppm. Measures of global intelligence, however, are designed to be resistant to the effects of acute and transient changes and are probably the least appropriate endpoints for such a study. More sensitive and suitable instruments are described later in the chapter.

Ozone's effects on the visual system were studied by Lagerwerff (1963), who documented decreases in visual acuity, changes in stereoscopic vision, and changes in ocular muscle balance in 22 subjects exposed for 3 and 6 hours to levels of ozone similar to that contained in the ozonosphere (16 ppm). Note that such a high level could produce lung damage. Subjective changes reported in this study (lethargy and difficulty concentrating) were similar to those previously reported in cases of ozone poisoning (Kleinfield & Giel, 1956; Griswold et al, 1957). Kleinfield and Giel (1956) state that "the delayed onset of symptoms (fatigability, weakness, exertional dyspnea) has not been sufficiently stressed in the literature."

Given the range of neurobehavioral questions evoked by ozone (particularly at current environmental concentrations) a focused program directed at these questions should be undertaken. It is especially pertinent to the public health issue of what constitutes toxicity. Also needing further study is ozone's potentiating effects as discussed by Spektor (1991). Several studies suggest that ozone serves as a potentiating factor for the immune system (Goldstein et al, 1974; 1979), with neoplastic drugs (Wenzel & Morgan, 1983), with thyroxine (Wong & Hochstein, 1981), and in inducing hyperactivity to histamine (Goldstein, 1978), but the implications of this research have not yet been investigated using neurobehavioral methods to assess cognitive changes and to quantify subjective experience.

BEHAVIORAL TOXICITY OF VOLATILE ORGANIC SOLVENTS

In the Federal Register of March 14, 1991, the U.S. Environmental Protection Agency (U.S. EPA, 1991) proposed a test rule governing 10 organic solvents. The agency noted that members of this class of chemicals posed significant neurotoxic risks that had not been adequately addressed and that additional information about these 10 agents (produced in high volumes) would be required to more fully determine such risks. The Multi-Chemical Test Rule requested more data on the following compounds:

acetone	n-amyl acetate
l-butanol	n-butyl acetate
diethyl ether	2-ethoxyethanol
ethyl acetate	isobutyl alcohol
methyl isobutyl ketone	tetrahydrofuran

EPA asked the manufacturers to supply four kinds of data based on inhalation exposures:

1   The results of a Functional Observation Battery (FOB) based on both acute and subchronic exposures. An FOB is a series of standardized observations (typically on rodents) and includes ratings of how the animal

responds to touch, sound, and other stimuli, of posture and movement, and general appearance.

2   Measurements of motor activity after both acute and subchronic exposure regimens. Locomotor activity is determined in an open field, or a maze in the shape of a figure-8, or in any similar device in which spontaneous movement can be recorded.

3   Neuropathology, consisting of a standard neuropathological assay following subchronic exposure.

4   Schedule-controlled operant behavior after subchronic exposure. Operant behavior is to be used here to answer other questions than those it was used to obtain in the ozone experiments described above.

The rationale for requiring operant behavior data was explained by EPA in these terms:

> Solvents may have neurotoxic effects on memory, learning, and performance which can be permanent. These effects are less well understood . . . The schedule-controlled operant behavior test has . . . typically been required as a second tier test . . . it is proposed as a first-tier test . . . because of EPA's desire to obtain data on the effects of solvents on learning, memory, and performance.

Although the proposed test rule has since deleted some of these requirements for some of the solvents, the rationale for their adoption remains a cogent one. Listed among the origins of the proposed rule is the experience of occupational toxicology. At least acutely, volatile organic solvents act primarily on the central nervous system. They possess potent narcotic properties. Threshold Limit Values (TLVs) for such agents published by the American Conference of Governmental Industrial Hygienists (ACGIH) are largely based on the air concentrations which produce adverse effects on behavioral function. The Short-Term Exposure Limit (STEL), for instance, is defined by the following criteria:

> The maximum concentration to which workers can be exposed for a period up to 15 minutes continuously without suffering from . . . narcosis of sufficient degree to increase accident proneness, impair self-rescue, or materially reduce work efficiency . . .

Experimental work confirms the sensitivity of behavioral measures as indices of acute solvent exposure. Controlled exposures with toluene provide a set of examples based on neurobehavioral endpoints. Neurobehavioral tests for humans encompass a broad range of endpoints, from traditional cognitive neuropsychological tests designed for diagnosis to occupationally related performance tests such as those designed to assay motor dexterity. This section will concentrate on acute exposures during experimental studies, using either traditionally administered neurobehavioral tests, or neuropsychological tests administered with computerized software. Subject selection in such studies is assumed to screen and exclude individuals with high current exposure or a history of such exposure. The tests and study designs are summarized in Table 1.

Generally, the results of studies using traditional neurobehavioral methods depended on exposure concentration rather than the type of test used, with the former TLV of 100 ppm serving as a divide. For example, Ogata et al. (1970) studied 23 male students exposed to 7 total hours of 200 ppm toluene, and found a longer latency in reaction time. Gamberale and Hultengren (1972) examined 12 young adult males exposed to 20 minutes each of escalating doses of toluene (100, 300, 500, and 700 ppm), and a control condition a week later. A significant lengthening of simple reaction time occurred after 300 ppm, in a task which required the subjects to respond to a light by pressing a button with a finger. A more complex task of perceptual speed, which required the subjects to underline numbers from a complex array that matched a sample provided on the top of the sheet, was not affected significantly until the 700 ppm dose. This study was flawed by its very short exposure periods and inadequate clearance time between doses.

Stewart et al. (1975) exposed adult women to 100 ppm of toluene for 7.5 hours finding reduced visual vigilance and tone detection, but there was no significant effect on time estimation, addition, or coordination. Cherry et al. (1983) exposed eight young adult males to 4 hours of toluene at 80 ppm. No effect was found for toluene alone on a battery of four tests (simple and choice reaction time, visual pursuit, and visual search). But when combined with alcohol, performance and mood tended to decline more than for either substance separately. Andersen et al. (1983) exposed 16 young male students for 6 hours to escalating doses of toluene (0, 10, 14, 100 ppm), giving them a battery of eight performance tests at two intervals (early and late). At the 100 ppm concentration, they found a performance decline of borderline significance in only three measures: multiplication errors, Landolt's rings, and the screw-plate test. No decline was seen on complex choice, rotary pursuit, sentence comprehension, word memory, or Bourdon-Wiersma. A significant interaction, however, was reported between exposure and time of test, with performance decrement enhanced later in the 6-hour period. The subject's subjective reports of effort required and self-perceived retarded reaction time were significant at the 100 ppm concentration level.

One report described the application of traditional neurobehavioral methods to a population hypothesized to be at greater risk for toluene sensitivity (i.e., subjects exposed previously to solvents in occupational settings). Baelum et al. (1985) examined 43 male rotogravure printers and 43 matched controls after 6-hour exposures to 100 ppm toluene or a room air control. When exposed to the toluene condition, both groups demonstrated decreased manual dexterity, decreased color discrimination, and decreased accuracy in visual perception, with printers showing a slightly greater sensitivity. As ex-

Table 1 Behavioral Tests Used with Toluene

Investigator	Test Name	Function	Measure
Gamberale and Hultengren, 1972	● Identical Numbers	visual-motor perceptual speed, attention	*time to complete
	● Spokes	visual-motor perceptual speed, coordination	time to complete
	● Simple Reaction Time	eye-hand reaction time	*mean for the last 20 stimul
	● Choice Reaction Time	3 light panel with 3 buttons, complex eye-hand reaction time	reaction time for the last 20 stimuli
Cherry et al. 1983	● Simple Reaction Time	eye-hand response	*reaction time
	● Choice Reaction Time	4-light panel with 4 buttons, speeded repetitive movement	*reaction time
	● Stressalayser	self-paced pursuit tracking task,	*reaction time, movement time, errors
	● Visual Search	matching of grid with target square	response time
Anderson et al. 1983	● Choice Reaction Time	5 random targets touched with metal probe	reaction time, aiming, vigilance
	● Rotary Pursuit	target on turntable, visual coordination	time on target
	● Screw Plate Test	32 screws and bolts, eye-finger coordination, manual dexterity	*time to completion
	● Landolt's Rings	card with 100 circles, 50 in need of closure, repetitive eye-hand coordination	*time to complete
	● Bourdon Wiersma	175 dot figures with 3, 4, or 5 dots	time to complete & errors
	● Multiplication Test	higher cortical function	*time & errors
	● Sentence Comprehension	higher cortical function	time & errors
	● Word Memory	recognition memory and discrimination	number correct, errors, omissions.

Table 1 (*Continued*)

Investigator	Test Name	Function	Measure
Baelum et al., 1985	● Peg Board Test	finger dexterity	*time to complete
	● Screw Plate Test	eye-hand coordination, finger dexterity	time to complete
	● Rotary Pursuit	visual motor coordination	time on target
	● Track Tracing Test	arm-hand steadiness	total time off track
	● Assembly Peg Test	30 minutes of peg placement	number of pegs placed & empty bars
	● Choice Reaction Test	5 choice (as in Anderson) response time	reaction time
	● Landolt's Ring	(as in Anderson)	*time to complete
	● Vigilance Clock	dual task to visually monitor rotating pointer and peripheral flashing light	*percent correct, errors, false reactions
	● Color Discrimination	20 samples matched from 210 choices	*deviations in color, density
	● Multiplication Test	higher cortical function	time & errors
	(Microcomputer Administered)		
Olson et al., 1985	● Simple Reaction Time	visual-motor response	reaction time
	● Memory Reproduction	type on keyboard letters and figures	number of correct
	● Choice Reaction Time	5 key response	reaction times for stimuli, combinations
Iregren et al., 1986	● Choice Reaction Time	(as in Olson)	reaction times for stimuli, combinations
	● Simple Reaction Time	(as in Olson)	reaction time
	● Color-word Vigilance	respond to matching word and color	time & errors
	● Memory Reproduction	(as in Olson)	number of correct series

Table 1 (Continued)

Investigator	Test Name	Function	Measure
Dick et al., 1984	• Visual Vigilance Test (Mackworth Clock Test)	critical signal response	*hits, false alarms
	• Choice Reaction Time	8 choice panel	reaction time
	• Pattern Recognition Test	dot patterns, visual discrimination	hits, false alarms, & latency correct
Echeverria et al., 1991	• Sternberg Memory	verbal short-term memory for letters	number correct, latency
	• Digit Span	verbal short-term memory for numbers	number correct, latency
	• Pattern Memory	visual memory	number correct, latency
	• Benton Visual Memory	visual memory	number correct, latency
	• Pattern Recognition	visual perception	number correct, latency
	• Simple Reaction Time	psychomotor skill	number correct, latency
	• Continuous Performance	psychomotor skill	number correct, latency
	• Symbol Digit Matching	psychomotor skill	number correct, latency
	• Critical Tracking	psychomotor skill	number correct, latency
	• One-hole Test	manual dexterity	number correct, latency
	• Hand-eye Coordination	manual dexterity	number correct, latency
	• Finger Tapping	manual dexterity	number correct, latency

*Indicates significant findings.

pected, even this older literature indicated that the detection of effects appears to be a function of the concentration and duration of exposure to toluene. Functional impairment first becomes noticeable in the form of increased latency to respond, with the suggestion that motor tasks may be especially sensitive to toluene.

The power of neurobehavioral testing to detect subtle performance changes has been enriched by the advent of computerized assessment methods, The advantages of computer-administered neuropsychological testing include the reproducibility of testing conditions, ease of data handling, and reduction of variation due to administrator error or systematic bias. Perhaps most intriguing is the ability to conduct detailed analyses of performance on a specific test. A test outcome, for example, is no longer restricted to one index such as the number of correct responses. It may also include the latency to emit both correct and incorrect responses, the change in performance over time, the rate (on self-paced tests) of responding, and other dimensions of performance. Such a richness of possibilities offers many perspectives from which to view the targeted behavior.

If the aim of neuropsychological testing is to assay performance, the subject should be adequately trained (usually accomplished with two training cycles) to eliminate the potentially misleading effects of practice. Sufficient practice should eliminate the possible confounding of performance with increased familiarity with task requirements. Most computerized test programs accommodate this need by making available multiple sets of randomly generated stimuli for each repetition of the test. A well-practiced subject can then yield a dependent measure of learning based on a novel set of stimuli within a familiar test requirement.

Olson et al. (1985) administered a computerized test battery to 16 men exposed for four hours to the Swedish occupational standard for toluene (80 ppm), xylene, the combination of toluene and xylene, and a control condition. No effect was found on computer-administered tests of simple and choice reaction time, and short-term memory. Iregren et al. (1986) also studied possible synergism between toluene and alcohol, exposing 12 men for 4.5 hours to 80 ppm toluene, ethanol, a combination of ethanol and toluene, and a control condition. Neither toluene alone nor in combination with ethanol produced reliable effects on tests of simple and choice reaction time, color-word vigilance, or memory reproduction. In a larger study than most, Dick et al. (1984), at NIOSH, divided 144 young adults (including 47 females) into eight groups, two of which received 4 hours of 100 ppm toluene or placebo. Results for this group indicated a significant effect on the accuracy of visual vigilance, with a nonsignificant trend toward performance decrement on pattern recognition, and no effect on choice reaction time. Note that although the TLVs of each country are set for a standard 8-hour exposure, testing for behavioral consequences of such exposure terminated at 4 hours. This may have been too brief a duration. Evidence from the traditional neurobehavioral tests (cited above) indicates an interaction with time and a trend toward increased impairment later in the exposure period.

Echeverria et al. (1991) conducted the most extensive study, examining 42 adult students to escalating doses of toluene (0, 75, and 150) for seven hours over three days. They then studied the same group using the same procedures but with escalating doses of alcohol (0, 0.33, and 0.66 g/kg body weight). Tests of seven performance functions were administered by computer: verbal skill, verbal short-term memory, visual memory, visual pattern recognition, and psychomotor skill as measured by a test which requires the matching of symbols to numbers. Tests aimed at manual dexterity, mood, and fatigue were also administered. The results indicated that response latency for pattern recognition was the most sensitive endpoint for both substances, while accuracy was affected only by toluene. Significant results for toluene exposure were also obtained for a test of manual dexterity and the symbol-digit matching test, although the authors recommend caution in interpreting the latter result.

Despite a variety of neurobehavioral studies conducted with humans exposed acutely to toluene, relatively few unambiguous conclusions emerge from the data. More general conclusions are hampered by the absence of a common protocol for exposure. The published studies have been conducted at various concentrations, with various durations of exposure, and with the exploratory use of tests which often are inappropriate to capture the effects of acute intoxication. Nonetheless, the past 25 years of neurobehavioral assays of toluene exposure has yielded some degree of consistency. Subjective complaints of headache, eye, and upper throat irritation are common. Several studies also suggest decreased performance efficiency based on measures of response latency, especially if visual (not linguistic) material is presented.

Like ozone, many opportunities exist for further exploration of research issues based on experimental procedures that are more analogous to typical human exposures. Even light work (50 W) performed during exposure, for example, can affect arterial blood concentration, raising it by as much as 100% (Arlien-Soberg, 1992). Yet, only one neurobehavioral study (Hyden et al. 1983) included light exercise. It exposed 15 workers for 70 minutes to 103–148 ppm toluene, then subjected them to a vestibulo-oculomotor test battery which was able to detect vestibular nystagmus. No experimental studies of even short-term subchronic toluene exposure has been reported, although toluene has been shown to accumulate in blood and subcutaneous fat if insufficient clearance time is experienced between exposures (Arlien-Soberg, 1992). Nor has the question of duration of effects been addressed; no tests have been performed following exposure to determine how long it takes to return to baseline. A few studies have attempted to address the interaction of toluene with other typical toxic human exposures, including alcohol ingestion, and previous occupational solvent exposure. But more work is necessary and might also include groups at risk from smoking or medications, particularly benzodiazapines (Arlien-Soberg, 1992).

Most importantly, the establishment of consistent standards of administration for computerized measures would allow greater confidence in the multiple endpoints (e.g., latency as well as accuracy) of a computerized neurobehavioral battery. With any assay, test selection is an important issue. When testing for behavioral (functional) impairment, test selection should be based on the known pharmacokinetics of the specific substances. For example, toluene in the rat brain is found at the highest concentrations in the brainstem and midbrain (Arlien-Soberg, 1992), so that tests of complex performance (based theoretically on intact cortical function) may be less likely to demonstrate an effect for toluene than for tests based on fine motor control which is largely controlled by the basal ganglia. It should not be a reflection on the test method if no effect is found, but perhaps it is a reflection of our limited appreciation of functional measures at this stage of test selection. Yet it is precisely for this reason that neurobehavioral methods, as they gain in power and achieve wider application, can be of great utility in exposing differential effects among toxicants.

Beginning in the early 1970s, another facet of solvent toxicity began to appear in the literature. Based mostly upon studies by Scandinavian investigators, the argument was made that chronic workplace exposure to volatile organic solvents caused enduring central nervous system dysfunction that could be exposed by psychological tests. Although exposure to high solvent levels, such as those experienced by abusers (glue-sniffers) could produce visible neuropathology, many observers derided these publications, pointing out the relatively small numbers of subjects in some of them, the absence of precise exposure data, the lack of adequate documentation about smoking and alcohol consumption, and other flaws from which retrospective epidemiological studies typically suffer. They also noted the generous compensation policies of Scandinavian governments for alleged victims of chronic solvent exposure. One of the widely cited critics acknowledged that a strong case could be made for carbon disulfide, recognized as a neurotoxicant since the 19th century, but not for the others (Grasso, 1984).

The earlier claims, however, found support in later, more stringent studies. The more recent studies have also uncovered additional signs of neurotoxicity. Mergler et al. (1987) discovered deficits in color vision in exposed workers and Schwartz et al. (1989) determined that workers in paint factories in which exposure monitoring indicated consistently precise control of workplace solvent concentrations displayed reductions in the ability to discriminate odors. The markers of chronic solvent toxicity are listed in Table 2. The Organic Solvent Syndrome (as it is sometimes called) markedly influenced the development of neurobehavioral toxicology because the challenges posed by critics of the earlier work led subsequent investigators to respond with

Table 2   Chronic Neurotoxic Effects of Solvents

Cognitive Variables	Motor Variables
• Intelligence	• Coordination
• Memory	• Response Speed Sensory
• Vigilance	• Color Vision
• Acquisition	• Odor Discrimination Personality
• Coding	• Mood Changes
• Concept Shifting	
• Spatial Relations	
• Categorization	

improved procedures and techniques. They also forced investigators to recognize that different solvents may induce different patterns of neurotoxicity, a principle that was not explicitly recognized at the beginning, in part because most workplaces contained solvent mixtures.

Developmental neurotoxicity is the engine that drives current exposure standards for substances such as methylmercury and lead. The Fetal Alcohol Syndrome has evoked concerns about the possible developmental effects of another alcohol—methanol—because of the role that it might be accorded as a replacement for gasoline. Widespread exposure to methanol vapor would result, and although the estimated ambient levels would tend to be low, the rudimentary level of knowledge now possessed about its actions on brain development has stirred efforts to expand the amount of information available. No such systematic program is visible for solvents, even though many women are employed in settings that expose them to such agents and despite some suggestive findings (Lipscomb et al. 1991). The literature remains scattered and relatively uninformative. Most of the animal studies are conducted with high doses and the epidemiologic data are sparse, although suggestive.

INHALATION AS A SOURCE OF EXPOSURE TO NEUROTOXIC METALS

Mercury, manganese, and lead are three metals of popular concern because of their neurotoxic properties. Inhalation is the only significant source of exposure in the workplace. For elemental mercury, inhalation is the predominant exposure medium in the communal environment as well, but even for lead (despite the emphasis on ingestion as the exposure source) inhalation is acknowledged as a major contributor to the body burden. Manganese is an essential dietary element, but its health risks have lately been posed in terms of neurotoxicity by the prospect of manganese fuel additives. The intersection of inhalation and neurobehavioral toxicity is recognized so implicitly for these metals that it hardly emerges as an issue for risk assessment. But those unaware of the intersections are prone to errors of interpretation.

Mercury

Mercury vapor, the form in which the metal is encountered in the workplace and often in the home, is currently the subject of intense public attention because of claims that amalgam tooth fillings, which contain about 50% mercury, are the origin of a vast galaxy of adverse health effects. Some of these claims are based on crude measures of mercury vapor in the mouth during chewing. Elemental mercury is clearly released in this way, but the science of inhalation toxicology tends to be disregarded in these assertions. The variables that inhalation toxicologists would deem important, such as the amount actually breathed in, its absorption via the lung, and its contribution to the target organs (brain and kidney) are spurned by those who advocate wholesale removal of amalgam fillings. Clarkson et al. (1988), on the basis of both measurements and calculations, demonstrated that only a slight proportion of the mercury vapor released by chewing reaches the lungs. The implications of such findings seem to escape even those advocates of amalgam removal with training in science.

Another source of skepticism about the validity of the amalgam allegations is the absence of any concrete health findings. Mercury vapor is primarily a central nervous system toxicant. Its dominant neurological expression is tremor and even in the absence of clinically detectable abnormalities in tremor, analytical techniques based upon Fourier analysis (reviewed in Newland, 1988) reveal shifts in the frequency spectrum of individuals exposed to relatively low ambient levels in the workplace. Wood et al. (1973) observed that tremor measurements in patients exposed to high levels of vapor showed two modes in a plot of frequency versus power. As blood concentrations decreased, the higher frequency mode gradually diminished in amplitude. Langolf et al. (1978) suggested that such an index might be used for decisions about reassignment of chlor-alkali workers whose duties expose them to mercury vapor in other work environments.

No tremor abnormalities have been detected in persons with many amalgam restorations. Nor were Swedish investigators able to detect any differences between individuals with many and individuals with few amalgam restorations on a lengthy inventory of subjective complaints (Ahlqwist et al. 1988). A specific collection of behavioral abnormalities known as erethism is characteristic of elemental mercury intoxication. It includes hyperirritability, labile temperament, excessive timidity, and depression. No measures of erethism, distinguishing subjects with many amalgam fillings from those with few such fillings, has appeared. Nor have any data been presented to support such differences on the basis of cognitive and performance tests which the published literature shows can discriminate between workers exposed to mercury and controls. Rejection or ignorance of the principles of both inhalation and behavioral toxicology seems to account for much of the current anti-amalgam movement.

Manganese

Manganese has claimed miners as its dominant victims. Inhalation of manganese dioxide dust is the exposure source. Although respiratory system damage has sometimes been reported, the central nervous system is the primary target organ. Manganese intoxication is characterized by two sets of signs. One is neurological. Clinically affected persons display disorders of posture and movement described by neurologists as dystonias, although the syndrome may share some commonalities with Parkinson's disease (Barbeau, 1984). The other is psychological. Victims complain of weakness and somnolence and are subject to episodes of abnormal laughing and weeping. At lower levels of exposure, seen in ore-crushing plants or in ferro-manganese processing plants, the manifestations are more subtle, but still detectable with appropriate psychological tests. For example, Roels et al. (1987) found higher levels of fatigue and irritability and a greater amplitude of finger tremor in those ore-crushing workers with the higher blood levels of manganese. In a later study (Roels et al. 1992), these investigators calculated a measure of total working lifetime exposure to respirable Mn dust and demonstrated a dose-response relationship with several indices of perceptual-motor performance.

Particle size and deposition—fundamental measures to inhalation toxicologists—have not been measured carefully in studies of manganese toxicity. Manganese chloride aerosols, however, yield some insights into the potential health risks of manganese additives to gasoline. Under such circumstances, combustion will expel Mn_3O_4, which is the form believed to exercise neurotoxicity. Experiments on the pharmacokinetics of inhaled manganese aerosols illustrate the potential environmental problems. Newland et al. (1987) exposed monkeys to radiolabeled $MnCl_2$. They discovered that the manganese lodged in the lung, providing a pool of slowly released manganese that eventually penetrated the brain. Calculations showed a half-life in the brain of about one year. Such data suggest that even small concentrations of atmospheric manganese (emitted by the combustion of treated gasoline) might accumulate in the lung, from where the manganese would be slowly and continuously released into the blood and stored in the brain. Further support for such a hypothesis was provided by Newland et al. (1989). They chronically exposed monkeys to a $MnCl_2$ aerosol and observed—by magnetic resonance imaging (MRI)—that manganese accumulated in the basal ganglia, which are brain structures critical for movement. MRI proved a useful tracer because manganese is paramagnetic. A succession of images indicated that manganese moves out of brain structures at a rather indolent pace, even when the manganese is administered intravenously. These experimental data suggest that even relatively modest environmental exposures to airborne manganese might accumulate to levels great enough to be discerned by neurobehavioral testing.

Lead

Regulations to remove lead pigments from interior paints (put in place 50 years ago) undoubtedly reduced the incidence of lead poisoning in children. However, many other exposure sources remain. At least 40 million households still retain lead pigment surfaces. Dust in urban areas, often ingested by children who explore the environment with their mouths, contains particles of airborne lead emitted by the combustion of leaded gasoline. Drinking water carried by lead pipe and soldered pipe joints also carries lead into the

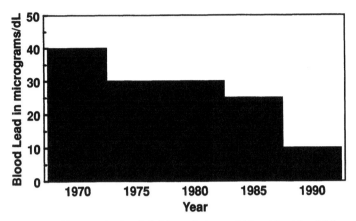

Figure 2 Changes in the definition of elevated blood lead in children by the Centers for Disease Control.

home. At the same time, inhalation remains a significant route of exposure. The U.S. Environmental Protection Agency showed a dramatic correlation between the fall in the amount of lead added to gasoline between the 1970s and 1980s and the fall in blood lead concentrations measured in the National Health and Nutrition Survey, NHANES II (EPA, 1986).

One reason for continuing to regard inhalation as a significant exposure medium for the developing brain are results such as those of Bellinger et al. (1987). This study compared three groups of children distinguished on the basis of cord blood levels at birth—low (mean of 1.8 μg/dl), medium (mean of 6.5μg/dl), and high (mean of 14.5 μg/dl). Tests of psychological development demonstrated a widening gulf between the scores achieved by high lead and the other two groups that amounted to about 8 percent at age 24 months. This is a disturbing result because it produces a sharp reduction in the proportion of children attaining superior scores and a concomitant rise in the proportion that fall into the retarded category. These results, when combined with numerous investigations at sites around the world, induced the Centers for Disease Control to once more reduce the blood lead levels it deems to pose a health risk as plotted in Fig. 2.

Inhalation is suspected as a significant exposure source in these studies because of the nature of the populations surveyed. The sample chosen by Bellinger et al. (1987) came from an upper middle class population that did not inhabit decrepit housing. Also, many of the studies abroad had not drawn from samples exposed to lead-based paints. EPA proposed a National Ambient Air Quality Standard (NAAQS) of 1.5 μg/m³ in 1978. At this level, the proportion of the lead body burden in children two years of age contributed directly by inhalation was estimated by EPA (1986) as 18%. The proportion contributed by ingested dust came to 56%.

Inhalation predominates in the workplace setting. The Permissible Exposure Level (PEL) for lead set by the Occupational Safety and Health Administration (OSHA) is now 40 $\mu g/m^3$. The associated blood lead maximum is 50 $\mu g/dl$. It strikes many observers (based on neurobehavioral testing) that these standards are excessively high. Stollery et al. (1991), Baker et al. (1985) and others have noted impaired cognitive performance, even in workers exhibiting significantly lower blood levels than 50 $\mu g/dl$. At least equally contentious for workplace environments is the exposure of women during their fertile years. If prenatal exposure levels greater than 10 $\mu g/dl$ are deemed as hazardous to fetal brain development, then the current PEL would have to be reduced sharply.

CHEMICAL SENSITIVITY SYNDROMES
Background

Today a new and challenging issue is facing the chemical industry, its regulators, the health care system, and even real estate businesses and transportation. It is claimed that many individuals suffer adverse health consequences because of an acquired sensitivity to low environmental levels of chemicals. Syndromes most closely identified with such conditions are Multiple Chemical Sensitivity Syndrome (MCS) and Sick Building Syndrome (SBS). An allied syndrome that is closely matched in its array of manifestations is the Chronic Fatigue Syndrome (CFS). A newer version has recently appeared in the form of health complaints by veterans of the Gulf War and has been called the Gulf War Syndrome (GWS).

The bulk of the adverse health effects associated with these syndromes emerge as subjective complaints. Most are referable to the central nervous system. Although the immune-system correlates have been investigated assiduously, only sporadic (but no compelling, replicable evidence for direct immunological dysfunction) has been revealed. Especially for MCS and SDS, odors have been often implicated as triggers for the emergence of symptoms. For most agents, odor thresholds lie far below the concentrations believed to evoke conventional neurotoxicity. This discrepancy is one of the sources of doubt about the validity of these syndromes.

The chemical industry and its customers (e.g., carpet manufacturers) are particularly concerned about these assertions as a result of a torrent of litigation. Most of the current personal injury and product liability cases are based on indoor air quality and involve MCS claims. Many are worker compensation cases. In addition, patient support groups have been formed for all four syndromes to press for recognition and research. A report by a committee of the National Academy of Sciences estimated a prevalence of 15%.

Air Freshener - Mothballs - Newsprint - Pesticides - Carpeting - Bug Spray - Fabric Softener - Tobacco Smoke - Particle Board - Paint & Varnish - Contaminated Water - Vinyl Shower Curtain - Chlorinated Cleanser - Perfume & Cologne - Woodsmoke - Detergents - Copy Machines - Disinfectants - Laser Printer - Landfill Emissions - Permanant Press Clothing - Carbonless Paper - Adhesives - Aftershave - Dry Cleaning - Mildewcide - Correction Fluid - Potpourri - Stainproofing - Felt Tip Marker - Nail Polish and Remover - Kerosene Heater Fuel - Veneered Wood - Plastic - Auto & Diesel Exhaust - Smokestack Emissions

Figure 3 Exposure sources asserted to induce reactions in patients claiming to suffer from Multiple Chemical Sensitivity.

Clinical Features

The most salient symptoms of MCS (the prototype) are dominated by central nervous system dysfunction. Included are lethargy, persistent fatigue, memory impairment, difficulty in concentrating, dizziness, lightheadedness, depression, nervousness, irritability, and disorientation. Respiratory and skin disorders and symptoms reminiscent of flu and allergic responses are sometimes reported, but far less frequently. MCS patients claim that these responses are triggered by a wide variety of chemicals, typically by inhalation exposure. Fig. 3 lists some of the sources of problems identified by MCS patients.

Commonly, patients report that the range of chemicals to which they are sensitive has progressively widened. Sometimes an initiating event (such as a chemical spill or the remodeling of a building) can be designated. Many patients report responsiveness to ambient levels significantly below those currently accepted as free of adverse effects—a feature of the syndrome which has contributed substantially to suspicion on the part of the biomedical community. Many physicians view MCS patients as malingerers or as suffering primarily from a psychiatric disorder. The bulk of the medical community regards clinical ecologists (a group of practitioners who specialize in MCS) with overwhelming distrust.

SBS differs from MCS in its presumed environmental triggers. It is typically aroused in a particular setting, most often the workplace. Frequently, the induction of symptoms is associated with a particular change in environmental conditions. For example, building renovation which is accompanied by painting, new carpet installation, and the consequent dispersal of volatile products by the ventilation system.

The Gulf War Syndrome is the subject of several current investigations. It has been linked hypothetically to low levels of poison gas, biological weapons, the oil well fires that spewed huge quantities of pollutants into the area, pesticides and repellants, insect vectors, and to a variety of other causes. CFS is most often ascribed to some postinfection response rather than chemical exposure, but no cogent immune system explanation has been forthcoming. Recognized hypersensitivity phenomena such as IgE-mediated reactions have not been detected with enough regularity to offer a convincing immune system explanation.

Characterization of Populations

Medical diagnoses rely on case definitions. A firmly entrenched case definition for MCS is lacking. One definition to which increasing numbers of investigators are beginning to subscribe was postulated by Cullen (1987) and appears in Table 3. Because of the lack of a consistent etiology, and the overwhelming preponderance of nervous system symptoms, the primary approach to MCS characterization is neuropsychological testing. Several studies have now uncovered differences between MCS, CFS, and control population samples. For example, one recent publication (Fiedler et al., 1992) reported a significant reduction in performance by MCS patients on a test of recognition memory. Similar data showing cognitive and memory impairment are beginning to appear.

Mechanistic Explanations

Because exposure is usually difficult to verify, and because odors are said by patients to play a major role in triggering reactions, the olfactory pathways have assumed a leading explanatory role. Some observers have pointed out that the olfactory receptors are linked directly to neural circuits connecting to the phylogenetically ancient structures of the brain known as the limbic system. This system is pivotal in the expression of emotional and sexual behavior. The hippocampus, which is crucial for short-term memory function, is a also limbic system structure.

On the basis of these neural connections, one set of hypotheses proposes that particular olfactory stimuli produce a nervous system response known

Table 3 Clinical Definition of MCS (Cullen, 1987)

- Initial symptoms acquired in relation to an identifiable environmental exposure
- Symptoms involve more than one organ system
- Symptoms recur and abate in response to predictable stimuli
- Symptoms are elicited by low-level exposures to chemicals or diverse chemical classes
- No standard test of organ system function can explain symptoms

as "kindling." It refers to a phenomenon in which repeated episodes of low-level stimulation (usually electrical) eventually produce a persistent hyperresponsiveness to stimulation. For example, low-amplitude electrical stimuli which originally evoked barely detectable responses may come to evoke massive electrical discharges and even epileptic-type convulsions.

Kindling is observed in most mammalian species. The anterior neocortex and the limbic system are among the most responsive and sensitive sites. The phenomenon is not, however, confined to electrical stimulation. Once convulsive episodes are induced, the response can generalize to other types of convulsants. For example, kindling can be induced by repeated stimulation of muscarinic cholinergic receptors by small doses of cholinergic agonists applied locally. Low-level chemical exposures by analogy can be argued to induce a similar state of hyperreactivity if the agents reach the brain.

A related phenomenon has been termed "time-dependent sensitization." A prototypical situation is one in which widely spaced intermittent drug administrations (or even a single administration weeks earlier) induce a nervous system state in which subsequent drug administrations elicit responses of much greater amplitude than that of the original response. Kindling is also often more readily induced by spacing successive stimulations by 1 or 2 days. Moreover, the "sensitized" subject may respond to a much wider array of drug classes than that originally administered. Even a stress experience (such as restraint) has been shown in animal experiments to sensitize subjects to subsequent drug injections. Finally, in another parallel with the reports of MCS patients, the spectrum of responsiveness seems to broaden with time.

A third explanation relies on learning. It asserts that the response evoked by an odor or odors represents a form of conditioning in which a previously neutral stimulus (by temporal association with one that elicits a physiological response) becomes able to elicit that response by itself. This form of learning is known as Classical or Pavlovian conditioning. For example, if a specific odor has been associated with an event (perhaps a chemical spill) that elicits aversive physiological reactions, the odor itself in the future might be sufficient to induce the same reaction. One form of such conditioning occurs with the phenomenon termed Conditioned Taste Aversion. Rats greatly prefer saccharine solutions to unflavored water. If, however, consumption of a saccharine solution is followed several hours later by administration of an event that produces illness (such as cyclophosphamide injection) subsequent preference trials will show a marked reduction in saccharine consumption and a rise in plain-water consumption.

One explanatory mechanism that has not yet received much attention invokes the immune system. The past two decades have uncovered a surprising degree of intimacy between the immune and nervous systems. In fact, the evidence is compelling that immune responses can be conditioned or learned.

The discipline of what is called psychoneuroimmunology grew out of this body of research. As a result, it is possible to conceive of a process in which a stressful or aversive experience that provokes an immune system response (perhaps mediated by neuroendocrine mechanisms) is paired with what was initially a neutral olfactory or other sensory stimulus. This previously neutral stimulus is then able to arouse the stress response. In turn, the stress response, which is neuroendocrine in character, generates a response by immune mechanisms which then elicit a variety of reactions translated by a subject or patient into the subjective complaints seen in the chemical sensitivity syndromes. Cytokines released by the immune system (such as interleukins) are clearly able to produce drastic behavioral responses. The tentative explanatory status of such a mechanism is due to the failure to detect unequivocal evidence of immune system dysfunction in these syndromes.

SUMMARY

Inhalation is the dominant route of exposure for many toxic materials whose primary actions influence central nervous system function. Volatile organic solvents such as toulene modify performance on a broad array of neuropsychological endpoints such as motor coordination, sensory acuity, reaction time, vigilance, mood, and complex cognitive discriminations. Neurotoxic metals also gain entrance to the central nervous system via inhalation. Metallic mercury and manganese provide the two most cogent examples, but airborne lead as a product of gasoline combustion was also a problem before its removal from gasoline. As with solvents, the most sensitive endpoints consists of neurobehavioral tests.

Agents such as ozone (classified primarily as a pulmonary toxicant) also exert effects on behavioral function or induce responses that can be measured behaviorally. One example is willingness to exercise, but other neurobehavioral endpoints have been explored only sporadically and ineffectively.

A new array of questions is now emerging that involves inhalation as the major source of exposure. They stem from chemical sensitivity syndromes such as Multiple Chemical Sensitivity (MCS) and Sick Building Syndrome (SBS). The emissions linked most closely to these syndromes by patients are odors and irritants, the predominant complaints are neurobehavioral in character, and the mechanisms proposed to account for them are also predominantly neurobehavioral.

ACKNOWLEDGMENTS

Preparation of this chapter was supported in part by grant ES-01247 from NIEHS, grant NAGW-2356 from NASA, and a fellowship to Alice Rahill from the Center for Indoor Air Research.

REFERENCES

Adams, WC, Schelegle, ES: Ozone and high ventilation effects on pulmonary function and endurance performance. J Appl Physiol (Respirat Environ Exercise Physiol) 55:805–812, 1983.

Ader, R., Cohen, N: CNS-immune system interactions: Conditioning phenomena. Behav Brain Sci 8:379–426, 1985.

Ahlqwist, M, Bengtsson C, Furunes, B, Hollender, L, Lapidus, L: Number of amalgam tooth fillings in relation to subjectively experienced symptoms in a study of Swedish women. Comm Dentist Oral Epidemiol 16:227–231, 1988.

Andersen, I, Lundquist, GR, Molhave, L, Pedersen, OF, Proctor, DF, Vaeth, M, Won, DP: Human response to controlled levels of toluene in six-hour exposures. Toxicol Appl Pharmacol 78:404, 1985.

Arlien-Soborg, P: Solvent Neurotoxicity. Boca Raton: CRC Press, 1992.

Ashford, NA, Miller, CS: Chemical Exposures: Low Levels and High Stakes. New York: Van Nostrand Reinhold, 1991.

Baelum, J, Andersen, I, Lundquist, GR, Molhave, L, Pedersen, OF, Vaeth, M, Wyon, DP: Response of solvent-exposed printers and unexposed controls to six-hour toluene exposure. Scand J Work Environ Hlth 11:271–280, 1985.

Baker, EL, White, RF, Pothier, IJ: Occupational lead neurotoxicity: improvement in behavioral effects after reduction of exposure. Br J Indust Med 42:507–516, 1985.

Barbeau, A: Manganese and extrapyramidal disorders. Neurotoxicology 5:13–36, 1984.

Bardana, EJ, Montanaro, A, O'Hollaran, MT: Building-related illness. Clin Rev Allergy 6:61–89, 1988.

Bell, IR, Schwartz, GE, Peterson, JM, Amend, D, Stini, WA: Possible time-dependent sensitization to xenobiotics: self-reported illness from chemical odors, foods, and opiate drugs in an older population. Arch Environ Hlth 48:315–327, 1993.

Bellinger, D, Leviton, A, Waternaux, C, Needleman, H, Rabinowitz, M: Longitudinal analyses of prenatal and postnatal lead exposure and early cognitive development. New Engl J Med 316:1037–1043, 1987.

Cherry, N, Johnston, JD, Venables, H, Waldron, HA: The effects of toluene and alcohol on psychomotor performance. Ergonomics 26:1081–1087, 1983.

Clarkson, TW, Friberg, L, Hursh, JB, Nylander, M: The prediction of intake of mercury vapor from amalgams. In: Clarkson, TW, Friberg, L, Nordberg, GG, Sager, PR, editors. Biological Monitoring of Toxic Metals. New York: Plenum Press: 247–264, 1988.

Council on Scientific Affairs, AMA: Clinical ecology. JAMA 268:3465–3467, 1992.

Cullen, MR: The worker with multiple chemical sensitivities: an overview. In: Occupational Medicine: State of the Art Reviews. Edited by MR Cullen. Philadelphia: Hanley and Belfus, 1987:655–662.

Dick, RB: Short duration exposures to organic solvents: The relationship between neurobehavioral test results and other indicators. Neurotoxicol Teratol 10:39–50, 1988.

Echeverria, D, Fine, L, Langolf, G, Schork, T, Sampaio, C: Acute behavioral comparisons of toluene and ethanol in human subjects. Br J of Indust Med 48:750–761, 1991.

Fiedler, N, Maccia, C, Kipen, H: Evaluation of chemically sensitive patients. J Occup Med 34:529–538, 1992.

Folinsbee, LJ, Drinkwater, BL, Bedi JF, Horvath, SM: The influence of exercise on the pulmonary function changes due to exposure to low concentrations of ozone. In: Environmental Stress. Edited by LJ Folinsbee, JA Wagner, JF Borgia, BL Drinkwater, JA Gliner, JF Bedi. New York: Academic Press, 1978:125–145.

Follinsbee, LJ, Silverman, F, Shepard, RJ: Decrease of maximum work performance following ozone exposure. J Appl Physiol (Respirat Environ Exercise Physiol) 42:531–536, 1977.

Gamberale, F, Hultengren M: Toluene exposure II. Psychophysiological functions. Scand J Work Environ Hlth 9:131–139, 1972.

Gilbert, ME: A characterization of chemical kindling with the pesticide endosulfan. Neurotoxicol Teratol 14:151–158, 1992.

Glick, HN: Hurricane intelligence. Science 91:450, 1940.

Goldstein, BD, Lai, LY, Cuzzi-Spada, R: Potentiation of complement-dependent membrane damage by ozone. Arch Environ Hlth 28:40–42, 1974.

Grasso, P, Sharratt, M, Davies, DM, Irvine, D: Neurophysiological and psychological disorders and occupational exposure to organic solvents. Food Chem Toxicol 22:819–852, 1984.

Hodgson, MJ, Frohliger, J, Permar, E, Tidwell, C, Traven, C, Olenchuk, S, Karpf, M: Symptoms and microenvironmental measures in non-problem buildings. J Occup Med 33:527–533, 1991.

Hore, T, Gibson, DE: Ozone exposure and intelligence tests. Arch Environ Hlth 17:77–79, 1968.

Hyden, D, Larsby, B, Andersson, H, Odkvist, LM, Liedgren SRC, Tham, R: Impairment of visuovestibular interaction in humans exposed to toluene. Otol Rhinol Laryngol 45:262–269, 1983.

Iregren, A, Akerstedt, T, Olson, BA, Gamberale, F: Experimental exposure to toluene in combination with ethanol intake. Scand J Work Environ Hlth 12:128–136, 1986.

Kalivas, PW, Richardson-Carlson, R, Van Orden, G: Cross-sensitization between food shock stress and enkephalin-induced motor activity. Biol Psychiat 21:939–950, 1986.

Kleinfeld, M, Giel, CP: Clinical manifestations of ozone poisoning: Report of a new source of exposure. AM J Med Sci 231:638–643, 1956.

Koningsberg, AS, Bachman, CH: Ozonized atmosphere and gross motor activity of rats. Int J Biometeor 14:261– , 1970.

Lagerwerff, JM: Prolonged ozone inhalation and its effects on visual parameters. Aerospace Med 34:479–486, 1963.

Langolf, GD, Chaffin, DB, Henderson, R, Whittle, HP: Evaluation of workers exposed to elemental mercury using quantitative tests of tremor and neuromuscular functions. Am Indust Hyg Assoc J 39:976–984, 1978.

Leach, LJ, Spiegl, CJ, Wilson, RH, Sylvester, GE, and Lauterbach, KE: A multiple chamber exposure unit designed for chronic inhalation studies. Am Ind Hyg Assoc J 20:13–22, 1959.

McDonnell, WF, Horstman, DH, Hazucka, MJ, Seal, E, Haak, ED, Salaam, SA, House, DE: Pulmonary effects of ozone exposure during exercise: Dose-response characteristics. J Appl Physiol (Respir Environ Exercise Physiol) 54:1345–1352, 1983.

Mergler, D, Blain, L: Assessing color vision loss among solvent-exposed workers. Am J Indust Med 12:195–203, 1987.

Murphy, SD, Ulrich, CE, Frankowitz, SH, Xintaras, C: Altered function in animals inhaling low concentrations of ozone and nitrogen dioxide. Am Indust Hyg Assoc J 25:246–253, 1964.

National Research Council: Multiple Chemical Sensitivities: Addendum to Biologic Markers in Immunotoxicity. Washington, DC: National Academy Press, 1992.

Newland, MC: Quantification of motor function in toxicology. Toxicol Lett 43:295–319, 1988.

Newland, MC, Ceckler, TL, Kordower, JH, Weiss, B: Visualizing manganese in the primate basal ganglia with magnetic resonance imaging. Exper Neurol 106:251–258, 1989.

Newland, MC, Cox, C, Hamada, R, Oberdorster, G, Weiss, B: The clearance of manganese chloride in the primate. Fund Appl Toxicol 9:314–328, 1987.

Ogata, M, Tomokuni, K, Takatsuka, Y: Urinary excretion of hippuric acid and m- or p-methylhippuric acid in the urine of persons exposed to vapours of toluene and m- or p-xylene as a test of exposure. Br J Indust Med 27:43–50, 1970.

Olson, BA, Gamberale, F, Iregren, A: Coexposure to toluene and p-xylene in man: central nervous functions. British J Ind Med 42:117–122, 1985.

Peters, CA: Ozone in the '38 hurricane. Science 90:491, 1939.

Premack, D, Schaeffer, RW: Distributional properties of operant-level locomotion in the rat. J Exp Anal Behav 5:89– , 1962.

Ritchie, CA: Ozone in the '38 hurricane. Cited by Peters. Science 90:491, 1939.

Roels, HA, Lauwerys, R, Buchet, JP, Genet, P, Sarhan, MJ, Hanotiau, I, deFays, M, Bernard, A, Stanescu, D: Epidemiological survey among workers exposed to manganese: Effects on lung, central nervous system, and some biological indices. Am J Indust Med 11:307–327, 1987.

Roels, HA, Ghyselen, P, Buchet, JP, Ceulemans, E, Lauwerys, RR: Assessment of the permissible exposure level to manganese in workers exposed to manganese dioxide dust. Br J Indust Med 49:25–34, 1992.

Schelegle ES, Adams, WC: Reduced exercise time in competitive simulations consequent to low level ozone exposure. Med Sci Sports Exercise 18:408–414, 1986.

Schwartz, BS, Doty RL, Monroe, C, Frye, R, Barker, S: Olfactory function in chemical workers exposed to acrylate and methacrylate vapors. Am J Publ Hlth 79:613–619, 1989.

Spektor, DM, Thurston, GD, Mao, J, He, D, Hayes, C, Lippman, M: Effects of single- and multiday ozone exposures on respiratory function in active normal children. Environ Res 55:107–122. 1991.

Stewart, J, Badiani, A: Tolerance and sensitization to the behavioral effects of drugs. Behav Pharmacol 4:289–312, 1993.

Stewart, RD, Hake, CL, Forster, HV, Labrum, AJ, Peterson, JE, Wu, A: Toluene: Development of a biologic standard for the industrial worker by breath analysis, NIOSH Contract Report No. 99-72-84. Cincinnati: National Institute for Occupational Safety and Health, 1975.

Stollery, BT, Broadbent, DE, Hanks, HA, Lee, WR: Short term prospective study of cognitive functioning in lead workers. Br J Indust Med 48:739–749, 1991.

Tepper, JL, Weiss, B, Cox, C: Microanalysis of ozone depression of motor activity. Toxicol Appl Pharmacol 64:317–326, 1982.

Tepper, JL, Weiss, B, Wood, RW: Alterations of behavior produced by inhaled ozone or ammonia. Fundamental and Applied Toxicology 5:1110–1118, 1985.

Tepper, JL, Weiss, B, Wood, RW: Behavioral indices of ozone exposure. In: Advances in Modern Toxicology, Vol. V. Edited by SD Lee and MG Mustafa. Princeton, NJ: Princeton Scientific, 1993:515–526.

Tepper, JL, Weiss, B: Determinants of behavioral response with ozone exposure. J Appl Physiol 60:868–875, 1986.

Tepper, JL, Wood, RW: Behavioral evaluation of the irritating properties of ozone. Toxicol Appl Pharmacol 78:404–411, 1985.

Umezu, T, Shimojo, N, Tsubone, H, Suzuki, AK, Kubota, K, Shimizu, A: Effect of ozone toxicity in the drinking behavior of rats. Arch Environ Hlth 42:58–61, 1987.

U.S. Environmental Protection Agency: Air Quality Criteria for Lead (EPA-6008-83-028). Research Triangle Park, NC: U.S. EPA, Environmental Criteria and Assessment Office, 1986.

U.S. Environmental Protection Agency: Multi-substance rule for the testing of neurotoxicity. Federal Register 56:9105–9119, 1991.

Warren, DL, Guth, DJ, Last, JA: Synergistic interaction of ozone and respirable aerosols on rat lungs. II. Synergy between ammonium sulfate aerosol and various concentrations of ozone. Toxicol Appl Pharmacol 84:470–479, 1986.

Wayne, WS, Wehrle, PF, Carroll, RE: Oxidant air pollution and athletic performance. JAMA 199:901–904, 1967.

Weiss, B, Ferin, J, Merigan, M, Stern, S, Cox, C: Modification of rat operant behavior by ozone exposure. Toxicol Appl Pharmacol 58:224–251, 1981.

Weiss, B, Cory-Slechta, DA: Assessment of behavioral toxicity. In: Principles and Methods of Toxicology. Edited by AW Hayes. New York: Raven Press, 1994. 1091–1155.

Wenzel, DG, Morgan, DL: Interactions of ozone and antineoplastic drugs on rat lung fibro-blasts and Walker rat carcinoma cells. Res Comm Chem Path Pharm 40:279–287, 1983.

Wong, DS, Hochstein, P: The potentiation of ozone toxicity by thyroxine. Res Comm Chem Path Pharm 31:483–92, 1981.

Wood, RW, Weiss, AB, Weiss, B: Hand tremor induced by industrial exposure to inorganic mercury. Arch Environ Hlth 26:249–252, 1973.

Wood, RW: Behavioral evaluation of sensory irritation evoked by ammonia, Toxicol Appl Pharmacol 50:157–162, 1979.

Chapter Eighteen

Noncarcinogenic Responses of the Respiratory Tract to Inhaled Toxicants

**Donald L. Dungworth, Fletcher F. Hahn,
and Kristen J. Nikula**

INTRODUCTION

The purpose of this chapter is to outline major concepts necessary to understand the complexity of noncarcinogenic pathologic responses of the respiratory tract to inhaled toxicants. Major emphasis will be on factors affecting
the topographic distribution of damage within the respiratory tract, on features of acute and chronic pathologic responses, and especially on time-
response profiles in laboratory animals. The time-response profiles have a
crucial bearing on the correlation between acute and chronic events and on
exposure concentration-time-response relationships. They also help to illustrate the complexity of the respiratory tract responses to injury and the dan-
gers involved in attempting to interpret or predict overall consequences by
measuring only one or a few parameters.

The nature and concentration of the inhaled toxicant (or mixture of toxicants) are obviously important determinants of the responses of the respiratory tract. A limited selection of vapors, gases, particles, or mixed gaseous
and particulate toxicants will be considered in this chapter. The basic concepts presented in the first edition have been confirmed by recent findings,
and they will be refined here in light of current knowledge. Effects of inhaled

toxicants on nasal passages and of particulate materials on the lung receive more attention in this revised chapter because of the rapid growth of information about these two topics.

FACTORS AFFECTING ANATOMIC LOCATION OF DAMAGE

The size and complexity of the respiratory tract lend to considerable variations in the anatomic location of the principal sites of damage according to the nature and concentration of the inhaled toxicant. Interspecies differences in structure and function of the respiratory tract cause additional complications.

The pattern of distribution of damage throughout the respiratory tract depends on the interplay among local dosimetry at various anatomic sites along the tract, effectiveness of the local defense mechanisms, and the relative susceptibility to damage of the cells at risk in each location. These factors differ among animal species, and the summation of these differences partially account for interspecies variations in response.

Factors Relating to the Toxicant

In general, highly water-soluble or reactive chemicals and large particles (> 10 μm aerodynamic equivalent diameter) react with or impact upon the nasal tissues. Poorly soluble chemicals can primarily affect the nose, for example, ferrocene (Nikula et al., 1993), or they may penetrate to and affect the deep lung. Irritant gases of low-water solubility such as ozone at low concentrations can damage the anterior nasal cavity and the deep lung. Small particles and fibers (< 3 μm aerodynamic equivalent diameter) principally affect the deep lung. Deposition and clearance of inhaled particles, regional disposition of inhaled reactive gases, and factors modifying the disposition of inhaled organic compounds are all discussed at greater length elsewhere in this volume and will not be specifically addressed here.

Factors Relating to the Exposed Animal

Interspecies differences in structure and function of the respiratory tract influence local dosimetric patterns and hence are considered in previous chapters. The importance of ventilatory parameters is also mentioned, notably the increased dose to deep lung associated with exercise or with oronasal or oral breathing in those species capable of it (e.g., dog and human). Additional considerations concerning interspecies differences can be found in the review by Pauluhn (1994) and in the symposium proceedings summarized by Dahl et al. (1991).

Differences in cell populations at risk at various anatomic locations in

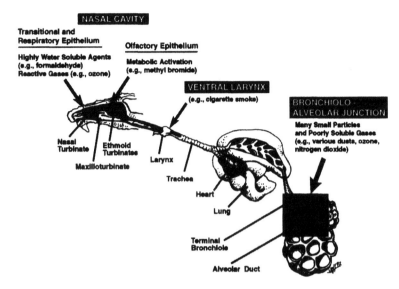

Figure 1 Stylized representation of the rat respiratory system illustrating three main regions vulnerable to toxicant-induced damage (drawing courtesy of Dr. J. R. Harkema).

the respiratory tract within any one species (Phalen et al., this volume; Plopper et al., 1994a; Harkema and Morgan, 1995a) have an important influence on susceptibility of the site to injury and its subsequent responses as will be discussed later in this chapter. The differences among species in cellular organization of the epithelium or mucosa at various anatomic sites, and the corresponding differences in metabolic capability of component cells are particularly important (Plopper et al., 1994a; Harkema, 1991; and Reed, 1993).

Observed Patterns of Anatomic Distribution of Damage

The net effect of integration of toxicant factors (affecting site-specific dosimetry) and host factors (affecting site-specific defense mechanisms and susceptibility of epithelial cells to injury) is that there are three main prediction sites for damage in the respiratory tract as illustrated in Fig. 1. Whether the nasal cavity, larynx, or deep lung is preferentially affected depends on the nature and dosimetry of the inhaled toxicant. As will become evident from more detailed discussion in subsequent sections, some toxicants essentially affect only one predilection site. Others such as ozone affect both the anterior nasal cavity and the deep lung (centriacinar region) at very low concentrations. High concentrations of most inhaled irritants cause severe, widespread damage which obscures any potential gradients of damage. At this juncture it is necessary to point out that nasal uptake of low-solubility gases and vapors through interaction with the surface-lining liquid can be considerable (Dahl, this volume; Miller, this volume). Deep lung injury will be nonetheless caused if a sufficient amount of the toxicant reaches this more vulnerable region.

Ozone has been the most extensively studied gas. Within the nasal cavity, at concentrations of 0.15 and 0.30 ppm, damage in monkeys occurs in the transitional and respiratory epithelium of the anterior and preturbinate portions of the septal and lateral walls of the cavity (Harkema et al., 1987a). Much more attention, however, has been focused on deep lung injury with ozone because, as with many inhaled toxicants, the principal location of damage is in the centriacinar region (bronchioloalveolar junction or "transitional" region). Damage at this site is both more evident and detrimental to pulmonary homeostasis and function.

With a reactive gas of low-water solubility such as ozone the closer the inhaled concentration is reduced to the no-observed-effect level, the more the detectable lesions are limited to the anterior nasal chamber and the proximal acinus. For short-term exposures, this occurs at approximately 0.15 ppm of ozone for the monkey (Castleman et al., 1977; Harkema et al., 1987a, 1993) and 0.08 (corrected to UV photometric standard) to 0.12 ppm for the rat (Plopper et al., 1979, Barry et al., 1985). Over long-term exposures (up to 20 months in the rat for instance) adaptation plays a role as described later. The centriacinus is the most vulnerable region of the lung to damage for ozone and many other inhaled irritants. It is, therefore, frequently the focus of attention for evaluation of the subtle effects of long-term experimental exposure to low levels of such toxicants. In this regard, it is important to note that although the proximal acinus is the site of most damage, the subgross and histologic structure of the site differs between primates and commonly used laboratory rodents (McLaughlin et al., 1961; Plopper et al., 1980, 1994a), and use of subtle lesions in experimental animals for predicting potentially harmful effects in humans must take this difference into account.

Reference has been made to the proximal acinar region as the principal site of damage of many inhaled irritants which by virtue of their physico-chemical properties can reach the deep lung and which are present as contaminants in the inhaled air. This was first emphasized by Gross et al. (1966) and has been confirmed with materials ranging from reactive gases such as ozone (Menzel, 1984) and nitrogen dioxide (Morrow, 1984) to particles and fibers such as asbestos (Brody et al., 1985; Warheit et al., 1986; Adamson and Bowden, 1986), and to mixed pollutants such as diesel exhaust (Mohr et al., 1986).

As with all generalizations, however, exceptions and special circumstances occur. One is that although the proximal acinar region is most vulnerable to damage by ozone, concentrations associated with evidence of airway hyperactivity in humans can damage the epithelium of conducting airways as well. This, together with interspecies differences, has resulted in conflicting information on the site of airway damage responsible for ozone-induced hyperreactivity in humans and on the pathogenic relationships involved

(O'Byrne et al., 1984; Lee and Murlas, 1985; Hazucha et al., 1986; Tepper et al., 1989; Folinsbee et al., 1994).

A second exception is that against the background of a general predilection for damage to be in proximal acinar regions, there are focal "hot spots" which can be of major importance as sites of predilection for carcinogenesis (Schlesinger, this volume), as exemplified by development of laryngeal tumors in Syrian golden hamsters chronically exposed to cigarette smoke (Bernfeld et al., 1979, 1983).

A third exception is that the location and nature of specific cellular damage caused by inhaled toxicants requiring metabolic activation are modified by the distribution of cells containing the necessary xenobiotic metabolizing enzyme systems. In the case of a compound such as 3-methylfuran (which is metabolized by cytochrome P-450 monooxygenases) there is preferential necrosis of olfactory epithelium and bronchiolar Clara cells at low concentrations, although high concentrations can obscure this pattern (Haschek et al., 1983). A more recent example is the initial selective sensitivity of Clara cells to damage by inhaled methylene chloride and the subsequent development of resistance associated with loss of cytochrome P-450 monooxygenase activity (Foster et al., 1992).

Implications of Patterns of Distribution of Lesions for Morphologic and Morphometric Examination

The size and structural complexity of the respiratory tract and the diversity of its responses to inhaled toxicants impose rigorous requirements for thorough examination. The anatomic variations in the tract necessitate wide and careful sampling. The inhomogeneity of response within any one anatomic compartment means that there must be deliberate sampling of specific sites, especially for ultrastructural and morphometric evaluations. Further, general consideration of this topic can be found in the reviews by Dungworth et al. (1985a), Pinkerton and Crapo (1985), and Schwartz (1987). The method for microdissecting airways to obtain precise sampling of specific-airway generations and path length is described by Plopper (1990) and Plopper et al. (1991). Introductions to morphometric methods for quantifying pulmonary damage are available in Hyde et al. (1992a) and Chang et al. (1991a). Special methods for examination of nasal cavities and of larynx are referred to in the subsequent sections on those anatomic sites.

NOSE
Patterns of Anatomic Distribution of Damage in the Nose

Examination of the nasal cavity of rodents requires multiple sections at standardized levels to assure adequate sampling of the various topographic sites and their mucosal lining. Most commonly, the fixed, decalcified nasal cavity

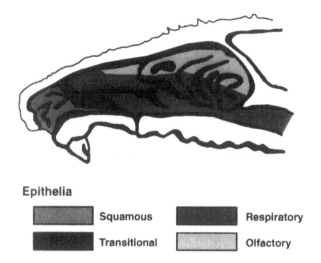

Epithelia

▨	Squamous	▰	Respiratory
▰	Transitional	▢	Olfactory

Figure 2 Diagram of the medial aspect of the right nasal passage of the rat showing the distribution of the four epithelial (mucosal) regions.

is sectioned transversely to yield four tissue blocks according to standardized anatomical landmarks as described by Young (1981). Sections from the anterior face of each block are then processed, usually for light microscopic examination. A series of nasal cavity maps for F344 rats and B6C3F$_1$ mice has been published to aid in documentation of the precise location of toxicant-induced damage in rodents (Mery et al., 1994a). A method of sectioning undecalcified, plastic embedded transverse sections of the nasal cavity of rhesus monkeys has been described by Randall et al. (1988).

Four distinct epithelia (mucosae) line the nasal cavity. In rodents, these epithelia are the stratified squamous epithelium that lines the nasal vestibule and the floor of the ventral meatus in the anterior portion of the nose; the nonciliated, pseudostratified, transitional epithelium that lies between the squamous epithelium and the respiratory epithelium and lines the lateral meatus; the ciliated, respiratory epithelium that lines the remainder of the nasal cavity anterior and ventral to the olfactory epithelium; and the olfactory epithelium (neuroepithelium) which lines the dorsal meatus and ethmoturbinates in the caudal portion of the nose (Fig. 2). These same epithelia line the nasal cavities of nonhuman primates and humans, but their relative abundance and exact locations differ between primates and rodents (Harkema, 1991; Negus 1958). Because of their differing regional distributions, the cells in these epithelia receive different doses of xenobiotics. These cells also differ in their inherent sensitivity to injury and their metabolic capabilities.

Generally, the olfactory epithelium has greater xenobiotic metabolizing activity than does the respiratory epithelium in most species (Dahl, this volume; Hadley and Dahl, 1983; Lewis and Dahl, 1994). When making cross-species extrapolations of metabolic capacities, however, the differing percentages of the nasal airway covered by olfactory epithelium must be considered. Approximately 50% and 37% of the nasal cavity surface area in F344 rats and B6C3F$_1$ mice, respectively, is lined by olfactory epithelium (Gross et al., 1982). In contrast, only 3% of the nasal cavity surface area of humans is lined by olfactory epithelium (Sorokin, 1988). The normal nasal morphology of rodents, including reference to cells with xenobiotic metabolizing capabilities and comparisons to primate species, is reviewed by Harkema and Morgan (1995a).

A variety of inhaled chemicals damage the nasal mucosa (Barrow, 1986; Buckley et al., 1984; Jiang et al., 1986; Monticello et al., 1990), and the observed lesions often are localized to particular nasal regions and cell types (Mery et al., 1994a; Morgan and Monticello, 1990a). As stated previously, the pattern of damage depends largely on the nature of the toxicant and local dosimetry, effectiveness of the local defense mechanisms, and the relative susceptibility to damage of the cells at risk in each location. Tissue- or cell-type-specific metabolism plays a role in dosimetry and defense through local generation of toxic metabolites or detoxication (Dahl and Hadley, 1991; Reed 1993) and by supplying other biochemical defenses. Airflow streams, mucus flow, and blood supply also influence dosimetry and defenses (reviewed in Morgan, 1994). Deposited particles are removed by the mucociliary apparatus of the transitional and respiratory regions of the nose (Phipps, 1981; Procter, 1982), and it has been proposed that mucus protects the airway epithelia from reactive oxygen species (Cross et al., 1984) and inhaled ozone (Pryor, 1992). Nasal uptake of nonreactive gases is strongly influenced by nasal blood perfusion rates (Morris et al., 1993), and nasal vessels may serve to remove toxic, nonreactive gases from the nose (Morris et al., 1986).

Morgan and his associates have shown that materials that are highly soluble in water and/or are highly reactive, and therefore are rapidly extracted from air streams passing over the nasal lining, cause a predictable pattern of lesions based on sites of highest local concentrations as a result of inspiratory airflow (Kimbell and Morgan, 1991). The epithelia lining the lateral, middle medial, and dorsal medial meatuses are most affected, and the lesions show a diminishing anterior-to-posterior severity gradient. Examples of inhaled toxicants affecting sites in this pattern include formaldehyde (Kimbell et al., 1993), dimethylamine (Gross et al., 1987), methyl isocyanate (Uriah et al., 1987), sulfur dioxide (Giddens and Fairchild, 1972), and methanol-fueled engine exhaust (Maejima et al., 1993).

The respiratory and olfactory epithelia of rats, other laboratory and nonlaboratory animals, and humans contain many xenobiotic metabolizing enzymes such as carboxylesterases, aldehyde and formaldehyde dehydrogenases, cytochromes P-450, epoxide hydrolase, rhodanese, UDP-glucuronyltransferase, and glutathione S-transferases (Dahl and Hadley, 1991; Reed, 1993). The nasal cell-type-specific distribution of these enzymes has been localized by histochemical (Bogdanffy, 1990) and immunohisto-chemical (Lewis et al., 1992; Lewis et al., 1994a) techniques. Nasal metabolism (Dahl, this volume) which can result in detoxification or activation of inhalants may render particular cell types or regions more susceptible or more resistant to chemically induced injury. An example of this is the predominant toxic effect of inhaled acetaldehyde on the olfactory epithelium which is essentially devoid of the detoxifying enzyme aldehyde dehydrogenase. Respiratory epithelium, which is rich in aldehyde dehydrogenase, is much less affected (Appelman et al., 1982; Bogdanffy et al., 1986).

Metabolism of acrylate esters by carboxylesterase provides an example not only of the potential for predicting sites of toxicity for inhaled substrates based on metabolism, but also of the potential for assessment of risks for mucosal damage in humans based on interspecies comparisons of enzyme activity and localization. Nasal carboxylesterases hydrolyze inhaled acetates and acrylate esters to their constituent carboxylic acids and alcohols which damage the nasal mucosa (Bogdanffy, 1990). Regional nasal carboxylesterase activity data for rats (Bogdanffy et al., 1987) and people (Mattes and Mattes, 1992) as well as enzyme localization data by region and cell types for both rats and people (Lewis et al., 1994a) are available that may allow predictions of toxicity after acute exposure to acrylate esters. Caution would be needed, however, when attempting to extrapolate expected results from the acute exposure scenario to repeated exposures or exposures to mixtures of inhalants. Predicted metabolism for an acute exposure may differ from that seen in these latter situations because (1) metabolites of carboxylesterase are often toxic to nasal tissue and nasal lesions affect carboxylesterase amounts and localization (Lewis et al., 1994a), and (2) because carboxylesterase can be induced by non-substrate exposures (Nikula et al., 1995) and by the common inhalant—tobacco smoke (Lewis et al., 1994b).

Responses of Nasal Mucosa to Inhaled Toxicants

The nature of toxicant-induced lesions varies with the test compound and the response of the nasal mucosa. The morphologic response changes or is modified by factors such as dose or degree of injury, duration of exposure, and time post-exposure. Within each mucosal region of the nose, certain morphologic responses to acute or chronic injury and certain metaplastic, potentially adaptive, or reparative responses are typical. These responses will be described in the following sections organized by the type of epithelium in each region of the nose. The descriptions focus on mucosal and luminal changes. Lesions of accessory structures such as the vomeronasal organ, the nasolacrimal duct, the maxillary recess, and the bone will not be addressed. Examples of lesions in these structures and many excellent photomicrographs of nasal lesions have been published (Toxicologic Pathology of the Upper Respiratory System, 1990).

Responses of Nasal Squamous Epithelium to Inhaled Toxicants Although the squamous epithelium is exposed to ambient concentrations of inhaled toxicants due to its proximal location in the nose, it is relatively resistant to injury so that lesions due to inhaled toxicants are not commonly reported. A few caustic or highly irritant chemicals have been shown to damage the squamous epithelium lining the nasal vestibule or the ventral meatus of rodents. Examples include dimethylamine (Buckley et al., 1985), glutaraldehyde (Gross et al., 1994), ammonia (Bolon et al., 1991), and hydrogen chloride (Jiang et al., 1986). Acute damage to the squamous epithelium (as visualized by light microscopy) usually consists of erosion or ulceration with or without inflammation. Chronic exposures may lead to hyperplasia or hyperplasia with hyperkeratosis.

Responses of Nasal Transitional Epithelium to Inhaled Toxicants Cells of the transitional epithelium are also exposed to near ambient concentrations of inhaled toxicants, and they are generally more susceptible to injury than those of the squamous epithelium. Toxicants that can damage the squamous epithelium usually cause more severe damage in the transitional epithelium, and less irritating toxicants that do not cause lesions in the squamous epithelium frequently cause morphologic alterations in the transitional epithelium. Examples include ozone (Johnson et al., 1990), chlorine gas (Jiang et al., 1983), tobacco smoke (Maples et al., 1993), and formaldehyde (Chang et al., 1983). Acute damage caused by caustic or highly irritating chemicals may consist of necrosis and erosion or ulceration, usually with an accompanying exudative inflammatory response. With less severe injury, acute inflammation may be the only acute response visible by light microscopy. Neutrophils may be found in the lamina propria, the epithelium, and the airway lumen. Repeated exposures lead to subacute or chronic alterations in the transitional epithelium which include epithelial hyperplasia, secretory cell hyperplasia or metaplasia, and squamous metaplasia. Repair of an acute, severe lesion can also lead to these same hyperplastic and/or metaplastic alterations, but the change in the epithelium is transient if the exposure is not repeated.

Ozone-induced alterations in the transitional epithelium have been extensively studied by Harkema and his colleagues. Ozone induces a mucous cell hyperplasia in the transitional epithelium of monkeys exposed to ozone (Harkema et al., 1987a,b). Exposure of rats to 0.8 ppm ozone (6 hours/day) induces an acute neutrophilic response (Hotchkiss et al., 1989) and an increase in epithelial DNA synthesis (Johnson et al., 1990; Hotchkiss and Harkema, 1992). Over a 7-day period, the inflammatory and epithelial replicative responses decrease dramatically as mucous cell metaplasia (transformation of an epithelium with no mucous cells into an epithelium with numerous mucous cells) develops (Harkema and Hotchkiss, 1994). Harkema and Hotchkiss (1994) have suggested that the inflammatory and replicative responses are transient because the increase in intraepithelial mucosubstances, and the expected corresponding increase in secreted mucins act as a protective, antioxidant shield for the nasal mucosa, thus preventing further injury by the continued ozone exposure. Chronic exposure of rats to 0.5 or 1.0 ppm ozone for 20 months (6 hours/day, 5 days/week) has been shown to cause marked mucous cell metaplasia in the transitional epithelium (Harkema et al., 1995).

The influence of exposure regimen on effects of reactive gases have not been as extensively studied in the nose as in the lung (described later in this chapter). However, Hotchkiss et al. (1991) assessed the effect of different cumulative exposure times on ozone-induced nasal epithelial hyperplasia and secretory metaplasia in the transitional epithelium. They exposed rats to 0.8 ppm O_3 for 7 days or for 3 days followed by a 4-day exposure to air. The hyperplastic and metaplastic changes within the transitional epithelium induced by the two exposure regimens were indistinguishable. These results suggest that O_3 rapidly induces hyperplasia and metaplasia within the nasal epithelium, and once initiated, development of these alterations does not require further ozone exposure.

The studies by Harkema and colleagues also provide information on comparative sensitivity of rat and monkey nasal tissue to ozone-induced alterations. Monkeys exposed to 0.15 ppm ozone (8 hours/day for 6 days) develop mucous cell hyperplasia in transitional and respiratory epithelium. In contrast, rats exposed to 0.8 ppm develop mucous cell metaplasia only in the transitional epithelium, and the transient epithelial cell replication is of lesser magnitude and shorter duration in respiratory than in transitional epithelium. Metaplasia did not occur in the transitional epithelium of rats exposed to 0.12 ppm ozone (6 hours/day for 7 days). Therefore, the monkey transitional and respiratory epithelia seem to be more susceptible to ozone-induced injury than the corresponding rat epithelia. Whether differing nasal airflow patterns, amounts of mucus overlying the epithelium, cell sensitivities, or some combination of these factors account for the differing epithelial sensitivities is currently unknown (Harkema and Morgan, 1995b).

Squamous metaplasia of the transitional epithelium has been induced by cigarette smoke exposure in rats (Maples et al., 1993), chlorine gas exposure in rodents (Jiang et al., 1983), and formaldehyde exposure in rodents (Chang et al., 1983) and nonhuman primates (Monticello et al., 1989). The sequence of chlorine-induced alterations, described by Jiang (1983), is acute epithelial degeneration with epithelial cell exfoliation, erosion, and ulceration following 1 to 3 days exposure, infiltration by neutrophils at 3 to 5 days, and squamous metaplasia after 5 days.

Responses of Nasal Respiratory Epithelium to Inhaled Toxicants   Acute injury to the respiratory epithelium may be manifest morphologically as attenuation of cilia or deciliation, degenerative cytoplasmic alterations (blebs, vacuoles), exfoliation of individual cells, erosion, or ulceration. Inflammation, varying from serocellular to fibrinocellular or purulent, may be present in the lamina propria, epithelium, and nasal lumen. The severity and extent of the inflammation usually correlate with the degree of epithelial damage.

Following epithelial ulceration and fibrinocellular exudative inflammation, regenerating epithelial cells may use the exudate as a scaffold, resulting in fusion of turbinates to each other, the septum, or the lateral wall. Subacute to chronic injury (usually the result of repeated exposures) may lead to the continued presence of the alterations described previously. Common morphologic manifestations of subacute to chronic injury include goblet cell hyperplasia, squamous metaplasia, and hyperplasia with folding of the epithelium. A hyaline change (referred to as eosinophilic globules in columnar epithelial cells) is sometimes observed in rats and mice exposed to a variety of toxicants. Similar globules are seen less frequently in control rodents. It is beyond the scope of this chapter to list all toxicants that cause the lesions listed above, but examples of toxicants causing various combinations of these lesions in the respiratory epithelium, depending on exposure concentration, exposure regimen, and the nature of the toxicant, include glutaraldehyde (Zissu et al., 1994; Gross et al., 1994), dimethylamine (Buckley et al., 1985; Gross et al., 1987), acrolein (Buckley et al., 1984), chlorine gas (Jiang et al., 1983), formaldehyde (Chang et al., 1983; Monticello et al., 1989), ozone (Harkema et al., 1987a), and tobacco smoke (Maples et al., 1993; Hotchkiss et al., 1995).

The responses of the respiratory and transitional epithelia to the toxicants listed above can be used to illustrate three important concepts. First, within a seemingly small region, a toxicant at a given exposure concentration may cause no morphologic response or varying responses, probably dependent on small differences in air flow (dosimetry) and the susceptibility of the local cell population. For example, rats exposed to tobacco smoke (250 mg/ m^3 total particulate matter, 6 hours/day, 5 days/week) for 2 weeks develop mucous cell hyperplasia in the respiratory epithelium lining the dorsal nasal septum, squamous metaplasia of the respiratory epithelium lining the mid septum, but no morphologic alterations in the respiratory epithelium of the ventral septum (Hotchkiss et al., 1995). Hotchkiss et al. (1995) used knowledge of regional differences in airflows (Kimbell et al., 1993) and in the amount of intraepithelial mucosubstances to suggest that the responses of the three septal regions could be explained on the basis of local dose and local protection/susceptibility. These correlations of local dose with morphology could not have been made without careful lesion mapping, and they emphasize the utility of this approach. Second, squamous metaplasia can be rapidly repaired following cessation of exposure. Rats exposed to tobacco smoke for 2 weeks (using the exposure regimen previously described) develop squamous metaplasia of the transitional epithelium and the respiratory epithelium of the midseptum. Two weeks after cessation of exposure, these epithelia return to a normal pseudostratified epithelium with no evidence of squamous metaplasia (Maples et al., 1993). Third, squamous metaplasia is induced by toxicants proven to be nasal carcinogens in chronic studies such as formaldehyde (Kerns et al., 1983) and by toxicants that are not nasal carcinogens such as dimethylamine (Gross et al., 1987) and chlorine gas (Wolf et al., 1995). Thus, this alteration may be an "adaptive" response (i.e., replacement of a more sensitive population by a more resistant one) (Burger et al., 1989; Monticello et al., 1990) or a "precancerous" lesion (Morgan and Monticello, 1990b; Klein-Szanto et al., 1987). Interpretation of observed squamous metaplasia, particularly when seen in a sub-chronic study, is difficult for the pathologist. Current evidence suggests that squamous metaplasia accompanied by dysplasia and cellular atypia should be considered precancerous (Morgan and Monticello, 1990b).

Responses of Nasal Olfactory Epithelium to Inhaled Toxicants Acute injury of the olfactory epithelium may result in loss of olfactory cilia, cell-type-specific or nonselective degeneration of epithelial cells, loss of sensory cells, atrophy, necrosis, erosion, and ulceration. As described for the respiratory epithelium, inflammation may accompany these lesions and can involve the lamina propria, epithelium, and lumen. Bowman's glands may also exhibit degenerative changes, atrophy, necrosis, or inflammation. These glandular lesions are commonly associated with toxicants that are activated by the xenobiotic metabolizing enzymes in acinar cells.

Subacute to chronic lesions observed in the olfactory mucosa may include the continued presence of any of the above alterations and/or replacement of the olfactory epithelium with ciliated respiratory epithelium (respiratory metaplasia) or squamous metaplasia. Other subacute to chronic alterations include eosinophilic globules in sustentacular cells, folding and/or the formation of intraepithelial gland-like structures in the epithelium, hypertrophy of Bowman's glands or ducts or (alternatively) loss of these structures, degeneration and atrophy of nerves, as well as fibrosis of the lamina propria. Organization of inflammatory exudate and remodeling associated with ulceration may lead to fusion of turbinates or partial occlusive fibrosis of the nasal cavity.

Following epithelial degeneration, necrosis, erosion, or ulceration, the epithelium may be repaired over a period of days to months, depending on the severity and extent of the damage (Lee et al., 1992). As in other epithelial regions, cell proliferation and regeneration can occur even in the face of continued exposure (Hurtt et al., 1988). Basal cells are considered the progenitors leading to the regeneration of the neuroepithelium (Gradziadei, 1973; Gradziadei and Gradziadei, 1979a, Lee et al., 1992). Remaining basal cells proliferate. They become hypertrophic and polyhedral and migrate across areas of ulceration (Genter et al., 1992). After forming an initial hypercellular zone of proliferating basal cells which frequently appear disorganized, they differentiate to form sustentacular and sensory cells. There is also earlier evidence that in some instances, proliferation and migration of cells from the ducts of Bowman's glands may contribute to the reepithelization (Mulvaney and Heist, 1971; Uriah et al., 1987). Thus, the source of the proliferating cells for regeneration of the olfactory epithelium is not unequivocally known. Damaged Bowman's glands and olfactory nerves may also repair at the same time as, or following, epithelial regeneration. Regeneration of the epithelium does not always result in a completely normal epithelium. Some degree of metaplasia or disorganization may remain. Occasionally, following epithelial regeneration, large bundles of axons are present within the epithelium (Gradziadei and Gradziadei, 1979b), and nests of sensory cells in the lamina propria have been reported after exposure of rats to 3-trifluoromethyl pyridine (Gaskell et al., 1988) and after exposure of hamsters to furfural (Feron et al., 1979).

As previously stated, the lesions frequently occur in specific foci. Local dosimetry and metabolic capacity of the cells are the major factors influencing the specific location of the lesions within the olfactory mucosa. A frequently affected site in rodents, particularly after exposure to inhaled irritants, is the anteriodorsal extension of the olfactory mucosa lining the dorsal meatus (Buckley et al., 1984). Because of its more anterior position compared to the remainder of the olfactory mucosa, this region is exposed to a relatively high concentration of inhaled toxicants. Other inhaled toxicants have a different pattern of lesion distribution, related frequently to sites of greater metabolic activity or a combination of metabolism and dosimetry factors. Examples of these toxicants include 3-methyl furan (Morse et al., 1984), methyl bromide (Hurtt et al., 1987), dibasic esters (Keenan et al., 1990), ferrocene (Nikula et al., 1993), chloroform (Mery et al., 1994b), and pyridine (Nikula and Lewis, 1994). Chlorthiamid is an example of a toxicant which initially causes necrosis of Bowman's glands, presumably due to local cytochrome P450-dependent activation, with degeneration and necrosis of the olfactory epithelium developing less rapidly (Brittebo et al., 1991).

The eosinophilic globules that develop in sustentacular cells following inhalation of a variety of toxicants are morphologically similar to those described earlier in the respiratory epithelium. Frequently, when globules develop in sustentacular cells, there is a concomitant sensory cell loss (neuronal atrophy). Until recently, the nature of the material comprising the globules was unknown. Eosinophilic globules develop in the respiratory and olfactory mucosae of rats exposed to cigarette smoke. Although morphologically similar globules can be found in control rat mucosae, the number and size of globules increases with increased cigarette smoke concentration and/or duration of exposure. The globules in smoke-exposed rats contain carboxylesterase, and the amount of carboxylesterase in the globules increases with exposure concentration and duration (Lewis et al., 1994b). These results suggest that the globule response to toxicants may represent induction of xenobiotic-metabolizing enzymes.

Epithelial lesions such as extensive erosion or ulceration, respiratory metaplasia, squamous metaplasia, or atrophy as well as obstruction of the airway lumen with inflammatory exudate would be expected to cause hyposmia or anosmia. Although olfactory dysfunction has been demonstrated after exposure to a variety of toxicants (Getchell et al., 1991), toxicology study designs rarely include tests of olfactory function.

LARYNX

The ventral larynx is one of the predilection sites in rodents for damage by a variety of inhaled irritants, cigarette smoke being a good example (Gopinath et al., 1987). More specifically, the transitional epithelium at the base of the epiglottis covering the region with submucosal seromucous glands is particularly vulnerable to alteration which is commonly manifest as squamous metaplasia. The ventral pouch and medial surfaces of the vocal processes are also susceptible locations. Detailed descriptions of the epithelium in these sites in rats and mice have been provided by Renne et al. (1992a). As in the nose, the transitional epithelium between squamous and respiratory epithelial zones is most susceptible to irritant-induced injury, and elevated local dosimetry which is determined by air flow and turbulence also plays a part. As with other portions of the respiratory tract, thorough examination of the larynx requires systematic and standardized processing (Lewis 1991; Renne et al., 1992a; and Sagartz et al., 1992).

Examples of recent inhalation studies where the larynx is the most vulnerable site for damage are those with glycerol (Renne et al., 1992b) and cobalt sulfate (Bucher et al., 1990). The latter is particularly interesting because of the development of extensive inflammatory polyps by the end of a 13-week exposure of rats to 3 mg/m³ or more of cobalt-sulfate heptahydrate aerosol.

ACUTE RESPONSES OF THE LUNG TO INHALED TOXICANTS

The general response to sufficiently severe acute injury is epithelial damage which triggers a large variety of acute inflammatory mediators and their modulators (see Cohn and Adler, 1992; Driscoll, this volume). This is accompanied by intraluminal and interstitial (mural) inflammation. Because epithelial, intraluminal, and interstitial events are assessed by different sets of parameters, and because their time-response profiles are different, each will be considered in turn. Vascular events, of course, are an integral part of interstitial inflammation but will not be a focus of attention in this chapter.

Acute Epithelial Responses

The common sequence of events in an acute epithelial response is death of sensitive cells, followed by a repair process consisting mainly of proliferation of resistant cells. Details depend on the anatomic location, the corresponding structural and functional properties of the epithelium at the affected site, as well as the properties and concentration of the toxicant. A common theme throughout the epithelium of the respiratory tract, however, is that cells with the most active secretory/metabolic capability are the least sensitive to damage by direct-acting toxicants and usually serve as the progenitors for reparative cell proliferation. In the hierarchy of sensitivity to direct-acting toxicants, alveolar type I cells and ciliated cells are most sensitive to damage, secretory bronchiolar (Clara) cells are of intermediate sensitivity, and mucous cells and type II alveolar cells are the most resistant. This pattern is changed for toxicants for which metabolic activation and deactivation play crucial roles as mentioned previously.

The response of respiratory epithelium to acute toxic injury has been studied most extensively at the bronchiolar and alveolar levels (Schwartz, 1987; Keenan, 1987; Plopper and Dungworth, 1987). In the alveolus, studies with oxidant gases (Kapanci et al., 1969; Evans et al., 1973; Adamson and Bowden, 1974) led to the recognition that the usual response to acute injury is rapid necrosis and sloughing of highly vulnerable type I cells. This stimulates proliferation of resistant type II cells. The peak of mitotic activity in type II cells is around 48 hours after onset of injury, and complete lining of the alveolus by type II cells or their type I daughter cells can occur within 7 days.

Acute injury to the terminal bronchioles causes damage and death of ciliated cells and degranulation of Clara cells. These reactions are followed by proliferation of Clara cells, with the peak at about 72 hours, and subsequent differentiation of a proportion of Clara cells to form new ciliated cells if the insult ceases or is relatively mild (Evans et al., 1976; Lum et al., 1978). An analogous situation occurs in species with respiratory bronchioles, but here the squamous (type I) cells lining the wall are replaced by proliferation of the nonciliated cuboidal cells (Castleman et al., 1980; Eustis et al., 1981).

The response to acute injury in the upper respiratory tract has been studied most extensively in the trachea (Keenan, 1987; Wilson et al., 1984). Here also, defects caused by necrosis and sloughing of epithelial cells are repaired by proliferation of secretory cells. The same holds true for bronchial epithelium (Evans et al., 1986). Whether the basal cell serves to replenish the "rapid-response" pool of secretory cells at steady state or has some entirely different function is not known. Hyde et al. (1992b) published the most comprehensive study to date on the sequence of epithelial changes in the lung following short-term exposure to a reactive gas. Microdissected airways were analyzed morphometrically and by labeling of replicating cells. Correlations were made with selected parameters of bronchoalveolar lavage fluid (BALF).

Hyde and associates exposed rhesus monkeys for 8 hours to 0.96 ppm of ozone and compared them to controls at post-exposure periods ranging from 1 to 168 hours. Their main findings were:

1 peak epithelial necrosis occurred earlier in distal trachea and respiratory bronchioles than in bronchi (1 hour post exposure versus 24), and

2 in bronchi the time course of intraepithelial appearance of polymorphonuclear cells paralleled that of necrosis, but in the trachea it lagged by about 12 hours.

Epithelial repair began by 12 hours after exposure (cell proliferation and/or ciliogenesis). It was essentially complete by 72 hours in the trachea and by 7 days in the bronchi and bronchioles. At 7 days, however, there was an increase in the numbers of intraepithelial inflammatory cells over those seen at 72 hours. This is not the place for detailed analysis of the results obtained by Hyde and associates but their paper is an excellent example both of the need for precise and well correlated investigations and the complexity in interpretation of resultant findings.

Returning to the main theme of this section, the most important consequence of the replacement of sensitive cellular components of epithelial cell populations by cells more resistant to injury is the effect this has on the response to chronic injury, as will be described later.

Acute Intraluminal Exudation

Acute injury to the respiratory tract results in an intraluminal exudate composed of cellular and glandular secretions, substances derived from degenerating or dead epithelial cells, and a large variety of cellular and humoral mediators and markers of acute inflammation (see Driscoll, this volume) which are mostly from the blood. The precise composition of the exudate depends on the cause of the injury and the site at which it occurs. The use of BALF as a technique for evaluating responses to inhaled toxicants has been responsible for greatly heightened interest in intraluminal events (Henderson, this volume).

Two important concepts must be kept in mind concerning the use of BALF in assessment of toxic injury of the respiratory tract. The first concept relates to the fact that BALF from acutely damaged lungs contains a large number of cellular and humoral components representing different time courses. The time courses will differ both according to the component measured and to the exposure protocol (e.g., species, strain, sex, toxicant, exposure concentration, and duration of exposure). These facts lead to the conclusion that care must be taken to ensure that erroneous comparisons are not made among different experiments, especially when only a small subset of BALF parameters has been measured.

The second concept also follows from the variety of components and time courses represented in BALF. When those are compared alongside the complexity of epithelial and inflammatory events *in situ* in different sites in the lungs, the difficulty in accurately estimating damage to the lung by use of BALF alone can easily be appreciated.

An example of the complexity of correlating BALF parameters and *in situ* pulmonary damage is given by the study of Hyde et al. (1992b) referred to earlier in the section on acute epithelial responses. Hyde and associates found that mucin and prostaglandins (mainly PGF_2 and PGE_2) were at a maximum in BALF of rhesus monkeys 1 hour after an 8-hour exposure to 0.96 ppm of ozone. Total protein, however, did not peak until 24 hours after exposure, and this was associated with maximal epithelial necrosis in bronchi, not trachea and bronchioles. A further, more clinical example is that of Schelegle et al. (1991) who concluded on the basis of human exposures to ozone that the time courses of symptoms, decrements in forced expiratory volume, and airway neutrophilia were different.

In spite of the need for care in interpretation, analysis of BALF is nevertheless an important tool because it both enables serial sampling and provides for the most direct comparison between responses in experimental animals and humans.

Acute Interstitial Inflammation

Components of the acute interstitial inflammation are essentially the same as those in any vascular connective tissue. The variety of cellular and humoral mediators and markers of acute inflammation is similar to that present in the intraluminal exudate, but the response is usually less intense, and the relative proportion of the various components is different. Special features of the interstitial response are those relating to the activities of mast cells (Nadel, 1992), fibroblasts (Dubaybo et al., 1992), and myofibroblasts (Zhang et al., 1994) and to the alteration of the connective tissue framework of elastin and collagen. The interstitial ground substance and its network of fibrils also modulate interaction between the normal cellular components and infiltrating inflammatory cells. Relatively little detailed information is available about acute interstitial responses, partly because they are not as amenable to study as for instance BALF, and partly because most attention has been on the chronic interstitial changes leading to either emphysema or pulmonary fibrosis (Last and Reiser, 1985; Snider et al., 1986; Hyde et al., 1992c; Cardoso et al., 1993).

Modifiers of Acute Responses

Major modifiers of the acute response to inhaled toxicants are the same as the set of factors that determine the anatomic location of damage—namely properties and dosimetry of the toxicant, effectiveness of local defense mechanisms, responsiveness of the various cell types comprising the population at risk, and other structural and functional properties that vary with the animal species under consideration.

There is no comprehensive review of the influence of interspecies differences in structure and function on the response to inhaled toxicants. The best sources of information for a general introduction to relevant subject matter are chapters in this volume and in the monograph edited by Parent (1992) entitled "Treatise on Pulmonary Toxicology: Comparative Biology of the Normal Lung." Age, breed, strain, and gender differences are beyond the scope of this review (see Pauluhn, 1994). Specifically on the topic of the influence of differences in cell biology among species of animals, it is important to note the developing knowledge about differences in cells comprising respiratory epithelium which is the first important target of inhaled toxicants (see Plopper et al., 1994a). Some differences in responsiveness among species of animals can now be explained on the basis of what is known about differences in regional dosimetry and susceptibility of cell populations at risk. Much remains to be elucidated, however, such as the mechanisms responsible for the greater inflammation in lungs of rats exposed to diesel exhaust compared to hamsters exposed to the same chamber concentrations (Stöber, 1986). Although differences in lung burden might account for part of the explanation, it appears that differences in amounts of inflammatory mediators generated by interaction between particles and alveolar macrophages or alveolar epithelial cells, or both, are also involved.

CHRONIC RESPONSES OF THE LUNG TO INHALED TOXICANTS

The following discussion relates mainly to effects of concentrations of toxicants at the low end of their respective dose-response ranges. The bronchioloalveolar region is emphasized because in general, this is the most vulnerable site for pulmonary damage by inhaled toxicants. As with acute events, epithelial, intraluminal, and interstitial responses will be considered separately. Although chronic vascular changes are of great importance in some pulmonary diseases, they are not a feature of primary importance in inhalation toxicology.

Chronic Epithelial Responses

The main types of noncarcinogenic responses of epithelial cells to chronic injury are hyperplasia, hypertrophy, and transdifferentiation (metaplasia), and usually a combination of these types occurs. Atrophy is not usually a significant feature, at least at the level of effects being discussed.

The early wave of proliferation of resistant cells to repopulate epithelium denuded of necrotic sensitive cells, which was described under acute responses, gives rise to a more steady-state population of the resistant epithelial cells. The size of the proliferating cell pool remains above control levels, however. The cells persisting during chronic injury vary with anatomic site, but they are predominantly of secretory or squamous types. Secretory and squamous metaplasias are the most obvious forms of cellular adaptation to injury, although more subtle adaptations are now being recognized. These will be discussed in the section on adaptation. Squamous metaplasia tends to be associated with the more severe forms of injury, although there is variation according to toxicant (Buckley et al., 1984). Other than the role of vitamin A, little is known about the mechanisms responsible for signaling squamous as opposed to secretory metaplasia (Keenan, 1987).

Secretory metaplasia in bronchioles is most evident when mucus-producing goblet cells replace the usual secretory bronchiolar (Clara) cells. This is well documented in chronic bronchiolitis in humans (Kilburn, 1974) and occurs experimentally in rats exposed to cigarette smoke (Reid, 1977; Reid and Jones, 1983), in guinea pigs exposed to cigarette smoke (Wright et al., 1992), and in hamsters with elastase-induced emphysema (Christensen et al., 1977). In species such as monkeys, dogs, and cats, which have well-developed respiratory bronchioles more analogous to humans (McLaughlin et al., 1961; Tyler, 1983), there are variations of bronchiolar epithelial hypertrophy and hyperplasia. In monkeys exposed to ozone for 3 months, the normal mixture of cuboidal and squamous (type 1) cells was replaced by a continuous lining of enlarged, hyperplastic bronchiolar cells mostly devoid of secretory organelles (Eustis et al., 1981; Harkema et al., 1993). In cats exposed to diesel exhaust for 27 months, there was hyperplasia of secretory bronchiolar (Clara) cells together with presence of ciliated and basal cells (Plopper et al., 1983). The extent to which the difference in these two responses was due to differences in species, toxicant, or length of exposure remains speculative. This is a good example, however, of the difficulty in making rigorous comparisons between studies because of the differences in experimental protocols.

A feature of the chronic response of bronchiolar epithelium to moderate or more severe damage by inhaled toxicants is the formation of hyperplastic nodules consisting of various proportions of ciliated and nonciliated cells. These have been produced by ozone in mice (Penha and Werthamer, 1974; Zitnik et al., 1978) and monkeys (Fujinaka et al., 1985), nitrogen dioxide in hamsters (Kleinerman et al., 1985), and automobile exhaust in dogs (Hyde et al., 1978). The response is therefore a general one, although in rodents exposed to oxidant gases, the mouse is more prone to develop hyperplastic nodules than the rat (Schwartz, 1987). An important characteristic of the nodules, to be discussed under time-response profiles, is that they persist after the exposure to the toxicant has ceased.

Chronic injury of alveolar epithelium results in one or more of persistence of alveolar type II epithelial cells, metaplasia into an epithelial cell type not normally found in the alveolus, or incomplete differentiation of type II cells toward type I epithelial cells (Plopper and Dungworth, 1987). Chronic exposure to inhaled toxicants causes these changes predominantly in centriacinar regions as previously stated. In these regions, chronic injury frequently causes persistent lining of alveolar spaces by enlarged cuboidal cells which are derived from either preexisting type II cells, nonciliated epithelial cells from adjacent bronchioles, or a mixture of the two. A feature of chronic centriacinar inflammation is the remodeling of alveolar ducts and their "bronchiolization" by becoming lined by bronchiolar epithelium (Nettesheim and Szakal, 1972; Pinkerton et al., 1993). This is generally believed to be the result of peripheral extension of bronchiolar cells, presumably because they have selective advantage for survival over type II cells. However, there is at least the possibility that some arise by metaplasia (transdifferentiation) from alveolar type II cells.

The phenotypic lability of respiratory epithelial cells has, in fact, been increasingly well recognized in recent years and is an important concept for the pathobiology of the respiratory tract. At the alveolar level, it has been shown for instance that type II epithelial cells are capable of transdifferentiation to ciliated cells and occasionally squamous cells as a result of bleomycin toxicity (Adamson and Bowden, 1979). In this study the authors inferred that these metaplastic changes were most likely to occur following damage to DNA. Generally, however, when squamous metaplasia or other manifestations of transdifferentiation are seen in bronchioloalveolar regions of animals chronically exposed to toxicants such as cigarette smoke (Leuchtenberger et al., 1963) or diesel exhaust (Mohr et al., 1986), neither the precise cell of origin nor the possible mechanisms involved are known. Recent evidence (Nolte et al., 1993) indicated that hyperplastic type II cells are an important source of squamous metaplasia. This is an important topic for future investigation because of its central role in both noncarcinogenic and carcinogenic responses of the lung to inhaled toxicants.

Chronic Intraluminal Exudate

Quantitative aspects of chronic intraluminal exudation are best evaluated by BALF, so reference should be made to the reviews by Henderson (this volume). Less information is available for chronic studies than for acute ones, and this is an important gap to be filled relative to modeling of time-response profiles (see below). In experimental situations, assessment of chronic responses by BALF should always be correlated with *in situ* observations. This is partly because the problems of sampling errors tend to increase when airways are narrowed and when exudate is lower in quantity, more cellular, and more adhesive. It is also partly because the pathogenetic events can become complicated by secondary events such as superimposed infection even in nominally specific-pathogen-free animals. One chronic sequel to acute necrosis of bronchiolar epithelium and exudation of fibrin-rich material is obliterative bronchiolitis in which the bronchioles become partially or completely occluded by fibrous plugs formed by organization of the fibrinous exudate. This is important at high concentrations of toxic deep-lung irritants.

Chronic Interstitial Responses

In the pulmonary parenchyma, chronic interstitial changes are the most important of the noncarcinogenic responses because they signify irreversibility of lesions which can seriously impair pulmonary function. The two types of lesions of most consequence are elastolysis leading to the rearrangement of the elastic fiber framework and emphysema (Snider et al., 1986; Cardoso et al., 1993) and excessive collagen deposition leading to pulmonary fibrosis (Last and Reiser, 1985; Zhang et al., 1994). Virtually any chronic injury capable of sustaining a significant amount of persistent damage to the centriacinar regions of the lung will produce some degree of bronchiolization of alveolar ducts and associated interstitial fibrosis (Barr et al., 1988, 1990). The most accurate assessment of these effects has been provided by the collaborative studies of NIEHS and HEI on the long-term inhalation of ozone in F344/N rats. Several detailed reports have been published (Harkema et al., 1995; Pinkerton et al., 1993; Plopper et al., 1994b,c; Dodge et al., 1994), but the one most precisely defining bronchiolization after 20 months of exposure to 1.0 ppm ozone (6 hours/day, 5 days/week) was Pinkerton et al. (1993). An important feature of this type of bronchiolization is not only is there distal extension of ciliated and nonciliated cells into the pulmonary acinus, but also remodeling of alveolar duct walls to produce structures with the characteristics of respiratory bronchioles (i.e., alveolar outpocketing separated by bronchiolar wall). Separate analysis of the lung material from the same study revealed morphometric evidence for significant increases in collagen, elastin, acellular space, and basement membrane components of interstitium after 20 months of exposure to 1.0 ppm ozone and lesser changes after exposure to 0.5 ppm. Only basement membrane and collagen were statistically significantly increased (Chang et al., 1995). This overall study indicated that effects were generally greater in male rats than females, possibly because of increased site-specific dosimetry in males. Other facets of the results from the NIEHS/HEI collaborative study will be discussed in the section on adaptation.

It is currently unclear what factors determine whether chronic bronchiolitis or bronchioloalveolar inflammation will lead predominantly to fibrosis and narrowing on the one hand or dilatation with retraction or destruction of septa (emphysema) on the other. At least in the experimental situation, it appears that the intensity of the initial burst of elastolysis is an important determinant for subsequent development of emphysema (Kuhn and Tavassoli, 1976; Kleinerman et al., 1985). Some degree of remodeling of the centriacinar region, which falls short of significant emphysema, occurs in rats exposed to 0.4 ppm ozone (corrected to UV photometric value) for 180 days (Moore and Schwartz, 1981), but progression into obvious emphysema of this type of bronchiolization with dilatation has not been documented. Several variations on the theme of remodeling of centriacinar regions, and their relevance to the development of emphysema have been discussed in connection with the effects of chronic exposure of dogs to components of automobile exhaust (Hyde et al., 1978).

Modifiers of Chronic Responses

Factors relating to the toxicant and the exposed species of animal modify chronic responses just as they do acute responses, but the emphases shift. As far as the toxicant is concerned, for instance, a major difference is between (1) inhalation of reactive gases where dosimetry is relatively little changed over time and ceases when exposure ceases, and (2) insoluble particles where burden and dosimetry increase with time and persistence of particles extends dosing well beyond the end of exposure. The role of inhaled particles in pulmonary response to injury will receive special consideration in the next section. With regard to the influence of animal species (because the main focus in most chronic studies is on the outcome of the smoldering proximal acinar inflammation) the role the structure of this region plays becomes an important consideration. It is a reasonable assumption that extrapolation from chronic respiratory bronchiolitis in monkeys to probable effects in humans is more valid than extrapolation from chronic bronchioloalveolar inflammation in rodents to humans (Castleman et al., 1977, 1980; Fujinaka et al., 1985; Harkema et al., 1993). Other important modifiers that have more time to be expressed in chronic studies are the tendency for superimposed infection (even in supposedly "clean" animals) and immunological responses to the inhaled materials (Bice, this volume).

   If administered in sufficient dose, insoluble particles or fibers may result in physiological or anatomical changes that significantly decrease the mechanical clearance of particles from the lung, thus increasing the amount of material (or dosage) in the lung. This phenomenon has been termed "overload" and is recognized as the outcome of excessive exposure to particles or fibers particularly occurring during chronic inhalation studies (Morrow, 1992). Particles widely regarded as being innocuous may induce fibrosis and proliferative keratinizing lesions similar to those induced by highly cytotoxic particles when excessive amounts of particles are retained in the lungs (Lee et al., 1985). This overload phenomenon with associated lesions has only been described in rats but not in other laboratory rodents or man. This issue confounds the toxicologic interpretation of these studies. The issue of overload has been discussed at length (Mauderly et al., 1990; Morrow, 1992; Mermelstein et al., 1994).

CORRELATION BETWEEN ACUTE AND CHRONIC RESPONSES

There are few quantitative data on which to base mathematical correlations between acute and chronic responses. Filling this gap is an important goal because until this is done, predictions of long-term effects based on short-term studies will not be possible, and many mechanistic aspects of the pathogenesis of pulmonary disease resulting from chronic, low-level injury will not be precisely defined.

Integration of Exposure Concentration and Response over Time

The response of the respiratory tract to chronic injury by a toxicant depends on the outcome of the integrated processes of injury, damage, and repair. The major factors influencing the interrelationship of injury, damage, and repair are the nature of the toxicant, the exposure regimen, the role of inflammatory amplifiers, and the possible effect of secondary factors such as superimposed infection. Only the first two of these will be considered further.

   Understanding of the time course of responses to prolonged exposure to inhaled toxicants (time-response profiles) is key to understanding exposure concentration-time-responses relationships over long time periods. Time-response profiles will therefore be discussed first, followed by a discussion of exposure concentration-time-response relationships.

Time-Response Profiles for Reactive Gases

The shape of the time-response profile differs according to whether epithelial, intraluminal, or interstitial responses are being evaluated. For each category of response, a dose-dependent family of curves can be constructed. Because the pattern of distribution of damage throughout the respiratory tract is not uniform for reasons outlined previously, it follows that the family of time-response curves also varies with anatomic site and with the nature of the inhaled toxicant.

   The concepts to be discussed are mainly derived from studies of the effects of relatively low concentrations of ozone on the lung. The quantitative data published since these concepts were presented in the first edition of this monograph have also been for ozone exposures, and they have confirmed the validity of the concepts (Chang et al., 1992; Plopper et al., 1994b,c). Refinement of these concepts and testing them for broader applicability will depend on generation of quantitative data from studies designed for that purpose.

   Time-Response Profile of Epithelial Changes   The acute death of cells sensitive to a reactive gas such as ozone is followed by proliferation of resistant cells and persistence of a more steady hyperplastic state as indicated

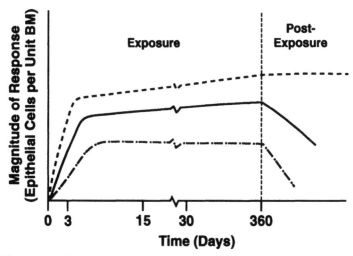

Figure 3 Schematic representation of the time-response profile of the hyperplastic change in bronchioloalveolar epithelium during and after exposure to a low concentration of a reactive gas such as ozone. The broken lines indicate variations in response across a narrow, low-dose range.

schematically in Fig. 3. Cell turnover measured by labeling index peaks as early as 2–3 days after onset of severe injury as denuded epithelium is repopulated. The peak can be delayed if injury is less severe, but the mitotic index subsequently returns closer to control levels as a reflection of the more steady state (Eustis et al., 1981; Moore and Schwartz, 1981; Schwartz, 1987; Rajni and Fritzer, 1994).

Chang et al. (1992) measured the changes in number, volume, size, and surface area of type I and type II cells in the proximal alveolar regions of rats chronically exposed to a pattern of ozone concentration simulating daily fluctuations in the South Coast (Los Angeles) air basin. They found that change in type I cell number was essentially the same as that shown in Fig. 4. The number of type II cells increased over age-matched controls over the 78-week exposure, but the percentage increase was greater because of an apparent reduction in number of type II cells in the 78-week control groups. Type I cells were also thicker and covered less surface area of basement membrane.

Two other important features of the epithelial response are the progressive bronchiolization referred to earlier (Barr et al., 1988; Pinkerton et al., 1993) and development of micronodular proliferations of cells in bronchioles after moderately severe chronic injury (Hyde et al., 1978; Harkema et al., 1984; Kleinerman et al., 1985).

Whether or not epithelial changes are reversible in post-exposure periods seems to depend heavily on duration of exposure. Short-term changes of a mild to moderate degree which do not involve significant alterations in basal lamina and interstitium are usually readily reversible. Epithelial changes

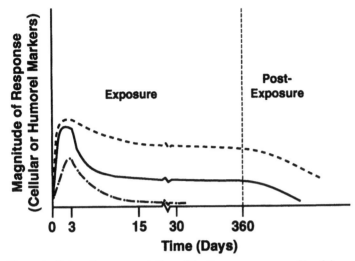

Figure 4 Schmatic representation of the time-response profile of the markers of intra-luminal exudation during and after exposure to a low concentration of a reactive gas such as ozone. The broken lines indicate variations in response across a narrow, low-dose range. The timing of the acute, peak response will vary according to the inflammatory marker measured.

present after long-term exposure persist, however, at least to some degree (Hyde et al., 1978; Harkema et al., 1984; Kleinerman et al., 1985; Tyler et al., 1987; Chang et al., 1992). The extent to which tissue remodeling has occurred appears to be an important factor. For instance, in bronchiolization there is remodeling which can permanently alter cell-to-cell and cell-to-matrix inter-actions. In less severe cases, thickening of basement membrane (possibly with altered collagen types) may be a factor in persistence of epithelial changes.

Time-Response Profile of Intraluminal Exudation The magnitude of in-traluminal exudation associated with injury by a reactive gas is greatest 2–3 days after onset of the injury. This correlates with acute epithelial necrosis and release of inflammatory mediators. The rate of generation and release of inflammatory mediators subsides as the epithelium becomes populated by more resistant cells. This is accompanied by a diminution in the amount of intraluminal exudate in the face of continued exposure (Boorman et al., 1980; Eustis et al., 1981). This is shown schematically in Fig. 4 which also indicates that in the post-exposure period, the intraluminal exudate decreases fairly rapidly in contrast to the epithelial and interstitial abnormalities.

Changes in characteristics of the intraluminal exudate with time as moni-tored by BALF are referred to by Henderson (this volume).

Time-Response of Interstitial Changes Interstitial fibrosis (with or with-out accompanying bronchiolization) and the type of remodeling leading to

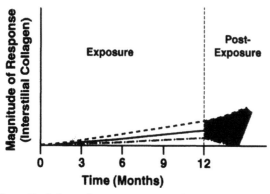

Figure 5 Schematic representation of the time-response profile of interstitial fibrosis during and after exposure to a low concentration of a reactive gas such as ozone. The broken lines indicate variations across a narrow, low-dose range. The shaded area containing a question mark (?) indicates the uncertainty as to when the rate of fibroplasia might increase in the post-exposure period or when it is at least partially reversible.

emphysema are the two most important chronic changes because of their irreversibility. Of these, interstitial fibrosis is the feature most often recorded as a response to inhaled toxicants. The time course of interstitial fibrosis during continuous exposure to a reactive gas is illustrated schematically in Fig. 5. Whereas the course of interstitial inflammation is similar to that of intraluminal changes (Fig. 4), there is a slowly progressive but not necessarily linear increase in interstitial collagen.

The issue of reversibility of interstitial fibrosis in post-exposure periods is far from straightforward. Very slight increase in interstitial matrix, including collagen, appears to be partially resolved by 17 weeks after a 78-week exposure to simulated ambient ozone (Chang et al., 1992). Thickening of basement membrane found in the study did not diminish, however, in the post-exposure period. In other situations, there is evidence that the rate of fibrosis (collagen deposition) can actually increase in the post-exposure period (Hyde et al., 1978, 1985; Last et al., 1984). The extent to which fibroplasia is either progressive or reversible in post-exposure periods (and the mechanisms responsible) are still important topics for investigation because their implications for enhanced recruitment of irreversible damage in intermittent exposure regimens such as occur with urban air pollution or occupational hazards.

The concept that the post-exposure period might be a period of worsening rather than recovery is also valid for some other forms of injury. A good example is the progressive development of pulmonary emphysema in hamsters following a pulse dose of elastase (Kuhn and Tavassoli, 1976).

A schematic comparison of the time-response profiles for epithelial, intraluminal, and interstitial changes is provided in Fig. 6.

Reversibility of Changes Caused by Reactive Gases These changes have already been mentioned under time-response profiles of epithelial, intraluminal, and interstitial compartments. They will be summarized here.

Intraluminal and interstitial inflammatory changes usually resolve rapidly and completely unless inhaled materials are highly antigenic.

The degree of reversibility for epithelial and interstitial changes is heavily influenced by duration of exposure. Especially where tissue remodeling has occurred, the epithelial and interstitial changes including persistent changes in basement membranes persist in the post-exposure period.

The situation is unclear with regard to interstitial fibrosis. There might be partial resolution in some instances. In others, there has been evidence for an increased rate of fibroplasia in the post-exposure period. As a corollary to this statement, it is important to differentiate between strongly cross-linked collagen (scar type) and matrix plus poorly cross-linked fibrils when evaluating studies on reversibility of fibrosis.

Considering all these factors, it is more appropriate to use the term "post-exposure period" than "recovery period."

EXPOSURE CONCENTRATION-TIME-RESPONSE RELATIONSHIPS FOR REACTIVE GASES

Most of the literature on this topic concerns evaluations of the validity of Haber's law ($C \times T = K$) for acute exposure situations. The broadest-based

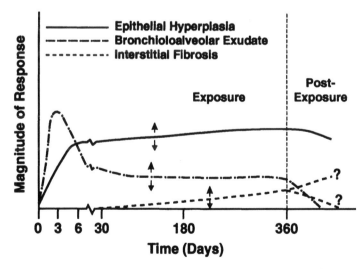

Figure 6 Comparison of the time-response profiles for epithelial hyperplasia, bronchioloalveolar exudate, and interstitial fibrosis in the centriacinar region of lung exposed to a constant dose rate of a low concentration of a reactive gas such as ozone. The arrows signify that the actual magnitude of the response varies according to dose.

mathematical approach is by Zwart et al. (1990, 1992). Review of several recent papers on ozone (Hazucha et al., 1992; Highfill et al., 1992) or ozone and nitrogen dioxide (Gelzleichter et al., 1992) reveals that Haber's law comes closest to being true when tested over a short period of time (up to 3 days) and when the response is measured by the amount of protein in BALF. This is not surprising because it comes closest to the response used by Haber— death from acute pulmonary edema. As soon as the time frame for exposure is extended and/or the response is measured by some parameter other than that of acute exudative damages, the concept of C × T no longer is valid for the reasons described below.

Influence of Exposure Regimen on Effects of Reactive Gases   Studies with ozone—the most widely investigated of the reactive gases (Menzel, 1984; Dungworth et al., 1985b)—have resulted in two important concepts. The first is that because the cycle of injury, damage, and repair proceeds once initiated, regardless of continued exposure, intermittent or periodic expo- sures can have a similar magnitude of effect to continuous exposure for any given exposure level. The second is that because cells most susceptible to injury are replaced by more resistant cells, the rate at which cumulative dam- age occurs is reduced during continued exposure. The latter concept will be discussed subsequently under adaptation. Both these concepts invalidate at- tempts to simply apply C × T in calculating equivalent damage over wide ranges of concentration and time. Although evidence is accumulating that in most cases C is more important than T, the mathematical functions to be applied cannot be determined from the available quantitative data.

The initiation of a sequence of injury, damage, and repair that proceeded largely independent of continued exposure was first demonstrated in short- term studies by Stephens et al. (1974) who found that bronchioloalveolar lesions produced in rats 48 hours after the start of exposure to 0.9 ppm ozone were similar regardless of whether the ozone exposure was limited to the first 8 hours or continued throughout the entire 48 hours. Similarly, pulmonary responses in rats after 7 days of exposure to ozone for 8 hours/day were only slightly reduced compared to responses after 7 days of continuous ex- posure (Schwartz et al., 1976). Concentrations of ozone evaluated were 0.16, 0.4, and 0.64 ppm (corrected from neutral-buffered KI to UV photo- metric standard). If the rats on the 8-hour-per-day schedule had been ex- posed during their nighttime activity when they probably would have re- ceived a higher effective dose of ozone (van Bree et al., 1992), the findings might have been closer. An analogous type of result was found for respiratory epithelium by Hotchkiss et al. (1991). On a more extensive timeframe, the effects of continuous and episodic exposure regimens were compared in rats using lung collagen content as the quantitative marker of response (Last et al., 1984). Rats exposed continuously (23.5 hours/day) for 90 days to 0.64 or 0.96 ozone were compared to rats exposed to the same concentrations for 8 hours/day for 5 consecutive days in each of seven, 2-week periods during a total of 90 days. Although the number of hours of exposure over the 90 days was 2115 compared to 280 (a 7.5-fold difference), the percentage increase in collagen content over control at the 0.96-ppm level of ozone was approximately 21% compared to 16% (a one-third difference which was not statistically significant). Thus, we have important evidence that episodic exposure might be as damaging as daily exposure, especially if effect is measured in terms of cumulative, largely irreversible changes. Similar conclusions concerning the effect of daily and episodic exposures were reached by Barr et al. (1990).

The second concept mentioned at the beginning of this section—the amount of epithelial damage and inflammation can lessen during continued, low-level exposure—will be considered further in the section on adaptation. A point to be emphasized here, however, is that comparison of intermittent and continuous exposures involves analysis of the sequence of cellular injury, damage, and repair whereas adaptation involves persistent phenotypic changes in cells either continuously exposed to the toxicant or exposed frequently enough (e.g., daily) for a continuing effect on the phenotypic expression of the cells.

Recognition of the importance of the sequence of injury, damage, and repair can resolve apparent paradoxes in the use of C × T. Swenberg et al. (1986) exposed rats and mice for 3 days to formaldehyde at a constant daily C × T provided by 12 ppm × 3 hours, 6 ppm × 6 hours, or 3 ppm × 12 hours, respectively. Using cell proliferation as the index of effect, they found that in the anterior nasal epithelium there was a similar magnitude of response in the three exposure groups, and they interpreted this to mean that use of cumulative exposure as measured by C × T was valid. In contrast, in more posterior and less susceptible regions of the nasal epithelium, the effect was strictly concentration-dependent. An alternative explanation for the findings in the anterior respiratory epithelium is that 3 ppm formaldehyde exerted maximum damage to the vulnerable cells and that higher concentrations (6 or 12 ppm) had no additional effect. In this type of situation, the use of C × T is misleading.

To summarize the complexity of the pathobiology of the respiratory tract with its shifting balance of injury, damage, adaptation and repair, means that effect is more concentration (dose-rate) dependent than time dependent. Even more importantly, experimental regimens designed to compare effects of different exposure profiles will give misleading results unless they are evaluated at the end of the same total elapsed time and unless they evaluate several parameters relevant to cumulative and most critical pulmonary damage. Using these criteria, for instance the study of Chang et al. (1991b) which relied on alveolar type I cell parameters measured after different total time intervals cannot be used to show a linear relationship between ozone effect and C × T.

Adaptation, Tolerance, and Attenuation

There is still confusion surrounding these terms when they are used in inhalation toxicology to refer to changes with time in the response to reactive gases, principally ozone. The confusion exists mainly because the terms are used globally instead of precisely. "Adaptation" has broad biologic meaning, but in pathobiology it refers to the response of cells or cell populations to a changed environment with the net result of the response being a minimization of harmful effects. "Tolerance" is a term used mainly in toxicology and is generally considered to be denoted by an absence or lessening of response with repeated exposure to a chemical agent. Because use of the term "tolerance" grew out of studies on protection against death caused by acute pulmonary edema, there is a tendency to view the state of tolerance as being one in which there is no ill effect of continued exposure once tolerance is established. This viewpoint is incorrect when for instance, attention is focused on changes in centriacinar regions of the lung during long-term exposure to ozone (discussed later in this section). "Attenuation" is a term used for lessening of some particular response without implication as to the mechanisms responsible. The term is especially applied to the lessening of pulmonary responses in humans exposed to ozone on consecutive days (Folinsbee et al., 1994).

To avoid confusion, it must be recognized that the specific manifestations of adaptation/tolerance/attenuation under investigation in any one study must be precisely defined, and findings from one study can be extrapolated to another for mechanistic interpretation only when the identical phenomenon is involved. The three terms represent different faces of any one of a variety of biologic responses and should not be used to imply specificity for any one response. With this is background, the findings of Tepper et al. (1989) and Folinsbee et al. (1994) are expected. There is no correspondence between the time course of attenuation of pulmonary functional changes in humans, rats, or monkeys exposed to ozone and the time course of injury, damage, and adaptation in the centriacinar regions of the lung.

The replacement of sensitive epithelial cells by a more resistant population (with corresponding reduction in inflammation) forms the basis for adaptation of airway epithelium as mentioned previously. This was documented initially by Boorman et al. (1980), Moore and Schwartz (1981), and Eustis et al. (1981). Adaptation can take place by population shift to a cell type inherently more resistant to damage by induction in cells of metabolic changes conferring a degree of protection (e.g., antioxidant enzymes), changes in the amount of secretory product present, or a combination of these. Recent evidence, in fact, indicates that the relative importance of these adaptive mechanisms in the case of ozone differ from one anatomic site to another. This evidence comes from a series of papers mainly derived from the combined NIEHS and HEI study on long-term effects of ozone (Plopper et al., 1994b,c; Dodge et al., 1994). This series of papers is the most detailed evaluation to date of *in situ* adaptive changes to ozone and relies heavily on airway microdissection for precise identification of site-specific differences in response. Plopper and associates found that in the distal trachea of rats exposed for 20 months to 1.0 ppm of ozone there was a decrease in intraepithelial mucin storage, an increase in the amount of superoxide dismutase and some decrease in epithelial thickness. In the centriacinar region there were bronchiolization (an increase in proportion of Clara cells to ciliated cells) increases in the three antioxidant enzymes measured (superoxide dismutase, glutathione peroxidase, and glutathione S-transferase), and an increase in the amount of stored Clara cell secretory protein (CC10). A dose response was observed between 0.5 and 1.0 ppm of ozone and for some parameters, from 0.12 to 0.5 to 1.0 ppm. The series of papers also reveals the complexity of differences between different levels of airway branching and different path lengths.

The same series of papers (Plopper et al., 1994b,c; Dodge et al., 1994) also presents an interesting example of the implications of using the term tolerance. The papers describe adaptive changes that are more pronounced at the highest level of ozone exposure. The authors' inference is that the cells are more tolerant to ozone at the highest level of exposure. While this might be true in one sense, there is more progressive damage at the highest level of exposure. Therefore, the degree of tolerance in the centriacinar region is insufficient to prevent continued damage at 1.0 ppm of ozone (see Catalano et al., 1995). That is why the term "adaptation" is less semantically confusing because it focuses on the real issue which is the extent to which the rate of cumulative damage is lessened.

TIME-RESPONSE RELATIONSHIPS FOR PARTICLES

The situation with regard to chronic injury by particles (and also fibers) is more complicated than is the case for reactive gases because of the persistence of particles and the influence of the dynamics of their deposition, retention, and clearance in the lung (Schlesinger, this volume). The responses to particles are modified by the dose of the material, the amount of material in the lungs, the toxicity of the particles, and in the case of particulate chemicals, metabolism of the particle. Solubility of the particle is a key characteristic governing the length of time the particle remains in the lung. Very soluble particles will leave the lung rapidly, resulting in a dose pattern to the lung that is similar to inhaled gases. The more usual situation, however, is a protracted dose to the lung after cessation of exposure because most particles and fibers of interest in pulmonary toxicology are relatively insoluble.

The intrinsic toxicity of the particle affects the damage to the pulmonary cells. Particles that are cytotoxic (such as silica) are generally more damaging to the lung than noncytotoxic compounds (such as titanium dioxide). The intrinsic toxicity may have another effect that should not be overlooked. The pulmonary lesions induced by toxic particles may alter clearance, thus increasing the retained dose to the lung or the distribution of that dose within the lung.

Chemicals inhaled in a particulate form will behave as particles in the lung in regard to deposition. If the chemical is metabolized in the lung, however, the clearance will be markedly affected. The issue of metabolism in the respiratory tract is covered elsewhere in this volume (Dahl).

Influence of Exposure Regimen on Effects of Insoluble Particles

Theoretical analyses and models of the overload phenomenon have predicted that a rapid rate of delivery of particles to the lung would overwhelm the normal clearance mechanisms of the lung and result in a higher lung burden of particles and greater inflammatory response than a slower rate of particle delivery (Morrow, 1992). This does appear to be the case with high concentrations of ultrafine (< 20 nm) particles (Ferin et al., 1992). With particle sizes and doses more commonly encountered, however, this increased effect with high delivery rates does not occur. A study specifically designed to determine the effect of exposure pattern on the accumulation of particles and the response of the lung to inhaled particles showed no increased effect of high, infrequent doses (Henderson et al., 1992). F344 rats were exposed over a 12-week period to the same weekly concentration × time product of carbon black particles, but at different exposure rates: 3.5 mg/m^3, 16 hours/day, 7 days/week; 13 mg/m^3, 6 hours/day, 5 days/week; or 98 mg/m^3, 4 hours/day, 1 day/week. The particles were in the respirable range (~2.1 μm MMAD). The results of this study showed no significant difference among exposure patterns on concentration of particles in the lung at the end of exposure or at the end of the 24-week recovery period. In addition, no effect was seen on inflammation as determined by BALF or histopathology. This study shows that the application of the concentration × time product is useful for some situations with inhaled relatively nontoxic particles.

Persistence of Pulmonary Lesions after Exposure

An important characteristic of toxic inhaled particles is their ability to initiate or result in lesions that persist after the number of particles in the lung are markedly reduced after cessation of exposure. With the cessation of exposure, the number of particles in the lung decreases with time, dependant upon mechanical clearance and solubility as noted earlier. Lesions induced in the lung such as emphysema (Hahn and Hobbs, 1979) or fibrosis (Lundgren et al., 1991) may retard the mechanical clearance of particles from the lung. This prolonged retention of particles may, in turn, result in a greater number or persistence of lesions. Silica and refractory ceramic fibers are good examples of this phenomenon (Hesterberg et al., 1993). In some cases, the lesions appear to progress after cessation of exposure. For example, gallium oxide causes scant lesions in the lung after a 4-week exposure, but after a 2 month post-exposure time, marked pulmonary lesions are present (Wolf et al., 1986). Lesions of alveolar histiocytosis, inflammatory infiltrates, or alveolar epithelial hyperplasia may regress after cessation of exposure; however, septal fibrosis does not. Regression is usually associated with less cytotoxic compounds such as polyolefin fibers (Hesterberg et al., 1992) or colloidal silica (Lee and Kelly, 1992). With highly toxic particles (such as beryllium oxide) one brief inhalation exposure is sufficient to achieve a dose to the lung that will result in both acute and chronic effects. The more usual case, however, is that repeated or chronic exposures are needed to induce significant lesions with inhaled particles or fibers. Repeated exposures are needed for such well-known toxic compounds as nickel subsulfide or asbestos.

CORRELATION BETWEEN *IN VITRO* AND *IN VIVO* FINDINGS

There is increasing use of *in vitro* methods for the investigation of mechanisms responsible for significant pathological and clinical endpoints observed in inhalation toxicology. Probably the most important question with regard to *in vitro* studies concerns their relevance to a significant disease endpoint in animals or humans. Individual mechanisms can be well explored in isolated *in vitro* systems (e.g., cell culture), but the relative importance in disease pathogenesis of the large array of factors which can be studied *in vitro* can only be determined by *in vivo* approaches. The challenge is to integrate findings from isolated cell systems, organ or organotypic cultures, and well-focused *in vivo* investigations.

Findings from isolated systems are not usually satisfying from the point of view of inhalation toxicology, especially if only one or a few parameters are measured. Relevance to *in vivo* events should clearly be established at the outset. In light of the evidence of the complexity of the pulmonary response to inhaled toxicants presented in this chapter, it should be clear that most progress will be made where multiple endpoints are assessed, preferably in the same laboratory, and where *in vitro* and *in vivo* findings are carefully integrated. A final point that follows from the complexity of the pulmonary response is that mechanistic questions must be posed against the background of the best available knowledge about the sequence of pathologic events leading *in vivo* to the disease endpoint of interest.

CONCLUSIONS

Relative to the concepts that have been presented, the critical needs are:

1   recognition of site-specific differences in response of the respiratory tract to inhaled toxicants;

2 recognition that time-response profiles differ according to the subsets of responses being measured;

3 generation of quantitative data for the various parameters of response to enable construction of mathematical models for interrelationships among the subsets of responses (i.e., epithelial, intraluminal, and interstitial);

4 use of the quantitative data to derive the mathematical functions for exposure concentration-time-response relationships while recognizing that the specific functions will be dependent on the particular effect being measured;

5 precision in the use of terms such as adaptation, tolerance, and attenuation with proper attention to the implications of their usage; and

6 need for integrated *in vitro* and *in vivo* studies in order to demonstrate the relevance of the former.

REFERENCES

Adamson, IYR, Bowden, DH: The type 2 cell as progenitor of alveolar epithelial regeneration. A cytodynamic study in mice after exposure to oxygen. Lab Invest 30:35–42, 1974.

Adamson, IYR, Bowden, DH: Bleomycin-induced injury and metaplasia of alveolar type 2 cells. Am J Pathol 96:531–544, 1979.

Adamson, IYR, Bowden, DH: Crocidolite-induced pulmonary fibrosis in mice: Cytokinetic and biochemical studies. Am J Pathol 122:261–267, 1986.

Appelman, LM, Woutersen, RA, Feron, VJ: Inhalation toxicity of acetaldehyde in rats. I. Acute and subacute studies. Toxicology 23:293–307, 1982.

Barr, BC, Hyde, DM, Plopper, CG, Dungworth, DL: Distal airway remodeling in rats chronically exposed to ozone. Am Rev Respir Dis 137:924–938, 1988.

Barr, BC, Hyde, DM, Plopper, CG, Dungworth, DL: A comparison of terminal airway remodeling in chronic daily versus episodic ozone exposure. Toxicol Appl Pharmacol 106(3):384–407, 1990.

Barrow, CS, (ed): Toxicology of the Nasal Passages. New York: Hemisphere, 1986.

Barry, BE, Wong, KC, Brody, AR, Crapo, JD: Reaction of rat lungs to inhaled chrysolite asbestos following acute and subchronic exposures. Exp Lung Res 5:1–21, 1983.

Barry, BE, Miller, FJ, Crapo, JD: Effects of inhalation of 0.12 and 0.25 parts per million ozone on the proximal alveolar region of juvenile and adult rats. Lab Invest 53:692–704, 1985.

Bernfeld, P, Homburger, F, Soto, E, Pai, KJ: Cigarette smoke inhalation studies in inbred Syrian golden hamsters. JNCI 63:675–689, 1979.

Bernfeld, P, Homburger, F, Soto, E: Subchronic cigarette smoke inhalation studies in inbred Syrian golden hamsters that develop laryngeal carcinoma upon chronic exposure. JNCI 71:619–623, 1983.

Bogdanffy, MS, Randall, HW, Morgan, KT: Histochemical localization of aldehyde dehydrogenase in the respiratory tract of the Fischer-344 rat. Toxicol Appl Pharmacol 82:560–567, 1986.

Bogdanffy, MS, Randall, HW, Morgan, KT: Biochemical quantitation and histochemical localization of carboxylesterase in the nasal passages of the Fischer-344 rat and B6C3F$_1$ mouse. Toxicol Appl Pharmacol 88:183–194, 1987.

Bogdanffy, MS: Biotransformation enzymes in the rodent nasal mucosa: The value of a histochemical approach. Environ Health Perspect 85:177–186, 1990.

Bolon, B, Bonnefoi, MS, Roberts, KC, Marshall, MW, Morgan, KT: Toxic interactions in the rat nose: Pollutants from soiled bedding and methyl bromide. Toxicol Pathol 19:571–579, 1991.

Boorman, GA, Schwartz, LW, Dungworth, DL: Pulmonary effects of prolonged ozone insult in rats: Morphometric evaluation of the central acinus. Lab Invest 43:108–115, 1980.

Brittebo, EB, Eriksson, C, Feil, V, Bakke, J, Brandt, I: Toxicity of 2,6-dichlorothiobenzamide (Chlorthiamid) and 2,6-dichlorobenzamide in the olfactory nasal mucosa of mice. Fundam Appl Toxicol 17:92–102, 1991.

Brody, AR, Hill, LH, Warheit, DB: Induction of early alveolar injury by inhaled asbestos and silica. Fed Proc 44:2596–2601, 1985.

Butcher, JR, Elwell, MR, Thompson, BM, Chou, BJ, Renne, R, Ragan, HA: Inhalation toxicity studies of cobalt sulfate in F344/N rats and B6C3F$_1$ mice. Fundam Appl Toxicol 15(2):357–372, 1990.

Buckley, LA, Jiang, XZ, James, RA, Morgan, KT, Barrow, CS: Respiratory tract lesions induced by sensory irritants at the RD$_{50}$ concentration. Toxicol Appl Pharmacol 74:417–429, 1984.

Buckley, LA, Morgan, KT, Swenberg, JA, James, RA, Hamm, TE, Jr, Barrow, CS: The toxicity of dimethylamine in F-344 rats and B6C3F$_1$ mice following a 1-year inhalation exposure. Fundam Appl Toxicol 5:341–352, 1985.

Burger, GT, Renne, RA, Sagartz, JW, Ayres, PH, Coggins, CRE, Mosberg, AT, Hayes, AW: Histologic changes in the respiratory tract induced by inhalation of xenobiotics, physiologic adaptation or toxicity? Toxicol Appl Pharmacol 101:521–542, 1989.

Cardoso, WV, Gekhon, HS, Hyde, DM, Thurlbook, WM: Collagen and elastin in human pulmonary emphysema. Am Rev Respir Dis 147(4):975–981, 1993.

Castleman, WL, Tyler, WS, Dungworth, DL: Lesions in respiratory bronchioles and conducting airways of monkeys exposed to ambient levels of ozone. Exp Mol Pathol 26:384–400, 1977.

Castleman, WL, Dungworth, DL, Schwartz, LW, Tyler, WS: Acute respiratory bronchiolitis: An ultrastructural and autoradiographic study of epithelial cell injury and renewal in rhesus monkeys exposed to ozone. Am J Pathol 98:811–840, 1980.

Catalano, PJ, Chang, L-Y, Harkema, JR, Kaden, DA, Last, JA, Mellick, PW, Parks, WC, Pinkerton, KE, Radhakrishnamurthy, B, Ryan, LM, and Szarek, JL: Consequences of prolonged inhalation of ozone on F344/N rats: Collaborative studies. Part XI: Integrative, summary. Research Report Number 65, pp. 1–137. Cambridge, MA: Health Effects Institute, 1995.

Chang, JCF, Gross, EA, Swenberg, JA, Barrow, CS: Nasal cavity deposition, histopathology and cell proliferation after single or repeated formaldehyde exposure in B6C3F$_1$ mice and F344 rats. Toxicol Appl Pharmacol 68:161–176, 1983.

Chang, LY, Mercer, RR, Pinkerton, KE, Crapo, JD: Quantifying lung structure. Experimental design and biological variation in various models of lung injury. Am Rev Respir Dis 143(3):625–634, 1991a.

Chang, LY, Miller, FJ, Ultman, J, Huang, Y, Stockstill, BL, Grose, E, Graham, JA, Ospital, JJ, Crapo, JD: Alveolar epithelial cell injuries by subchronic exposure to low concentrations of ozone correlate with cumulative exposure. Toxicol Appl Pharmacol 109(2):219–234, 1991b.

Chang, LY, Huang, Y, Stockstill, BL, Graham, JA, Grose, EC, Monacho, MC, Miller, FJ, Costa, DL, Crapo, JD: Epithelial injury and interstitial fibrosis in the proximal alveolar regions of rats chronically exposed to a simulated pattern of urban ambient ozone. Toxicol Appl Pharmacol 115(2):241–252, 1992.

Chang, LY, Stockstill, BL, Mánache, MG, Mercer, RR, Crapo, D: Consequences of prolonged inhalation of ozone on F344/N rats: Collaborative studies. Part VIII. Morphometric analysis of structural alterations in alveolar regions. Research report number 65, Health Effects Institute, Cambridge, MA, 1995.

Christensen, TG, Korthy, AL, Snider, GL, Hayes, JA: Irreversible bronchial goblet cell metaplasia in hamsters with elastase-induced panacinar emphysema. J Clin Invest 59:397–404, 1977.

Cohn, LA, Adler, KB: Interactions between airway epithelium and mediators of inflammation. Exp Lung Res 18:299–322, 1992.

Cross, AE, Halliwell, B, Allen, A: Antioxidant protection: A function of tracheobronchial and gastrointestinal mucus. Lancet 1:1328–1330, 1984.

Dahl, AR, Hadley, WH: Nasal cavity enzymes involved in xenobiotic metabolism: Effects on the toxicity of inhalants. CRC Crit Rev Toxicol 21:345–372, 1991.

Dahl, AR, Schlesinger, RB, D'A Heck, H, Medinsky, MA, and Lucier, GW: Comparative dosimetry of inhaled materials: Differences among animal species and extrapolation to man. Fundam Appl Toxicol 16:1–13, 1991.

Davis, IMG, Beckett, ST, Bolton, RE, Collings, P, Middleton, AP: Mass and number of fibers in the pathogenesis of asbestos-related lung disease in rats. Br J Cancer 37:673–688, 1978.

Dodge, DE, Rucker, RB, Pinkerton, KE, Haselton, CJ, Plopper, CG: Dose-dependent tolerance to ozone. III. Elevation of intracellular Clara cell 10-kDa protein in central acini of rats exposed for 20 months. Toxicol Appl Pharmacol 127(1):109–123, 1994.

Dubaybo, BA, Rubeiz, GJ, Fligiol, SE: Dynamic changes in the functional characteristics of the interstitial fibroblast during lung repair. Exp Lung Res 18(4):461–477, 1992.

Dungworth, DL, Tyler, WS, Plopper, CG: Morphological methods for gross and microscopic pathology. In: Toxicology of Inhaled Materials, edited by HP Witschi, JD Brain, pp. 229–258. New York: Springer-Verlag, 1985a.

Dungworth, DL, Goldstein, E, Ricci, PF: Photochemical air pollution—part II (specialty conference). West J Med 142:523–531, 1985b.

Eustis, SL, Schwartz, LW, Kosch, PC, Dungworth, DL: Chronic bronchiolitis in non-human primates after prolonged ozone exposure. Am J Pathol 105:121–137, 1981.

Evans, MJ, Cabral, LJ, Stephens, RJ, Freeman, G: Renewal of alveolar epithelium in the rat following exposure to NO_2. Am J Pathol 70:175–198, 1973.

Evans, MJ, Johnson, LV, Stephens, RJ, Freeman, G: Renewal of the terminal bronchiolar epithelium in the rat following exposure to NO_2 or O_3. Lab Invest 35:246–257, 1976.

Evans, MJ, Shami, SG, Cabral-Anderson, LJ, Dekker, NP: Role of non-ciliated cells in renewal of the bronchial epithelium of rats exposed to NO_2. Am J Pathol 123:126–133, 1986.

Ferin, J, Oberdörster, G, Penney, DP: Pulmonary retention of ultrafine and fine particles in rats. Am J Respir Cell Mol Biol 6:535–542, 1992.

Feron, VJ, Kruysse, A, Dreef-Van der Meulen, HC: Repeated exposure to furfural vapor: 13 weeks study in Syrian golden hamsters. Zentralbl Bakteriol Parasitenkd Infektionskr Hyg Abt 1 Orig Reihe B 168:442–451, 1979.

Folinsbee, LJ, Horstman, DH, Kehrl, HR, Harder, S, Abdul-Salaam, S, Ives, PJ: Respiratory responses to repeated prolonged exposure to 0.12 ppm ozone. Am J Respir Crit Carc Med 149(1):98–105, 1994.

Foster, JR, Green, T, Smith, LL, Lewis, RW, Hext, PM, Wyatt, I: Methylene chloride—an inhalation study to investigate pathological and biochemical events occurring in the lungs of mice over an exposure period of 90 days. Fundam Appl Toxicol 18(3):376–388, 1992.

Fujinaka, LE, Hyde, DM, Plopper, CG, Tyler, WS, Dungworth, DL, Lollini, W: Respiratory bronchiolitis following long-term ozone exposure in bonnet monkeys: A morphometric study. Exp Lung Res 8:167–190, 1985.

Gaskell, BA, Hext, PM, Pigott, GH, Hodge, MCH, Tinston, DJ: Olfactory and hepatic changes following inhalation of 3-trifluoromethyl pyridine in rats. Toxicology 50:57–68, 1988.

Gelzleichter, TR, Witschi, H, Last, JA: Concentration-response relationships of rat lungs to exposure to oxidant air pollutants: A critical test of Haber's Law for ozone and nitrogen dioxide. Toxicol Appl Pharmacol 112(1):73–80, 1992.

Genter, MB, Llorens, J, O'Callaghan, JP, Peele, DB, Morgan KT, Crofton, KM: Olfactory toxicity of B,B'-iminodipropionitrile in the rat. J Pharm Exp Ther 263:1432–1439, 1992.

Getchell, TV, Doty, RL, Bartoshuk, LM, Snow, JBJ: Smell and Taste in Health and Disease. New York: Raven Press, 1991.

Giddens, WEJ, Fairchild, GA: Effects of sulfur dioxide on the nasal mucosa of mice. Arch Environ Health 25:166–173, 1972.

Gopinath, C, Prentice, DE, Lewis, DJ: The respiratory system. In: Atlas of Experimental Toxicologic Pathology. Volume 13, Current Histopathology, edited by GA Gynesham, pp. 22–42. Norwell, MA: MTP Press, 1987.

Gradziadei, PPC: Cell dynamics in the olfactory mucosa. Tissue and Cell 5:113–131, 1973.

Gradziadei, PPC, Gradziadei, GAM: Neurogenesis and neuron regeneration in the olfactory system of mammals. I. Morphological aspect of differentiation and structural organization of the olfactory sensory neurons. J Neurocytol 8:1–8, 1979a.

Gradziadei, GAM, Gradziadei, PPC: Neurogenesis and neuron regeneration in the olfactory system of mammals. III. Degeneration and reconstitution of the olfactory sensory neurons after axotomy. J Neurocytol 8:197–213, 1979b.

Gross, EA, Swenberg, JA, Field, S, Popp, JA: Comparative morphometry of the nasal cavity in rats and mice. J Anat 135:83–88, 1982.

Gross, EA, Patterson, DL, Morgan, KT: Effects of acute and chronic dimethylamine exposure on the nasal mucociliary apparatus of F344 rats. Toxicol Appl Pharmacol 90:359–376, 1987.

Gross, EA, Mellick, PW, Kari, FW, Miller, FJ, Morgan, KT: Histopathology and cell replication responses in the respiratory tract of rats and mice exposed by inhalation to glutaraldehyde for up to thirteen weeks. Fundam Appl Toxicol 23:348–362, 1994.

Gross, P, Pfitzer, EA, Hatch, TF: Alveolar clearance: Its relation to lesions of the respiratory bronchiole. Am Rev Respir Dis 94:10–19, 1966.

Hadley, WM, Dahl, AR: Cytochrome P-450-dependent monooxygenases in nasal membranes of six species. Drug Metab Dispos 11:275–276, 1983.

Hahn, FF, Hobbs, CH: The effect of enzyme-induced pulmonary emphysema in Syrian hamsters on the deposition and long-term retention of inhaled particles. Arch Environ Health 34:302–311, 1979.

Harkema, JR, Plopper, C, St. George, J, Fujinaka, L, Lollini, L, Hyde, D, Tyler, W, Dungworth, DL: Persistent centriacinar lesions in bonnet monkeys following cessation of exposures to ambient levels of ozone. Am Rev Respir Dis 129:A137, 1984.

Harkema, JR, Plopper, CG, Hyde, DM, St. George, JA, Wilson, DW, Dungworth, DL: Response of the macaque nasal epithelium to ambient levels of ozone: A morphologic and morphometric study of the transitional and respiratory epithelium. Am J Pathol 128:29–44, 1987a.

Harkema, JR, Plopper, CG, Hyde, DM, St. George, JA, Dungworth, DL: Effects of an ambient level of ozone on primate nasal epithelial mucosubstances: Quantitative histochemistry. Am J Pathol 127:90–96, 1987b.

Harkema, JR: Comparative aspects of nasal airway anatomy: Relevance to inhalation toxicology. Toxicol Pathol: 19:321–336, 1991.

Harkema, JR, Plopper, CG, Hyde, DM, St. George, JA, Wilson, DW, Dungworth, DL: Response of macaque bronchiolar epithelium to ambient concentrations of ozone. Am J Pathol 143:857–866, 1993.

Harkema, JR, Hotchkiss, JA: Ozone-induced proliferative and metaplastic lesions in nasal transitional and respiratory epithelium: Comparative pathology. Inhal Toxicol 6:187–204, 1994.

Harkema, JR, Morgan, KT: Normal morphology of the nasal passages in laboratory rodents. In: International Life Sciences Monographs on Pathology of Laboratory Animals, Respiratory System. Edited by PC Jones and U Mohr, Berlin: Springer-Verlag, 1995a, (in press).

Harkema, JR, Morgan, KT: Proliferative and metaplastic lesions in nonolfactory nasal epithelia induced by inhaled chemicals. In: International Life Sciences Monographs on Pathology of Laboratory Animals, Respiratory System. Edited by PC Jones and U Mohr, Berlin: Springer-Verlag, 1995b, (in press).

Harkema, JR, Morgan KT, Gross, EA, Catalano, PJ, Griffith, WC: Consequences of prolonged inhalation of ozone on F344/N rats: Collaborative studies. Part VII. Effects on the nasal mucociliary apparatus. Research Report Number 65, Health Effects Institute, Cambridge, MA, 1995.

Haschek, WM, Morse, CC, Boyd, MR, Hakkinen, PJ, Witschi, HP: Pathology of acute inhalation exposure to 3-methylfuran in the rat and hamster. Exp Mol Pathol 39:342–354, 1983.

Hazucha, MJ, Bates, DV, Bromberg, PA: Mechanism of action of ozone on the human lung. Am Rev Respir Dis 133:A214, 1986.

Henderson, RF, Barr, EB, Cheng, YS, Griffith, WC, Hahn, FF: The effect of exposure pattern on the accumulation of particles and the response of the lung to inhaled particles. Fund Appl Toxicol 19:367–374, 1992.

Hesterberg, TW, McConnell, EE, Miller, WC, Hamilton, R, Bunn, WB: Pulmonary toxicity of inhaled polypropylene fibers in rats. Fund Appl Toxicol 19:358–366, 1992.

Hesterberg, TW, Miller, WC, McConnell, EE, Chevalier, J, Hadley, JG, Bernstein, DM, Thevenaz, Anderson, R: Chronic inhalation toxicity of size-separated glass fibers in Fischer 344 rats. Fund Appl Toxicol 20:464–476, 1993.

Highfill, JW, Hatch, GE, Slade, R, Krissman, KM, Norwood, J, Dovlin, RB, Costa, DL: Concentration-time models for the effects of ozone on bronchoalveolar lavage fluid protein from rats and guinea pigs. Inhal Toxicol 4:1–16, 1992.

Hotchkiss, JA, Harkema, JR, Sun, JP, Henderson, RF: Comparison of acute ozone-induced nasal and pulmonary inflammatory responses in rats. Toxicol Appl Pharmacol 98:289–302, 1989.

Hotchkiss, JA, Harkema, JR, Henderson, RF: Effects of cumulative ozone exposure on ozone-induced nasal epithelial hyperplasia and secretory metaplasia in rats. Exp Lung Res 17(3):589–600, 1991.

Hotchkiss, JA, Harkema, JR: Endotoxin and cytokins attenuate ozone-induced DNA synthesis in rat transitional epithelium. Toxicol Appl Pharmacol 114:182–187, 1992.

Hotchkiss, JA, Evans, WA, Chen, BT, Finch, GL, Harkema, JR: Regional differences in the effects of mainstream cigarette smoke on stored mucosubstances and DNA synthesis in F344 rat nasal respiratory epithelium. Toxicol Appl Pharmacol, 1995, (in press).

Hurtt, ME, Morgan, KT, Working, PK: Histopathology of acute toxic responses in selected tissues from rats exposed by inhalation to methyl bromide. Fundam Appl Toxicol 9:352–365, 1987.

Hurtt, ME, Thoma, DA, Working, PK, Monticello, TM, and Morgan, KT: Degeneration and regeneration of the olfactory epithelium following inhalation exposure to methyl bromide: Pathology, cell kinetics, and olfactory function. Toxicol Appl Pharmacol 94:311–328, 1988.

Hyde, DM, Orthoefer, J, Dungworth, DL, Tyler, W, Carter, R, Lum, H: Morphometric and morphologic evaluation of pulmonary lesions in beagle dogs chronically exposed to high ambient levels of air pollutants. Lab Invest 38:455–469, 1978.

Hyde, DM, Plopper, CG, Weir, AJ, Murnane, RD, Warren, DL, Last, JA, Pepelko, WE: Peribronchiolar fibrosis in lungs of cats chronically exposed to diesel exhaust. Lab Invest 52:195–206, 1985.

Hyde, DM, Magliano, DJ, Plopper, CG: Morphometric assessment of pulmonary toxicity in the rodent lung. Toxicol Pathol 19:428–466, 1992a.

Hyde, DM, Hubbard, WC, Wong, V, Wu, R, Pinkerton, K, Plopper, CG: Ozone-induced acute tracheobronchial epithelial injury. Relationship to granulocyte emigration in the lung. Am J Respir Cell Mol Biol 6:481–497, 1992b.

Hyde, DM, King, TE, Jr, McDermott, T, Waldron, JA, Jr, Colby, TV, Thurlbeck, WM, Fling, WM, Ackerson, T, Cherniack, RM: Idiopathic pulmonary fibrosis. Quantitative assessment of lung pathology. Comparison of a semiquantitative and morphometric histopathologic scoring system. Am Rev Respir Dis 146(4):1042–1047, 1992c.

Jiang, XZ, Buckley, LA, Morgan, KT: Pathology of toxic responses to the RD_{50} concentration of chlorine gas in the nasal passages of rats and mice. Toxicol Appl Pharmacol 71:225–236, 1983.

Jiang, XZ, Morgan, KT, Beauchamp, RO, Jr: Histopathology of acute and subacute nasal toxicity. In: Toxicology of the Nasal Passages. Edited by CS Barrow. Washington, DC: Hemisphere, pp. 51–66, 1986.

Johnson, NF, Hotchkiss, JA, Harkema, JR, Henderson, RF: Proliferative responses of rat nasal epithelia to ozone. Toxicol Appl Pharmacol, 103:143–155, 1990.

Kapanci, Y, Weibel, ER, Kaplan, GP, Robinson, FR: Pathogenesis and reversibility of the pulmonary lesions of oxygen toxicity in monkeys: Ultrastructural and morphometric studies. Lab Invest 20:101–118, 1969.

Keenan, KP: Cell injury and repair of tracheobronchial epithelium. In: Current Problems in Tumour Pathology, Vol. 3: Lung Carcinomas. Edited by EM McDowell, Edinburgh: Churchill Livingstone, pp. 74–93, 1987.

Keenan, CM, Kelly, DP, Bogdanffy, MS: Degeneration and recovery of rat olfactory epithelium following inhalation of dibasic esters. Fundam Appl Toxicol 15:381–393, 1990.

Kerns, WD, Pavkov, KL, Donofrio, DJ, Gralla, EJ, Swenberg, JA: Carcinogenicity of formaldehyde in rats and mice after long-term inhalation exposure. Cancer Res 43:4382–4392, 1983.

Kilburn, KH: Functional morphology of the distal lung. Int Rev Cytol 37:153–270, 1974.

Kimbell, JS, Morgan, KT: Airflow effects on regional disposition of particles and gasses in the upper respiratory tract. Radiat Protect Dosim 38:213–219, 1991.

Kimbell, JS, Gross, EA, Joyner, DR, Godo, MN, Morgan, KT: Application of computational fluid dynamics to regional dosimetry of inhaled chemicals in the upper respiratory tract of the rat. Toxicol Appl Pharmacol 121:253–263, 1993.

Klein-Szanto, AJP, Boysen, M, Reith, A: Keratin and involucrin in preneoplastic and neoplastic lesions, distribution in the nasal mucosa of nickel workers. Arch Pathol Lab Med 111:1057–1061, 1987.

Kleinerman, J, Ip, MPC, Gordon, RE: The reaction of the respiratory tract to chronic NO_2 exposure. Monogr Pathol 26:200–210, 1985.

Kuhn, C, Tavassoli, F: The scanning electron microscopy of elastase-induced emphysema: A comparison with emphysema in man. Lab Invest 34:2–9, 1976.

Last, JA, Reiser, KM, Tyler, WS, Rucker, RB: Long-term consequences of exposure to ozone. I. Lung collagen content. Toxicol Appl Pharmacol 72:111–118, 1984.

Last, JA, Reiser, KM: Effects of pneumotoxins on lung connective tissue. In: Toxicology of Inhaled Materials. Edited by HP Witschi, JD Brain, New York: Springer-Verlag, pp. 503–535, 1985.

Lee, HK, Murlas, C: Ozone-induced bronchial hyperreactivity in guinea pigs is abolished by BW 755C or FPL 55712 but not by indomethacin. Am Rev Respir Dis 132:1005–1009, 1985.

Lee, KP, Trochimowicz, HJ, Reinhard, CF: Pulmonary response of rats exposed to titanium dioxide (TiO_2) by inhalation for two years. Toxicol Appl Pharmacol 79:179–192, 1985.

Lee, KP, Valentine, R, Bogdanffy, MS: Nasal lesion development and reversibility in rats exposed to aerosols of dibasic esters. Toxicol Pathol 3:376–393, 1992.

Lee, KP, Kelly, DP: The pulmonary response and clearance of ludox colloidal silica after a 4-week inhalation exposure in rats. Fundam Appl Toxicol 19:399–410, 1992.

Leuchtenberger, C, Leuchtenberger, R, Ruch, R, Tanaka, K, Tanaka, T: Cytological and cytochemical alterations in the respiratory tract of mice after exposure to cigarette smoke, influenza virus, and both. Cancer Res 23:555–565, 1963.

Lewis, DJ: Morphological assessment of pathological changes within the rat larynx. Toxicol Pathol 19(4 pt 1):352–357, 1991.

Lewis, JL, Rhoades, CE, Bice, DE, Harkema, JR, Hotchkiss, JA, Sylvester, DM, Dahl, AR: Interspecies comparison of cellular localization of the cyanide metabolizing enzyme rhodanese within olfactory mucosa. Anat Rec 232:620–627, 1992.

Lewis, JL, Dahl, AR: Olfactory mucosa: Composition, enzymatic localization, and metabolism. In: Handbook of Clinical Olfaction and Gustation. Edited by RL Doty, New York: Marcel Dekker, pp. 33–52, 1994.

Lewis, JL, Nikula, KJ, Novak, R, Dahl, AR: Comparative localization of carboxylesterase in F344 rat, Beagle dog, and human nasal tissue. Anat Rec 239:55–64, 1994a.

Lewis, JL, Nikula, KJ, Sachetti, LA: Induced xenobiotic-metabolizing enzymes localized to eosinophilic globules in olfactory epithelium of toxicant-exposed F344 rats. Inhal Toxicol 6:422–425, 1994b.

Lum, H, Schwartz, LW, Dungworth, DL, Tyler, WS: A comparative study of cell renewal after exposure to ozone or oxygen. Response of terminal bronchiolar epithelium in the rat. Am Rev Respir Dis 118:335–345, 1978.

Lundgren, DL, Mauderly, JL, Rebar, AH, Gillett, NA, Hahn, FF: Modifying effects of preexisting pulmonary fibrosis on biological responses of rats to inhaled ^{239}PuO$_2$. Health Phys 60(3):353–363, 1991.

Maejima, K, Suzuki, T, Numata, H, Maekawa, A, Nagase, S, Ishinishi, N: Recovery from changes in the blood and nasal cavity and/or lungs of rats caused by exposure to methanolfueled engine exhaust. J Toxicol Environ Health 39(3):323–340, 1993.

Maples, KR, Nikula, KJ, Chen, BT, Finch, GL, Griffith, WC, Jr, Harkema, JR: Effects of cigarette smoke on the glutathione status of the upper and lower respiratory tract of rats. Inhal Toxicol 5:389–401, 1993.

Mattes, PM, Mattes, WB: μ-Naphthyl butyrate carboxylesterase activity in human and rat nasal tissue. Toxicol Appl Pharmacol 114:71–76, 1992.

Mauderly, JL, Cheng, YS, Snipes, MB: Particle overload in toxicological studies: Friend or foe? J Aerosol Med 3(1):S-169, 1990.

McLaughlin, RF, Tyler, WS, Canada, RO: A study of the subgross pulmonary anatomy in various mammals. Am J Anat 108:149–164, 1961.

Menzel, DB: Ozone: An overview of its toxicity in man and animals. J Toxicol Environ Health 13:183–204, 1984.

Mermelstein, R, Kilpper, RW, Morrow, PE, Muhle, H: Lung overload, dosimetry of lung fibrosis and their implications to the respiratory dust standard. Ann Occup Hyg 38(1):313–322, 1994.

Mery, S, Gross, EA, Joyner, DR, Godo, M, Morgan, KT: Nasal diagrams: A tool for recording the distribution of nasal lesions in rats and mice. Toxicol Pathol 22:353–372, 1994a.

Mery, S, Larson, JL, Butterworth, BE, Wolf, DC, Harden, R, Morgan, KT: Nasal toxicity of chloroform in male Fischer-344 rats and female B6C3F$_1$ mice following a 1-week inhalation exposure. Toxicol Appl Pharmacol 125:214–227, 1994b.

Mohr, U, Takenaka, S, Dungworth, DL: Morphologic effects of inhaled diesel engine exhaust on lungs of rats: Comparison with effects of coal oven flue gas mixed with pyrolized pitch. In: Carcinogenic and Mutagenic Effects of Diesel Engine Exhaust. Edited by N Ishinishi, A Koizumi, RO McClellan, W Stöber. New York: Elsevier, pp. 459–470, 1986.

Monticello, TM, Morgan, KT, Everitt, JI, Popp, JA: Effects of formaldehyde gas on the respiratory tract of rhesus monkeys: Pathology and cell proliferation. Am J Pathol 134:515–527, 1989.

Monticello, TM, Morgan, KT, Uriah, L: Nonneoplastic nasal lesions in rats and mice. Environ Health Perspect 85:249–274, 1990.

Moore, PF, Schwartz, LW: Morphological effects of prolonged exposure to ozone and sulfuric acid aerosol on the rat lung. Exp Mol Pathol 35:108–123, 1981.

Morgan, KT, Monticello, TM: Airflow, gas deposition, and lesion distribution in the nasal passages. Environ Health Perspect 85:209–218, 1990a.

Morgan, KT, Monticello, TM: Formaldehyde toxicity: Respiratory epithelial injury and repair. In: Biology, Toxicology, and Carcinogenesis of Respiratory Epithelium. Edited by DG Thomassen and P Nettesheim. New York: Hemisphere, pp. 155–171, 1990b.

Morgan, KT: Nasal dosimetry, lesion distribution, and the toxicologic pathologist: A brief review. Inhal Toxicol 6:41–57, 1994.

Morris, JB, Clay, RJ, Cavanagh, DG: Species differences in upper respiratory tract deposition of acetone and ethanol vapors. Fundam Appl Toxicol 7:671–680, 1986.

Morris, JB, Hassett, DN, Blanchard, KT: A physiologically based pharmacokinetic model for nasal uptake and metabolism of nonreactive vapors. Toxicol Appl Pharmacol 123:120–129, 1993.

Morrow, PE: Toxicological data on NO$_x$: An overview. J Toxicol Environ Health 13:205–227, 1984.

Morrow, PE: Dust overloading of the lungs: Update and appraisal. Toxicol Appl Pharmacol 113:1–12, 1992.

Morse, CC, Boyd, MR, Witschi, H: The effect of 3-methylfuran inhalation exposure on the rat nasal cavity. Toxicology 30:195–204, 1984.

Mulvaney, BD, Heist, HE: Regeneration of rabbit olfactory epithelium. Am J Anat 131:241–252, 1971.

Nadel, JA: Biological effects of mast cell enzymes. Am Rev Respir Dis 145(2 pt 2):S37–S41, 1992.

Negus, VE (ed): The Comparative Anatomy and Physiology of the Nose and Paranasal Sinuses. Edinburgh, Scotland: E and S Livingstone, 1958.

Nettesheim, DE, Szakal, AK: Morphogenesis of alveolar bronchiolarization. Lab Invest 26:210–219, 1972.

Nikula, KJ, Sun, JD, Barr, EB, Bechtold, WE, Haley, PJ, Benson, JM, Eidson, AF, Burt, DG, Dahl, AR, Henderson, RF, Chang, IY, Mauderly, JL, Dieter, MP, Hobbs, CH: Thirteen-week, repeated inhalation exposure of F344/N rats and B6C3F₁ mice to ferrocene. Fundam Appl Toxicol 21:127–139, 1993.

Nikula, KJ, Lewis, JL: Olfactory mucosal lesions in F344 rats following inhalation exposure to pyridine at threshold limit value concentrations. Fundam Appl Toxicol 23:510–517, 1994.

Nikula, KJ, Novak, RF, Chang, IY, Dahl, AR, Kracko, DA, Zangar, RC, Kim, SG, Lewis, JL: Induction of nasal carboxylesterase in F344 rats following inhalation exposure to pyridine. Drug Metab Dispos 23:529–535, 1995.

Nolte, T, Thiedemann, KU, Dungworth, DL, Ernst, H, Pomlini, I, Heinrich, U, Dosenbrock, C, Peters, L, Ueberschär, S, Mohr, U: Morphology and histogenesis of squamous cell metaplasia of the rat lung after chronic exposure to pyrolized pitch condensate and/or carbon black or to a combination of pyrolized pitch condensate, carbon black and irritant gases. Exp Toxicol Pathol 45:135–144, 1993.

O'Byrne, PM, Walters, EH, Gold, BD, Aizawa, HA, Fabbri, LM, Alpert, SE, Nadel, JA, Holtzman, MJ: Neutrophil depletion inhibits airway hyperresponsiveness induced by ozone exposure. Am Rev Respir Dis 130:214–219, 1984.

Parent, RA, Editor: Treatise on Pulmonary Toxicology: Comparative Biology of the Normal Lung. Boca Raton, FL: CRC Press, 1992.

Pauluhn, J: Species differences impact on testing. In: Respiratory Toxicology and Risk Assessment. Edited by PG Jenkins. Stuttgart: Wis. Vert. Ges. (IPCS Joint Series No. 18), pp. 145–173, 1994.

Penha, PD, Werthamer, S: Pulmonary lesions induced by long-term exposure to ozone. II. Ultrastructural observations of proliferative and regressive lesions. Arch Environ Health 29:282–289, 1974.

Phipps, RJ: The airway mucociliary system. In: Respiratory Physiology III. Edited by JG Widdicombe. Baltimore: University Park Press, pp. 213–260, 1981.

Pinkerton, KE, Pratt, PC, Brody, AR, Crapo, JD: Fiber localization and its relationship to lung reaction in rats after chronic inhalation of chrysolite asbestos. Am J Pathol 117:484–498, 1984.

Pinkerton, KE, Crapo, JD: Morphometry of the alveolar region of the lung. In: Toxicology of Inhaled Materials. Edited by HP Witschi, JD Brain. New York: Springer-Verlag, pp. 259–285, 1985.

Pinkerton, KE, Dodge, DE, Cederdahl-Demmler, J, Wong, VJ, Peake, J, Haselton, CJ, Mellick, PW, Singh, G, Plopper, CG: Differentiated bronchiolar epithelium in alveolar ducts of rats exposed to ozone for 20 months. Am J Pathol 142(3):947–956, 1993.

Plopper, CG, Dungworth, DL, Tyler, WS, Chow, CK: Pulmonary alterations in rats exposed to 0.2 and 0.1 ppm ozone: A correlated morphological and biochemical study. Arch Environ Health 34:390–395, 1979.

Plopper, CG, Hill, LH, Marrassy, AT: Ultrastructure of the nonciliated bronchiolar epithelial

(Clara) cell of mammalian lung. III. A study of man with comparison of 15 mammalian species. Exp Lung Res 1:171–180, 1980.

Plopper, CG, Hyde, DM, Weir, AJ: Centriacinar alterations in lungs of cats chronically exposed to diesel exhaust. Lab Invest 49:391–399, 1983.

Plopper, CG, Dungworth, DL: Structure, function, cell injury and cell renewal of bronchiolaralveolar epithelium. In: Current Problems in Tumour Pathology, Vol. 3: Lung Carcinomas. Edited by EM McDowell. Edinburgh: Churchill Livingstone, pp. 94–128, 1987.

Plopper, CG: Structural methods for studying bronchiolar epithelial cells. In: Models of Lung Disease: Microscopy and Structural Methods. Edited by J Gil, New York: Marcel Dekker, Inc., pp. 537–559, 1990.

Plopper, CG, Chang, AM, Pang, A, Buckpitt, AR: Use of microdissected airways to define metabolism and cytotoxicity in murine bronchiolar epithelium. Exp Lung Res 17:197–212, 1991.

Plopper, CG, Overby, LH, Nishio, SJ, Weir, AJ, Serabjit-Singh, CJ, Philpot, RM, Buckpitt, AR: Differences in cellular organization and metabolism among species used as models for responses to inhaled substances in the human respiratory system. In: Respiratory Toxicology and Risk Assessment. Edited by PG Jenkins. Stuttgart: Wis. Vert. Ges. (IPCS Joint Series No. 18), pp. 207–221, 1994a.

Plopper, CG, Chu, FP, Haselton, CJ, Peake, J, Wu, J, Pinkerton, KE: Dose-dependent tolerance to ozone. I. Tracheobronchial epithelial reorganization in rats after 20 months' exposure. Am J Pathol 144(2):404–420, 1994b.

Plopper, CG, Duan, X, Buckpitt, AR, Pinkerton, KE: Dose-dependent tolerance to ozone. IV. Site-specific elevation in antioxidant enzymes in the lungs of rats exposed for 90 days or 20 months. Toxicol Appl Pharmacol 127(1):124–131, 1994c.

Proctor, DF: The mucociliary system. In: The Nose, Upper Airway Physiology and the Atmospheric Environment. Edited by DF Proctor and I Andersen. New York: Elsevier, pp. 245–278, 1982.

Pryor, WA: How far does ozone penetrate into the pulmonary air/tissue boundary before it reacts? Free Radical Biol Med 12:83–88, 1992.

Rajni, P, Fritcher, D: Cell proliferation in the respiratory tract of hamsters continuously exposed during 4 weeks to 0.8 ppm ozone. Inhal Toxicol. 6:335–346, 1994.

Randall, HW, Monticello, TM, Morgan, KT: Large area sectioning for morphologic studies of nonhuman primate nasal cavities. Stain Technol, 63:355–362, 1988.

Reed, CJ: Drug metabolism in the nasal cavity: Relevance to toxicology. Drug Metab Rev 25:173–205, 1993.

Reid, L: Secretory cells. Fed Proc 36:2703–2707, 1977.

Reid, L, Jones, R: Experimental chronic bronchitis. Int Rev Exp Pathol 24:335–382, 1983.

Renne, RA, Gideon, KM, Miller, RA, Mellick, PW, Grumbein, SL: Histological methods and interspecies variations in the laryngeal histology of F344/N rats and B6C3F₁ mice. Toxicol Pathol 20(1):44–51, 1992a.

Renne, RA, Wehner, AP, Greenspan, BJ, DeFord, HS, Ragan, HA, Westerberg, RB, Buschbom, RL: Two-week and 13-week inhalation studies of aerosolized glycerol in rats. Inhal Toxicol 4:95–111, 1992b.

Sagartz, JW, Madarasz, AJ, Forsell, MA, Burger, GT, Ayres, PH, Coggins, CR: Histological sectioning of the rodent larynx for inhalation toxicity testing. Toxicol Pathol 20(1):118–121, 1992.

Schelegle, ES, Siefkin, AD, McDonald, RJ: Time course of ozone-induced neutrophilia in normal humans. Am Rev Respir Dis 143(6):1353–1358, 1991.

Schwartz, LW, Dungworth, DL, Mustafa, MG, Tarkington, BK, Tyler, WS: Pulmonary responses of rats to ambient levels of ozone. Effects of 7-day intermittent or continuous exposure. Lab Invest 34:565–578, 1976.

Schwartz, LW: Pulmonary responses to inhaled irritants and the morphological evaluation of those responses. In: Inhalation Toxicology. Research Methods, Applications, and Evaluation. Edited by H Salem. New York: Marcel Dekker, pp. 293–348, 1987.

Snider, GL, Lucey, EC, Stone, PJ: Animal models of emphysema. Am Rev Respir Dis 133:149–169, 1986.

Snipes, MB, Boecker, BB, McClellan, RO: Respiratory tract clearance of inhaled particles in laboratory animals. In: Lung Modelling for Inhalation of Radioactive Materials. Edited by H Smith, G Gerber. Luxembourg: Commission of European Communities and National Radiological Protection Board, EUR 938 4 EN, pp. 63–71, 1984.

Sorokin, SP: The respiratory system. In: Cell and Tissue Biology: A Textbook of Histology. Edited by L Weiss. Urban and Schwarzenberg, Inc., Baltimore, MA, pp. 751–814, 1988.

Stephens, RJ, Sloan, MF, Evans, MJ, Freeman, G: Early response of lung to low levels of ozone. Am J Pathol 74:31–58, 1974.

Stöber, W: Experimental induction of tumors in hamsters, mice and rats after long-term inhalation of filtered and unfiltered diesel engine exhaust. In: Carcinogenic and Mutagenic Effects of Diesel Engine Exhaust. Edited by N Ishinishi, A Koizumi, RO McClellan, W Stöber. New York: Elsevier, pp. 421–439, 1986.

Strom, KA, Chan, TL, Soderholm, SC: Modeling diesel particulate retention in the rat lung. GMR-4735, pp. 1–22. Warren, MI: General Motors Research Laboratories, 1985.

Swenberg, JA, Gross, EA, Randall, HW: Localization and quantitation of cell proliferation following exposure to nasal irritants. In: Toxicology of the Nasal Passages. Edited by CS Barrow. Washington, DC: Hemisphere, pp. 291–300, 1986.

Tepper, JS, Costa, DL, Lehmann, JR, Weber, MF, Hatch, GE: Unattenuated structural and biochemical alterations in the rat lung during functional adaptation to ozone. Am Rev Resp Dis 140:493–501, 1989.

Toxicologic Pathology of the Upper Respiratory System, edited by LC Uriah, KT Morgan, KR Maranpot. Environ Health Perspect 85:163–352, 1990.

Tyler, WS: Small airways and terminal units. Comparative subgross anatomy of lungs. Pleuras, interlobular septa and distal airways. Am Rev Respir Dis 128:S32–S36, 1983.

Tyler, WS, Tyler, NK, Last, JA, Barstow, TJ, Magliano, DJ, Herds, DM: Effects of ozone on lung and somatic growth. Pair fed rats after ozone exposure and recovery periods. Toxicology 46:1–20, 1987.

Uriah, LC, Talley, FA, Mitsumori, K, Gupta, BN, Bucher, JR, Boorman, GA: Ultrastructural changes in the nasal mucosa of Fischer 344 rats and B6C3F₁ mice following an acute exposure to methyl isocyanate. Environ Health Perspect 72:77–88, 1987.

van Bree, L, Marra, M, Rombout, PJ: Differences in pulmonary biochemical and inflammatory responses of rats and guinea pigs resulting from daytime or nighttime, single and repeated exposure to ozone. Toxicol Appl Pharmacol 116(2):209–216, 1992.

Wagner, JC, Berry, B, Skidmore, JW, Pooley, FC: The comparative effects of three chrysolites by injection and inhalation in rats. In: Biological Effects of Mineral Fibers. Edited by JC Wagner. Lyon, France: IARC Scientific Publications, pp. 363–372, 1980.

Warheit, DB, Hill, LH, George, G, Brody, AR: Time course of chemotactic factor generation and the corresponding macrophage response to asbestos inhalation. Am Rev Respir Dis 134:128–133, 1986.

Wilson, DW, Plopper, CG, Dungworth, DL: The response of the macaque tracheobronchial epithelium to acute ozone injury. A quantitative ultrastructural and autoradiographic study. Am J Pathol 116:193–206, 1984.

Wolf, DC, Morgan, KT, Gross, EA, Barrow, C, Moss, OR, James, RA, Popp, JA: Two-year inhalation exposure of female and male B6C3F₁ mice and F344 rats to chlorine gas induces lesions confined to the nose. Fundam Appl Toxicol 24:111–131, 1995.

Wolff, RK, Henderson, RF, Gray, RH, Carpenter, RL, Hahn, FF: Effects of sulfuric acid mist inhalation on mucous clearance and on airway fluids of rats and guinea pigs. J Toxicol Environ Health 17:129–142, 1986.

Wright, JL, Ngai, T, Churg, A: Effect of long term exposure to cigarette smoke on the small airways of the guinea pig. Exp Lung Res 18(1):105–114, 1992.

Young, JT: Histopathologic examination of the rat nasal cavity. Fundam Appl Toxicol 1:309–312, 1981.

Zhang, K, Rekhter, MD, Gordon, D, Phau, SH: Myofibroblasts and their role in lung collagen gene expression during pulmonary fibrosis. A combined immunohistochemical and *in-situ* hybridization study. Am J Pathol 145(1):114–125, 1994.

Zissu, D, Gagnaire, F, Bonnet, P: Nasal and pulmonary toxicity of glutaraldehyde in mice. Toxicol Lett 71:53–62, 1994.

Zitnik, LA, Schwartz, LW, McQuillen, NK, Zee, YC, Osebold, JW: Pulmonary changes induced by low-level ozone: Morphological observations. J Environ Pathol Toxicol Oncol 1:365–378, 1978.

Zwart, A, Arts, JH, Klokman-Houweling, JM: Determination of concentration-time-mortality relationships to replace LC50 values. Inhal Toxicol 2:105–117, 1990.

Zwart, A, Arts, JH, Ten Bergc, WF, Appelman, LM: Alternative acute inhalation toxicity testing by determination of the concentration-time-mortality relationship: Experimental comparison with standard LC50 testing. Regul Toxicol Pharmacol 15(3):278–290, 1992.

Part Six

Risk Assessment

Chapter Nineteen

Risk Assessment for Inhaled Toxicants

Roger O. McClellan

INTRODUCTION

Risk assessment is the process of characterizing and quantifying potential adverse effects on humans of exposure to physical agents, chemicals, or situations that pose a health hazard. The concept of risk assessment is not new: we have always had to make decisions about risk as a part of our daily lives. Relatively new, however, is the shift from concern for immediate hazards with readily discernible linkages between a specific agent and an adverse outcome to situations where there are only probabilistic linkages between exposure to an agent and the occurrence of an adverse effect over a long period of time. The legendary coal miner's canary illustrates a readily discernible linkage between poor air quality and acute asphyxiation. The need to set standards for protection against lung cancer risks of 1 in 10^6 over a lifetime represents the other extreme. The heavy cigarette smoker who knows that the odds of developing lung cancer late in life are 1 in 10 and yet continues to smoke represents an intermediate paradox.

Inhalation toxicology and risk assessment have been closely linked for many years. The link between poor air quality in coal mines and acute asphyxiation and that between cigarette smoking and lung cancer have already been noted. Other examples are the development of silicosis in individuals occupationally exposed to quartz and the acute morbidity and mortality in residents of London, the Neuse Valley in Belgium, and Donora, Pennsylvania, following relatively brief episodes (days) of severe air pollution. More recently, we have seen the occurrence of respiratory tract cancers in workers exposed to bischloromethylether, liver cancer in occupationally exposed vinyl chloride workers, leukemia in individuals exposed to high levels of benzene, and lung cancer in uranium miners. The acute morbidity and mortality in residents of Bhopal, India, following a catastrophic accident with release of a large quantity of methyl isocyanate are reminders of the need to be vigilant to accidents as well as the occurrence of acute respiratory effects and to avoid excessive emphasis on concern for long-term exposures and cancer. In each case involving occupational or environmental exposure, steps were taken based on experience to minimize future risk. Cigarette smoking, which is a matter of personal choice, remains an unresolved problem with consequences that far outweigh all other inhalation risks taken in aggregate.

These examples of obvious inhalation risks have paved the way for the current proactive linkage of toxicology and risk assessment, which strives to minimize the potential for occurrence of adverse effects in humans attributable to airborne materials. This proactive approach places an emphasis on obtaining toxicological information from *in-vitro* systems and laboratory animals to complement limited or in some cases nonexistent human data. It also places an emphasis on prospectively estimating levels of risk for various situations in which the estimates can never be validated with actual data unless an extraordinary accident occurs. These risk assessments are then used as input in establishing limits for control of air pollutants.

One of the major roots of prospective, probabilistic risk assessment is in the nuclear field. Early attention was focused on risks related to the production and use of nuclear weapons and later to nuclear power generation. In both situations, the need to mine uranium ore resulted in the exposure of miners to radon and progeny. Unfortunately, the high risk of lung cancer was not recognized early on, perhaps because mining was such a familiar occupation, and attention focused on immediate traumatic injury and related frequent deaths. However, concern did develop for occupational exposure during the fabrication of fuel elements containing uranium and later on plutonium as well as during the processing of irradiated fuel elements containing fission products and plutonium. Exposure of workers to pure uranium and plutonium during the fabrication of nuclear weapons was of special concern. An overriding concern was for accidents that would result in a breach of containment systems and exposure of populations downwind, a concern that became a reality with the Chernobyl accident.

Among the important concepts that grew out of nuclear risk assessment was the concept of dose, which is central to all radiation risk assessment activities. With radioactive materials, a relatively clear distinction exists between exposure (a product of the duration of intake time and concentration of a material in the intake medium, in this case, air) and dose (the time integral of the material in the target tissue and the transfer of energy to the tissue). The concept of dose extends from airborne radioactivity to all airborne toxicants. With radiation, there is concern for limiting the total radiation dose to the critical tissue, whether the source is penetrating external radiation or internal dose arising from inhalation or ingestion. The establishment of dose limits or prediction of exposure-related outcomes is dependent upon knowing both the relationship between exposure and dose and between dose and health response. The concept of the linkage between exposure and dose is now quite widely accepted for chemicals as well as for radioactivity. Because our understanding of the myriad of complex mechanisms involved in interactions of different chemicals with biological systems is limited, we have only recently begun to explore the quantitative linkage between dose and health response for chemical toxicants.

Early risk assessment in protection from radiation was based on the assumption of a threshold relationship between dose and response. No response is expected below a given dose, while responses are expected above that dose. After World War II, two new concepts emerged: (1) radiation as a genetic toxicant and (2) DNA as genetic material that transfers encoded information for cell structure and function from generation to generation of both cells and organisms. Scientists soon recognized that genetic damage in the form of mutations in both germ and somatic cells could occur. Changes resulting from mutations in germ cells could be transmitted to subsequent generations, while mutations in somatic cells could be transmitted to cell progeny and ultimately result in cancer. Observations relating dose to response were soon made, and various empirical and biologically based models were developed to describe the dose-response relationship over the range of observations. Most importantly, linear extrapolations to lower doses were made based on the theoretical assumption that the smallest quantity of dose had an associated level of calculable risk for induction of mutations and cancer in an individual and that the risk for a population could be estimated by summing the risk for the individuals in the population. The calculated levels of risk were frequently for readily measurable exposures that nonetheless had associated levels of risk well below those that could actually be observed and differentiated from the background level of cancer. Because of statistical limitations imposed on the development of experimental observations, the possibility could not be excluded that the actual risk was zero at some low level of dose.

Quantitation of exposure-dose-response relationships in the field of chemical toxicology and carcinogenesis generally lagged behind those in the radiation field. However, the concept soon emerged that certain chemicals had electrophilic characteristics and associated interactions with DNA that mimicked those of radiation and were therefore capable of causing genetic mutations and cancer in a manner analogous to that of radiation. In risk assessments, a rather natural default assumption was that these chemicals, like radiation, could also produce calculable levels of risk at the lowest levels of chemical exposure.

Two other innovations in toxicology that affected risk assessment are also noteworthy, both involving the screening of chemicals for their potential human hazards. First was the development of life-span bioassays using rodent species (in a sense as surrogates for humans) to assess the toxicological and carcinogenic hazard of chemicals. Widespread use of the life-span bioassay system in government laboratories first occurred in the Cancer Bioassay Program of the National Cancer Institute, which focused on evaluating synthetic chemicals, including many pesticides. In this program, which later evolved into the National Toxicology Program, groups of rats and mice were administered graded doses of chemicals for up to two years. The highest dose level was intended to be the maximum tolerated dose (MTD) of the chemical, a quantity that would produce some evidence of toxicity but only minimal mortality in short-term studies of up to 90 days. The MTD was selected without considering the relevance of the exposure level to human exposures, which are frequently many orders of magnitude lower than the MTD. Since the objective of the program was to detect carcinogens without concern for exposure-response issues, the experimental design using the MTD was intended to maximize statistical power when only about 50 animals per sex and species could be used at each dose level. Legacies of this program are the many cancer bioassay results that have been qualitatively evaluated as no, equivocal, some, or clear evidence of carcinogenicity, as will be discussed later. The significance for human risk assessment of the carcinogenic responses in rodents at the MTD has been the subject of much debate.

The second innovation in toxicology affecting human risk assessment was the development of cell and tissue systems for assessing the toxic and mutagenic potential of chemicals. This development was stimulated by the recognition of mutagenic changes as a key step in the complex process of carcinogenesis and the need for rapid and inexpensive systems to screen chemicals for their toxic and mutagenic properties. Of the large number of systems developed, the Ames assay pioneered by Bruce Ames and associates is probably the best known. It uses strains of bacterial species with various kinds of mutations to assay for the test agent's potential to reverse-mutate the cells. The results of the Ames assay are typically reported in qualitative fashion as positive or negative for mutagenicity. Many individuals view a mutagenicity finding as indicative of the cancer-causing potential of the chemical.

With this as background, let us examine how toxicological and carcinogenic data are used to assess human risks for airborne materials by first considering the sources of information.

SOURCES OF INFORMATION
FOR ASSESSING RISK

As illustrated in Fig. 1, information for assessing the health risks of exposure
to airborne materials can come from multiple sources such as epidemiological, clinical, laboratory animal, and *in-vitro* studies. Each approach has advantages and disadvantages and yields data that are often complementary to
the other approaches. All four approaches have benefited from the remarkable advances in investigational techniques of recent years. Especially notable
are the advances in cellular and molecular biology and computational
science.

Since our ultimate interest is in assessing human health risks, it follows
that human data should be used when available. The use of human data
avoids the uncertain and often controversial extrapolations from laboratory
animals to humans or from human cells or tissues to the whole human organism. Epidemiological studies also have the advantage of involving real-world
situations. If extrapolations to lower levels of exposure are necessary (which
is usually the case), the starting point at least can be actual human exposures.
However, a related problem is that exposure conditions are certainly not controlled and are rarely quantified. Frequently, employment status is used as a
qualitative indicator of exposure. In exceptional cases, data from a few area
monitors in the workplace may be used to reconstruct the exposure of a

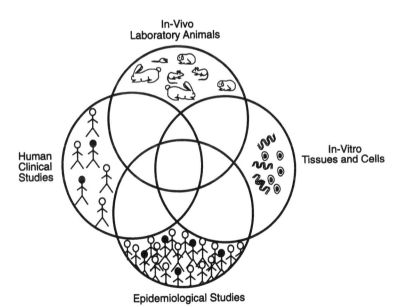

Figure 1 Risk assessments for airborne materials require the synthesis and integration of information from multiple sources.

population of workers. The procedures that can be used to evaluate the health status of the workers can range from symptom questionnaires to sophisticated measurements of pulmonary function or pulmonary exfoliative cytology to morality records. The development of quantitative exposure-response relationships from epidemiological data is extremely difficult since exposure conditions are rarely quantifiable, especially over the long periods of time necessary for chronic diseases like cancer or emphysema. At best, these relationships are semiquantitative. More frequently, they demonstrate an association between exposure conditions and disease outcome (e.g., sufficient evidence of exposure to bischloromethylether causing respiratory tract cancer). Perhaps the best quantitation of an exposure-response relationship for an airborne toxicant is that for radon and its progeny and the development of lung cancer (Whittemore and McMillan, 1983; NRC, 1988; Lubin et al., 1994).

A problem almost universally encountered in the conduct and interpretation of epidemiological studies of respiratory disease (including lung cancer) is the overwhelming impact of cigarette smoking (discussed in Chapter 1). Determining the attributable risk for the agent in question versus smoking is frequently difficult (Lubin and Steindorf, 1995). The question frequently arises as to whether low levels of exposure to the toxicant alone, in the absence of cigarette smoking, would result in an increase in lung cancer incidence.

Clinical studies in which exposures of human subjects are carried out under carefully controlled laboratory conditions allow the test atmosphere to be precisely defined and the exposure concentration and duration to be experimentally controlled. A drawback in such studies is the extent to which the range of exposure conditions under study must be limited to those that the clinician feels confident will not produce irreversible effects since the use of higher exposures would be unethical. A wide range of procedures, including symptom questionnaires and all the tools of modern diagnostic medicine, can be used to evaluate changes related to exposure conditions. The resulting data can be readily evaluated to quantitatively define relationships between the concentration and duration of exposure and response. Because of the expense, such studies rarely involve exposures of more than a few hours. An example of a controlled clinical study is that in which the effects of ozone were evaluated by Horstmann et al. (1990) and Devlin et al. (1991).

Humans can also be studied under natural exposure conditions rather than using the carefully controlled conditions of the laboratory. In these studies, comparisons can be made between responses evaluated under conditions of low and high pollutant exposure. Exposure gradients and quality of the exposure characterization are likely to be best for short-term observations (days) rather than long-term (months). A broad range of procedures, which are usually conducted in the field, can evaluate the functional status of the individuals being studied. The results of such studies can be evaluated to provide a semiquantitative or perhaps in some cases a quantitative relationship between exposure and response. The relevance of the exposure is not open to debate since it occurred naturally. The studies by Spektor et al. (1988a) of the effects of air pollution on lung function in children attending camp and of exercising healthy adults (Spektor et al., 1988b) are examples of field studies using natural exposure conditions.

As with controlled human exposure studies, laboratory animal studies have the advantage of carefully controlled exposure conditions matched to experimental needs. Moreover, the range of procedures used to evaluate the dose of pollutant received by the experimental subjects can include invasive procedures as well as end-of-life observations that cannot be used with humans. The study of intact mammals is advantageous in that all the body functions are subject to the complex, integrated physiology that occurs in intact humans. Because exposure conditions and evaluation procedures can be rigorously controlled, quantitative assessments of exposure-response relationships can be developed. Approaches to the conduct of animal bioassays with inhaled materials are discussed in the chapter by Hahn. Of course, laboratory animal studies have the major disadvantage of requiring extrapolation of data from animals to humans. However, we are rapidly approaching the point where such extrapolations can be greatly facilitated by taking into account species characteristics (described in several earlier chapters). There is almost always a need to extrapolate from the high exposure level used in laboratory animal studies to the frequently much lower levels of exposure likely to be encountered by humans. The validity of such extrapolations is substantially enhanced by conducting studies that use multiple levels of exposure and incorporate serial observations into the study. Such studies provide insight into the complex interaction among exposure level and time and biological response. An example of a well-conducted, long-term animal study is that of Mauderly et al. (1987), who reported that chronic exposure of rats to high levels of diesel exhaust produced lung cancer. The interpretation of the results was greatly facilitated by the detailed attention given to the lung burden of diesel soot, which increased at a rate disproportionate to (i.e., greater than) the exposure rate (Wolff et al., 1987). Guidelines for conducting adequate long-term carcinogenicity experiments have been outlined by Montesano et al. (1986).

In-vitro studies using cells and tissues from humans and laboratory animals represent the ultimate reductionist approach to defining pollutant effects. Advantages of cell and tissue studies are precise control of exposure conditions and the availability of a wide range of procedures of various complexities for evaluating responses. Observations such as the influence of pollutants on the production and release of specific cellular mediators from defined cell populations can be made at a level of detail that is impossible in intact laboratory animals or humans. The related disadvantage is that the observations must be extrapolated to the intact mammal, with its complex array of feedback mechanisms that modulate interactions at all levels of organization in the body.

In the final analysis, risk assessment for an air pollutant should be based on the available data irrespective of the particular methodological approach used to acquire it. Maximum use should always be made of any human data. Results from other systems can be used to complement and extend the utility of the available human data, frequently by providing insight into the mechanisms that may be operative in humans. In the absence of human data, priority should be given to data obtained in controlled studies of laboratory animals, which preferably include life-span studies. If the material is an air pollutant with inhalation as the primary route of entry to humans, then laboratory animal studies using inhalation exposures are appropriate. This obviates the need for making extrapolations between routes such as extrapolations from oral intake data (if that is the only source of data) to the inhalation route of intake. Studies with isolated cells or tissues can be used as screening systems to identify and rank potential toxicants and, of course, give insight into mechanisms of action. In the absence of data from biological systems, some insight into the potential toxicity of new materials can be gained by evaluating structure-activity relationships for the new material relative to materials that have been extensively studied in biological systems.

APPROACHES TO RISK ASSESSMENT

Approaches to evaluating the health risks of airborne toxicants can be conveniently divided into those for which the primary concern is adverse functional changes and those for which the primary concern is cancer. Traditionally, the exposure-response relationship of a functional change has been viewed as exhibiting a threshold level of exposure. Below the threshold, the adverse effect is considered not to occur or to be at a minimal level. Above the threshold, the severity of the adverse effect is considered to increase with increasing exposure. With functional changes, the response variable is continuous. In contrast, exposure-response relationships between air toxicants and cancer have been assumed to lack a threshold. Here, the relationship between exposure and incidence of cancer in the population is probabilistic. The disease outcome, cancer, is dichotomous in the individual: cancer or no cancer.

Threshold Limit Values

The earliest risk assessments for airborne toxicants, which were much less formal that those conducted today, focused on alterations in function and structure. A threshold was assumed to exist for the exposure-response relationship so that a level of exposure could be defined below which no effects would be observed: hence, the term *threshold limit values* (TLV), which are usually expressed as a *time-weighted average* (TWA) for a normal 8-hour work day and a 40-hour work week.

TLV are set by the American Conference of Governmental Industrial Hygienists (ACGIH) for a number of compounds. TLV are routinely reevaluated and current values published in a regularly updated document (*1995– 1996 Threshold Limit Values for Chemical Substances and Physical Agents and Biological Exposure Indices*) (ACGIH, 1995). As defined in the introduction, "Threshold Limit Values (TLV) refer to airborne concentrations of substances and represent conditions under which it is believed that nearly all workers may be repeatedly exposed day after day without adverse health effects." The document goes on to note that individuals vary in their susceptibility and that a small portion of individuals may be affected at levels below the TLV. It also emphasizes the harmful effects of smoking. Since TLV are specifically developed to provide guidance for occupational exposures, the document contains appropriate admonishments about not using TLV for other purposes such as setting community air pollution indices or estimating toxic potential. The document also emphasizes that the line is not a fine one between safe and dangerous concentrations of a compound. As is implicit in the term threshold limit values, the TLV are developed within a structure that implies a threshold relationship between exposure and health effect. Therefore, it is feasible to establish values that accord relative safety in contrast to values for higher exposure, which have a higher degree of hazard.

In addition, the ACGIH establishes Biological Exposure Indices (BEI) to complement the TLV. BEI are reference values used in assessing the concentration of appropriate determinants in biological specimens collected from workers at specified times relative to chemical exposures. The measured value (such as methyl chloroform concentration in exhaled air or trichloroethanol concentration in blood or urine) relative to the BEI gives an indication of the extent to which compliance with TLV for the compound has been achieved. Because TLV involve threshold relationships, the establishment of a TLV is quite naturally based on the use of safety factors. A level for some identified hazard and for the lowest observed adverse effect may be established and then safety factors added to extrapolate to lower levels with an associated lower probability for hazard. The basis for TLV will not be covered in detail. However, the basic methodology used for the derivation of the closely related inhalation reference concentration will be discussed. The ACGIH also classifies compounds according to their potential for human carcinogenicity. This classification system will be discussed later.

In the light of the Clean Air Act as amended in 1990 (Clean Air Act Amendments, 1990), risk assessments of air pollutants require extensive scientific information, ranging from sources of air pollution to effects on health and ecology. The Clean Air Act classifies air pollutants into two broad categories: criteria air pollutants and hazardous air pollutants. The Act requires the EPA administrator to set National Ambient Air Quality Standards for criteria air pollutants, such as ozone, carbon monoxide, particulate matter, sulfur dioxide, nitrogen dioxide, and lead, whose occurrence and potential health consequences are of national concern. The process by which standards are set for the criteria pollutants has previously been reviewed (Grant and Jordan, 1988; Padgett and Richmond, 1983; Lippmann, 1987) and will be briefly reviewed later in this chapter. The Clean Air Act also requires the Administrator to regulate hazardous air pollutants, although this regulatory process more specifically relates to emission of pollutants by specific sources.

Cancer Classification Schemes

Concern for cancer causation by chemicals has been a dominant factor in the risk assessment arena, especially during the last quarter century. Therefore, it is appropriate to spend some time addressing risk assessment practices for carcinogens even though the general approach is not exclusive to the inhalation route of entry or to respiratory tract cancers. Early cancer risk assessments focused on whether or not a chemical was carcinogenic or whether a particular industrial occupation posed a carcinogenic hazard based on epidemiological evidence. Later, the assessments were broadened to include evidence for carcinogenic hazard based on laboratory animal studies. As multiple sources of data were considered and the process became more complex, the need arose for a more formalized approach.

On the international front, the International Agency for Research on Cancer (IARC) initiated such a program in 1969 to evaluate the carcinogenic hazards of agents and occupations. Results of the evaluation process are published in a series of IARC monographs (IARC, 1994). The IARC program has had major international impact because of the prominence of the Agency (an arm of the World Health Organization) and the authoritative nature of its reviews. Moreover, the IARC classifications are widely used as a basis of action by federal and state regulatory agencies around the world.

The IARC approach to evaluation of carcinogenic hazards is briefly described in a preamble to each monograph. The IARC approach uses international working groups of experts with input from the IARC staff to carry out five tasks: "(i) to ascertain that all appropriate references have been collected; (ii) to select the data relevant for the evaluation on the basis of scientific merit; (iii) to prepare accurate summaries of the data to enable the reader to follow the reasoning of the Working Group; (iv) to evaluate the results of experimental and epidemiological studies; and (v) to make an overall evaluation of the carcinogenicity of the agent to humans" (IARC, 1994, p. 15).

In the monographs, the term *carcinogenic* is used to denote an agent that is capable of increasing the incidence of malignant neoplasms. Traditionally, IARC evaluations were based on direct evidence of cancer observed in humans or laboratory animals without regard to the mechanism(s) involved. In 1991, IARC convened a group of experts to advise the Agency on the use of mechanistic data in the classification process (IARC, 1991). The group suggested a greater use of mechanistic data relevant to the extrapolation of information from laboratory animals to humans and interpretation of the evidence of carcinogenicity in humans. As will be noted later, this may result in either upgrading or downgrading the cancer classification for specific chemicals.

The IARC evaluation process considers four types of data: (1) exposure data, (2) human carcinogenicity data, (3) experimental animal carcinogenicity data, and (4) other data relevant to an evaluation of carcinogenicity and its mechanisms. Epidemiological studies of three types can contribute to the assessment of human carcinogenicity: cohort studies, case-control studies, and correlation (or ecological) studies. The monographs are not intended to summarize all published studies; rather the emphasis is given to reviewing and citing those that contribute to the evaluation. The evaluation process considers the quality of the studies, inferences about mechanisms of action, and evidence of causality. Since the 1991 meeting on use of mechanistic data, the reports include the statement that "Special attention is given to measurements of biological markers of carcinogen exposure or action, such as DNA or protein adducts, as well as markers of early steps in the carcinogenic process, such as protooncogene mutation, when these are incorporated into epidemiological studies focused on cancer incidence or mortality." Such measurements may allow inferences to be made about putative mechanisms of action (IARC, 1991; Vainio et al., 1992a; Vainio et al., 1992b).

The epidemiological evidence is classified into four categories:

1 *Sufficient evidence of carcinogenicity* is used when a causal relationship has been established between exposure to the agent and human cancer.

2 *Limited evidence of carcinogenicity* is used when a positive association between exposure to an agent and human cancer is considered to be credible, but chance, bias, or confounding could not be ruled out with reasonable confidence.

3 *Inadequate evidence of carcinogenicity* is used when the available studies are of insufficient quality, consistency, or statistical power to permit a conclusion regarding the presence or absence of a causal association.

4 *Evidence suggesting lack of carcinogenicity* is used when there are several adequate studies covering the full range of doses to which human beings are known to be exposed. These studies must be mutually consistent in not showing a positive association between exposure and any studied cancer at any observed level of exposure.

The IARC evaluation process gives substantial weight to carcinogenicity data from laboratory animals. The preamble to the IARC report notes that all known human carcinogens that have been studied adequately in experimental animals have produced positive results in one or more animal species (Wilbourn et al., 1986; Tomatis et al., 1989). The preamble also notes that for some 24 agents or exposure situations (including cyclophosphamide, estrogens, and vinyl chloride), the evidence of carcinogenicity was first established or became highly suspected from studies in laboratory animals and then was confirmed in human studies (Vainio et al. 1994). The preamble to the IARC report goes on to state that "Although this association cannot establish that all agents and mixtures that cause cancer in experimental animals also cause cancer in humans, nevertheless, in the absence of adequate data on humans, it is biologically plausible and prudent to regard agents and mixtures for which there is sufficient evidence (p. 27) of carcinogenicity in experimental animals as if they presented a carcinogenic risk to humans. The possibility that a given agent may cause cancer through a species-specific mechanism which does not operate in humans (p. 28) should also be taken into consideration" (IARC, 1994, p. 21).

Thus the IARC classifies the strength of the carcinogenicity evidence in experimental animals in a fashion analogous to that used for human data. As in the case of human data, particular attention is given to mechanistic information relevant to the understanding of the process of carcinogenesis in humans that may strengthen the plausibility of a conclusion that the agent being evaluated is carcinogenic to humans.

The evidence of laboratory animal carcinogenicity is classified into four categories.

1 *Sufficient evidence of carcinogenicity* is used when a working group considers that a causal relationship has been established between the agent and an increased incidence of malignant neoplasms or an appropriate combination of benign and malignant neoplasms in (a) two or more species of animals or (b) two or more independent studies in one species carried out at different times, in different laboratories, or under different protocols. A single study in one species might be considered under exceptional circumstances to provide sufficient evidence when malignant neoplasms occur to an unusual degree with regard to incidence, site, type of tumor, or age at onset.

2 *Limited evidence of carcinogenicity* is used when the data suggest a carcinogenic effect but are limited for making a definitive evaluation.

3 *Inadequate evidence of carcinogenicity* is used when studies cannot be interpreted as showing either the presence or absence of a carcinogenic effect because of major qualitative or quantitative limitations.

4 *Evidence suggesting lack of carcinogenicity* is used when adequate studies involving at least two species show that the agent is not carcinogenic within the limits of the tests used. Such a conclusion is inevitably limited to the species, tumors, and doses of exposure studied.

Supporting evidence includes a range of information such as structure-activity correlations, toxicological information, and data on kinetics, metabolism, and genotoxicity from laboratory animals, humans, and lower levels of biological organization such as tissues and cells. In short, any information that may provide a clue as to the cancer-causing potential of an agent will be reviewed and presented.

Finally, all relevant data are integrated, and the agent is categorized on the strength of the evidence derived from studies in humans and experimental animals and from other studies, as shown in Table 1 (IARC, 1994, pp. 29–30).

As noted above, the overall evaluation criteria were changed after the meeting on the use of mechanisms of carcinogenesis in risk identification (IARC, 1991). The first major impact of this change came in the 1994 reevaluation of the evidence on ethylene oxide, which had earlier been classified as 2A based on limited human evidence and sufficient animal evidence (IARC, 1987). As a result of the 1994 review, ethylene oxide was reclassified: "Ethylene Oxide is carcinogenic to humans (Group 1)." Since the evaluation is highly informative, it is quoted below (IARC, 1994, p. 139).

There is *limited evidence* in humans for the carcinogenicity of ethylene oxide.

There is *sufficient evidence* in experimental animals for the carcinogenicity of ethylene oxide.

In making the overall evaluation, the Working Group took into consideration the following supporting evidence. Ethylene oxide is a directly acting alkylating agent that:
  (i) induces a sensitive, persistent dose-related increase in the frequency of chromosomal aberrations and sister chromatid exchange in peripheral lymphocytes and micronuclei in bone marrow cells of exposed workers;
 (ii) has been associated with malignancies of the lymphatic and haematopoietic system in both humans and experimental animals;
(iii) induces a does-related increase in the frequency of haemoglobin adducts in exposed humans and dose-related increases in the numbers of adducts in both DNA and haemoglobin in exposed rodents;
 (iv) induces gene mutations and heritable translocations in germ cells of exposed rodents; and
  (v) is a powerful mutagen and clastogen at all phylogenetic levels.

This change in classification results in a rather awkward situation: a Group 1 that in reality has two subgroups, the largest of which comprises agents for which there is *sufficient* evidence of human carcinogenicity and a single agent (ethylene oxide) subgroup for which there is limited evidence of carcinogenicity.

It will be of interest to learn how future monograph working groups deal with the numerous other chemicals that are currently classified as Group 2A, probably carcinogenic to humans. Recognizing the current rapid increase in the use of new techniques for evaluating molecular and cellular changes that may be linked to carcinogenesis, it is reasonable to expect that evidence may be found for many chemicals to have exposure-related increases in various parameters causally associated with cancer. Future working groups may conceivably use such evidence to upgrade other current Group 2A chemicals to Group 1 even if sufficient evidence of carcinogenicity in humans is absent.

As may be noted, the IARC classification scheme does not address the potency of carcinogens. In short, a carcinogen is a carcinogen irrespective of

Table 1   Classification Scheme for Cancer Causation by Agents, International Agency for Research on Cancer

Group 1   The agent (mixture) is carcinogenic to humans. The exposure circumstance entails exposures that are carcinogenic to humans. This category is used when there is sufficient evidence of carcinogenicity in humans. Exceptionally, an agent (mixture) may be placed in this category when evidence in humans is less than sufficient but there is sufficient evidence of carcinogenicity in experimental animals and strong evidence in exposed humans that the agent (mixture) acts through a relevant mechanism of carcinogenicity. [This last sentence was added to the Preamble after the meeting on the use of mechanisms of carcinogenesis in risk identification (IARC, 1991).]
Group 2   This category includes agents, mixtures, and exposure circumstances for which, at one extreme, the degree of evidence of carcinogenicity in humans is almost sufficient, as well as those for which, at the other extreme, there are no human data, but for which there is evidence of carcinogenicity in experimental animals. Agents, mixtures, and exposure circumstances are assigned to either Group 2A (probably carcinogenic to humans) or Group 2B (possibly carcinogenic to humans) on the basis of epidemiological and experimental evidence of carcinogenicity and other relevant data.
Group 2A   The agent (mixture) is probably carcinogenic to humans. The exposure circumstance entails exposures that are probably carcinogenic to humans. This category is used when there is limited evidence of carcinogenicity in humans and sufficient evidence of carcinogenicity in experimental animals. In some cases, an agent (mixture) may be classified in this category when there is inadequate evidence of carcinogenicity in humans and sufficient evidence of carcinogenicity in experimental animals and strong evidence that the carcinogenesis is mediated by a mechanism that also operates in humans. Exceptionally, an agent, mixture, or exposure circumstance may be classified in this category solely on the basis of limited evidence of carcinogenicity in humans.

its potency. This poses serious constraints on the utility of IARC classification data for use beyond hazard identification. This lumping of carcinogens irrespective of potency can be misleading to the nonspecialist, including the lay public. Because of the authoritative role of IARC, it would be helpful if its guidance could be extended to include cancer potency so that others could use this information in conjunction with exposure assessment to derive estimates of risk, which require information on both exposure and potency. Estimates of risk would be of great value since it is risk that we ultimately wish to control.

The National Toxicology Program (NTP) uses a classification scheme that is somewhat similar to that of IARC (Table 2). However, the scheme is clearly cast as evidence of carcinogenicity, as reflected in the wording for the various classifications. The NTP scheme has a particularly important role since NTP uses it to provide input into the *Annual Report on Carcinogens* (U.S. DHHS, 1994) This report is frequently used by other agencies as a basis for regulatory decisions on the listed chemicals.

The United States Environmental Protection Agency (EPA) has also developed a similar cancer classification that is described in the 1986 guidelines

Table 1 *(Continued)*

Group 2B The agent (mixture) is possibly carcinogenic to humans. The exposure circumstance entails exposures that are possibly carcinogenic to humans. This category is used for agents, mixtures, and exposure circumstances for which there is limited evidence of carcinogenicity in humans and less than sufficient evidence of carcinogenicity in experimental animals. It may also be used when there is inadequate evidence of carcinogenicity in humans, but there is sufficient evidence of carcinogenicity in experimental animals. In some instances, an agent, mixture, or exposure circumstance for which there is inadequate evidence of carcinogenicity in humans but limited evidence of carcinogenicity in experimental animals together with supporting evidence from other relevant data may be placed in this group.

Group 3 The agent (mixture or exposure circumstance) is not classifiable as to its carcinogenicity to humans. This category is used most commonly for agents, mixtures, and exposure circumstances for which the evidence of carcinogenicity is inadequate in humans and inadequate or limited in experimental animals. Exceptionally, agents (mixtures) for which the evidence of carcinogenicity is inadequate in humans but sufficient in experimental animals may be placed in this category when there is strong evidence that the mechanism of carcinogenicity in experimental animals does not operate in humans. Agents, mixtures, and exposure circumstances that do not fall into any other group are also placed in this category.

Group 4 The agent (mixture) is probably not carcinogenic to humans. This category is used for agents or mixtures for which there is evidence suggesting lack of carcinogenicity in humans and in experimental animals. In some instances, agents or mixtures for which there is inadequate evidence of carcinogenicity in humans but evidence suggesting lack of carcinogenicity in experimental animals, consistently and strongly supported by a broad range of other relevant data, may be classified in this group.

(*From:* IARC, 1994, pp. 29–30.)

for cancer risk assessment (U.S. EPA, 1986b). The guidelines build on a policy document on chemical carcinogenesis prepared by the United States Office of Science and Technology Policy (U.S. OSTP, 1985). The EPA scheme is shown in Table 3. Although the EPA approach is quite similar to that of IARC, it uses a weight-of-evidence approach (contrasted with the IARC strength-of-evidence approach), and the resulting classification has alpha categorization rather than numeric captioning. In the strength-of-evidence approach, one or two positive studies are taken as sufficient or limited evidence irrespective of how many well-conducted negative studies exist. The weight-of-evidence approach weighs all the evidence and takes into account the strengths and weaknesses of all studies.

In 1994, an NRC Committee released a report (NRC, 1994) that reviewed EPA risk assessment practices with special reference to hazardous air pollutants and offered several recommendations for improvements. One of the recommendations was that EPA should consider modifying its cancer classification scheme to include incorporation of a narrative statement. This statement would be expected to more adequately describe the evidence for carcinogenic risk, including conditions under which the studies were done

Table 2   National Toxicology Program (NTP) Classification Scheme for Animal Bioassays

---

Clear Evidence of Carcinogenic Activity   Demonstrated by studies that are interpreted as showing a dose-related (i) increase of malignant neoplasms, (ii) increase of a combination of malignant and benign neoplasms, or (iii) marked increase of benign neoplasms if there is an indication from this or other studies of the ability of such tumors to progress to malignancy.
Some Evidence of Carcinogenic Activity   Demonstrated by studies that are interpreted as showing a chemical-related increased incidence of neoplasms (malignant, benign, or combined) in which the strength of the response is less than that required for clear evidence.
Equivocal Evidence of Carcinogenic Activity   Describes studies that are interpreted as showing a marginal increase of neoplasms that may be chemical-related.
No Evidence of Carcinogenic Activity   Demonstrated by studies that are interpreted as showing no chemical-related increases in malignant or benign neoplasms.
Inadequate Study of Carcinogenic Activity   Demonstrated by studies that because of major qualitative or quantitative limitations cannot be interpreted as valid for showing either the presence or absence of carcinogenic activity.

From: NTP, 1992.

and the relevance to human exposure conditions. For example, any nonphysiological method of administering a toxicant such as intraperitoneal or intrapleural instillation would be noted in the narrative. Likewise, the relationship between dosage or exposure levels and likely human exposures would be noted. EPA (U.S. EPA, 1994) has proposed the use of such a narrative statement in its draft revision of the guidelines for cancer risk assessment. This narrative approach could be more informative than a single alphanumeric classification for a material such as fiberglass, where the primary animal evidence of carcinogenicity is from studies with rats given very large quantities of fiberglass as a series of intraperitoneal injections (McClellan, 1994).

Risk Paradigm

Approaches to risk assessment have become more formalized in recent years, largely as a result of demands created by environmental and occupational health legislation and public awareness. Some useful insights into how risk assessment procedures evolved within the EPA are provided in a review by Albert (1994). Recognizing that needs for risk assessment and risk management extend across the federal government and indeed well beyond federal agencies, several agencies asked the National Research Council (NRC) of the National Academy of Sciences to review the process and offer recommendations on how it could be improved. This request was addressed by the NRC Committee on the Institutional Means for Assessment of Risks to Public Health. The Committee's now-classic report, *Risk Assessment in the Federal*

Table 3 U.S. Environmental Protection Agency (EPA) Cancer Categorization Scheme

Human Evidence	Animal Evidence				
	Sufficient	Limited	Inadequate	No data	No evidence
Sufficient	A	A	A	A	A
Limited	B1	B1	B1	B1	B1
Inadequate	B2	C	D	D	D
No data	B2	C	D	D	E
No evidence	B2	C	D	D	E

Group A: Human carcinogen.
Group B: Probable human carcinogen. B1: Limited evidence of carcinogenicity from epidemiology studies. B2: Inadequate human evidence but positive animal evidence.
Group C: Possible human carcinogen.
Group D: Not classifiable as to human carcinogenicity.
Group E: Evidence of noncarcinogenicity for humans.
(*From:* U.S. EPA, 1986b.)

Government: Managing the Process (NRC, 1983), has profoundly influenced the conduct of risk assessment by providing a structure that is now widely used (Fig. 2).

During the 1980s, there was considerable debate over the relationship between risk assessment and risk management. Some individuals argued for a rigid separation, with science involved only in risk assessment and judgment entering only into risk management. Others argued that judgment would inevitably enter the process even in the conduct of risk assessments. Unfortunately, in my opinion, arguments over the relationship between two components of the risk paradigm (risk assessment and risk management) probably obscured consideration of the third component (research) and its role in the total process. Failure to include research in the risk paradigm has resulted in a lost opportunity for creating new knowledge that would be uniquely useful in the risk assessment process. Inadequacies in scientific knowledge as well as endless debate over the risk assessment process and the application of risk assessments to meet legislative intent have slowed the process of developing new standards such as standards for hazardous air pollutants required by the Clean Air Act. Impatient with the slow progress, Congress modified the Act in 1990, shifting from standards based on health risk to a phased process in which technology-based standards are used in the first phase and standards based on health risk are used in the second. In addition, the Clean Air Act Amendments of 1990 called on the National Academy of Sciences to "review the methods used by EPA to determine the carcinogenic risk associated with exposure to hazardous air pollutants. . . ." The requested review was to focus on "methods used for estimating carcinogenic potency of hazardous air pollutants and for estimating human exposure to these pollutants."

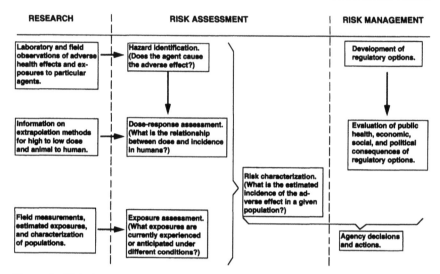

Figure 2 Risk Assessment—Risk Management Paradigm of the National Research Council (NRC), National Academy of Sciences (NAS). (*From:* NRC, 1983).

In response to the Congressional request, the National Research Council formed the Committee on Risk Assessment of Hazardous Air Pollutants. The Committee began its deliberations in 1991 and published its report, *Science and Judgment in Risk Assessment,* in 1994 (NRC, 1994). The report and its appendixes provide a wealth of information on current risk assessment practices of the EPA, especially in regard to hazardous air pollutants and numerous recommendations for improvements in the process. The report gives special attention to the separate but interrelated issues of uncertainty (a lack of precise knowledge about what the truth is) and variability (differences in quantities among individuals in time or across space).

The report also addresses the contentious issue of the role of default options in risk assessment and especially decisions about when default options should be replaced with specific scientific information. Default options were defined in the 1983 NRC report as "the option chosen on the basis of risk assessment policy that appears to be the choice in the absence of data to the contrary" (NRC, 1983, p. 63). Some of the key default options are as follows:

1 Humans are as sensitive as the most sensitive laboratory animal species, strain, or sex evaluated.

2 Chemicals act like radiation at low doses in inducing cancer, with a linearized multistage model appropriate for estimating dose-response relationships below the range of experimental observations.

3 The biology of humans and laboratory animals, including the rate of metabolism of chemicals, is a function of body surface area.

4   A given unit of intake of chemical has the same effect irrespective of the intake time or duration.

5   Laboratory animals are a surrogate for humans in assessing cancer risks with positive cancer bioassay results in laboratory animals taken on evidence of the chemical's potential to cause cancer in people (U.S. EPA, 1986b).

To avoid underestimating risk, EPA has selected default options that are scientifically plausible and conservative. This is not an unreasonable position to take in the absence of scientific data to the contrary. The critical issue becomes one of determining the kind and strength of scientific evidence required to depart from a default. Because the NRC Committee could not reach a consensus on this issue, it chose to publish two appendixes prepared by Committee members that address the issue of default options.

In one of the two appendixes to *Science and Judgment in Risk Assessment,* Finkel (1994) advocates the use of "plausible conservatism" in choosing and altering defaults. He argues that the EPA should assemble all the available data and models that are deemed plausible by knowledgeable scientists and then "from this 'plausible set,' EPA should adopt (or should reaffirm) as a generic default that model or assumption which tends to yield risk estimates more conservative than the other plausible choices" (p. 613). In contrast, McClellan and North (1994) argue that "plausible conservatism" has a role in the initial selection of default options but is not an appropriate criterion to use for determining when to depart from a default option and use specific scientific information. They argue that such a test places an excessively high hurdle on the introduction of new science and will tend to freeze risk characterizations at the level determined by conservative default options.

In conducting science-based risk assessments, McClellan and North (1994) go on to advocate building on the general principles outlined in the 1983 NRC report. They argue for full use of scientific information, including a closer linkage between the components of research and risk assessment, in the science-based risk paradigm (Fig. 3). In this approach, the risk assessment process is envisioned as yielding both risk characterizations and identification of research needs. The latter is derived from considering the major uncertainties in the science undergirding the risk characterization. If these uncertainties can be reduced through further research, they serve as a basis for creating the agenda and priorities for subsequent research. In this manner, a feedback loop is created from the risk assessment to the research component. As conservative default options are replaced over time with specific scientific information, it follows that there should be a general reduction in the estimated risk levels and a tightening of the bounds of uncertainty (Fig. 4). This is illustrated later with the formaldehyde cancer risk assessment example.

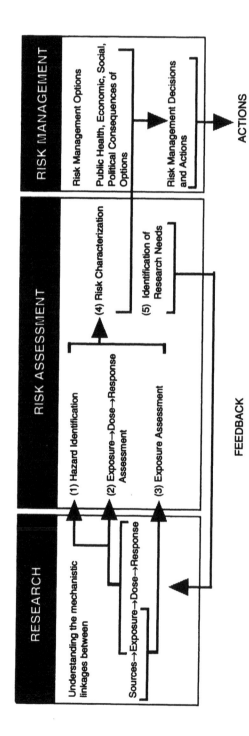

Figure 3 The Risk Research–Risk Assessment Paradigm as modified by McClellan and North (1994) from the original NRC-NAS Paradigm (NRC, 1983).

Hazard Identification   Let us briefly consider the four major components of risk assessment. The hazard identification phase is a qualitative process of determining whether a physical agent, chemical, or situation *might* pose a threat or potential harm to the health of individuals or populations. The 1983 NRC report put it simply: "Does the agent cause the adverse effect?" It entails synthesis and integration of all available information gathered from multiple sources, including studies of humans, laboratory animals, cells, and tissues. It includes consideration of whether a chemical may cause cancer under some exposure conditions. One of the often-cited shortcomings of the present hazard identification process is that the hazard identification phase is not constrained to any actual exposure situations or, for that matter, even plausible exposure scenarios. In my view, the use of hazard identification information without considering actual or plausible exposures may frequently mislead individuals. The answer is to make certain that hazard identification is always joined with consideration of exposure-response relationships and plausible exposures.

Exposure-Dose-Response Assessment   In the 1983 NRC report, this step in risk assessment was identified as "dose-response assessment" with a straightforward question: "What is the relationship between dose and incidence in humans?" In Fig. 3, the term is expanded to "exposure-dose-response assessment" in recognition of the fact that there is a clear distinction between exposure and dose in the field of inhalation toxicology, perhaps more so than in other subspecialties of toxicology. Exposure entails consideration of the concentration and time dimensions of the availability of the agent for intake by humans. As is apparent from information presented in earlier

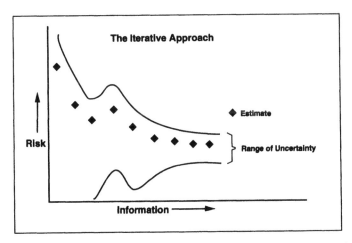

Figure 4   As scientific information is used in risk assessments to replace conservative default options, the risk estimate and bounds of uncertainty should be reduced.

chapters, the physical form (liquid droplet size, particle size, vapor, or gas) and chemical characteristics of the agent in the air may have a profound influence on the actual dose received by the individual, that is, the quantity deposited in the body. Moreover, the quantity of material deposited is obviously only one dose metric. For example, the formaldehyde example cited later in this chapter emphasizes the use of DNA-protein cross-links as a measure of effective dose for inhaled formaldehyde.

Most human risk assessments ultimately involve developing some measure of risk per unit of exposure or identifying a level of exposure below which the risk is considered not to be important. Frequently, such expressions of "exposure or dose-response" relationship use exposure as a surrogate for dose. This approach may overlook important factors that can modify exposure-dose relationships such as nonlinearities in exposure-dose relationships and species differences that must be accounted for in extrapolating to humans. Moreover, I believe that risk is better related to the actual doses received by the cells and tissues exhibiting the biological effects of concern. These considerations argue for using the expanded exposure-dose-response assessment.

The importance of knowing the actual doses to the tissues at risk is greatest when (1) information derived from studies using laboratory animals is applied to human health risk assessments, and (2) information derived from one exposed human population is used to assess the risks to a second population that has a different level or route of exposure. In the first case, laboratory animals and humans may have significant differences in the manner in which they deposit, clear, and metabolize a toxic substance (see chapters by Schlesinger, Snipes, Miller and Kimbell, and Bond and Medinsky). In the second case, the two populations may have dissimilar target organ burdens per unit of exposure because of differences in the form of the toxic substance, route of exposure, exposure conditions, or the related pathways for metabolism.

Perhaps the most significant obstacle to developing exposure-dose and, ultimately, exposure-dose-response relationships in toxicology and risk assessment is the difficulty in defining the most appropriate expressions of dose. For studies using ionizing radiation, there is general agreement that the total energy deposited in the tissues at risk represents the dose. This can be measured or calculated with confidence for penetrating radiations, but the concept becomes less clear for types of radiation that have very short ranges in tissue. In the latter case, the dose distribution can be highly nonuniform, reaching cells immediately adjacent to the sources but not reaching cells that are more than a few 10s of micrometers away.

Identifying the most appropriate expression of dose is even more difficult when the exposures involve chemically toxic agents. The tissue concentration of the chemical, the time-integrated concentration, and the peak concentration may all represent appropriate dose parameters. Also, the tissue concentrations of metabolites of the chemical agent or the amount of a chemical or one of its metabolites bound to cell DNA may also be important dose parameters. Even greater uncertainty in describing dose arises when the exposure involves complex mixtures of toxic chemicals such as with inhaled cigarette smoke or atmospheric pollutants. Derivation of appropriate exposure-dose relationships in these cases may never be possible, although risk assessments must still be attempted.

Dosimetry models used to calculate dose to organs, tissues, or cells at risk for a given level of exposure provide typical, or median, values. However, substantial variability occurs in the doses received by individuals within a group of people or laboratory animals, even when all members of the group are exposed to the same inhaled material. This is due to anatomical and physiological differences in factors that determine deposition, absorption, and internal organ redistribution.

Analyses of well-controlled laboratory studies with animals that received a single brief inhalation exposure indicate that about 2% of an exposed group can be expected to receive more than three times the average organ doses to the group (Cuddihy et al., 1979). Further analyses of inhalation exposures of people to substances distributed uniformly in the environment showed that a small percentage of the individuals receive more than five times the average dose. Typical predicted and measured distributions of internal organ burdens resulting from inhalation of relatively insoluble material are shown in Fig. 5. The distribution predicted from the results of laboratory studies represents a single inhalation exposure. The individual values would be expected to show less spread if the exposures were repeated or chronic. The measured distribution from the human exposures represents chronic inhalation of material in the environment so that variability in the exposure or contact with the environment is also included in the data. However, this is the level of uncertainty that may be expected in predicting organ doses to individuals exposed by inhalation of toxic substances released to the environment.

A key step in the risk assessment process is to develop an exposure-dose-risk relationship that applies to the physical or chemical agent of concern and the mode and level of exposure in the population at risk. The mathematical form of the relationship used has generally depended on whether or not the agent is a carcinogen. For carcinogens, it is most frequently assumed as a default option that a linear or curvilinear relationship should be applied and that the relationship should pass through the point of zero risk only at zero exposure. This will be illustrated later with the formaldehyde cancer risk assessment. An important implication of this type of relationship is that all levels of exposure, however small, add to the background risk. On the other hand, for low levels of exposure to carcinogens having low potency, the experimental data are never usually adequate to exclude the possibility of no added risk to the exposed population. For toxicants that are not carcinogens, a threshold type of relationship is frequently used. This implies that the exposure must exceed a threshold level before its effects become apparent. This will be illustrated later with the ozone and bromomethane risk assessments.

Several types of dose-effect relationships are illustrated in Fig. 6. The use of each type of relationship can be supported by theoretical arguments concerning the mechanisms of the injury process. However, accepting such arguments always requires assumptions. Perhaps the most significant difficulty faced when selecting a dose-effect of comparable dose-risk relationship lies in extrapolating from the high-dose region, where effects data are available, to the lower-dose region, where most human exposures are apt to occur. Seldom are data available covering a sufficiently broad range of doses to justify the use of one particular form of dose-effect relationship over all others. Thus in most cases it is difficult to predict human risk at appropriately low doses with a high degree of confidence. This difficulty is readily apparent from consideration of Fig. 6. Because the toxicity or carcinogenicity of some chemicals results from their metabolic products rather than from the chemicals per se, dose-risk relationships based on the concentrations of the chemicals in the affected tissues may not be suitable. These risks are better related to the tissue concentrations or rates of formation of the metabolites. One of the best illustrations of the importance of metabolites in deriving a dose-effect relationship is seen with vinyl chloride (Gehring et al., 1978). Exposure to high concentrations of vinyl chloride can cause hepatic angiosarcomas, but this has been related to its metabolic products chloroethylene oxide and chloroacetaldehyde.

Maltoni and Lefemine (1975) determined the incidence of hepatic angiosarcomas in studies with rats exposed to vinyl chloride by chronic inhalation. Analysis of these data by Gehring et al. (1978) resulted in the relationships

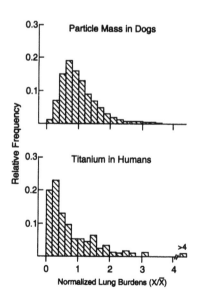

Figure 5 Lung burdens projected for dogs (above) exposed once by inhalation to an insoluble aerosol and measured in humans (below) for titanium inhaled in atmospheric particles. (*Adapted from:* Cuddihy et al., 1979.)

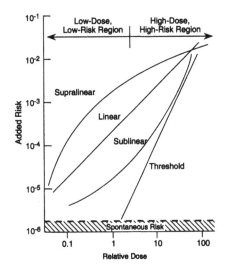

Figure 6 Typical shapes of dose-risk relationships used for assessing health effects in exposed populations of humans and laboratory animals.

illustrated in Fig. 7. When related to the exposure concentrations of vinyl chloride, the incidence of liver tumors remained constant or actually decreased with increasing dose in the three highest exposure groups. When related to the rate of metabolism of vinyl chloride, the incidence of liver tumors increased uniformly over the entire range of exposures. The rate of formation of metabolites of vinyl chloride was assumed to follow Michaelis-Menton-type kinetics. This analysis was then used to predict liver cancer risk in human exposure situations by adjusting for differences between rats and humans in exposure level and body surface area. The latter parameter was used as an index of the relative rates of metabolism of vinyl chloride in rats and humans.

As additional information becomes available on the mechanisms of action of toxic agents, it should be incorporated into the risk assessment process. In doing this with certain substances, it may become apparent that different mechanisms of injury are involved to various extents at different exposure or dose levels. The research of greatest interest to the risk assessment process, including that on mechanisms of action, is that which is most relevant to the exposure levels encountered in the human exposure situations being evaluated. This includes studies done *in vitro* using tissues, cells, or biochemical preparations. In such studies, it is useful to always place the concentrations (or doses) of toxicant into perspective relative to those expected under various exposure scenarios for humans or laboratory animals.

Exposure Assessment The third step in the risk assessment process is to fully characterize the exposures of people to potentially harmful substances. This should include a determination of

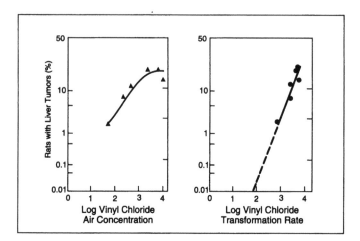

Figure 7 Relationships between the incidence of liver tumors in rats that inhaled vinyl chloride and the air concentrations inhaled (left) or rate of formation of metabolites (right). (*Adapted from:* Gehring et al., 1978.)

   1 The physical and chemical characteristics of each toxic substance, especially those characteristics that influence its disposition in the environment and in exposed people.
   2 The amount of the toxic substance present and its potential for release into the atmosphere.
   3 The probable dispersion of material released into the atmosphere and ambient air concentrations near people.
   4 The pattern and duration of the exposure and the magnitude of the exposed population.

   The most important characteristics of toxic substances that affect their dispersion in the environment and their internal deposition after inhalation by people are chemical composition, physical form, and the size, shape, and density of particles. It is also important to know how and where the substances may be released in relation to the exposed population, especially if the substances can undergo transformations in the environment.
   A full discussion of the subject of exposure assessment is beyond the scope of this book. The interested reader is referred to EPA exposure assessment guidelines (U.S. EPA, 1992b), a report of the NRC on exposure assessments for airborne toxicants (NRC, 1990), and the NRC risk assessment report (NRC, 1994), which includes numerous references to the role of the exposure assessment component in assessing human risk. The NRC report contains a very informative appendix that considers the role of various factors, including several related to exposure, in estimating human cancer risk to individuals living in the vicinity of facilities releasing 1,3-butadiene (Del Pup et al., 1994). Changes in the assumptions used in assessing exposure had a major impact on the resultant estimates of risk.

Risk Characterization   The fourth step of the risk assessment process is estimating the incidence of health effects in a human population assuming specific conditions of exposure. In addition to providing the best estimate of this risk, it is important to indicate the amount of uncertainty that is likely to have resulted considering the quality of the input data and the method of analysis. The risk characterization should be presented in as robust a manner as possible based on the input data. The report *Science and Judgment in Risk Assessment* (NRC, 1994) emphasized the importance of distinguishing between uncertainty and variability in the risk assessment process.

Uncertainty can be defined as a lack of precise knowledge as to what the truth is, whether qualitative or quantitative. An example is knowledge of the shape of the exposure-cancer response relationship at low levels of exposure. This kind of uncertainty may be classified as model uncertainty. Variability can be described as the difference in values of a parameter across space, in time, or among individuals. Examples include wind speed and direction around a pollutant source; interindividual differences in breathing rate, food consumption, and activity; and interindividual differences in metabolism of a toxicant or sensitivity to injury.

In the past, regulatory agencies such as the EPA have frequently reported only upbound values (95% confidence limit) for risk. When the data are sufficient, every possible effort should be made to provide a best or central estimate of risk. Guidance developed internal to EPA (U.S. EPA, 1992a) has called for describing the range of risk, including central tendency and high-end positions of the risk distribution.

Stochastic analyses of risk provide a distribution of estimated risk based on use of probability density functions for input parameters instead of single point estimates. Monte Carlo simulation techniques can be used to calculate a distribution of risks through numerous iterations using randomly generated values from the defined probability functions for input variables. Del Pup et al. (1994) used this approach in estimating risks for emissions from a facility handling large quantities of butadiene. Using a conservative screening approach, the EPA estimated a maximum individual cancer risk of 1 in 10. After plant modernization, the estimate was reduced to a maximum individual cancer risk of 5–10 in 1000. Using Monte Carlo techniques, the corresponding cancer risk estimates for the 5th and 95th percentile at the nearest residence were 4 in 100,000,000 to 2 in 10,000. The key factors influencing the uncertainty were (1) the meteorological data used in estimating transport and fate of butadiene; (2) butadiene decay factors; (3) exposure time, frequency, and duration; and (4) the cancer potency slope factor for butadiene. The cancer potency slope factor contributed almost 3 orders of magnitude to the estimates separating different scenarios. Information on uncertainty (such as just described) should be conveyed to the public and to regulatory agencies whenever the results of a risk assessment are to be applied in formulating decisions related to the health of workers or the public. When risks of different activities are compared, comparable output parameters should be compared (i.e., the maximum likelihood values), and making comparisons between an upper 95% confidence limit value for one activity and a central tendency estimate for another activity should be avoided. Concern for the nature of the comparison extends to consideration of common risks such as the occurrence of deaths in automobile accidents or cancer mortality. These are most frequently actuarially determined and presented as best estimates. Hence the use of these values as a basis of comparison with upper bound estimates of risk for air pollutants can be very misleading.

RISK ASSESSMENTS FOR REGULATED AIR POLLUTANTS

During the 1960s and 1970s, the public became increasingly concerned about the potential toxic and carcinogenic effects of chemicals. Public concern was soon translated into legislative concern, and a number of laws directing executive actions were passed. In the United States, two principal federal agencies regulate air pollutants and assess risks: the Occupational Safety and Health Administration (OSHA), which regulates occupational exposures, and the Environmental Protection Agency, which is primarily concerned with environmental exposures to airborne materials. The EPA does have some authority over occupational exposures under the Toxic Substances Control Act (TSCA) and has also ventured indoors with recently activity on environmental tobacco smoke and radon in homes. The Mine Safety and Health Administration has some specific authority over exposure of miners.

Beyond government involvement, many responsible companies whose workers are potentially exposed to air pollutants maintain internal risk management systems in addition to those mandated by law. Using risk assessment approaches similar to those used by government agencies, these companies have developed and use workplace exposure guidance values to avoid exposure and adverse effects in workers.

As amended in 1990, the Clean Air Act (Clean Air Act Amendments, 1990; Mayer and Francis, 1991) is a dominant piece of legislation driving the risk assessment process for airborne materials. Within the context of the Clean Air Act, airborne materials may be broadly viewed as falling into two categories: the criteria pollutants and hazardous air pollutants. The regulations for criteria pollution are heavily focused on setting National Ambient Air Quality Standards (NAASQS) for these pollutants based primarily on concern for limiting adverse effects other than cancer. In contrast, the hazardous air pollutants include a diverse range of individual chemicals for which there is concern for both cancer and noncancer functional decrements. The following sections describe the basic approach used for assessing human health risks of criteria pollutants and hazardous air pollutants.

Criteria Air Pollutants

The Clean Air Act requires that the primary standards for the criteria pollutants "shall be ambient air quality standards the attainment and maintenance of which in the judgment of the administrator, based on such criteria and allowing an adequate margin of safety, are required to protect the public health" (Section 109[b]1, Clean Air Act Amendments of 1970). The intent is to have NAAQS that are uniform nationwide. In addition to the primary (or health-based) standard, the Clean Air Act specifies the need for a secondary (or welfare-based) standard for effects of air pollutants on vegetation, buildings, visibility, and other aspects of the general welfare. Implementation of the standards is primarily the responsibility of state and regional authorities.

The primary standards are intended to protect against adverse effects and not necessarily against all identifiable effects of air pollutants. The question of what constitutes an adverse effect has been the subject of extensive debate. Although Congress did not rigorously define an adverse effect, the debate on the Clean Air Act did provide some general guidance in its concern with effects ranging from cancer, metabolic and respiratory disease, and impairment of mental processes to headaches, dizziness, and nausea. An American Thoracic Society (ATS) committee (Andrews et al., 1985) has developed guidelines for defining adverse respiratory health effects as identified in epidemiological studies. The ATS guidelines call attention to such issues as reversible versus irreversible changes, variability among individuals, and changes in an individual or population over time and note that the percentage of ‹change indicative of an adverse effect can vary. ATS gives heavy weight to irreversible changes and departures from normal trends as revealed by serial measurements associated with growth or aging. With children, for example, the failure to maintain a normal predicted lung-growth curve as a result of exposure to air pollution is regarded as clearly indicative of an adverse effect.

Congress also took into consideration sensitive population groups in setting the NAAQS. In particular, Congress noted that the standard should protect "particularly sensitive citizens as bronchial asthmatics and emphysematics who in the normal course of daily activity are exposed to the ambient environment." This statement has been interpreted to exclude individuals who are not performing normal activities such as the hospitalized. Congress gave further guidance in noting that the standard is statutorily sufficient whenever there is "an absence of adverse effect on the health of a statistically related sample of persons in sensitive groups from exposure to the ambient air." A statistically related sample is defined as "the number of persons necessary to test in order to detect a deviation in the health of any persons within such sensitive group which is attributable to the condition of the ambient air." The administrator of EPA was also directed to specify a primary NAAQS for each pollutant that includes an "adequate margin of safety." The margin of safety is intended to protect against effects that have not yet been uncovered by research and effects whose medical significance is a matter of disagreement. In this chapter, ozone will be used as a case study to illustrate how standards for criteria air pollutants are set.

The criteria air pollutants are listed in Table 4, along with key information on their effects and the current primary standards, which are set to protect against a broad range of effects. Some effects, such as decrements in pulmonary function, occur indirectly in the respiratory tract. Thus the respiratory tract is both the portal of entry and the target tissue for pollutants such as ozone, oxides of nitrogen, sulfer dioxide, and particulate material. With other pollutants, the respiratory tract serves as the portal of entry, but the effect of concern involves another system. For example, carbon monoxide affects the cardiovascular system and lead the nervous system.

Under the Clean Air Act, EPA is charged with reviewing and updating the criteria every five years. In reality, the review interval has been closer to every ten years. Lippmann (1987) has described the process for preparing and reviewing the documents. It includes preparation of a detailed compilation of everything that is published in the peer-reviewed literature about the criteria pollutant under review. The 1986 ozone criteria document (U.S. EPA, 1986a) and a supplement updating the earlier material (U.S. EPA, 1988) are examples. Following preparation of the criteria document and its acceptance by the EPA Clean Air Scientific Advisory Committee (CASAC), a staff position paper is prepared that summarizes the crucial studies (reviewed in the criteria document) to be used in setting or revising the standard. An example is the last staff position paper on ozone (U.S. EPA, 1989). I highly recommend use of the criteria documents and staff position papers as authoritative review documents if the reader desires information on the criteria pollutants.

As this chapter is being written in early 1995, much is happening with respect to reaffirmation or revision of the NAAQS. CASAC has provided closure letters to the Administrator of EPA on oxides of nitrogen and sulfur dioxide. The closure letters indicate that the staff position papers provide a scientifically adequate basis for regulatory decisions on these substances and, in addition, provide some specific comments for the Administrator to consider in setting the standards. A revised criteria document on ozone is in the final stages of review, and the associated staff position paper is being finalized. A revised criteria document on particulate material is in the final stages of preparation, and it is expected that both the criteria document and staff position papers on particulate material will be reviewed by the CASAC on an accelerated basis in 1995–1996.

In the next section, the approach to setting a NAAQS will be illustrated using ozone as an example.

Ozone: A Criteria Pollutant Example   Ozone, a criteria pollutant, is a highly reactive gas that is ubiquitous in its distribution. For purposes of this presentation, attention will be focused on trophospheric ozone that poses a

Table 4   National Ambient Air Quality Standards

Pollutant	Sensitive population	Health effects	Averaging time	Primary standard*
Particles (PM$_{10}$)	Individuals with preexisting respiratory disease.	Changes in mortality in sensitive populations, increase in respiratory symptoms, reduced pulmonary function.	24 hour Annual	150 μg/m³ 50 μg/m³
Sulfur dioxide	Asthmatics.	Increased respiratory symptoms, reduced pulmonary function.	24 hour Annual	0.140 ppm 0.003 ppm
Carbon monoxide	Individuals with heart disease.	Aggravation of angina pectoris.	8 hour 1 hour	9.000 ppm 35.000 ppm
Nitrogen dioxide	Young children and asthmatics. Individuals with preexisting respiratory disease.	Increased pulmonary symptoms, reduced pulmonary function.	Annual	0.053 ppm
Ozone	None identified.	Increased respiratory symptoms, reduced pulmonary function.	1 hour	0.120 ppm
Lead	Fetuses and young children.	Neurobehavioral development, impaired heme synthesis.	Quarterly	1.5 μg/m³

*μg/m³, micrograms per cubic meter; ppm, parts per million.

potential inhalation hazard. For completeness, it should be noted that concern also exists for depletion of stratospheric ozone, whose concentration influences the amount of ultraviolet light reaching the earth. Ironically, our concern in one situation is for too much ozone and in the other situation for too little.

Trophospheric ozone is produced by a complex series of chemical reactions involving hydrocarbons and oxides of nitrogen. Since the reactions are strongly activated by both temperature and sunlight, the highest concentrations of ozone are observed in the daytime and the summer. Because it is very reactive, ozone is rapidly depleted from the atmosphere when it is no longer formed because precursors are no longer available or when the sun goes down. Since both hydrocarbons and oxides of nitrogen are ubiquitous in the atmosphere, ozone is found everywhere, including areas remote from human activity. Nonetheless, human activities that give rise to the precursors certainly increase atmospheric concentrations. Since the reactions involved in both formation and degradation of ozone are complex, ozone concentrations are not a simple proportion of the concentrations of either hydrocarbons or oxides of nitrogen (NRC, 1991). This makes control of ozone an especially vexing problem.

Ozone toxicity has been the subject of intensive investigation for many years, resulting in numerous published papers. Lippmann (1992) has prepared an authoritative review on ozone toxicity. Research has included controlled human exposure studies, epidemiological investigations, animal studies, and studies with cell and tissue preparations. The human data, obviously, play the central role in evaluating the risks of human exposure to ozone.

A number of carefully controlled human exposure studies have been done using ozone concentrations that approximate levels ranging from current ambient ozone levels in the most highly ozone-polluted areas such as Los Angeles to levels somewhat higher. The exposures have extended typically from 1 to 6.6 hours and usually used exercising young men. A range of responses has been evaluated, including changes in Forced Expiratory Volume in 1 second (FEV_1), cough reflex, airway reactivity, airway permeability, tracheobronichial clearance of particles, lung lavage biomarkers, and respiratory tract symptoms. The results of these studies have been reviewed in the last criteria documents (U.S. EPA, 1986a and 1988) and staff position paper (U.S. EPA, 1989) for ozone.

The key responses to acute ozone exposures are summarized in Table 5. The changes in FEV_1 are shown graphically in Figs. 8 and 9. In summary, decrements in pulmonary function related to exposure concentration, exposure duration, and level of exercise are observed. These changes are observed with conditions similar to that of the current standard, which is 120 ppb averaged over 1 hour.

Clinical studies on individuals exposed to ambient air have also been attempted. Spektor et al. (1988a and 1988b) report studies conducted on children in a camp setting and adults carrying out a normal exercise regime. The interpretation of the results of such studies is complicated by the fact that the ambient air typically contains multiple pollutants. It therefore becomes difficult to ascribe the effects to any single pollutant. Nonetheless, these studies conducted in the subjects of concern (humans) must be utilized to the maximum extent possible to gain insight into the key causative factors in air pollution that influence human health.

Table 5 Gradation of Individual Physiological Response to Acute O₃ Exposure

Physiological response	Gradation of response			
	Mild	Moderate	Severe	Incapacitating
Change in spirometry FEV₁, FVC	5–10%	10–20%	20–40%	> 40%
Duration of effect	Complete recovery in < 30 min.	Complete recovery in < 6 h.	Complete recovery in 24 h.	Recovery in > 24 h.
Symptoms	Mild to moderate cough.	Mild to moderate cough, pain on deep inspiration, shortness of breath.	Repeated cough, moderate to severe pain on deep inspiration, shortness of breath, breathing distress.	Severe cough, pain on deep inspiration, shortness of breath, obvious distress.
Limitation of activity	None	Few individuals choose to discontinue activity.	Some individuals choose to discontinue activity.	Many individuals choose to discontinue activity.

(*From:* U.S. EPA, 1989.)

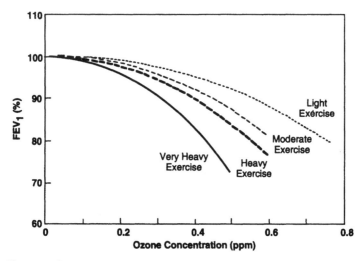

Figure 8 Changes in pulmonary function (FEV,) related to ozone exposure concentration and level of exercise. (*Adapted from:* U.S. EPA, 1986a.)

Epidemiological studies with ozone have yielded limited information, largely for the same reason that there are difficulties in interpreting clinical studies on individuals exposed to ambient pollution. Moreover, one is always faced with accounting for the impact of the primary inhaled toxicant—cigarette smoking (U.S. DHHS, 1989). Indeed, when assessing subtle respiratory effects, it is also necessary to consider the role of environmental tobacco smoke (Samet, 1992). The most notable of the epidemiological studies relevant to assessing ozone effects is the study of Detels et al. (1987). They reported lower baseline pulmonary function in a high oxidant community and a more rapid, age-related decline in a high-oxidant compared with a low-oxidant air pollution community.

Numerous studies of ozone have been conducted in laboratory animals. The acute effects are generally similar to those observed in humans. The range of response factors that can be evaluated in laboratory animals is much greater than can be carried out in humans since it is possible to kill animals and make serial observations using a wide array of techniques. Thus, changes are observed at the ultrastructural, biochemical, and functional levels (U.S. EPA, 1988; Lippmann, 1992). Unfortunately, the interpretation of many of the studies is complicated by the frequent use of a single exposure level and in most cases short observation periods. An exception is a major recent study sponsored by the National Toxicology Program (NTP) in which mice and rats were exposed daily to three concentration of ozone (0.12, 0.50, and 1.0 ppm) for 6 hours/day, 5 days/week up to two years (Boorman et al., 1995). A large series of detailed studies to evaluate structural, biochemical, and functional changes near the end of the exposure period (20 months) were sup-

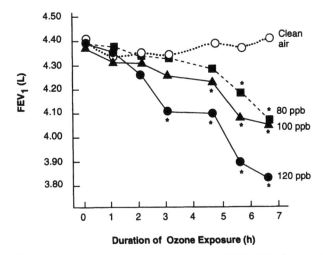

Figure 9 Changes in pulmonary function (FEV₁) related to ozone exposure concentration and duration of exposure for individuals undergoing moderate exercise. (*From:* Horstman et al., 1990.)

ported by the Health Effects Institute and should provide greatly improved insight into the effects of chronic exposure to ozone (Boorman et al., 1995; Catalano et al., 1995). The results were reassuring with regard to the risk from chronic inhalation exposure to ozone.

The clearest changes induced by ozone were seen in the evaluation of chronic rhinitis, where a statistically significant trend of increased response was noted with increased ozone concentration. The differences between measurements in the controls and the 0.5 ppm and 1.0 ppm rats were statistically significant. Although marginally significant or significant trends toward increased responses were seen for centriacinar fibrosis and airway disease, the differences between control rats and rats exposed to ozone were not statistically significant. When interpreting the results of laboratory animal studies for purposes of human risk assessment, it is always important to keep in mind species differences in the linkage between both exposure and dose and dose and response. The issue of interspecies extrapolation of exposure-dose relationships for ozone was addressed by Miller and Kimbell in an earlier chapter.

A major issue in interpreting the significance of laboratory animal studies for quantitative human risk assessment relates to interspecies differences in the exposure-dose linkage. With the approaches outlined in the chapter by Miller and Kimbell, it should be possible to develop scaling factors (exposure concentration-local tissue dose) for the various laboratory animal species that will allow consideration of dose-response relationships across the species to humans.

And finally, it should be noted that *in-vitro* studies have been conducted with ozone to assess its effects on lung macrophages and tracheal epithelium (Valentine, 1985; Leikauf et al., 1988). These studies are difficult to interpret because of uncertainties in extrapolating the dosimetry from the *in-vitro* to *in-vivo* situation for the highly reactive ozone. This complicating factor is one that must always be kept in mind when making *in-vitro* observations that ultimately are desirable to extrapolate to humans *in vivo.*

Data of the kind described above were used in 1971 to set the standard at 80 ppb for total oxidants. This was revised in 1979 with the exposure metric defined as ozone and the standard set at 120 ppb with a 1-hour maximum not to be exceeded more than once a year. In practice, a community is considered to be out of compliance if the 1-hour maximum concentration is exceeded a fourth time in a 3-year period. The ozone criteria documents (U.S. EPA, 1986a, 1988) and staff position paper (U.S. EPA, 1988) were reviewed by the Clean Air Scientific Advisory Committee in 1988. The staff recommended that the next NAAQS for ozone should be based on a 1-hour maximum concentration with the range of 80–120 ppb. Half the CASAC endorsed the staff proposal, and half recommended a reduced upper limit, arguing that if the 1-hour standard were reaffirmed at 120 ppb, the standard would not provide an ample margin of safety. Both the EPA staff and CASAC recommendations were to be based on the information included in the criteria document, which was only current as of 1986. Perhaps the EPA staff and CASAC would have offered a different opinion if the new information more recently summarized by EPA (U.S. EPA, 1992c) had been available.

The EPA administrator has since reaffirmed the previous standard and initiated preparation of a new criteria document and staff position paper on an accelerated schedule. The observation of changes in pulmonary function and accompanying cellular and biochemical changes, as evaluated using lavage fluid in exercising individuals exposed to 80 ppb of ozone for 6.6 hours (McDonnell et al., 1991; Devlin et al., 1991), will likely have a strong influence on the standard in regard to both concentration and averaging time. Other factors that will be considered in setting the standard are patterns of ozone concentration in parts of the country that are not characterized by the classical 1- to 2-hour peak observed in California. This and the experimental data on effects have been used to argue for a standard with a 6- to 8-hour averaging time. One approach to resolution of the issue of the level and duration of the standard may be to establish two interrelated standards, as I previously suggested (McClellan, 1989). A short-term standard with a 6-hour averaging time could be set to be protective of acute responses. A second, long-term standard protective of potential human chronic pulmonary disease could be based on measurements averaged over an extended period of time, perhaps up to 1 year. I would argue that the highest priority should be accorded to achieving compliance with the long-term standard.

Hazardous Air Pollutants

The other broad category of pollutants regulated by the Clean Air Act consists of the hazardous air pollutants. In some portions of the Act, some of these chemicals are also identified as air toxics. The 1970 Clean Air Act specified that standards for hazardous air pollutants should be established to protect public health with an ample margin of safety. Unfortunately, this section of the Clean Air Act was very difficult to implement, especially with known or suspected cancer-causing agents for which there was no clear evidence of a threshold for cancer causation. As a result, only seven National Emission Standards for Hazardous Air Pollutants were promulgated from 1970 to 1990, when the Clean Air Act was amended. These standards were for arsenic, asbestos, benzene, beryllium, mercury, vinyl chloride, and radionuclides.

Largely because of frustration with the slow pace of regulating hazardous air pollutants, Congress took a radically different approach in amending the Clean Air Act in 1990. The approach uses two phases. The first phase is technology-based and requires the use of Maximum Achievable Control Technology (MACT) with regulation by source categories. This phase depends on installation of control equipment, process changes, operator training, and substitution of materials to reduce emissions of 189 hazardous air pollutants. A second phase, which is based on health risk, will be implemented some years after the MACT source category control phase. The health risk phase requires that more stringent standards be placed in effect if necessary to protect the public health with an ample margin of safety. Implementation of this phase will require quantitative exposure-dose-response data on each of the 189 hazardous air pollutants as well as quantitative exposure assessments for sources to ascertain whether the residual health risk exceeds 10^{-6}. If so, further actions will be required to achieve additional reductions in risk.

Table 6 lists 48 of the 189 chemicals (or mixtures) identified in the Clean Air Act Amendments of 1990. A complete list appears as an appendix in *Science and Judgment in Risk Assessment* (NRC, 1994). The chemicals listed are those for which the 1991 Toxic Release Inventory data base identified emissions in excess of 500 tons per year. The emissions data are used here as a criterion for listing the compounds because it is only a crude index as to the potential hazard of the chemicals. Many factors influence the extent to which emissions become actual sources of individual exposure. In addition, as emphasized in this chapter, exposure must be joined with considerations of hazard identification and exposure-dose-response assessment to characterize risk.

For each of the chemicals listed in Table 6, current information (as of November 1, 1993) is provided on various indices of toxicity and carcinogenicity. For many of the chemicals, summary data are not shown, indicating

Table 6 Data on Selected Hazardous Air Pollutants (from a total of 189 listed in the Clean Air Act Amendments of 1990)

Genetic toxicity data[g] columns — Mammalian: In vivo (S, G), In vitro (M, C); Bacterial (S, E).

Chemical name	1991 TRI emissions[a] (tons/yr)	IUR[b] per µg/m³	OUR[c] per µg/L	EPA[d] WOE	RfC[e] (mg/m³)	IARC[f] WOE	In vivo S	In vivo G	In vitro M	In vitro C	Bacterial S	Bacterial E	R-D data[h]
1,3-Butadiene	1975.2	2.8E-4		B2		2B	+				+		X
2,4-Toluene diisocyanate	661.9				V	2B					−		
Acetaldehyde	3540.5	2.2E-6		B2	9.0E-3	2B	+		+		+	+	X
Acetonitrile	683.9				UR								X
Acrylonitrile	1094.4	6.8E-5	1.5E-5	B1	2.0E-3	2A	−		+	+/−	+	+	X
Benzene	8737.2	8.3E-6	8.3E-7	A	UR	1	+/−		−/+	−	−	−	X*
Bis(2-ethylhexyl)phthalate (DEHP)	521.7		4.0E-7	B2		2	−				−	−	X
Carbon disulfide	44669.6				UR		+	+	−		−		X*
Carbon tetrachloride	773.4	1.5E-5	3.7E-6	B2		2B	+	+		−		+	X
Carbonyl sulfide	8362.6				NV			−					X*
Chlorine	38804.7			D									
Chlorobenzene	1198.1			D	UR		+		+	−	−	−	X
Chloroform	9541.4	2.3E-5	1.7E-7	B2	UR	2B	−		−	−	−	−	X
Chloroprene (2-chloro-1,3-butadiene)	735.3				7.0E-3	3	+	+	−		+	−	X*
Cumene (isopropylbenzene)	1638.8				UR								
Ethyl benzene	4320.5			D	1.0E+0		−		+		−		X
Ethyl chloride (chloroethane)	1431.6				1.3E+1								X
Ethylene dichloride (1,2-dichloroethane)	1997.7	2.6E-5	2.6E-6	B2		2B	−	−	+		+		X
Ethylene glycol	5330.1												
Formaldehyde	5109.2	1.3E-5		B1		2A	−/+		+	+	+	+	X*
Glycol ethers	21957.1												
Hydrochloric acid	41460.7				7.0E-3								X
Hydrogen fluoride (hydrofluoric acid)	4590.6				UR								

Compound	TRI[a]	UR[b]	OUR[c]	EPA/WOE[d]	RfC (mg/m³)[e]	IARC/WOE[f]	Mammalian in vivo S[g]	G	Mammalian in vitro M	C	Bacterial S	E	Repro-Dev[h]
Lead compounds	703.8			B2	4.03-4	2B	+/-	-/+			-	-	
Manganese compounds	623.2			D	UR		+	+			+	+	
Methanol	99841.5												X
Methyl bromide (bromomethane)	1222.8			D	5.0E-3	3	+	+			+	+	X
Methyl chloride (chloromethane)	2849.4				UR	3		+		+			X
Methyl chloroform (1,1,1-trichloroethane)	68753.1			D	UR	3				+			X
Methyl ethyl ketone (2-butanone)	51710.9			D	1.0E+0								X
Methyl isobutyl ketone (hexone)	13599.3				UR								X
Methyl methacrylate	1278.7					3							X
Methyl tert butyl ether	1519.1				5.0E-1	3							X
Methylene chloride (dichloromethane)	39669.2	4.7E-7	2.1E-7	B2	UR	2B	-	-		-	+	-	X
m-Xylene	718.1				NV		-		+		-		
Naphthalene	1335.9	4.2E-6		C			-				-		
o-Xylene	864.9				NV		-		+		+		X
Phenol	3165.6			D	NV		+	+		+		V	
Propionaldehyde	694.1				NV								
Propylene oxide	533.3	3.7E-6	6.8E-6	B2	3.0E-2	2A	+	+	+	+	+	+	
p-Xylene	2639.2				NV								
Styrene	14238.2			UR	1.0E+0	2B	+/-	+	+	+	+	+	X*
Tetrachloroethylene (perchloroethylene)	8343.7				UR	2B			+		-		X
Toluene	99260.1			D	4.0E-1	3	+	-	+/-	+	+	-	X*
Trichloroethylene	17529.2			UR	UR	3	+	+/-	+/-	+	+	+	X
Vinyl acetate	2743.2			UR	2.0E-1		+	+	+		-		X
Vinyl chloride	523.7	8.4E-5	5.4E-5	A		1	+	+	+	+	+	+	X*
Xylenes (isomers and mixture)	57776.5			D	NV		-	-	-	-	-	-	X

[a] TRI = 1991 Toxic Release Inventory data in tons/year.

[b] UR = Inhalation unit risk estimate per µg/m³. Source is the EPA Integrated Risk Information System (IRIS) database.

[c] OUR = Oral unit risk per ug/L. Source is the EPA IRIS database.

[d] EPA/WOE = EPA Weight-of-Evidence Cancer Classification. Source is the EPA IRIS.

[e] RfC Workgroup; V = verified, on IRIS = concentration given in mg/m³; NV = not verified; UR = under review.

[f] IARC/WOE = International Agency for Research on Cancer Classification.

[g] Genetic Toxicity Data; mammalian in vivo, S = somatic, G = germ cell; mammalian in vitro, M = mutation, C = chromosome aberration; bacterial, S = *Salmonella typhi*, E = *Escherichia coli* (EPA Genetic Activity Profile database provided by Dr. Michael Waters, EPA, current as of 1992).

[h] Reproductive-Developmental Toxicity Data provided by Dr. John Vandenburg. EPA; X = data available, X* = some human data available.

(Adapted from: National Research Council (NRC), 1994, Table 6, pp. 334–343.)

that the chemical has not been evaluated by EPA or IARC for carcinogenic potential or by EPA for noncancer end points. The lack of evaluation in some cases relates to a lack of information on toxicity or carcinogenicity for the chemical in question.

A useful reference source for information on the toxicity and carcinogenicity of hazardous air pollutants and other chemicals is the U.S. EPA Integrated Risk Information System (IRIS), an on-line database (U.S. EPA, 1993a and 1993b; Griffin, 1994). IRIS, originally developed for internal EPA use, contains the rationale for the risk status of a chemical, including its reference dose and reference concentration values and cancer classification if available and provides background bibliographic documentation for over 500 chemicals. The IRIS database is one of the databases available through the National Library of Medicine's Toxicology Data Network (TOXNET), which may be accessed via the Internet (toxnet.nlm.nih.gov).

A number of the hazardous air pollutants have been classified by IARC or EPA as to their carcinogenicity. In this chapter, formaldehyde will be used as a case study to illustrate some of the issues involved in performing risk assessments for hazardous air pollutants with potential carcinogenic effects.

Numerous pollutants may produce adverse effects not involving the development of cancer. For these pollutants, it is generally accepted that a threshold exists in the development of a toxic response as the result of exposure. The EPA (U.S. EPA, 1992b) developed a procedure that integrates available toxicity information in laboratory animals and humans to establish an exposure level that will protect both sensitive and nonsensitive individuals from adverse health effects resulting from lifetime exposure. For inhaled materials, this level is referred to as the inhalation reference concentration (RfC) and takes into account the unique dosimetric considerations in the regional deposition and retention of inhaled toxicants. The RfC methodology is discussed in more detail later in the chapter, and examples demonstrating methods for estimating the RfC for inhaled toxicants are presented.

Formaldehyde: A Cancer Risk Assessment Example   Formaldehyde is one of the chemicals listed in the Clean Air Act Amendments of 1990 as a hazardous air pollutant (Table 6). The database for formaldehyde will be summarized here to illustrate the risk assessment process for hazardous air pollutants.

Formaldehyde is a widely used commodity chemical and also a normal body constituent as an essential intermediate in cellular metabolism. It is widely used in the manufacture of resins, plywood, particle board, textiles, leather goods, paper, and pharmaceuticals. It has also been used widely as a tissue preservative and as an embalming agent.

The irritant properties of formaldehyde are well known by anyone who has had contact with it. These irritant properties were the basis for early health concerns about formaldehyde and also the basis for establishing the TLV by the ACGIH. The TLV-TWA ceiling for formaldehyde is currently set at 0.3 ppm (ACGIH, 1995).

Formaldehyde was one of the first commodity chemicals selected for evaluation in the late 1970s by the newly established Chemical Industry Institute of Toxicology (CIIT). The results of the long-term bioassay were reported by Kerns et al. (1983), who showed that chronic exposure of rats to a high concentration of formaldehyde (14.3 ppm) caused about a 50% incidence of nasal cancer, a relatively rare tumor in control rats (Table 7). The results of this study shifted attention from the irritant properties of formaldehyde to its carcinogenic potential for humans. Thus, the focus in this case study is on assessment of carcinogenic risk. The reader interested in more detail on the toxicity and carcinogenicity of formaldehyde is referred to the

Table 7   Incidence of Nasal Tumors in F-344 Rats Exposed to Formaldehyde and Comparison of EPA Estimates of Human Cancer Risk Associated with Continuous Exposure to Formaldehyde

Exposure rate (ppm)[a]	Incidence of rat nasal tumors		
14.3	94/140		
5.6	2/153		
2.0	0/159		
0	0/156		
	1987 risk estimates[b]	1991 risk estimates[c]	
	Rat-based	Rat-based	Monkey-based
Exposure concentration (ppm)	Upper 95% confidence limit estimates		
1.0	2×10^{-2}	1×10^{-2}	7×10^{-4}
0.5	8×10^{-3}	3×10^{-3}	2×10^{-4}
0.1	2×10^{-3}	3×10^{-4}	3×10^{-5}
	Maximum likelihood estimates		
1.0	1×10^{-2}	1×10^{-2}	1×10^{-4}
0.5	5×10^{-4}	1×10^{-3}	1×10^{-5}
0.1	5×10^{-7}	3×10^{-5}	4×10^{-7}

[a]Exposed 6 h/day, 5 days/week for 2 years.
[b]Estimated with 1987 inhalation cancer unit risk of 1.6×10^{-2} per ppm, which used airborne concentration as measure of exposure.
[c]Estimated with 1991 inhalation cancer unit risks of 2.8×10^{-3} per ppm (rat) and 3.3×10^{-4} per ppm (monkey), which used DNA-protein cross-links as measure of exposure.
(*Adapted from:* U.S. EPA, 1991.)

review of Heck et al. (1990), documents of the U.S. EPA (1991) and California Air Resources Board (CARB, 1992), and an IARC monograph published in 1995 (IARC, 1995).

In assessing the carcinogenic risk of a chemical, the epidemiological evidence appropriately received first consideration. Since formaldehyde has been widely used as a commodity chemical, a relatively large body of epidemiological data is available on it. Heck et al. (1990) reviewed 36 papers reporting on epidemiological studies of formaldehyde-exposed populations, including retrospective cohort mortality and case-control studies. They noted that deficits of sinonasal cancer (nasal cavity or nasal sinuses) were observed in two large retrospective cohort mortality studies of chemical workers and in three of five case-control studies. Two case-control studies that showed an excess of sinonasal cancer involved concurrent exposure to wood dust, which is known to produce an excess of nasal cancer. Thus the epidemiological data taken as a whole do not indicate a role for formaldehyde in causing sinonasal cancer. This is reassuring since the nasal cavity and nasal sinuses are sites of excess cancers in rats exposed to high concentrations of formaldehyde.

The evaluation of nasopharyngeal and oropharyngeal cancer as a group has yielded equivocal results. Lung cancer deaths were examined in 12 cohort studies reviewed by Heck et al. (1990). Nine studies showed a deficit relative to the number expected. In one study, the positive finding may have related to the absence of control for smoking habits. In another study, the increase in lung cancer was not statistically significant and occurred in the workers with the most recent and presumably lowest exposures. In the third study, the increase reported among formaldehyde-exposed workers was exceeded by a similar increase in a group not exposed to formaldehyde. Since the review of Heck et al. (1990), Andjelkovich et al. (1990, 1992, 1994, and 1995) have published their analyses of the causes of mortality of iron foundry workers. Some of the workers were exposed to high concentrations of formaldehyde over many years arising from the use of formaldehyde resins as binders in molds used in the metal casting process. The workers were also exposed to other noxious agents, including silica. And as is always the case, a significant number of workers were cigarette smokers. The bottom-line conclusion from the studies was most evident in the most recent publication (Andjelkovich et al., 1995), which reported on a subset cohort for which cigarette smoking history was ascertained. They reported that "no association between formaldehyde exposure and deaths from malignant or nonmalignant diseases of the respiratory tract was found. Cigarette smoking and silica exposures were found to be significantly associated with deaths attributed to lung cancer and disease of the respiratory tract." A Working Group Panel of IARC (1995) reviewed the epidemiological data on formaldehyde in 1994 and concluded that there was limited evidence in humans for the carcinogenicity of formaldehyde.

Let us now turn to the laboratory animal evidence of carcinogenicity.

The key study in this area comprises the original bioassays sponsored by CIIT. The studies involved F-344 rats and B6C3F1 mice exposed to 0.0, 2.0, 5.6, or 14.3 ppm formaldehyde for 6 hours/day, 5 days/week, for 24 months (Kerns et al., 1983) (Table 7). During the course of the study, it became apparent that an excess incidence of nasal carcinoma, a rare cancer, was being induced in the rats at the highest level of exposure. Because of their potential significance, the preliminary results were promptly made public. Ultimately, the results of the study were documented in the peer-reviewed literature (Kerns et al., 1983). The difference in nasal cancer incidences in the rats at the 5.6 ppm level compared to the controls was not statistically significant, although it should be noted that nasal tumors are rare in nonexposed rats. The difference at the 14.3 ppm level was clearly significant. In mice, nasal carcinomas were only observed at the 14.3 ppm level, where the incidence was 3.3%. This was not statistically significant. Formaldehyde-induced nasal cancers have also been reported in Sprague-Dawley and Wistar rats (Albert et al., 1982; Feron et al., 1988; Woutersen et al., 1989).

Subsequent to the determination of the nasal carcinogenicity of formaldehyde in the rat, a great deal of research has been conducted to elucidate the mechanisms of nasal cancer induction, with particular attention directed to the relevance of the finding for estimating the cancer risk to individuals exposed to formaldehyde. The resulting large body of literature has been reviewed by Heck et al. (1990) and Leikauf (1992). Two major lines of research will be discussed here to illustrate some key points and concepts: the molecular dosimetry of formaldehyde and the role of cell proliferation in formaldehyde-induced nasal cancer.

The molecular dosimetry studies have been based on the quantitation of DNA-protein cross-links (DPX) in the living tissue of the respiratory tract. Measurements have been made in both monkeys and rats exposed to various concentrations of formaldehyde (Casanova et al., 1989, 1991). The data are summarized in Fig. 10. The key features to be noted are the nonlinear relationships between exposure concentration and DPX formation and the approximately 10-fold greater cross-link formation rate in rats compared with monkeys at all exposure concentrations studied. Casanova (1991) used the rat data as input into a pharmacokinetic model using physiological parameters for rats and monkeys. The model yielded predictions for DPX formation in the monkey that were in excellent agreement with the experimentally derived values. This gives confidence in the use of both the rat and monkey data for predicting the human situation using human physiological parameters.

The regional distribution of DPX was qualitatively different in rats and monkeys. These differences also correlated with differences in the distribution of lesions in the respiratory tract. Both species had lesions in the nasal cavity. In the monkey, however, lesions were also evident in the more distal portions of the respiratory tract extending to the carina (Monticello et al., 1989; 1991).

A number of studies have focused on the mechanisms of formaldehyde

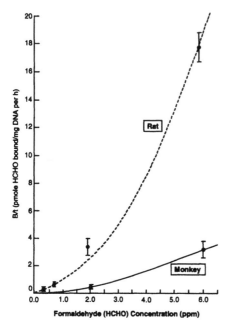

Figure 10 Formation of DNA-protein cross-links in nasal tissue of Fischer-344 rats and rhesus monkeys versus the airborne formaldehyde concentration, showing nonlinear exposure concentration-delivered dose curves and species differences. (*From:* Casanova et al., 1989 and 1991.)

carcinogenesis, with particular attention directed to the role of cell proliferation. Swenberg et al. (1983) reported that in the rat there was a marked increase in cell proliferation at high exposure concentrations as a compensatory response to cytotoxicity. These studies have been extended to consider a range of exposure situations for rats (Wilmer et al., 1987, 1989; Zwart et al., 1988; Monticello et al., 1991) and also to observations in monkeys (Monticello et al., 1989).

Conolly et al. (1992) proposed a risk assessment strategy for formaldehyde that would incorporate data on DNA-protein cross-links, cell proliferation, and putative preneoplastic changes. The kinds of biologically based models advocated by Moolgavkar (1986) would be built into such a strategy.

The role of formaldehyde in inducing mutations was reviewed by Heck et al. (1990) (see also Table 6). In both bacterial and mammalian *in-vitro* systems, it is a weak mutagen. The *in-vivo* data on somatic cell mutations are equivocal. This latter finding can soon be evaluated using the transgenic animals that are being developed. To date, clear linkages have not been established between DPX and the mutations. However, this is an area worthy of exploration.

The data summarized above have been used to classify formaldehyde with regard to both carcinogenicity and its cancer-causing potency (Table 7). Both EPA (U.S. EPA, 1987 and 1991) and IARC (1995) have classified formaldehyde as a probable human carcinogen, Group B1 by EPA and Group 2A by IARC. The classification is based on (1) sufficient animal evidence, (2) limited human evidence, and (3) supporting information, including structure-activity relationships that associate formaldehyde with genotoxic activity and carcinogenicity, including formation of DNA adducts and cyto-toxicity.

The quantitative estimates of formaldehyde cancer risk are of special interest (Table 7). All the quantitative estimates use the rat cancer data reported by Kerns et al. (1983). The observed data have been fit with a linearized multistage model and extrapolated to lower levels of exposure (GLOBAL 83 or 86, ICF Clement Associates, Rushton, LA). This is the approach advocated within EPA guidelines for carcinogenic risk assessment (U.S. EPA, 1986b). The earliest cancer risk estimate of EPA (U.S. EPA, 1987) used air concentration as the dose metric and yielded 95% confidence limits on risk of 1.6×10^{-2} per ppm or 1.3×10^{-5} per $\mu g/m^3$. More recently, EPA (U.S. EPA, 1991) has developed risk estimates using the DPX data as a measure of the delivered or effective dose while continuing to use the cancer response data from Kerns et al. (1983). With this approach, the upper 95% confidence limit is reduced to 2.8×10^{-3} per ppm using the rat DPX data and further reduced to 3.3×10^{-4} per ppm using the monkey DPX data. In my opinion, the latter value is the most scientifically sound of the estimates and is compatible with the equivocal-negative epidemiological observations. Unfortunately, the EPA has not yet formally adopted the revised estimate based on the monkey DPX data.

Table 7 is informative with respect to several aspects of the cancer risk assessment process typically used by EPA. The EPA routinely reports its estimates as unit risks expressed as cancer risk per $\mu g/m^3$. This value is the slope of the linear component produced using the linearized multistage model. This linear slope is apparent at the 0.1 ppm concentration for all the risk estimates. However, the 1991 risk estimates for 0.5 and 1.0 ppm do not scale proportional to the 0.1 ppm values. The values shown are greater than proportionality, indicating that the 0.5 and 1.0 ppm estimates for these data are in the curvilinear portion of the exposure-response relationship where an increase in exposure yields a disproportionate increase in risk. It is also informative to consider the maximum likelihood estimates. As expected, they diverge markedly from the upper 95% confidence limits, especially for the risk estimates at the lowest exposure concentrations. The several risk estimates illustrate the schematic relationship shown in Fig. 4, where the risk estimate decreases and the bounds of uncertainty are reduced as more scientific information is acquired and used to estimate risk.

Inhalation Reference Concentration Methodology

To provide guidance for evaluating noncancer health effects, the EPA has developed an inhalation reference concentration (RfC) methodology (Barnes et al., 1988; U.S. EPA, 1990; Jarabek et al., 1990; Shoaf, 1991; Jarabek, 1995). Jarabek (1995) has reviewed the methodology in detail and defined an RfC as "an estimate (with uncertainty spanning perhaps an order of magnitude) of a continuous inhalation exposure to the human population (including sensitive subgroups) that is likely to be without appreciable risk of deleterious noncancer health effects during a lifetime." The EPA RfC methodology is compared with several of the other approaches used for assessing and managing cancer risks of airborne materials (Table 8). All five approaches have in common the use of safety factors or uncertainty factors for making extrapolations to humans from data developed in laboratory animals when human data are not available. Of the five approaches, only the EPA RfC methodology has provision for using dosimetry data to make extrapolations. Recognizing the extent to which there are marked species differences in exposure-dose relationships, the dosimetry adjustment provision in the RfC methodology is a significant advance over the other approaches that do not have provision for such adjustments.

The RfC methodology is intended to provide guidance for noncancer toxicity, that is, adverse health effects or toxic end points other than cancer and gene mutations due to effects of environmental agents on the structure or function of various organ systems. This includes those effects observed in the respiratory tract as well as extrarespiratory effects that occur when the respiratory tract serves as a portal of entry. An assumption inherent in the approach is that the relationship between dose and noncancer responses manifests a threshold, which contrasts with the nonthreshold relationship between dose and cancer incidence that is a key default assumption of the EPA Cancer Risk Assessment Guidelines (U.S. EPA, 1986b). The noncancer responses are defined by both continuous incidence and severity scales, which contrast with the quantal or dichotomous nature of cancer response. The models used for the continuous effects estimate expected changes in individuals or shifts in population means.

In developing and implementing the RfC methodology, special attention has been directed to the nature of the database available, with provision made for various levels of confidence related to the quantity of data available (Table 9). The system also has provision for using a range of effects levels (Table 10). Jarabek (1995) provides an excellent discussion of the difficulties involved in distinguishing between adverse and nonadverse effects and assigning levels of adversity.

The RfC is intended to estimate a benchmark level for continuous exposure. Thus normalization procedures are used to adjust less than continuous exposure data to 24 hours/day for a lifetime of 70 years.

A wide range of dosimetric adjustments are accommodated within the RfC methodology to take account of differences in exposure-dose relationships among species. Regional differences (extrathoracic, tracheobronchial, and pulmonary) are taken into account as adjustments for particles versus gases and within gases for three categories based on reactivity (including both dissociation and local metabolism) and water solubility. Provision is made for using more detailed, experimentally derived models when they are available.

Table 8 Comparison of Risk Assessment and Risk Management Estimates

Estimate	NAS paradigm	Exposure scenario	Effect level	Population	Database	Dosimetry	SF or UF[a]
ACGIH TLV-TWA	Management	8 h/day 40 h/wk 40 yr	Impairment of health or freedom from irritation. Narcosis. Nuisance. Stress.	Healthy worker.	Industrial experience. Experimental human and animal.	No	SF
NIOSH REL	Characterization	10 h/day 40 h/wk "Working lifetime"	Impairment of health or functional capacity and technical feasibility.	Healthy worker.	Medical. Biological. Chemical. Trade.	No	SF
OSHA PEL	Management	8 h/day 40 h/wk 45 yr	Impairment of health or functional capacity and technical feasibility.	Healthy worker.	Medical. Biological. Chemical. Trade.	No	SF
EPA RfC	Dose-response	24 h/day 70 yr	NOAEL.	General population, including susceptible.	Occupational. Experimental human and animal.	Yes	UF
ATSDR MRL	Dose-response	24 h/day 70 yr	NOAEL.	General population, including susceptible.	Occupational. Experimental human and animal.	No	UF

[a]SF = safety factor; UF = uncertainty factor for explicit extrapolations applied to data.
(*From:* Jarabek, 1995.)

Table 9 Minimum Animal Bioassay Database for Various Levels of Confidence in the Inhalation Reference Concentration (RfC)

Mammalian database[a]	Confidence	Comments

1. A. Two inhalation bioassays[b] in different species.
    B. One two-generation reproductive study.
    C. Two developmental toxicity studies in different species.
	High	Minimum database for high confidence

2. 1A and 1B, as above.	Medium to high	
3. Two of three studies, as above in 1A and 1B; one or two developmental toxicity studies.	Medium to high	
4. Two of three studies, as above in 1A and 1B.	Medium	
5. One of three studies, as above in 1A and 1B; one or two developmental toxicity studies.	Medium to low	
6. One inhalation bioassay.[c]	Low	Minimum database for estimation of an RfC

[a]Composed of studies published in refereed journals, final quality assured and quality checked and approved contract laboratory studies, or core minimum Office of Pesticide Programs rated studies. It is understood that adequate toxicity data in humans can form the basis of an RfC and yield high confidence in the RfC without this database. Pharmacokinetic data indicating insignificant distribution occurring remote to the respiratory tract may decrease requirements for reproductive and developmental data.

[b]Chronic data.

[c]Chronic data preferred but subchronic acceptable.

(*From:* Jarabek, 1995.)

A key step in arriving at the RfC is the development of the entire toxicity profile or data array, which is examined to select the prominent toxic effect. The toxicity profile is defined as the critical effect pertinent to the mechanism of action of the chemical that is at or just below the threshold for more serious effects. The study that best characterizes the critical effect is identified as the principal study. The critical effect chosen is generally characterized by the lowest NOAEL, adjusted to the Human Equivalent Concentration (HEC), that is representative of the threshold region for the entire data array.

The RfC is then derived from $NOAEL_{(HEC)}$ for the critical effect by application of uncertainty factors (UF). A UF may be included when effects are observed that are considered to be related to exposure duration, contrasted to those that are concentration-related (Fig. 11). In addition, a modifying factor (MF) may also be applied when scientific uncertainties in the principal study are not explicitly addressed by the standard UF. Guidelines for the use of UF and MF are shown in Table 11.

The RfC methodology just discussed focuses on the establishment of either a lowest-observed-adverse-effect level (LOAEL) or a no-observed-adverse-effect level (NOAEL) as the starting point for the derivation of expo-

Table 10   Effect Levels Considered in Deriving RfC in Relationship to Empirical Severity Rating Values

Effect or no-effect level	Rank	General effect
NOEL	0	No observed effects
NOAEL	1	Enzyme induction or other biochemical change, consistent with possible mechanism of action, with no pathologic changes and no change in order weights
NOAEL	2	Enzyme induction and subcellular proliferation or other changes in organelles, consistent with possible mechanism of action, but no other apparent effects
NOAEL	3	Hyperplasia, hypertrophy, or atrophy, but no change in organ weights
NOAEL-LOAEL	4	Hyperplasia, hypertrophy, or atrophy, with changes in organ weights
LOAEL	5	Reversible cellular changes, including cloudy swelling, hydropic change, or fatty changes
(LO)AEL*	6	Degenerative or necrotic tissue changes with no apparent decrement in organ function
(LO)AEL-FEL	7	Reversible slight changes in organ function
FEL	8	Pathological changes with definite organ dysfunction that are unlikely to be fully reversible
FEL	9	Pronounced pathologic changes with severe organ dysfunction with long-term sequelae
FEL	10	Death or pronounced life-shortening

Note: Adapted from DeRosa et al. (1985) and Hartung (1986). Ranks are from lowest to highest severity.
*The parentheses around the "LO" in the acronym "LOAEL" refer to the fact that any study may have a series of doses that evoke toxic effects of rank 5 through 7. All such doses are referred to as adverse effect levels (AEL). The lowest AEL is the (LO)AEL.
(From: Jarabek, 1995.)

sure limits. One criticism of this approach is that it does not make use of all the data available on a chemical. This has led to a proposal for an alternative procedure in which the lower 95% confidence limit for a response function is used to establish the benchmark dose (Crump, 1984).

Bromomethane: A Noncancer Risk Assessment Example   The derivation of an inhalation reference concentration (RfC) for bromomethane (also called methyl bromide) will be used to illustrate the methodology currently used by EPA for developing quantitative estimates of risk for noncancer end points. Bromomethane is widely used as a fumigant and pesticide. It is listed in the Clean Air Act Amendments of 1990 as a hazardous air pollutant. The derivation of the bromomethane RfC was reviewed in detail by Guth and Jarabek (1995).

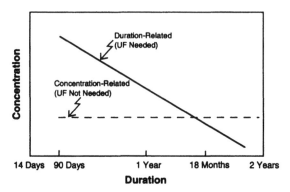

Figure 11 A schematic illustrating the relationship of a fixed effect (toxicity indexed as either an increased incidence or an increase in severity, or both) as a function of concentration and duration. Different lines indicate the relationship where toxicity is predominantly explained either by concentration or by duration. (*From:* Jarabek, 1995.)

The basic biological data for the derivation are from the study of Reuzel et al. (1991), who exposed Wistar rats to 0, 3, 30, or 90 ppm bromomethane for 6 hours/day, 5 days/week for up to 29 months. They observed changes in the nasal cavity, hyperplasia of basal cells, and degeneration of the olfactory epithelium that were exposure concentration-dependent in both incidence and severity and were statistically significant at 29 months (Fig. 12). The effect was not statistically significant in small groups observed after 12 or 24 months of exposure. The lowest-observed-adverse-effect level (LOAEL) in the rat study was 3.0 ppm.

The NTP (1992) conducted a study in which B6C3F1 mice were exposed to 0, 10, 33, or 100 ppm bromomethane for 6 hours/day, 5 days/week, for 6, 15, or 24 months. The incidence of olfactory epithelial necrosis and metaplasia in the nasal cavity was increased in a statistically significant manner at the 100 ppm level. Mice exposed to the lower concentrations did not exhibit significant increases in any of the lesions (Fig. 13). For the mouse study, the LOAEL was 100 ppm, and the no-observed-adverse-effect level (NOAEL) was 33 ppm.

The RfC was based on the more sensitive outcome in the rat study. The rat LOAEL of 3 ppm (11.7 mg/m³) was adjusted for continuous exposure to yield a duration-adjusted LOAEL of 2.08 mg/m³ (11.7 mg/m³ × 6 hours/24 hours × 5 days/7 days = 2.08 mg/m³). Then using the default dosimetric adjustments (U.S. EPA, 1990; Jarabek et al., 1990), the human equivalent concentration (HEC) was calculated using a regional gas deposition ratio (RGDR). The RGDR is the ratio of the animal ventilation rate over animal surface area of the extrathoracic region of the respiratory tract to human ventilation rate over human surface area of the extrathoracic region:

$$[(0.30 \text{ m}^3/\text{day})/11.6 \text{ cm}^2/(20 \text{ m}^3/\text{day})/177 \text{ cm}^2] = 0.23.$$

Table 11 Guidelines for the Use of Uncertainty Factors in Deriving Information Reference Concentration (RfC)

Standard uncertainty factors (UF)	Processes considered in UF purview
H = Human to sensitive human. Use a 10-fold factor when extrapolating from valid experimental results from studies using prolonged exposure to average healthy humans. This factor is intended to account for the variation in sensitivity among the members of the human population.	Pharmacokinetics-pharmacodynamics. Sensitivity. Differences in mass (children, obese). Concomitant exposures. Activity pattern. Does not account for idiosyncrasies.
A = Animal to human. Use a 3-fold factor when extrapolating from valid results of long-term studies on experimental animals when results of studies of human exposure are not available or are inadequate. This factor is intended to account for the uncertainty in extrapolating animal data to the case of average healthy humans. Use of a 3 is recommended with default dosimetric adjustments. More rigorous adjustments may allow additional reduction. Conversely, judgment that the default may not be appropriate could result in an application of a 10-fold factor.	Pharmacokinetics-pharmacodynamics. Relevance of laboratory animals model. Species sensitivity.
S = Subchronic to chronic. Use a 10-fold factor when extrapolating from less than chronic results on experimental animals or humans when there are no useful long-term human data. This factor is intended to account for the uncertainty in extrapolating from less than chronic NOAEL to chronic NOAEL.	Accumulation-cumulative damage. Pharmacokinetics-pharmacodynamics. Severity of effect. Recovery. Duration of study. Consistency of effect with duration.

Then the

$$\text{LOAEL}_{(HEC)} = 2.08 \text{ mg/m}^3 \times 0.23 = 0.48 \text{ mg/m}^3$$

For comparison, the HEC from the NTP mouse study was calculated in the same way, and the resulting RGDR for mice was 0.19. The follow-on calculations for the mice resulted in a $\text{LOAEL}_{(HEC)}$ of 13 mg/m^3 and a $\text{NOAEL}_{(HEC)}$ of 4.4 mg/m^3.

The next step in RfC methodology involves introduction of uncertainty

Table 11   (Continued)

Standard uncertainty factors (UF)	Processes considered in UF purview
L = LOAEL$_{[HEC]}$ to NOAEL$_{[HEC]}$. Use a 10-fold factor when deriving an RfC from a LOAEL$_{[HEC]}$, instead of a NOAEL$_{[HEC]}$. This factor is intended to account for the uncertainty in extrapolating from LOAEL$_{[HEC]}$ to NOAEL$_{[HEC]}$.	Severity. Pharmacokinetics-pharmacodynamics. Slope of dose-response curve. Trend, consistency of effect. Relationship of end points. Functional vs. histopathological evidence. Exposure uncertainties.
D = Incomplete to complete database. Use up to a 10-fold factor when extrapolating from valid results in experimental animals when the data are "incomplete." This factor is intended to account for the inability of any single animal study to adequately address all possible adverse outcomes in humans.	Quality of critical study. Data gaps. Power of critical study and supporting studies. Exposure uncertainties.
Modifying Factor (MF). Use professional judgment to determine whether another uncertainty factor (MF) that is ≤10 is needed. The magnitude of the MF depends upon the professional assessment of scientific uncertainties of the study and database not explicitly treated above (e.g., the number of animals tested or quality of exposure characterization). The default value of the MF is 1.	

Note: Assuming the range of the UF is distributed log-normally, reduction of a standard 10-fold UF by half (i.e., 10^{-8}) results in a UF of 3. Composite UF for derivation involving 4 areas of uncertainty is 3000 in recognition of the lack of independence of these factors. Inhalation reference concentrations are not derived if all five areas of uncertainty are invoked.

(From: Jarabek, 1995.)

factors (UF). A UF of 10 for intraspecies differences in sensitivity was introduced because there were no human data indicating that such a UF was not warranted. A UF of 3 was used to adjust from a LOAEL to a NOAEL. This value was reduced from 10 due to the mildness of the effect. A UF for interspecies scaling of 3 rather than 10 was used due to the application of default dosimetry. Because the factors of 3 represent operational application of a geometric half of the standard factor of 10 rounded to a single significant figure, multiplication of the two factors of 3 results in a composite factor of 10; hence $3 \times 3 \times 10 = 100$. When the NOAEL$_{(HEC)}$ from the rat study of 0.48 mg/m^3 is divided by the total uncertainty factor of 100 and the result is rounded to one significant factor, the resultant RfC for bromomethane is 5 μg/m^3.

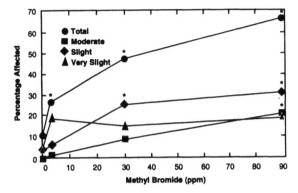

Figure 12 Basal cell hyperplasia of the olfactory epithelium in male rats exposed to methyl bromide (bromomethane) for 29 months, percentage of rats affected; $n = 46$, 48, 49, and 48 in control, and 3, 30, and 90 ppm groups, respectively; $*p < 0.05$. (*From:* Guth and Jarabek, 1995; based on data from Reuzel et al., 1991.)

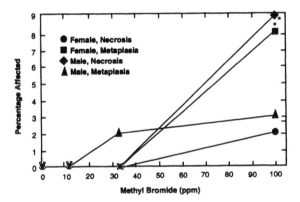

Figure 13 Lesions of the olfactory epithelium in mice exposed to methyl bromide (bromomethane) for 2 years, percentage of rats affected; $*p < 0.05$. (*From:* Guth and Jarabek, 1995; based on data from NTP, 1992.)

Guth and Jarabek (1995) also provided a qualitative summary of the robustness of the RfC derivation based on the principal study and the data base. Because it did not identify a NOAEL, they accorded the Reuzel et al. (1991) study a medium confidence rating even though it was well-conducted, used an adequate number of animals, and included a thorough histopatho-logical examination of the respiratory tract. The database was given a high confidence rating because there were chronic inhalation studies conducted in two species that were supported by subchronic studies, because data were available on developmental and reproductive effects (less sensitive indicators of effect), and because pharmacokinetic data following inhalation exposure were available. Taking into account their confidence in the principal study as well as in the database, Guth and Jarabek (1995) accorded high confidence to the RfC of 5 μg/m^3.

In considering the RfC derivation for bromomethane, the difference in the exposure-response relationship between the rats and mice is striking. This carries over into the RfC calculation, with a rat $LOAEL_{(HEC)}$ of 0.48 mg/m³ compared with a mouse $LOAEL_{(HEC)}$ of 13 mg/m³ and a mouse $NOAEL_{(HEC)}$ of 4.4 mg/m³. If one were to have confidence in the mouse data over the rat data to estimate human risk, then it would be appropriate to base an RfC on it. Because the starting point would be a $NOAEL_{(HEC)}$, it could be argued that a UF of 3 could be removed from the calculation. Thus, the RfC based on mice could be 4.4 mg/m³/30 = 147 μg/m³, a value 30 times higher than that based on the rat data. This difference emphasizes the importance of understanding interspecies differences as a basis for assessing human health risks from laboratory animal studies.

SUMMARY

In this chapter, I have reviewed some of the key concepts involved in assessing the human health risks of airborne toxicants. The chapter began with a review of some of the historical aspects in assessing risks of airborne toxicants and then proceeded to a discussion of the sources of information used for assessing human health risks. It emphasized the preeminent role of human data if they are available. For a variety of reasons, however, human data are now limited. Looking to the future, it is hoped that positive epidemiological data will only rarely become available because such data can only be developed if our past exposure control practices were inadequate. For many chemicals, it is not ethical to conduct controlled human exposures. Thus, for the majority of new materials, it will be necessary to rely on risk assessments developed with information from studies of cells, tissues, and laboratory animals. The use of such data typically involves two extrapolations: from laboratory animals to humans and from high-level exposures that can be studied with statistical confidence to the low levels of exposure likely to be encountered by people in situations other than accidents. The validity of these extrapolations and all human risk assessment depends on the quality of the underlying scientific information and especially on our knowledge of interspecies similarities and differences in exposure-dose-response relationships for airborne toxicants. The scientific basis for human risk assessments can be enhanced when the risk assessment process is used to identify uncertainties in risk characterization that, if addressable by investigation, serve as feedback in the form of research needs to the research arena. If these research needs are addressed, the resulting information should serve to reduce the uncertainty in future risk assessments. Thus, in the final analysis, there is major societal benefit to be gained from a joining of risk research, risk assessment, and risk management as an integrated approach to minimizing human health risks effectively and efficiently.

REFERENCES

Albert, RE: Carcinogen risk assessment in the U.S. Environmental Protection Agency. Crit Rev Toxicol 24:75–85, 1994.

Albert, RE, Sellakumar, AR, Laskin, S, Kuschner, M, Nelson, N, Snyder, CA: Gaseous formaldehyde and hydrogen chloride induction of nasal cancer in the rat. J Natl Cancer Inst 68:597–603, 1982.

American Conference of Governmental Industrial Hygienists (ACGIH): Threshold limit values and biological exposure indices for 1994–1995. ACGIH, Cincinnati, OH, 1994.

Andjelkovich, DA, Janszen, DB, Brown, MH, Richardson, RB, Miller, FJ: Mortality of iron foundry workers. IV. Analysis of a subcohort exposed to formaldehyde. J Occup Med, 1995. In press.

Andjelkovich, DA, Mathew, RM, Richardson, RB, Levine, RJ: Mortality of iron foundry workers; I. Overall findings. J Occup Med 32:529–540, 1990.

Andjelkovich, DA, Mathew, RM, Yu, RC, Richardson, RB, Levine, RJ: Mortality of iron foundry workers; II. Analysis by work area. J Occup Med 34:391–401, 1992.

Andjelkovich, DA, Shy, CM, Brown, MH, Janszen, DB, Levine, RJ, Richardson, RB: Mortality of iron foundry workers. III. Lung cancer case-control study. J Occup Med 36:1301–1309, 1994.

Andrews, C, Buist, S, Ferris, BG, Hackney, J, Rom, W, Samet, J, Schenker, M, Shy, C, Strieder, D: Guidelines as to what constitutes an adverse respiratory health effect, with special reference to epidemiologic studies of air pollution. Am Rev Respir Dis 131:666–668, 1985.

Barnes, DG, Dourson, M: Reference dose (RfD): Description and use in health risk assessments. Regul Toxicol Pharmacol 8:471–486, 1988.

Boorman, GA, Catalano, PJ, Jacobson, BJ, Kaden, DA, Mellick, PW, Nauss, KM, Ryan, LM: Consequences of prolonged inhalation of ozone on F344/N rats: Collaborative studies. Part VI. Background and study design. [Research Report Number 65]. Health Effects Institute, Cambridge, MA, 1995. In press.

California Air Resources Board (CARB): Formaldehyde Identification. Office of Environmental Health Hazard Assessment, Air Toxicology and Epidemiology Section, 1992.

Casanova, M, Deyo, DF, Heck, Hd'A: Covalent binding of inhaled formaldehyde to DNA in the nasal mucosa of Fischer-344 rats: Analysis of formaldehyde and DNA by high-performance liquid chromatography and provisional pharmacokinetic interpretation. Fundam Appl Toxicol 12:397–417, 1989.

Casanova, M, Morgan KT, Steinhagen, WH, Everitt, JI, Popp, JA, Heck, Hd'A: Covalent binding of inhaled formaldehyde to DNA in the respiratory tract of rhesus monkeys: Pharmacokinetics, rat-to-monkey interspecies scaling, and extrapolation to man. Fundam Appl Toxicol 17:409–428, 1991.

Catalano, PJ, Rogus, J, Ryan, LM: Consequences of prolonged inhalation of ozone on F344/N rats: Collaborative studies. Part X. Robust composite scores based on median Polish analysis. [Research Report Number 65]. Health Effects Institute, Cambridge, MA, 1995.

Clean Air Act Amendments, Public Law no. 101–549; 104 STAT.2399, 1990.

Clean Air Act Amendments of 1970, Public Law no. 91–604; 84 STAT.1676, 1970.

Conolly, RB, Morgan, KT, Andersen, ME, Monticello, TM, Clewell, HJ: A biologically based risk assessment strategy for inhaled formaldehyde. Comments Toxicol 4(4):269–293, 1992.

Crump, KS: An improved procedure for low-dose carcinogenic risk assessment from animal data. JEPTO 5:339–348, 1984.

Cuddihy, RG, McClellan, RO, Griffith, WC: Variability in target organ deposition among individuals exposed to toxic substances. Toxicol Appl Pharmacol 49:179–187, 1979.

Del Pup, J, Kmiecik, J, Smith, S, Reitman, F: Improvement in human health risk assessment utilizing site- and chemical-specific information: A case study. Appendix G, pp. 479–502. In: Science and Judgment in Risk Assessment, National Research Council, Committee on Risk Assessment of Hazardous Air Pollutants. National Academy Press, Washington, DC, 1994.

DeRosa, CT, Stara JF, Durkin, PR: Ranking chemicals based on chronic toxicity data. Toxicol Ind Health 1(4):177–191, 1985.

Detels, R, Tashkin, DP, Sayre, JW, Rokaw, SN, Coulson, AH, Massey, FJ, Jr, Wegman, DH: The UCLA population studies of chronic obstructive respiratory disease. 9. Lung function changes associated with chronic exposure to photochemical oxidants; a cohort study among never-smokers. Chest 92:594–603, 1987.

Devlin, RB, McDonnell, WF, Mann, R, Becker, S, House, DE, Schreinemachers, D, Koren, HS: Exposure of humans to ambient levels of ozone for 6.6 hours causes cellular and biochemical changes in the lung. Am J Respir Cell Mol Biol 4:72–81, 1991.

Feron, VJ, Bruyntjes, JP, Woutersen, RA, Immel, HR, Appelman, LM: Nasal tumours in rats after short-term exposure to a cytotoxic concentration of formaldehyde. Cancer Lett 39:101–111, 1988.

Finkel, AM: The case for "plausible conservatism" in choosing and altering defaults. Appendix N-1, pp. 601–627. In: Science and Judgment in Risk Assessment, National Research Council, Committee on Risk Assessment of Hazardous Air Pollutants. National Academy Press, Washington, DC, 1994.

Gehring, PJ, Watanabe, PG, Park, CN: Resolution of dose-response toxicity data for chemicals requiring metabolic activation: Example—vinyl chloride. Toxicol Appl Pharmacol 44:581–591, 1978.

Grant, LD, Jordan, BC: Basis for primary air quality criteria and standards. U.S. Environmental Protection Agency, Environmental Criteria and Assessment Office. Research Triangle Park, NC, 1988.

Griffin, WA: Toward an improved information resource for risk assessment: EPA's integrated risk information system (IRIS). CIIT Activities 14(10):1–7, 1994.

Guth, DJ, Jarabek, AM: U.S. EPA inhalation reference concentration for bromomethane. In: Nasal Toxicity and Dosimetry of Inhaled Xenobiotics: Implications for Human Health. Edited by FJ Miller, pp. 327–339. Taylor & Francis, Washington, DC, 1995.

Hartung, R: Ranking the severity of toxic effects. In: Trace Substances in Environmental Health—XX: Proceedings of the University of Missouri's 20th Annual Conference on Trace Substances in Environmental Health. Edited by PD Hemphill, pp. 204–211. University of Missouri, Columbia, MO, 1986.

Heck, Hd'A, Casanova, M, Starr, T: Formaldehyde toxicity—new understanding. Crit Rev Toxicol 20:397–426, 1990.

Horstman, DH, Folinsbee, LJ, Ives, PJ, Abdul-Salaam, S, McDonnell, WF: Ozone concentration and pulmonary response relationships for 6.6-hour exposures with five hours of moderate exercise to 0.08, 0.10, and 0.12 ppm. Am Rev Respir Dis 142:1158–1163, 1990.

International Agency for Research on Cancer (IARC): A Consensus Report of an IARC Monographs Working Group on the Use of Mechanisms of Carcinogenesis in Risk Identification (IARC intern. tech. Rep. No. 91/002). IARC, Lyon, France, 1991.

International Agency for Research on Cancer (IARC): Occupational Exposures to Wood Dusts and Formaldehyde [IARC Monographs on the Evaluation of Carcinogenic Risks to Humans, Volume 62]. IARC, Lyon, France, 1995.

International Agency for Research on Cancer (IARC): Overall Evaluations of Carcinogenicity: An Updating of IARC Monographs [IARC Monographs on the Evaluation of Carcinogenic Risks to Humans, Supplement 70, Vol. 1–42]. IARC, Lyon, France, 1987.

International Agency for Research on Cancer (IARC): Some Industrial Chemicals [IARC Monographs on the Evaluation of Carcinogenic Risks to Humans, Vol. 60]. IARC, Lyon, France, 1994.

Jarabek, AM: Inhalation RfC methodology: Dosimetric adjustments and dose-response estimation of noncancer toxicity in the upper respiratory tract. In: Nasal Toxicity and Dosimetry of Inhaled Xenobiotics: Implications for Human Health. Edited by FJ Miller, pp. 301–325. Taylor & Francis, Washington, DC, 1995.

Jarabek, AM, Menache, MG, Overton, JH, Jr, Dourson, ML, Miller, FJ: The U.S. Environmental Protection Agency's inhalation RfD methodology: Risk assessment for air toxics. Toxicol Ind Health 6(5):279–301, 1990.

Kerns, WD, Pavkov, KL, Donofrio, DJ, Gralla, EJ, Swenberg, JA: Carcinogenicity of formaldehyde in rats and mice after long-term inhalation exposure. Cancer Res 43:4382–4392, 1983.

Leikauf, GD: Formaldehyde and other aldehydes. In: Environmental Toxicants: Human Exposures and Their Health Effects. Edited by M Lippmann, pp. 299–330. Van Nostrand Reinhold, New York, NY, 1992.

Leikauf, GD, Driscoll, KE, Wey, HE: Ozone-induced augmentation of eicosanoid metabolism in epithelial cells from bovine trachea. Am Rev Respir Dis 137:435–442, 1988.

Lippmann, M: Ozone. In: Environmental Toxicants: Human Exposures and their Health Effects. Edited by M Lippmann, pp. 465–519. Van Nostrand Reinhold, New York, NY, 1992.

Lippmann, M: Criteria and standards for occupational exposures to airborne chemicals. Clin Podiatr Med Surg 4:619–628, 1987.

Lubin, JH, Boice, JD, Edling, C, Hornung, R, Howe, G, Kunz, E., Kusaik, RA, Morrison, HI, Radford, EP, Samet, JM, Trimarche, M, Woodward, A, Xiang, YS, Pierce, DA: Radon and lung cancer risk: A joint analysis of 11 underground miners studies [NIH Publication No. 94–3644]. National Institutes of Health, Bethesda, MD, 1994.

Lubin, JH, Steindorf, K: Cigarette use and the estimation of lung cancer attributable to radon in the United States. Radiat Res 141:79–85, 1995.

Maltoni, C, Lefemine, G: Carcinogenicity bioassays of vinyl chloride: Current results. Ann NY Acad Sci 246:195–218, 1975.

Mauderly, JL, Jones, RK, Griffith, WC, Henderson, RF, McClellan, RO: Diesel exhaust is a pulmonary carcinogen in rats exposed chronically by inhalation. Fundam Appl Toxicol 9:208–221, 1987.

Mayer, CA, Francis, MA: Clean Air Act Handbook: A Practical Guide to Compliance. Clark Boardman Company, New York, NY, 1991.

McClellan, RO: Assessing health risks of synthetic vitreous fibers: An integrative approach. Regul Toxicol Pharmacol 20(3):S121-S134, 1994.

McClellan, RO: Prepared discussion—Health effects of ozone. J Air Pollut Control Assoc 39:1186–1188, 1989.

McClellan, RO, North, DW: Making full use of scientific information in risk assessment. Appendix N-2, pp. 629–640. In: Science and Judgment in Risk Assessment. National Research Council, Committee on Risk Assessment of Hazardous Air Pollutants. National Academy Press, Washington, DC, 1994.

McDonnell, WF, Kehrl, HR, Abdul-Salaam, S, Ives, PJ, Folinsbee, LJ, Devlin, RB, O'Neil, JJ, Horstman, DH: Respiratory response of humans exposed to low levels of ozone for 6.6 hours. Arch Environ Health 46:145–150, 1991.

Montesano, R, Bartsch, H, Vainio, H, Wilbourn, J, Yamasaki, H, editors: Long-term and Short-term Assays for Carcinogenesis—A Critical Appraisal [IARC Scientific Publications No. 83]. IARC, Lyon, France, 1986.

Monticello, TM, Miller, FJ, Morgan, KT: Regional increases in rat nasal epithelial cell proliferation following acute and subchronic inhalation of formaldehyde. Toxicol Appl Pharmacol 111:409–421, 1991.

Monticello, TM, Morgan, KT, Everitt, JI, Popp, JA: Effects of formaldehyde gas on the respiratory tract of rhesus monkeys. Pathology and cell proliferation. Am J Pathol 134:515–527, 1989.

Moolgavkar, SH: Carcinogenesis modeling: from molecular biology to epidemiology. Annu Rev Public Health 7:151–169, 1986.

National Research Council (NRC): Health risks of radon and other internally deposited alpha-emitters [National Research Council Report BEIR IV]. National Academy Press, Washington, DC, 1988.

National Research Council (NRC): Human exposure assessment for airborne pollutants: Advances and opportunities: National Academy Press, Washington, DC, 1990.

National Research Council (NRC): Rethinking the ozone problem in urban and regional air pollution. National Academy Press, Washington, DC, 1991.

National Research Council (NRC), Committee on the Institutional Means for Assessment of Risks to Public Health: Risk Assessment in the Federal Government: Managing the Process. National Academy Press, Washington, DC, 1983.

National Research Council (NRC), Committee on Risk Assessment of Hazardous Air Pollutants: Science and Judgment in Risk Assessment. National Academy Press, Washington, DC, 1994.

National Toxicology Program (NTP): Toxicology and carcinogenesis studies of methyl bromide (CAS No. 74–83–9) in B6C3F1 Mice (Inhalation Studies) [NTP TR 385]. National Toxicology Program, Research Triangle Park, NC, 1992.

Padgett, J, Richmond, H: The process of establishing and revising National Ambient Air Quality Standards. J Air Pollut Control Assoc 33:13–16, 1983.

Reuzel, PGJ, Dreef-van der Meulen, HC, Hollanders, VMH, Kuper, CF, Feron, VJ, van der Heijden, CA: Chronic inhalation toxicity and carcinogenicity study of methyl bromide in Wistar rats. Food Chem Toxicol 29:31–39, 1991.

Samet, JM: Environmental tobacco smoke. In: Environmental Toxicants: Human Exposures and their Health Effects. Edited by M Lippmann, pp. 231–265. Van Nostrand Reinhold, New York, 1992.

Shoaf, CR: Current assessment practices for noncancer end points. Environ Health Perspect 95:111–119, 1991.

Spektor, DM, Lippmann, M, Lioy, PJ, Thurston, GD, Citak, K, James, DJ, Bock, N, Speizer, FE, Hayes, C: Effects of ambient ozone on respiratory function in active, normal children. Am Rev Respir Dis 137:313–320, 1988a.

Spektor, DM, Lippmann, M, Thurston, GD, Lioy, PJ, Stecko, J, O'Connor, G, Garshick, E, Speizer, FE, Hayes, C: Effects of ambient ozone on respiratory function in healthy adults exercising outdoors. Am Rev Respir Dis 138:821–828, 1988b.

Swenberg, JA, Gross, EA, Randall, HW, Barrow, CS: The effect of formaldehyde exposure on cytotoxicity and cell proliferation. In: Formaldehyde: Toxicology Epidemiology, Mechanisms. Edited by JJ Clary, JE Gibson, RS Waritz, pp. 225–236. Marcel Dekker, New York, 1983.

Tomatis, L, Aitio, A, Wilbourn, J, Shuker, L: Human carcinogens so far identified. Jpn J Cancer Res 80:795–807, 1989.

U.S. Department of Health and Human Services (DHHS): Reducing the health consequences of smoking. 25 years of progress. A report of the Surgeon General. U.S. Government Printing Office, Washington, DC, 1989.

U.S. Department of Health and Human Services (DHHS): Seventh Annual Report on Carcinogens. Summary 1994. Technical Resources, Inc., Rockville, MD, 1994.

U.S. Environmental Protection Agency (EPA): Air Quality Criteria for Ozone and Other Photochemical Oxidants [EPA-600/8-84-020aF]. U.S. EPA, Research Triangle Park, NC, 1986a.

U.S. Environmental Protection Agency (EPA): Guidance on Risk Characterization for Risk Managers and Risk Assessors. Internal Memorandum by F. Henry Habicht III, February 26, 1992. USEPA Office of the Administrator, Washington, DC, 1992a.

U.S. Environmental Protection Agency (EPA): Guidelines for Carcinogen Risk Assessment. Fed Reg 51:33992–34003, 1986b.

U.S. Environmental Protection Agency (EPA): Guidelines for exposure assessment. Fed Reg 57:22888–22938, 1992b.

U.S. Environmental Protection Agency (EPA): Integrated Risk Information System; Announcement of Availability of Background Paper; Notice. Fed Reg 58(36):11490–11495, 1993a.

U.S. Environmental Protection Agency (EPA): Interim methods for development of inhalation reference concentrations [EPA/600/8–90/066A]. Office of Research and Development, Washington, DC, August, 1990.

U.S. Environmental Protection Agency (EPA): IRIS Background Paper. Integrated Risk Information System, Office of Health and Environmental Assessment, Office of Research and Development, 1993b.

U.S. Environmental Protection Agency (EPA): Review of the National Ambient Air Quality Standards for Ozone. Assessment of Scientific and Technical Information[EPA-450/2-92-001]. Office of Air Quality Planning and Standards Staff Paper, 1989.

U.S. Environmental Protection Agency (EPA): Summary of selected new information on effects of ozone on health and vegetation: Draft supplement to air quality criteria for ozone and other photochemical oxidants [EPA/600/8–88/105A]. U.S. EPA, Research Triangle Park, NC, 1988.

U.S. Environmental Protection Agency (EPA): Summary of selected new information on effects of ozone on health and vegetation: Supplement to 1986 air quality criteria for ozone and other photochemical oxidants [EPA/600/8–88/105F]. U.S. EPA, Washington, DC, 1992c.

U.S. Environmental Protection Agency (U.S. EPA): Working paper for considering draft revisions to the U.S. EPA guidelines for cancer risk assessment. Appendix D. In: Science and Judgment in Risk Assessment. National Research Council, Committee on Risk Assessment of Hazardous Air Pollutants, National Academy Press, Washington, DC: U.S. Environmental Protection Agency, pp. 383–447, 1994.

U.S. Environmental Protection Agency (EPA), Office of Pesticides and Toxic Substances: Assessment of health risks to garment workers and certain home residents from exposure to formaldehyde. U.S. EPA, Washington, DC, 1987.

U.S. Environmental Protection Agency (EPA), Office of Pesticides and Toxic Substances: Formaldehyde risk assessment update, 1991.

U.S. Office of Science and Technology Policy (OSTP): Chemical carcinogens: A review of the science and its associated principles, 1985.

Vainio, H, Heseltine, E, McGregor, D, Tomatis, L, Wilbourn, J: Working group on mechanisms of carcinogenesis and the evaluation of carcinogenic risks. Cancer Res 52:2357–2361, 1992a.

Vainio, H, Magee, P, McGregor, D, McMichael, A, editors: Mechanisms of Carcinogenesis in Risk Identification [IARC Scientific Publications No. 116]. IARC, Lyon, France, 1992b.

Vainio, H, Wilbourn, J, Tomatis, L: Identification of environmental carcinogens: the first step in risk assessment. In: The Identification and Control of Environmental and Occupational Diseases: Asbestos and Cancers. Edited by MA Mehlman, A Puton. Princeton Scientific Publishing Company, Princeton, 1994.

Valentine, R: An in vitro system for exposure of lung cells to gases: Effects of ozone on rat macrophages. J Toxicol Environ Health 16:115–126, 1985.

Whittemore, AS, McMillan, A: Lung cancer mortality among U.S. uranium miners: A reappraisal. J Natl Cancer Inst. 71:489–499, 1983.

Wilbourn, J, Haroun, L, Heseltine, E, Kaldor, J, Partensky, C, Vainio, H: Response of experimental animals to human carcinogens: An analysis based upon the IARC Monographs Programme. Carcinogenesis 7:1853–1863, 1986.

Wilmer, JWGM, Woutersen, RA, Appelman, LM, Leeman, WR, Feron, VJ: Subacute (4-week) inhalation toxicity study of formaldehyde in male rats: 8-hour intermittent versus 8-hour continuous exposures. J Appl Toxicol 7:15–16, 1987.

Wilmer, JWGM, Woutersen, RA, Appelman, LM, Leeman, WR, Feron, VJ: Subchronic (13-week) inhalation toxicity study of formaldehyde in male rats: 8-hour continuous exposures. Toxicol Lett 47:287–293, 1989.

Wolff, RK, Henderson, RF, Snipes, MB, Griffith, WC, Mauderly, JL, Cuddihy, RG, McClellan, RO: Alterations in particle accumulation and clearance in lungs of rats chronically exposed to diesel exhaust. Fundam Appl Toxicol 9:154–166, 1987.

Woutersen, RA, van Garderen-Hoetner, A, Bruijntjes, JP, Zwart, A, Feron, VJ: Nasal tumors in rats after severe injury to the nasal mucosa and prolonged exposure to 10 ppm formaldehyde. J Appl Toxicol 9:39–46, 1989.

Zwart, A, Woutersen, RA, Wilmer, JWGM, Spit, BJ, Feron, VJ: Cytotoxic and adaptive effects in rat nasal epithelium after 3-day and 13-week exposure to low concentrations of formaldehyde vapour. Toxicology 51:87–99, 1988.

Index

Milton Keynes UK
Ingram Content Group UK Ltd.
UKHW020003071024
449327UK00031B/2636